TIME SERIES ECONOMETRICS

Unit Roots and Trend Breaks

Volume 1

TIME SERIES ECONOMETRICS

Unit Roots and Trend Breaks

Volume 1

Editor

Pierre Perron

Boston University, USA

World Scientific

NEW JERSEY • LONDON • SINGAPORE • BEIJING • SHANGHAI • HONG KONG • TAIPEI • CHENNAI • TOKYO

Published by

World Scientific Publishing Co. Pte. Ltd.

5 Toh Tuck Link, Singapore 596224

USA office: 27 Warren Street, Suite 401-402, Hackensack, NJ 07601

UK office: 57 Shelton Street, Covent Garden, London WC2H 9HE

Library of Congress Cataloging-in-Publication Data
Names: Perron, Pierre, 1959– editor.
Title: Time series econometrics / edited by Pierre Perron (Boston University, USA).
Description: New Jersey : World Scientific, [2018]
Contents: Part I. Unit roots and trend breaks -- Part II. Structural change.
Identifiers: LCCN 2018002771| ISBN 9789813237858 (hc : set : alk. paper) |
ISBN 9789813237865 (hc: v. 1 : alk. paper) | ISBN 9789813237896 (hc: v. 2 : alk. paper)
Subjects: LCSH: Econometrics. | Time-series analysis.
Classification: LCC HB139 .T534 2018 | DDC 330.01/51955--dc23
LC record available at https://lccn.loc.gov/2018002771

British Library Cataloguing-in-Publication Data
A catalogue record for this book is available from the British Library.

For any available supplementary material, please visit
https://www.worldscientific.com/worldscibooks/10.1142/10930#t=suppl

Desk Editors: Anthony Alexander/Alisha Nguyen

Typeset by Stallion Press
Email: enquiries@stallionpress.com

About the Editor

Professor Pierre Perron received his B.A. in Economics from McGill University, Montréal, Canada, M.A. in Economics from Queens University, Kingston, Canada, and Ph.D. in Economics from Yale University, Connecticut, USA. He has been a Professor at the Department of Economics at Boston University since 1997, prior to which he was a full Professor at the Université de Montréal and an Assistant Professor at Princeton University. He was a Visiting Professor at Pontifícia Universidade Católica, Brazil, Universidade de São Paulo, Brazil, the Université de Bourgogne, Dijon, France and the Université de Lausanne, Switzerland as well as a Visiting Scholar at Centre de Recherche en Économie et Statistique, Institut Nationale de la Statistique et des Études Économiques, Paris, France, and Département de Statistiques, Université Libre de Bruxelles, Belgium. Professor Perron was a co-editor for the *Econometrics Journal* and on the editorial board of numerous journals such as *Econometrica*, the *Journal of Econometrics*, *Econometric Reviews*, *Econometric Theory*, *Economics Letters*, *Review of Economics and Statistics*. He was a referee for numerous other journals and grant agencies. He is an elected fellow of the Econometric Society, a fellow of the *Journal of Econometrics* and has been listed in the book *Who's Who in Economics: A Biographical Dictionary of Major Economists 1700 to 1995*, published by Edwin

Elgar Publishing. Professor Perron has published nearly 100 papers in many referred journals such as *Econometrica*, the *Journal of Econometrics*, the *Journal of Business and Economic Statistics*, the *Journal of the American Statistical Association*, *Econometric Theory*, *Biometrika*, *Nature Geoscience* to name a few. He has also contributed many book chapters and is the recipient of various grants from the National Science Foundation, the Fonds pour la Formation de Chercheurs et l'Aide à la Recherche du Québec, the Social Sciences and Humanities Research Council of Canada and the Natural Sciences and Engineering Council of Canada. Professor Perron has presented his work at numerous conferences, both national and international. He also acted as the main Ph.D. supervisor for over 25 students and numerous others as secondary or external advisors. His work is very highly cited in economics as well as in a variety of other fields.

Acknowledgments

I am especially grateful to all the talented students I had over many years who have ventured working with me on various projects. Those associated with the papers included in this volume are Dukpa Kim, Serena Ng, Zhongjun Qu, Gabriel Rodriguez and Timothy Vogelsang. Many students contributed to research, unfortunately not included, which helped make possible the work that was included or extended it: Francisco Estrada, Silvie Mallet, Cosme Vodounou, Tatsuma Wada, Tomoyoshi Yabu, and Xiaokang Zhu. Special thanks to Robert Shiller for giving me a head start on research. Thanks also to my other co-authors: Josep Lluís Carrion-i-Silvestre, Marcio Garcia, Seiji Nabeya and Peter C.B. Phillips. This book is dedicated to my wife, and the best life companion I could have hoped for, Regina Celia Cati (also a co-author on Chapter 9) and my children (Fanny, Laurie and Rafael) and grand-child (Phoebe).

I would like to thank the following publishers and journals for granting their permission to reproduce the articles in this volume:

Publishers
American Statistical Association
Cambridge University Press
Elsevier Science
John Wiley & Sons
Oxford University Press

Journals

Biometrika
Econometric Theory
Econometrica
Economics Letters
International Economic Review
Journal of the American Statistical Association
Journal of Applied Econometrics
Journal of Business and Economic Statistics
Journal of Econometrics
Review of Economic Studies

Contents

Introduction

Pierre Perron

My adventure in research about problems related to unit root testing started with Robert Shiller, to whom I was and remain very grateful for the opportunity he gave me in getting going on this and other research topics. We published Chapter 1 while I was a graduate student. It deals with a very interesting issue related to the power of unit root tests when considering that a larger sample size can be associated with a different sampling interval. For instance, 100 observations over 100 years versus 400 observations over 25 years. One would think that a larger number of observations would lead to more powerful tests. This proved wrong. What matters is the span of data. Indeed, if you keep the span fixed, then the power does not converge to 1 as the sample size increases. It levels off at some value that depends mostly on the span of the data. This paper involved results based on simulations. I set upon a dissertation topic to try to get theoretical results, in particular about the limit value of the power as the sample size increases while the span is kept fixed (what I used to call the infinity column). Because of personal reasons, I went on the job market (a limited one in Québec) in the middle of my third year. My job talk was something like 'here is what I will do for my thesis'. I got a job at the Université de Montréal and started the following fall but without solving what I said I would. It would take more skills and more time, eventually published as Chapter 5. At the time, nobody at Yale was working on such time series topics; so no help was available and I obviously lacked the necessary mathematical background. I later found out that Peter Phillips had gotten interested in unit root issues. I joined him in writing the *Biometrika* paper (Chapter 2), which has become a landmark paper, with the resulting tests now referred to as the Phillips–Perron tests. This was a part of my thesis; the rest involved, among other things, preliminary versions of material published in Perron (1989, 1990). An early application of these new unit root tests appears in Perron (1988).

I started to investigate in more detail issues about limit distributions related to unit root tests. One was Chapter 3, about how to derive the appropriate limit moment-generating function to be used via an inversion theorem to compute the quantiles of the limit distribution of the normalized least-squares bias for a unit root process with independent and identically distributed errors. An extension to the seasonal unit root case was provided in Perron (1992). This was closely followed by Chapter 4, which provided the limit distribution under a continuous-time asymptotic framework, closely related to my initial goal of getting the 'infinity column' in the paper with Robert Shiller (Chapter 1). See also Perron (1991) for a similar analysis for the strictly stationary case and Perron and Vodounou (2001) for extensions. The issue of testing for a unit root with an arbitrary path for the sample size with respect to the sampling interval was finally solved in Chapter 5.

I then set out to work on issues that would be of more practical relevance, an approach which I continued throughout my career. It was by then well known that the Phillips–Perron tests have very poor finite sample properties, especially with a negative moving average component in the noise. To understand why, consider a process $y_t = \gamma z_t + v_t$, $v_t = v_{t-1} + u_t$ and z_t a vector of deterministic components (e.g., a linear trend, $z_t = (1, t)$), the Phillips–Perron test Z_α and its limit distribution is given by

$$
\begin{aligned}
Z_\alpha &= T(\hat{\alpha} - 1) - X_T \\
&\Rightarrow \frac{\int_0^1 W^*(r)dW(r) + \delta}{\int_0^1 W^*(r)^2 dr} - \frac{\delta}{\int_0^1 W^*(r)^2 dr} = \frac{\int_0^1 W^*(r)dW(r)}{\int_0^1 W^*(r)^2 dr},
\end{aligned}
$$

where $\hat{\alpha}$ is the least-squares estimate in the regression $y_t = \gamma z_t + \alpha y_{t-1} + v_t$, $X_T = (\hat{\sigma}^2 - \hat{\sigma}_u^2)/2T^{-2}\sum_{t=1}^{T}(y_{t-1}^*)^2$, $\delta = (\sigma^2 - \sigma_u^2)/2\sigma^2$, σ^2 is the so-called long-run variance of u_t (2π time the spectral density at frequency zero), σ_u^2 is the variance of u_t, with $\hat{\sigma}^2$ and $\hat{\sigma}_u^2$ consistent estimates, y_{t-1}^* is the detrended version of y_{t-1}, while $W^*(r)$ is a detrended Weiner process. The first step is to investigate the adequacy of the limit distribution. This was analyzed in Chapter 7, which showed that $(\int_0^1 W^*(r)dW(r) + \delta)/\int_0^1 W^*(r)^2 dr$ is a very bad approximation to $T(\hat{\alpha} - 1)$ when there is a negative moving-average component in u_t, while $\delta/\int_0^1 W^*(r)^2 dr$ is a good approximation to X_T. Hence, the overall limit distribution provides a poor approximation which results in severe size distortions. The problem is that the standard functional central limit theorem underlying the derivations provides a poor approximation with a negative moving average component; see Perron and Mallet (2000). For alternative asymptotic frameworks that provide better approximations, see Nabeya and Perron (1994, 1995).

The problem is that $\hat{\alpha}$ is severely biased below one with a negative moving-average component. So, one would like to minimize the impact of this bias in devising alternative unit root tests. This was analyzed in Chapter 8. Suppose that one uses an estimate of the variance of the noise imposing the null hypothesis, i.e., $\tilde{\sigma}_u^2 = T^{-1}\sum_{t=1}^{T}(\Delta y_t^*)^2$, then the so-called modified PP test is (also suggested earlier as a test by Stock, 1999):

$$
\begin{aligned}
MZ_\alpha &= T(\hat{\alpha} - 1) - \frac{\hat{\sigma}^2 - \tilde{\sigma}_u^2}{2T^{-2}\sum_{t=1}^{T}(y_{t-1}^*)^2} \\
&\approx Z_\alpha + T(\hat{\alpha} - 1)^2 \approx \frac{T^{-1}(y_T^*)^2 - \hat{\sigma}^2}{2T^{-2}\sum_{t=1}^{T}(y_{t-1}^*)^2}.
\end{aligned}
$$

One could try the same thing for the estimate of the long-run variance, but the test is then inconsistent. To solve this problem, the idea is to use an autoregressive spectral density estimator that includes an additional lagged dependent variable, i.e., $s_{AR}^2 = \hat{\sigma}_{ek}^2/(1 - \sum_{i=1}^{k}\hat{b}_i)^2$, where $\hat{\sigma}_{ek}^2 = T^{-1}\sum_{t=k+1}^{T}\hat{e}_{tk}^2$ with $\hat{b}_i$ and $\hat{e}_{tk}$ obtained from the following OLS regression:

$$\Delta y_t^* = \hat{b}_0 y_{t-1}^* + \sum_{i=1}^{k} \hat{b}_i \Delta y_{t-i}^* + \hat{e}_{tk}.$$

This last equation is simply the Augmented Dickey–Fuller regression (Dickey and Fuller, 1979). Introducing the lagged level is necessary to ensure a consistent test. More details can be found in Perron and Ng (1998). Other modified tests, which perform similarly were proposed in Chapter 8. The remaining issue to obtain tests with good sizes in finite samples is to appropriately select k, the order of the autoregression. On this issue, substantial but incomplete progress was made in Chapter 6, which advocated a general to specific method. A method which delivers very good results was proposed in Chapter 10 and refined in Chapter 11; see also Ng and Perron (2005). This involves the so-called modified information criteria. This work also uses quasi-GLS detrended series as suggested by Elliott *et al.* (1996) to increase power. Hence, the class of tests suggested in Chapter 10 have good size properties, even in the presence of a negative moving average component, and good power overall. They have become extensively used in practice and available in popular software.

The fact that the tests are robust to the presence of a negative moving-average component makes them robust to outliers as well. Another type of external disturbance, which can affect the properties of unit root tests are 'inliers' when a series is temporarily brought down to low levels before returning to its previous trend path. This issue was analyzed in Chapter 9 in the context of Brazilian inflation series.

I never believed that unit root processes made sense for the majority of economic time series, and I still don't. Yet, most tests failed to reject the null hypothesis of a unit root for most series. An influential paper when I started was Nelson and Plosser (1982). I then set out an agenda to understand why, which led me to the issue of trend breaks. To understand the main idea, consider a series generated by the following popular process: $y_t = \mu_t + c_t$;

$\mu_t = \mu_{t-1} + \beta_{t-1} + \eta_t$; $\beta_t = \beta_{t-1} + v_t$, where c_t is the stationary or cyclical component, μ_t is the trend and β_t is the slope of the trend (the noise components η_t and v_t are stationary). This implies a series y_t that is integrated of order 2, $I(2)$, but typically the variance of v_t is small so that the series is close to being integrated of order 1, $I(1)$. See Perron and Wada (2009, 2016) for more details. On the other hand, a trend-stationary process is specified by $y_t = a + \delta t + c_t$. Unit root tests aim to distinguish between the two. It is useful to see the implications of what is being tested. For $I(1)$ processes, using a standard Beveridge and Nelson (1981) decomposition, the trend changes every period, induced by level shifts. For $I(2)$ processes and performing a unit root test in first-differences (e.g., growth rates), the slope changes every period. For trend-stationary processes, the level and slope of the trend never change. So unit root tests are basically asking the question: is the level (or slope) of the trend function changing every period or never? Why not ask changing every period or 'sometimes'. My argument was that 'sometimes' is most often a very small number meaning that influential changes in level and slope are quite rare. For historical GDP series, say, a level shift occurs during the Great Crash of 1929, or World War II for Europe; changes in slope occur after World War II for Europe and after the Oil Price shock in 1973 for most industrialized countries. This was the argument put forward in Chapter 12. Since such changes are rare, it is difficult to cast them in a probabilistic framework, and hence I treated them as exogenous. I showed that standard unit root tests are biased toward a nonrejection of a unit root when a change in level or slope is present in an otherwise stationary time series. Appropriate unit root tests that allow for such changes were developed. The results were quite striking. It was now easy to reject the null of a unit root. This work was extended to nontrending series in Chapters 13 and 14.

This work initially treated the dates of the changes as known. This drew a lot of criticisms, the argument being that I had peeked at the data prior to selecting the break dates (even though I really only tried a few around the stated ones; e.g., 1929 and 1973) and this induced a preselection bias. For good or bad reasons (better size versus reduced power), the suggested solution was to be completely agnostic about where the break dates could be. Important contributions along those lines are Zivot and Andrews (1992) and Chapters 15–18 (see also Perron, 1994). The problem with this work is that the changes in level and/or slope were only allowed under the alternative hypothesis of a broken-trend stationary process. Hence, a rejection can occur if a break is present in an otherwise nonstationary series. This applies especially to Zivot and Andrews's (1992) suggested test. Though Chapters 15–18 adopted the same setup, they also consider different methods to select the break date, which were shown to imply less size distortions. The problem was finally solved in Chapter 19. This built upon work by Perron and Zhu (2005) and Perron and Yabu (2009a, 2009b), which allowed the construction of a pretest for a break in trend allowing the noise to be integrated or stationary. Upon a rejection, the limit distribution pertaining to the known break date case can then be used; see the extension of Chapter 12 by Perron and Vogelsang (1993a, 1993b), otherwise the standard limit distribution applies. This has now become the leading method to test for unit roots in the presence of a break in trend. For an interesting application to climate change data, see Estrada *et al.* (2013). This method was extended to the multiple breaks case in Chapter 20.

References

Beveridge, S. and Nelson, C.R. (1981). A new approach to decomposition of economic time series into permanent and transitory components with particular attention to measurement of the business cycle. *Journal of Monetary Economics* 7, 151–174.

Dickey, D.A. and Fuller, W.A. (1979). Distribution of the estimator for autoregressive time series with a unit root. *Journal of the American Statistical Association* 74, 427–431.

Elliott, G., Rothenberg, T. and Stock, J.H. (1996). Efficient tests for an autoregressive unit root. *Econometrica* 64, 813–836.

Estrada, F., Perron, P. and Martinez-López, B. (2013). Statistically-derived contributions of diverse human influences to 20th century temperature changes. *Nature Geoscience* 6, 1050–1055.

Nabeya, S. and Perron, P. (1994). Local asymptotic distribution related to the AR(1) model with dependent errors. *Journal of Econometrics* 62, 229–264.

Nabeya, S. and Perron, P. (1995). Approximations to some exact distributions in the first order autoregressive model with dependent errors. *Econometric Reviews* 14, 421–457.

Nelson, C.R. and Plosser, C.I. (1982). Trends and random walks in macroeconomic time series. *Journal of Monetary Economics* 10, 139–162.

Ng, S. and Perron, P. (2005). A note on the selection of time series models. *Oxford Bulletin of Economics and Statistics* 67, 115–134.

Perron, P. (1988). Trends and random walks in macroeconomic time series: Further evidence from a new approach. *Journal of Economic Dynamics and Control* 12, 297–332.

Perron, P. (1989). Testing for a Random Walk: A Simulation Experiment of Power When the Sampling Interval is Varied. In *Advances in Econometrics and Modeling*, Baldev Raj (ed.). Kluwer Academic Publisher, pp. 47–68.

Perron, P. (1990). Tests of Joint Hypotheses in Time Series Regression with a Unit Root. In *Advances in Econometrics: Co-integration, Spurious Regression and Unit Roots*, G.F. Rhodes and T.B. Fomby (eds.) Vol. 8. JAI Press, pp. 135–159.

Perron, P. (1991). A continuous time approximation to the stationary first-order autoregressive model. *Econometric Theory* 7, 236–252.

Perron, P. (1992). The limiting distribution of the least-squares estimator in nearly integrated seasonal models. *Canadian Journal of Statistics* 20, 121–134.

Perron, P. (1994). Trend, Unit Root and Structural Change in Macroeconomic Time Series. In *Cointegration for the Applied Economist*, B.B. Rao (ed.). Macmillan Press, Basingstoke, pp. 113–146.

Perron, P. and Mallet, S. (2000). A look at the quality of the approximation of the functional central limit theorem. *Economics Letters* 68, 225–234.

Perron, P. and Ng, S. (1998). An autoregressive spectral density estimator at frequency zero for nonstationarity tests. *Econometric Theory* 14, 560–603.

Perron, P. and Vodounou, C. (2001). Asymptotic approximations in the near-integrated model with a non-zero initial condition. *Econometrics Journal* 4, 143–169.

Perron, P. and Vogelsang, T.J. (1993a). The great crash, the oil price shock and the unit root hypothesis: Erratum. *Econometrica* 61, 248–249.

Perron, P. and Vogelsang, T.J. (1993b). A note on the additive outlier model with breaks. *Revista de Econometria* 13, 181–201.

Perron, P. and Wada, T. (2009). Let's take a break: Trends and cycles in U.S. real GDP. *Journal of Monetary Economics* 56, 749–765.

Perron, P. and Wada, T. (2016). Measuring business cycles with structural breaks and outliers: Applications to international data. *Research in Economics* 70, 281–303.

Perron, P. and Yabu, T. (2009a). Estimating deterministic trends with an integrated or stationary noise component. *Journal of Econometrics* 151, 56–69.

Perron, P. and Yabu, T. (2009b). Testing for shifts in trend with an integrated or stationary noise component. *Journal of Business and Economic Statistics* 27, 369–396.

Perron, P. and Zhu, X. (2005). Structural breaks with stochastic and deterministic trends. *Journal of Econometrics* 129, 65–119.

Stock, J.H. (1999). A Class of Tests for Integration and Cointegration. In *Cointegration, Causality and Forecasting. A Festschrift in Honour of Clive W.J. Granger*, Engle, R.F., and H. White (eds.), Oxford University Press, Oxford, pp. 137–167.

Zivot, E. and Andrews, D.W.K. (1992). Further evidence on the great crash, the oil price shock and the unit root hypothesis. *Journal of Business and Economic Statistics* 10, 251–270.

Part A

Unit Root

Chapter 1

Testing the Random Walk Hypothesis: Power versus Frequency of Observation*

Robert J. Shiller and Pierre Perron

Yale University, New Haven, CT 06520, USA

Power functions of tests of the random walk hypothesis versus stationary first-order autoregressive alternatives are tabulated for samples of fixed span but various frequencies of observation.

Keywords: Unit root, sampling interval, power function, test of market efficiency, Ornstein–Uhlenbeck process, autoregressive process.

1. Introduction

Those who test the random walk model for prices of speculative assets often use very many observations, as data may be readily available on monthly, weekly or even daily basis. It is often casually asserted that, even though the data may come from a relatively short sample period, with so many observations power of tests ought to be very high.

To evaluate such an assertion, power functions for an important class of alternatives will be tabulated here where the span of the data (measured in years, say, whether observations are annual, monthly or daily) is held fixed and the number of observations is varied by changing the frequency of observation. We shall see that over a

*Originally Published in *Economics Letters* 18, pp. 381–386.

substantial range of parameter values it is more useful to think of power of a t-test or normalized beta test as depending on the span of the data rather than the number of observations, and that the power of a runs tests will be destroyed by too many observations.

2. Null and Alternative Hypotheses

The null hypothesis and alternative hypothesis to be considered are given by the stochastic differential equations:

$$\mathrm{H}_0 : \mathrm{d}p_t = \sigma_0 \mathrm{d}w_t, \quad t > 0, \tag{1}$$

$$\mathrm{H}_1 : \mathrm{d}p_t = -\tau(p_t - \mu)\mathrm{d}t + \sigma_1 \mathrm{d}w_t, \quad -\infty < t < \infty, \ \ \tau > 0, \tag{2}$$

where w_t is a unit Wiener process. In the finance context, the random walk null hypothesis means that price p_t can never be described as 'too high' (i.e., that it can be expected to fall in the future) or 'too low' (i.e., that it can be expected to rise in the future). Many other, more complicated, variations on the null hypothesis exist of course in the finance literature, and it is a matter for further research to see to what extent the results obtained here extend to these other hypotheses.

The alternative hypothesis here asserts price p_t is mean-reverting. In Shiller (1981), this alternative model was referred to as a 'fads' model; stock prices move because of repeated investor fads (represented by the innovations $\sigma_1 dw_t$) but these fads are gradually forgotten. Forgetting takes place with an exponential decay pattern as commonly modeled by mathematical psychologists, so that τ might be thought of as a parameter reflecting human memory retention. If the public information set at time t consists of the history of p so that there is steady information flow, then this hypothesis shows the 'martingale-like behavior' that Sims (1984) argues, from arbitrage considerations, ought to obtain.

The alternative hypothesis H_1 is equivalent for $t \geq 0$ (all observations will have $t \geq 0$) to specifying that Equation (2) holds for $t > 0$ and that p_0 is normally distributed with mean μ and variance $\sigma_t^2/2\tau$.

If sampled at discrete points of time at intervals of length h the processes are

$$\mathrm{H}_0 : p_{ht} = p_{h(t-1)} + u_{ht}^{(h)}, \quad t = 1, 2, \ldots, \tag{3}$$

$$\mathrm{H}_1 : p_{ht} = \mu + \beta_h(p_{h(t-1)} - \mu) + \upsilon_{ht}^{(h)}, \quad t = \ldots, -2, -1, 0, 1, 2, \ldots, \tag{4}$$

where $u_{ht}^{(h)}$ is normally distributed with mean zero and variance $h\sigma_0^2$ while $\upsilon_{ht}^{(h)}$ is normally distributed with mean zero and variance $\sigma_1^2(l - \exp(-2\tau h))/(2\tau h)$. The discrete time autoregressive parameter β_h is equal to $\exp(-\tau h)$. Regardless of the sampling interval h. the alternative hypothesis is that the process p_{ht} is first-order autoregressive around the mean μ.

3. Test Statistics

We are given $1 + S/h$ observations $p_0, p_h, p_{2h}, \ldots, p_S$ where h is the sampling interval and S is the span of the data.

The t-test whose power we shall investigate employs the conventional t-statistic to test whether the slope coefficient equals one in a regression of p_{ht} on a constant term and $p_{h(t-1)}$ with $T = S/h$ observations,

$$t = \left[\sum_{t=1}^{T} (p_{h(t-1)} - \bar{p}_{-h})^2\right]^{0.5} (\hat{\beta} - 1)/s, \tag{5}$$

$$\hat{\beta} = \sum_{t=1}^{T} (p_{h(t-1)} - \bar{p}_{-h}) p_{ht} \left[\sum_{t=1}^{T} (p_{h(t-1)} - \bar{p}_{-h})^2\right]^{-1}, \tag{6}$$

$$\hat{\mu} = \bar{p}_0 - \hat{\beta}\bar{p}_{-h}, \tag{7}$$

$$\bar{p}_{-j} = T^{-1} \sum_{t=1}^{T} p_{ht-j}, \quad j = 0, h, \tag{8}$$

$$s^2 = (T-2)^{-1} \sum_{t=1}^{T} (p_{ht} - \hat{\mu} - \hat{\beta} p_{h(t-1)})^2. \tag{9}$$

This t-statistic is not distributed as student's t under the null hypothesis, is not asymptotically normal, and is indeed very badly approximated by the normal (see Nankervis and Savin, 1983). The normalized beta test statistic B as used in Evans and Savin (1981, 1984) is proportional to the $\hat{\beta}$ from (6) above,

$$B = (T(T-1)/2)^{0.5}(\hat{\beta} - 1). \tag{10}$$

The runs test uses the number of runs r in the sequence of T observed price changes; r is one plus the number of times a price change is followed by a price change of the opposite sign. Under the assumption that the total number n_1 of positive price changes and n_2 of negative price changes ($n_1 + n_2 = T$) are given, and that all arrangements of these price changes occur with equal probability, then the mean m_r and standard deviation σ_r of the number r of runs are

$$m_r = 2n_1n_2/T + 1, \tag{11}$$

$$\sigma_r = \sqrt{2n_1n_2(2n_1n_2 - T)/(T^2(T-1))}. \tag{12}$$

The test statistic as used for example by Fama (1965) to test the random walk hypothesis for stock prices is given by[1]

$$K = (r + 0.5 - m_r)\sigma_r. \tag{13}$$

It can be shown using a theorem in Mood (1940) that the ratio K is (under H_0) asymptotically (where n_1/T is kept constant as T is increased) normal with zero mean and unit variance.

Under our hypotheses all test statistics are similar; we can therefore carry the simulations under the assumptions that $\mu = 0$ and $\sigma_0^2 = 1/h$ and $\sigma_1^2 = 2\tau/(1 - \exp(-2\tau h))$.

---

[1]Fama (1965) used a formula for K which allows separately for +, − and 0 price changes and which reduces to this formula if there are no zero price changes.

4. Results

Power functions for a size 0.05 test were tabulated with 40,000 replications for number of observations $T = 8$ to 512 and with 10,000 replications for $T = 1024$.[2]

In Tables 1–3, power is shown only for a single value of $\tau (\tau = 0.2)$ but powers of tests for $\tau = 0.2 * 2^j$ can be found by reading j rows down from the power for $\tau = 0.2$. In each table, we can read how power depends on the number of observations for a fixed span (reading across rows), how power depends on the span for a fixed number of observations (reading down columns) or how power depends on the number of observations when, as is usually assumed, span is proportional to observations (reading along diagonals from upper left to lower right).[3]

For the t-test and normalized beta test, Tables 1 and 2, we see that with short spans there is never much power.[4] For intermediate

---

[2]All simulations were carried out on a CDC Cyber 915 at the Université de Montréal. The $N(0, 1)$ random deviates $u_t^{(h)}$ in (3) and (4) were obtained from the subroutine GGNML of the International Mathematical and Statistical Library (IMSL) package, version 9.1. In the first step the empirical distributions under the null hypothesis of the t-statistic and of the normalized β-statistic were computed with 20,000 replications. The series p_{ht} were generated from (3) for $t = 1, \ldots, T$ starting from the initial condition $p_0 = 0$. The significance points were taken from the sorted arrays. Critical values were taken as the 2.5% and 97.5% points of the empirical distributions. Critical values for the runs test were taken from the assumption that the statistic K was normally distributed. In the second step the empirical power functions of the tests using the significance points from step one were computed with 20,000 replications (except for $T = 1024$ which used 10,000). Under the alternative the series p_{ht} were generated from (4) for $t = 1, \ldots, T$ starting from the initial condition $p_0 = u_0/(1 - \exp(-2\tau h))^{0.5}$. This procedure was repeated twice (except for $T = 1024$) and the tables show the averages of the two power computations.

[3]Some of the numbers in the tables correspond approximately with power computations in the literature (Dickey, 1976; Dickey and Fuller, 1979; Evans and Savin, 1984; Fuller, 1976; Nankervis and Savin, 1983) and in these cases our results were roughly confirmed.

[4]When the span of the data is very short, the test is slightly biased (the power is less than the size of the test of 0.05). Such a bias in the case of a fixed startup at the mean alternative was noted before by Nankervis and Savin (1983).

Table 1. Power of a two-tailed t-test, number of observations: $T = S/h$; S: span of the data.[a]

S	8	16	32	64	128	256	512	1024
8	0.040	0.038	0.037	0.039	0.039	0.041	0.037	0.039
16	0.042	0.042	0.043	0.044	0.045	0.045	0.044	0.043
32	0.059	0.071	0.079	0.085	0.088	0.089	0.088	0.086
64	0.111	0.191	0.246	0.284	0.296	0.304	0.307	0.307
128	0.175	0.470	0.701	0.824	0.872	0.892	0.901	0.907
256	0.195	0.681	0.973	0.999	1.000	1.000	1.000	1.000
512	0.196	0.726	0.997	1.00	1.00	1.00	1.00	1.00
1024	0.196	0.728	0.998	1.00	1.00	1.00	1.00	1.00
Lower bound	−3.893	−3.421	−3.271	−3.181	−3.155	−3.137	−3.135	−3.135
Higher bound	0.562	0.379	0.312	0.275	0.240	0.217	0.225	0.246

Notes: [a]Table gives power of two-tailed t-test at 0.05 level against alternative with $\gamma = 0.2$. Thus, for example, if $\gamma = 0.2$ where time is measured in years (so that H_1 would predict a rate of decline for p of $0.2(p_i - \mu)$ per year) and we have 128 quarterly observations covering a span of 32 years, the probability of rejecting is 0.088. If $\gamma = 0.05$ where time is measured in years and we have 128 quarterly observations covering a span on 32 years, then the probability of rejecting is given by 0.039. The final rows give the critical values of the t-statistic, Equation (5).

spans (32 or 64) power rises gradually as observations are increased and then levels off substantially below one. Over a substantial range, power depends more on the span of the data than on the number of observations.

For the two-tailed runs test, Table 3, we see that if span is held fixed as the number of observations is increased, power tends to rise and then fall back towards the size of the test. If there are many observations, there is never much power, for any spans shown in the table.

It is thus wrong to presume that power of tests of the random walk hypothesis must be high just because there are very many observations (see also Summers, 1982). In all the tables shown here, power does not approach one in the limit as we either move to the right along rows or move down along columns, but approaches on as a

Table 2. Power of a two-tailed test with the normalized estimator of beta; number of observations: $T = S/h$; S: span of the data.[a]

S	8	16	32	64	128	256	512	1024
8	0.050	0.047	0.044	0.045	0.045	0.046	0.046	0.046
16	0.058	0.059	0.057	0.058	0.060	0.060	0.061	0.058
32	0.094	0.116	0.123	0.126	0.128	0.130	0.133	0.132
64	0.185	0.309	0.371	0.411	0.427	0.434	0.445	0.439
128	0.283	0.644	0.843	0.925	0.953	0.964	0.969	0.971
256	0.312	0.828	0.993	1.00	1.00	1.00	1.00	1.00
512	0.313	0.860	1.00	1.00	1.00	1.00	1.00	1.00
1024	0.313	0.862	1.00	1.00	1.00	1.00	1.00	1.00
Lower bound	−6.837	−8.717	−10.174	−10.988	−11.472	−11.693	−11.784	−11.897
Higher bound	0.775	0.486	0.384	0.350	0.297	0.276	0.283	0.315

Notes: [a]Table gives power of a two-tailed normalized beta test of the null hypothesis against alternative with $\gamma = 0.2$. The final rows give the critical values for the statistic B, Equation (10).

Table 3. Power of a two-tailed runs test; number of observations; $T = S/h$; S: Span of the data.[a]

S	8	16	32	64	128	256	512	1024
8	0.048	0.045	0.062	0.050	0.050	0.048	0.050	0.054
16	0.056	0.044	0.065	0.050	0.049	0.049	0.050	0.055
32	0.076	0.053	0.074	0.055	0.052	0.048	0.051	0.056
64	0.110	0.080	0.105	0.068	0.061	0.052	0.053	0.055
128	0.141	0.139	0.189	0.123	0.094	0.072	0.063	0.059
256	0.150	0.200	0.358	0.280	0.214	0.141	0.104	0.082
512	0.151	0.217	0.504	0.566	0.532	0.385	0.256	0.170
1024	0.151	0.218	0.542	0.756	0.877	0.833	0.682	0.464
Lower bound	−1.96	−1.96	−1.96	−1.96	−1.96	−1.96	−1.96	−1.96
Higher bound	1.96	1.96	1.96	1.96	1.96	1.96	1.96	1.96

Notes: [a]Table gives power of a two-tailed runs test at the 0.05 level against alternative with $\gamma = 0.2$. The final rows give the critical values for the runs test statistic K, Equation (13).

limit only if we move diagonally down from upper left to lower right, i.e., increasing both span and number of observations.[5] Some of the relevant asymptotic distributions are developed in Perron (1984).

Acknowledgment

This research was supported by the U.S. National Science Foundation (SES-8408565) and the Université de Montréal. The authors are indebted to P.C.B. Phillips, Stephen A. Ross and N.E. Savin for helpful discussions.

References

Dickey, D.A. (1976). Estimation and hypothesis testing in non-stationary time series, unpublished Ph.D. dissertation, Iowa State University, Ames, IA.

Dickey, D.A. and Fuller, W.A. (1979). Distribution of the estimates for the autoregressive time series with a unit root, *Journal of the American Statistical Association* 74(366), 427–431.

[5]For some values of S or T in the tables the power is given as 1.000. This reflects the fact that the power is very dose to one for these values, but not that the power has the limit of one along either rows or columns.

Dickey, D.A. and Fuller, W.A. (1981). Likelihood ratio statistics for autoregressive time series with a unit root, *Econometrica* 49(4), 1957–1972.

Evans, G.B.A. and Savin, N.E. (1981). Testing for unit roots: 1, *Econometrica* 49(3), 753–780.

Evans, G.B.A. and Savin, N.E. (1984). Testing for unit roots: 2, *Econometrica* 52(5), 1241–1270.

Fama, F. (1965). The behavior of stock market prices. *Journal of Business* 38, 34–105.

Fuller, W.A. (1976). *Introduction to Statistical Time Series* (Wiley, New York).

Mood, A.M. (1940). The distribution theory of runs, *Annals of Mathematical Statistics* 11, 367–392.

Nankervis, J.C. and Savin, N.E. (1983). *Testing the Autoregressive Parameter with the t Statistic*, unpublished (Trinity College, Cambridge).

Perron, P. (1984). *The Distribution of the Least Squares Estimator of the Autoregressive Coefficient with a Continuum of Data*, unpublished (Yale University, New Haven, CT).

Shiller, R.J. (1981). The use of volatility measures in assessing market efficiency, *The Journal of Finance* 36(2), 291–304.

Sims, C.A. (1984). Martingale-like behavior of prices and interest rates. Discussion paper no. 205, University of Minnesota, Minneapolis, MN.

Summers, L.H. (1982). Do we really know that financial markets are efficient? Working paper no. 994, NBER, Cambridge, MA.

Chapter 2

Testing for a Unit Root in Time Series Regression*

Peter C. B. Phillips[†] and Pierre Perron[‡]

[†] *Cowles Foundation for Research in Economics,*
Yale University, Yale Station,
New Haven, Connecticut 06520, USA

[‡] *Département des Sciences Economiques,*
Université de Montréal, Montréal, Québec,
Canada H3C 3J7

This paper proposes new tests for detecting the presence of a unit root in quite general time series models. Our approach is nonparametric with respect to nuisance parameters and thereby allows for a very wide class of weakly dependent and possibly heterogeneously distributed data. The tests accommodate models with a fitted drift and a time trend so that they may be used to discriminate between unit root nonstationarity and stationarity about a deterministic trend. The limiting distributions of the statistics are obtained under both the unit root null and a sequence of local alternatives. The latter noncentral distribution theory yields local asymptotic power functions for the tests and facilitates comparisons with alternative procedures due to Dickey and Fuller. Simulations are reported on the performance of the new tests in finite samples.

Keywords: Brownian motion; noncentral distribution; time series; unit root; weak convergence.

*Originally Published in *Biometrika* 75, pp. 335–346.

1. Introduction

Methods for detecting the presence of a unit root in parametric time series models have lately attracted a good deal of interest in both statistical theory and application. Fuller (1984) and Dickey *et al.* (1986) review much of the literature. The latter article provides a helpful practical guide to the use of some of the formal tests.

One major field of application where the hypothesis of a unit root has important implications is economics. This is because a unit root is often a theoretical implication of models which postulate the rational use of information that is available to economic agents. Examples include various financial market variables such as futures contracts (Samuelson, 1965), stock prices (Samuelson, 1973), dividends (Kleidon, 1986), spot and forward exchange rates (Meese and Singleton, 1983), and even aggregate variables like real consumption (Hall, 1978). Formal statistical tests of the unit root hypothesis are of additional interest to economists because they can help to evaluate the nature of the nonstationarity that most macroeconomic data exhibit. In particular, they help in determining whether the trend is stochastic, through the presence of a unit root, or deterministic, through the presence of a polynomial time trend. A recent examination of historical economic time series by Nelson and Plosser (1982), for example, found strong evidence in favor of unit root nonstationarity using the testing procedure of Dickey and Fuller (1979).

Recently, Said and Dickey (1984) have shown that the Dickey–Fuller procedure, which was originally developed for autoregressive representations of known order, remains valid asymptotically for a general ARIMA $(p, 1, q)$ process in which p and q are of unknown orders, that is for an autoregressive integrated moving average process of the indicated order. More specifically, Said and Dickey (1984) show that the Dickey–Fuller regression t test for a unit root may still be used in an ARIMA $(p, 1, q)$ model provided the lag length in the autoregression increases with the sample size, T, at a controlled rate less than $T^{1/3}$.

An alternative procedure for testing the presence of a unit root in a general time series setting has recently been proposed by Phillips

(1987a). This approach is nonparametric with respect to nuisance parameters and thereby allows for a very wide class of time series models in which there is a unit root. This includes ARIMA models with heterogeneously as well as identically distributed innovations. The method seems to have significant advantages when there are moving average components in the time series and, at least in this respect, offers a promising alternative to the Dickey–Fuller and Said–Dickey procedures.

The present paper extends the study of Phillips (1987a) to the cases where (a) a drift, and (b) a drift and a linear trend are included in the specification. These extensions are important for practical applications, where the presence of a nonzero drift is very common. Moreover, in many cases and, particularly, with economic time series, the main competing alternative to the presence of a unit root is a deterministic linear time trend. It is therefore important that regression tests for unit roots allow for this possibility.

The methods of the paper are asymptotic and rely on the theory of functional weak convergence. The limit distributions of the new test statistics developed here are expressed as functionals of standard Brownian motion and are the same as those tabulated by Fuller (1976). This means that our tests may be used with existing tabulations even though they allow for much more general time series specifications. The asymptotic local power properties of our tests are studied using the theory of near-integrated processes (Phillips, 1987b). Some simulation evidence on the finite sample performance of the new tests is also provided.

2. Preliminaries

The models we consider are driven by a sequence of innovations denoted by $\{u_t\}$. Throughout we assume that $\{u_t\}$ satisfies the following general conditions:

(i) $E(u_t) = 0$ for all t;
(ii) $\sup_t E|u_t|^{\beta+\varepsilon} < \infty$ for some $\beta > 2$ and $\varepsilon > 0$;
(iii) as $T \to \infty$, $\sigma^2 = \lim E(T^{-1}S_T^2)$ exists and $\sigma^2 > 0$, where $S_t = u_1 + \cdots + u_t$;

(iv) $\{u_t\}$ is strong mixing with mixing coefficients α_m that satisfy $\sum \alpha_m^{1-2/\beta} < \infty$, where the sum is over $m = 1, \ldots, \infty$.

The conditions allow many weakly dependent and heterogeneously distributed time series. They include a wide variety of possible data generating mechanisms such as finite order ARMA models under very general conditions on the underlying errors (Withers, 1981). Condition (ii) controls the allowable heterogeneity of the process, whereas (iv) controls the extent of permissible temporal dependence in relation to the probability of outlier occurrences; see Phillips (1987a) for more discussion of these conditions, and Hall and Heyde (1980, p. 132) for the definition of strong mixing and the mixing coefficients. If $\{u_t\}$ is weakly stationary with spectral density $f_u(\lambda)$ then condition (iii) is a consequence of (ii) and (iv). In this case we have

$$\sigma^2 = E(u_1^2) + 2\sum_{k=2}^{\infty} E(u_1 u_k) = 2\pi f_u(0).$$

From the sequence of partial sums $\{S_t\}$ we construct the random element

$$X_T(r) = T^{-\frac{1}{2}}\sigma^{-1}S_{[Tr]} = T^{\frac{1}{2}}\sigma^{-1}S_{j-1}$$

for $(j-1)/T \le r < j/T$ $(j = 1, \ldots, T)$, where $[Tr]$ denotes the integral part of Tr. Then $X_T(r)$ lies in $D = D[0,1]$, the space of real valued functions on the interval [0, 1] that are right continuous and have finite left limits. Under very general conditions the random element $X_T(r)$ obeys a central limit theory on the function space D. In particular, under (i)–(iv) above we have (Herrndorf, 1984) that, as $T \to \infty$,

$$X_T(r) \Rightarrow W(r). \tag{1}$$

The symbol '$\Rightarrow$' signifies weak convergence of the associated probability measures. In this case, the probability measure of $X_T(r)$ converges weakly to the probability measure of the standard Brownian motion $W(r)$; see Billingsley (1968, Section 16) and Pollard (1984, Chapter 5) for further discussion.

Using (1) it is simple to deduce the asymptotic behavior of the sample moments of the process $\{S_t\}$ and the innovations $\{u_t\}$. For a full development, see Phillips (1987a). We state below those results which are most useful subsequently. Unless otherwise indicated sums are over $t = 1, \ldots, T$. The limit distributions are expressed as functions of standard Brownian motion $W(r)$. To simplify formulae all integrals are understood to be taken over the interval [0, 1], integrals such as $\int W, \int W^2, \int rW$ are understood to be taken with respect to Lebesgue measure and we write $W_1 = W(1)$. Then, as $T \to \infty$,

$$
\begin{aligned}
&T^{-3/2} \sum S_t \Rightarrow \sigma \int W, \quad T^{-2} \sum S_t^2 \Rightarrow \sigma^2 \int W^2, \\
&T^{-3/2} \sum t u_t \Rightarrow \sigma \int r dW = \sigma(W_1 - \int W), \\
&T^{-5/2} \sum t S_t \Rightarrow \sigma \int rW, \quad T^{-1} \sum u_t S_{t-1} \Rightarrow \sigma^2 \int W dW + \lambda \\
&\qquad\qquad = \frac{1}{2}(\sigma^2 W_1^2 - \sigma_u^2),
\end{aligned}
$$

where

$$
\lambda = \frac{1}{2}(\sigma^2 - \sigma_u^2), \quad \sigma_u^2 = \lim_{T \to \infty} \frac{1}{T} \sum E(u_t^2).
$$

Joint weak convergence for the sample moments given above to their respective limits is also easily established and will be used below.

3. The Models and Estimators

Let $\{y_t\}$ be a time series generated by

$$
y_t = \alpha y_{t-1} + u_t \quad (t = 1, 2, \ldots), \tag{2}
$$

$$
\alpha = 1. \tag{3}
$$

Initial conditions for (2) are set at $t = 0$ and y_0 may be any random variable, including a constant, whose distribution is fixed and independent of the sample size T. The innovation sequence $\{u_t\}$ satisfies conditions (i)–(iv).

We consider the two least-squares regression equations

$$
y_t = \hat{\mu} + \hat{\alpha} y_{t-1} + \hat{u}_t, \tag{4}
$$

$$
y_t = \tilde{\mu} + \tilde{\beta}\left(t - \frac{1}{2}T\right) + \tilde{\alpha} y_{t-1} + \tilde{u}_t, \tag{5}
$$

where $(\hat{\mu}, \hat{\alpha})$ and $(\tilde{\mu}, \tilde{\beta}, \tilde{\alpha})$ are the conventional least-squares regression coefficients. We use X to denote the $T \times 3$ matrix of explanatory variables in (5). We define also the following regression t statistics:

$$t_{\hat{\alpha}} = (\hat{\alpha} - \alpha)\left\{\sum (y_{t-1} - \bar{y}_{-1})^2\right\}^{\frac{1}{2}} / \hat{s},$$

$$t_{\hat{\mu}} = (\hat{\mu} - \mu)\left\{\sum (y_{t-1} - \bar{y}_{-1})^2 / \sum y_{t-1}^2\right\}^{\frac{1}{2}} / \hat{s},$$

$$t_{\tilde{\mu}} = (\tilde{\mu} - \mu)/(\tilde{s}^2 c_1)^{\frac{1}{2}}, \quad t_{\tilde{\beta}} = (\tilde{\beta} - \beta)/(\tilde{s}^2 c_2)^{\frac{1}{2}},$$

$$t_{\tilde{\alpha}} = (\tilde{\alpha} - \alpha)/(\tilde{s}^2 c_3)^{\frac{1}{2}},$$

where $\hat{s}$ and $\tilde{s}$ are the standard errors of regressions (4) and (5), c_i is the ith diagonal element of the matrix $(X'X)^{-1}$, and $\bar{y}_{-1} = T^{-1}\sum y_{t-1}$.

Following Dickey and Fuller (1979), we shall be concerned with the limiting distributions of the regression coefficients in (4) and (5) and their t statistics under the hypothesis that the data are generated by (2) and (3). Thus the null values of the coefficients in the above tests become $\alpha = 1$, $\mu = \beta = 0$. However, the coefficient $\tilde{\alpha}$ and its t statistic in (5) are invariant with respect to the introduction of a nonzero drift $\mu \neq 0$ in the generating process. Thus we may replace (2) with the generating mechanism

$$y_t = \mu + \alpha y_{t-1} + u_t \quad (t = 1, 2, \ldots), \tag{2'}$$

and the distributions and asymptotic distributions of the above-mentioned statistics are unchanged. Thus there is no loss in generality by assuming $\mu = 0$.

Under (2) and (3), $y_t = S_t + y_0$ and the asymptotic behavior of sample moments of y_t and u_t follows directly from that of S_t and u_t given earlier. Thus, as $T \to \infty$,

$$T^{-3/2}\sum y_t \Rightarrow \sigma \int W, \quad T^{-2}\sum y_t^2 \Rightarrow \sigma^2 \int W^2,$$

$$T^{-5/2}\sum t y_{t-1} \Rightarrow \sigma \int rW, \quad T^{-1}\sum y_{t-1}u_t \Rightarrow \sigma^2 \int W dW + \lambda.$$

Again joint weak convergence to the stated limits applies.

4. Limiting Distributions of the Statistics

In this section we characterize the limiting distributions of the standardized coefficient estimators $T(\hat{\alpha}-1), T(\tilde{\alpha}-1)$ and the t statistics $t_{\hat{\alpha}}$ and $t_{\tilde{\alpha}}$ $(\alpha=1), t_{\hat{\mu}}$ and $t_{\tilde{\mu}}$ $(\mu=0)$ and $t_{\tilde{\beta}}$ $(\beta=0)$ under the maintained hypothesis that the time series $\{y_t\}$ is generated by (2) and (3).

Theorem 1. *For the regression model* (4), *as* $T \to \infty$,

(a) $T(\hat{\alpha}-1) \Rightarrow (\int W_*^2)^{-1}(\int W_* dW + \lambda')$,
(b) $t_{\hat{\alpha}} \Rightarrow (\sigma/\sigma_u)(\int W_*^2)^{-\frac{1}{2}}(\int W_* dW + \lambda')$,
(c) $t_{\hat{\mu}} \Rightarrow (\sigma/\sigma_u)(\int W^2 \int W_*^2)^{-\frac{1}{2}}\{W_1 \int W^2 - (\int W dW + \lambda')\int W\}$.

For the regression model (5), *as* $T \to \infty$,

(d) $T(\tilde{\alpha}-1) \Rightarrow (\int W dW + \lambda' + A_1)/D$,
(e) $t_{\tilde{\alpha}} \Rightarrow (\sigma/\sigma_u)(\int W dW + \lambda' + A_1)/D^{\frac{1}{2}}$,
(f) $t_{\tilde{\mu}} \Rightarrow (\sigma/\sigma_u)\{A_2 - (\int W dW + \lambda')\int W\}/D^{\frac{1}{2}} A_3$,
(g) $t_{\tilde{\beta}} \Rightarrow (\sigma/\sigma_u)\{A_4 + (\int W dW + \lambda')(\frac{1}{2}\int W - \int rW)\}/(D/12)^{\frac{1}{2}}$ $(\int W_*^2)^{\frac{1}{2}}$,

where

$$D = \int W^2 - 12\left(\int rW\right)^2 + 12\int W \int rW - 4\left(\int W\right)^2,$$

$$A_1 = 12\left(\int rW - \frac{1}{2}\int W\right)\left(\int W - \frac{1}{2}W_1\right) - W_1\int W,$$

$$A_2 = W_1\left\{\int W^2 - 12\left(\int rW\right)^2 + 18\int W\int rW - 6\left(\int W\right)^2\right\}$$
$$+6\int W\left\{\left(\int W\right)^2 - 2\int rW\int W\right\},$$

$$A_3 = \left\{D + \left(\int W\right)^2\right\}^{\frac{1}{2}},$$

$$A_4 = W_1 \left\{ \frac{1}{2} \int W^2 + \int W \int rW - \left(\int W \right)^2 \right\} - \int W \int W_*^2,$$

$$W_*(r) = W(r) - \int W, \lambda' = \lambda/\sigma^2.$$

Note that $W_*(r)$ in the above formulae can be called a demeaned Brownian motion. The stated results follow in a direct way from the asymptotic behavior of the sample moments given earlier. In some cases the calculations are lengthy and full details will be supplied on request.

When the innovation sequence $\{u_t\}$ is independent and identically distributed, we have $\sigma_u^2 = \sigma^2$ and $\lambda = \lambda' = 0$. In this case, the limiting distributions of the statistics given in Theorem 1 are independent of nuisance parameters, as is readily seen by inspection; and percentage points of the asymptotic distributions have been calculated by Monte Carlo methods by Dickey and Fuller. Specifically, critical values of $T(\hat{\alpha}-1), T(\tilde{\alpha}-1), t_{\hat{\alpha}}$ and $t_{\tilde{\alpha}}$ are tabulated by Fuller (1976, Tables 8.5.1 and 8.5.2); and tabulated critical values of $t_{\hat{\mu}}, t_{\tilde{\mu}}$ and $t_{\tilde{\beta}}$ are also given by Dickey and Fuller (1981, Tables I–III).

Theorem 1 extends the results of Dickey and Fuller to the general case of weakly dependent and heterogeneously distributed data. Interestingly, our result shows that the limiting distributions of these statistics have the same general form for a very wide class of innovation processes $\{u_t\}$. This feature enables us to derive transformations of the statistics that opens the way to hypothesis testing in the general case. This is the approach that we pursue in Section 5.

We remark that independent and identically distributed innovations $\{u_t\}$ are not necessary for the equivalence $\sigma^2 = \sigma_u^2$. The equivalence also holds for innovations that are martingale differences under mild additional moment conditions. Thus, unmodified versions of the Dickey–Fuller tests are valid asymptotically in the presence of some heterogeneity in the innovation sequence provided the innovations are martingale differences and (1) holds. However, when the innovations are nonorthogonal and $\sigma^2 \neq \sigma_u^2$, the Dickey–Fuller tests do not have the correct asymptotic size.

5. Statistical Inference in the Presence of a Unit Root

The limiting distributions of the regression coefficients and associated t statistics given in Section 4 all depend upon the nuisance parameters σ^2 and σ_u^2. This presents an obstacle to conventional procedures of inference in the general case where $\sigma^2 \neq \sigma_u^2$. However, since σ^2 and σ_u^2 may be consistently estimated there exist simple transformations of the test statistics which eliminate the nuisance parameters asymptotically. This idea was first developed by Phillips (1987a) in the context of tests for a unit root. Here we show how the procedure may be extended to apply quite generally to statistical tests in regressions with a fitted drift and time trend.

Consistent estimates of σ_u^2 are provided by $\hat{s}^2, \tilde{s}^2$ and $s^2 = T^{-1}\sum(y_t - y_{t-1})^2$ for data generated by (2) and (3). When there is a nonzero drift in the model, as in (2′), both $\hat{s}^2$ and $\tilde{s}^2$ are consistent. Since we often wish to allow for a nonzero drift in regressions such as (5) we use $\tilde{s}^2$ as our preferred estimator of σ_u^2 in this regression.

Consistent estimation of σ^2 is discussed by Phillips (1987a). When $\{u_t\}$ is weakly stationary with spectral density $f_u(\lambda)$ we have $\sigma^2 = 2\pi f_u(0)$. In this case, estimation of σ^2 is equivalent to estimating the spectral density of $\{u_t\}$ at the origin. Many consistent estimates are available. Consider, for example, the simple estimate based on truncated sample autocovariances, namely

$$s_{Tl}^2 = T^{-1}\sum_{t=1}^{T} u_t^2 + 2T^{-1}\sum_{s=1}^{l}\sum_{t=s+1}^{T} u_t u_{t-s}, \tag{6}$$

where $u_t = y_t - y_{t-1}$. Conditions for the consistency of s_{Tl}^2 are explored by Phillips (1987a). It is shown there that $s_{Tl}^2 \to \sigma^2$ in probability as $T \to \infty$ provided the moment condition

(ii′) $\sup_t E|u_t|^{2\beta} < \infty$ for some $\beta > 2$

holds in place of (ii) of Section 2, and provided

(v) $l \to \infty$ as $T \to \infty$ and $l^4/T \to 0$.

According to this, if we allow the number of estimated autocovariances in (6) to increase as $T \to \infty$ but control the rate of increase so that $l^4/T \to 0$, then s^2_{Tl} yields a consistent estimator of σ^2. Of course, faster rates of increase in l are allowable if we make stronger assumptions on $\{u_t\}$ as, for example, in the case of spectral estimation for weakly stationary processes. In what follows we shall assume that conditions (ii′) and (v) hold, in addition to (i), (iii) and (iv) given earlier.

Rather than using first differences $u_t = y_t - y_{t-1}$ in the construction of s^2_{Tl} we could use the residuals from the regression equations (4) and (5). Since the coefficients in these regressions are consistent, it is easy to show that these modifications to s^2_{Tl}, which we denote by $\hat{s}^2_{Tl}$ and $\tilde{s}^2_{Tl}$ respectively, are also consistent estimates of σ^2 under the same conditions. Once again $\tilde{s}^2_{Tl}$ will be the preferred estimator when we wish to allow for a nonzero drift as in (2′).

Note that (6) is not constrained to be nonnegative as it is presently defined; it can be negative when there are large negative sample serial covariances. Simple modifications to (6) overcome this difficulty. For example, the weighted variance estimators

$$\hat{\sigma}^2_{Tl} = T^{-1}\sum_{t=1}^{T} \hat{u}_t^2 + 2T^{-1}\sum_{s=1}^{l} W_{sl} \sum_{t=s+1}^{T} \hat{u}_t\hat{u}_{t-s}, \tag{7}$$

$$\tilde{\sigma}^2_{Tl} = T^{-1}\sum_{t=1}^{T} \tilde{u}_t^2 + 2T^{-1}\sum_{s=1}^{l} W_{sl} \sum_{t=s+1}^{T} \tilde{u}_t\tilde{u}_{t-s}, \tag{8}$$

where

$$w_{sl} = 1 - s/(l+1), \tag{9}$$

are non-negative and, for stationary $\{u_t\}$, are simply 2π times the corresponding Bartlett estimates of $f_u(0)$. Other choices of lag window besides the triangular window (9) are possible. We use the Parzen window in the simulations reported in Section 7. The estimator (7) was recently suggested in the context of variance estimates by Newey and West (1987).

We now define some simple transformations of conventional test statistics from the regressions (4) and (5) which eliminate the nuisance parameter dependencies asymptotically. Specifically, we define

$$
\begin{aligned}
Z(\hat{\alpha}) &= T(\hat{\alpha}-1) - \hat{\lambda}/\bar{m}_{yy},\\
Z(t_{\hat{\alpha}}) &= (\hat{s}/\hat{\sigma}_{Tl})t_{\hat{\alpha}} - \hat{\lambda}'\hat{\sigma}_{Tl}/\bar{m}_{yy}^{\frac{1}{2}},\\
Z(t_{\hat{\mu}}) &= (\hat{s}/\hat{\sigma}_{Tl})t_{\hat{\mu}} + \hat{\lambda}'\hat{\sigma}_{Tl}m_y/\bar{m}_{yy}^{\frac{1}{2}}m_{yy}^{\frac{1}{2}},\\
Z(\tilde{\alpha}) &= T(\tilde{\alpha}-1) - \tilde{\lambda}/M,\\
Z(t_{\tilde{\alpha}}) &= (\tilde{s}/\tilde{\sigma}_{Tl})t_{\tilde{\alpha}} - \tilde{\lambda}'\tilde{\sigma}_{Tl}/M^{\frac{1}{2}},\\
Z(t_{\tilde{\mu}}) &= (\tilde{s}/\tilde{\sigma}_{Tl})t_{\tilde{\mu}} - \tilde{\lambda}'\tilde{\sigma}_{Tl}m_y/M^{\frac{1}{2}}(M+m_y^2)^{\frac{1}{2}},\\
Z(t_{\tilde{\beta}}) &= (\tilde{s}/\tilde{\sigma}_{Tl})t_{\tilde{\beta}} - \tilde{\lambda}'\tilde{\sigma}_{Tl}\left(\frac{1}{2}m_y - m_{ty}\right)/(M/12)^{\frac{1}{2}}\bar{m}_{yy}^{\frac{1}{2}},
\end{aligned}
$$

where

$$
\begin{aligned}
m_{yy} &= T^{-2}\sum y_t^2, \quad \bar{m}_{yy} = T^{-2}\sum (y_t-\bar{y})^2,\\
m_y &= T^{-3/2}\sum y_t, \quad m_{ty} = T^{-5/2}\sum t y_t,\\
M &= (1-T^{-2})m_{yy} - 12m_{ty}^2 + 12(1+T^{-1})m_{ty}m_y\\
&\quad - (4+6T^{-1}+2T^{-2})m_y^2,\\
\hat{\lambda} &= \frac{1}{2}(\hat{\sigma}_{Tl}^2 - \hat{s}^2), \quad \hat{\lambda}' = \hat{\lambda}/\hat{\sigma}_{Tl}^2, \quad \tilde{\lambda} = \frac{1}{2}(\tilde{\sigma}_{Tl}^2 - \tilde{s}^2), \quad \tilde{\lambda}' = \tilde{\lambda}/\tilde{\sigma}_{Tl}^2.
\end{aligned}
$$

These Z-statistics extend those of Phillips (1987a) for the case of an autoregression with no fitted constant or time trend. The idea behind their construction is to correct the conventional regression statistics so that they allow for the effects of serially correlated and heterogeneously distributed innovations. Thus, the standard errors of regression $\hat{s}$ and $\tilde{s}$ which measure scale effects in the conventional t ratios are now replaced by the general standard error estimates $\hat{\sigma}_{Tl}$ and $\tilde{\sigma}_{Tl}$ which allow for serial covariation as well as variance. Each Z-statistic also involves an additive correction term whose magnitude depends on the difference between the corresponding variance estimates $\hat{\sigma}^2_{Tl} - \hat{s}^2$ or $\tilde{\sigma}^2_{Tl} - \tilde{s}^2$. Once again these differences capture the effects of serial correlation and the transformations are designed to remove these effects asymptotically. The limiting distributions of the Z-statistics are given in our next main result, which follows simply from Theorem 1.

Theorem 2. (a) *For the regression model* (4) *the statistics* $Z(\hat{\alpha})$, $Z(t_{\hat{\alpha}})$, *and* $Z(t_{\hat{\mu}})$ *have limit distributions given by those of* $T(\hat{\alpha}-1), t_{\hat{\alpha}}$ *and* $t_{\hat{\mu}}$, *respectively in Theorem* 1 *with* $\sigma^2 = \sigma_u^2$.

(b) *For the regression model* (5) *the statistics* $Z(\tilde{\alpha}), Z(t_{\tilde{\alpha}})Z(t_{\tilde{\mu}})$ *and* $Z(t_{\tilde{\beta}})$ *have limit distributions given by those of* $T(\tilde{\alpha}-1), t_{\tilde{\alpha}}$, $t_{\tilde{\mu}}$ *and* $t_{\tilde{\beta}}$, *respectively, in Theorem* 1 *with* $\sigma^2 = \sigma_u^2$. *The stated results for* $Z(\tilde{\alpha})$ *and* $Z(t_{\tilde{\alpha}})$ *remain valid if the generating mechanism of* $\{y_t\}$ *is* (2′) *rather than* (2).

Theorem 2 shows that the limiting distributions of the Z-statistics are invariant within a wide class of weakly dependent and possibly heterogeneously distributed innovations $\{u_t\}$. Furthermore, the limiting distributions of the Z-statistics are identical to the limiting distributions of the original untransformed statistics considered in Section 4, when $\sigma_u^2 = \sigma^2$. Thus, the critical values derived in the studies of Dickey and Fuller under the assumption of independent and identically distributed errors $\{u_t\}$ may be used with the new tests proposed here, which are valid under much more general conditions.

6. Power Functions for Unit Root Tests

We may develop asymptotic power functions for unit root tests by considering the sequence of local alternatives to (3) given by

$$\alpha = e^{c/T} \sim 1 + c/T. \tag{10}$$

When $c = 0$, (10) reduces to the null hypothesis (3); $c > 0$ gives local explosive alternatives to (3); and $c < 0$ corresponds to local stationary alternatives. The idea of developing a noncentral asymptotic distribution theory using the specification (10) was explored by Phillips (1987b). Time series generated by models such as (2) or (2′) with a coefficient α of the form (10) were called near-integrated in that paper. The asymptotic theory developed there showed that the sample moments of a near-integrated time series converge weakly to corresponding functionals of a diffusion process rather than standard Brownian motion. Specifically, we have, as $T \to \infty$,

$$T^{-3/2}\sum y_t \Rightarrow \sigma \int J_c, \quad T^{-2}\sum y_t^2 \Rightarrow \sigma^2 \int J_c^2,$$

$$T^{-5/2}\sum t y_t \Rightarrow \sigma \int r J_c, \quad T^{-1}\sum y_{t-1}u_t \Rightarrow \sigma^2 \int J_c dW + \lambda,$$

where

$$J_c(r) = \int_0^r e^{(r-s)c} dW(s)$$

is the Ornstein–Uhlenbeck process generated in continuous time by the stochastic differential equation $dJ_c(r) = cJ_c(r)dr + dW(r)$, with initial condition $J_c(0) = 0$. As before, joint weak convergence of these sample moments to their respective limits also applies. Note that $\int J_c dW$ is interpreted as a stochastic integral in the above formulae.

Using these results for sample moments we may now develop an asymptotic theory for the regression coefficients and t-statistics in (4) and (5). Moreover, it is a simple matter to find the noncentral asymptotic distributions of the new unit root test statistics developed in Section 6. The main results of interest are contained in the following theorem which concentrates on estimates of the autoregressive coefficient α and its associated t-ratio. The derivations follow those of Phillips (1987b) and will be supplied on request.

Theorem 3. *If* $\{y_t\}$ *is a near-integrated time series generated by* (2) *and* (10), *then as* $T \to \infty$:

(a) $Z(\hat{\alpha}) \Rightarrow c + (\int J_c^{*2})^{-1} \int J_c^* dW$,
(b) $Z(t_{\hat{\alpha}}) \Rightarrow c(\int J_c^{*2})^{\frac{1}{2}} + (\int J_c^{*2})^{-\frac{1}{2}} \int J_c^* dW$,
(c) $Z(\tilde{\alpha}) \Rightarrow c + (\int J_c dW + A_{1c})/D_c$,
(d) $Z(t_{\tilde{\alpha}}) \Rightarrow cD_c^{\frac{1}{2}} + (\int J_c dW + A_{1c})/D_c^{\frac{1}{2}}$,

where

$$D_c = \int J_c^2 - 12\left(\int rJ_c\right)^2 + 12\int rJ_c \int J_c - 4\left(\int J_c\right)^2,$$

$$A_{1c} = 6W_1 \int rJ_c - 4W_1 \int J_c - 12\int rdW \int rJ_c + 6\int rdW \int J_c,$$

$$J_c^*(r) = J_c(r) - \int J_c.$$

Results (c) *and* (d) *remain valid if the generating mechanism of* $\{y_t\}$ *is* (2′) *rather than* (2).

Theorem 3 gives the noncentral limiting distributions of the Z-statistics for testing $\alpha = 1$ under the sequence of local alternatives (10) to the unit root hypothesis (3). It therefore delivers asymptotic local power functions for the new unit root tests.

It is interesting to compare these asymptotic local power functions with those of the conventional Dickey–Fuller tests. The latter are based on the statistics $T(\hat{\alpha} - 1)$ and $t_{\hat{\alpha}}$ for regression (4) and $T(\tilde{\alpha} - 1)$ and $t_{\tilde{\alpha}}$ for regression (5). When the innovation sequence $\{u_t\}$ is independent and identically distributed these statistics have identical limiting distributions under the local alternative hypothesis (10) as the Z-statistics given above. We deduce that the new tests based on $Z(\hat{\alpha}), Z(t_{\hat{\alpha}})$ $Z(\tilde{\alpha})$ and $Z(t_{\tilde{\alpha}})$ have the same asymptotic local power properties for a wide class of possible time series innovations $\{u_t\}$ as the regression-based tests $(T(\hat{\alpha}-1), t_{\hat{\alpha}})$ and $(T(\tilde{\alpha}-1), t_{\tilde{\alpha}})$ do in the case of independent and identically distributed errors. Thus, there is no loss in asymptotic power in the use of the new tests over the Dickey–Fuller procedure in spite of the fact that they allow for a more general class of error processes.

7. Experimental Evidence

Simulations were run to assess the adequacy of the new tests and to evaluate their performance in comparison with the procedure suggested by Said and Dickey (1984). As explained earlier, Said and Dickey recommend the use of the Dickey–Fuller regression t test for a unit root in the autoregression

$$\Delta y_t = \hat{\mu} + \hat{\alpha}_* y_{t-1} + \sum_{i=1}^{l} \hat{\phi}_i \Delta y_{t-i} + \hat{v}_t. \tag{11}$$

We denote this test statistic $t(\hat{\alpha}_*)$. Said and Dickey show that when the lag length $l \to \infty$ in (11) as $T \to \infty$ then $t(\hat{\alpha}_*)$ has the same limit distribution as the conventional Dickey–Fuller t test. This corresponds with our statistic $Z(t_{\hat{\alpha}})$ given above. Note that Said and Dickey do not suggest a statistic based on the coefficient $\hat{\alpha}_*$ in (11), since the limit distribution of $T\hat{\alpha}_*$ depends on nuisance parameters. Thus, there is no analogue of our $Z(\hat{\alpha})$ test in the paper by Said and Dickey (1984).

Data were generated by the model (2) with moving average errors

$$u_t = e_t + \theta e_{t-1} \tag{12}$$

and the e_t independent and identically distributed $N(0,1)$. We set $y_0 = 0$, and used various lag lengths l in (11) and lag truncations l in (7) to evaluate the effects of these choices on test performance. A Parzen window and fitted residuals $\hat{u}_t$ from (4) were used in the construction of the variance estimate $\hat{\sigma}^2_{Tl}$. The simulations reported in Table 1 are based on 2000 replications and give results for one-sided tests under the null hypothesis $\alpha = 1$ and for the alternative $\alpha = 0.85$.

The results show size and power computations for six different values of θ in (12). When $\theta = 0$ there is no need to employ the transformations leading to $Z(\hat{\alpha})$ and $Z(t_{\hat{\alpha}})$ or the long autoregression (11). However, we gather from Table 1 that there is little loss in accuracy with respect to the size of the $Z(\hat{\alpha})$ and $Z(t_{\hat{\alpha}})$ tests. In fact, the $Z(\hat{\alpha})$ test is conservative and at the same time has greater power than either $Z(t_{\hat{\alpha}})$ or $t(\hat{\alpha}_*)$ for all choices of l. The $Z(t_{\hat{\alpha}})$ and $t(\hat{\alpha}_*)$ tests are both liberal in terms of size at $T = 100$ and the size distortions of $t(\hat{\alpha}_*)$ increase appreciably with the length of the autoregression. At the same time, the power of the $t(\hat{\alpha}_*)$ tests decreases as l increases.

Table 1. Simulations based on 2000 replications; $T = 100$, nominal size 5%.

θ	Size					Power $(\alpha = 0 \cdot 85)$				
	$l = 2$	$l = 4$	$l = 6$	$l = 8$	$l = 12$	$l = 2$	$l = 4$	$l = 6$	$l = 8$	$l = 12$
	(a) Said–Dickey $t(\hat{\alpha}_*)$ test									
0.0	0.068	0.064	0.078	0.086	0.106	0.557	0.472	0.406	0.354	0.303
0.5	0.052	0.064	0.071	0.085	0.102	0.378	0.411	0.377	0.348	0.308
0.8	0.037	0.051	0.072	0.082	0.109	0.267	0.302	0.313	0.308	0.273
−0.2	0.063	0.065	0.084	0.092	0.111	0.591	0.490	0.421	0.358	0.304
−0.5	0.116	0.076	0.073	0.086	0.105	0.868	0.626	0.512	0.434	0.350
−0.8	0.677	0.343	0.201	0.142	0.120	1.00	0.988	0.900	0.772	0.523
	(*b*)$Z(\hat{\alpha})$ test									
0.0	0.044	0.049	0.049	0.049	0.055	0.772	0.776	0.784	0.794	0.805
0.5	0.010	0.021	0.022	0.021	0.014	0.301	0.524	0.534	0.500	0.378
0.8	0.003	0.020	0.024	0.022	0.015	0.247	0.513	0.526	0.474	0.330
−0.2	0.134	0.105	0.110	0.115	0.133	0.960	0.940	0.946	0.952	0.967
−0.5	0.537	0.417	0.428	0.451	0.516	1.00	1.00	1.00	1.00	1.00
−0.8	0.997	0.988	0.988	0.991	0.995	1.00	1.00	1.00	1.00	1.00
	(*c*)$Z(t_{\hat{\alpha}})$ test									
0.0	0.063	0.062	0.066	0.069	0.069	0.669	0.688	0.696	0.705	0.705
0.5	0.028	0.036	0.038	0.035	0.030	0.191	0.367	0.371	0.333	0.214
0.8	0.026	0.034	0.035	0.031	0.026	0.144	0.353	0.357	0.305	0.167
−0.2	0.135	0.115	0.119	0.127	0.140	0.933	0.910	0.918	0.930	0.945
−0.5	0.534	0.438	0.445	0.460	0.518	1.00	1.00	1.00	1.00	1.00
−0.8	0.995	0.988	0.990	0.993	0.995	1.00	1.00	1.00	1.00	1.00

Thus, the cost of using $t(\hat{\alpha}_*)$ when $\theta = 0$ is appreciably greater than that of using $Z(t_{\hat{\alpha}})$ and $Z(\hat{\alpha})$ is the preferred test.

When $\theta > 0$ similar results apply. Since $T = 100$ moderate choices of l around $l = 8$ seem appropriate. We observe that the $Z(\hat{\alpha})$ test is again very conservative and has higher power than the $Z(t_{\hat{\alpha}})$ and $t(\hat{\alpha}_*)$ tests for all choices of $l \geq 4$. The $Z(t_{\hat{\alpha}})$ test is also conservative and has similar power to $t(\hat{\alpha}_*)$ for $l = 4, 6, 8$. The $t(\hat{\alpha}_*)$ test is liberal for $l \geq 4$ and the size distortions increase with the length of the autoregression. These results again suggest that $Z(\hat{\alpha})$ is the preferred test.

When $\theta < 0$ the results are very different. Both $Z(\hat{\alpha})$ and $Z(t_{\hat{\alpha}})$ have significant size distortions and are too liberal to be useful for $\theta = -0.5, -0.8$. The $t(\hat{\alpha}_*)$ test also suffers size distortions but these are attenuated as the lag length in the autoregression (11) increases. However, as l increases the $t(\hat{\alpha}_*)$ test also suffers appreciable loss in power. None of the tests therefore has accurate size and good power properties when $\theta < 0$. But, of the three tests, $t(\hat{\alpha}_*)$ seems preferable in this case.

Asymptotic expansions recently obtained by Phillips (1987c) may be used to shed light on this simulation evidence. In particular, formula (34) of that paper may be extended to apply to the regression (4) with fitted intercept giving

$$T(\hat{\alpha}-1) \equiv \left(\int W_*^2\right)^{-1} \left\{\int W_* dW - k_0 - (2T)^{-\frac{1}{2}} k_1 \eta\right\} + O_p(T^{-1}),$$

where $\equiv$ signifies equivalence in distribution,

$$k_0 = \frac{1}{2}\sigma_u^2/\sigma^2 = \frac{1}{2}(1+\theta^2)/(1+\theta)^2, \quad k_1 = (1+4\theta^2+\theta^4)^{\frac{1}{2}}/(1+\theta)^2,$$

and η is standard $N(0,1)$ and independent of $W(r)$. The correction term on the asymptotic depends on k_1. For $\theta \geq 0$, we have $0 < k_1 \leq 1$. When $\theta < 0$, k_1 is unbounded and rapidly becomes large as $\theta \to -1$. These results show clearly how the quality of the asymptotic approximation given in Theorem 1(a) depends on the size and the magnitude of θ. When $\theta > 0$ we have $k_1 < 1$ and the asymptotic distribution may be expected to deliver at least as good an approximation to the finite sample distribution of $T(\hat{\alpha}-1)$ as it does for the case where $\theta = 0$ and $k_1 = 1$. When $\theta < 0$ this is not the case and the asymptotic distribution may be expected to be poor, particularly as $\theta \to -1$. Similar behavior may be inferred for the test statistics $Z(\hat{\alpha})$ and $Z(t_{\hat{\alpha}})$, which are both based on $T_{(\hat{\alpha}-1)}$. This helps to explain the size distortions in these tests that were evident in the simulations when $\theta < 0$.

8. Concluding Comments

The present approach gives a simple test for a unit root in univariate time series against stationary and trend alternatives. One needs only to estimate a first-order autoregression with a constant and possibly a time trend and to calculate the appropriate transformed Z-statistic.

The distribution theory underlying this procedure is asymptotic and critical values already provided by Fuller (1976) may be used.

As we have seen in Section 6, there is no loss in asymptotic local power in the use of the $Z(\hat{\alpha})$ tests for a unit root. But the simulations reported in Section 7 indicate that test performance can differ substantially in finite samples among asymptotically equivalent tests. For models with positive moving average errors the $Z(\hat{\alpha})$ test is conservative and has better power properties than the other tests. For models with independent and identically distributed errors where the transformations that lead to the Z tests are not strictly needed, $Z(\hat{\alpha})$ again seems to be the preferred test. For models with moving average errors and negative serial correlation the Z tests suffer appreciable size distortions and are not recommended. In such cases the Said–Dickey procedure of using a long autoregression seems preferable.

Acknowledgments

The authors thank the referees, Joon Park and Gene Savin for helpful comments on an earlier draft, the National Science Foundation for research support at Yale University, and the Université de Montréal for computing facilities.

References

Billingsley, P. (1968). *Convergence of Probability Measures*, Wiley, New York.

Dickey, D.A., Bell, W.R. and Miller, R.B. (1986). Unit roots series model: Tests and implications. *American Statistician* 40, 12–26.

Dickey, D.A. and Fuller, W.A. (1979). Distribution of the estimators for autoregressive time series with a unit root. *Journal of American Statistical Association* 74, 427–431.

Dickey, D.A. and Fuller, W.A. (1981). Likelihood ratio statistics for autoregressive time series with a unit root. *Econometrica* 49, 1057–1072.

Fuller, W.A. (1976). *Introduction to Statistical Time Series*, Wiley, New York.

Fuller, W.A. (1984). Nonstationary autoregressive time series. In *Handbook of Statistics* 5, *Time Series in the Time Domain*, E.J. Hannan, P.R. Krishnaiah and M.M. Rao (eds.), Ch. 1, North Holland, Amsterdam, pp. 1–23.

Hall, R.E. (1978). Stochastic implications of the life cycle-permanent income hypothesis: Theory and evidence. *Journal of Political Economy* 86, 971–987.

Hall, P. and Heyde, C.C. (1980). *Martingale Limit Theory and its Applications*, Academic Press, New York.

Herrndorf, N. (1984). A functional central limit theorem for weakly dependent sequences of random variables. *Annals of Probability* 12, 141–153.

Kleidon, A.W. (1986). Variance bounds tests and stock price valuation models. *Journal of Political Economy* 94, 953–1001.

Meese, R.A. and Singleton, K.J. (1983). On unit roots and the empirical modeling of exchange rates. *International Economical Review* 24, 1029–1035.

Nelson, C.R. and Plosser, C.I. (1982). Trends and random walks in macroeconomic time series: Some evidence and implications. *Journal of Monetary Economics* 10, 139–162.

Newey, W.K. and West, K.D. (1987). A simple, positive definite, heteroskedasticity and autocorrelation consistent covariance matrix. *Econometrica* 55, 703–708.

Phillips, P.C.B. (1987a). Time series regression with a unit root. *Econometrica* 55, 277–301.

Phillips, P.C.B. (1987b). Towards a unified asymptotic theory of autoregression. *Biometrika* 74, 535–548.

Phillips, P.C.B. (1987c). Asymptotic expansions in nonstationary vector autoregressions. *Econometric Theory* 2, 45–68.

Pollard, D. (1984). *Convergence of Stochastic Processes,* Springer-Verlag, New York.

Said, S.E. and Dickey, D.A. (1984). Testing for unit roots in autoregressive-moving average of unknown order. *Biometrika* 71, 599–607.

Samuelson, P.A. (1965). Proof that properly anticipated prices fluctuate randomly. *Industrial Management Review* 6, 41–50.

Samuelson, P.A. (1973). Proof that properly discounted present values of assets vibrate randomly. *Bell Journal of Economics Management Science* 4, 369–374.

Withers, C.S. (1981). Conditions for linear processes to be strong mixing. *Z. Wahr. verw. Get.* 57, 477–480.

Chapter 3

The Calculation of the Limiting Distribution of the Least-Squares Estimator in a Near-Integrated Model*

Pierre Perron

Department of Economics, Princeton University, Princeton, NJ 08544, USA
Centre de Recherche et Développement en Economique, Université de Montréal, C.P. 6128, Succ. A, Montréal, Canada H3C-3J7

We tabulate the limiting cumulative distribution and probability density functions of the least-squares estimator in a first-order autoregressive regression when the true model is near-integrated in the sense of Phillips [19]. The results are obtained using an exact numerical method which integrates the appropriate limiting moment generating function. The adequacy of the approximation is examined for various first-order autoregressive processes with a root close to unity.

Keywords: Density distribution, unit root, moment-generating function, autoregressive process, numerical integration.

1. Introduction

Consider the first-order autoregressive model:

$$y_t = \alpha y_{t-1} + u_t, \quad t = 1, 2, \ldots, T \tag{1}$$

where y_0 is fixed and $\{u_t\}$ is a sequence of independently and identically distributed random variables with mean zero and finite variance σ^2. The properties of the least-squares estimator $\hat{\alpha}$ have

*Originally Published in *Econometric Theory* 5, pp. 241–255.

been extensively studied. Mann and Wald (1978) and Rubin (1950) showed that $T^{1/2}(\hat{\alpha} - \alpha)$ has a limiting normal distribution when $|\alpha| < 1$. When $|\alpha| > 1$, White (1958) showed that the limiting distribution of $|\alpha|^T(\alpha^2 - 1)^{-1}(\hat{\alpha} - \alpha)$ is Cauchy and that for $|\alpha| = 1$ the limiting distribution of $T(\hat{\alpha} - \alpha)$ can be expressed in terms of the ratio of two functionals of a Wiener process. In the latter case, Rao (1978) obtained a computable expression.

The case of the unit root $\alpha = 1$ has attracted a great deal of attention. The critical values of the asymptotic distribution of $T(\hat{\alpha} - 1)$ have been derived by Dickey (1976) by simulation methods. Evans and Savin (1981) obtained tabulated values of both the cumulative distribution function (c.d.f.) and the probability density function (p.d.f.) of the limiting distribution of $(T/\sqrt{2})(\hat{\alpha} - 1)$ by numerically integrating the limiting moment generating function derived by White (1958). Although Evans and Savin and White imposed the normality assumption in their derivation, their result holds under more general conditions. Furthermore, a variety of procedures have also been suggested which permit a considerable weakening of the conditions imposed on the innovation sequence $\{u_t\}$ while at the same time using the same critical values derived in the earlier studies; see, for example, Dickey and Fuller (1979, 1981), Said and Dickey (1984) and Phillips (1987).

The major point which emerges from these studies is that the asymptotic distributions have a discontinuity at one. Because the exact distribution of $\hat{\alpha}$ is continuous for all values of α, this suggests that the limiting distribution inadequately approximates the finite-sample distribution for α near the discontinuity point $\alpha = 1$. This inadequacy has been well documented by Evans and Savin (1981, 1984) and by Phillips (1977) who also showed that Edgeworth expansions also perform poorly for α less than but close to one.

Recently, a new class of models which specifically deal with the presence of a root close to, but not necessarily equal to, one have been studied. Phillips (1988) introduced the concept of a near-integrated random process where the autoregressive parameter is defined by

$$\alpha = \exp(c/T). \tag{2}$$

Here c, a real-valued constant, is a measure of the deviation from the unit root case. The models (1) and (2) may also be described as having a root local to unity (see Cavanagh, 1986): as the sample size increases, the autoregressive parameter converges to unity. When c is negative, the process $\{y_t\}$ is said to be (locally) stationary and when c is positive, it is said to be (locally) explosive. An expression for the limiting distribution of $T(\hat{\alpha} - \alpha)$ under (2) has been derived by Phillips (1988), Cavanagh (1986), and Chan and Wei (1987). This class of models has been quite useful in studying various problems such as the power of tests of a unit root under local alternatives (Phillips, 1987a; Phillips and Perron, 1988; Perron 1989), the derivation of confidence intervals when α is near unity (Cavanagh, 1986), and the calculation of the power of tests of a unit root with a continuum of observations (Perron, 1987).

Phillips (1987b) suggested that percentage points of the limiting distribution of $T(\hat{\alpha} - \alpha)$ under (2) could be obtained by numerical integration of the limiting moment generating function obtained in Phillips (1987). This paper is aimed at this task. We evaluate both the limiting c.d.f. and p.d.f. of $T(\hat{\alpha} - \alpha)$ when α is of the form (2), for various values of c. These are obtained in Section 2 and use exact numerical methods integrating the limiting moment generating function. Relevant percentage points are tabulated. In Section 3, we also provide some assessment of the adequacy of the approximation to the finite-sample distribution.

Recent work related to our study includes Chan (1988), Cavanagh (1986), and Nabeya and Tanaka (1987). Cavanagh extends the expansion given in Dickey and Fuller (1979) for $T(\hat{\alpha}-a)$ to this nearly nonstationary situation and evaluates the asymptotic distribution using a simulation method. Chan is concerned with the t-statistic on $\hat{\alpha}$ in a regression with or without a constant term included. He derives three alternate expressions for the asymptotic distribution and evaluates them by simulation. The study of Nabeya and Tanaka (1987) is more closely related to ours. They calculate the asymptotic cumulative and probability density functions for the normalized least-squares estimator $T(\hat{\alpha} - 1)$ under (2) for both cases where a constant is included or not in the regression. These are obtained by numerical integration of the limiting moment generating function derived using the theory of Fredholm determinants. Our results agree with theirs when suitably normalized.

2. The Limiting Distribution of $T(\hat{\alpha} - \alpha)$

The normalized estimator to be considered can be expressed as follows:

$$T(\hat{\alpha} - \alpha) = \sum_{1}^{T} y_{t-1}u_t \Big/ \sum_{1}^{T} y_{t-1}^2.$$

When the errors $\{u_t\}$ are martingale differences (with some additional restrictions on their moments), the limiting distribution of $T(\hat{\alpha} - \alpha)$ under (2) is given by (see Phillips, 1987, Theorem 3.2)

$$T(\hat{\alpha} - \alpha) \Rightarrow \int_0^1 J_c(r)dw(r) \Big/ \int_0^1 J_c(r)^2 dr,$$

where $J_c(r) = \int_0^r e^{(r-s)c} dw(s)$, $w(r)$ is a Wiener process defined on $C[0,1]$, and $\Rightarrow$ denotes weak convergence. Note that when $c = 0$, $J_c(r) = w(r)$ and the limiting distribution reduces to $(\frac{1}{2})[w(1)^2 - 1]/\int_0^1 w(r)^2 dr$. Our results will contain, as a special case, those derived by Evans and Savin (1981a).

The limiting characteristic function of $(T^{-1}\sum_1^T y_{t-1}u_t, T^{-2}\sum_1^T y_{t-1}^2)$ as $T \to \infty$ is the characteristic function of $(\int_0^1 J_c(r)dw(r), \int_0^1 J_c(r)^2 dr)$ and has the form (see Phillips, 1987, Equation (A.1) corrected for a misprint):

$$M_c(\mu, v) = \left\{ \left(\frac{1}{2}\right) h^{-1} e^{c+i\mu} [\{h - (c + i\mu)\} e^{h} + \{h + (c + i\mu)\} e^{-h}] \right\}^{-1/2}, \tag{3}$$

where $h = h(\mu, v) = (c^2 + 2ci\mu - 2iv)^{1/2}$.

Since $\Pr[\int_0^1 J_c(r)^2 dr \leq 0] = 0$, we can use Gurland's (1948) theorem to derive the following characterization of the limiting distribution and density functions:

$$\begin{aligned} F(x,c) &= \lim_{T\to\infty} \Pr[T(\hat{\alpha}-\alpha) < x], \\ &= \Pr\left[\int_0^1 J_c(r)dw(r) \Big/ \int_0^1 J_c(r)^2 dr < x\right], \\ &= \frac{1}{2} - \frac{1}{2\pi i}\int_{-\infty}^{\infty} \frac{M_c(\mu,-\mu x)}{\mu} d\mu, \\ &= \frac{1}{2} - \frac{1}{\pi}\int_0^{\infty} \frac{\text{Imag}[M_c(\mu,-\mu x)]}{\mu} d\mu, \end{aligned} \tag{4}$$

Similarly, the limiting probability density function can be derived as follows:

$$\begin{aligned} f(x,c) &= \frac{d}{dx}F(x,c) = \frac{1}{2\pi i}\int_{-\infty}^{\infty} \left[\frac{\partial M_c(\mu,v)}{\partial v}\right]_{v=-\mu x} d\mu, \\ &= \frac{1}{\pi}\int_0^{\infty} \text{Real}[\{M_c(\mu,-\mu x)\}^3 B] d\mu, \end{aligned} \tag{5}$$

where

$$\begin{aligned} B = \left(\frac{1}{4}\right) & [e^{c+i\mu}/h^2(\mu,-\mu x)]\{[(e^{h(\mu,-\mu x)} - e^{-h(\mu,-\mu x)}] \\ & \cdot [h(\mu,-\mu x) + (c+i\mu)/h(\mu,-\mu x)] \\ & - (c+i\mu)[e^{-h(\mu,-\mu x)} + e^{h(\mu,-\mu x)}]\}. \end{aligned}$$

An alternative expression for the moment generating function may be given in the notation of Evans and Savin (1981a).

$$M_c(\mu,v) = \exp(-w/2)(\cos\theta - (w/\theta)\sin\theta)^{-1/2},$$

where

$$w = (c+i\mu), \quad \theta = (2iv - 2ci\mu - c^2)^{1/2}.$$

Letting $A = (\cos\theta - (w/\theta)\sin\theta)$ and $v = -\mu x$, we have

$$F(x,c) = \frac{1}{2} - \frac{1}{\pi}\int_0^{\infty} \frac{\text{Imag}[e^{-w/2}A^{-1/2}]}{\mu} d\mu, \tag{6}$$

and

$$f(x,c) = \frac{1}{\pi}\int_0^\infty \text{Real}[e^{-w/2}A^{-3/2}B]d\mu, \tag{7}$$

where

$$B = \frac{1}{2}[\sin(\theta)/\theta + (w/\theta^2)(\cos\theta - \sin(\theta)/\theta)].$$

When $c = 0$, and allowing for the fact that their results apply to the limiting distribution of $(T/\sqrt{2})(\hat{\alpha}-1)$, it is easy to verify that (6) and (7) correspond to the expressions derived by Evans and Savin (1981).

The expressions (4) and (5), or equivalently (6) and (7), can be numerically integrated using an appropriate upper bound of integration. Care is needed, however, in evaluating the integrand of both $F(x,c)$ and $f(x,c)$ since each involves the square root of a complex-valued quantity. The automatic use of the principal value of the argument does not necessarily preserve the continuity of the integrand. Details of the method used are given in the Appendix.

Table 1 presents the limiting cumulative distribution function for values of c given by $c = 0.0, \pm 0.5, \pm 1.0, \pm 2.0$, and ± 5.0. Note that for negative values of c, $\{y_t\}$ is a (locally) stationary process, whereas for positive values, $\{y_t\}$ is a (locally) explosive process. Figure 1 presents a graph of these functions for $c = 0.0, \pm 1.0$, and ± 5.0. Figure 2 presents the corresponding results for the probability density function. Table 2 presents the critical values for selected percentage points.

A striking feature is that the limiting density function of $T(\hat{\alpha}-\alpha)$ becomes more concentrated as c increases. For example, when $c = 5.0$, 90% of the density lies within the range (–1.0, 0.25). When c is negative, the density is flatter and the tails are more important. As in the case of a (strict) unit root process, the distribution is skewed to the left. Although the median of the distribution is between -1.0 and 0.0 (closer to 0.0 as c increases), the peak in the density occurs between 0.0 and 1.0 (closer to 0.0 as c increases). Also to be noticed is the sharp nonnormality of the distribution, especially when $c = -5.0$. This feature makes apparent the difference between the usual asymptotic theory and the local asymptotic theory described here.

Table 1. Cumulative distribution function.

x	$c = -5.0$	$c = -2.0$	$c = -1.0$	$c = -0.5$	$c = 0.0$	$c = 0.5$	$c = 1.0$	$c = 2.0$	$c = 5.0$
−14.000	0.0202	0.0132	0.0111	0.0101	0.0092	0.0083	0.0074	0.0058	0.0021
−13.000	0.0261	0.0174	0.0147	0.0140	0.0121	0.0109	0.0098	0.0077	0.0028
−12.000	0.0338	0.0228	0.0193	0.0177	0.0161	0.0145	0.0130	0.0102	0.0036
−11.000	0.0436	0.0300	0.0256	0.0234	0.0213	0.0193	0.0173	0.0135	0.0047
−10.000	0.0563	0.0394	0.0338	0.0311	0.0284	0.0257	0.0230	0.0181	0.0060
−9.000	0.0725	0.0519	0.0449	0.0413	0.0378	0.0343	0.0308	0.0241	0.0076
−8.000	0.0933	0.0685	0.0596	0.0551	0.0501	0.0460	0.0414	0.0324	0.0096
−7.000	0.1199	0.0904	0.0795	0.0738	0.0679	0.0620	0.0559	0.0437	0.0118
−6.500	0.1358	0.1040	0.0918	0.0855	0.0789	0.0721	0.0651	0.0509	0.0130
−6.000	0.1537	0.1196	0.1062	0.0991	0.0917	0.0840	0.0760	0.0594	0.0142
−5.500	0.1739	0.1377	0.1230	0.1151	0.1067	0.0980	0.0888	0.0693	0.0156
−5.000	0.1965	0.1585	0.1425	0.1338	0.1245	0.1146	0.1040	0.0811	—
−4.500	0.2218	0.1825	0.1653	0.1558	0.1455	0.1343	0.1222	0.0951	0.0186
−4.000	0.2502	0.2102	0.1921	0.1817	0.1703	0.1578	0.1439	0.1117	0.0205
−3.500	0.2819	0.2422	0.2234	0.2124	0.2000	0.1860	0.1702	0.1315	0.0227
−3.000	0.3170	0.2791	0.2602	0.2487	0.2356	0.2202	0.2021	0.1548	0.0254
−2.750	0.3360	0.2997	0.2810	0.2695	0.2560	0.2400	0.2206	0.1680	0.0270
−2.500	0.3560	0.3217	0.3035	0.2921	0.2785	0.2619	0.2412	0.1822	0.0288
−2.250	0.3769	0.3454	0.3280	0.3169	0.3032	0.2862	0.2641	0.1978	0.0309
−2.000	0.3989	0.3708	0.3547	0.3440	0.3306	0.3132	0.2895	—	0.0334
−1.750	0.4218	0.3981	0.3837	0.3738	0.3609	0.3433	0.3178	0.2354	0.0365
−1.500	0.4458	0.4273	0.4152	0.4065	0.3946	0.3770	0.3492	0.2591	0.0403
−1.250	0.4708	0.4585	0.4495	0.4425	0.4319	0.4146	0.3842	0.2879	0.0426
−1.000	0.4968	0.4919	0.4868	0.4820	0.4735	0.4563	—	0.3239	0.0523

(*Continued*)

Table 1. (*Continued*)

x	$c = -5.0$	$c = -2.0$	$c = -1.0$	$c = -0.5$	$c = 0.0$	$c = 0.5$	$c = 1.0$	$c = 2.0$	$c = 5.0$
−0.875	0.5101	0.5094	0.5067	0.5032	0.4959	0.4788	0.4467	0.3457	0.0570
−0.750	0.5237	0.5274	0.5273	0.5255	0.5195	0.5025	0.4710	0.3708	0.0629
−0.625	0.5375	0.5460	0.5489	0.5488	0.5441	0.5277	0.4977	0.4001	0.0709
−0.500	0.5516	0.5652	0.5712	0.5731	0.5698	—	0.5269	0.4348	0.0822
−0.375	0.5658	0.5849	0.5945	0.5984	0.5965	0.5831	0.5590	0.4760	0.0997
−0.250	0.5803	0.6052	0.6185	0.6246	0.6242	0.6136	0.5941	0.5252	0.1317
−0.125	0.5950	0.6259	0.6434	0.6517	0.6530	0.6460	0.6325	0.5836	0.2114
0.000	0.6098	0.6472	0.6689	0.6793	—	0.6803	0.6737	0.6504	0.5520
0.125	0.6248	0.6689	0.6951	0.7073	0.7134	0.7160	0.7172	0.7219	0.8937
0.250	0.6400	0.6911	0.7217	0.7354	0.7447	0.7526	0.7615	0.7907	0.9584
0.375	0.6552	0.7135	0.7483	0.7636	0.7762	0.7892	0.8047	0.8493	0.9789
0.500	0.6706	0.7362	0.7748	—	0.8072	0.8245	0.8447	0.8943	0.9879
0.625	0.6861	0.7591	0.8001	0.8189	0.8371	0.8572	0.8796	0.9265	0.9926
0.750	0.7016	0.7819	0.8256	0.8450	0.8650	0.8864	0.9086	0.9488	0.9952
0.875	0.7172	0.8047	0.8493	0.8696	0.8903	0.9113	0.9315	0.9641	0.9968
1.000	0.7333	0.8267	—	0.8921	0.9123	0.9317	0.9490	0.9745	0.9978
1.250	0.7637	0.8686	0.9109	0.9294	0.9462	0.9606	0.9720	0.9869	0.9990
1.500	0.7943	0.9051	0.9414	0.9561	0.9682	0.9776	0.9846	0.9931	0.9995
1.750	0.8240	0.9344	0.9630	0.9735	0.9815	0.9873	0.9914	0.9962	0.9997
2.000	0.8525	—	0.9774	0.9843	0.9893	0.9928	0.9952	0.9979	0.9999
2.500	0.9042	0.9826	0.9919	0.9946	0.9964	0.9976	0.9985	0.9994	1.0000
3.000	0.9456	0.9935	0.9971	0.9981	0.9988	0.9992	0.9995	0.9998	1.0000
3.500	0.9740	0.9976	0.9990	0.9993	0.9996	0.9997	0.9998	0.9999	1.0000
4.000	0.9893	0.9991	0.9996	0.9998	0.9999	0.9999	1.0000	1.0000	1.0000
4.500	0.9969	0.9997	0.9999	0.9999	1.0000	1.0000	1.0000	1.0000	1.0000
5.000	1.0000	1.0000	1.0000	1.0000	1.0000	1.0000	1.0000	1.0000	1.0000

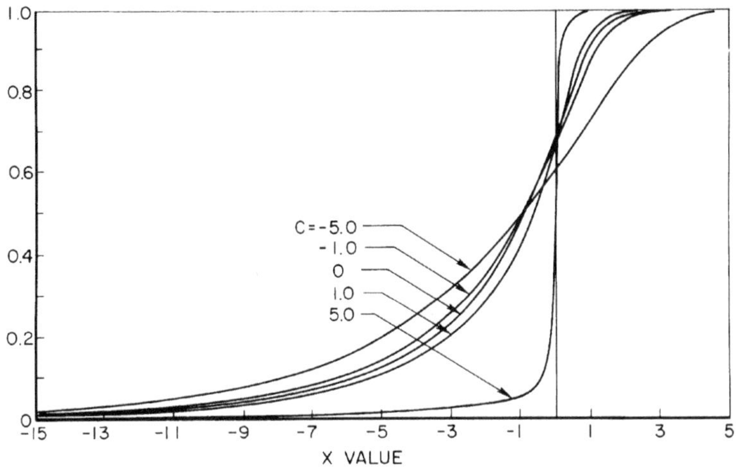

Fig. 1. Cumulative distribution function of the asymptotic distribution ($c =$ -5.0; -1.0; 0.0; 1.0; 5.0; $T(\hat{\alpha} - \alpha)$).

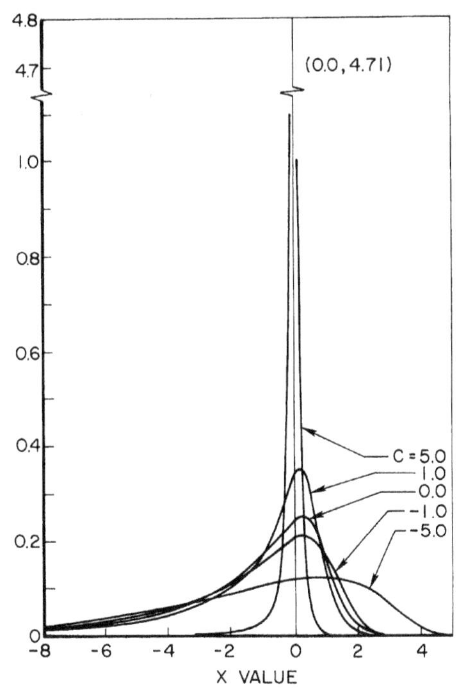

Fig. 2. Probability density function of the asymptotic distribution ($c = -5.0$; -1.0; 0.0; 1.0; 5.0; $T(\hat{\alpha} - \alpha)$).

Table 2. Selected percentage points of the limiting distribution of $T(\hat{\alpha} - \alpha)$; $\alpha = \exp(c/T)$.

Prob.	$c = -5.0$	$c = -2.0$	$c = -1.0$	$c = -0.5$	$c = 0.0$	$c = 0.5$	$c = 1.0$	$c = 2.0$	$c = 5.0$
0.01	−16.738	−15.034	−14.390	−14.048	−13.692	−13.317	−12.923	−12.056	−7.807
0.025	−13.174	−11.660	−11.081	−10.769	−10.440	−10.091	−9.718	−8.876	−3.064
0.05	−10.463	−9.138	−8.618	−8.337	−8.038	−7.718	−7.371	−6.559	−1.071
0.10	−7.724	−6.641	−6.207	−5.970	−5.700	−5.434	−5.124	−4.343	−0.373
0.90	2.456	1.462	1.175	1.048	0.928	0.816	0.710	0.519	0.132
0.95	3.064	1.901	1.589	1.434	1.285	1.143	1.008	0.758	0.221
0.975	3.524	2.310	1.950	1.779	1.612	1.451	1.297	1.007	0.341
0.99	4.034	2.785	2.401	2.214	2.033	1.855	1.683	1.355	0.547

The general pattern of these results points to the fact that $\hat{\alpha}$ converges faster to its true value α when the latter is greater than one. The greater concentration of the local asymptotic distribution around zero is a reflection of the greater precision of the estimates.

Finally, note that when $c = 0.0$, our results agree with Evans and Savin (1981): since they evaluate the distribution of $(T/\sqrt{2})(\hat{\alpha} - 1)$, $F(0, x) \equiv H(x/\sqrt{2})$ (in their notation) and $f(0, x) = h(x/\sqrt{2})/\sqrt{2}$.

3. The Adequacy of the Approximation to the Exact Distribution

A number of experiments were performed to assess the adequacy of the asymptotic distribution for nearly integrated series to the exact distribution of $T(\hat{\alpha} - \alpha)$ for various values of T and α. A more extensive description is contained in the working paper version of this study (available upon request). We present here a number of comparisons which are suggestive of the approximation. The exact distribution of the statistic $T(\hat{\alpha} - \alpha)$ presented here was obtained using the numerical procedure of Imhof (1961) as programmed by Koerts and Abrahamse (1969). The distributions obtained therefore pertain to model (1) with $N(0, \sigma^2)$ innovations.

Table 3 presents the exact probability of obtaining a smaller value of $T(\hat{\alpha} - \alpha)$ when using the asymptotic critical values of Table 2. Results are presented for the cases where $c = \pm 5.0, \pm 1.0$, and $T = 10$, 25, and 50. The values obtained can therefore be used to assess the adequacy of the approximation in the tails and evaluate the exact error in finite samples that would occur when performing a test using the asymptotic critical values.

Several interesting features emerge from the results of Table 3. First, the approximation is better as the noncentrality parameter **c** increases. It is not adequate for $T = 10$ or $T = 25$ when $c = -5.0$, but is very accurate when $c = 5.0$ for a sample size as small as 10, even in the extreme tail of the distribution. Second, the approximation is better for the right tail than the left (see, for example, the case with $c = -1.0$ and $T = 25$). Third, even if the approximation is not very adequate when $T = 10$ or $T = 25$ with negative values

Table 3. Exact size of tests using asymptotic critical values.

	Nominal size							
	0.01	0.025	0.05	0.10	0.90	0.95	0.975	0.99
$c = -5.0$								
$T = 10$	0.000	0.000	0.002	0.018	0.943	0.986	0.998	1.000
$T = 25$	0.002	0.008	0.023	0.061	0.917	0.966	0.987	0.997
$T = 50$	0.005	0.015	0.035	0.080	0.908	0.958	0.981	0.994
$c = -1.0$								
$T = 10$	0.000	0.003	0.015	0.051	0.907	0.968	0.993	1.000
$T = 25$	0.004	0.014	0.035	0.080	0.902	0.956	0.981	0.995
$T = 50$	0.007	0.019	0.042	0.090	0.901	0.953	0.978	0.992
$c = 1.0$								
$T = 10$	0.001	0.009	0.029	0.078	0.907	0.965	0.990	0.999
$T = 25$	0.006	0.018	0.042	0.091	0.902	0.955	0.980	0.994
$T = 50$	0.008	0.022	0.046	0.096	0.901	0.952	0.977	0.992
$c = 5.0$								
$T = 10$	0.009*	0.033*	0.058	0.113	0.907	0.958	0.982	0.995
$T = 25$	0.009*	0.028*	0.053	0.104	0.904	0.953	0.978	0.992
$T = 50$	0.009*	0.028*	0.052	0.102	0.902	0.952	0.977	0.991

Note: * entry indicates that the Imhof routine did not perform adequately in these cases due to ill-conditioned eigenvalues. The entry was then obtained by Monte-Carlo simulations using 20,000 replications.

of c (stationary processes), the adequacy improves markedly as T increases to 50 suggesting a rapid convergence to the asymptotic distribution.

To give an idea of the fit for the whole distribution, Fig. 3 compares the asymptotic and exact distributions for the case where $c = -5.0$ and $T = 10$, 25, and 50. This figure illustrates the rapid convergence to the asymptotic distribution over the full range of support. This convergence is even more rapid for larger values of c.

The results presented in this section provide evidence of the usefulness of such a local approximation. In this sense, the type of asymptotic theory for nearly integrated series studied here provides an interesting alternative to the standard limiting normal approximation or Edgeworth expansions in the case of stationary processes,

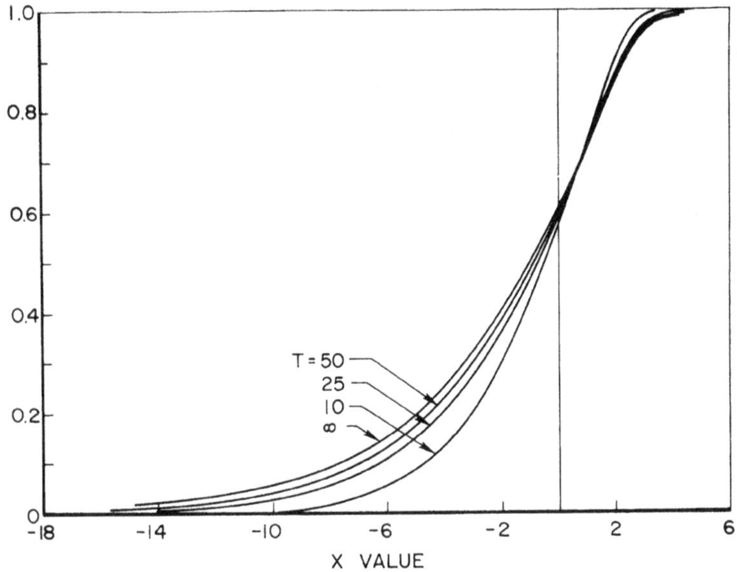

Fig. 3. The cumulative distribution functions of $T(\hat{\alpha} - \alpha)$; $\alpha = \exp(c/T)$; $c = -5.0$.

or to the limiting Cauchy or Satchell's (1984) approximation in the case of explosive processes.

4. Discussion

The results derived in this study suggest that the asymptotic distribution of near-integrated processes provides a satisfactory approximation to the exact distribution of processes with a root in the vicinity of one. The class of processes considered is characterized by a family of asymptotic distribution functions indexed by a noncentrality parameter c which measure the deviation of the root from unity, i.e., $\alpha = \exp(c/T) \approx 1 + c/T$.

The main advantage of this approach is that it yields a family of asymptotic distribution functions which is continuous with respect to the true value of the autoregressive parameter α, contrary to the usual asymptotic distributions which have a discontinuity at one. This permits the derivation of asymptotically valid confidence intervals for an autoregressive parameter in the neighborhood of one (see Cavanagh, 1986, for a description of the method).

The asymptotic distributions of near-integrated processes are also useful for other statistical problems. First, they can be used to evaluate the asymptotic power of tests for a unit root against a local alternative either stationary or explosive (see Phillips, 1988, for details). Second, they can be used to evaluate the exact power of tests for a unit root when a continuum of data is available. In this case, the noncentrality parameter c is a function of the span of the data (see Perron, 1987, for details). In this sense, Table 1 can be used to provide an upper bound to the power function of tests for a unit root with a given span when using the statistic $T(\hat{\alpha}-1)$. To see this, consider a continuous time Ornstein–Uhlenbeck process

$$dy_t = \gamma y_t dt + \sigma^2 dw_t, \tag{8}$$

where w_t is a unit Wiener process. The null hypothesis of a random walk is given by $\gamma = 0$. The discrete time representation of the process (8) is given by

$$y_{th} = \exp(\gamma h) y_{(t-1)h} + u_{th}, \tag{9}$$

with

$$u_{th} \sim N(0, \sigma^2(\exp(2\gamma h) - 1)/2\gamma),$$

where $h = S/T$ is the sampling interval and S the available span of data. Under this setting, we have that the normalized least-squares estimator in (1) has the following limiting distribution as $h \to 0$, or equivalently as $T \to \infty$ for a given fixed span S:

$$T(\hat{\alpha}-1) \to \gamma S + \left\{\int_0^1 J_c(r)^2 dr\right\}^{-1} \left\{\int_0^1 J_c(r) dw(r)\right\},$$

where $c = \gamma S$. Therefore, the power function of the statistic $T(\hat{\alpha}-1)$ for the null hypothesis that $\gamma = 0$ against the alternative that $\gamma \neq 0$

with a continuum of observation is given by (for a size α-test):

$$\text{Prob}\left[\left\{\int_0^1 J_c(r)^2 dr\right\}^{-1}\left\{\int_0^1 J_c(r)dw(r)\right\} \leq \delta_\alpha - \gamma S\right], \tag{10}$$

where δ_α is the α-percentage point of the distribution of $(\frac{1}{2})[w(1)^2 - 1]\{\int_0^1 w(r)drt\}^{-1}$ tabulated in Fuller (1976). The relation (10) can easily be evaluated using Table 1, letting $x = \delta_\alpha - \gamma S$, and provides an upper bound to the power function of $T(\hat{\alpha} - 1)$ with a given fixed span S.

Possible extensions to this line of research include the derivation of the local asymptotic distribution of the least-squares estimator when a constant and/or a trend are included in the regression and the consideration of more general error processes.

Appendix

This appendix describes how the numerical integrations were performed. First, let us note that Figs. 1 and 2 were drawn using the c.d.f. and p.d.f. values calculated for 155 values of x in the range (–15.0 to 5.0). The interval between each abscissa was 0.125 and we excluded values where $x = -c$ since at these points the integral is ill-defined.

The integral (4) was evaluated in the range of integration $(0 + \epsilon, \overline{U})$, where $\overline{U}$ is such that $|\text{Im}[M_c^2(\overline{U}, -\overline{U}x)]|/\overline{U} \leq \epsilon; \epsilon$ was set at 1.0×10^{-08}. The integral was evaluated using the subroutine DCADRE of IMSL with an absolute error criterion set at ϵ. The same procedure applies to evaluating the integral (5) except that $\overline{U}$ was chosen such that $|\text{Re}[M_c^6(\overline{U}, -\overline{U}x)B(\overline{U})]| \leq \epsilon$ and ϵ was set at 1.0×10^{-07}. We performed some sensitivity checks by evaluating the integrals using a greater value of $\overline{U}$ than the one given by the above error criterion. The results were essentially the same.

As mentioned in the text, care must be taken in evaluating these integrals since the integrand involves the square root of a complex quantity. The Fortran function CSQRT gives the principal value of the square root. Let A be the complex-valued quantity for which we need the square root. Now A can be represented in polar form as

$R = e^{i\theta}$ and $A^{1/2} = R^{1/2}\exp(i[\theta/2 + k\pi])$, $k = 0,1$. For $k = 0$, we get the principal value; note that $A^{1/2}(k = 0) = -A^{1/2}(k = 1)$.

The problem is that as we integrate over the range $(0 + \epsilon, \overline{U})$, the value of $A^{1/2}$ which ensures the continuity of the integrand may change. We must therefore integrate over the Riemann surface consisting here of two planes. As we change planes, we must change the value of $A^{1/2}$. To determine when to change planes, we proceed as follows: let $A = a + bi$ and $\delta = \arg(A(\mu)) = \arctan(a/b)$. For a given value of μ, the Fortran function ATAN gives ARG$(\cdot)$ in the range $(-\pi, \pi)$. Therefore, the points at which we change planes on the Riemann surface are the points $-\pi$ and π joined in the complex circle. For each values of μ in $(0+\epsilon, \overline{U})$, we must keep track of which plane we are on. Each time ARG$(A(\mu))$ crosses the points $-\pi$ or π, we must change the value of the square root used. Accordingly, an array for a grid of values of μ in $(0 + \epsilon, \overline{U})$ must be given to the subroutine DCADRE to indicate the position on the Riemann surface corresponding to the particular μ where the integrand is evaluated.

This procedure effectively ensures the continuity of the integrand by appropriately integrating in the complex domain over the Riemann surface. It was used to evaluate all points of both the c.d.f. and the p.d.f.

Acknowledgments

The author would like to thank Bryan Campbell and two anonymous referees for their comments and especially Peter C.B. Phillips for his help in designing the program for the numerical integration described in the Appendix. This research was supported by a grant from the Université de Montréal (CAFIR).

References

Cavanagh, C. (1986). Roots local to unity. Mimeo, Harvard University.

Chan, N.H. (1988). The parameter inference for nearly nonstationary time series. *Journal of the American Statistical Association* 83, 857–862.

Chan, N.H. and Wei, C.Z. (1987). Asymptotic inference for nearly nonstationary AR(1) processes. *The Annals of Statistics* 15, 1050–1063.

Dickey, D.A. (1976). Estimation and hypothesis testing for nonstationary time series. Ph.D. Thesis, Iowa State University, Ames, IA.

Dickey, D.A. and Fuller, W.A. (1979). Distribution of the estimators for autoregressive time series with a unit root. *Journal of the American Statistical Association* 74, 427–431.

Dickey, D.A. and Fuller, W.A. (1981). Likelihood ratio statistics for autoregressive time series with a unit root. *Econometrica* 49, 1057–1072.

Evans, G.B.A. and Savin, N.E. (1981a). The calculation of the limiting distribution of the least-squares estimator of the parameter in a random walk model. *Annals of Mathematical Statistics* 5, 114–118.

Evans, G.B.A. and Savin, N.E. (1981b). Testing for unit roots: 1. *Econometrica* 49, 753–779.

Evans, G.B.A. and Savin, N.E. (1984). Testing for unit roots: 2. *Econometrica* 52, 1241–1269.

Fuller, W.A. (1976). *Introduction to Statistical Time Series.* Wiley: New York.

Gurland, J. (1948). Inversion formulae for the distribution of ratios. *Annals of Mathematical Statistics* 19, 228–237.

Imhof, P.J. (1961). Computing the distribution of quadratic forms in normal variables. *Biometrika* 48, 419–426.

Koerts, J. and Abrahamse, A.P.J. (1969). *On the Theory and Application of the General Linear Model.* Rotterdam University Press: Rotterdam.

Mann, H.B. and Wald, A. (1943). On the statistical treatment of linear stochastic difference equations. *Econometrica* 11, 173–220.

Nabeya, S. and Tanaka, K. (1987). Asymptotic distribution of the estimator for the nonstable first-order autoregressive model. Mimeo, Hitotsubashi University.

Perron, P. (1987). Test consistency with varying sampling frequency. Cahier de recherche no 4187, C.R.D.E., Université de Montréal.

Perron, P. (1989). Tests of joint hypotheses in time series regression with a unit root. In *Advances in Econometrics: Cointegration, Spurious Regression, and Unit Roots*, T.B. Fomby & G.F. Rhodes, (eds.), JAI Press: Greenwich, CT.

Phillips, P.C.B. (1977). Approximations to some finite-sample distributions associated with a first-order stochastic difference equation. *Econometrica* 45, 463–485.

Phillips, P.C.B. (1987). Towards a unified asymptotic theory for autoregression. *Biometrika* 74(3), 535–547.

Phillips, P.C.B. (1987a). Time series regression with a unit root. *Econometrica* 55, 277–301.

Phillips, P.C.B. (1987b). Multiple regression with integrated time series. Cowles Foundation Discussion Paper No. 852, Yale University.

Phillips, P.C.B. (1988). Regression theory for near-integrated time series. *Econometrica*, 56, 1021–1044.

Phillips, P.C.B. and Perron, P. (1988). Testing for a unit root in time series regression. *Biometrika* 75, 335–346.

Rao, M.M. (1978). Asymptotic distribution of an estimator of the boundary parameter of an unstable process. *Annals of Statistics* 6, 185–190.

Rubin, H. (1950). Consistency of maximum likelihood estimates in the explosive case. In *Statistical Inference in Dynamic Economic Models*, T.C. Koopmans (ed.), Wiley: New York, pp. 356–364.

Said, S.E. and Dickey, D.A. (1984). Testing for unit roots in autoregressive moving average of unknown order. *Biometrika* 71, 599–607.

Satchell, S. (1984). Approximation to the finite-sample distribution for nonstable first-order stochastic difference equations. *Econometrica* 52, 1271–1289.

White, J.S. (1958). The limiting distribution of the serial correlation coefficient in the explosive case. *Annals of Mathematical Statistics* 29, 1188–1197.

Chapter 4

A Continuous Time Approximation to the Unstable First-Order Autoregressive Process: The Case Without an Intercept*

Pierre Perron

Department of Economics, Princeton University, Princeton, NJ 08544, USA
Centre de Recherche et Développement en Economique, Université de Montréal, C.P. 6128, Succ. A, Montréal, Canada H3C-3J7

Consider the first-order autoregressive process $y_t = \alpha y_{t-1} + e_t, y_0$ a fixed constant, $e_t \sim$ i.i.d. $(0, \sigma^2)$, and let $\hat{\alpha}$ be the least-squares estimator of α based on a sample of size $(T+1)$ sampled at frequency h. Consider also the continuous time Ornstein–Uhlenbeck process $dy_t = \theta y_t dt + \sigma dw_t$ where w_t is a Wiener process and let $\hat{\theta}$ be the continuous time maximum likelihood (conditional upon y_0) estimator of θ based upon a single path of data of length N. We first show that the exact distribution of $N(\hat{\theta} - \theta)$ is the same as the asymptotic distribution of $T(\hat{\alpha} - \alpha)$ as the sampling interval converges to zero. This asymptotic distribution permits explicit consideration of the effect of the initial condition y_0 upon the distribution of $\hat{\alpha}$. We use this fact to provide an approximation to the finite sample distribution of $\hat{\alpha}$ for arbitrary fixed y_0. The moment-generating function of $N(\hat{\theta} - \theta)$ is derived and used to tabulate the distribution and probability density functions. We also consider the moment of $\hat{\theta}$ and the power function of test statistics associated with it. In each case, the adequacy of the approximation to the finite sample distribution of $\hat{\alpha}$ is assessed for values of α in the vicinity of one. The approximations are, in general, found to be excellent.

Keywords: Ornstein–Uhlenbeck process; moment-generating function; near-integrated processes; unit root; distribution theory.

*Originally Published in *Econometrica* 59, pp. 211–236.

1. Introduction

Consider the following first-order stochastic difference equation:

$$y_t = \alpha y_{t-1} + e_t \quad (t = 1, \ldots, T), \tag{1}$$

where $y_0 = b$, a fixed constant, and $e_t \sim$ i.i.d. $\mathcal{N}(0, \sigma^2)$ where $\mathcal{N}$ denotes the normal distribution. The unrestricted maximum likelihood estimate of α (conditional upon the initial observation y_0) based upon a sequence of observations $\{y_t\}_0^T$, is the least-squares estimator:

$$\hat{\alpha} = \sum_{t=1}^{T} y_t y_{t-1} \left(\sum_{t=1}^{T} y_{t-1}^2 \right)^{-1}.$$

The distribution of $\hat{\alpha}$ has been extensively studied. One topic of concern has been the adequacy of various asymptotic approximations to the finite sample distribution. Mann and Wald (1943) and Rubin (1950) showed that $T^{1/2}(\hat{\alpha} - \alpha)(1 - \alpha^2)^{-1/2}$ has a limiting $\mathcal{N}(0, 1)$ distribution when $|\alpha| < 1$. White (1958) showed that when $|\alpha| > 1$, the limiting distribution of $|\alpha|^T(\alpha^2 - 1)^{-1}(\hat{\alpha} - \alpha)$ is Cauchy provided that $y_0 = 0$. White also considered the case $|\alpha| = 1$ and showed that the limiting distribution of $T(\hat{\alpha} - 1)$ can be expressed in terms of the ratio of two functionals of a Wiener process (see also Phillips, 1987a). In the latter case Rao (1978) obtained a computable expression.

The case of the unit root $\alpha = 1$ has attracted a great deal of attention. The critical values of the asymptotic distribution of $T(\hat{\alpha} - 1)$ have been derived by Dickey (1976) by simulation methods. Evans and Savin (1981a) obtained tabulated values of both the cumulative distribution function and the probability density function of the limiting distribution of $(T/\sqrt{2})(\hat{\alpha} - 1)$ by numerically integrating the limiting moment-generating function derived by White (1958). A variety of procedures have also been suggested which permit a considerable weakening of the conditions imposed on the innovation sequence $\{e_t\}$ while at the same time using the same critical values derived in the earlier studies; see, for example, Dickey and Fuller (1979), Said and Dickey (1984), and Phillips (1987a).

The major point which emerges from these studies is that the asymptotic distributions have a discontinuity at one. Because the exact distribution of $\hat{\alpha}$ is continuous for all values of α, this suggests that the limiting distribution inadequately approximates the finite sample distribution for values of α near the discontinuity point $\alpha = 1$. This inadequacy has been well-documented by Evans and Savin (1981b, 1984) and by Phillips (1977) who also showed that Edgeworth expansions perform poorly for α less than but close to one.

Recently, a new class of models which specifically deal with the presence of a root close to, but not necessarily equal to one, has been studied. Phillips (1988) considered a near-integrated random process where the autoregressive parameter is defined by:

$$\alpha = \exp(c/T). \tag{2}$$

Here, the constant c is a measure of the deviation from the unit root case. The model (1) and (2) may also be described as having a root local to unity (see Cavanagh, 1986): as the sample size increases, the autoregressive parameter converges to unity. When $c < 0$, the process $\{y_t\}$ is said to be (locally) stationary and when $c > 0$, it is said to be (locally) explosive. An expression for the limiting distribution of $T(\hat{\alpha} - \alpha)$ under (2) has been derived by Cavanagh (1986), Chan and Wei (1987), and Phillips (1987b). This framework has been quite useful in studying various problems such as the power of tests of a unit root under local alternatives (Phillips, 1987b; Phillips and Perron, 1988; Perron, 1990), the derivation of confidence intervals when α is near unity (Cavanagh, 1986), and the limiting power of tests of a unit root with varying sampling intervals (Perron, 1989b).

Tabulations of the limiting distribution of $T(\hat{\alpha} - \alpha)$ under (2) have been obtained by Cavanagh (1986), Nabeya and Tanaka (1990) and Perron (1989a) using different procedures. These studies also provide measures of the adequacy of this limiting distribution as an approximation to the finite sample distribution of $\hat{\alpha}$ when α is in the vicinity of 1. They show the approximation to be quite good in the case where $y_0 = 0$. A feature of substantial interest is that the limiting distribution of $T(\hat{\alpha} - \alpha)$ under the near-integrated process (2) is invariant to the value of the initial observation y_0. As thoroughly documented by Evans and Savin (1981b), the finite sample distribution of $\hat{\alpha}$ is very sensitive to the value of y_0 when α is near 1 even for quite large sample sizes. Hence, the approximation provided by the near-integrated limiting distribution is adequate only in the special case where $y_0 = 0$. When $y_0 \neq 0$, it fails to capture the substantial effect a nonzero initial condition has on the finite sample distribution.

The purpose of this paper is to present a rather different approach in deriving an asymptotic approximation which permits the explicit consideration of the effects of different initial conditions on the distribution of $T(\hat{\alpha} - \alpha)$. Instead of the usual framework where the sample size increases to infinity keeping a fixed sampling interval, we use the so-called continuous records asymptotic analysis where T is increased to infinity keeping the span of the data fixed, i.e., by letting the sampling interval converge to zero at the same rate as T increases to infinity. This method is closely related to the exact distribution of estimators in continuous time.

Consider the following continuous time Ornstein–Uhlenbeck diffusion process:

$$dy_t = \theta y_t \, dt + \sigma \, dw_t; \quad y_0 = b, \ t > 0. \tag{3}$$

θ and $\sigma > 0$ are unknown parameters and w_t is the standard Wiener process. The unique solution to $\{y_t\}$ (in the mean squared sense) is given by (see, e.g., Arnold (1974)):

$$y_t = \exp(\theta t)b + \sigma \int_0^t \exp(\theta(t-s))dw_s = \exp(\theta t)b + \sigma J_\theta(t) \tag{4}$$

where, following the notation of Phillips (1987b), $J_\theta(t)$ is a stochastic integral defined as $J_\theta(t) = \int_0^t \exp(\theta(t-s))\, dw_s$. Note that $J_\theta(t) \sim \mathcal{N}(0, (e^{2\theta t} - 1)/2\theta)$. Our concern is the estimation of the unknown parameter θ given a single sample path of observations $\{y_t, \ 0 \leq t \leq N\}$, where N is the span of the data. The analog to the least-squares procedure in continuous time yields the following estimator:

$$\hat{\theta}_N(y) = \int_0^N y_t\, dy_t \bigg/ \int_0^N y_t^2\, dt. \tag{5}$$

$\hat{\theta}_N(y)$ is also the unrestricted maximum likelihood estimator when $b = 0$ (see, e.g., Liptser and Shiryayev, 1978, 17.1.1). For simplicity of notation, we shall simply write $\hat{\theta} \equiv \hat{\theta}_N(y)$ and analyze the distributional properties of the standardized estimator $N(\hat{\theta} - \theta)$. The discrete-time representation of the process y_t is easily shown to be given by:

$$y_{th} = \exp(\theta h) y_{(t-1)h} + u_{th}, \quad y_0 = b, \quad t \geq 0, \tag{6}$$

where $u_{th} \sim \mathcal{N}(0, \sigma^2(e^{2\theta h} - 1)/2\theta)$ and h is the sampling interval. In this discrete-time framework the goal is to estimate the unknown quantity $\alpha_h = \exp(\theta h)$ given a sequence of observations $\{y_{th}, t = 0, 1, 2, \ldots, T\}$ where $T = N/h$ is the total number of observations available (minus one given the initial condition y_0). The least squares estimator (and maximum likelihood estimator conditional upon y_0) is:

$$\hat{\alpha}_h = \sum_{t=1}^{T} y_{th}\, y_{(t-1)h} \bigg/ \sum_{t=1}^{T} y_{(t-1)h}^2. \tag{7}$$

We derive the limiting distribution of $T(\hat{\alpha}_h - \alpha_h)$ under (6) as $h \to 0$ with N fixed and show that it is identical to the exact distribution of $N(\hat{\theta} - \theta)$ under (3). Furthermore, this distribution contains an explicit dependence upon the initial condition y_0 and reduces to the near-integrated asymptotic distribution in the special case where $y_0 = 0$. These distributional results are considered in Section 2 which also discusses some interesting corollaries. Section 3 derives the joint moment-generating function of $(N^{-1} \int_0^N y_t\, dw_t, N^{-2} \int_0^N y_t^2\, dt)$. The expression obtained contains as a special case the limiting joint moment-generating function derived by Phillips (1987b) for the near-integrated framework by setting $y_0 = 0$. The result of White (1958) and Evans and Savin (1981a) is a special case obtained by setting $y_0 = 0$ and $\theta = 0$.

In Section 3, we also derive and tabulate the cumulative distribution and probability density functions for selected combinations of the parameters θ and y_0. Section 4 considers the moments of $N(\hat{\theta}-\theta)$ and discusses their use in approximating the exact moments of $\hat{\alpha}$. Power functions of specific hypotheses are also discussed in that Section with special emphasis on the power of tests for a unit root against alternatives close to one. Finally, Section 5 offers remarks about the usefulness of our method to other frameworks. A mathematical appendix contains the proofs of some theorems. Throughout the paper, we provide a measure of the adequacy of the asymptotic approximation to the finite sample counterpart. The various percentage points of the distribution, the moments of the distribution, and the power functions are considered in turn. The results are quite striking, showing a very close approximation.

2. Some Distributional Results

The main result of this section ties together the exact distribution of $N(\hat{\theta}-\theta)$, given a sample path of observations $\{y_t\}_0^N$, and the asymptotic distribution of $T(\hat{\alpha}_h - \alpha_h)$ with the continuous records asymptotic ($h \to 0$ with a fixed span N) given a discrete set of observations sampled at interval h, $\{y_{th};\ t = 0, 1, \ldots, T = N/h)$. For simplicity, we consider a limiting sequence $\{h = h_1, h_2, \ldots, h_n\}$ such that $T_n = N/h_n$ is integer-valued with $h_n \to 0$ and $T_n \to \infty$, as $n \to \infty$. Throughout the paper we denote by '$\Rightarrow$' weak convergence in distribution, and by '$\stackrel{d}{=}$' equality in distribution.

Theorem 1. *Let $J_c(r) = \int_0^r \exp(c(r-s))dw_s$ and w_r be the standard Wiener process on $C(0,1)$; define $\gamma = b/\sigma N^{1/2}$ and $c = \theta N$, and the following variables*:

$$A(\gamma, c) = \gamma \int_0^1 \exp(cr)\, dw_r + \int_0^1 J_c(r)\, dw_r,$$

$$B(\gamma, c) = \gamma^2(\exp(2c) - 1)/2c + 2\gamma \int_0^1 \exp(cr) J_c(r)\, dr + \int_0^1 J_c(r)^2\, dr.$$

(i) *Let* $\{y_t;\ 0 \le t \le N\}$ *be a continuous-time stochastic process generated by* (3) *and let* $\hat{\theta}$ *be the estimator of* θ *defined by* (5); *then*

$$N(\hat{\theta}-\theta) \stackrel{d}{=} A(\gamma,c)/B(\gamma,c) \equiv Z(\gamma,c).$$

(ii) *Let* $\hat{\alpha}_h$ *be defined by* (7) *with* $\{y_{th};\ t=0,\ldots,T\}$ *a discrete time stochastic process generated by* (6); *then as* $h \to 0$ *with* $T \to \infty$ *and* N *fixed,*

$$T(\hat{\alpha}_h - \alpha_h) \Rightarrow Z(\gamma,c).$$

Remark. The result given under (ii) is valid under more general conditions than those implied by the process (6). Indeed the same result holds if the innovations $\{u_{ht}\}$ are martingale differences with a variance function, say $\mathrm{var}(u_{th}) = a(h)$, such that $a(h)/h \to \sigma^2$ as $h \to 0$, and the stochastic process $\{e_{th} = u_{th}/a(h)^{1/2}\}$ satisfies conditions permitting the application of a functional central limit theorem to their partial sums, i.e., $T^{-1/2}\sum_1^{[\mathrm{Tr}]} e_{th} \Rightarrow W(r)$ as $T \to \infty$ (see, e.g., Phillips, 1987a). This precludes, among other things, processes which, in continuous time, are characterized by a Poisson process. In particular, the normality assumption is not necessary. The discretization of the process (3), given by (6), provides a natural instance where the relevant condition is satisfied given that the variance of this process is such that $h^{-1}\sigma^2(\exp(2\theta h)-1)/2\theta \to \sigma^2$ as $h \to 0$. Alternatively, the same result can be obtained using the usual $(T \to \infty)$ asymptotic framework by specifying an initial condition such that $T^{-1/2}y_0 \to \gamma$ as $T \to \infty$.

Proof. The proof of part (ii) is closely related to the proof of Theorem 3.1 in Phillips (1987b) and is presented in the Appendix. To prove part (i), we first express $N(\hat{\theta}-\theta)$ as a function of a transformed diffusion process. Consider a scale transformation from $t \in (0,N)$ to $r \in (0,1)$ such that $t \to Nr = t$. Then $J_\theta(t) = \int_0^t \exp(\theta(t-s))dw_s \sim \mathcal{N}(0,\ (e^{2\theta t}-1)/2\theta)$ can be expressed as $J_\theta(t) = N^{1/2}J_c(r)$ where $c = \theta N$. Using (4), we obtain the solution:

$$y_r = \exp(cr)b + \sigma N^{1/2}J_c(r), \quad 0 \le r \le 1, \tag{8}$$

with initial condition $y_0 = b$. Using (3), the normalized estimator of $\hat{\theta}$ is:

$$\begin{aligned} N(\hat{\theta} - \theta) &= \left[\int_0^N y_t\, dy_t - \theta \int_0^N y_t^2\, dt\right]\left[N^{-1}\int_0^N y_t^2\, dt\right]^{-1} \\ &= \sigma \int_0^N y_t\, dw_t \left[N^{-1}\int_0^N y_t^2\, dt\right]^{-1} \\ &= \sigma N^{-1/2}\int_0^1 y_r\, dw_r \left[N^{-1}\int_0^1 y_r^2\, dr\right]^{-1} \end{aligned} \tag{9}$$

where y_r is defined by (8). Given that $\hat{\theta}$ is invariant to a scale transformation of the data, let $x_r = y_r/\sigma N^{1/2}$. Then,

$$N(\hat{\theta} - \theta) = \int_0^1 x_r\, dw_r \bigg/ \int_0^1 x_r^2\, dr \tag{10}$$

where

$$x_r = \exp(cr)\gamma + J_c(r). \tag{11}$$

The expression (11) is the solution x_r of the following stochastic differential equation:

$$dx_r = cx_r\, dr + dw_r, \quad x_0 = \gamma, \quad 0 < r \leq 1, \tag{12}$$

where $c = \theta N$ and $\gamma = b/\sigma N^{1/2}$. Consider first the numerator of (10). Using (11) we have $\int_0^1 x_r\, dw_r = \gamma \int_0^1 \exp(cr)\, dw_r + \int_0^1 J_c(r)\, dw_r \equiv A(\gamma, c)$. Similarly, we can express the denominator of (9) as $\int_0^1 x_r^2\, dr = \int_0^1 [\exp(cr)\gamma + J_c(r)]^2\, dr = \gamma^2 \int_0^1 \exp(2cr)\, dr + 2\gamma \int_0^1 \exp(cr) J_c(r)\, dr + \int_0^1 J_c(r)^2\, dr \equiv B(\gamma, c)$ given that $\int_0^1 \exp(2cr)\, dr = (\exp(2c) - 1)/2c$. □

Theorem 1 suggests several avenues of investigation. First, the small-h (or continuous records) asymptotic distribution of $T(\hat{\alpha}_h - \alpha_h)$ is the same as the exact distribution of the normalized continuous time estimator $N(\hat{\theta} - \theta)$. This justifies studying the distributional properties of $N(\hat{\theta} - \theta)$ as an approximation to the exact distribution of the discrete-time least-squares estimator $T(\hat{\alpha} - \alpha)$. Since the stochastic representation of the variable $Z(\gamma, c)$ is explicitly affected by the value of the initial condition $y_0 = b$, we can hope to capture the effects of different values of b on the finite sample distribution of the discrete-time least-squares estimator. More precisely, the distribution of the random variable $Z(\gamma, c)$ depends only upon two parameters, namely $c = \theta N$ and $\gamma = b/\sigma N^{1/2}$. We shall therefore characterize the distributional properties of $N(\hat{\theta} - \theta)$, in subsequent discussions, directly as functions of these two quantities.

The dependence of the asymptotic distribution upon the nuisance parameter γ implies that this quantity must be consistently estimated in order to use our distributional results in practice. Consider the normalized sum of squared residuals, $\hat{\sigma}^2 = T^{-1}\sum_{t=1}^{T}(y_{th} - \hat{\alpha}_h\, y_{ht-1})^2$, as an estimator of σ^2. Using Lemma A.1 in the appendix, it can be shown that $\hat{\sigma}^2$ is an inconsistent estimator of $\hat{\sigma}^2$ as $h \to 0$; indeed $\hat{\sigma}^2 \to 0$ in probability as $h \to 0$. However, $T\hat{\sigma}^2 \to N\sigma^2$ in probability as $h \to 0$ keeping N fixed. Hence, a consistent estimator of γ is given by $y_0/T^{1/2}\hat{\sigma}$.

Several known distributional results can be obtained as special cases of the exact distribution of $N(\hat{\theta} - \theta)$. First, we obtain from Theorem 1 that the asymptotic distribution of $T(\hat{\alpha} - \alpha)$ in the near-integrated case is a special case of the exact distribution of $N(\hat{\theta} - \theta)$ obtained by setting $y_0 = b = 0$. In this case:

$$N(\hat{\theta} - \theta) \stackrel{d}{=} Z(0, c) = \int_0^1 J_c(r)\, dw_r \bigg/ \int_0^1 J_c(r)^2\, dr,$$

which is the asymptotic distribution of $T(\hat{\alpha} - \alpha)$ in the near-integrated case as derived by Phillips (1987b). Hence, it appears that when $b \neq 0$, we can expect a better approximation to the exact distribution of $\hat{\alpha}$ using the continuous record asymptotic rather than the near-integrated framework since in the latter case the effect of b vanishes asymptotically. Secondly, when $b = \theta = 0$, we have:

$$N(\hat{\theta} - \theta) \stackrel{d}{=} (1/2)(w_1^2 - 1) \bigg/ \int_0^1 w_r^2\, dr$$

which corresponds to the limiting distribution of $T(\hat{\alpha} - \alpha)$ in the unit root case where $\alpha = 1$ (see White, 1958; Phillips, 1987a, among others). It is also straightforward to obtain different asymptotic results. Consider first the behavior of $N(\hat{\theta} - \theta)$ as $N \to \infty$, documented in the following Corollary whose proof is given in the Appendix.

Corollary 1. *For any fixed* $y_0 = b$, (i) *if* $\theta < 0$, $N^{1/2}(\hat{\theta} - \theta) \Rightarrow \mathcal{N}(0, -2\theta)$, *as* $N \to \infty$; (ii) *if* $\theta > 0$, $e^{\theta N}(\hat{\theta} - \theta)2\theta \Rightarrow [d\eta + \xi\eta]/[d+\eta]^2$, *as* $N \to \infty$ *where* η *and* ξ *are independent* $\mathcal{N}(0,1)$ *variates and* $d = b(2\theta)^{1/2}/\sigma$.

Part (i) of Corollary 1 is a standard result in the literature (e.g., Brown and Hewitt, 1975; Basawa and Rao, 1980). Our framework provides a simple alternative derivation. Part (ii) appears not to have been derived previously. Note, however, that when $y_0 = b = 0$, the limiting distribution of $e^{\theta N}(\hat{\theta} - \theta)/2\theta$ as $N \to \infty$ is the same as the limiting distribution of $|\alpha|^T(\alpha^2 - 1)^{-1}(\hat{\alpha} - \alpha)$ when $|\alpha| > 1$ as $T \to \infty$, namely a Cauchy variate.

It is also possible to study the behavior of $N(\hat{\theta} - \theta)$ as $b \to \infty$ or, equivalently as $\sigma \to 0$, the so-called 'small-σ asymptotics.' It is a straightforward consequence of Theorem 1 that $N(\hat{\theta} - \theta) \to 0$ as $b \to \infty$. Hence, one can expect the bias and variance of $\hat{\theta}$ to decrease as b increases. The same behavior should hold concerning the discrete-time estimator $\hat{\alpha}$. These conjectures are verified in Section 4. The approximate distribution of $N(\hat{\theta} - \theta)$ for large γ (i.e., large b) is easily shown to be:

$$N(\hat{\theta} - \theta) \overset{a}{\sim} \int_0^1 \exp(cr)\, dw_r [\gamma(\exp(2c) - 1)/2c]^{-1},$$

where $\overset{a}{\sim}$ denotes 'approximately distributed as'. Now $\int_0^1 \exp(cr)\, dw_r \sim \mathcal{N}(0, (e^{2c} - 1)/2c)$ (see the Appendix); hence, $N(\hat{\theta} - \theta) \overset{a}{\sim} \mathcal{N}(0, 2c/\gamma^2(e^{2c} - 1))$. Given that $N(\hat{\theta} - \theta)$ approximates $T(\hat{\alpha}_h - \alpha_h)$, we have the following approximation to the discrete-time estimator $\hat{\alpha}_h$ for large b (or equivalently small-σ asymptotic):

$$T(\hat{\alpha}_h - \alpha_h) \overset{a}{\sim} \mathcal{N}(0, 2c/\gamma^2(e^{2c} - 1)). \tag{13}$$

The approximation (13) generalizes a result derived by Phillips (1987a) who considered the unit root case. Indeed, by taking the limit of (13) as $c \to 0$, we have $T(\hat{\alpha}_h - 1) \overset{a}{\sim} \mathcal{N}(0,\, 1/\gamma^2)$ which corresponds to his expression (27). As he remarked for the unit root case, the usual ($T \to \infty$) asymptotic distribution obscures the dependence of the distribution of $\hat{\alpha}$ on the parameter y_0/σ. The same feature emerges in the ($T \to \infty$) asymptotic distribution for the near-integrated case. The ($h \to 0$) asymptotic distribution highlights the effects of the initial condition and may be a better guide to the finite sample behavior of $T(\hat{\alpha} - \alpha)$ for values of α near 1.

In order to use these distributional results as approximations to the finite sample behavior, we need a method to compute the distribution function of the random variable $Z(\gamma,\, c)$ or equivalently of $N(\hat{\theta} - \theta)$. To this effect, the next section derives a computable expression for this distribution.

3. The Exact Distribution of $N(\hat{\theta} - \theta)$

To study the exact distribution of $N(\hat{\theta} - \theta)$, we derive the joint moment generating function of $(\int_0^1 x_r\, dw_r,\, \int_0^1 x_r^2\, dr)$ given that x_r is a random variable in a probability space $(\Omega,\, F,\, \mu_x^c)$, say, where μ_x^c is the measure induced by the diffusion process (12). We denote this joint moment-generating function as

$$M_{c,\gamma}(v, u) = E\left[\exp\left(v \int_0^1 x_r\, dw_r + u \int_0^1 x_r^2\, dr\right)\right],$$

where the expectation is taken with respect to the measure μ_x^c. The main result of this section is an expression for $M_{c,\gamma}(v, u)$ which is contained in the following theorem.

Theorem 2. *Let* $\lambda = (c^2 + 2cv - 2u)^{1/2}$ *and*

$$\begin{aligned}\psi_c(v, u) = \{&2\lambda \exp(-(v + c))/[(\lambda + (v + c)) \exp(-\lambda)\\ &+ (\lambda - (v + c)) \exp(\lambda)]\}^{1/2};\end{aligned}$$

then

$$\begin{aligned}M_{c,\gamma}(v, u) = \psi_c(v, u)\ &\exp\{-(\gamma^2/2)(v + c - \lambda)\\ &\times (1 - \exp(v + c + \lambda)\psi_c^2(v, u))\}.\end{aligned}$$

Remark. When $\gamma = 0$, $M_{c,\gamma}(v,u) = M_{c,0}(v,u) = \psi_c(v,u)$, which is the joint moment-generating function of $(\int_0^1 J_c(r)\,dw_r, \int_0^1 J_c(r)^2\,dr)$ derived by Phillips (1987b). When $\gamma = c = 0$, $M_{c,\gamma}(v,u) = M_{0,0}(v,u) = \{[(\lambda+v)\exp(-\lambda)+(\lambda-v)\exp(\lambda)]/2\lambda\exp(-v)\}^{-1/2}$, which is the limiting joint moment-generating function of $(\int_0^1 w_r\,dw_r, \int_0^1 w_r^2\,dr)$ derived by White (1958) and analyzed by Evans and Savin (1981a).

Proof. The proof of Theorem 2 is of interest by itself. It proposes a new, and we believe, quite simple way of deriving limiting distributions involving functionals of Wiener processes. It offers an interesting alternative to the approach of Nabeya and Tanaka (1990) based on Fredholm determinants, the approach used by White (1958) and Phillips (1987b) which deals with the limit of the finite sample moment-generating function, and the methodology proposed by Chan (1988) based on simulations. In the simple framework considered, it also allows greater generality given that we recover earlier results as special cases (i.e., when setting γ and θ equal to 0). We believe that the methodology proposed below will also allow the derivation of similar results in more complex models.

The proof is inspired by the development in Lemma 17.3 of Liptser and Shiryayev (1978). Denote by μ_x^c and μ_x^λ the measures induced by the processes x^c and x^λ (defined on the same probability space) generated by the following stochastic differential equations:

$$dx_r^c = cx_r^c\,dr + dw_r, \quad x_0^c = \gamma, \tag{14}$$

$$dx_r^\lambda = \lambda x_r^\lambda\,dr + dw_r, \quad x_0^\lambda = \gamma. \tag{15}$$

The measures μ_x^c and μ_x^λ are equivalent (Liptser and Shiryayev, 1978, Theorem 7.19) and the density or Radon–Nikodym derivative $d\mu_x^c/d\mu_x^\lambda$ evaluated with the random process x^λ is given by

$$d\mu_x^c/d\mu_x^\lambda(x^\lambda) = \exp\left\{(c-\lambda)\int_0^1 x_r^\lambda\,dx_r^\lambda - (1/2)(c^2-\lambda^2)\int_0^1 (x_r^\lambda)^2\,dr\right\}.$$

Denoting by E_c the expectation operator taken with respect to the measure μ_x^c, we have

$$M_{c,\gamma}(v,u) = E_c\left[\exp\left\{v\int_0^1 x_r\,dw_r + u\int_0^1 x_r^2\,dr\right\}\right]$$
$$= E\left[\exp\left\{v\int_0^1 x_r^c\,dw_r + u\int_0^1 (x_r^c)^2\,dr\right\}\right].$$

Using (14), $\int_0^1 x_r^c\,dw_r = \int_0^1 x_r^c\,dx_r^c - c\int_0^1 (x_r^c)^2\,dr$, and upon substitution:

$$M_{c,\gamma}(v,u) = E\left[\exp\left\{v\int_0^1 x_r^c\,dx_r^c + (u - vc)\int_0^1 (x_r^c)^2\,dr\right\}\right]$$
$$= E\left[\exp\left\{v\int_0^1 x_r^\lambda\,dx_r^\lambda + (u - vc)\right.\right.$$
$$\left.\left.\times\int_0^1 (x_r^\lambda)^2\,dr\right\} \times d\mu_x^c/d\mu_x^\lambda(x^\lambda)\right]$$
$$= E\left[\exp\left\{v\int_0^1 x_r^\lambda\,dx_r^\lambda + (u - vc)\int_0^1 (x_r^\lambda)^2\,dr\right.\right.$$
$$\left.\left.+ (c-\lambda)\int_0^1 x_r^\lambda\,dx_r^\lambda - (1/2)(c^2-\lambda^2)\int_0^1 (x_r^\lambda)^2\,dr\right\}\right]$$
$$= E\left[\exp\left\{(v + c - \lambda)\int_0^1 x_r^\lambda\,dx_r^\lambda\right.\right.$$
$$\left.\left.+ [u - vc - (c^2-\lambda^2)/2]\int_0^1 (x_r^\lambda)^2\,dr\right\}\right].$$

Now, let $\lambda = (c^2 + 2cv - 2u)^{1/2}$ and $a = v + c - \lambda$. Then, $[u - vc - (c^2-\lambda^2)/2] = 0$ and

$$M_{c,\gamma}(v,u) = E\left[\exp\left\{a\int_0^1 x_r^\lambda\,dx_r^\lambda\right\}\right].$$

By Ito's Lemma applied to the function $(x_r^\lambda)^2$, $\int_0^1 (x_r^\lambda)\,dx_r^\lambda = [(x_1^\lambda)^2 - (x_0^\lambda)^2 - 1]/2$. Hence,

$$\begin{aligned} M_{c,\gamma}(v,u) &= E\left[\exp\left\{(a/2)\left((x_1^\lambda)^2 - (x_0^\lambda)^2 - 1\right)\right\}\right] \\ &= \exp\{(-a/2)(\gamma^2+1)\}E\left[\exp\left\{(a/2)(x_1^\lambda)^2\right\}\right], \quad (16) \end{aligned}$$

since $x_0^\lambda = \gamma$, a fixed constant. Now, the solution to (15) is given by

$$x_r^\lambda = \exp(\lambda r)\gamma + \int_0^r \exp(\lambda(r-s))\,dw_s = \exp(\lambda r)\,\gamma + J_\lambda(r)$$

and hence, $x_1^\lambda = \exp(\lambda)\gamma + J_\lambda(1)$. Now $J_\lambda(1) \sim \mathcal{N}(0, (e^{2\lambda}-1)/2\lambda)$ and therefore x_1^λ is $\mathcal{N}(\exp(\lambda)\gamma, (e^{2\lambda}-1)/2\lambda)$. Let $s^2 = (e^{2\lambda} - 1)/2\lambda$ and $q = \exp(\lambda)\gamma$. Then $(x_1^\lambda)^2/s^2$ is distributed as a noncentral chi-square variate with noncentrality parameter q. Hence:

$$\begin{aligned} E\left[\exp\left\{(a/2)(x_1^\lambda)^2\right\}\right] &= E\left[\exp\left\{(a/2)s^2(x_1^\lambda/\mathrm{s})^2\right\}\right] \\ &= (1-as^2)^{-1/2}\exp\left[(aq^2/2)(1-as^2)^{-1}\right] \end{aligned}$$

and

$$\begin{aligned} M_{c,\gamma}(v,u) = \exp&\left\{(-a/2)(\gamma^2+1)\right\}(1-as^2)^{-1/2} \\ &\times\exp\left\{(aq^2/2)(1-as^2)^{-1}\right\}. \end{aligned}$$

Upon substitution for $s^2 = (e^{2\lambda}-1)/2\lambda$ and $a = v + c - \lambda$, it is straightforward to verify that $\psi_c(v,u) = (1-as^2)^{-1/2}\exp(-a/2)$, and it follows that:

$$\begin{aligned} M_{c,\gamma}(v,u) &= \psi_c(v,u)\exp\left\{(-a\gamma^2/2) + (aq^2/2)(1-as^2)^{-1}\right\} \\ &= \psi_c(v,c)\exp\left\{(-a\gamma^2/2)(1-\exp(2\lambda+a)\psi_c(v,u)^2)\right\}, \end{aligned}$$

which reduces to the expression in Theorem 2 upon substitution for $a = v + c - \lambda$. □

The moment-generating function derived in Theorem 2 allows us to compute the cumulative distribution function as well as other distributional quantities using numerical integration. Denote the joint characteristic function of $(\int_0^1 x_r\, dw_r, \int_0^1 x_r^2\, dr)$ by $cf_{c,\gamma}(v,u)$. With i denoting the imaginary number, we have

$$cf_{c,\gamma}(v,u) = M_{c,\gamma}(iv,\, iu) = E\left[\exp\left(iv\int_0^1 x_r\, dw_r + iu\int_0^1 x_r^2\, dr\right)\right]. \tag{17}$$

The distribution function of $N(\hat{\theta}-\theta)$ can be obtained as follows. Let $F_{c,\gamma}(z) = P[N(\hat{\theta}-\theta) \le z]$ and recall that $N(\hat{\theta}-\theta) = \int_0^1 x_r\, dw_r / \int_0^1 x_r^r\, dr$. By Theorem 1 of Gurland (1948):

$$\begin{aligned} F_{c,\gamma}(z) &= (1/2) - (1/2\pi i) \lim_{\varepsilon_1\to 0,\, \varepsilon_2\to\infty} \int_{\varepsilon_1<|v|<\varepsilon_2} [cf_{c,\gamma}(v,-vz)/v]\, dv \\ &= (1/2) - (1/2\pi)\int_0^\infty \mathrm{AIMAG}[cf_{c,\gamma}(v,-vz)/v]\, dv \end{aligned} \tag{18}$$

where AIMAG$(\cdot)$ denotes the imaginary part of the complex number. Further, the density function is derived as follows:

$$\begin{aligned} f_{c,\gamma}(z) &= dF_{c,\gamma}(z)/dz \\ &= (1/2\pi i) \lim_{\varepsilon_1\to 0,\varepsilon_2\to\infty} \int_{\varepsilon_1<|v|<\varepsilon_2} \{\partial cf_{c,\gamma}(v,u)/\partial u\}_{u=-vz} dv \\ &= (1/\pi)\int_0^\infty \mathrm{REAL}[cf_{c,\gamma}(v,-vz)(A1 + A2/2)]\, dv, \end{aligned} \tag{19}$$

where

$$\begin{aligned} A1 = (-\gamma^2/2)\{&(1-\exp(iv+c+\lambda))\,[\psi_c(iv,-ivz)^2] \\ &+ (iv+c-\lambda)\exp(iv+c+\lambda)\psi_c(iv,-ivz)(-\lambda^{-1}+A2)\} \end{aligned}$$

and

$$\begin{aligned} A2 &= -\lambda^{-2} + D2/D1; \\ D1 &= [\lambda + (iv+c)]e^{-\lambda} + [\lambda - (iv+c)]e^{\lambda}, \\ D2 &= D1/\lambda + (e^{\lambda} + e^{-\lambda})/\lambda, \\ \lambda &= (c^2 + 2iv(c+z))^{1/2}, \end{aligned}$$

and REAL$(\cdot)$ denotes the real part of the complex number.

The expressions (18) and (19) can be numerically integrated to obtain values for the cumulative distribution function and the probability density function. When calculating these functions, we evaluate the integrals in the range $(0 + \varepsilon, \bar{V})$ where $\bar{V}$ is an upper bound set such that the integrand evaluated at $\bar{V}$ is less than ε (ε was set at $1.0E{-}07$ in each integration). The integrals are then evaluated in this range using the subroutine DCADRE of the International Mathematical and Statistical Library (IMSL) (the error criterion for this routine integration was also set at $1.0E{-}07$). Special care, however, must be applied to the integration since it involves the square root of a complex valued function. The use of the principal value of the square root may not ensure the continuity of the integrand. We must therefore integrate over the Riemann surface consisting here of two planes. The method used is described in more detail in the appendix to Perron (1989a).

Figures 1–3 present the probability density functions of $N(\hat{\theta} - \theta)$ for $c = -5$, 0, and 2, respectively. Each figure contains four curves corresponding to the following values for γ: 0, 0.5, 1.0, and 2.0. Each curve was constructed by evaluating the p.d.f. at 100 equidistant points in the relevant range.

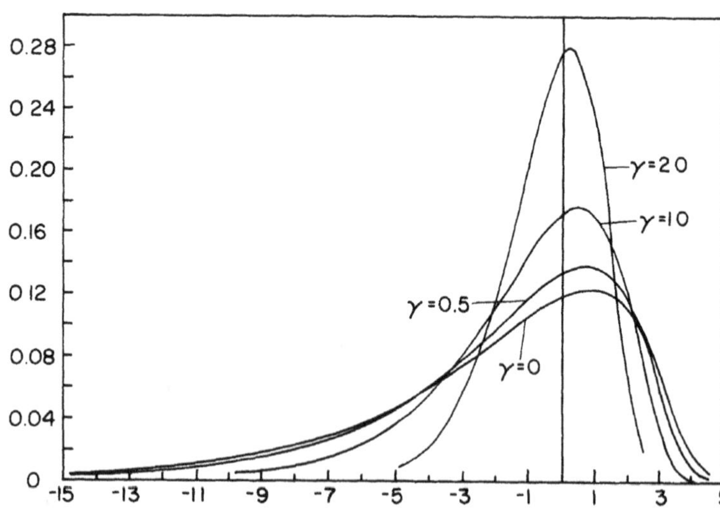

Fig. 1. Probability density function of $N(\hat{\theta} - \theta)$; $c = -5.0$.

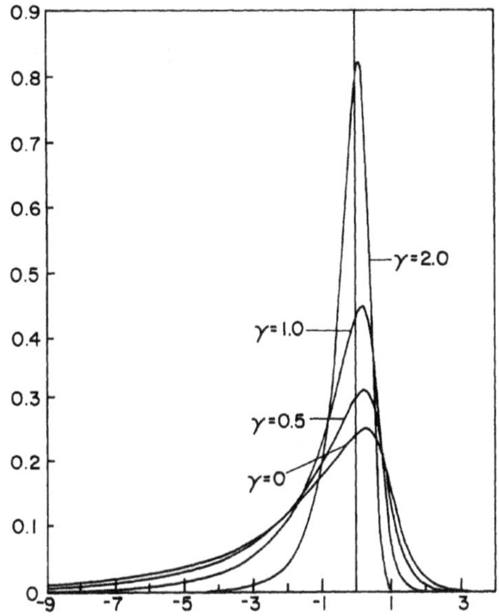

Fig. 2. Probability density function of $N(\hat{\theta} - \theta)$; $c = 0.0$.

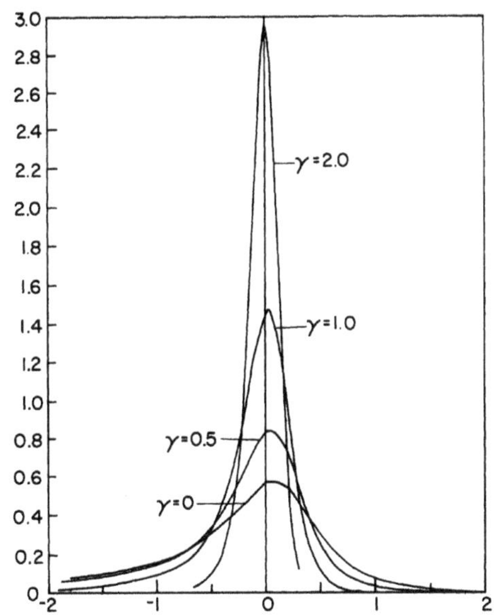

Fig. 3. Probability density function of $N(\hat{\theta} - \theta)$; $c = 2.0$.

Several striking features emerge from these graphs. First, the distributions are sharply nonnormal. Second, the shapes are significantly affected by changes in the values of γ and c. The most important feature is that the distribution becomes more closely centered and more concentrated near zero as γ and c increase. As c increases, the distribution becomes less skewed to the left and shows much less variability around zero. A similar behavior occurs as γ increases. Another interesting feature is the effect of increasing γ for different values of c. For a given change in γ, the effect on the shape of the distribution is more pronounced the larger c is. In other words, the effects of different values of γ are smaller when θ is smaller (i.e., 'more stationary') or when N is larger, provided $\theta < 0$. When $\theta > 0$, the effect of the initial condition is considerable. To summarize the main features of the various distributions, Table 1 presents selected percentage points of the distribution of $N(\hat{\theta} - \theta)$ for various values of c and γ.

In this framework, what is an 'exceptionally large' initial condition γ depends upon the value of c. To gain some insights into this matter it is useful to consider an asymptotically stationary process ($c < 0$). In this case, the stationary distribution of $\{x_r\}$ is $\mathcal{N}(0, -1/2c)$. Using this benchmark distribution, we can calculate how far a specific value of γ is in the tail of the distribution. For the cases reported in Table 1, the probabilities of obtaining an absolute value of x_r less than the initial condition γ are as follows: for $c = -5$, 0.8858 ($\gamma = 0.5$), 0.9984 ($\gamma = 1.0$), ≈ 1.0 ($\gamma = 2.0$); for $c = -2.0$, 0.6826 ($\gamma = 0.5$), 0.9544 ($\gamma = 1.0$), ≈ 1.0 ($\gamma = 2.0$); and for $c = -1.0$, 0.5222 ($\gamma = 0.5$), 0.8414 ($\gamma = 1.0$) and 0.9954 ($\gamma = 2.0$). Hence $\gamma = 2.0$ can be viewed as an exceptionally large initial condition unless c is very close to 0. On the other hand, a value $\gamma = 0.5$ could be considered a standard draw while a value $\gamma = 1.0$ can be considered large only if c is quite small.

To assess the adequacy of the asymptotic approximation to the finite sample distribution, we proceed as follows. We simulated the least-squares estimator $\hat{\alpha}$ from an AR(1) model with $\mathcal{N}(0,1)$ errors. The number of replications was set at 20,000. We obtained the percentage points of the distribution of $T(\hat{\alpha}-\alpha)$ for selected values of T, α, and $y_0 = b$. The correspondence between the exact distribution

Table 1. Selected percentage points of the distribution of $N(\hat{\theta} - \theta)$.

	1%	2.5%	5%	10%	90%	95%	97.5%	99%
$c = -5.0$								
$\gamma = 0.0$	−16.7378	−13.1743	−10.4633	−7.7242	2.4561	3.0637	3.5239	4.0342
$= 0.5$	−13.8955	−10.9832	−8.7550	−6.4899	2.2695	2.8257	3.2413	3.6788
$= 1.0$	−9.5405	−7.6095	−6.1140	−4.5703	1.9279	2.4017	2.7578	3.1215
$= 2.0$	−4.8739	−3.9531	−3.2206	−2.4413	1.3836	1.7357	2.0123	2.3035
$c = -2.0$								
$\gamma = 0.0$	−15.0342	−11.6602	−9.1377	−6.6406	1.4622	1.926*	2.3095	2.7849
$= 0.5$	−12.2749	−9.5498	−7.5038	−5.4726	1.3047	1.6839	2.005*	2.3921
$= 1.0$	−8.1023	−6.3512	−5.0258	−3.6952	1.0808	1.3688	1.5996	1.8594
$= 2.0$	−3.8162	−3.0438	−2.4456	−1.8281	0.7943	0.9940	1.1599	1.3310
$c = -1.0$								
$\gamma = 0.0$	−14.3901	−11.0813	−8.6182	−6.2072	1.1752	1.5887	1.9502	2.4005
$= 0.5$	−11.6484	−8.9854	−7.0027	−5.0566	1.007*	1.3498	1.6426	2.0031
$= 1.0$	−7.5197	−5.8323	−4.5689	−3.3203	0.8076	1.056*	1.2535	1.4819
$= 2.0$	−3.3477	−2.6348	−2.0925	−1.5445	0.5908	0.7512	0.8784	1.012*
$c = 0.0$								
$\gamma = 0.0$	−13.6919	−10.4399	−8.0383	−5.7133	0.9280	1.2854	1.6122	2.0325
$= 0.5$	−10.9570	−8.3535	−6.4315	−4.5709	0.7643	1.0481	1.3042	1.6338
$= 1.0$	−6.8480	−5.2202	−4.0196	−2.8568	0.5876	0.7780	0.9384	1.1302
$= 2.0$	−2.7392	−2.0886	−1.6093	−1.1454	0.4116	0.5271	0.6211	0.7260
$c = 1.0$								
$\gamma = 0.0$	−12.9229	−9.7183	−7.3710	−5.1243	0.7101	1.0079	1.2966	1.6832
$= 0.5$	−10.1753	−7.6177	−5.7469	−3.9627	0.5544	0.7771	0.9932	1.2871
$= 1.0$	−6.0273	−4.4386	−3.2793	−2.1696	0.4017	0.5366	0.6571	0.8093
$= 2.0$	−1.7104	−1.1842	−0.8640	−0.5948	0.2610	0.3360	0.3994	0.4725
$c = 2.0$								
$\gamma = 0.0$	−12.0557.	−8.8762	−6.5587	−4.3431	0.5191	0.7585	1.0067	1.3550
$= 0.5$	−9.2554	−6.7080	−4.8444	−3.0206	0.3756	0.5401	0.7143	0.9680
$= 1.0$	−4.8763	−3.1253	−1.7404	−0.8782	0.2493	0.3357	0.4182	0.5294
$= 2.0$	−0.5897	−0.4307	−0.3294	−0.2374	0.1466	0.1896	0.2272	0.2712
$c = 5.0$								
$\gamma = 0.0$	−7.8066	−3.0640	−1.0714	−0.3734	0.1316	0.2212	0.3410	0.5473
$= 0.5$	−2.2400	−0.5435	−0.2103	−0.0969	0.0611	0.0974	0.1505	0.2571
$= 1.0$	−0.0826	−0.0579	−0.0438	−0.0315	0.0283	0.0385	0.0491	0.0650
$= 2.0$	−0.0275	−0.0225	−0.0186	−0.0142	0.0137	0.0177	0.0214	0.0259

Note: The integral (15) which evaluates $P[N(\hat{\theta} - \theta) \leq z]$ is undefined for $z = -c$ and the numerical integration is accordingly ill-behaved for values of z around $-c$. A * indicates where such a situation occurs. While all digits reported are accurate for the other entries, in these cases the accuracy is not as precise.

Table 2. Selected percentage points of the distribution of $T(\hat{\alpha} - \alpha)y_t = \alpha y_{t-1} + e_t$, $y_0/\sigma = T^{1/2}\gamma$, $\alpha = \exp(c/T)$.

	1%	2.5%	5%	10%	90%	95%	97.5%	99%
$c = -5.0$								
$\gamma = 0.5$								
$T = 10$	−7.750	−6.555	−5.442	−4.225	2.065	2.556	2.962	3.436
$T = 25$	−10.539	−8.711	−7.123	−5.421	2.165	2.722	3.143	3.578
$T = 50$	−12.061	−9.530	−7.840	−5.935	2.213	2.772	3.206	3.646
$T = 100$	−13.305	−10.487	−8.314	−6.266	2.212	2.777	3.193	3.614
$\gamma = 1.0$								
$T = 10$	−5.642	−4.756	−3.941	−3.050	1.693	2.079	2.395	2.713
$T = 25$	−7.407	−6.053	−5.002	−3.874	1.815	2.260	2.610	2.930
$T = 50$	−8.488	−6.930	−5.616	−4.228	1.862	2.335	2.677	3.038
$T = 100$	−8.969	−7.193	−5.956	−4.464	1.869	2.341	2.713	3.056
$\gamma = 2.0$								
$T = 10$	−3.282	−2.708	−2.245	−1.727	1.167	1.464	1.694	1.939
$T = 25$	−4.048	−3.308	−2.708	−2.099	1.278	1.594	1.869	2.152
$T = 50$	−4.462	−3.683	−3.031	−2.317	1.323	1.656	1.943	2.232
$T = 100$	−4.696	−3.831	−3.117	−2.385	1.333	1.675	1.956	2.251
$c = 0.0$								
$\gamma = 0.5$								
$T = 10$	−7.892	−6.392	−5.230	−3.897	0.864	1.188	1.508	1.915
$T = 25$	−9.589	−7.578	−5.897	−4.347	0.791	1.098	1.372	1.765
$T = 50$	−9.954	−7.873	−6.128	−4.454	0.797	1.092	1.348	1.688
$T = 100$	−10.457	−8.159	−6.417	−4.587	0.772	1.052	1.313	1.608
$\gamma = 1.0$								
$T = 10$	−5.253	−4.209	−3.364	−2.469	0.634	0.843	1.001	1.221
$T = 25$	−5.892	−4.638	−3.677	−2.653	0.597	0.793	0.961	1.167
$T = 50$	−6.557	−5.128	−3.995	−2.788	0.598	0.798	0.955	1.156
$T = 100$	−6.629	−5.178	−4.063	−2.891	0.588	0.777	0.935	1.125
$\gamma = 2.0$								
$T = 10$	−2.433	−1.873	−1.462	−1.057	0.430	0.549	0.648	0.758
$T = 25$	−2.589	−1.996	−1.569	−1.102	0.417	0.531	0.630	0.737
$T = 50$	−2.715	−2.055	−1.582	−1.131	0.416	0.534	0.629	0.738
$T = 100$	−2.721	−2.107	−1.613	−1.151	0.411	0.527	0.617	0.718
$c = 2.0$								
$\gamma = 0.5$								
$T = 10$	−7.598	−5.796	−4.461	−2.832	0.459	0.655	0.867	1.163
$T = 25$	−8.423	−6.302	−4.686	−2.962	0.406	0.583	0.770	1.054
$T = 50$	−8.953	−6.493	−4.751	−3.016	0.401	0.565	0.745	1.008
$T = 100$	−9.207	−6.595	−4.801	−3.000	0.377	0.550	0.718	0.957

(Continued)

Table 2. (*Continued*)

	1%	2.5%	5%	10%	90%	95%	97.5%	99%
$\gamma = 1.0$								
$T = 10$	−4.359	−2.564	−1.421	−0.788	0.290	0.386	0.481	0.611
$T = 25$	−4.306	−2.870	−1.617	−0.833	0.263	0.356	0.445	0.558
$T = 50$	−4.703	−3.036	−1.676	−0.860	0.262	0.348	0.432	0.543
$T = 100$	−5.037	−3.163	−1.735	−0.887	0.252	0.340	0.418	0.526
$\gamma = 2.0$								
$T = 10$	−0.562	−0.437	−0.338	−0.247	0.165	0.214	0.257	0.311
$T = 25$	−0.585	−0.432	−0.334	−0.241	0.154	0.200	0.237	0.288
$T = 50$	−0.596	−0.445	−0.335	−0.242	0.151	0.196	0.235	0.279
$T = 100$	−0.597	−0.424	−0.325	−0.238	0.149	0.192	0.226	0.272

and the asymptotic approximation is obtained by setting $c = T\ln(\alpha)$ (since $\alpha = \exp(\theta N/T)$) and $\gamma = b/T^{1/2}$ since $\sigma^2 = 1$. Table 2 presents the results for $T = 10$, 25, 50, and 100 with $c = -5$, 0, and 2 and $\gamma = 0.5$, 1.0, 2.0. We omit the case where $\gamma = 0.0$ since it was thoroughly analyzed in Cavanagh (1986), Nabeya and Tanaka (1990), and Perron (1989a). These studies found the approximation to be adequate for various values of c. The approximation deteriorates as c becomes smaller.

In our more general context, the approximation remains quite good especially for values of T at 50 or 100. The approximation gets more accurate as (i) c increases, (ii) γ increases, and (iii) if we consider the right tail rather than the left. It is also relatively less accurate in the extreme tails. For example, consider the case with $c = 2$, $\gamma = 2$, and the 5th percentage point. The asymptotic approximation is −0.3294 and the exact values range from −0.338 for $T = 10$ to −0.325 for $T = 100$, showing a high degree of accuracy even for a very small sample size. On the other hand, when c is small, the convergence to the asymptotic value is slower. Consider $c = -5$, $\gamma = 0.5$, and the 5th percentage point. The asymptotic critical value is −8.755 while the exact values range from −5.442 ($T = 10$) to −8.314 ($T = 100$). However, even for small values of c, it gets more accurate as γ increases. For example, when $c = -5.0$, $\gamma = 2.0$, the asymptotic critical value for the 95th percentage point is 1.7357, and the exact values range from 1.464 ($T = 10$) to 1.675 ($T = 100$).

Overall, the asymptotic distribution approximates rather well the major changes in the distribution as c and γ are varied and for sample sizes typically available the approximation is very good.

4. The Moments of $\hat{\theta}$ and the Power Function of Tests

The moment-generating function derived in Section 3 can be used to derive the moments of $N(\hat{\theta}-\theta)$. Using Mehta and Swamy's (1978) result, we have

$$E[N(\hat{\theta}-\theta)]^r = \Gamma(r)^{-1}\int_0^\infty u^{r-1}\{\partial^r M_{c,\gamma}(v,-u)/\partial v^r\}_{v=0}\,du, \tag{20}$$

where $M_{c,\gamma}(v,u) = E[\exp(v\int_0^1 x_r\,dw_r + u\int_0^1 x_r^2\,dr)]$ is the joint moment-generating function of $(\int_0^1 x_r\,dw_r, \int_0^1 x_r^2\,dr)$ derived in Theorem 2.

Figure 4 presents the normalized bias of $\hat{\theta}$ (i.e., the first moment of $N(\hat{\theta}-\theta)$) as a function of $c=\theta N$ for various values of $\gamma = b/\sigma N^{1/2}$ (a similar graph for the second moment is available on request; it also displays the feature that the second moment decreases

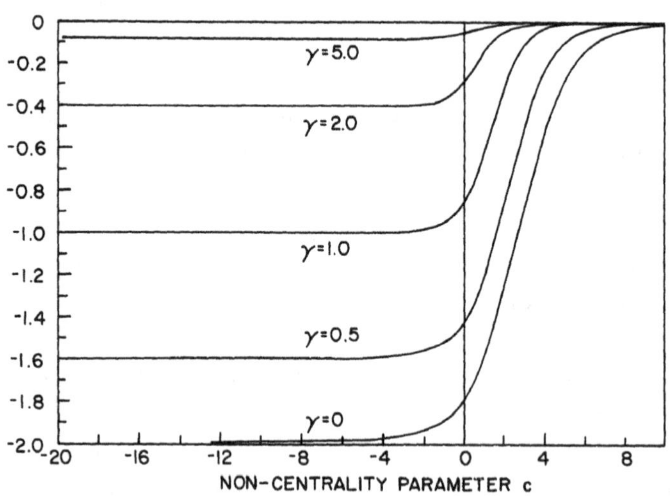

Fig. 4. Normalized bias of $N(\hat{\theta}-\theta)$ as a function of $c=\theta N$; $E[N(\hat{\theta}-\theta)]$.

Table 3. Asymptotic approximation to the expectation of $g(T)(\hat{\alpha} - \alpha)$.

T	$\alpha = 0.90$	$\alpha = 0.99$	$\alpha = 1.00$	$\alpha = 1.01$	$\alpha = 1.05$
			$y(0)/\sigma = 0.0$		
25	−0.8933 (−0.7339)	−2.5718 (−2.2980)	−1.2597 (−1.1347)	−4.4482 (−4.0382)	−2.0080 (−1.8770)
50	−0.6416 (−0.5528)	−1.8457 (−1.7374)	−1.2597 (−1.1955)	−2.7778 (−2.6566)	−2.4096 (−2.3709)
100	−0.4570 (−0.4027)	−1.3342 (−1.2890)	−1.2597 (−1.2271)	−2.1295 (−2.0902)	−3.6704 (−3.7018)
200	−0.3240 (−0.2886)	−0.9675 (−0.9467)	−1.2597 (−1.2433)	−2.2841 (−2.2710)	−5.2877 (−5.3839)
400	−0.2293 (−0.2054)	−0.6975 (−0.6867)	−1.2597 (−1.2514)	−3.3706 (−3.3726)	— —
			$y(0)/\sigma = 4.0$		
25	−0.5509 (−0.4622)	−1.5492 (−1.4152)	−0.7480 (−0.6882)	−2.5895 (−2.3978)	−0.9968 (−0.9368)
50	−0.4878 (−0.4222)	−1.4021 (−1.3279)	−0.9467 (−0.9038)	−2.0429 (−1.9642)	−1.3231 (−1.2855)
100	−0.3943 (−0.3477)	−1.1555 (−1.1179)	−1.0835 (−1.0571)	−1.7800 (−1.7490)	−1.8444 (−1.8091)
200	−0.3001 (−0.2673)	−0.8981 (−0.8791)	−1.1657 (−1.1509)	−2.0108 (−1.9989)	−2.3509 (−2.39)*
400	−0.2205 (−0.1975)	−0.6712 (−0.6608)	−1.2110 (−1.2032)	−2.9477 (−2.9467)	— —
			$y(0)/\sigma = 16.0$		
25	−0.0805 (−0.0713)	−0.1677 (−0.1592)	−0.0732 (−0.0697)	−0.2244 (−0.2145)	−0.0455 (−0.0439)
50	−0.1060 (−0.0942)	−0.2605 (−0.2523)	−0.1509 (−0.1468)	−0.2556 (−0.2493)	−0.0308 (−0.0302)
100	−0.1288 (−0.1148)	−0.3678 (−0.3598)	−0.2968 (−0.2920)	−0.3150 (−0.3102)	−0.0060 (−0.0059)
200	−0.1423 (−0.1272)	−0.4311 (−0.4236)	−0.5120 (−0.5072)	−0.3955 (−0.3910)	−0.0001 (−0.0001)
400	−0.1399 (−0.1253)	−0.4287 (−0.4225)	−0.7480 (−0.7440)	−0.4107 (−0.4045)	— —

Note: The values in parentheses are the exact values as derived by Evans and Savin (1981b) except for the starred entry. See footnote 2 for a discussion about the value reported in this case.

markedly as either c or γ increases). The conjectures discussed in Section 2 are verified. The bias and mean-squared error (MSE) are smaller (in absolute value) as c increases. Furthermore, both decrease (in absolute value) as γ increases. An increase in γ has a larger reduction effect on both the bias and the mean-squared error when c is small. This is mainly due to the fact that the bias and MSE are quite small for large c when $\gamma = 0$.

Both the bias and MSE converge quite rapidly to zero as c and/or γ increase. On the other hand, the bias appears to attain an upper limit (in absolute value) when $c \to -\infty$ for any fixed γ and it approaches this limit fairly rapidly. The limit of the bias as $c \to -\infty$ is a decreasing (in absolute value) function of γ.

To assess the relevance of the asymptotic approximation to the exact moments of $\hat{\alpha}$ we consider applying our approximation to the experimental setting in Table 3 of Evans and Savin (1981b) which presents the exact mean and variance of a normalized estimator of $\hat{\alpha}$; $\alpha = 0.90$, 0.99, 1.00, 1.01, and 1.05; $y_0/\sigma = 0$, 4, 16; and $T = 25$, 50, 100, 200, and 400. The normalization is defined as $g(T)(\hat{\alpha} - \alpha)$ where: $g(T) = [T(1-\alpha^2)^{-1}]^{1/2}$ for $|\alpha| < 1; g(T) = T/\sqrt{2}$ for $|\alpha| = 1$; and $g(T) = |\alpha|^T(\alpha^2 - 1)^{-1}$ for $|\alpha| > 1$. To approximate the moments of $g(T)(\hat{\alpha} - \alpha)$, we consider the correspondence $c = T\ \mathrm{In}(\alpha)$ and $y = y_0/\sigma T^{1/2}$ and use (20).

The results concerning the expectation of $g(T)(\hat{\alpha} - \alpha)$ are presented in Table 3. The values in parentheses are taken from Table 3 of Evans and Savin (1981b). The approximation is, in general, very good and certainly captures most of the changes in the distribution as $\alpha, y_0/\sigma$, and T are varied.[1] As expected, the

[1]The exact value reported in Table 3 for the entry ($\alpha = 1.05$, $T = 200$, and $y_0/\sigma = 4$) does correspond to the result reported in Evans and Savin's (1981b, Table 3) which is –0.0688. This latter result is erroneous, probably due to a numerical integration error that occurred because of an inadequate choice for the bounds of integration. In a case such as this one, the numerical integrations are very sensitive to the proper choice of these bounds. The value reported here was obtained by a simulation experiment using 1,000,000 replications. Evans and Savin's result for the standard error corresponding to this case is also erroneous; our simulations

approximation is better as α is closer to 1 and as T is larger. But it is also better as y_0/σ increases. The same experiment was performed for the standard deviation of $g(T)(\hat{\alpha} - \alpha)$. The results showed an approximation with similar features and hence are not reported. Overall, our procedure appears to provide a very adequate approximation to the exact moments of $\hat{\alpha}$ when α is in the vicinity of one, allowing the initial condition y_0/σ to be an arbitrary fixed value.

Using the statistic $N(\hat{\theta} - \theta_0)$, the results of Section 3 can also be used to analyze the power function for tests of the null hypothesis H_0: $\theta = \theta_0$ against various alternatives. Denote by z^* the value such that $P_{\theta_0}[N(\hat{\theta}-\theta_0) \leq z^*] = \beta$. Then, the power function of a one-sided test with size β for testing $\theta = \theta_0$ against $\theta < \theta_0$ is given by:

$$\begin{aligned} P_\theta[N(\hat{\theta} - \theta_0) < z^*] &= P_\theta[N(\hat{\theta} - \theta) < z^* + N(\theta - \theta_0)] \\ &= P_\theta[Z(\gamma, c) < z^* + (c - c_0)] \end{aligned} \tag{21}$$

with $c = N\theta$ and $c_0 = N\theta_0$. Expression (18) can be used to evaluate the power function (21) for various values of c, γ, c_0, and significance level β.

Much attention has recently been given to testing the null hypothesis of a unit root ($\alpha = 1$, $\theta = 0$). Letting $\theta_0 = 0$ and using (21), the power function of a one-sided test of the unit root hypothesis using the statistic $N\hat{\theta}$ is given by

$$P_\theta[N\hat{\theta} < z_0^*] = P_\theta[N(\hat{\theta} - \theta) < z_0^* - c] = P_\theta[Z(\gamma, c) < z_0^* - c], \tag{22}$$

which can be evaluated using (18). Here, z_0^* is the critical value such that $P[Z(\gamma, 0) < z_0^*] = \beta$ where β is the size of the test.

Interestingly, (22) gives the local asymptotic power function of a test for a unit root using the statistic $T(\hat{\alpha} - 1)$ (in discrete-time) where the sequence of local alternatives is given by $\alpha = \exp(\theta h)$ and the asymptotic is derived as $h \to 0$, i.e., using the continuous records

---

yielded a value of 158.35 instead of their reported 5.6014 figure. Our asymptotic approximation for the standard deviation this case was 161.275, indeed quite close to the exact value obtained by our simulations.

asymptotic instead of the usual ($T \to \infty$) asymptotic. To elaborate, note that

$$\lim_{h \to 0} P_\alpha[T(\hat{\alpha} - 1) < z_0^*] = \lim_{h \to 0} P_\alpha[T(\hat{\alpha} - \alpha) < z_0^* + T(\alpha - 1)]$$
$$= P[Z(\gamma, c) < z_0^* - c]$$

since $T(\hat{\alpha} - \alpha) \Rightarrow Z(\gamma, c)$ as $h \to 0$ (Theorem 1) and $T(\alpha - 1) = Nh^{-1}(\exp(\theta h) - 1) \to \theta N \equiv c$ as $h \to 0$.

This framework generalizes the analysis of Phillips (1987b) who considers the local asymptotic power in the usual ($T \to \infty$) asymptotic context. Here, the continuous records asymptotic ($h \to 0$) permits explicit consideration of the effect of a nonzero initial condition. Our results specialize to those of Phillips by setting $\gamma = 0$. Note finally, that when $\gamma = 0$, the critical values z_0^* correspond to those derived by Dickey and Fuller (1979); see Fuller (1976).

Figure 5 presents the power function of a one-sided 5% size test that $\theta = 0$ against both positive and negative alternative values of $c = \theta N$. Four curves are drawn corresponding to values of γ at 0, 0.5, 1.0, and 2.0. Several features emerge. First, as expected from the studies of Evans and Savin (1981b) and others,

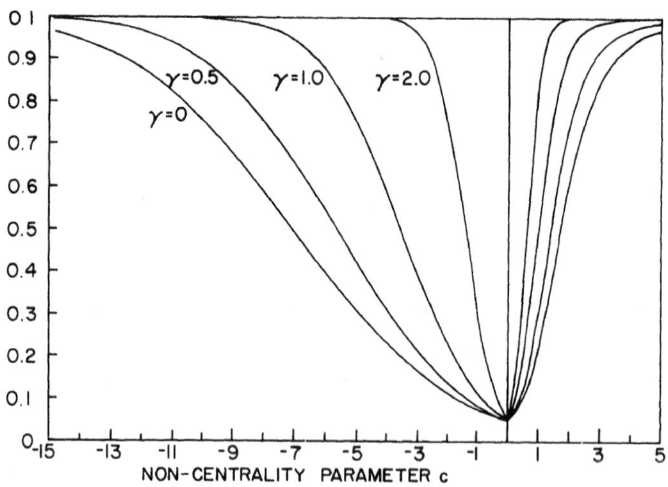

Fig. 5. Power function of 5% one-tailed tests of $\theta = 0$ as a function of $c = \theta N$.

the power is much higher against explosive alternatives ($c > 0$). Secondly, the power function is significantly influenced by different values of the startup condition γ. Indeed, the power increases as γ increases. There is an interesting analogy with the bias function (see Fig. 4) in that changes in γ have a larger effect on the power function for stationary alternatives. This is due to the fact that when $\gamma = 0$ the power function against explosive alternatives is already quite high compared to the power function under stationary alternatives.

To assess the adequacy of our asymptotic power function in the context of finite sample tests for a unit root against alternatives close to one, we considered the experimental framework adopted in Evans and Savin (1981b). We used our asymptotic framework to approximate each entry in their Table IV. The results (available upon request) are quite striking. For almost all combinations of parameters the approximation is indeed excellent. The worst fit occurs for a small sample size ($T = 25$) and large values of α (1.05). The conclusions reached by Evans and Savin (1981b) are well-captured by the asymptotic approximations.

5. Concluding Comments

This paper has considered an alternative asymptotic approximation to the least-squares estimator of the parameter in a first-order autoregressive model with a fixed initial condition. Our approach is different in that we consider a framework using the continuous records asymptotics ($h \to 0$) instead of the usual ($T \to \infty$) asymptotic theory. The main advantage over the ($T \to \infty$) near integrated framework is that our method allows for explicit consideration of the effects of nonzero initial conditions.

The continuous record asymptotic distribution for the discrete-time estimator is shown to be equivalent to the distribution of the continuous time estimator of the parameter in an Ornstein–Uhlenbeck process. The exact distribution of this continuous time estimator was derived. It was found to yield interesting theoretical results for continuous time estimation. More importantly, it has proved to be a good approximation to the finite sample distribution of the discrete-time least squares estimator.

This study indicates the usefulness of the continuous records asymptotic distributional theory which can be used in other contexts. One such possible extension is the possibility to approximate the finite sample distribution of the least-squares estimator $\hat{\alpha}$ when a random initial condition is specified. Indeed, Equation (16) can be used to derive the moment-generating function of $N(\hat{\theta} - \theta)$ with a random initial condition as well. A study of the case where the stochastic process is stationary is presented in Perron (1988). Another avenue of research, using the proposed framework, involves models where the innovations are dependent. We hope to report results on this topic in the near future.

Further possible extensions include the case where a fitted mean is included in the regression. Here, the parameter affecting the shape of the distribution is a normalization of the mean of the variables instead of the initial condition *per se* (see Evans and Savin (1984)). Again, the analysis can distinguish between a fixed and random initial condition. Such an analysis will be presented in future work. It remains to be seen whether analogs to Theorem 2 can be obtained in the context of more general models and different statistics such as the t ratio.

Mathematical Appendix

Proof of Theorem 1, Part (ii): Let $a(h) = \sigma^2(\exp(2\theta h) - 1)/2\theta$ denote the variance of the innovations $\{u_{th}\}$ in (6). Since $\hat{\alpha}$ is invariant to a scale transformation of the data, we can, without loss of generality, divide the data by $a(h)^{1/2}$. Denote this transformation by $x_{th} = y_{th}/a(h)^{1/2}$. Note that $\lim_{h\to 0} h^{-1}a(h) = \sigma^2$. We consider a triangular array of random variables $\{\{x_{nt}\}_{t=1}^{T_n}\}_{n=1}^{\infty}$. We have $T_n = N/h_n$ and $T_n \to \infty$, $h_n \to 0$ as $n \to \infty$. For a given n, the sequence $\{x_{nt}\}_{n=1}^{T_n}$ is generated by

$$x_{nt} = \exp(\theta h_n)x_{n(t-1)} + \varepsilon_{nt}, \quad x_{n0} = y_0/a(h_n)^{1/2} \quad (t = 1, \ldots, T_n), \tag{A.1}$$

where the innovation sequence $\{\varepsilon_{nt}\}_{t=1}^{T_n}$ is i.i.d. normal with mean 0 and variance 1. From (A.l) we deduce that

$$x_{nt} = \sum_{j=1}^{t} \exp\left((t-j)\theta h_n\right)\varepsilon_{nj} + \exp\left(t\theta h_n\right)x_{n0}. \tag{A.2}$$

We consider a sequence of statistics $\{T_n(\hat{\alpha}_n - \alpha_n)\}_{n=1}^{\infty}$ where, for a fixed n, we have

$$T_n(\hat{\alpha}_n - \alpha_n) = T_n^{-1}\sum_{t=1}^{T_n} x_{nt-1}\varepsilon_{nt} \Big/ T_n^{-2}\sum_{t=1}^{T_n} x_{nt-1}^2, \tag{A.3}$$

with x_{nt} defined by (A.l). Lemma A.l characterizes the limit of the sample moments of the sequence $\{x_{nt}\}$ necessary to derive the asymptotic distribution of $T_n(\hat{\alpha}_n - \alpha_n)$ as $n \to \infty$.

Lemma A.1. *If* $\{\{x_{nt}\}_{t=1}^{T_n}\}_{n=1}^{\infty}$ *is a triangular array of random variables defined by* (A.l), *then as* $n \to \infty$ ($h_n \to 0$ *and* $T_n \to \infty$ *with* N *fixed*) :

(a) $T_n^{-1/2}x_{T_n} \Rightarrow J_c(1) + \exp(c)\gamma;$

(b) $T_n^{-3/2}\sum_{t=1}^{T_n} x_{nt} \Rightarrow \int_0^1 J_c(r)\,dr + \gamma(\exp(c) - 1)/c;$

(c) $$T_n^{-2}\sum_{t=1}^{T_n} x_{nt}^2 \Rightarrow \int_0^1 J_c(r)^2\,dr + 2\gamma\int_0^1 \exp(cr)J_c(r)\,dr + \gamma^2(\exp(2c) - 1)/2c;$$

(d) $$\begin{aligned} T_n^{-1}\sum_{t=1}^{T_n} x_{nt-1}\varepsilon_{nt} &\Rightarrow \int_0^1 J_c(r)\,dw_r + \gamma\Big[\exp(c)J_c(1) - 2c\int_0^1 \exp(cr)J_c(r)\,dr\Big] \\ &\overset{d}{=} \int_0^1 J_c(r)\,dw_r + \gamma\int_0^1 \exp(cr)\,dw_r; \end{aligned}$$

where $J_c(r) = \int_0^r \exp(c(r-s))\,dw_s$, $c = \theta N$, *and* $\gamma = y_0/(\sigma N^{1/2})$.

To prove part (a) use (A.2) to obtain:

$$T_n^{-1/2} x_{T_n} = T_n^{-1/2} \sum_{j=1}^{T_n} \exp\left((T_n - j)\theta h_n\right) \varepsilon_{nj} + T_n^{-1/2} \exp\left(T_n \theta h_n\right) x_{n0}$$

$$= T_n^{-1/2} \sum_{j=1}^{T_n} \exp\left((T_n - j)\theta N / T_n\right) \varepsilon_{nj}$$

$$+ N^{-1/2} h_n^{1/2} a(h_n)^{-1/2} \exp\left(\theta N\right) y_0.$$

The first term converges to $J_c(1)$ by Lemma 1(a) of Phillips (1987b) and the second term converges to $\gamma \exp(c)$ given that $h_n^{-1} a(h_n) \to \sigma^2$ as $h_n \to 0$. For part (b) we have:

$$T_n^{-3/2} \sum_{t=1}^{T_n} x_{nt} = T_n^{-3/2} \sum_{t=1}^{T_n} \sum_{j=1}^{t} \exp\left((t - j)\theta h_n\right) \varepsilon_{nj}$$

$$+ T_n^{-3/2} \sum_{t=1}^{T_n} \exp(t\theta h_n) x_{n0}$$

$$= T_n^{-3/2} \sum_{t=1}^{T_n} \sum_{j=1}^{t} \exp((t - j)\theta N / T_n) \varepsilon_{nj} + y_0 N^{-1/2} h_n^{-1/2}$$

$$\times a(h_n)^{1/2} T_n^{-1} (\exp\left(\theta N\right) - 1) / (\exp(\theta N / T_n) - 1).$$

The first term converges to $\int_0^1 J_c(r)\, dr$ using Lemma 1(b) of Phillips (1987b) and the second term converges to $\gamma(\exp(c) - 1)/c$ given that $h_n^{-1} a(h_n) \to \sigma^2$ as $h_n \to 0$, and $T_n(\exp(\theta N / T_n) - 1) \to \theta N \equiv c$, as $T_n \to \infty$. To prove part (c) consider

$$T_n^{-2} \sum_{t=1}^{T_n} x_{nt}^2 = T_n^{-2} \sum_{t=1}^{T_n} \left[\sum_{j=1}^{t} \exp\left((t - j)\theta h_n\right) \varepsilon_{nj} + \exp\left(t\theta h_n\right) x_{n0} \right]^2$$

$$= T_n^{-2} \sum_{t=1}^{T_n} \left(\sum_{j=1}^{t} \exp\left((t - j)\theta N / T_n\right) \varepsilon_{nj} \right)^2$$

$$+\,2T_n^{-1/2}x_{n0}T_n^{-3/2}\sum_{t=1}^{T_n}\exp\,(t\theta N/T_n)$$

$$\times\sum_{j=1}^{t}\exp\,((t-j)\theta N/T_n)\varepsilon_{nj}$$

$$+\,T_n^{-1}x_{n0}^2\,T_n^{-1}\sum_{t=1}^{T_n}\exp\,(2t\theta N/T_n).$$

The first term converges to $\int_0^1 J_c(r)^2\,dr$ using Lemma 1(c) of Phillips (1987b). Using arguments similar to those found in the proof of this lemma, it can be shown that the second term converges to $2\gamma\int_0^1\exp\,(cr)J_c(r)\,dr$. The third term converges to $\gamma^2(\exp(2c)-1)/2c$ in a manner similar to part (b). To prove part (d), note that by squaring (A.l), subtracting x_{nt-1}^2 from both sides, and summing over $t=1,\ldots,T_n$, we obtain

$$x_{T_n}^2-x_{nt=0}^2(\exp(2\theta h_n)-1)\sum_{t=1}^{T_n}x_{nt-1}^2+2\exp(\theta h_n)\sum_{t=1}^{T_n}x_{nt-1}\varepsilon_{nt}+\sum_{t=1}^{T_n}\varepsilon_{nt}^2$$

and

$$T_n^{-1}\sum_{t=1}^{T_n}x_{nt-1}\varepsilon_{nt}=(1/2)\exp(-\theta h_n)$$

$$\times\left[T_n^{-1}x_{T_n}^2-T_n^{-1}x_{n0}^2-T_n(\exp(2\theta N/T_n)-1)\right.$$

$$\left.\times\,T_n^{-2}\sum_{t=1}^{T_n}x_{nt-1}^2-T_n^{-1}\sum_{t=1}^{T_n}\varepsilon_{nt}^2\right].$$

Now, $T_n^{-1}\sum_{t=1}^{T_n}\varepsilon_{nt}^2\to 1$ as $T_n\to\infty$, $T_n(\exp\,(2\theta N/T_n)-1)\to 2\theta N\equiv 2c$ as $T_n\to\infty$, $\exp(-\theta h_n)\to 1$ as $h_n\to 0$, $T_n^{-1}x_{n0}^2\to y_0^2/\sigma^2N\equiv\gamma^2$, and using parts (a), (b), and (c) we obtain

$$T_n^{-1}\sum_{t=1}^{T_n}x_{nt-1}\varepsilon_{nt}\Rightarrow(1/2)\left[(J_c(1)+\gamma\,\exp\,(c))^2-\gamma^2-2c[\psi]-1\right],$$

where

$$\psi = \int_0^1 J_c(r)^2\, dr + 2\gamma \int_0^1 \exp{(cr)} J_c(r)\, dr + \gamma^2 (\exp{(2c)} - 1)/2c.$$

Rearranging and using the fact that

$$\int_0^1 J_c(r)\, dw_r = (1/2)\left[J_c(1)^2 - 2c \int_0^1 J_c(r)^2\, dr - 1\right]$$

(see Phillips, 1987b, Equation (8)), we have

$$T_n^{-1} \sum_{t=1}^{T_n} x_{nt-1}\varepsilon_{nt} \Rightarrow \int_0^1 J_c(r)\, dw_r + \gamma \left[\exp{(c)} J_c(1) - 2c \int_0^1 \exp{(cr)} J_c(r)\, dr\right].$$

What remains to be shown is that $\exp(c)J_c(1) - 2c\int_0^1 \exp{(cr)} J_c(r)\, dr \overset{d}{=} \int_0^1 \exp{(cr)}\, dw_r$. Now, from the proof of Lemma A.2 below: $\int_0^1 \exp{(cr)}\, dw_r \sim \mathcal{N}(0, (\exp(2c){-}1)/2c)$. Hence, we must show that $\exp{(c)} J_c(1) - 2c \int_0^1 \exp{(cr)}\, J_c(r)\, dr \sim \mathcal{N}(0, (\exp{(2c)} - 1)/2c)$.

Note that

$$J_c(1) \sim \mathcal{N}(0, (\exp{(2c)} - 1)/2c)$$

and

$$\int_0^1 \exp{(cr)} J_c(r)\, dr \sim \mathcal{N}(0, (\exp{(4c)} - 1 - \exp{(2c)}4c)/8c^3)$$

(see the proof of Lemma A.2). Therefore,

$$\left[\exp{(c)} J_c(1) - 2c \int_0^1 \exp{(cr)} J_c(r)\, dr\right]$$

is Gaussian with mean 0 and variance given by

$$\exp{(2c)}\ \mathrm{Var}\ [J_c(1)] + 4c^2\ \mathrm{Var} \left[\int_0^1 \exp{(cr)} J_c(r)\, dr\right] - 4c \exp{(c)}\, E\left[J_c(1) \int_0^1 \exp{(cr)} J_c(r)\, dr\right].$$

Now,

$$\begin{aligned} E\left[J_c(1)\int_0^1 \exp(cr)J_c(r)\,dr\right] &= \int_0^1 \exp(cr)E[J_c(1)J_c(r)]\,dr \\ &= \int_0^1 \{\exp(cr)[\exp(c(r+1)) \\ &\quad - \exp(c(1-r))]/2c\}\,dr \\ &= \exp(c)[\exp(2c) - 1 - 2c]/(2c)^2. \end{aligned}$$

Therefore $[\exp(c)J_c(1) - 2c\int_0^1 \exp(cr)J_c(r)\,dr]$ has variance given by $(\exp(2c)-1)/2c$. This proves part (d). The proof of Theorem 1 follows by using (A.3) and parts (c) and (d) of Lemma A.l.

Proof of Corollary 1. We first prove the following Lemma concerning the limit as $c \to \pm\infty$ of the various terms involved in the definition of $A(\gamma, c)$ and $B(\gamma, c)$.

Lemma A.2. *Let* $J_c(r) = \int_0^r \exp(c(r-s))dw_s$, *and* η *and* ξ *be independent* $\mathcal{N}(0,1)$ *variates; then:*

(i) $(-2c)\int_0^1 J_c(r)^2\,dr \to 1$ *in probability, as* $c \to -\infty$;
(ii) $(2c)^2 e^{-2c}\int_0^1 J_c(r)^2\,dr \Rightarrow \eta^2$, *as* $c \to +\infty$;
(iii) $(-2c)^{1/2}\int_0^1 J_c(r)\,dw_r \Rightarrow \mathcal{N}(0,1)$, *as* $c \to -\infty$;
(iv) $(2c)e^{-c}\int_0^1 J_c(r)\,dw_r \Rightarrow \xi\eta$, *as* $c \to +\infty$;
(v) $(-2c)^{3/2}\int_0^1 \exp(cr)J_c(r)\,dr \Rightarrow \mathcal{N}(0,1)$, *as* $c \to -\infty$;
(vi) $(2c)^{3/2} e^{-2c}\int_0^1 \exp(cr)J_c(r)\,dr \Rightarrow \eta$, *as* $c \to +\infty$;
(vii) $(-2c)^{1/2}\int_0^1 \exp(cr)\,dw_r \Rightarrow \mathcal{N}(0,1)$, *as* $c \to -\infty$;
(viii) $(2c)^{1/2} e^{-c}\int_0^1 \exp(cr)\,dw_r \to \eta$, *as* $c \to +\infty$.

Parts (i) *through* (iv) *of Lemma* A.2 *are proved in* Phillips (1987b), *Lemma* 2. *To prove parts* (*v*) *and* (*vi*), *wc first note that* $w(s)$ *is Gaussian with mean zero and variance* s; *then* $J_c(r)$ *and* $\int_0^1 \exp(cr)J_c(r)\,dr$ *are also Gaussian with mean zero for any* c. *Now* $J_c(r) \sim \mathcal{N}(0, (\exp(2cr) - 1)/2c)$ *and using the fact that* $E(w(s)w(t)) = \min(t,s)$, *simple calculations yield* $E[J_c(r)J_c(k)] = [\exp(c(r+k)) - \exp(c(M-m))]/2c$, *where* $M = \max(r,k)$ *and* $m = \min(r,k)$. *This result can be used to show that* $\int_0^1 \exp(cr) J_c(r)\, dr \sim \mathcal{N}(0, \upsilon)$, *where* $\upsilon = (\exp(4c) - 1 - 4c \exp(2c))/8c^3$. *Parts* ($\upsilon$) *and* ($\upsilon i$) *follow by noting, respectively, that* $(-2c)^3\upsilon \to 1$ *as* $c \to -\infty$, *and that* $(2c)^3 \exp(-4c)\upsilon \to 1$ *as* $c \to \infty$.

To prove parts (vii) *and* (viii), *note that* $\int_0^1 \exp(cr)\, dw_r = \exp(c) \int_0^1 \exp(-c(1-r))\, dw_r = \exp(c) J_{-c}(1)$. *Now,* $J_{-c}(1) \sim \mathcal{N}(0, (\exp(-2c) - 1)/(-2c))$; *hence* $\int_0^1 \exp(cr)\, dw_r \sim \mathcal{N}(0, (\exp(2c) - 1)/2c) \equiv \mathcal{N}(0, q)$, *say. Parts* (vii) *and* (viii) *follow by noting, respectively, that* $(-2c)q \to 1$ *as* $c \to -\infty$, *and that* $(2c)\exp(-2c)q \to 1$ *as* $c \to +\infty$.

The proof of Corollary 1 (i) follows by writing

$$N^{1/2}(\hat{\theta} - \theta) = (-2\theta)^{1/2}[(-2c)^{1/2} A(\gamma, c)/(-2c)B(\gamma, c)],$$

and noting that $(-2c)^{1/2} A(\gamma, c) \Rightarrow \mathcal{N}(0, 1)$ as $c \to -\infty$, from Lemma A.2 (iii, vii), and $(-2c)B(\gamma, c) \to 1$ in probability as $c \to -\infty$, from Lemma A.2 (i,v) (also noting that $\gamma \to 0$ as $N \to \infty$). The proof of Corollary 1 (ii) follows by writing

$$\begin{aligned} \exp(\theta N)(\hat{\theta} - \theta)/2\theta &= (2c)^{-1} \exp(c) Z(\gamma, c) \\ &= [(2c) \exp(-c) A(\gamma, c)]/[(2c)^2 \exp(-2c) B(\gamma, c)]. \end{aligned}$$

Now $(2c)\exp(-c)A(\gamma, c) \Rightarrow d\eta + \xi\eta$, using Lemma A.2 (iv, viii) and the fact that $(2c)^{1/2}\gamma = d$. Furthermore, $(2c)^2 \exp(-2c) B(\gamma, c) \Rightarrow d^2 + 2d\eta + \eta^2$ using Lemma A.2 (ii, vi) and the fact that $(2c)^2 \exp(-2c)\gamma^2(\exp(2c) - 1)/2c \to d^2$ as $c \to +\infty$.

Acknowledgments

Author has benefited from comments by Daniel Nelson, Peter Phillips, Eugene Savin, the referees, and especially a co-editor. Research support is acknowledged from the Social Sciences and Humanities Research Council of Canada, the Fonds pour la Formation de Chercheurs et l'Aide à la Recherche du Québec, and the Natural Sciences and Engineering Research Council of Canada.

References

Arnold, L. (1974). *Stochastic Differential Equations: Theory and Applications.* John Wiley: New York.

Basawa, I.V. and Rao, B.P. (1980). *Statistical Inference for Stochastic Processes.* Academic Press: New York.

Brown, B.M. and Hewitt, J.I. (1975). Asymptotic likelihood theory for diffusion processes. *Journal of Applied Probability* 12, 228–238.

Cavanagh, C. (1986). Roots local to unity. Mimeo, Department of Economics, Harvard University.

Chan, N.H. (1988). The parameter inference for nearly nonstationary time series. *Journal of the American Statistical Association* 83, 857–862.

Chan, N.H. and Wei, C.Z. (1987). Asymptotic inference for nearly nonstationary AR(1) processes. *Annals of Statistics* 15, 1050–1063.

Dickey, D.A. (1976). Estimation and hypothesis testing for nonstationary time series. Unpublished Ph.D. Dissertation, Iowa State University, Ames.

Dickey, D.A. and Fuller, W.A. (1979). Distribution of the estimators for autoregressive time series with a unit root. *Journal of the American Statistical Association* 74, 427–431.

Evans, G.B.A. and Savin, N.E. (1981a). The calculation of the limiting distribution of the least squares estimator of the parameter in a random walk model. *Annals of Statistics* 9, 1114–1118.

Evans, G.B.A. and Savin, N.E. (1981b). Testing for unit roots: 1. *Econometrica* 49, 753–779.

Evans, G.B.A. and Savin, N.E. (1984). Testing for unit roots: 2. *Econometrica* 52, 1241–1269.

Fuller, W.A. (1976). *Introduction to Statistical Time Series.* John Wiley: New York.

Gurland, J. (1948). Inversion formulae for the distribution of ratios. *Annals of Mathematical Statistics* 19, 228–237.

Liptser, R.S. and Shiryayev, A.N. (1978). *Statistics of Random Processes II: Applications.* Springer-Verlag, New York.

Mann, H.B. and Wald, A. (1943). On the statistical treatment of linear stochastic difference equations. *Econometrica* 11, 173–220.

Mehta, J.S. and Swamy, P.A.V.B. (1978). The existence of moments of some simple bayes estimators of coefficients in a simultaneous equation model. *Journal of Econometrics* 7, 1–14.

Nabeya, S. and Tanaka, K. (1990). A general approach to the limiting distribution for estimators in time series regression with nonstable autoregressive errors. *Econometrica* 58, 145–163.

Perron, P. (1988). A continuous-time approximation to the stationary first-order autoregressive model. Econometric research program memorandum No. 337, Princeton University.

Perron, P. (1989a). The calculation of the limiting distribution of the least squares estimator in a near-integrated model. *Econometric Theory* 5, 241–255.

Perron, P. (1989b). Test consistency with varying sampling frequency. Econometric research program memorandum No. 345, Princeton University.

Perron, P. (1990). Tests of joint hypotheses for time series regression with a unit root. in *Advances in Econometrics: Cointegration, Spurious Regression, and Unit Roots,* ed. by T. B. Fomby and G. F. Rhodes. Vol. 8, JAI Press: Greenwich, CT, pp. 135–159.

Phillips, P.C.B. (1977). Approximations to some finite sample distributions associated with a first-order stochastic difference equation. *Econometrica* 45, 463–485.

Phillips, P.C.B. (1987a). Time series regression with unit roots. *Econometrica* 55, 277–302.

Phillips, P.C.B. (1987b). Towards a unified asymptotic theory for autoregression. *Biometrika* 74, 535–547.

Phillips, P.C.B. (1988). Regression theory for near-integrated time series. *Econometrica* 56, 1021–1044.

Phillips, P.C.B. and Perron, P. (1988). Testing for a unit root in time series regression. *Biometrika* 75, 335–346.

Rao, M.M. (1978). Asymptotic distribution of an estimator of the boundary parameter of an unstable process. *Annals of Statistics* 6, 185–190.

Rubin, H. (1950). Consistency of maximum likelihood estimates in the explosive case. In *Statistical Inference in Dynamic Economic Models,* T.C. Koopmans. ed. by Wiley: New York.

Said, S.E. and Dickey, D.A. (1984). Testing for unit roots in autoregressive-moving average models of unknown order. *Biometrika* 71, 599–607.

White, J.S. (1958). The limiting distribution of the serial correlation coefficient in the explosive case. *Annals of Mathematical Statistics* 29, 1188–1197.

Chapter 5

Test Consistency with Varying Sampling Frequency*

Pierre Perron

Department of Economics, Princeton University,
Princeton, NJ 08544, USA
Centre de Recherche et Développement en Economique,
Université de Montreal, Montreal QC H3T 134, Canada

This paper considers the consistency property of some test statistics based on a time series of data. While the usual consistency criterion is based on keeping the sampling interval fixed, we let the sampling interval take any equispaced path as the sample size increases to infinity. We consider tests of the null hypotheses of the random walk and randomness against positive autocorrelation (stationary or explosive). We show that tests of the unit root hypothesis based on the first-order correlation coefficient of the original data are consistent as long as the span of the data is increasing. Tests of the same hypothesis based on the first-order correlation coefficient of the first-differenced data are consistent against stationary alternatives only if the span is increasing at a rate greater than $T^{1/2}$, where T is the sample size. On the other hand, tests of the randomness hypothesis based on the first-order correlation coefficient applied to the original data are consistent as long as the span is not increasing too fast. We provide Monte Carlo evidence on the power, in finite samples, of the tests studied allowing various combinations of span and sampling frequencies. It is found that the consistency properties summarize well the behavior of the power in finite samples. The power of tests for a unit root is more influenced by the span than the number of observations while tests of randomness are more powerful when a small sampling frequency is available.

Keywords: Hypothesis testing, unit root, randomness, near-integrated processes, time-series, continuous-time asymptotic.

---

*Originally Published in *Econometric Theory* 7, pp. 341–368.

1. Introduction

Foremost among the various criteria which have been proposed to assess the adequacy of a test statistic is the consistency property. It is generally considered a useful necessary condition for a test to possess since it ensures that a statistic would distinguish a fixed false null hypothesis from a disjoint true alternative if the sample size is sufficiently large (at least for an interesting part of the possible parameter space which is complement to the null space). More formally, consider a parameterized testing procedure. Let θ_0 be the null value of θ which we want to test and $P_T(\theta)$ the power function of the test considered based on a sample of size T and evaluated at a fixed alternative $\theta \neq \theta_0$. A test statistic is said to be consistent if its power function $P_T(\theta)$, evaluated at any *fixed* alternative $\theta \neq \theta_0$, converges to one as the sample size increases to infinity when the true value is θ (e.g., Rao, 1973, p. 464). Consistency justifies our belief that a larger sample leads to a more powerful testing procedure. Of course how large 'sufficiently large' is depends on the particular circumstances.

As stated, the consistency criterion appears straightforward and desirable but some ambiguity arises when considering a test statistic in the context of a time series of data. In this case the total number of observations depends upon both the total span (S) of the available data as well as the sampling frequency (h) via the relation $T = S/h$. An increase in the sample size T can be achieved by any of the following schemes: increasing S (with h fixed), decreasing h (with S fixed), increasing S and decreasing h, increasing S faster than increasing h, or decreasing h faster than decreasing S.

The requirement of a *fixed* alternative in the definition of consistency has usually implied that h is treated as fixed since in this case the parametric discrete time representation of the process is unchanged. Consider, for example, a first-order autoregressive process with a parameter $\alpha_Q = \overline{\alpha}$, say, when sampled quarterly. Sampled annually, the parameter becomes $\alpha_A = \overline{\alpha}^4$. Therefore, the autoregressive parameter is not fixed as the sampling frequency is changed. It is argued, accordingly, that to maintain the requirement of a fixed alternative in the definition of consistency one should only consider increasing the sample size without changing the sampling interval, by increasing the span one for one with the sample size.

This argument is not without ambiguity. Under both the annual and quarterly schemes the stochastic processes are the same. We simply use a different framework to represent them. Hence, these processes could be represented using a common fixed parameter. In general, one can represent this fixed parameter by considering a 'basic' sampling interval. This could be the sampling interval believed to represent the actual timing of the process. Usually, however, this choice is arbitrary since there is generally no basis suggesting such an interval. In order to avoid this type of arbitrariness one can view the basic process as one occurring in continuous time and consider the discrete time representation of the process in terms of the parameters of this continuous-time process and the actual sampling interval.

In this context, there are no reasons to view the consistency criterion solely as a sequence of tests with the sampling interval fixed as the sample size increases. It is useful to consider a more general power function, say $P_{T,h(T)}(\theta)$, indexed by both the sample size T and the sampling frequency h (or equivalently by T and $S(T)$ with $P_{T,S(T)}(\theta)$, say; or by h and S with $P_{h(T),S(T)}(\theta)$, say, given that, by definition, $T = S/h$). The sampling frequency, $h(T)$, is indexed by T to highlight the fact that as T increases h may vary. The criterion of consistency is the same as before, namely that $\lim_{T\to\infty} P_{T,h(T)}(\theta) = 1$ given any *fixed* alternative $\theta \neq \theta_0$.

A number of interesting questions can be posed in this more general context. Do our test statistics have more power if we increase the sample size by decreasing the sampling interval keeping a fixed span; for example, by going from annual to quarterly data? If the test is consistent, considering a sequence of sampling intervals that goes to zero as T tends to infinity, then we should have some reasons to answer yes. Since in economics the sample size is often increased this way this aspect of the consistency criterion may provide useful insight. Or consider the following problem: is the power greater with 100 years of annual data or 20 years of monthly data? If the test is not consistent as h decreases faster than T increases (i.e., S decreases) we may have reasons to expect the power to be greater with 100 years of annual data.

In this paper, we analyze in detail the general consistency properties of some statistics for testing the null hypotheses (i) of a random walk and (ii) of randomness, against stationary or explosive first-order correlation. We study the behavior of the power function allowing any path for the sampling interval as the sample size increases to infinity. We also restrict our analysis to the case of point sampled data, that is, we do not consider the effect of aggregation as the sampling interval changes.

Section 2 considers the null hypothesis of a random walk against the alternative hypothesis of either stationary or explosive positive autocorrelation. We first consider the Dickey and Fuller (1979) statistics based on the normalized least-squares estimator in a first-order autoregression. It is shown to be consistent against the specified alternative processes if and only if the span of the data increases as the sample size increases, that is, if the sampling interval decreases to zero at a rate slower than T. Hence, the power function can be said to be influenced more strongly by the span of the data than by the number of observations *per se*. A simulation experiment illustrates the usefulness of the asymptotic results in assessing the behavior of the power function in finite samples. In this section, extensive use is made of results originally developed by Perron (1991) and Phillips (1987).

We also consider in Section 2 the same null and alternative hypotheses but this time by applying a test of randomness on the first differences of the data. Here the power function converges to unity if and only if the span increases at least at a rate greater than $T^{1/2}$. This result is interesting because it suggests that, in some instances, it may be possible to increase the power by simply deleting observations, for example, by using the same span but a longer sampling interval. Furthermore, it can be said that this procedure is dominated by the Dickey and Fuller (1979) procedure in the sense that the latter is consistent over a wider range of possible paths for the sampling interval as the sample size increases. Again we provide simulation evidence to support the usefulness of our asymptotic framework in finite samples.

In Section 3, we consider the null hypothesis of randomness against the alternative hypothesis of positive serial correlation. The statistic investigated is simply the normalized least-squares estimator in a first-order autoregression. Here the behavior of the power function is rather different. The test statistic is consistent against stationary positive autocorrelation if and only if the span of the data does not increase too fast (in a sense to be made precise) relative to the number of observations. This behavior is unlike tests for the random walk hypothesis. Here a smaller sampling interval is preferable. Again, simulation evidence supports our asymptotic results. Section 4 presents a discussion of some issues and proposes simplified consistency criteria to assess the behavior of power functions in finite samples. Section 5 offers some concluding comments. A mathematical appendix contains the proof of some theorems.

The results obtained are reasonable from an intuitive perspective. Indeed, a random walk process is essentially characterized by long sustained deviations from its initial starting point but with an eventual return. Such a feature can be distinguished from stationary or explosive alternatives only with a long span of data and the longer the span the higher the chances of depicting departures from such a pattern. On the other hand, when dealing with a test of randomness against stationary positive autocorrelation we should expect higher power with a data set sampled frequently given that an increase in the sampling interval makes the dependence between adjacent observation smaller.

2. Testing for a Random Walk

Consider the simple Ornstein–Uhlenbeck diffusion process:

$$dy_t = \gamma y_t \, dt + \sigma dw_t; \quad -\infty < \gamma < \infty; \quad 0 < t \leq S; \quad y_0 = y(0); \tag{1}$$

where w_t is a unit Wiener process and γ, σ, and y (0) are fixed constants. S is the span of the data from which the process is sampled, that is, the length, in units of time, of the observed record of data. We consider testing for a random walk, hence our null hypothesis is H_0: $\gamma = 0$, and the alternative hypothesis is H_1 : $\gamma = \bar{\gamma}(-\infty < \bar{\gamma} < \infty, \bar{\gamma} \neq 0)$. The class of alternative hypotheses encompasses both stationary ($\gamma < 0$) and explosive ($\gamma > 0$) processes.

In this context, it is straightforward to derive the following discrete time representation of the process (1) in terms of the sampling interval parameter h:

$$y_{th} = \alpha_h y_{(t-1)h} + v_{th}, \quad y_0 = y(0); \quad t = 1, \dots, T = S/h, \tag{2}$$

where $\alpha_h = \exp(\gamma h)$ and $v_{th} \sim N(0, \sigma^2(\exp(2\gamma h) - 1)/2\gamma)$. This solution exists and is unique in a mean-squared sense (see, e.g., Bergstrom, 1984). Under the null hypothesis we have $\gamma = 0$ and $\alpha_h = 1$ for all h. It is interesting to remark that for any given fixed alternative $\gamma \neq 0$, α_h converges to 1 as h converges to zero. This implies that when testing H_0, the discrete time parameter α_h converges to the null value as the sampling interval decreases. On the other hand, if h increases to infinity, α_h converges to 0 when $\gamma < 0$ and to infinity when $\gamma > 0$, hence it moves away from the null value of the parameter.

A wide variety of statistics have been proposed to test H_0. We shall concentrate on two simple ones that are representative of broader classes of statistics. Consider first the following regression estimated by ordinary least-squares:

$$y_{th} = \hat{\alpha}_h y_{(t-1)h} + \hat{u}_{th}.$$

The statistic $T(\hat{\alpha}_h - 1)$ has been frequently used to test H_0 and the critical values under the null hypothesis have been tabulated by Dickey (1976) (see also Fuller, 1976). Under the null hypothesis of a random walk the sequence of first-differences $\{x_{th} = y_{th} - y_{(t-1)h}\}$ is i.i.d., while under the alternative hypothesis it is correlated. A commonly used statistic to test for randomness is the usual first-order autocorrelation coefficient denoted by

$$R_h = \frac{\sum_{t=2}^{T} x_{th} x_{(t-1)h}}{\sum_1^T x_{th}^2}.$$

We shall be concerned with the asymptotic behavior of the normalized statistic $T^{1/2}R_h$ as a test of the random walk hypothesis.

These statistics have been chosen to represent the following class of tests: (a) tests of a random walk based on the original series $\{y_{th}\}$; (b) tests of a random walk based on a test of randomness applied to the differenced series $\{x_{th}\}$. Many other statistics in these classes are available. An extensive Monte Carlo experiment with a wide class of statistics is presented in Perron (1989); their behavior corresponds to that of the statistics analyzed here.

2.1. *Testing for a random walk using* $T(\hat{\alpha}_h - 1)$

In order to analyze the limiting distribution of the statistics under the null hypothesis, we specify a triangular array of random variables $\{\{y_{nt}\}_{t=1}^{T_n}\}_{n=1}^{\infty}$. First consider the process under the null hypothesis of a random walk. For a given n, the sequence $\{y_{nt}\}_1^{T_n}$ is generated by (2) with $\gamma = 0$:

$$y_{nt} = y_{nt-1} + u_{nt}; \quad y_0 = y(0); \quad t = 1, \ldots, T_n, \tag{3}$$

where the innovation sequence $\{u_{nt}\}_{t=1}^{T_n}$ is i.i.d. $N(0, \sigma^2 h_n)$. T_n and h_n are related as $T_n h_n = S_n$ with $T_n \in Z$. We require $T_n \to \infty$ as $n \to \infty$ but do not impose any prescribed path for h_n and S_n. The sequence $\{\{u_{nt}\}_1^{T_n}\}_1^{\infty}$ is a triangular array of i.i.d. N $(0, \sigma^2 h_n)$ variates and $\{\{y_{nt}\}_1^{T_n}\}_1^{\infty}$ a triangular array of random walks.

Each row of the triangular array $\{\{y_{nt}\}_1^{T_n}\}_1^{\infty}$ is a sequence of random variables generated by a random walk in discrete time with a sampling interval h_n. In this framework, we can formally analyze the limiting behavior of the statistic $T(\hat{\alpha}_h - 1)$ as T increases but without imposing any conditions on h and S. For any given n, we let $T_n(\hat{\alpha}_n - 1)$ be the standardized least-squares estimator in a first-order autoregression based on a sample of size T_n sampled at frequency h_n. The asymptotic distribution of the statistic is contained in the following lemma, where, here and elsewhere in this article, $\Rightarrow$ denotes weak convergence in distribution and $w(r)$ is a standard Wiener process defined on C [0, 1], the space of all real valued continuous functions on the interval [0, 1]. As a matter of notation, we also let

$J_c(r) = \int_0^r \exp(c(r-s))dw(s)$ and define the following variables:

$$A(\delta, c) = \int_0^1 J_c(r)dw(r) + \delta \int_0^1 \exp(cr)dw(r);$$

$$B(\delta, c) = \int_0^1 J_c(r)^2 dr + 2\delta \int_0^1 \exp(cr)J_c(r)dr + \frac{\delta^2(\exp(2c)-1)}{2c};$$

and

$$Z(\delta, c) = \frac{A(\delta, c)}{B(\delta, c)}. \tag{4}$$

Note that when $\delta = c = 0$, $Z(\delta,c)$ reduces to the usual limit of the normalized least-squares estimator $T(\hat{\alpha} - 1)$ in the random walk case, that is, $Z(0,0) = \int_0^1 w(r)\,dw(r) / \int_0^1 w(r)^2 dr$.

Lemma 1. *Let* $\{\{y_{nt}\}_1^{T_n}\}_1^\infty$ *be a triangular array of random walks defined by* (3) *and let the variable* $Z(\delta, c)$ *be as defined in* (4), *then as* $n \to \infty$:

Case (1): $y(0) = 0$.

$T_n(\hat{\alpha}_n - 1) \Rightarrow Z(0,0)$; *for any possible path for* h_n *as* $n \to \infty$.

Case (2): $y(0) \neq 0$.

(a) *If* $S_n \to 0$ *as* $n \to \infty$:

$$S_n^{-1/2} T_n(\hat{\alpha}_n - 1) \Rightarrow N(0, \sigma^2/y(0)^2);$$

(b) *If* $S_n \to S$ *as* $n \to \infty$:

$$T_n(\hat{\alpha}_n - 1) \Rightarrow Z(\delta, 0), \quad \textit{with } \delta = y(0)/\sigma S^{1/2}.$$

(c) *If* $S_n \to \infty$ *as* $n \to \infty$: *The result in case* (1) *still applies.*

This lemma (proved in the appendix) generalizes a result obtained by Dickey and Fuller (1979) which considered the limiting distribution of $T_n(\hat{\alpha}_n - 1)$ when h is fixed (Phillips, 1987a also considered the case where S is fixed). Note that this result holds under the more general specifications where, for instance, the errors $\{u_{th}\}$ in (3) are martingale differences with a variance function $a(h)$, say, such that $h^{-1}a(h) \to \sigma^2$ and the stochastic process $\{\epsilon_{th} \equiv u_{th}/a(h)^{1/2}\}$ satisfies conditions permitting the application of a functional central limit theorem to their partial sums, that is, $T^{-1/2}\sum_1^{[Tr]}\epsilon_{th} \Rightarrow w(r)$ as $T \to \infty$ (see, e.g., Phillips, 1987). This precludes, among other things, processes which in continuous time are characterized by a Poisson process.

This lemma implies that, when $y(0) = 0$, the limiting distribution of $T_n(\hat{\alpha}_n - 1)$ is invariant with respect to possible paths for h_n and S_n; it is the same as long as T increases to infinity. When $y(0) \neq 0$, things are different unless $S_n \to \infty$ as $n \to \infty$. In particular, if $S_n \to 0$ as $n \to \infty$, a suitably normalized function of $\hat{\alpha}$ has an asymptotic normal distribution.

To analyze the limiting distribution of the statistic under the alternative hypothesis we shall again consider a triangular array of random variables $\{\{y_{nt}\}_{t=1}^{T_n}\}_{n=1}^{\infty}$. For a given n, the sequence $\{y_{nt}\}_{t=1}^{T_n}$ is generated by:

$$y_{nt} = \exp(\gamma h_n) y_{nt-1} + u_{nt}, \quad y_0 = y(0), \quad t = 1, \ldots, T_n; \qquad (5)$$

where the innovation sequence $\{u_{nt}\}_{t=1}^{T_n}$ is i. i. d. normal with mean 0 and variance $\sigma^2(\exp(2\gamma h_n) - 1)/2\gamma$. By analogy with the random walk case we refer to the sequence $\{\{y_{nt}\}_1^{T_n}\}_1^{\infty}$ as a triangular array of first-order autoregressive processes. This class of processes can accommodate stationary sequences ($\gamma < 0$) as well as explosive ones ($\gamma > 0$). We do not consider the case $\gamma = 0$ since it was analyzed in Lemma 1. Our next lemma concerns the limiting distribution of $T_n(\hat{\alpha}_n - 1)$ under (5). Its proof is presented in the appendix.

Lemma 2. *Let $\{\{y_{nt}\}_1^{T_n}\}_1^{\infty}$ be a triangular array of random variables defined by* (5) *and let $Z\ (\delta,\ c)$ be the variable defined by* (4); *then as* $n \to \infty$:

Case (1): $y(0) = 0$.

(a) *If $S_n \to 0$ as $n \to \infty : T_n(\hat{\alpha}_n - 1) \Rightarrow Z(0,0)$;*
(b) *If $S_n \to S$ as $n \to \infty : T_n(\hat{\alpha}_n - 1) \Rightarrow c + Z(0,c)$, with $c = \gamma S$;*
(c) *If $S_n \to \infty$ and $h_n \to 0$ as $n \to \infty$, and $\gamma < 0$:*

$$S_n^{-1/2} T_n(\hat{\alpha}_n - \alpha_n) \Rightarrow N(0, -2\gamma\sigma^2),$$

$$T_n(\hat{\alpha}_n - 1) \to -\infty \quad \textit{and} \quad S_n^{-1} T_n(\hat{\alpha}_n - 1) \to \gamma;$$

(d) *If* $S_n \to \infty$ *and* $h_n \to 0$ *as* $n \to \infty$, *and* $\gamma > 0$:

$$(2\gamma S_n)^{-1} \exp(\gamma S_n) T_n(\hat{\alpha}_n - \alpha_n) \Rightarrow Cauchy,$$
$$T_n(\hat{\alpha}_n - 1) \to +\infty \quad and \quad S_n^{-1} T_n(\hat{\alpha}_n - 1) \to \gamma.$$

Case (2): y *(0)* $\neq 0$.

(a) *If* $S_n \to 0$ *as* $n \to \infty$: $S_n^{-1/2} T_n(\hat{\alpha}_n - \alpha_n) \Rightarrow N(0, \sigma^2/y(0)^2)$;
(b) *If* $S_n \to S$ *as* $n \to \infty$: $T_n(\hat{\alpha}_n - 1) \Rightarrow c + Z(\delta, c)$, *with* $c = \gamma S$; *and* $\delta = y(0)/\sigma S^{1/2}$.
(c) *If* $S_n \to \infty$ *as* $n \to \infty$: *If* $\gamma < 0$, *the result of case* $(1, (c))$ *still applies and if* $\gamma > 0$:

$$(2\gamma S_n)^{-1} \exp(\gamma S_n) T_n(\hat{\alpha}_n - \alpha_n) \Rightarrow [d\eta + \xi\eta]/[d + \eta]^2;$$
$$S_n^{-1} T_n(\hat{\alpha}_n - 1) \to \gamma;$$

and $T_n(\hat{\alpha}_n - \alpha_n) \to +\infty$; *where* $d = y(0)\,(2\gamma)^{1/2}/\sigma$, *and* η *and* ξ *are independent* $N(0,1)$ *variates.*

Lemma 2 provides the limiting distribution of the standardized least-squares estimator under the alternative hypothesis, allowing the sampling interval to decrease as the sample size increases to infinity. Parts (a) consider the case where both the sampling interval and the span decrease to zero as the sample size increases to infinity. By comparing this result with that of Lemma 1, we observe that the limiting distribution is the same under both the null and the alternative hypotheses irrespective of the value of $y(0)$. This result holds for any alternative $\gamma \neq 0$ whether negative or positive. Therefore, in this case, the power of a test of the random walk hypothesis ($\gamma_0 = 0$) converges to the size of the test as the sample size increases. The test statistic $T_n(\hat{\alpha}_n - 1)$ is therefore not consistent if the span decreases as the sample size increases.

Parts (b) contain the theoretically interesting special case of a fixed span. Here the sampling interval h converges to zero at the same rate as the sample size increases to infinity. The limiting distribution derived in this case is equivalent to the so-called 'continuous records' asymptotics (see, e.g., Perron, 1991). In this case the limiting distributions under the null and the alternative hypotheses are different and so the limiting power of the tests does not converge to the size of the test as the sample size increases. However, $T_n(\hat{\alpha}_n - 1)$ is bounded in probability under both the null and the alternative hypotheses; hence, the test is not consistent. This result agrees with the fact that estimating γ with a continuum of observations and a fixed span yields an estimator with a nonzero variance; see, for example, Perron (1991) for the case analyzed here.

With $\gamma < 0$, the limiting power function of a size-β one-tailed test is:

$$\begin{aligned}\lim_{T\to\infty} P_{T,h(T)}(\gamma) &= \lim_{n\to\infty} \text{Prob}\left[T_n(\hat{\alpha}_n - 1) < \lambda_\beta\right] \\ &= \text{Prob}\left[\gamma S + Z\left(\delta, c\right) < \lambda_\beta\right] \\ &= \text{Prob}\left[Z\left(\delta, c\right) < \lambda_\beta - \gamma S\right]\end{aligned}$$

where $\delta = y(0)/\sigma S^{1/2}$ and λ_β is the $\beta\%$ point of the distribution of Z $(\delta, 0)$. When $y\ (0) = 0$, $\delta = 0$ and $Z\ (0{,}0)$ reduces to $(1/2)(w\,(1)^2 - 1)(\int_0^1 w(r)^2 dr)^{-1}$ whose percentage points can be obtained from the work of Dickey (1976) (see also Fuller, 1976; Evans and Savin, 1981). The same principle holds when $\gamma > 0$ except that the inequality sign is reversed and δ_β is the $(1 - \beta)\%$ point. A calculation of the limiting power function of the test in the general case can be obtained by numerically integrating the characteristic function of $Z(\delta, c)$ (see Perron, 1991, 1989).

Parts (c) and (d) consider the case where the sampling interval converges to zero but the span increases to infinity. In that case, $T_n(\hat{\alpha}_n - 1)$ converges to $-\infty$ if the alternative is stationary $(\gamma < 0)$ and to $+\infty$ if the alternative is explosive. The power of the test therefore converges to one in each case.

When the sampling interval of the data, h_n, increases with n, we adopt a rather different approach to prove the consistency of the test statistic $T_n(\hat{\alpha}_n - 1)$. We start with the statement of an auxiliary lemma proved in Mikulski and Monsour (1989).

Lemma 3. *Let $\hat{\alpha}$ be the least-squares estimator in a first-order autoregression with i.i.d. Normal errors with finite variance estimated* *from a sample of size T. Then for any* α *and* $\delta > 0$, *and for all* $T \geq T_0$ (δ) :

$$E|\hat{\alpha} - \alpha| \leq \{(1+\delta)6T/[(T-3)(T-5)]\}^{1/2} = k/T^{1/2} + o(T^{-1/2})$$

where $k = [6(1+\delta)]^{1/2}$.

It is useful to note that, in Lemma 3, the stated bound does not depend upon α. This result enables us to analyze the limiting behavior of the power function when $h_n \to \infty$ given that, for a fixed n, (5) satisfies the relevant conditions. Consider first the case where $\gamma < 0$ (stationary alternatives). We can write:

$$\begin{aligned} \Pr_{\alpha_n}\{T_n(\hat{\alpha}_n - 1) < \lambda_\beta\} &= 1 - \Pr_{\alpha_n}\{T_n(\hat{\alpha}_n - 1) > \lambda_\beta\} \\ &= 1 - \Pr_{\alpha_n}\{(\hat{\alpha}_n - \alpha_n) > \lambda_\beta/T_n + (1-\alpha_n)\}. \end{aligned} \tag{6}$$

Let $\epsilon_n = \lambda_\beta/T_n + (1-\alpha_n)$ and note that there will exist an n^* such that $\epsilon_n > 0$ for all $n > n^*$. Hence, for all $n > n^*$:

$$\begin{aligned} \Pr_{\alpha_n}\{(\hat{\alpha}_n - \alpha_n) > \epsilon_n\} &\leq \Pr_{\alpha_n}\{|\hat{\alpha}_n - \alpha_n| > \epsilon_n\} \\ &\leq (1/\epsilon_n)E|\hat{\alpha}_n - \alpha_n|, \text{ by Markov's inequality,} \\ &\leq (1/\epsilon_n)[k/T_n^{1/2} + o(T_n^{-1/2})], \end{aligned}$$

for all $n > \max(n^*, n(\delta))$, by Lemma 3, where $n(\delta)$ is such that $T_{n(\delta)} = T_0(\delta)$. Since $\alpha_n \to 0$ as $n \to \infty$, $\varepsilon_n \to 1$ as $n \to \infty$ and, using (6) $\lim_{n\to\infty} \Pr_{\alpha_n}\{T_n(\hat{\alpha}_n - 1) < \lambda_\beta\} = 1$, showing consistency. When $\gamma > 0$ (explosive alternatives), the development of the proof is similar and we obtain

$$\Pr_{\alpha_n}\{T_n(\hat{\alpha}_n - 1) > \lambda_{1-\beta}\} \geq 1 - (1/\epsilon_n)[k/T_n^{1/2} + o(T_n^{-1/2})],$$

for all n sufficiently large, where $\epsilon_n = (\alpha_n - 1) - \lambda_{1-\beta}/T_n$. Since $\alpha_n \to \infty$ as $n \to \infty$, $\epsilon_n \to \infty$ as $n \to \infty$ and $\lim_{n\to\infty} \Pr_{\alpha_n}\{T_n (\hat{\alpha}_n - 1) > \lambda_{1-\beta}\} = 1$, showing consistency.

The results of Lemmas 1 and 2 and the above discussion permit us to state the following result about the limiting power of the statistics $T_n(\hat{\alpha}_n - 1)$ as $n \to \infty$ when testing $\gamma = 0$ against an alternative $\gamma \neq 0$.

Theorem 1. *Let $\{y_{nt}\}$ be generated by (5) with $y(0)$ any fixed value. The statistic $T_n(\hat{\alpha}_n - 1)$ for testing the null hypothesis $\gamma = 0$ is consistent against an alternative hypothesis $\gamma \neq 0$ if and only if the span of the data increases as the sample size increases; that is,*

$$\lim_{n\to\infty} P_{T_n,S_n}(\gamma) = 1 \quad \text{iff} \quad S_n \to \infty \quad \text{as } n \to \infty;$$

When the statistic is not consistent we have

(i) *If $S_n \to 0$ as $n \to \infty$: $\lim_{n\to\infty} P_{T_n,S_n}(\gamma) = \beta$, where β is the size of the test, and*

(ii) *If $S_n \to S$ as $n \to \infty$:*

  (a) *if $\gamma < 0$: $\lim_{n\to\infty} P_{T_n,S_n}(\gamma) = Prob[Z(\delta, c) < \lambda_\beta - \gamma S]$, where $Z(\delta, c)$ is defined in (4) with $c = \gamma S$ and $\delta = y(0)/\sigma S^{1/2}$; and λ_β is the β-percentage point of the distribution of $Z\ (\delta, 0)$;*

  (b) *if $\gamma > 0$: $\lim_{n\to\infty} P_{T_n,S_n}(\gamma) = Prob[Z(\delta, c) < \lambda_{(1-\beta)} - \gamma S]$.*

Theorem 1 illustrates in a formal way the idea that the power of a test of the random walk hypothesis depends much more on the span of the data than on the number of observations *per se*. Indeed, given these theoretical results one can expect several features related to the behavior of the power function in finite samples: (a) the increase in power should eventually decrease and become marginal as the number of observations is increased, keeping a fixed span; (b) the power should eventually decrease if more observations are added while reducing the span of the data; (c) a longer span of data given a fixed number of observations should yield a higher power; and (d) it may be possible that the power function be higher with less observations when these fewer observations are spaced at longer intervals.

Some intuition behind the results of Theorem 1 can be obtained by considering the power of tests for a unit root under local alternatives. An implication of Theorem 1 in Phillips and Ouliaris (1990) is that, in the fixed sampling interval case, $T(\hat{\alpha} - 1)$ diverges at rate T under the alternative hypothesis of stationary fluctuations. This fact, in the unit root context, is easily deduced given that $T(\hat{\alpha}-1) = T^{1/2}[T^{1/2}(\hat{\alpha}-\alpha)] + T(\alpha-1)$ which diverges at rate T since, for any fixed α ($|\alpha| < 1$), $T^{1/2}(\hat{\alpha}-\alpha)$ has a nondegenerate asymptotic distribution and $T(\alpha-1)$ diverges at rate T. In the context of varying sampling interval, α is not fixed. Indeed given $\alpha = \exp(\gamma S/T)$ we have that $T(\alpha - 1)$ is bounded as $T \to \infty$ with S fixed. Hence, the interpretation is that as $h \to 0$ (at rate T), the discrete time parameter α approaches the null value 1 at rate T, that is, at the same rate at which $T(\hat{\alpha}-1)$ diverges under the alternative hypothesis (in the fixed interval framework). The nondegenerate power function in the fixed S case occurs because the two rates are the same.

To assess the significance of these theoretical results we conducted a simulation experiment similar to the one performed in Shiller and Perron (1985). We simulated the power of the statistic $T(\hat{\alpha} - 1)$ for the null hypothesis of a random walk ($\gamma = 0$) against a stationary alternative $\gamma = -0.2$. Each series was generated by the following process:

$$y_t = \exp(\gamma S/T)y_{t-1} + u_t, \quad t = 1, \ldots, T \tag{7}$$

with $y_0 = 0$. We specify $u_t \sim N(0, 1)$, without loss of generality, since $\hat{\alpha}$ is invariant with respect to the variance of the innovation when $y_0 = 0$. We simulated 10,000 replications for each of the following values of S and T: 8, 16, 32, 64, 128, 256, 512. The critical values of $T(\hat{\alpha} - 1)$ under the null hypothesis were also simulated using 10,000 replications of a random walk model. The size of the tests was set at 5%, and we considered only one-tailed tests against stationary alternatives. Table 1 presents the results. The values obtained clearly show the relevance of the asymptotic properties in assessing the behavior of the power function in finite samples. The infinity row in Table 1 is the limiting power of the test as the span increases to infinity keeping T fixed. As can be verified from Equation (7), this is simply the power of a test of the random walk hypothesis against the alternative hypothesis of randomness, since $\lim_{S\to\infty} \exp(\gamma S/T) = 0$.

In Table 1, the power values obtained for the statistic $T(\hat{\alpha} - 1)$ show that the power increases with the sample size keeping the span fixed but that it tends to level off as T increases. When both S and h decrease the power decreases. The power is always higher for a given

Table 1. Power of a one-tailed test of the random walk hypothesis with the statistic $T(\hat{\alpha} - 1)$; H_0:$\gamma = 0$; $H_1 : \gamma = -0.2$.

	T						
S	8	16	32	64	128	256	512
8	0.101	0.095	0.101	0.105	0.107	0.108	0.111
16	0.169	0.171	0.182	0.174	0.183	0.187	0.197
32	0.320	0.366	0.398	0.416	0.436	0.423	0.439
64	0.555	0.731	0.825	0.869	0.899	0.896	0.891
128	0.700	0.947	0.996	1.000	1.000	1.000	1.000
256	0.724	0.988	1.000	1.000	1.000	1.000	1.000
512	0.728	0.991	1.000	1.000	1.000	1.000	1.000
Inf	0.737	0.991	1.000	1.000	1.000	1.000	1.000

Note: Number of observations: $T = S/h$.

sample size if the span is larger. Finally they are instances where fewer observations lead to higher power if the span is correspondingly higher. Consider, for example, the case with 16 observations and a span of 64 units; the power is 0.731, while it is 0.174 with 64 observations and a span of 16 units. Other, more extreme, examples can be obtained from Table 1.

2.2. *Using a test of randomness on the first-differenced data*

We consider a sequence of statistics $\{R_n\}_1^\infty$ where, for a fixed n, R_n is the first-order correlation coefficient of the first differences of the data:

$$R_n = \frac{\sum_{t=2}^{T_n} x_{nt} x_{n(t-1)}}{\sum_{t=1}^{T_n} x_{nt}^2},$$

where $x_{nt} = y_{nt} - y_{n(t-1)}$, and y_{nt} is given by (5). In this section, we consider the case where $\gamma < 0$ and study the limiting power function of the statistic R_n which tests for a random walk against the alternative that the process $\{y_{nt}\}$ is stationary and positively correlated. Since R_n is invariant to a scale transformation of the data, we may, without loss of any generality, set σ^2 in (5) such that $\{u_{nt}\} \sim$ i.i.d. $N(0, 1)$. This affects the value of the initial condition of the process but this is of no consequence given that the results to be presented below do not depend on the initial condition. Hence, to ease presentation, we set $y(0) = 0$ without loss of generality. These simplifications leave us with $\{y_{nt}\}$ defined by

$$y_{nt} = \exp(\gamma h_n) y_{nt-1} + u_{nt} \tag{8}$$

with $y_{n0} = 0$ and $u_{nt} \sim$ i.i.d. N (0, 1). If $\{y_{nt}\}$ follows the stationary process (8) then the first differences $\{x_{nt}\}$ are an ARMA (1,1) with a moving-average parameter on the unit circle, that is,

$$x_{nt} = \alpha_n x_{nt-1} + u_{nt} - u_{nt-1}.$$

Now $\{x_{nt}\}$ has the infinite moving average representation $x_{nt} = \sum_{i=0}^{\infty} g_{ni} u_{nt-i}$ with $g_{n0} = 1$ and $g_{ni} = (\alpha_n - 1)\alpha_n^{i-1}$ for all $i \geq 1$. Since $\sum_{i=0}^{\infty} |g_{ni}| = 2$ and $\sum_{i=0}^{\infty} i g_{ni}^2 = (1 + \alpha_n)^{-2} < \infty$, we have (e.g., Priestely, 1981, p. 339):

$$T_n^{1/2}(R_n - \rho_n(1))/\sigma_n(R_n) \Rightarrow N(0,1) \tag{9}$$

for any fixed α_n in the range [0, 1]. This result holds irrespective of the initial condition $y(0)$. Here $\rho_n(1)$ is the true first-order autocorrelation coefficient of the x_{nt}'s and $\sigma_n^2(R_n)$ is the variance of the statistic $T_n^{1/2} R_n$. Simple calculations using results in Priestely, (1981, p. 332), yield

$$\rho_n(1) = (\alpha_n - 1)/2, \tag{10}$$

$$\sigma_n^2(R_n) \equiv var\,(T_n^{1/2} R_n) \approx 1 - (\alpha_n - 1)^2(\alpha_n - 2)/\{2(1 + \alpha_n)\}. \tag{11}$$

The derivation of (10) and (11) is valid for any fixed $|\alpha_n| < 1$. Note, however, that when $\alpha_n = 1$, we have $T_n^{1/2} R_n \Rightarrow N(0,1)$ (since, in this case, x_{nt} is simply an i.i.d. sequence and invariant with respect to $y(0)$). Therefore, the limiting distribution (9) is also valid with $\alpha_n = 1$ since (10) and (11) yield $\rho_n(1) = 0$ and $var\,(T_n^{1/2} R_n) \approx 1$, when evaluated at $\alpha_n = 1$. Therefore, (9) is valid for all fixed α_n in the range (0, 1) including both endpoints. To analyze the limiting behavior of the power function for various paths for the sampling interval, h_n, our analysis uses different methods of proof according to whether $T_n^{1/2}h_n \to \infty$ or not.

Consider first the case where $T_n^{1/2}h_n \to \infty$. Denoting by b the β percentage point of the $N(0,\,1)$ distribution, the power function of a size β test is given by

$$\Pr_{\alpha_n}[T_n^{1/2}R_n < b] = 1 - \Pr_{\alpha_n}[R_n - \rho_n(1) \geq \epsilon_n], \tag{12}$$

where $\rho_n(1)$ is as defined in (10) and $\epsilon_n \equiv b/T_n^{1/2} - \rho_n(1)$. Note that there will exist an n^* such that $\epsilon_n > 0$ for all $n \geq n^*$. Hence, for all $n \geq n^*$ we have

$$\begin{aligned}\Pr_{\alpha_n}[R_n - \rho_n(1) \geq \epsilon_n] &\leq \Pr_{\alpha_n}[|R_n - \rho_n(1)| \geq \epsilon_n] \\ &\leq E(R_n - \rho_n(1))^2/\epsilon_n^2 \end{aligned} \tag{13}$$

by Chebyshev's inequality. We need to show that this upper bound converges to 0 as $n \to \infty$. We make use of the following approximations described in, for example, Priestely (1981, p. 332): (a) $E(R_n) \approx (1 - T_n^{-1})\rho_n(1)$; and (b) $var\,(R_n) \approx \sigma_n^2(R_n)/T_n$ where $\rho_n(1)$ and $\sigma_n^2(R_n)$ are defined by (10) and (11). Using these approximations, straightforward manipulations show that $T_nE(R_n - \rho_n(1))^2 = \sigma_n^2(R_n) + o(1)$. Noting that $\alpha_n = \exp(\gamma h_n)$ converges to 0 if $h_n \to \infty$ and converges to 1 if $h_n \to 0$, we deduce (using (11)) that $T_nE(R_n - \rho_n(1))^2$ is bounded as $n \to \infty$ for any $\gamma < 0$ and any path for the sampling interval. Also, $T_n^{1/2}\epsilon_n = b - T_n^{1/2}\rho_n(1) = b - T_n^{1/2}(\exp(\gamma h_n) - 1)/2 \to \infty$ provided $T_n^{1/2}h_n \to \infty$ as $n \to \infty$. Hence, in this case, the bound in (13) converges to zero and the statistic is shown to be consistent using (12).

When $T_n^{1/2}h_n \to k$, a (possibly zero) fixed constant, we adopt a different approach to characterize the limiting behavior of the power function. Without loss of generality, set $h_n = k/T_n^a$ with $a \geq 1/2$. The following lemma proved in the appendix characterizes the relevant limiting behavior of sample moments of the process $\{y_{nt}\}$ as $T_n \to \infty$.

Lemma 4. *Let $\{y_{nt}\}$ be generated by (8) with $h_n = k/T_n^a$ for any $a \geq \frac{1}{2}$. Let $\alpha_n = \exp(\gamma h_n)$ and denote convergence in probability by '$\to$'; then as $n \to \infty$:*

(i) $T_n^{-1/2}(\alpha_n - 1)y_{T_n}^2 \to 0$ *for any* $a \geq \frac{1}{2}$;

(ii) $T_n^{-1/2}(\alpha_n - 1)^2 \sum_1^{T_n} y_{nt}^2 \to 0$ *for any* $a \geq \frac{1}{2}$, *and* $\to -\gamma k/2$ *if* $a = \frac{1}{2}$;

(iii) $T_n^{-1/2}(\alpha_n - 1)^2 \sum_1^{T_n} y_{nt-1}u_{nt} \to 0$ *for any* $a \geq \frac{1}{2}$.

To derive the limiting distribution, first note that

$$T_n^{1/2} R_n = \frac{T_n^{-1/2} \sum_1^{T_n} x_{nt}x_{nt-1}}{T_n^{-1} \sum_1^{T_n} x_{nt-1}^2}.$$

Straightforward manipulations show that

$$\begin{aligned} T_n^{-1/2} \sum_1^{T_n} x_{nt}x_{nt-1} &= T_n^{-1/2}(\alpha_n - 1)^2 \alpha_n \sum_1^{T_n} y_{nt-2}^2 + (2\alpha_n - 1 + \alpha_n^{-1}) \\ &\quad \times (\alpha_n - 1) T_n^{-1/2} \sum_1^{T_n} y_{nt-2}u_{nt-1} + T_n^{1/2}(\alpha_n - 1) \\ &\quad \times T_n^{-1} \sum_1^{T_n} u_{nt-1}^2 + T_n^{-1/2}(\alpha_n - 1)\alpha_n y_{T_n-1}u_{T_n} \\ &\quad + \alpha_n^{-1} T_n^{-1/2} \sum_1^{T_n} u_{nt-1}u_{nt}, \end{aligned} \tag{14}$$

and

$$\begin{aligned} T_n^{-1} \sum_1^{T_n} x_{nt-1}^2 &= T_n^{-1}(\alpha_n - 1)^2 \alpha_n \sum_1^{T_n} y_{nt-2}^2 \\ &\quad + \alpha_n(\alpha_n - 1) T_n^{-1} \sum_1^{T_n} y_{nt-2}u_{nt-1} + T_n^{-1} \sum_1^{T_n} u_{nt-1}^2. \end{aligned} \tag{15}$$

Using Lemma 4, (15) and the fact that $T_n^{-1} \sum_1^{T_n} u_{nt-1}^2 \to 1$, we deduce that $T_n^{-1} \sum_1^{T_n} x_{nt-1}^2 \to 1$ for any $a \geq \frac{1}{2}$. On the other hand, from (14), Lemma 4 and the fact that $T_n^{1/2}(\alpha_n - 1)T_n^{-1} \sum_1^{T_n} u_{nt-1}^2 \to \gamma k$ and $T_n^{-1/2}\sum_1^{T_n} u_{nt-1}u_{nt} \Rightarrow N(0,1)$, we have $T_n^{-1/2}\sum_1^{T_n} x_{nt}x_{nt-1} \Rightarrow N(0,1)$ if $a > \frac{1}{2}$, and $T_n^{-1/2}\sum_1^{T_n} x_{nt}x_{nt-1} \Rightarrow N(\gamma k/2, 1)$ if $a = \frac{1}{2}$. Therefore, $T_n^{1/2}R_n \Rightarrow N(\gamma k/2, 1)$ and the limiting power function of the test when $T_n^{1/2}h_n \to k$ is given by $\lim_{n\to\infty} \Pr_{\alpha_n}[T_n^{1/2}R_n < b] = \Phi(b - \gamma k/2)$. When $T_n^{1/2}h_n \to 0$ we deduce that $\lim_{n\to\infty} \Pr_{\alpha_n}[T_n^{1/2}R_n < b] = \Phi(b) = \beta$, the size of the test. Note that the results hold with a nonzero initial condition as well. However, the proof relies on the normality assumption (though it could probably be relaxed).

The implication of these results for the behavior of the power function as the sample size increases allowing any path for the sampling interval are summarized in the following theorem.

Theorem 2. *Consider $\{\{y_{nt}\}_1^{T_n}\}_1^\infty$ a triangular array of random variables defined by (5) with $\gamma < 0$, and $\{R_n\}_1^\infty$ a sequence of test statistics, where R_n is the first-order correlation coefficient of the triangular array of first differences of y_{nt}. Suppose that we test the null hypothesis that $\gamma = 0$, then $T_n^{1/2}R_n$ is a consistent test, as $n \to \infty$, under any alternative hypothesis that $\gamma < 0$ if, and only if $T_n^{1/2}h_n \to \infty$ as $T_n \to \infty$. If $T_n^{1/2}h_n \to 0$, the power of the test converges to its size. Finally, if $T_n^{1/2}h_n \to k(k > 0)$, the power of the test converges to $\Phi(b - \gamma k/2)$ where Φ is the standardized normal distribution and b its associated β percentage point, such that $\Phi(b) = \beta$, the size of the test.*

Theorem 2 states that the statistic $T_n^{1/2}R_n$ is a consistent test of the random walk hypothesis against stationary positive autocorrelation if, and only if, the sampling interval decreases at a rate less than $T^{1/2}$ as T, the sample size, increases. Alternatively, it is consistent if the span of the data increases at a rate higher than $T^{1/2}$ as the sample size increases.

These results are to be contrasted with those obtained for the statistic $T(\hat{\alpha} - 1)$ which was consistent as long as the span was at least increasing, and this, at any rate in relation to the sample size. This implies that a test based on the original series $\{y_t\}$, such as

$T(\hat{\alpha}-1)$, is consistent over a much wider range of possible paths for the span or the sampling interval than is the case for a test based on the first-differenced series such as $T^{1/2}R$. According to a more general consistency criterion, the statistic $T(\hat{\alpha}-1)$ can be said to dominate the statistic $T^{1/2}R$. An explanation in terms of local asymptotic power function in the fixed sampling interval case is also helpful here. Unlike the statistic $T(\hat{\alpha}-1)$ applied to the level of the series, $T^{1/2}R$ is not $O_p(T)$ under the alternative hypothesis that the process is stationary but is $O_p(T^{1/2})$. Given that α approaches one at rate T if T increases with a fixed span, the rate of approach of α to the null value 1 is too fast relative to the rate at which the statistic diverges under the alternative hypothesis. Hence, one can expect the power to equal the size of the test in the limit.

A most interesting feature of our result concerning tests of the random walk hypothesis based on first-differenced data is that more observations need not lead to a higher power if the span is kept fixed. Indeed as more observations are added the power will eventually decrease towards the size of the test. This implies that in some instances it may be possible to increase the power of the test by simply deleting some of the available data keeping the observations associated with a longer sampling interval.

We also conducted a simulation experiment to assess the relevance of these asymptotic results as a guide to the finite sample behavior. The framework is the same as in Section 2.1 and the results are presented in Table 2. The results clearly show the relevance in finite samples of the fact that R has a power which converges to the size of the test as $h \to 0$. Indeed, for any given span, the power initially increases with the sample size until it reaches a maximum at a value of T between 16 and 64 (increasing with S) and then declines steadily as T is further increased.

Our results also show that tests of the random walk hypothesis based on first-differenced data have lower power than tests based on the level of the data (compare Tables 1 and 2). A similar observation has been made by Sargan and Bhargava (1983) in the context of tests of random walk residuals against stationary AR(1) residuals in the linear regression model. They documented

Table 2. Power of a one-tailed test of the random walk hypothesis with the statistic $T^{1/2}R$; $H_0 : \gamma = 0$; $H_1 : \gamma = -0.2$.

	T						
S	8	16	32	64	128	256	512
8	0.035	0.055	0.058	0.058	0.058	0.059	0.061
16	0.059	0.072	0.070	0.071	0.068	0.061	0.057
32	0.090	0.116	0.109	0.096	0.080	0.068	0.062
64	0.169	0.231	0.204	0.160	0.127	0.104	0.086
128	0.231	0.429	0.447	0.364	0.260	0.186	0.139
256	0.254	0.588	0.758	0.728	0.595	0.420	0.274
512	0.261	0.630	0.905	0.970	0.953	0.858	0.656
Inf	0.254	0.631	0.919	0.998	1.000	1.000	1.000

Note: Number of observations: $T = S/h$.

the fact that using the Durbin–Watson ratio in a regression equation specified in first-differences leads to a test with significantly lower power than using the same Durbin–Watson ratio in a regression specified in levels, unless the alternative hypothesis is close to the null hypothesis of a random walk. Our analysis has, however, an interesting implication for their results, namely that the figures reported in their Appendix C about the power of the Durbin–Watson ratio applied to first-differenced data may not represent the maximal power possible with the given data set used. It is indeed possible that using their same experiment and deleting observations such as to obtain a longer sampling interval may lead to a test with higher power.

3. Testing for Randomness

Consider the following continuous-time diffusion process:

$$dy_t = \gamma y_t dt + \sigma(1 + |2\gamma|)^{1/2} dw_t; \quad 0 < t \leq S; \quad y_0 = y(0), \tag{16}$$

where w_t is again a unit Wiener process and $\gamma(-\infty \leq \gamma < \infty)$, $\sigma(> 0)$, and $y(0)$ are constants. The discrete-time representation of y_t sampled at an interval h is

$$y_{th} = \alpha_h y_{(t-1)h} + v_{th}; \quad t = 1, 2, \ldots, T = S/h; \quad y_0 = y(0), \tag{17}$$

with $\alpha_h = \exp(\gamma h)$ and $v_{th} \sim N(0, \sigma^2(1 + |2\gamma|)(\exp(2\gamma h) - 1)/2\gamma)$. As $\gamma \to -\infty$ (17) becomes

$$y_{th} = e_{th}; \quad t = 1, 2, \ldots, T = S/h; \tag{18}$$

with $e_{th} \sim N\ (0,\ \sigma^2)$. Hence (18) represents a process where the series y_{th} is random. In this framework, the null and alternative hypotheses are nested within model (17) in such a way that H_0: $\gamma = -\infty$ and $H_1 : \gamma = \bar{\gamma}(-\infty < \bar{\gamma} < \infty)$. The class of alternative hypotheses permitted includes stationary first-order processes, a unit root process, and explosive first-order difference equations.

To test the null hypothesis of randomness we consider the OLS estimator in a first-order autoregression denoted as $\hat{\alpha} = \sum_1^T y_{th} y_{(t-1)h} / \sum_1^T y_{(t-1)h}^2$. It is easy to see that $\hat{\alpha}$ is invariant to a scale transformation of the data. Hence, we can, without loss of generality, choose σ^2 such that $v_{th} \sim N(0, 1)$ for all h in (17). In this setting, we consider a triangular array of random variables $\{\{y_{nt}\}_1^{T_n}\}_1^\infty$ defined under the null hypothesis by

$$y_{nt} = v_{nt}; \quad v_{nt} \sim N(0, 1) \tag{19}$$

and under the alternative hypothesis by

$$y_{nt} = \exp(\gamma h_n) y_{nt-1} + v_{nt}; \quad v_{nt} \sim N(0, 1) \tag{20}$$

with $y_{n0} = 2\gamma y(0)/[(1 + |2\gamma|)(\exp(2\gamma h_n) - 1)]$. We consider an array of test statistics $\{T_n^{1/2} \hat{\alpha}_n\}_1^\infty$ defined for a fixed n by $T_n^{1/2} \hat{\alpha}_n = T_n^{-1/2} \sum_{t=1}^{T_n} y_{nt} y_{nt-1} / T_n^{-1} \sum_{t=1}^{T_n} y_{nt-1}^2$. Under the null hypothesis $T_n^{1/2} \hat{\alpha}_n \Rightarrow N(0, 1)$ as $n \to \infty$, no matter what the path of h or S is as n increases. This result holds for any fixed $y(0)$ given the invariance of y_{nt} with respect to $y(0)$ under the null hypothesis.

To prove consistency, we proceed in a manner similar to that of Section 2.1. Let b^* be the $(1-\beta)$ percentage point of the standardized normal distribution. The power of the test is given by

$$\Pr_{\alpha_n}[T_n^{1/2} \hat{\alpha}_n > b^*] = 1 - \Pr_{\alpha_n}[(\alpha_n - \hat{\alpha}_n) > \epsilon_n],$$

where $\epsilon_n = \alpha_n - b^*/T_n^{1/2}$. Note that there will exist an n^* such that $\epsilon_n > 0$ for all $n \geq n^*$. Hence, for all $n \geq n^*$,

$$\begin{aligned}\Pr_{\alpha_n}[(\alpha_n - \hat{\alpha}_n) > \epsilon_n] &\leq \Pr_{\alpha n}[|\hat{\alpha}_n - \alpha_n| > \epsilon_n] \\ &\leq (1/\epsilon_n)E|\hat{\alpha}_n - \alpha|, \quad \text{by Markov's inequality} \\ &\leq (1/\epsilon_n)[k/T_n^{1/2} + o(T_n^{-1/2})], \quad \text{by Lemma 3.}\end{aligned}$$

for all $n \geq \max(n^*, n_o(\delta))$ where $n_o(\delta)$ is such that $T_{n_o(\delta)} = T_0(\delta)$ in Lemma 3.

Consider first the case where $h_n \to 0$ as $n \to \infty$. Then $\alpha_n \to 1$ for any γ and $\epsilon_n = \alpha_n - b^*/T_n^{1/2} \to 1$. Hence, $T_n^{1/2}\epsilon_n \to \infty$ and $\lim_{n\to\infty} \Pr_{\alpha n}[T_n^{1/2}\hat{\alpha}_n > b^*] = 1$, showing consistency. If h_n converges to some constant, h say, $\alpha_n \to \exp(\gamma h)$, a constant, and $T_n^{1/2}\epsilon_n \to \infty$, showing consistency again.

Things are different when $h_n \to \infty$. Consider first the case where $\gamma = 0$ and $\alpha_n = 1$ for all n. We have $\epsilon_n \to 1$ and, hence, $T_n^{1/2}\epsilon_n \to \infty$, showing consistency. If $\gamma > 0$, $\alpha_n \to \infty$ and $T_n^{1/2}\epsilon_n \to \infty$, showing consistency again. Hence, a test of randomness is consistent against an alternative that the process is a random walk or is explosive under any path for the sampling interval, h_n, as n increases.

Such is not the case when $\gamma < 0$ (stationary alternatives). In that case, we have $\alpha_n \to 0$ and $\epsilon_n \to 0$ as $n \to \infty$. The test will be consistent if $T_n^{1/2}\epsilon_n \to \infty$ as $n \to \infty$. This will be the case whenever $\lim\inf_{n\to\infty} h_n^{-1}\ln(T_n^{-1/2\gamma}) > 1$ using the fact that $\alpha_n = \exp(\gamma h_n)$. To provide limiting results when $\lim\sup_{n\to\infty} h_n^{-1}\ln(T_n^{-1/2\gamma}) \leq 1$ we need a different approach based on sequences of local alternatives.

Consider the following problem of limiting distributions under a sequence of local alternatives $H_n : \alpha = \alpha_n = \alpha_0 T_n^{-1/2} + o(T_n^{-1/2})$. These alternatives are local to the null hypothesis H_0: $\alpha = 0$. Let $\{Q_n\}$ be the sequence of probability measures of the data $\{y_1, \ldots, y_{T_n}\}$ generated by (20) under H_n and let $\{P_n\}$ be the sequence of probability measures of the data $\{y_1, \ldots, y_{T_n}\}$ under H_0. From Kreiss (1987, Corollary 3.2) or Aiyar (1981, Lemma 2(i)), the probability measures $\{P_n\}$ and $\{Q_n\}$ are contiguous. Using this result it is shown in the appendix that under the sequence of measures $\{Q_n\}$,

$$T_n^{1/2}\hat{\alpha}_n \Rightarrow N(\alpha_0, 1). \tag{21}$$

When $\lim_{n\to\infty} h_n^{-1}\ln(T_n^{-1/2\gamma}) = 1$ (or $T_n^{1/2}\alpha_n \to 1$) as $n \to \infty$, we have $\alpha_0 = 1$. Hence, using (21), we deduce that, in this case, $\lim_{n\to\infty} \Pr_{\alpha_n}[T_n^{1/2}\hat{\alpha}_n > b^*] = 1 - \Phi(b^* - 1)$, where Φ is the standardized normal distribution function. When $\lim\sup_{n\to\infty} h_n^{-1}\ln(T_n^{-1/2\gamma}) < 1$ (or $T_n^{1/2}\alpha_n \to 0$) as $n \to \infty$, the process $\{y_1, \ldots, y_{T_n}\}$ can be specified under H_n with $\alpha_0 = 0$. Hence, using (21), we have $\lim_{n\to\infty} \Pr_{\alpha_n}[T_n^{1/2}\hat{\alpha}_n > b^*] = 1 - \Phi(b^*) = \beta$ and the power function converges to the size of the test.

We summarize our results on the consistency of a test of randomness against correlated alternatives in the following theorem.

Theorem 3. *Let $\{y_{nt}\}$ be generated according to (20).*

(a) *The statistic $T_n^{1/2}\hat{\alpha}_n$, for testing the null hypothesis of randomness ($\gamma = -\infty$), is consistent against the alternative hypothesis of stationary positive autocorrelation ($-\infty < \gamma < 0$), if and only if the sampling interval is such that $\lim\inf_{n\to\infty} h_n^{-1}\ln(T_n^{-1/2\gamma}) > 1$. More specifically the limiting behavior of the power function is as follows for different paths for the sampling interval as n increases:*

$$\lim_{n\to\infty} P_{T_n,h_n}(\gamma) = 1 \quad \textit{iff } \liminf_{n\to\infty} h_n^{-1}\ln(T_n^{-1/2\gamma}) > 1;$$

$$\lim_{n\to\infty} P_{T_n,h_n}(\gamma) = \beta \quad \textit{if } \limsup_{n\to\infty} h_n^{-1}\ln(T_n^{-1/2\gamma}) < 1;$$

and

$$\lim_{n\to\infty} P_{T_n,h_n}(\gamma) = 1 - \Phi(b^* - 1) \quad \textit{if } \lim_{n\to\infty} h_n^{-1}\ln(T_n^{-1/2\gamma}) = 1;$$

where Φ is the standardized normal distribution and $1 - \Phi(b^*) = \beta$, the size of the test.

(b) *The same statistic $T_n^{1/2}\hat{\alpha}_n$ is consistent against the alternative hypothesis of explosive positive autocorrelation or against the* *random walk hypothesis* ($\gamma \geq 0$), *regardless of the path taken by the sampling interval as the sample size increases.*

The practical implications of these results are as follows. In a test of randomness against stationary positive autocorrelation, the power function is greater if the sample size is increased with a reduction in the sampling interval rather than with an increase in both the span and the sampling interval. It may even be the case that a smaller sample size yields a greater power than a larger one if the former is associated with a small sampling interval.

To assess the significance of these theoretical results we conducted a simulation experiment similar to those in Section 2. We generated 10,000 replications of (7) with $\gamma = -2.0$, hence analyzing the case where the alternative is a stationary process. The critical values under the null hypothesis were taken from the asymptotic standard $N(0, 1)$ distribution since it provides an adequate approximation even if the sample size is quite small.

The results in Table 3 are quite striking. Clearly, the power of the test converges to 1 as $h \to 0$, whenever S is fixed or decreases. Overall, it is apparent that the power increases whenever h decreases. On the other hand, the decrease in the power is quite dramatic when h increases. The infinity row corresponds to the theoretical

Table 3. Power of a one-tailed test of the randomness hypothesis with the statistic $T^{1/2}\hat{\alpha}$; $H_0 : \gamma = -\infty$; $H_1 : \gamma = -2.0$.

	T						
S	8	16	32	64	128	256	512
8	0.083	0.392	0.956	1.000	1.000	1.000	1.000
16	0.050	0.115	0.651	0.999	1.000	1.000	1.000
32	0.047	0.057	0.185	0.898	1.000	1.000	1.000
64	0.050	0.047	0.064	0.286	0.994	1.000	1.000
128	0.049	0.048	0.047	0.065	0.455	1.000	1.000
256	0.051	0.045	0.049	0.052	0.074	0.703	1.000
512	0.048	0.046	0.048	0.051	0.048	0.092	0.921
1024	0.049	0.048	0.044	0.046	0.048	0.047	0.110

Note: Number of observations: $T = S/h$.

case where an infinite span would be available with a fixed number of observations. As the span increases to infinity with a fixed T and $\gamma(< 0)$, the process (7) is a random one, hence the power is equal to the size of the test. This is basically the intuitive reason why the power decreases as the sampling interval increases. In that case the discrete-time representation of the process under the alternative hypothesis gets closer and closer to the discrete-time representation of the process under the null hypothesis.

4. Simplified Consistency Criteria

In general, deriving the limiting power function for any possible path for the sampling interval as the sample size increases can be a complex task. Accordingly, it may be desirable to consider special cases of the general consistency criterion that are easier to derive and yet can still provide interesting information about issues concerning frequency versus number of observations. Consider the following cases:

Definitions.

(a) Small-h consistency: $\lim_{S\to\infty} [\lim_{h\to 0} P_{S,h}(\theta)] = 1$;
(b) Large-S consistency: $\lim_{T\to\infty} [\lim_{S\to\infty} P_{T,S}(\theta)] = 1$;
(c) Small-S consistency: $\lim_{S\to 0} [\lim_{h\to 0} P_{S,h}(\theta)] = 1$;

where in each case the limiting operations are taken sequentially.

Determining whether a test statistic satisfies one or more of these criteria is relatively easier than analyzing the consistency properties for any possible paths for the sampling interval as the sample size increases. Yet these simple criteria summarize quite well the behavior of the power function in finite samples when considering various sampling intervals as the number of observations is changed. For example, if a test is small-h consistent and large-S consistent but not small-S consistent, we can infer that a larger span of data is preferable in terms of power, and that the power eventually decreases as the span decreases. On the other hand, if a test is small-S consistent but not large-S consistent, a smaller span is preferable.

Let us consider the behavior of these simplified criteria with respect to the statistics and models considered in this paper.

Consider first the statistic $T(\hat{\alpha} - 1)$ when testing for a random walk. The limit of the power function as $S \to \infty$ when T is fixed is the power function of a test of the random walk hypothesis against the alternative that the process is random (since $\alpha = \exp(\gamma S/T) \to 0$ as $S \to \infty$ if $\gamma < 0$). This power function is increasing with T and converges to one as $T \to \infty$. Hence, the statistic is large-S consistent. Consider now the other two criteria. Both involve first the limit of the power function as $h \to 0$ keeping S fixed. It can be shown that this power function converges to one as $S \to \infty$ and converges to the size of the test as $S \to 0$. Hence, $T(\hat{\alpha} - 1)$ is small-h but not small-S consistent. The simulations in Section 2.1 show the usefulness of these criteria as guides to the behavior of the power function in finite samples.

Consider now using a test of randomness on the first differences of the data to test for a random walk against stationary alternatives. Since the limit of the power function is the size of the test as $h \to 0$ keeping the span fixed, it is the case that here the test is neither small-h nor small-S consistent. Hence, using these simple criteria, one would expect this test to have a power function declining as the sample size increases if the span does not increase sufficiently rapidly. For the same reason as before the test is, however, large-S consistent; hence, we can expect increasing power if the sampling interval increases with the number of observations. These implications are confirmed by our simulation experiment.

Finally, let us analyze the behavior of the first-order autocorrelation coefficient when testing for randomness. Consider first the case of nonstationary alternatives ($\gamma \geq 0$). The statistic satisfies all the criteria: large-S, small-h, and small-S consistency. The test is small-h and small-S consistent because its power converges to one as the sampling interval tends to zero for any fixed span. Now the limit of the power function as the span increases keeping a fixed sample size, when $\gamma = 0$, is simply the power function (in finite samples) of a test of randomness against a random walk hypothesis. This power function converges to one as the sample size increases, so we have large-S consistency against an alternative that $\gamma = 0$. When $\gamma > 0$, the discrete time value of the autoregressive parameter diverges to infinity as the span increases. Hence, the test is also large-S consistent in this case.

Consider now the case of stationary alternatives ($\gamma < 0$). The test is again small-S and small-h consistent since the power function converges to one as the sampling interval decreases (for any given fixed span). However, the test is not large-S consistent. As the span increases with a fixed sample size the discrete time value of the autoregressive parameter tends to zero (i.e., $\alpha = \exp(\gamma S/T) \to 0$ as $S \to \infty$ if $\gamma < 0$). Therefore, the power is given by the size of the test for any sample size and the test is not large-S consistent. Again, our simulation experiment reveals the usefulness of these simplified criteria.

5. Concluding Comments

This paper considered the consistency criterion for a test statistic in a more general framework than usual. Standard practice considers a test statistic consistent if its power against fixed alternatives converges to one as the sample size increases but constrains the sampling interval to remain unchanged as the sample size increases. While this criterion is useful and can be regarded to be a minimal property of a satisfactory test, it cannot shed light on some important features of the test when the sampling interval is changed at the same time as the sample size increases.

To answer some of these interesting practical questions we considered the consistency criterion under all possible paths of the sampling interval as the sample size increases. We analyzed in detail the properties of three commonly used test statistics: (a) the normalized least-squares estimator in a first-order autoregression as a test of the random walk hypothesis; (b) the first-order correlation coefficient based test of the random walk hypothesis applied to the first differences of the data; and (c) the normalized least-squares estimator in a first-order autoregression as a test of randomness. Some of the results obtained are: (1) the normalized least-squares estimator,

$T(\hat{\alpha} - 1)$, is a consistent test of the random walk hypothesis as long as the span increases with the sample size (whether the alternative is stationary or explosive); (2) the first-order correlation coefficient based test using the first differences of the data, $T^{1/2}R$, is a consistent test of the random walk hypothesis against stationary alternatives as long as the span of the data increases at a rate greater than $T^{1/2}$. In this sense the normalized least-squares estimator dominates the latter test statistic since it is consistent over a wider range of possible paths for the sampling interval; and (3) the normalized least-squares estimator in a first-order autoregression, $T^{1/2}\hat{\alpha}$, is a consistent test of the hypothesis of randomness against stationary alternatives as long as the sampling interval is not increasing too fast. If the alternative is explosive, then the test is consistent whatever the path for the span or the sampling interval.

We introduced three summary measures of the consistency properties as well: large-S, small-S, and small-h consistency. These were found to adequately capture both the asymptotic and finite sample properties of the statistics considered. The normalized least-squares estimator, as a test of the random walk hypothesis, is both large-S and small-h consistent but not small-S consistent. Therefore, when testing the random walk hypothesis with this statistic, a large span of data is to be preferred (in some cases), even if this entails less observations than would be possible at a smaller sampling frequency. The normalized least-squares estimator in a first-order autoregression as a test of randomness is both small-h and small-S consistent but not large-S consistent (against stationary alternatives). In this case, a smaller sampling interval is to be preferred, in some instances, even if this entails a smaller sample size. The first-order correlation coefficient based test applied to the first differences is only large-S consistent but not small-h nor small-S consistent as a test of the random walk hypothesis. The result implies that a greater number of observations may lead to a test with smaller power if the span is not increased sufficiently.

The paper has considered only a few (though important in practice) test statistics in a specialized framework. The results, however, can be extended in several directions. First, it is relatively straightforward to show that the results of Section 2.1 apply to the t-statistic on the coefficient in a first-order autoregression as well, though the distributional results of Lemmas 1 and 2 would need to be modified. The analysis could also be extended to regression models with an estimated intercept and trend function; again the same consistency results would apply though the limiting distributions would change. Perron (1989) also analyzes the consistency properties of some 13 different test statistics through Monte Carlo methods. What emerges from these sets of additional results is that the consistency properties of $T(\hat{\alpha} - 1)$ derived in Section 2 also apply to any test of the random walk hypothesis using the original series (as opposed to a test of randomness on the first differences). With respect to tests of randomness on the first-differenced data, the simulations suggest that the results obtained in Section 2.2 also apply to a wide class of statistics. For these reasons, the results of the present study indicate that, under the framework and hypotheses considered, tests of the random walk hypothesis should be based on a statistic using the undifferenced series since they dominate, in terms of a more general consistency criterion, those based upon the differenced series. Of course, this result is contingent upon the class of alternative hypotheses specified.

Finally, there is ample scope for further research by analyzing more complex models based on a time series of data; for example, models with lagged dependent variables and exogenous regressors as well as models with time-averaged data. The major practical problem to be addressed in this context is to see whether it is worth, in terms of power and efficiency, increasing the sample size if this entails a fixed or, even more realistically, a reduced span. For example, is it better to have 100 annual observations, 150 postwar quarterly observations, or, say, 250 monthly observations? Some answers to these questions can be given using the approach suggested in this paper.

Appendix

To prove Lemma 1, case (1) with $y(0) = 0$, we first note that, in that case, $\hat{\alpha}_n$ is invariant with respect to the variance of the innovations

$\{u_{nt}\}$ in (3). Hence, without loss of generality, we can set $u_{nt} \sim N(0,1)$ for all n. $\{y_{nt}\}$ is then not a function of h_n and the usual result carries over; see Phillips (1987a). To see this, note that $Z(0,\, 0)$ reduces to $\frac{1}{2}(w\,(1)^2 - 1)/\int_0^1 w(r)^2 dr$. We postpone the proof of case (2) with $y(0) \neq 0$ as part of the proof of Lemma 2.

Proof of Lemma 2. From Perron (1991, Theorem 1) we have, if $T_n \to \infty$ with S fixed:

$$T_n(\hat{\alpha}_n - \alpha_n) \Rightarrow Z(\delta, c) \tag{A.1}$$

with $c = \gamma S$ and $\delta = y(0)/\sigma S^{1/2}$. A slight modification of the proof of that theorem allows the more general case where $S_n \to S$ as $n \to \infty$. Given that

$$T_n(\hat{\alpha}_n - 1) = T_n(\hat{\alpha}_n - \alpha_n) + T_n(\alpha_n - 1) \tag{A.2}$$

and $T_n(\alpha_n - 1) = T_n(\exp{(\gamma S_n/T_n)} - 1) \to \gamma S$ as $n \to \infty$, we obtain

$$T_n(\hat{\alpha}_n - 1) \Rightarrow c + Z\,(\delta, c) \tag{A.3}$$

proving part 2(b). Part 1(b) follows directly given that $\delta = 0$ when $y(0) = 0$. Note from (A.3) that when $\gamma = 0$, $c = 0$, and $Z(\delta, 0)$ depends on S only through $\delta = y(0)/\sigma S^{1/2}$. Hence as $S \to \infty$, $Z(\delta, 0) \to Z(0,0)$, proving Lemma 1, part 2(c). To prove part (a) of Lemma 2 when $y(0) = 0$, note that in this case $Z(0,c) = \{\int_0^1 J_c(r)^2 dr\}^{-1} \times \{\int_0^1 J_c(r)\, dw\,(r)\}$. As $S \to 0$, $c \to 0$, $J_c(r) \to w(r)$ and $Z(0,\, c) \to Z(0,0)$. □

Proof of Part 2(c). From Perron (1991, Corollary 1) we have $(-2c)^{-1/2} Z\,(\delta,\, c) \Rightarrow N(0,\, 1)$ as $c \to -\infty$ for any value of $y(0)$. Using $c = \gamma S_n$, we deduce that $S_n^{-1/2} T_n(\hat{\alpha}_n - \alpha_n) \Rightarrow N(0, -2\gamma\sigma^2)$. Since $h_n = S_n/T_n$ is decreasing as $n \to \infty$, we have

$$S_n^{-1} T_n(\exp(\gamma S_n/T_n) - 1) \to \gamma \quad \text{as } S_n/T_n \to 0. \tag{A.4}$$

Using (A.2) and (A.4) we deduce that $S_n^{-1} T_n(\hat{\alpha}_n - 1) \to \gamma$ and, given that $\gamma < 0$, $T_n(\hat{\alpha} - 1) \to -\infty$ as $n \to \infty$. □

Proof of Part 2(d). From Perron (1991, Corollary 1) we have $(2c)^{-1} \exp(c) Z(\gamma, c) \Rightarrow [d\eta + \xi\eta]/[d + \eta]^2$ as $c \to +\infty$. Replacing

c by γS_n we have

$$(2\gamma S_n)^{-1}\exp(\gamma S_n)T_n(\hat{\alpha}_n - \alpha_n) \Rightarrow [d\eta + \xi\eta]/[d + \eta]^2. \qquad \text{(A.5)}$$

Using (A.2), (A.4), and (A.5) and the fact that $S_n^{-1}(2\gamma S_n)$ $\exp(-\gamma S_n) \rightarrow 0$ as $S_n \rightarrow \infty$, we have $S_n^{-1}T_n(\hat{\alpha}_n - 1) \rightarrow \gamma$ and $T_n(\hat{\alpha}_n - 1) \rightarrow +\infty$ given that $\gamma > 0$. Part 1(d) is easily deduced from (A.5) upon evaluation at $d = 0$. □

Proof of Cases where $S_n \rightarrow 0$. Let $a(h) = \sigma^2(\exp(2\gamma h) - 1)/2\gamma$ denote the variance of the innovations $\{u_{th}\}$ in (5). Since $\hat{\alpha}$ is invariant to a scale transformation of the data, we can, without loss of generality, divide the data by $a(h)^{1/2}$. Denote this transformation by $x_{th} = y_{th}/a(h)^{1/2}$. Note that $\lim_{h\rightarrow 0}h^{-1}a(h) = \sigma^2$. We consider a triangular array of random variables $\{\{x_{nt}\}_{t=1}^{T_n}\}_{n=1}^{\infty}$. For a given n, the sequence $\{x_{nt}\}_{n=1}^{T_n}$ is generated by

$$x_{nt} = \exp(\gamma h_n)x_{n(t-1)} + \epsilon_{nt},$$
$$x_{n0} = y(0)/a(h_n)^{1/2}, \quad (t = 1, \ldots, T_n) \qquad \text{(A.6)}$$

where the innovation sequence $\{\epsilon_{nt}\}_{t=1}^{T_n}$ is i.i.d. normal with mean 0 and variance 1. From (A.6) we deduce that

$$x_{nt} = \sum_{j=1}^{t} \exp\left((t-j)\gamma h_n\right)\epsilon_{nj} + \exp\left(t\gamma h_n\right)x_{n0}. \qquad \text{(A.7)}$$

In this framework, the sequence of statistics $\{T_n(\hat{\alpha}_n - \alpha_n)\}_{n=1}^{\infty}$ can equivalently be represented as

$$T_n(\hat{\alpha}_n - \alpha_n) = T_n^{-1}\sum_{t=1}^{T_n} x_{nt-1}\epsilon_{nt}/T_n^{-2}\sum_{t=1}^{T_n} x_{nt-1}^2, \qquad \text{(A.8)}$$

with x_{nt} defined by (A.6). Lemma A.1 characterizes the limit of the sample moments of the sequence $\{x_{nt}\}$ necessary to derive the asymptotic distribution of $T_n(\hat{\alpha}_n - \alpha_n)$ as $S_n \rightarrow 0$.

Lemma A.1. *If* $\{\{x_{nt}\}_{t=1}^{T_n}\}_{n=1}^{\infty}$ *is a triangular array of random variables defined by* (A.6), *then as* $n \rightarrow \infty$ *with* $h_n \rightarrow 0$ *and* $S_n \rightarrow 0$:

(i) $S_n^{1/2}T_n^{-1/2}x_{T_n} \Rightarrow y(0)/\sigma$;

(ii) $S_n^{1/2}T_n^{-3/2}\sum_{t=1}^{T_n} x_{nt} \Rightarrow y(0)/\sigma$;
(iii) $S_nT_n^{-2}\sum_{t=1}^{T_n} x_{nt}^2 \Rightarrow y(0)^2/\sigma^2$;
(iv) $S_n^{1/2}T_n^{-1}\sum_{t=1}^{T_n} x_{nt-1}\epsilon_{nt} \Rightarrow N(0, y(0)^2/\sigma^2)$.

To prove part (i), use (A.6) to obtain

$$\begin{aligned} S_n^{1/2}T_n^{-1/2}x_{T_n} &= S_n^{1/2}T_n^{-1/2}\sum_{j=1}^{T_n}\exp((T_n-j)\gamma h_n)\epsilon_{nj} \\ &\quad + S_n^{1/2}T_n^{-1/2}\exp(T_n\gamma h_n)x_{n0} \\ &= S_n^{1/2}T_n^{-1/2}\sum_{j=1}^{T_n}\exp((T_n-j)\gamma S_n/T_n)\epsilon_{nj} \\ &\quad + h_n^{1/2}a(h_n)^{-1/2}\exp(\gamma S_n)y(0). \end{aligned} \tag{A.9}$$

The first term converges to 0 (since $T_n^{-1/2}\sum_{j=1}^{T_n}\exp((T_n-j)\gamma S_n/T_n)\epsilon_{nj} \Rightarrow w(1)$) and the second term converges to $y(0)/\sigma$ given that $h_n^{-1}a(h_n) \to \sigma^2$ as $h_n \to 0$. The proof of parts (ii) and (iii) are analogous and omitted. To prove part (iv), note that we can write (see Perron, 1991)

$$\begin{aligned} S_n^{1/2}T_n^{-1}\sum_{t=1}^{T_n} x_{nt-1}\epsilon_{nt} &= \left(\frac{1}{2}\right)\exp(-\gamma h_n)\left[S_n^{1/2}T_n^{-1}(x_{T_n}^2 - x_{n0}^2)\right. \\ &\quad -S_n^{1/2}\left[S_n^{-1}T_n\left(\exp\left(\frac{2\gamma S_n}{T_n}\right)-1\right)\right] \\ &\quad \left.\times S_nT_n^{-2}\sum_{t=1}^{T_n} x_{nt-1}^2 - S_n^{1/2}T_n^{-1}\sum_{t=1}^{T_n}\epsilon_{nt}^2\right]. \end{aligned}$$

Now, $T_n^{-1}\sum_{t=1}^{T_n}\epsilon_{nt}^2 \to 1$ as $T_n \to \infty$, $S_n^{-1}T_n(\exp(2\gamma S_n/T_n)-1) \to 2\gamma$ as $S_n/T_n \to 0, \exp(-\gamma h_n) \to 1$ as $h_n \to 0$, and using parts (a), (b), and (c) we obtain

$$S_n^{1/2}T_n^{-1}\sum_{t=1}^{T_n} x_{nt-1}\epsilon_{nt} = \left(\frac{1}{2}\right)S_n^{1/2}T_n^{-1}(x_{T_n}^2 - x_{n0}^2) + o_p(1). \tag{A.10}$$

From (A.9),

$$
\begin{aligned}
S_n^{1/2}T_n^{-1}(x_{T_n}^2 - x_{n0}^2) = S_n^{1/2}T_n^{-1}\left(\sum_{j=1}^{T_n}\exp((T_n-j)\gamma h_n)\epsilon_{nj}\right)^2 \\
+2S_n^{1/2}T_n^{-1}\exp(T_n\gamma h_n)x_{n0} \\
\times\left(\sum_{j=1}^{T_n}\exp((T_n-j)\gamma h_n)\epsilon_{nj}\right) \\
+S_n^{1/2}T_n^{-1}(\exp(2T_n\gamma h_n)-1)x_{n0}^2.
\end{aligned}
$$

It is straightforward to show that the first and last terms of this expression converge to 0. Hence,

$$
\begin{aligned}
&S_n^{1/2}T_n^{-1}(x_{T_n}^2 - x_{n0}^2) \\
&= 2S_n^{1/2}T_n^{-1}\exp(T_n\gamma h_n)x_{n0}\left(\sum_{j=1}^{T_n}\exp((T_n-j)\gamma h_n)\epsilon_{nj}\right)+o(1) \\
&= 2\exp(\gamma S_n)h_n^{1/2}a(h_n)^{-1/2}y(0)T_n^{-1/2} \\
&\quad\times\sum_{j=1}^{T_n}\exp((T_n-j)\gamma h_n)\epsilon_{nj}+o(1) \Rightarrow \frac{2y(0)w(1)}{\sigma}
\end{aligned}
$$

given $\exp(\gamma S_n) \to 1$ as $S_n \to 0$, $h_n^{-1}a(h_n) \to \sigma^2$, and $T_n^{-1/2}\sum_{j=1}^{T_n} \exp((T_n - j)\gamma h_n)\epsilon_{nj} \Rightarrow w(1)$. The proof of part (d) follows using (A.10) and the fact that w(l) $\sim N(0,1)$. The proof of Lemma 2, part 2(a) follows using Lemma A.1 (iii, iv), (A.8), and (A.2) noting that $S_n^{-1/2}T_n(\alpha_n - 1) \to 0$ as $n \to \infty$. Since the result does not depend upon the value of γ, this also proves Lemma 1, part 2(a).

□

Proof of Lemma 4. Consider first the case where $a = 1$ (i.e., $S_n \to S$ as $n \to \infty$). Using arguments similar to those in Phillips (1987b) and in the proof of Lemma 2, we have: $T_n^{-1}y_{T_n}^2 \Rightarrow J_c(1)^2, T_n^{-2}\sum_1^{T_n} y_{nt}^2 \Rightarrow \int_0^1 J_c(r)^2dr$ and $T_n^{-1}\sum_1^{T_n} y_{nt-1}u_{nt} \Rightarrow \int_0^1 J_c(r)\,dw(r)$, where $J_c(r)$ is as defined above (4) with $c = \gamma S$. Similar results hold if $a < 1$ (i.e., $S_n \to 0$ as $n \to \infty$) with $J_c(r)$ replaced by $w(r)$. Note that $h_n^{-1}(\alpha_n - 1) = h_n^{-1}(\exp(\gamma h_n) - 1) \to \gamma$ as $n \to \infty$. Hence, $T_n^{-1/2}(\alpha_n - 1)y_{T_n}^2 = S_n T_n^{-1/2}[h_n^{-1}(\alpha_n - 1)] \times T_n^{-1} y_{T_n}^2 \to 0$, proving part (i). Parts (ii) and (iii) follow analogously. Using similar arguments and the results of Lemma A. l, it can be shown that the results remain valid if $y(0) \neq 0$.

Further developments are needed when $\frac{1}{2} \leq a < 1$. Consider first the behavior of $y_{T_n}^2$. Using $y(0) = 0$ and $h_n = k/T_n^a$, we have $y_{T_n} = \sum_{t=1}^{T_n} \exp[(T_n - t)\gamma k/T_n^a]u_{nt}$. First note that $E[y_{T_n}] = 0$. Simple algebra yields

$$T_n^{-a}\operatorname{var}(y_{T_n}) = [\exp(-2\gamma k/T_n^a) - \exp((T_n - 1)2\gamma k/T_n^a)]/ \\ T_n^a[\exp(-2\gamma k/T_n^a) - 1].$$

Using the fact that $a<1$ and $\gamma<0$, we have $T_n^{-a} var(y_{T_n}) \to -1/2\gamma k$ as $n \to \infty$. Using the normality assumption, $T_n^{-a/2} y_{T_n} \Rightarrow N(0, -1/2\gamma k)$ and $2(\alpha_n - 1)y_{T_n}^2 \Rightarrow \chi_1^2$, using the fact that $T_n^a(\alpha_n - 1) = T_n^a(\exp(\gamma k/T_n^a) - 1) \to \gamma k$ as $n \to \infty$. Hence, $T_n^{-1/2}(\alpha_n - 1)y_{T_n}^2 \to 0$, proving part (i). To prove part (ii), note that (given our assumptions) $T_n^{-1/2}(\alpha_n - 1)^2 \sum_1^{T_n} y_{nt}^2 \to \lim_{T_n \to \infty} T_n^{-1/2}(\alpha_n - 1)^2 \sum_1^{T_n} E[y_{nt}^2]$. We have $y_{nt} = \sum_{j=1}^{t} \exp[(t - j)\gamma k/T_n^a]u_{nj}$. Simple algebra (using var $(u_{nj}) = 1$) yields $E(y_{nt}^2) = var(y_{nt}) = [1 - \exp(2t\gamma k/T_n^a)]/[1 - \exp(2\gamma k/T_n^a)]$ and

$$\sum_1^{T_n} \operatorname{var}(y_{nt}) = \frac{[T_n - \{\exp(2\gamma k T_n^{1-a}) - 1\}/\{1 - \exp(-2\gamma k/T_n^a)\}]}{1 - \exp(2\gamma k/T_n^a)}.$$

Using standard limiting arguments, $T_n^{-2a-1/2} \sum_1^{T_n} var(y_{nt}) = (-1/2\gamma k)T_n^{-a+1/2} + o(1)$. Hence, $T_n^{-2a-1/2} \sum_1^{T_n} \operatorname{var}(y_{nt}) \to 0$ if $a > \frac{1}{2}$, and $T_n^{-3/2} \sum_1^{T_n} \operatorname{var}(y_{nt}) \to -1/2\gamma k$ $(if\ a = \frac{1}{2})$. Part (ii) follows using the fact that $T_n^{-1/2}(\alpha_n - 1)^2 \sum_1^{T_n} y_{nt}^2 = T_n^{2a}(\alpha_n - 1)^2 T_n^{-2a-1/2} \sum_1^{T_n} y_{nt}^2$ which converges to 0 if $a > \frac{1}{2}$ and converges to $-\gamma k/2$ if $a = \frac{1}{2}$ using the fact that $T_n^{2a}(\alpha_n - 1)^2 \to (\gamma k)^2$.

To prove part (iii), first note that

$$T_n^{-1/2}(\alpha_n - 1)\sum_1^{T_n} y_{nt-1}u_{nt} = \left(\frac{1}{2\alpha_n}\right)\left[T_n^{-1/2}(\alpha_n - 1)y_{T_n}^2 - T_n^{-1/2}(\alpha_n - 1)^2(\alpha_n + 1) \times \sum_1^{T_n} y_{nt}^2 - T_n^{-1/2}(\alpha_n - 1)\sum_1^{T_n} u_{nt}^2\right].$$

Now $\alpha_n \to 1$ and $T_n^{-1}\sum_1^{T_n} u_{nt}^2 \to 1$. Hence, $T_n^{-1/2}(\alpha_n - 1)\sum_1^{T_n} u_{nt}^2 = T_n^a(\alpha_n - 1) \times T_n^{-1/2-a}\sum_1^{T_n} u_{nt}^2$ which converges to 0 if $a > \frac{1}{2}$ and converges to γk if $a = \frac{1}{2}$ (since $T_n^a(\alpha_n - 1) \to \gamma k$). Parts (i) and (ii) show the other terms to vanish as $T_n \to \infty$ if $a > \frac{1}{2}$ When $a = \frac{1}{2}$, first term converges to 0, the second term converges to $-\gamma k$, and the result of part (iii) follows.

Proof of (21). Define the sequence of log likelihood ratios as $\Lambda_n = \log\,(dQ_n/dP_n)$ where $\{Q_n\}$ is the sequence of probability measures under H_n and $\{P_n\}$ is the sequence of probability measures under H_0. Both measures are defined over a common σ-algebra $\mathscr{G}_n$. To prove the result we make use of the following lemma stated in Roussas (1972, Theorem 7.2).

Lemma A.2. *Let J_n be a random sequence ($\mathscr{G}_n$ measurable). Suppose the sequences of probability measures $\{P_n\}$ and $\{Q_n\}$ are contiguous and that $\mathscr{L}(J_n|P_n) \Rightarrow N(0,\Gamma)(\Gamma \neq 0)$ as $n \to \infty$, where $\mathscr{L}(J_n|P_n)$ denotes the distribution of J_n under the measure P_n. Suppose also that $\Lambda_n - \delta J_n \to -\delta^2\Gamma/2$ in P_n-probability. Then $\mathscr{L}(J_n|Q_n) \Rightarrow N(\delta\Gamma,\Gamma)$ as $n \to \infty$.*

In the present context $J_n = T_n^{1/2}\hat{\alpha}_n$ and we have $\mathscr{L}(T_n^{1/2}\hat{\alpha}_n|P_n) \Rightarrow N(0,1)$. Hence, $\Gamma = 1$ and we need to show (using $\delta \equiv \alpha_0$) that $\Lambda_n - \alpha_0 T_n^{1/2}\hat{\alpha}_n \to -\alpha_0^2/2$ in P_n-probability. Using normality, we deduce that $\Lambda_n = -(1/2)\sum_1^{T_n}(y_{nt} - \alpha_n y_{nt-1})^2 + (1/2)\sum_1^{T_n} y_{nt}^2 = \alpha_0 T_n^{-1/2} \sum_1^{T_n} y_{nt}y_{nt-1} - (\alpha_0^2/2)T_n^{-1}\sum_1^{T_n} y_{nt-1}^2 + o_p(1)$, using the fact that $\alpha_n = \alpha_0 T_n^{-1/2} + o(T_n^{-1/2})$. Hence, $\Lambda_n - \alpha_0 T_n^{1/2}\hat{\alpha}_n = \alpha_0 T_n^{-1/2}\sum_1^{T_n} y_{nt}y_{nt-1} - (\alpha_0^2/2)T_n^{-1}\sum_1^{T_n} y_{nt-1}^2 - \alpha_0 T_n^{-1/2}\sum_1^{T_n} y_{nt}y_{nt-1}/T_n^{-1}\sum_1^{T_n}$

$y_{nt-1}^2 + o_p(1)$. Evaluating under the measure P_n (i.e., under the null that $y_{nt} = \upsilon_{nt}$ where $\upsilon_{nt} \sim$ i.i.d. N (0, 1), see (19)), we obtain

$$(\Lambda_n - \alpha_0 T_n^{1/2}\hat{\alpha}_n|P_n) = \alpha_0 T_n^{-1/2}\sum_1^{T_n}\upsilon_{nt}\upsilon_{nt-1} - (\alpha_0^2/2)T_n^{-1}\sum_1^{T_n}\upsilon_{nt-1}^2$$

$$-\alpha_0 T_n^{-1/2}\sum_1^{T_n}\upsilon_{nt}\upsilon_{nt-1}/T_n^{-1}\sum_1^{T_n}\upsilon_{nt-1}^2 + o_p(1).$$

Using $T_n^{-1/2}\sum_1^{T_n}\upsilon_{nt}\upsilon_{nt-1} \Rightarrow N(0,1)$ and $T_n^{-1}\sum_1^{T_n}\upsilon_{nt-1}^2 \to 1$ as $n \to \infty$, we deduce that $(\Lambda_n - \alpha_0 T_n^{1/2}\hat{\alpha}_n|P_n) \to -\alpha_0^2/2$. This shows the conditions of Lemma. A.2 to be satisfied and we therefore deduce from its conclusion that $\mathscr{L}(T_n^{1/2}\hat{\alpha}_n|Q_n) \Rightarrow N(\alpha_0, 1)$, proving (21).

Acknowledgments

The author wishes to thank Jean-Marie Dufour, Whitney Newey, anonymous referees, and Peter Phillips (the editor) for comments. This research was supported by the Social Sciences and Humanities Research Council of Canada, the Natural Sciences and Engineering Council of Canada, and the Fonds pour la Formation de Chercheurs et l'Aide à la Recherche du Québec.

References

Aiyar, R.J. (1981). Asymptotic efficiency of rank tests of randomness against autocorrelation. *Annals of the Institute of Statistical Mathematics* 33, 255–262.

Bergstrom, A.R. (1984). Continuous time stochastic models and issues of aggregation over time. In *Handbook of Econometrics*, M.D. Intriligator and Z. Griliches (eds.), vol. 2, North-Holland: Amsterdam, pp. 1145–1212.

Dickey, D.A. (1976). Estimation and hypothesis testing for nonstationary time series. PhD Thesis, Iowa State University, Ames, IA,

Dickey, D.A. and Fuller, W.A. (1979). Distribution of the estimators for autoregressive time series with a unit root. *Journal of the American Statistical Association* 74, 427–431.

Evans, G.B.A. and Savin, N.E. (1981). The calculation of the limiting distribution of the least squares estimator of the parameter in a random walk model. *Annals of Statistics* 9, 1114–1118.

Fuller, W.A. (1976). *Introduction to Statistical Time Series.* Wiley: New York.

Kreiss, J.P. (1987). On adaptive estimation in stationary ARMA processes. *Annals of Statistics* 15, 112–133.

Mikulski, P.W. and Monsour, M.J. (1991). Optimality of the maximum likelihood estimator in first-order autoregressive processes. *Journal of Time Series Analysis* 12(3), 237–253.

Perron, P. (1989a). The calculation of the limiting distribution of the least squares estimator in a near-integrated model. *Econometric Theory* 5, 241–255.

Perron, P. (1989b). Testing for a random walk: A simulation experiment when the sampling interval is varied. In *Advances in Econometrics and Modeling*, B. Raj (ed.), Kluwer Academic Publisher: Dordrecht, pp. 47–68.

Perron, P. (1991). A continuous time approximation to the unstable first-order autoregressive model: The case without an intercept. *Econometrica* 59, 211–256.

Phillips, P.C.B. (1987a). Time series regression with unit roots. *Econometrica* 55, 277–301.

Phillips, P.C.B. (1987b). Towards a unified asymptotic theory for autoregression. *Biometrika* 74, 535–547.

Phillips, P.C.B. and Ouliaris, S. (1990). Asymptotic properties of residual based tests for cointegration. *Econometrica* 58, 165–193.

Priestley, M.B. (1981). *Spectral Analysis and Time Series.* Academic Press: New York.

Rao, C.R. (1973). *Linear Statistical Inference and Its Applications*, 2nd ed. Wiley: New York.

Roussas, G.G. (1972). *Contiguity of Probability Measures.* Cambridge University Press: Cambridge.

Sargan, J.D. and Bhargava, A. (1983). Testing residuals from least squares regression for being generated by the Gaussian random walk. *Econometrica* 51, 153–174.

Shiller, R.J. and Perron, P. (1985). Testing the random walk hypothesis: Power versus frequency of observations. *Economics Letters* 18, 381–386.

Chapter 6

Unit Root Tests in ARMA Models with Data-Dependent Methods for the Selection of the Truncation Lag*

Serena Ng and Pierre Perron

Départment de Sciences Economiques et C.R.D.E., Université de Montréal, Québec H3C 3J7, Canada

We analyze the choice of the truncation lag in the context of the Said–Dickey test for the presence of a unit root in a general autoregressive moving average model. It is shown that a deterministic relationship between the truncation lag and the sample size is dominated by data-dependent rules that take sample information into account. In particular, we study data-dependent rules that are not constrained to satisfy the lower bound condition imposed by Said–Dickey. Akaike's information criterion falls into this category. The analytical properties of the truncation lag selected according to a class of information criteria are compared to those based on sequential testing for the significance of coefficients on additional lags. The asymptotic properties of the unit root test under various methods for selecting the truncation lag are analyzed, and simulations are used to show their distinctive behavior in finite samples. Our results favor methods based on sequential tests over those based on information criteria, because the former show less size distortions and have comparable power.

Keywords: Dickey–Fuller t test; general to specific; information criteria; order selection; Said–Dickey test.

*Originally Published in *Journal of the American Statistical Association* 90, pp. 268–281.

1. Introduction

Testing for the presence of a unit root in a time series of data has become a common starting point of applied work in macroeconomics. Except in very special cases, one often assumes that the series to be tested is driven by serially correlated innovations and tests for the presence of a unit root using statistics that take serial dependence into account. One such statistic that has become very popular is the augmented Dickey–Fuller t test due to Dickey and Fuller (1979) and Said and Dickey (1984). Their test, hereafter referred to as t_ρ, is based on estimates from an augmented autoregression. The test is valid for stationary and invertible autoregressive moving-average (ARMA) noise functions of unknown order provided that the truncation lag, k, is chosen in relationship to the sample size, T, to satisfy lower and upper bound conditions.

An issue that arises with the implementation of t_ρ is the choice of k. Work of Schwert (1989), Agiakloglou and Newbold (1991), and Harris (1992) have found the order of the autoregression to have important size and power implications. This article provides a formal analysis of the relevance of k in the test procedure. One of our objectives is to show, via simulations, that a deterministic rule that relates k to T is inferior to a data-dependent rule that takes sample information into account. Another objective is to clarify the role of the lower bound and the upper bound on k in the limiting behavior of the statistic t_ρ. We study the asymptotic properties of t_ρ and of the estimates from the augmented autoregression with k chosen using different data-dependent rules. Among these are information-based model selection rules, such as the Akaike information criterion (AIC) and the Schwartz criterion, and sequential testing for the significance of the coefficients on lags, such as F or t tests. We show that with parameter values for which size problems surface, information-based rules tend to select values of k that are consistently smaller than those chosen through sequential testing for the significance of coefficients on additional lags, and the size distortions associated with the former method are correspondingly larger. Thus the choice of the data-dependent rule has bearing on the size and power of the test. These issues are of particular relevance in finite samples.

The article is structured as follows. Section 2 puts forth the Said and Dickey framework, the role of the upper and lower bound conditions on k, and the implications for t_ρ with and without the lower bound. Section 3 provides a discussion of procedures typically used to select k. Formal definitions of 'deterministic' and 'adaptive' rules are given. Sections 4 and 5 analyze the properties of t_ρ with k chosen according to information criteria and sequential testing for the significance of coefficients on lags. Section 6 presents implications of these results. We conclude with suggestions for procedures to select k and directions for future research. Proofs of theorems are given in the Appendix.

2. The Said–Dickey Approach

2.1. *The test statistic*

Suppose the data-generating process (DGP) for $\{y_t\}$ is given by

$$y_t = \rho y_{t-1} + u_t, \tag{1}$$

$$u_t = \sum_{i=1}^{p} \alpha_i u_{t-i} + e_t + \sum_{j=1}^{q} \theta_j e_{t-j},$$

where $e_t \sim \text{iid}(0, \sigma_e^2)$ with bounded fourth moment. Assuming that $\{u_t\}$ is stationary and invertible with autoregressive and moving-average polynomials that do not share common roots, $\{y_t\}$ evolves according to

$$\Delta y_t = (\rho - 1)y_{t-1} + \sum_{i=1}^{\infty} d_i u_{t-i} + e_t, \tag{2}$$

where the coefficients $d_i (i = 1, \ldots, \infty)$ are functions of the parameters $\{\alpha_i, \theta_j; i = 1, \ldots, p, j = 1, \ldots, q\}$. The true order of the autoregression is infinity when $q > 0$. The null hypothesis of interest is $\rho = 1$, in which case a unit root is said to exist and the DGP is an ARIMA $(p, 1, q)$. Because $\Delta y_t = u_t$ under the null hypothesis, (2) can also be seen as an autoregression in Δy_t augmented by y_{t-1}, namely

$$\Delta y_t = (\rho - 1)y_{t-1} + \sum_{i=1}^{\infty} d_i \Delta y_{t-i} + e_t. \tag{3}$$

When the orders p and q are unknown, as is often the case in practice, Said and Dickey (1984) suggested approximating the infinite autoregression by a truncated version whose order is a function of the number of observations, T:

$$\Delta y_t = d_0 y_{t-1} + \sum_{i=1}^{k} d_i \Delta y_{t-i} + e_{tk}, \tag{4}$$

where $d_0 = \rho - 1$, and for future reference, we denote $\mathbf{d}(k) = (d_1, \ldots, d_k)$. The ordinary least squares (OLS) estimates are similarly defined as $\hat{d}_0 = \hat{\rho} - 1$ and $\hat{\mathbf{d}}(k) = (\hat{d}_1, \ldots, \hat{d}_k)$. The order of truncation, k, is assumed to satisfy some conditions to ensure consistency of the least squares estimates. More precisely, Said and Dickey (1984) assumed the following:

Assumption A1. k is chosen as a function of T such that

$$k^3/T \to 0 \quad \text{and} \quad k \to \infty \quad \text{as} \quad T \to \infty.$$

Assumption A2. There exist $c > 0$ and

$$r > 0 \quad \text{such that } ck > T^{1/r}.$$

Assumption A1 is based on the work of Berk (1974) who showed consistency of the parameter estimates in an autoregression of the form (4) but without the level regressor, y_{t-1}, and when the process is stationary. The assumption is imposed to ensure that the number of regressors does not increase so fast as to induce excess variability in the estimators. Assumption A2, often an overlooked condition, is a lower-bound condition that restricts k to be at least a polynomial rate in T. It rules out values of k that are proportional to $\log T$. Intuitively, A2 prohibits k from being so small as to provide an inadequate approximation to the true model. It is more restrictive than the following assumption:

Assumption A2′. k satisfies $k^{1/2} \sum_{i=k+1}^{\infty} |d_i| \to 0$ and

$$k \to \infty \quad \text{as } T \to \infty.$$

Assumption A2′ was used by Berk (1974), and in related work by Lewis and Reinsel (1985), to show consistency of the OLS estimates in an autoregression applied to a stationary process. Note that A2$'$ is satisfied for any $\{u_t\}$ that is a stationary and invertible ARMA process as long as $k \to \infty$ as $T \to \infty$, irrespective of the rate at which k increases. Of particular importance is the fact that unlike A2, A2$'$ allows k to grow at a logarithmic rate. Berk (1974) and Lewis and Reinsel (1985) strengthened Assumption A2$'$ to the following:

Assumption A2$''$. k satisfies $T^{1/2}\sum_{i=k+1}^{\infty}|d_i| \to 0$ as

$$k \to \infty \quad \text{and} \quad T \to \infty,$$

to ensure $\sqrt{T}$ consistency of $\hat{\mathbf{d}}(k)$. Note that A2$''$ implicitly rules out k growing at a $\log(T)$ rate and is basically equivalent to A2. Consistency of $\hat{\mathbf{d}}(k)$ may be achieved at a rate slower than $\sqrt{T}$ if A2$'$ is satisfied but not A2$''$.

The foregoing discussion applies when the DGP is an infinite autoregression, as would be the case if moving-average components were present. When dealing with a finite autoregression, A2$''$ is automatically satisfied. In fact, k need not grow to infinity as long as it is selected to be larger than the true order. Hence most of the results that follow also apply to the case of a finite autoregression. (For a more specific treatment of this case, see Hall, 1994.)

Said and Dickey's result states that when k satisfies A1 and A2, the least squares estimates $\hat{\mathbf{d}}(k)$ are $\sqrt{T}$-consistent, and the coefficient on y_{t-1} provides a basis for testing the unit root hypothesis. The limiting distribution for the t statistic on $\hat{d}_0 = (\hat{\rho} - 1)$ for testing $\rho = 1$ is such that

$$t_\rho \Rightarrow \left(\int_0^1 W(r)\,dW(r)\right)\left(\int_0^1 W(r)^2 dr\right)^{-1/2}, \tag{5}$$

where $W(r)$ is a standard Brownian motion in the space $C[0, 1]$. Percentiles of this distribution were given by Fuller (1976). The result stated in (5) extends naturally to the inclusion of deterministic components in (4). In that case the Wiener process is replaced by its detrended counterpart.

2.2. *A useful result*

Of interest are the properties of the test statistic when k is chosen as a function of T to satisfy A1 but not necessarily A2, because such procedures are commonly used in applied work. The following lemma considers the validity of Said and Dickey's (1984) result when the lower-bound condition A2 is not imposed.

Lemma 2.1. *Let $\{y_t\}$ be given by (1). Let t_ρ be obtained from the truncated autoregression (4) with k chosen such that A1 is satisfied. Then (a) the asymptotic distribution of t_ρ continues to be given by (5) without A2, and (b) $\hat{\mathbf{d}}(k) = (\hat{d}_1, \ldots, \hat{d}_k)$ is not in general $\sqrt{T}$-consistent for $d(k) = (d_1, \ldots, d_k)$ if A2 or A2′ does not hold. In that case there exists a λ, with $|d_j| \leq C_1\lambda^j$ for some constant C_1 and $0 < \lambda < 1$, such that $\lambda^{-k}(\hat{d}_i - d_i) = O_p(1), (i = 1, \ldots, k)$.*

Lemma 2.1 states that although $\sqrt{T}$ consistency of the coefficients on Δy_{t-i} is not assured without A2, $\hat{d}_0$ is still consistent for d_0 at rate T, and t_ρ attains the same limiting distribution as defined in (5) with Assumption A1 alone. The proof of consistency of $\hat{d}_0$ and $\hat{\mathbf{d}}(k)$ under A1 and A2 was given by Said and Dickey (1984). The lower-bound condition enters the analysis only when considering the properties of coefficients pertaining to Δy_{t-i}. Specifically, $\sqrt{T}$ consistency of $\hat{\mathbf{d}}(k)$ requires, from Berk (1974, Lemma 2), that

$$E\left((T-k)^{-1}\sum_{j=1}^{k}\left\{\sum_{t=k+1}^{T} u_{t-j}(e_{tk}-e_t)\right\}^2\right) \leq k(T-k)\sum_{i=k+1}^{\infty} d_i^2 \rightarrow 0. \tag{6}$$

Because $(e_{tk} - e_t)$ is the error in approximating an infinite autoregression by a truncated autoregression, it is larger the smaller is k; the role of the lower bound is thus intuitive in this context. Sufficient conditions for (6) to hold are provided by either A2, as invoked by Said and Dickey (1984), or A2″, as used by Berk (1974) and by Lewis and Reinsel (1985). Recall, however, that k growing at a logarithmic rate is ruled out by either A2 or A2″.

To see the ramifications of this condition, suppose that $\{u_t\}$ is an MA(1) with coefficient θ, and hence $d_i = -(-\theta)^i$. The condition (6) is equivalent to requiring that $\log(k) + \log(T - k) + k \log \theta^2$ diverges to $-\infty$. Now take $k = b\log(T)$ for some constant $b > 0$. Clearly, $k^3/T \to 0$ and A1 is satisfied, but the condition for $\sqrt{T}$ consistency is (approximately) $1 + b\log(\theta^2) < 0$. This condition fails when $|\theta| > exp(-1/2b)$. Hence for any fixed rule satisfying $k = b\log(T)$, there will exist a range of values of θ such that (6) does not hold. In that case $\hat{\mathbf{d}}(k)$ will achieve consistency not at rate $\sqrt{T}$ but rather at the slower rate of $T^{(1-a)/2}$, with $a = 1+b\log(\theta^2)$ in the case of an MA(1) (or, equivalently, $|\theta|^{-k}(\hat{d}_i - d_i) = O_p(1)$ as stated in Lemma 2.1). As we will see in subsequent sections, this logarithmic rate is of special interest.

The result that the estimates for the coefficients on Δy_{t-i} might achieve consistency at a rate slower than $\sqrt{T}$ extends to the case when $\{u_t\}$ satisfies a general ARMA (p, q) model, using the fact that the coefficients d_i are such that $|d_i| < C_1\lambda^i, 0 < \lambda < 1$ for some constant C_1 (see, for example, Fuller, 1976). The important point is that $(\hat{\rho} - 1)$ will continue to be order T-consistent even without the lower-bound condition. The asymptotic equivalence of t_ρ with and without A2 follows from this result and the result that consistency of the least squares estimates is enough to ensure the consistency of $\hat{\sigma}_k^2$ for σ_e^2.

Although all estimates from the regression (4) will be consistent whether or not A2 holds, the lower-bound condition on k is important. The coefficients on the stationary regressors will converge at a rate slower than $\sqrt{T}$ when the lower-bound condition is not satisfied. Therefore, choices of k that satisfy A2 will yield coefficient estimates on the stationary differences that achieve consistency at a faster rate and can be expected to lead to unit root tests having better finite-sample properties than those choices that do not satisfy A2.

3. Selection of *k*

This section consists of three parts. First, in Section 3.1 we use simulations to show that any *a priori* rule that presets the value of k is likely to result in size distortions and/or power loss, unless that value of k happens to be chosen appropriately. This is so, even if k is chosen to be a fixed function of T. In Section 3.2, we discuss the specifics of two data-dependent rules whereby the relationship between k and T depends on the given sample of data. In Section 3.3 we further restrict our analysis to data-dependent rules that satisfy A1 only and analyze the limiting distribution of t_ρ when such data-dependent rules are used.

3.1. *Rules of thumb*

3.1.1. *Fixing k*

Although the asymptotic distribution of t_ρ is derived under the assumption that k increases at an appropriate rate with T, the theoretical conditions A1 and A2 provide little practical guidance for choosing k. The common practice is to fix k at a value independent of T. Using (1) as the DGP, we considered numerous parameterizations of α_i and θ_j, with k fixed to be 1–10. As the results reported in Table 1 (moving-average case) indicate, the properties of the statistic can be quite different, depending on the chosen value of k. For example, when $\theta = -0.8$, fixing k to be 4 yields an exact size of 28% instead of the 5% nominal size, noting that the exact size worsens to 0.939 when θ is -0.95. But size distortions are much smaller as k becomes larger. Although size distortions are much smaller when θ is positive, t_ρ is oversized when k is odd but undersized when k is even.

Although in autoregressive models (see Table 2) the exact size of the test for all choices of k is generally close to the nominal size (provided that k is larger than the true order), the choice of k has implications for power. As is evident from Table 2, an overparameterized model is associated with lower power. Thus, although a liberal choice of k will reduce size distortions in moving-average models, it will generally yield lower power.

We also performed similar simulations for $T = 200$ and $T = 500$. As expected, power increases for every value of k in both the moving-average and the autoregressive cases. With respect to the size of the test, the results for the autoregressive case are qualitatively the same as when $T = 100$. For positive moving-average models, the zig-zag pattern of size distortions as k alternates between odd and

Table 1. Size and power of unit root tests, moving-average case, $T = 100$ (5,000 replications).

ρ	θ	$k=1$	$k=2$	$k=3$	$k=4$	$k=5$	$k=6$	$k=7$	$k=8$	$k=9$	$k=10$
1.0	0.8	0.123	0.031	0.075	0.037	0.062	0.041	0.062	0.041	0.051	0.041
1.0	0.5	0.103	0.047	0.065	0.052	0.057	0.054	0.053	0.053	0.051	0.048
1.0	0.3	0.073	0.051	0.055	0.056	0.053	0.047	0.050	0.048	0.045	0.046
1.0	0.0	0.049	0.048	0.046	0.044	0.046	0.044	0.045	0.044	0.044	0.041
1.0	−0.3	0.091	0.062	0.057	0.052	0.052	0.048	0.049	0.048	0.048	0.045
1.0	−0.5	0.214	0.099	0.068	0.055	0.051	0.051	0.052	0.050	0.049	0.048
1.0	−0.8	0.880	0.640	0.434	0.283	0.200	0.132	0.110	0.082	0.074	0.060
0.95	0.8	0.307	0.044	0.176	0.059	0.129	0.064	0.106	0.067	0.085	0.064
0.95	0.5	0.237	0.074	0.130	0.089	0.103	0.086	0.086	0.079	0.077	0.072
0.95	0.3	0.162	0.088	0.101	0.092	0.092	0.085	0.087	0.081	0.075	0.069
0.95	0.0	0.117	0.111	0.108	0.102	0.099	0.093	0.083	0.082	0.080	0.068
0.95	−0.3	0.219	0.139	0.122	0.108	0.100	0.091	0.094	0.089	0.087	0.081
0.95	−0.5	0.477	0.246	0.157	0.115	0.104	0.087	0.085	0.081	0.078	0.069
0.95	−0.8	0.997	0.941	0.782	0.584	0.444	0.329	0.255	0.203	0.164	0.130
0.85	0.8	0.788	0.201	0.524	0.212	0.379	0.203	0.278	0.166	0.207	0.149
0.85	0.5	0.708	0.316	0.425	0.297	0.308	0.252	0.239	0.203	0.191	0.169
0.85	0.3	0.598	0.393	0.395	0.334	0.306	0.267	0.241	0.214	0.196	0.171
0.85	0.0	0.510	0.436	0.399	0.343	0.316	0.271	0.251	0.218	0.193	0.166
0.85	−0.3	0.746	0.540	0.452	0.376	0.344	0.287	0.266	0.233	0.209	0.176
0.85	−0.5	0.961	0.779	0.614	0.482	0.423	0.353	0.315	0.275	0.242	0.208
0.85	−0.8	1.000	1.000	0.996	0.956	0.886	0.759	0.671	0.565	0.483	0.399

Note: DGP: $y_t = \rho y_{t-1} + u_t, u_t = e_t + \theta e_{t-1}$. Regression: $\Delta y_t = \delta_0 y_{t-1} + \sum_{i=1}^{k} \delta_i \Delta y_{t-1} + v_t$.

even persists even when T is 500. But for negative moving-average models, size distortions increase with T for a given value of k. For example, with $\theta = -0.8$ and $k = 3$, the exact size increases from 0.455 to 0.598 as T increases from 100 to 500.

3.1.2. *Choosing k as a fixed function of T*

Any rule that defines k as a deterministic function of T fits into this category. A rule often used in unit root tests is due to Schwert (1989). For given constants c and d, the truncation lag, k, is chosen according to

$$k = \text{int}\{c(T/100)^{1/d}\}.$$

Table 2. Size and power of unit root tests, autoregressive case, $T = 100$ (5,000 replications).

ρ	ϕ_1	ϕ_2	ϕ_3	ϕ_4	$k = 1$	$k = 2$	$k = 3$	$k = 4$	$k = 5$	$k = 6$	$k = 7$	$k = 8$	$k = 9$	$k = 10$
1.0	0.6	0.0	0.0	0.0	0.058	0.055	0.057	0.053	0.053	0.053	0.058	0.055	0.055	0.052
1.0	−0.6	0.0	0.0	0.0	0.054	0.054	0.054	0.052	0.052	0.047	0.045	0.043	0.044	0.043
1.0	0.4	0.2	0.0	0.0	0.033	0.052	0.052	0.051	0.050	0.052	0.050	0.048	0.047	0.046
1.0	0.3	0.3	0.25	0.14	0.062	0.034	0.038	0.051	0.054	0.052	0.050	0.047	0.048	0.052
0.95	0.6	0.0	0.0	0.0	0.391	0.346	0.320	0.281	0.260	0.225	0.210	0.191	0.178	0.154
0.95	−0.6	0.0	0.0	0.0	0.078	0.075	0.073	0.068	0.070	0.061	0.061	0.058	0.056	0.055
0.95	0.4	0.2	0.0	0.0	0.125	0.354	0.328	0.288	0.275	0.237	0.225	0.204	0.194	0.165
0.95	0.3	0.3	0.25	0.14	0.137	0.650	0.865	0.903	0.876	0.814	0.763	0.675	0.608	0.522
0.85	0.6	0.0	0.0	0.0	0.976	0.938	0.883	0.805	0.720	0.626	0.560	0.479	0.435	0.357
0.85	−0.6	0.0	0.0	0.0	0.252	0.224	0.207	0.188	0.171	0.159	0.154	0.134	0.129	0.122
0.85	0.4	0.2	0.0	0.0	0.818	0.933	0.889	0.799	0.718	0.626	0.560	0.479	0.435	0.357
0.85	0.3	0.3	0.25	0.14	0.688	0.964	0.991	0.993	0.984	0.960	0.908	0.827	0.748	0.647

Note: DGP: $y_t = \rho y_{t-1} + \sum_{i=1}^{4} \phi_i \Delta y_{t-1} + u_t$. Regression: $\Delta y_t = \delta_0 y_{t-1} + \sum_{i=1}^{k} \delta_i \Delta y_{t-1} + v_t$.

Values of $c = 4$ and 12 and $d = 4$ were used in Schwert's extensive Monte Carlo analysis. He found that the size of the test is significantly better with $c = 12$ the closer the moving-average coefficient, θ, is to -1. Problems encountered in fixing k arbitrarily will also arise if k is chosen as a deterministic function of T, because one is faced with a given sample size in practice. In general, there is no way to assure that arbitrarily chosen values of c and d are adequate for a given data series unless c and d happen to be chosen correctly.

The simulations highlight the fact that conditions on k appropriate for asymptotic inference are not necessarily good practical guidelines for selecting k. Indeed, the value of k that ensures an exact size close to the nominal size and also produces high power is highly dependent on the actual DGP; that is, the values of the AR and MA parameters. Rules of thumb ignore such sample information and are the main reason why fixing k is to be avoided as a matter of practice.

3.2. *Data-dependent rules*

3.2.1. *Information-based rules*

The order of an autoregressive process is often chosen by minimizing an objective function that trades off parsimony against reductions in the sum of squared residuals. Following Hannan and Deistler (1988), we consider an objective function of the general form

$$I_k = \log \hat{\sigma}_k^2 + kC_T/T, \tag{7}$$

where C_T is a sequence that satisfies $C_T > 0, C_T/T \to 0$. The familiar Akaike information criterion (AIC) (Akaike, 1974) is obtained as a special case with $C_T = 2$. Another popular criterion is that of Schwartz (1978) with $C_T = \log T$. Other criteria, such as the Bayesian information criterion (BIC), can be shown to fall within the class of I_k. For econometric applications, the AIC and the Schwartz criteria are more common and will be considered in subsequent sections.

3.2.2. *Sequential tests for the significance of the coefficients on lags*

The premise of a sequential test is a general-to-specific modeling strategy that chooses between a model with m lags and a model with $r = m + n$ lags. Let $\hat{\mathbf{d}}(m,r)$ denote the vector of coefficients $(\hat{d}_{m+1}, \ldots, \hat{d}_r)$ obtained by applying OLS to (4), with $\hat{\sigma}_r^2 = (T-r)^{-1}\sum_{t=r+1}^{T} \hat{e}_{tr}^2$ and

$$\mathbf{M}_r = \sum_{t=r+1}^{T} (y_{t-1}, \Delta y_{t-1}, \ldots, \Delta y_{t-r})' \times (y_{t-1}, \Delta y_{t-1}, \ldots, \Delta y_{t-r}).$$

Let $\mathbf{M}_r^{-1}(n)$ be the lower-right $(n \times n)$ block of $\mathbf{M}_r^{-1}$. We define the Wald test for the null hypothesis that the coefficients on the last n lags are jointly equal to 0 as

$$J(m,\ r) = \hat{\mathbf{d}}(m,\ r)'(\mathbf{M}_r^{-1}(n))^{-1}\hat{\mathbf{d}}(m,\ r)/\hat{\sigma}_r^2. \tag{8}$$

We now provide a formal definition of the procedure for choosing $\hat{k}$ from a set of possible values $\{0, 1, \ldots, k\max\}$, where $k\max$ is selected *a priori*.

Definition 3.1. The general-to-specific modeling strategy chooses $\hat{k}$ to be either (i) $m+1$ if, at significance level α, $J(m,r)$ is the first statistic in the sequence $J(i, i+n), \{i = k\max - 1, \ldots, 1\}$, which is significantly different from zero, or (ii) 0 if $J(i, i+n)$ is not significantly different from zero for all $i = k\max - 1, k\max - 2, \ldots, 1$.

The idea is to start with the most general model with $k\max + n - 1$ lags and test whether the coefficients of the last lags are significant. If they are, then $\hat{k} = k\max$; otherwise, the next step is to estimate an autoregression of order $k\max - 2 + n$ and perform the joint test again. This procedure is repeated until a rejection occurs or the sequential testing leads to the boundary of zero lags. This procedure has been analyzed by Hall (1994) in the case of a pure autoregressive process.

The $J(m,r)$ statistic specializes to a t statistic on the last lag if the test is performed with $n = 1$. This special case of the general-to-specific procedure has been used by Perron (1989). (See also Perron,

1990; Perron and Vogelsang, 1992 for simulation results for unit root tests allowing a break in the trend function and the noise function assumed to be an ARMA process.) Although in principle one can start with k min lags and gradually increase k until the next included lag is insignificant, Hall (1994) found that a specific-to-general approach is not generally asymptotically valid in the pure AR case. He also found the finite sample properties of statistics associated with a specific-to-general approach to be inferior to those based on a general-to-specific scheme in more general ARMA models. In subsequent analyses only the general-to-specific approach will be analyzed.

3.3. *Rules satisfying the upper-bound condition*

We now restrict our attention to deterministic and data-dependent rules that satisfy Assumption A1. Formal definitions for the rules considered are as follows.

Definition 3.2. *Deterministic Rules.* Let $\tilde{K} = (\tilde{k}_1, \tilde{k}_2, \ldots)$ be the set of points in $\tilde{K}_\infty = \prod_{T=1}^{\infty} S_T$, where $S_T = \{0, 1, \ldots, [T/2]\}$, with $\tilde{k}_T \to \infty$ and $\tilde{k}_T^3/T \to 0$ as $T \to \infty$.

Simply put, $\tilde{K}$ is the collection of deterministic rules that satisfy the conditions of Lemma 2.1. Our definition of deterministic rules is adapted from Eastwood and Gallant (1991), who studied the selection of the truncation point in a univariate Fourier series expansion fitted by least squares. In our context Schwert's rule of thumb is, for example, an element of $\tilde{K}$.

Definition 3.3. *Adaptive Rules.* An adaptive truncation rule is a sequence of random variables $\hat{K}_\infty = (\hat{k}_1, \hat{k}_2, \ldots)$. We say that an adaptive truncation rule maps into the set of deterministic rules $\tilde{K}$ if there exists a deterministic rule $\tilde{k}_T$ such that $\tilde{K} = (\tilde{k}_1, \tilde{k}_2, \ldots)$ is a subset of $\tilde{K}_\infty$ and $\hat{k}_T - \tilde{k}_T \to_p 0$.

The following lemma considers the limiting distribution of t_ρ when it is based on adaptive rules that map into the set of deterministic rules $\tilde{K}$.

Lemma 3.4. *Suppose that we have an adaptive truncation rule* $\hat{K}_\infty = (\hat{k}_1, \hat{k}_2, \ldots)$ *that maps into the set of deterministic rules* $\tilde{K}$ *stated in Definition* 3.2, *and let* $t_\rho(\hat{k}_T)$ *be the* t *statistic for testing* $\rho = 1$ *in regression* (4) *estimated with* $\hat{k}_T$ *lags. Then* $t_\rho(\hat{k}_T) \Rightarrow \int_0^1 W(r)dW(r)(\int_0^1 W(r)^2 dr)^{-1/2}$.

The proof is analogous to Theorem 5 of Eastwood and Gallant (1991) and thus is omitted. The importance of Lemma 3.4 is that the limiting distribution of t_ρ is the same whether one uses a deterministic rule in $\tilde{K}$ or an adaptive rule that maps into $\tilde{K}$. The issue then becomes which of the selection procedures delivers better finite-sample properties in testing for the presence of a unit root.

Deterministic rules are useful for analytical purposes, because they help establish the properties of t_ρ under adaptive rules. But as seen from the results reported earlier, size and power will be affected whenever k is fixed in a deterministic way unless the rule happens to be chosen correctly. Adaptive rules take sample information into account and are thus likely to dominate deterministic rules. In the next two sections our analysis will be further restricted to adaptive rules only.

4. Adaptive Rule 1: Information Criteria

This section presents properties of $\hat{k}$ and t_ρ when an information criterion as defined in (7) is used to select the truncation lag in regression (4). A related issue has been studied by Hannan and Deistler (1988) in the context of stationary variables with the autoregression

$$x_t = \sum_{i=1}^{k} \delta_i x_{t-i} + e_{tk}. \tag{9}$$

The next lemma summarizes a result of theirs that is relevant to our analysis.

Lemma 4.1. *Let* x_t *be a stationary and invertible ARMA process with finite fourth moment and* $\tilde{\sigma}_k^2 = (T-k)^{-1}\sum_{t=k+1}^{T} \tilde{e}_{tk}^2$ *with* $\tilde{e}_{tk}$ *the OLS residuals from regression* (9). *Let* C_T *be a function of* T *such* *that* $C_T > 0$ *and* $C_T/T \to 0$, *and* $\tilde{k}_T = \arg\min_k(\log(\tilde{\sigma}_k^2) + kC_T/T)$. *Then* $\lim_{T\to\infty} \tilde{k}_T/(b\log T) = 1$ *for some constant* b.

The result that the AIC with $C_T = 2$ chooses a value of k that is proportional to $\log T$ in a univariate Gaussian ARMA model is due to Shibata (1980). Hannan and Deistler (1988) provided a unified asymptotic framework to show that the feature of $\log T$ proportionality is generic to information-based rules applied, in particular, to stationary and invertible ARMA models. The logarithmic rule also extends to multivariate and/or autoregressive moving-average with (lagged) exogenous terms (ARMAX) models, as Hannan and Deistler (1988) have shown. Their result is useful in studying the properties of $\hat{k}_T$ within the context of an augmented autoregression of the form (4) derived for an autoregressive integrated moving-average (ARIMA) $(p, 1, q)$ process. The following lemma shows that their result extends to this latter case.

Lemma 4.2. *Let* y_t *satisfy* (1) *and define* $\hat{\sigma}_k^2 = (T-k)^{-1}\sum_{t=k+1}^{T}\hat{e}_{tk}^2$, *where* $\hat{e}_{tk}$ *are the least squares residuals from the augmented autoregression* (4). *Let* $\tilde{\sigma}_k^2 = (T-k)^{-1}\sum_{t=k+1}^{T}\tilde{e}_{tk}^2$, *where* $\tilde{e}_{tk}$ *are the OLS residuals from the restricted regression* (9) *with* $x_t = \Delta y_t$. *Then* $\tilde{\sigma}_k^2 = \hat{\sigma}_k^2 + o_p(T^{-1/2})$, *provided* k *satisfies A1.*

Lemma 4.2 implies that the difference between the residual sum of squares from an augmented autoregression and a restricted one is $o_p(T^{-1/2})$ uniformly in k. Hence the information criteria and the corresponding values of k that minimize such criteria are asymptotically the same in both cases. Thus, the AIC and Schwartz criteria, when applied to the augmented autoregression defined in (4), also select truncation lags proportional to $\log T$ under the null hypothesis of a unit root. The implication for the unit root test is summarized in the following theorem.

Theorem 4.3. *If* k *is selected using an information criterion in the class* I_k *as defined in* (7), *then* t_ρ *has a limiting distribution defined by* (5) *under the null hypothesis of a unit root.*

The order of truncation selected by the AIC or the Schwartz criteria is proportional to $\log T$. Because such a rule satisfies A1 that $k^3/T \to 0$ and $k \to \infty$ as $T \to \infty$, it is also an adaptive rule that maps into the set of deterministic rules $\tilde{K}$. Thus the result follows directly from Lemmas 2.1 and 3.4. Theorem 4.3 still holds when the DGP is a finite instead of an infinite autoregression, provided that the information criterion does not asymptotically underparameterize the model (see Hall, 1994). Note, however, that the information criterion is not an adaptive rule that maps into the set of deterministic rules that satisfy both A1 and A2.

4.1. *A special case: An MA(1)*

Because the truncation lag selected from regression (4) when the series is an ARIMA$(p, 1, q)$ and the truncation lag selected on the basis of (9) when the series is a stationary and invertible ARMA process have the same asymptotic properties, we can, for simplicity, use the restricted framework to provide more insight about the properties of the truncation lag selected using information criteria. Specifically, we consider an MA(1) process defined as

$$x_t = e_t + \theta e_{t-1} = \sum_{i=1}^{\infty} \phi_i x_{t-i} + e_t,$$

where $\phi_i = -(-\theta)^i$. The true order of the autoregression is infinity for all values of $\theta \neq 0$. The estimated regression is

$$x_t = \sum_{i=1}^{k} \phi_i x_{t-i} + e_{tk}.$$

It is straightforward to show that $\tilde{\sigma}_k^2$ is approximately related to k by

$$\tilde{\sigma}_k^2 \simeq \sigma_e^2 (1 - \theta^{2(k+2)})(1 - \theta^{2(k+1)})^{-1}.$$

Minimizing the AIC, $\log \tilde{\sigma}_k^2 + 2k/T$, the solution is asymptotically equivalent to

$$\tilde{k}(AIC) \approx (\log(T) + \log[(\theta^2 - 1)\log\theta^2] - \log 2) \times (|\log\theta^2|)^{-1}. \tag{10}$$

Table 3. Approximation to the selected truncation lag using AIC in the MA(1) model.

$\|\theta\|$	0.2	0.4	0.6	0.8
$T = 100$	1	2	3	3
$T = 10{,}000$	3	4	7	13
$T = 1{,}000{,}000$	4	7	12	23

Note: DGP: $x_t = e_t + \theta e_{t-1}$. Regression: $x_t = \sum_{i=1}^{k} \phi_i x_{t-i} + e_{tk}$.

Table 3 presents the approximation to $\tilde{k}(\mathrm{AIC})$ provided by (10) for various values of $|\theta|$ and T. For small θ, ϕ_i is small and declines geometrically as i increases. One might then expect the AIC to choose a low order, because extra parameters have little information content but reduce the degrees of freedom. Table 3 shows that indeed for $|\theta| \leq 0.4$, low values of k are selected by AIC. But as $|\theta|$ gets large, ϕ_i will remain nonnegligible even for i quite large. Increasing the length of the autoregression should, in principle, improve the approximation to the DGP. But the k selected by AIC increases only at a logarithmic rate. Except when T becomes impracticably large, the AIC will abandon information at large lags in favor of a very parsimonious model. Hence, in practice one can expect the chosen k to be no higher than 5 even with T as large as 500 when $|\theta|$ is close to 1.

5. Adaptive Rule 2: Testing for the Significance of Coefficients on Lags

This section analyzes the properties of $\hat{k}$ and t_ρ when $\hat{k}$ is chosen by the $J(m, r)$ statistic described in Section 3.2 to sequentially test for the significance of coefficients on additional lags. The following lemma is useful in establishing the limiting distribution of t_ρ.

Lemma 5.1. *Let $\{y_t\}$ be generated by* (1) *and suppose that Assumptions* A1 *and* A2$''$ *hold. Let $\hat{\mathbf{d}}(k)$ be obtained from the augmented autoregression* (4), *and let $J(k-n, k)$ be as defined in* (8). *Then $J(k-n, k)$ is asymptotically distributed as χ^2 with n degrees of freedom.*

Berk (1974) proved consistency and asymptotic normality of the coefficients in the restricted regression under A1 and A2″ (see also Lewis and Reinsel, 1985). The crucial element in the proof of Lemma 5.1 is the fact that when $\{\Delta y_t\}$ is a stationary and invertible ARMA process, the coefficients $\mathbf{d}(k)$ converge to 0 at a rate that yields an asymptotic equivalence between the Wald test that $\hat{\mathbf{d}}(k) = \mathbf{d}(k)$ and the Wald test that $\hat{\mathbf{d}}(k) = 0$. Indeed, Lemma 5.1 requires Assumption A2″ to ensure that $\sqrt{T}\mathbf{d}(k) \to 0$, which in turn ensures asymptotic normality of $\sqrt{T}\hat{\mathbf{d}}(k)$.

5.1. *A special case: An MA(1) and the t test*

We now specialize the sequential procedure described in Section 3.2.2 to the case where $n = 1$. The square root of the statistic $J(k-1, k)$ then simplifies to a t test for the significance of the coefficient on the last lag in an autoregression of order k:

$$t_{\hat{d}_k} = \sqrt{T}\hat{d}_k(\hat{\sigma}_k^2 T\mathbf{M}_k^{-1}(1))^{-1/2}.$$

The sequential procedure chooses a value of $\hat{k}$ if $t_{\hat{d}_k}$ is significant at some prespecified level α in an estimated autoregression of order $\hat{k}$, whereas the t statistics $t_{\hat{d}_k}$ are insignificant in estimated autoregressions of order k for all k in the range $(\hat{k}, k\max]$. We can show that if Δy_t is an MA(1) (i.e., $\Delta y_t = e_t + \theta e_{t-1}$), then

$$\hat{d}_k \simeq \theta^{k-1}(1-\theta^2)(1-\theta^{2(k+1)})^{-1}$$

and

$$t_{\hat{d}_k} \simeq \sqrt{T}\theta^{k-1}(1-\theta^2)((1-\theta^{2k})(1-\theta^{2(k+2)}))^{-1/2}.$$

These results show that both $t_{\hat{d}_k}$ and $\sqrt{T}\hat{d}_k$ will converge to zero if k increases at a polynomial rate. Given the result of Lewis and Reinsel (1985, Theorem 4) that $t_{\hat{d}_k = d_k} \equiv (\hat{d}_k - d_k)/(\hat{\sigma}_k^2\mathbf{M}_k^{-1}(1))^{1/2}$ is asymptotically distributed $N(0, 1)$ if k increases at a polynomial rate satisfying A1 and A2″, $t_{\hat{d}_k}$ can be shown to have the same asymptotic distribution under these restrictions on the rate of increase of k.

It is of some interest to note that the foregoing results also imply that a specific-to-general procedure starting from any lower bound $k\,\text{min}$ that tests for the significance of the coefficient on the last lag would select a $\hat{k}$ that increases to infinity at a logarithmic rate when $\{\Delta y_t\}$ contains a moving-average component. Hence such a specific-to-general procedure would have the same asymptotic properties as a selection rule based on an information criterion.

Note that the asymptotic normality result of Berk (1974) and Lewis and Reinsel (1985) we used to prove Lemma 5.1 requires that $\hat{k}$ increase at some polynomial rate, or at least at a rate that ensures A2$''$ is satisfied. A logarithmic rate is not sufficient. We now show that the truncation lag selected by a general-to-specific procedure will be of an order higher than $\log T$ provided that $k\,\text{max}$ increases at a rate faster than $\log T$. In fact the selected truncation lag will grow at the same rate as $k\,\text{max}$.

Lemma 5.2. *If $\hat{k}$ is selected by means of the general-to-specific strategy described in Definition 3.1 and $k\,\text{max}$ increases at a rate such that* A1 *and* A2$''$ *are satisfied, then $\hat{k}$ increases at the same rate as $k\,\text{max}$.*

The intuition behind the result stated in Lemma 5.2 is as follows. Under the assumptions of Lemma 5.1, $J(k-n,k)$ is asymptotically distributed as a χ^2 random variable with n degrees of freedom. Thus the limiting probability that $J(k-n,k)$ is statistically significant is α, the size of the test. For a given $\hat{k} < k\,\text{max}$ to be chosen, it must be the case that all prior statistics in the sequential procedure ($J(i-n,i)$; $i = k\,\text{max}-1,\dots,\hat{k}-1$) are statistically insignificant. This event occurs for large samples with probability $\alpha(1-\alpha)^{k\,\text{max}-\hat{k}}$. Because $\hat{k} \leq k\,\text{max}$ and $k\,\text{max} \to \infty$, this probability vanishes as $T \to \infty$ unless $\hat{k}$ increases at the same rate as $k\,\text{max}$.

The importance of Lemma 5.2 is that if $k\,\text{max}$ is chosen to increase at a polynomial rate, then $\hat{k}$ will also increase at a polynomial rate. This implies that Assumption A2 or A2$''$ can be satisfied with judicious choice of $k\,\text{max}$, thereby ensuring that the results of Lemma 5.1 hold. Lemma 5.2 allows us to state the following theorem concerning the limiting behavior of the unit root test under this truncation lag selection rule.

Theorem 5.3. *If* $k\max$ *satisfies* A1 *and* A2″ *and* $\hat{k}$ *is chosen from the general-to-specific sequential procedure stated in Definition* 3.1, *then* $t_\rho(\hat{k})$ *has the same limiting distribution as* (5).

Because $\hat{k}$ maps into a deterministic rule in the set $\tilde{K}$ by Lemma 5.2, the result follows from Lemma 3.4. In fact $\hat{k}$ maps into the set of deterministic rules that satisfies A1 and A2, because k increases at a polynomial rate under the conditions of Theorem 5.3.

6. Finite-Sample Simulations

The results of the preceding sections can be summarized as follows. An information criterion will choose values of k that are proportional to $\log T$, a rate ruled out by A2. But the k selected using the $J(m,r)$ statistic to test for the significance of lags will increase at the same rate as the prespecified $k\max$, itself increasing at a polynomial rate. Because a logarithmic rate of increase is slow compared to a polynomial rate, an information criterion will choose values of k that are generally much smaller than those chosen by a general-to-specific t test, for example. Although the log proportionality rule might fail the lower-bound condition, the limiting distribution of t_ρ is unaffected. In such a case the estimates of the coefficients on Δy_{t-i} in the augmented autoregression will be consistent at a rate slower than $\sqrt{T}$ for some DGP's. In the MA(1) case, a large value of $|\theta|$ is more likely to be associated with a slower rate of consistency for $\hat{\mathbf{d}}(k)$. We now examine the implications of these results in finite samples.

The results we report are based on 5,000 simulations for different values of θ_j and α_i. For each parameterization, the selected values of k and the corresponding values of t_ρ are recorded. The simulations were performed on a 486/25 MHz PC with code compiled using the Borland C (Version 3) compiler. Random numbers were generated using the ran1( ) function from Press, Teukolsky, Vetterling, and Flannery (1988), with time (in seconds) as seed. We considered $T = 100$, 200, and 500. For a given T, different values for $k\max$ and $k\min$ were examined. We focus on results for $T = 100$ with $k\max = 10$ and $k\min = 0$ without loss of generality and discuss results for other configurations where appropriate. The complete set of results is available on request.

We select, for presentation, results based on two information criteria: the AIC and Schwartz. The results for the BIC and the Hannan–Quinn criteria show no appreciable difference. For the general-to-specific strategy, we considered the t as well as the F test, but only present results for the t test at the 5% and 10% levels. In general, a tighter model is selected using a lower significance level.

6.1. *Frequency distribution of $\hat{k}$*

We first examine the number of times that $k = i$ ($i = 1, \ldots, 10$) is being selected by each procedure during the 5,000 simulations. Tables 4 and 5 are the frequency counts. As we can see, both information criteria consistently select values of k less than 3.

Table 4. Frequency count of selected lag lengths k, moving-average case, $T = 100$.

	$k=1$	$k=2$	$k=3$	$k=4$	$k=5$	$k=6$	$k=7$	$k=8$	$k=9$	$k=10$
					$\theta = 0.8$					
$t_{\text{sig}}(10)$	0.001	0.028	0.115	0.131	0.149	0.113	0.135	0.104	0.121	0.102
$t_{\text{sig}}(5)$	0.013	0.103	0.210	0.157	0.147	0.092	0.097	0.063	0.065	0.053
AIC	0.016	0.149	0.325	0.219	0.150	0.069	0.039	0.020	0.009	0.005
Schwartz	0.151	0.381	0.326	0.100	0.035	0.004	0.001	0.001	0.000	0.000
					$\theta = 0.3$					
$t_{\text{sig}}(10)$	0.304	0.081	0.067	0.055	0.059	0.062	0.074	0.079	0.086	0.086
$t_{\text{sig}}(5)$	0.468	0.071	0.047	0.040	0.038	0.039	0.043	0.031	0.049	0.039
AIC	0.676	0.131	0.037	0.013	0.007	0.002	0.002	0.001	0.000	0.000
Schwartz	0.611	0.037	0.004	0.001	0.000	0.000	0.000	0.000	0.000	0.000
					$\theta = -0.5$					
$t_{\text{sig}}(10)$	0.220	0.189	0.075	0.061	0.058	0.068	0.069	0.077	0.077	0.083
$t_{\text{sig}}(5)$	0.387	0.212	0.054	0.041	0.039	0.041	0.041	0.041	0.041	0.039
AIC	0.484	0.331	0.073	0.025	0.011	0.007	0.003	0.001	0.000	0.000
Schwartz	0.621	0.169	0.016	0.003	0.000	0.000	0.000	0.000	0.000	0.000
					$\theta = -0.8$					
$t_{\text{sig}}(10)$	0.074	0.113	0.109	0.116	0.076	0.097	0.075	0.092	0.080	0.085
$t_{\text{sig}}(5)$	0.123	0.162	0.121	0.100	0.061	0.065	0.045	0.047	0.045	0.041
AIC	0.197	0.225	0.146	0.091	0.033	0.019	0.008	0.003	0.002	0.001
Schwartz	0.264	0.172	0.056	0.019	0.002	0.000	0.000	0.000	0.000	0.000

Note: DGP: $y_t = y_{t-1} + u_t$; $u_t = e_t + \theta e_{t-1}$. Regression: $\Delta y_t = \delta_0 y_{t-1} + \sum_{i=1}^{k} \delta_i \Delta y_{t-i} + v_t$.

Table 5. Frequency count of selected lag lengths k, autoregressive case, $T = 100$.

	$k=1$	$k=2$	$k=3$	$k=4$	$k=5$	$k=6$	$k=7$	$k=8$	$k=9$	$k=10$
				$\phi_1 = 0.6, \phi_2 = \phi_3 = \phi_4 = 0.0$						
$t_{\text{sig}}(10)$	0.407	0.045	0.047	0.051	0.057	0.064	0.071	0.078	0.091	0.089
$t_{\text{sig}}(5)$	0.644	0.037	0.033	0.033	0.037	0.042	0.040	0.043	0.047	0.044
AIC	0.878	0.075	0.027	0.010	0.004	0.002	0.001	0.001	0.001	0.000
Schwartz	0.976	0.020	0.003	0.000	0.000	0.000	0.000	0.000	0.000	0.000
				$\phi_1 = -0.6, \phi_2 = \phi_3 = \phi_4 = 0.0$						
$t_{\text{sig}}(10)$	0.407	0.049	0.051	0.050	0.060	0.069	0.068	0.076	0.081	0.090
$t_{\text{sig}}(5)$	0.652	0.036	0.037	0.035	0.041	0.040	0.036	0.040	0.040	0.044
AIC	0.866	0.086	0.027	0.011	0.006	0.003	0.000	0.000	0.000	0.000
Schwartz	0.978	0.018	0.002	0.001	0.000	0.000	0.000	0.000	0.000	0.000
				$\phi_1 = 0.4, \phi_2 = 0.2, \phi_3 = \phi_4 = 0.0$						
$t_{\text{sig}}(10)$	0.200	0.242	0.055	0.057	0.060	0.058	0.073	0.070	0.092	0.089
$t_{\text{sig}}(5)$	0.391	0.285	0.038	0.039	0.038	0.038	0.043	0.035	0.042	0.044
AIC	0.455	0.443	0.057	0.021	0.008	0.003	0.002	0.001	0.001	0.000
Schwartz	0.680	0.277	0.008	0.001	0.000	0.000	0.000	0.000	0.000	0.000
				$\phi_1 = 0.30, \phi_2 = 0.30, \phi_3 = 0.25, \phi_4 = 0.14$						
$t_{\text{sig}}(10)$	0.008	0.095	0.299	0.149	0.059	0.063	0.067	0.078	0.084	0.092
$t_{\text{sig}}(10)$	0.023	0.192	0.388	0.137	0.034	0.037	0.042	0.047	0.043	0.046
AIC	0.026	0.196	0.494	0.204	0.035	0.016	0.007	0.003	0.001	0.001
Schwartz	0.082	0.369	0.419	0.076	0.005	0.002	0.000	0.000	0.000	0.000

Note: DGP: $y_t = y_{t-1} + \sum_{i=1}^{4} \phi_i \Delta y_{t-i} + u_t$. Regression: $\Delta y_t = \delta_0 y_{t-1} + \sum_{i=1}^{k} \delta_i \Delta y_{t-i} + v_t$.

Although the k's selected for autoregressive models seem appropriate given that the DGP's considered are of order no higher than 4, the information criteria yield very parsimonious models when the DGP is driven by a moving-average process. Although the true order of autoregression is infinity in those cases, the AIC and Schwartz criteria continue to choose values of 2 and 3 for k. When θ is large, the coefficients in the autoregression die off only slowly. Truncating the autoregression at a low order will yield a more parsimonious model but with a loss of information. The cost of parsimony will be judged in terms of the size and power of t_ρ in the next subsection.

In the moving-average case, the values of k selected by a general-to-specific modeling strategy are quite evenly distributed over the range $[2, k\max = 10]$, with some mass concentrated at $k = 1$. This result follows directly from Lemma 5.1. A further implication of this lemma is that the chosen value of k will be closer to $k\max$ the more liberal the size of the test. Thus the frequency of k chosen to be 5 and above is higher under the 10% t test than under the 5% t test.

6.2. *Size and power*

Having confirmed that information criteria choose values of k that tend to be small, we now proceed to show that in many cases, the method used to choose k can have size and power implications. The results are reported in Tables 6 and 7 for $T = 100$, with the power of the test evaluated at $\rho = 0.95$ and 0.85. Turning first to moving-average models (Table 6), we see that for positive values of θ, the size of the test is similar for all methods of selecting k. When $\theta = 0.8$,

Table 6. Size and power of unit root tests; moving average case, $T = 100$, $k\max = 10$.

ρ	θ	$t_{\text{sig}}(10)$	$t_{\text{sig}}(5)$	AIC	Schwartz
1.0	0.80	0.069	0.073	0.068	0.071
1.0	0.50	0.083	0.087	0.082	0.088
1.0	0.30	0.075	0.077	0.070	0.069
1.0	0.00	0.063	0.059	0.052	0.046
1.0	−0.30	0.097	0.126	0.127	0.174
1.0	−0.50	0.116	0.158	0.167	0.244
1.0	−0.80	0.304	0.424	0.561	0.733
0.95	0.80	0.136	0.151	0.146	0.158
0.95	0.50	0.164	0.184	0.170	0.196
0.95	0.30	0.158	0.162	0.152	0.144
0.95	0.00	0.153	0.151	0.140	0.126
0.95	−0.30	0.228	0.292	0.294	0.393
0.95	−0.50	0.254	0.336	0.377	0.510
0.95	−0.80	0.534	0.704	0.877	0.963
0.85	0.80	0.347	0.387	0.405	0.451
0.85	0.50	0.445	0.510	0.520	0.586
0.85	0.30	0.465	0.513	0.536	0.505
0.85	0.00	0.486	0.540	0.580	0.575
0.85	−0.30	0.555	0.682	0.758	0.859
0.85	−0.50	0.627	0.753	0.860	0.936
0.85	−0.80	0.825	0.908	0.996	1.000

Note: DGP: $y_t = \rho y_{t-1} + u_t$, $u_t = e_t + \theta e_{t-1}$. Regression: $\Delta y_t = \delta_0 y_{t-1} + \sum_{i=1}^{k} \delta_i \Delta y_{t-i} + v_t$.

Table 7. Size and power of unit root tests; autoregressive case, $T = 100$, $k\,\text{max} = 10$.

ρ	ϕ_1	ϕ_2	ϕ_3	ϕ_4	$t_{\text{sig}}(10)$	$t_{\text{sig}}(5)$	AIC	Schwartz
1.0	0.6	0.0	0.0	0.0	0.078	0.075	0.066	0.060
1.0	−0.6	0.0	0.0	0.0	0.068	0.066	0.066	0.060
1.0	0.4	0.2	0.0	0.0	0.066	0.062	0.055	0.047
1.0	0.3	0.3	0.25	0.14	0.066	0.062	0.058	0.052
0.95	0.6	0.0	0.0	0.0	0.371	0.399	0.404	0.394
0.95	−0.6	0.0	0.0	0.0	0.101	0.099	0.087	0.080
0.95	0.4	0.2	0.0	0.0	0.346	0.336	0.338	0.267
0.95	0.3	0.3	0.25	0.14	0.822	0.840	0.886	0.837
0.85	0.6	0.0	0.0	0.0	0.782	0.870	0.960	0.972
0.85	−0.6	0.0	0.0	0.0	0.269	0.274	0.268	0.256
0.85	0.4	0.2	0.0	0.0	0.763	0.824	0.899	0.867
0.85	0.3	0.3	0.25	0.14	0.901	0.937	0.976	0.947

Note: DGP: $y_t = \rho y_{t-1} + \sum_{i=1}^{4} \phi_i \Delta y_{t-i} + u_t$. Regression: $\Delta y_t = \delta_0 y_{t-1} + \sum_{i=1}^{k} \delta_i \Delta y_{t-i} + v_t$.

the 10% t test picks k to be 5 or smaller 40% of the time, whereas the AIC picks k to be in the same range twice as often (see Table 4). Although such variations in the choice of k appear to yield small size differences, power is slightly higher the more parsimonious the model. It is well known that the Schwartz criterion imposes a heavy penalty for overparameterization. Thus for positive moving-average models, the Schwartz criterion tends to yield higher power for a given size.

The result that stands out in Table 6 is the large size distortions when θ is negative. The problem of size distortion with unit root tests in the presence of negative moving errors is well documented (e.g., Schwert, 1989). Although Schwert used deterministic rules to select k, he also noted that the exact size depends on the choice of k. Our results confirm that the more conservative the criterion for selecting the truncation order, the larger the size distortions associated with t_ρ. For example, size distortions associated with the conservative Schwartz criterion are significantly larger than those associated with the 10% t test, the most liberal criterion considered. From the frequency counts, we see that the Schwartz criterion chooses values of k less than 3 90% of the time, whereas the 10% t test chooses values of k *greater* than 3 with a probability of 0.9.

Table 7 indicates that for autoregressive models, all methods produce estimates of k that are as large as the true order with high probability. Accordingly, all selection procedures produce an exact size close to the nominal size. The 10% t test tends to have lower power, however. According to the frequency counts, the t test tends to overparameterize autoregressive models. For example, the 10% t test selects k greater than 4 more than 40% of the time when the DGP is a fourth-order autoregression. Thus underparameterization is associated with larger size distortions, and overparameterization with power loss when $T = 100$.

The size of the test for moving-average models with $T = 200$ is reported in Table 8. Note that size distortions in negative moving-average models persist as T increases. The Schwartz criterion continues to be associated with significantly larger size distortions than the 10% t test. But in cases for which size distortion is not an issue, as in autoregressive models, the discrepancies in power across selection procedures vanish almost completely when $T = 500$. We report in Table 9 the size and power for autoregressive models at $T = 200$. Compared to the results for $T = 100$, power is higher throughout, and the differences in power across selection procedures are smaller. Thus, discrepancies in power across selection

Table 8. Size of unit root tests; moving-average case, $T = 200$, $k\,\max = 12$.

ρ	θ	$t_{\text{sig}}(10)$	$t_{\text{sig}}(5)$	AIC	Schwartz
1.0	0.80	0.056	0.060	0.059	0.063
1.0	0.50	0.061	0.064	0.056	0.064
1.0	0.30	0.061	0.064	0.061	0.066
1.0	0.00	0.064	0.066	0.059	0.057
1.0	−0.30	0.067	0.076	0.076	0.102
1.0	−0.50	0.085	0.110	0.121	0.168
1.0	−0.80	0.177	0.250	0.366	0.557

Note: DGP: $y_t = \rho y_{t-1} + u_t$, $u_t = e_t + \theta e_{t-1}$. Regression: $\Delta y_t = \delta_0 y_{t-1} + \sum_{i=1}^{k} \delta_i \Delta y_{t-i} + v_t$.

Table 9. Size and power of unit root tests; autoregressive case, $T = 200$, $k\max = 12$.

ρ	ϕ_1	ϕ_2	ϕ_3	ϕ_4	$t_{\text{sig}}(10)$	$t_{\text{sig}}(5)$	AIC	Schwartz
1.0	0.6	0.0	0.0	0.0	0.063	0.060	0.057	0.054
1.0	−0.6	0.0	0.0	0.0	0.063	0.064	0.058	0.056
1.0	0.4	0.2	0.0	0.0	0.062	0.061	0.056	0.048
1.0	0.3	0.3	0.25	0.14	0.076	0.072	0.070	0.059
0.95	0.6	0.0	0.0	0.0	0.738	0.815	0.897	0.908
0.95	−0.6	0.0	0.0	0.0	0.166	0.168	0.160	0.153
0.95	0.4	0.2	0.0	0.0	0.712	0.709	0.837	0.784
0.95	0.3	0.3	0.25	0.14	0.979	0.988	1.000	0.998
0.85	0.6	0.0	0.0	0.0	0.955	0.974	1.000	1.000
0.85	−0.6	0.0	0.0	0.0	0.608	0.603	0.738	0.749
0.85	0.4	0.2	0.0	0.0	0.954	0.975	1.000	1.000
0.85	0.3	0.3	0.25	0.14	0.994	0.997	1.000	1.000

Note: DGP: $y_t = \rho y_{t-1} + \sum_{i=1}^{4} \phi_i \Delta y_{t-i} + u_t$. Regression: $\Delta y_t = \delta_0 y_{t-1} + \sum_{i=1}^{k} \delta_i \Delta y_{t-i} + v_t$.

procedures are small for typical sample sizes encountered in economic analyses, but size distortions are not. A t or an F test, therefore, has an advantage over information criterion in that it produces tests with more accurate size without much loss of power.

6.3. *The choice of k and size distortions*

When θ in the noise function is large and negative, y_t is close to having a common factor and behaves more like a white noise than an integrated process. The asymptotic properties of the normalized least squares estimator in this case have been shown by Nabeya and Perron (1994) and Perron (1992) to be different from those derived under standard assumptions. In view of those results, one would conjecture that there is also a discrepancy between the finite-sample distribution of t_ρ and its approximate distribution as defined by (5). But as we can see, the extent of size distortions varies with k. This suggests that k affects the adequacy of (5) as an approximating distribution. The question is, how?

Using a local asymptotic framework, Pantula (1991) parameterized θ as $-(1 - T^{-\eta})$ and showed that the limit of t_ρ is given by (5) only if $0 < \eta < 0.25$, but diverges to $-\infty$ at rate T^η/k if $0.25 < \eta < 0.625$, with limiting distribution given by

$$kT^{-\eta}t_\rho \Rightarrow -\left(\int_0^1 W(r)^2 dr\right)^{-1}. \tag{11}$$

Because $k = O(T^{1/4})$ by assumption, and (11) is valid for $\eta > 0.25$, the limiting distribution of t_ρ will always tend to $-\infty$. But the larger the rate of increase of k, the slower the rate of divergence and the smaller the discrepancies between the exact and the approximate distributions of t_ρ. Consequently, even though η is 0.35 when $T = 100$ and $\theta = -0.8$, size distortions are noticeably smaller at larger values of k when critical values from (5) are used for hypothesis testing.

To reinforce the importance of a large k when θ is large and negative, we report in Table 10 the size of the test at selected parameter values for $T = 200$ and $T = 500$ when a different lower bound, $k\min$, is prescribed. We set $k\min$ and $k\max$ to 4 and 12 when $T = 200$, and to 6 and 14 when $T = 500$. Evidently, the larger the $k\min$ the larger the k and the smaller the size distortions.

The importance of $k\min$ and $k\max$ in all selection procedures must be emphasized. If we raise the value of $k\min$ and let the information criteria select k from the range $[k\min, k\max]$ and $k\min > \log T$, then the criteria will choose $k\min$ in large samples, because $\log T$ is outside the permissible range. Loosely speaking, the choice of $k\min$ can be seen as a practical way of imposing the lower-bound condition A2. On the other hand, the choice of $k\max$ is more important in a general-to-specific model selection strategy. By Lemma 5.1, the test statistic will choose $k \in [k\min, k\max]$ with declining probability as k moves away from $k\max$. Thus, the larger the $k\max$, the higher the probability that a larger k will be chosen. The larger the k, the better the size — at the expense, however, of power losses.

Table 10. Size of unit root tests moving-average case; different choices of T, $k\max$, and $k\min$.

ρ	θ	$t_{\text{sig}}(10)$	$t_{\text{sig}}(5)$	AIC	Schwartz
		$T = 200,\ k\max = 12,\ k\min = 4$			
1.0	0.80	0.066	0.068	0.066	0.056
1.0	0.50	0.057	0.055	0.050	0.048
1.0	0.30	0.062	0.060	0.057	0.054
1.0	0.00	0.056	0.056	0.053	0.052
1.0	−0.30	0.061	0.059	0.051	0.050
1.0	−0.50	0.061	0.061	0.060	0.059
1.0	−0.80	0.168	0.228	0.281	0.343
		$T = 500,\ k\max = 14,\ k\min = 6$			
1.0	0.80	0.050	0.048	0.050	0.046
1.0	0.50	0.053	0.052	0.048	0.047
1.0	0.30	0.057	0.057	0.060	0.060
1.0	0.00	0.052	0.052	0.053	0.052
1.0	−0.30	0.065	0.064	0.062	0.059
1.0	−0.50	0.059	0.060	0.058	0.056
1.0	−0.80	0.104	0.134	0.167	0.213
		$T = 500,\ k\max = 14,\ k\min = 9$			
1.0	0.80	0.059	0.062	0.057	0.058
1.0	0.50	0.059	0.059	0.058	0.057
1.0	0.30	0.058	0.058	0.059	0.056
1.0	0.00	0.056	0.056	0.056	0.056
1.0	−0.30	0.052	0.053	0.052	0.052
1.0	−0.50	0.058	0.056	0.056	0.056
1.0	−0.80	0.093	0.103	0.108	0.116

Note: DGP: $y_t = \rho y_{t-1} + u_t$, $u_t = e_t + \theta e_{t-1}$. Regression: $\Delta y_t = \delta_0 y_{t-1} + \sum_{i=1}^{k} \delta_i \Delta y_{t-i} + v_t$.

7. Conclusions

This article has analyzed issues related to the selection of the truncation lag in unit root tests of the type proposed by Dickey and Fuller (1979) and Said and Dickey (1984). We have focused on the implications of the lower-bound condition on t_ρ used by Said and Dickey (1984). Procedures that do not satisfy this condition tend to select truncation lags that are too small for some parameter values. Information-based rules such as AIC and Schwartz fit into this category.

A general feature of our results is that an overly parsimonious model can have large size distortions, but an overparameterized model may have low power. But the size problem is more severe than power loss in the sense that discrepancies in power across selection procedures diminish as T increases, but size distortions persist even for large sample sizes for some methods of selecting k. In this regard, a t or an F test for the significance of lags will have an advantage over information-based rules such as the AIC, because the former produces tests with more robust size properties across models.

There remain, of course, several avenues for further research that follow from the framework used in this article. First, given the problems associated with approximating a general ARMA process by a finite autoregression, one might be tempted to construct unit root tests from an estimated ARMA(p, q) process whose order is selected using a consistent procedure, such as the one discussed by Dickey and Said (1981). But in view of the problems associated with maximum likelihood estimates of processes with moving-average components, it is not evident that this method can provide statistical improvement. A comparison of the various estimation methods in the context of unit root tests would be useful.

The second avenue is an extension of the results to the multivariate case whereby vector autoregressive processes are used to approximate more general multivariate linear processes. Although one expects, and preliminary work suggests, that the same qualitative results would hold, the analysis is not a straightforward extension because of possible cointegration among the variables.

The third topic concerns the issue of optimal lag selection. Our analysis has concentrated on two particular classes of lag length selection that are widely used in practice. None of these needs to be optimal. The difficulty, however, lies in finding the proper way to assess the procedures for selecting k, because the purpose of estimating these autoregressions is not in obtaining a particular estimate that is as precise as possible, but rather in obtaining the unit root test itself. The optimality criterion thus needs to be based on an appropriate trade-off between type I and type II errors in the application of the unit root test.

Appendix: Proofs

The following notation will be used in this Appendix. Unless otherwise stated, we shall let C_1 be an arbitrary constant (not necessarily the same throughout). Let $\mathbf{D}_T = \text{diag}\{(T-k)^{-1}, (T-k)^{-1/2}, \ldots, (T-k)^{-1/2}\}$, $\mathbf{U}'_t = (y_{t-1}, \mathbf{X}'_t)$, and $\mathbf{X}'_t = (\Delta y_{t-1}, \ldots, \Delta y_{t-k})$. Let $\mathbf{M}_k = \sum_{t=k+1}^{T} \mathbf{U}_t\mathbf{U}'_t$ and $\mathbf{R}_k = \sum_{t=k+1}^{T} \mathbf{X}_t\mathbf{X}'_t$. Thus

$$\mathbf{R}_k = \sum_{t=k+1}^{T} (\Delta y_{t-1}, \Delta y_{t-2}, \ldots, \Delta y_{t-k})'(\Delta y_{t-1}, \Delta y_{t-2}, \ldots, \Delta y_{t-k})$$

and

$$\mathbf{D}_T\mathbf{M}_k\mathbf{D}_T = \begin{bmatrix} (T-k)^{-2} \sum_{t=k+1}^{T} y_{t-1}^2 & (T-k)^{-3/2} \sum_{t=k+1}^{T} y_{t-1}\mathbf{X}'_t \\ (T-k)^{-3/2} \sum_{t=k+1}^{T} y_{t-1}\mathbf{X}_t & (T-k)^{-1}\mathbf{R}_k \end{bmatrix}.$$

Note that from Said and Dickey (1984), the limit of $\mathbf{D}_T\mathbf{M}_k\mathbf{D}_T$ is block diagonal with the two blocks corresponding to the limits of $(T-k)^{-2}\sum_{t=k+1}^{T} y_{t-1}^2$ and $(T-k)^{-1}\mathbf{R}_k$. We also let $\mathbf{M}_k^{-1}(1)$ be the first diagonal element of the matrix $\mathbf{M}_k^{-1}$ and let $\mathbf{M}_k^{-1}(k)$ be the lower right $k \times k$ block of $\mathbf{M}_k^{-1}$.

For a matrix $\mathbf{C}$, the matrix norm is defined by $\|\mathbf{C}\| = \sup_{\|X\|<1} \|\mathbf{Cx}\|$, where for a vector $\mathbf{z}$, $\|\mathbf{z}\| = (\mathbf{z}'\mathbf{z})^{1/2}$. Using this norm, lemma 3 of Berk (1974) showed that $k^{1/2}\|(T-k)\mathbf{R}_k^{-1} - \mathbf{\Gamma}^{-1}\| \to 0$, where $\mathbf{\Gamma}$ is a $k\times k$ matrix with typical elements $\mathbf{\Gamma}_{ij} = E(\Delta y_{t-i}\Delta y_{t-j})$.

Proof of Lemma 2.1

The proof for consistency of the least squares estimates in the augmented autoregression (4) was given by Said and Dickey (1984) and will not be repeated here. Nevertheless, it is important to point out the two steps involved in the proof. The first step is to show that $k^{1/2}\|(\mathbf{D}_T\mathbf{M}_k\mathbf{D}_T)^{-1} - \mathbf{M}^{-1}\|$ converges to zero for some limiting block-diagonal matrix $\mathbf{M}$. For this step, Assumption A1 is sufficient and the argument follows from lemma 3 of Berk (1974). The second step is to show that $\|\mathbf{D}_T \sum_{t=k+1}^{T} \mathbf{U}_t e_{tk}\| = O_p(k^{1/2})$. The combination of the two steps implies that $Td_0 = O_p(1)$, $\sqrt{T}(\hat{d}_i - d_i) = O_p(1)$ $(i = 1, \ldots, k)$, and $\hat{\sigma}_k^2 \to \sigma_e^2$. Assumption A2 is used only in this second step and, more specifically, to ensure that

$$E\left((T-k)^{-1}\sum_{j=1}^{k}\left(\sum_{t=k+1}^{T}\Delta y_{t-j}(e_{tk}-e_t)\right)^2\right) \leq C_1 \cdot k(T-k)\sum_{i=k+1}^{\infty} d_i^2 \to 0, \tag{A.1}$$

as $T \to \infty$ for some constant C_1. Note that A2″ is also sufficient to guarantee that A1 holds; however, A1 is sufficient but not necessary to ensure that t_ρ has the limiting distribution given by (5). To see this, we first express t_ρ as

$$t_\rho = \left((T-k)^{-1}\sum_{t=k+1}^{T} y_{t-1}e_{tk}\right)(\hat{\sigma}_k^2(T-k)^{-2}[\mathbf{M}_k^{-1}(1)]^{-1})^{-1/2}.$$

From Said and Dickey (1984), $T^{-2}[\mathbf{M}_k^{-1}(1)]^{-1} \Rightarrow \sigma_e^2 \int_0^1 W(r)^2 dr$ provided that A1 holds. Consider now the numerator,

$$(T-k)^{-1}\sum_{t=k+1}^{T} y_{t-1}e_{tk} = (T-k)^{-1}\sum_{t=k+1}^{T} y_{t-1}e_t + (T-k)^{-1}\sum_{t=k+1}^{T} y_{t-1}\sum_{i=k+1}^{\infty} d_i \Delta y_{t-i}. \tag{A.2}$$

It is straightforward to show that $(T-k)^{-1}\sum_{t=k+1}^{T} y_{t-1}e_t \Rightarrow \sigma_e^2 \int_0^1 W(r)\, dW(r)$, provided that $k \to \infty$ and $k/T \to 0$ as $T \to \infty$.

Consider now the second term in A2. We have

$$E\left[(T-k)^{-1}\sum_{t=k+1}^{T} y_{t-1}\sum_{i=k+1}^{\infty} d_i\Delta y_{t-i}\right]^2$$

$$= \sum_{i=k+1}^{\infty}\sum_{j=k+1}^{\infty} d_i d_j (T-k)^{-2}$$

$$\times \sum_{t=k+1}^{T}\sum_{s=k+1}^{T} E\left[y_{t-1}\Delta y_{t-i}\, y_{s-1}\Delta y_{s-j}\right]$$

$$\leq C_1 \sum_{i=k+1}^{\infty} d_i \sum_{j=k+1}^{\infty} d_j \leq C_1 \sum_{i=k+1}^{\infty} \lambda^i \sum_{j=k+1}^{\infty} \lambda^j$$

$$= C_1 \left(\sum_{i=k+1}^{\infty} \lambda^i\right)^2 = C_1\lambda^{2k}/(1-\lambda)^2 \to 0 \text{ as } k \to \infty.$$

The first inequality follows from Said and Dickey (1984, p. 601) who stated that there exists a constant C_1 such that $(T-k)^{-2}\sum_{t=k+1}^{T}\sum_{s=k+1}^{T} E\left[y_{t-1}\Delta y_{t-i} y_{s-1}\Delta y_{s-j}\right] \leq C_1$. The second inequality uses the fact that Δy_t is a stationary and invertible ARMA process, and hence there exists λ, $0<\lambda<1$, such that $|d_i| < C_1\lambda^i$ for a different constant C_1. Therefore, $(T-k)^{-1}\sum_{t=k+1}^{T} y_{t-1}e_{tk} \Rightarrow \sigma_e^2\int_0^1 W(r)\,dW(r)$ under the sole condition that $k/T \to 0$ and $k\to\infty$ as $T\to\infty$. Neither A2 nor A2$''$ is needed to establish the limiting distribution of (5).

To consider the properties of $\hat{\mathbf{d}}(k)$ without the lower-bound condition, it can be seen, from Berk (1974, Lemma 2) or Said and Dickey (1984, Lemma 5.2), that consistency of $\hat{\mathbf{d}}(k)$ still holds if

$$E\left((T-k)^{-2}\sum_{j=1}^{k}\left(\sum_{t=k+1}^{T} u_{t-j}(e_{tk}-e_t)\right)^2\right)$$

$$\leq C_1 k \sum_{i=k+1}^{\infty} d_i^2 \leq C_1 k\lambda^{2k}/(1-\lambda^2) \to 0. \tag{A.3}$$

The condition (A.3) is satisfied for any stationary and invertible ARMA process provided that $k \to \infty$, which is assured under A1. More generally, the rate at which $\hat{\mathbf{d}}(k)$ converges to $\mathbf{d}(k)$ can be found by writing

$$\begin{aligned}\hat{\mathbf{d}}(k) - \mathbf{d}(k) &= ((T-k)\mathbf{M}_k^{-1}(k) - \mathbf{\Gamma}^{-1})(T-k)^{-1}\sum_{t=k+1}^{T}\mathbf{X}_t e_{tk}\\ &\quad + \mathbf{\Gamma}^{-1}(T-k)^{-1}\sum_{t=k+1}^{T}\mathbf{X}_t e_t\\ &\quad + \mathbf{\Gamma}^{-1}(T-k)^{-1}\sum_{t=k+1}^{T}\mathbf{X}_t(e_{tk}-e_t). \qquad (A.4)\end{aligned}$$

Taking norms, the first term is $o_p(T^{-1/2})$ and the second is $O_p(k^{1/2}T^{-1/2})$ by the results of Said and Dickey (1984), whether or not A2 is satisfied. Using (A.3), the third term is $O_p(k^{1/2}\lambda^k)$ for some λ such that $|d_i| \le C_1\lambda^i$. If A2 or A2$''$ is satisfied, then the second term in (A.4) dominates, because the third term is $o_p(1)$. In that case, $\|\sqrt{T}(\hat{\mathbf{d}}(k)-\mathbf{d}(k))\| = O_p(k^{1/2})$ and $\sqrt{T}(\hat{d}_i-d_i) = O_p(1)$, $i = 1,\dots,k$. If A2$''$ is not satisfied, then the third term in (A.4) dominates and $\|\lambda^{-k}(\hat{\mathbf{d}}(k)-\mathbf{d}(k))\| = O_p(k^{-1/2})$ or $\lambda^{-k}(\hat{d}_i - d_i) = O_p(1)$ $(i = 1,\dots,k)$.

The proof of Lemma 2.1 is completed by showing $\hat{\sigma}_k^2 \to \sigma_e^2$ without any lower-bound condition. The result follows from consistency of the least squares estimates. The proof is standard and is omitted.

Proof of Lemma 4.2

Let $\hat{\mathbf{d}}(k) = (\hat{d}_1,\dots,\hat{d}_k)$ be obtained by applying OLS to the augmented autoregression (4), let $\tilde{\mathbf{d}}(k) = (\tilde{d}_1,\dots,\tilde{d}_k)$ be obtained by applying OLS to (9) with $x_t = \Delta y_t$. We have $\hat{\mathbf{d}}(k) - \mathbf{d}(k) = \mathbf{M}_k^{-1}(k)\sum_{t=k+1}^{T}\mathbf{X}'_t e_{tk}$ and $\tilde{\mathbf{d}}(k) - \mathbf{d}(k) = \mathbf{R}_k^{-1}\sum_{t=k+1}^{T}\mathbf{X}'_t(e_{tk} + d_0 y_{t-1}) \equiv \mathbf{R}_k^{-1}\sum_{t=k+1}^{T}\mathbf{X}'_t e_{tk}$, because $d_0 = 0$ under the null hypothesis of a unit root. Hence

$$\hat{\mathbf{d}}(k) - \tilde{\mathbf{d}}(k) = (\mathbf{M}_k^{-1}(k) - \mathbf{R}_k^{-1})\sum_{t=k+1}^{T}\mathbf{X}'_t e_{tk}.$$

Note from Said and Dickey (1984, Lemma 5.2) that $\|(T-k)^{-1}\sum_{t=k+1}^{T}\mathbf{X}'_t e_{tk}\| = O_p(k^{1/2}T^{-1/2})$. By partition inversion, $(T-k)\mathbf{M}_k^{-1}(k) = ((T-k)^{-1}\mathbf{R}_k - \mathbf{A})^{-1}$, where

$$\begin{aligned}\mathbf{A} &= (T-k)^{-1}\left(\sum_{t=k+1}^{T} y_{t-1}\mathbf{X}_t\right)\left(\sum_{s=k+1}^{T} y_{s-1}\mathbf{X}'_s\right)\left(\sum_{t=k+1}^{T} y_{t-1}^2\right)^{-1}\\ &= (T-k)^{-3}\left(\sum_{t=k+1}^{T}\sum_{s=k+1}^{T} y_{t-1}y_{s-1}\mathbf{X}_t\mathbf{X}'_s\right)\Big/(T-k)^{-2}\sum_{t=k+1}^{T} y_{t-1}^2.\end{aligned}$$

Note that $(T-k)^{-2}\sum_{t=k+1}^{T} y_{t-1}^2 = O_p(1)$, and by Said and Dickey (1984, p. 601), each element of the numerator of $\mathbf{A}$ is bounded by $C_1/(T-k)$ for some constant C_1. Because $\mathbf{A}$ is a $k \times k$ matrix and $E(\|\mathbf{A}\|^2) \leq C_1k^2/(T-k)$, we have that $k^{1/2}\|\mathbf{A}\|$ converges to zero provided that $k^3/T \to 0$. Thus

$$\begin{aligned}&\|(T-k)\mathbf{M}_k^{-1}(k) - (T-k)\mathbf{R}_k^{-1}\|\\ &\quad = \|(T-k)\mathbf{M}_k^{-1}(k)((T-k)^{-1}\mathbf{R}_k\\ &\qquad -((T-k)\mathbf{M}_k^{-1}(k))^{-1})(T-k)\mathbf{R}_k^{-1}\|\\ &\quad = \|(T-k)\mathbf{M}_k^{-1}(k)\mathbf{A}(T-k)\mathbf{R}_k^{-1}\|\\ &\quad \leq \|(T-k)\mathbf{M}_k^{-1}(k)\|\,\|\mathbf{A}\|\,\|(T-k)\mathbf{R}_k^{-1}\|.\end{aligned}$$

Because $\|(T-k)\mathbf{M}_k^{-1}(k)\|$ and $\|(T-k)\mathbf{R}_k^{-1}\|$ are O_p (1) (see Said and Dickey, 1984) and $k^{1/2}\|\mathbf{A}\| \to 0$, $k^{1/2}\|(T-k)\mathbf{M}_k^{-1}(k) - (T-k)\mathbf{R}_k^{-1}\| \to 0$. Combining these results, we have

$$\begin{aligned}T^{1/2}\|\hat{\mathbf{d}}(k) - \tilde{\mathbf{d}}(k)\| &\leq k^{1/2}\|(T-k)\mathbf{M}_k^{-1}(k) - (T-k)\mathbf{R}_k^{-1}\|\\ &\quad \times k^{-1/2}T^{1/2}\|(T-k)^{-1}\sum_{t=k+1}^{T}\mathbf{X}'_t e_{tk}\| \qquad \text{(A.5)}\end{aligned}$$

$\to 0$ as $T \to \infty$, provided that $k^3/T \to 0$.

We are now in a position to prove Lemma 4.2. Using the definitions of $\hat{e}_{tk}$ and $\tilde{e}_{tk}$, we have

$$\hat{\sigma}_k^2 = (T-k)^{-1}\sum_{t=k+1}^{T}(\Delta y_t - \hat{d}_0 y_{t-1} - \hat{\mathbf{d}}(k)'\mathbf{X}_t)^2$$

$$= \tilde{\sigma}_k^2 + (T-k)^{-1}\hat{d}_0^2\sum_{t=k+1}^{T} y_{t-1}^2 - 2(T-k)^{-1}\hat{d}_0$$

$$\times\sum_{t=k+1}^{T} y_{t-1}\tilde{e}_{tk} + (\hat{\mathbf{d}}(k)-\tilde{\mathbf{d}}(k))'\left[(T-k)^{-1}\sum_{t=k+1}^{T}\mathbf{X}_t\mathbf{X}'_t\right]$$

$$\times(\hat{\mathbf{d}}(k)-\tilde{\mathbf{d}}(k)) - 2(\hat{\mathbf{d}}(k)-\tilde{\mathbf{d}}(k))'(T-k)^{-1}$$

$$\times\sum_{t=k+1}^{T}\mathbf{X}_t\tilde{\mathbf{e}}_{tk} + 2(T-k)^{-1}\hat{d}_0(\hat{\mathbf{d}}(k)-\tilde{\mathbf{d}}(k))'\sum_{t=k+1}^{T}\mathbf{X}_t y_{t-1}.$$

We now consider each term individually.

1. $(T-k)^{-1}(T-k)^2\hat{d}_0^2(T-k)^{-2}\sum_{t=k+1}^{T} y_{t-1}^2 = O_p(T^{-1})$, because $(T-k)^2\hat{d}_0^2 = O_p(1)$ and $(T-k)^{-2}\sum_{t=k+1}^{T} y_{t-1}^2 = O_p(1)$.
2. $(T-k)^{-1}(T-k)\hat{d}_0(T-k)^{-1}\sum_{t=k+1}^{T} y_{t-1}\tilde{e}_{tk} = O_p(T^{-1})$. Because $T\hat{d}_0 = O_p(1)$, we need to show that $(T-k)^{-1}\sum_{t=k+1}^{T} y_{t-1}\tilde{e}_{tk}$ is $O_p(1)$. Using the fact that $\tilde{e}_{tk} = e_{tk} + (\tilde{\mathbf{d}}(k)-\mathbf{d}(k))'\mathbf{X}_t$, we have

$$(T-k)^{-1}\sum_{t=k+1}^{T} y_{t-1}\tilde{e}_{tk}$$

$$= (T-k)^{-1}\sum_{t=k+1}^{T} y_{t-1}e_{tk} + (T-k)^{-1}(\tilde{\mathbf{d}}(k)$$

$$-\mathbf{d}(k))'\sum_{t=k+1}^{T} y_{t-1}\mathbf{X}_t.$$

The first term is $O_p(1)$ (see the proof of Lemma 2.1). We now show that the second term vanishes. We have for $|d_i| \le C_1\lambda^i$, with $0 < \lambda < 1$,

$$\|\tilde{\mathbf{d}}(k) - \mathbf{d}(k)\| \left\| (T-k)^{-1} \sum_{t=k+1}^{T} y_{t-1}\mathbf{X}_t \right\|$$

$$= O_p(k^{1/2}\lambda^k)O_p(k^{1/2}) \text{ if A2 is not satisfied;}$$

$$= O_p(k^{1/2}T^{-1/2})O_p(k^{1/2}) \text{ if A2 is satisfied;}$$

and is $o_p(1)$, because $k^2/T \to 0$ in the latter case and $k \to \infty$ with $\lambda < 1$ in the former case.

3. Taking norms, for the third term we have

$$\left\| (\hat{\mathbf{d}}(k) - \tilde{\mathbf{d}}(k))' \left((T-k)^{-1} \sum_{t=k+1}^{T} \mathbf{X}_t\mathbf{X}'_t \right) (\hat{\mathbf{d}}(k) - \tilde{\mathbf{d}}(k)) \right\|$$

$$\leq \|(\hat{\mathbf{d}}(k) - \tilde{\mathbf{d}}(k))\| \left\| (T-k)^{-1} \sum_{t=k+1}^{T} \mathbf{X}_t\mathbf{X}'_t \right\| \|(\hat{\mathbf{d}}(k) - \tilde{\mathbf{d}}(k))\|$$

$$= o_p(T^{-1/2}) \cdot O_p(1) \cdot o_p(T^{-1/2}) = o_p(T^{-1})$$

using (A.5). Hence the third term is $o_p(T^{-1})$.

4.

$$(\hat{\mathbf{d}}(k) - \tilde{\mathbf{d}}(k))'(T-k)^{-1} \sum_{t=k+1}^{T} \mathbf{X}_t\tilde{e}_{tk}$$

$$= (\hat{\mathbf{d}}(k) - \tilde{\mathbf{d}}(k))'(T-k)^{-1} \sum_{t=k+1}^{T} \mathbf{X}_t e_{tk}$$

$$+ (\hat{\mathbf{d}}(k) - \tilde{\mathbf{d}}(k))' \left((T-k)^{-1} \sum_{t=k+1}^{T} \mathbf{X}_t\mathbf{X}'_t \right) (\hat{\mathbf{d}}(k) - \mathbf{d}(k)).$$

Taking norms, for the first term we have

$$\left\|(\hat{\mathbf{d}}(k)-\tilde{\mathbf{d}}(k))'(T-k)^{-1}\sum_{t=k+1}^{T}\mathbf{X}_t e_{tk}\right\|$$

$$\leq (\hat{\mathbf{d}}(k)-\tilde{\mathbf{d}}(k))\|\left\|(T-k)^{-1}\sum_{t=k+1}^{T}\mathbf{X}_t e_{tk}\right\|$$

$$= o_p(T^{-1/2})\cdot O_p(k^{1/2}T^{-1/2}) = o_p(k^{1/2}T^{-1}).$$

For the second term we have

$$\left\|(\hat{\mathbf{d}}(k)-\tilde{\mathbf{d}}(k))'\left((T-k)^{-1}\sum_{t=k+1}^{T}\mathbf{X}_t\mathbf{X}'_t\right)(\hat{\mathbf{d}}(k)-\mathbf{d}(k))\right\|$$

$$\leq \|\hat{\mathbf{d}}(k)-\tilde{\mathbf{d}}(k))\|\left\|(T-k)^{-1}\sum_{t=k+1}^{T}\mathbf{X}_t\mathbf{X}'_t\right\|\|(\hat{\mathbf{d}}(k)-\mathbf{d}(k))\|$$

$$= o_p(T^{-1/2})\cdot O_p(1)\cdot O_p(k^{1/2}T^{-1/2})$$

$$= o_p(k^{1/2}T^{-1}) \text{ if A2 is satisfied;}$$

$$= o_p(T^{-1/2})\cdot O_p(1)\cdot O_p(k^{1/2}\lambda^k)$$

$$= o_p(k^{1/2}T^{-1/2}\lambda^k) \text{ if A2 is not satisfied.}$$

5. Because $T\hat{d}_0 = O_p(1)$, we consider

$$\left\|(\hat{\mathbf{d}}(k)-\tilde{\mathbf{d}}(k))(T-k)^{-2}\sum_{t=k+1}^{T}\mathbf{X}_t y_{t-1}\right\|$$

$$\leq \|\hat{\mathbf{d}}(k)-\tilde{\mathbf{d}}(k)\|\left\|(T-k)^{-2}\sum_{t=k+1}^{T}\mathbf{X}_t y_{t-1}\right\|$$

$$= o_p(T^{-1/2})O_p(k^{1/2}T^{-1}) = o_p(k^{1/2}T^{-3/2}).$$

Collecting results from 1 – 5, we have

$$\hat{\sigma}_k^2 = \tilde{\sigma}_k^2 + o_p(k^{1/2}T^{-1}) \text{ if A2 is satisfied;}$$
$$\hat{\sigma}_k^2 = \tilde{\sigma}_k^2 + o_p(k^{1/2}T^{-1/2}\lambda^k) \text{ if A2 is not satisfied.}$$

Because $k/T \to 0$ and $k^{1/2}\lambda^k \to 0$ as $k \to \infty$ and $T \to \infty$, we have, whether or not A2 is satisfied,

$$\hat{\sigma}_k^2 = \tilde{\sigma}_k^2 + o_p(T^{-1/2}).$$

Proof of Lemma 5.1

We first note (from the proof of Lemma 4.1) that $\hat{\mathbf{d}}(n) = \tilde{\mathbf{d}}(n) + o_p(1)$, where $\tilde{\mathbf{d}}(n)$ corresponds to the OLS estimates from the restricted regression without the lagged dependent variable. Using the block diagonality of $\mathbf{M}_k$, we have the following asymptotic relation:

$$J(k-n,k) = (T-k)\tilde{\mathbf{d}}(n)'((T-k)\hat{\sigma}_r^2\mathbf{R}_k^{-1}(n))^{-1}\tilde{\mathbf{d}}(n) + o_p(1),$$

where $\mathbf{R}_k^{-1}(n)$ is the lower $n \times n$ block of $\mathbf{R}_k^{-1}$. We now apply the following decomposition:

$$\begin{aligned} J(k-n,k) &= \sqrt{T}(\tilde{\mathbf{d}}(n) - \mathbf{d}(n))'((T-k)^{-1}\hat{\sigma}_r^2\mathbf{R}_k^{-1}(n))^{-1}\sqrt{T}(\tilde{\mathbf{d}}(n) - \mathbf{d}(n)) \\ &\quad + 2\sqrt{T}(\tilde{\mathbf{d}}(n) - \mathbf{d}(n))'((T-k)\hat{\sigma}_r^2\mathbf{R}_k^{-1}(n))^{-1}\sqrt{T}\mathbf{d}(n)' \\ &\quad + \sqrt{T}\mathbf{d}(n)'((T-k)\hat{\sigma}_r^2\mathbf{R}_k^{-1}(n))^{-1}\sqrt{T}\mathbf{d}(n) + o_p(1). \end{aligned}$$

By Lewis and Reinsel (1984, Theorem 4), the first term is asymptotically distributed as χ^2 with n degrees of freedom. To complete the proof, it remains to show that the other terms vanish as $T \to \infty$. We first note that $[(T-k)\hat{\sigma}_r^2\mathbf{R}_k^{-1}(n)]^{-1} \to_p \mathbf{R}$, say. Given that Δy_t is a stationary and invertible ARMA process, a typical element of $\sqrt{T}\mathbf{d}(n)$, say $\sqrt{T}d_{k+i}$ $(i = 1, \ldots, n)$, is such that $|\sqrt{T}d_{k+i}| \leq C_1\sqrt{T}\lambda^{k+i}$ for some C_1 and $0 < \lambda < 1$. Hence, under the conditions of A2$''$, $\sqrt{T}\mathbf{d}(n) \to 0$. It follows that the last term converges to zero. Finally, to show that the second term also vanishes, we simply note that under the conditions of A2$''$, $\sqrt{T(\tilde{\mathbf{d}}(n) - \mathbf{d}(n))} = O_p(1)$.

Proof of Lemma 5.2

Because $k\max$ is assumed to increase in such a way that Assumption A2$''$ is satisfied, the conditions of Lemma 5.1 hold and $J(k\max, k\max+n)$ is asymptotically distributed as a χ^2 random variable with n degrees of freedom. Let $\hat{k}_T$ be the estimate of k selected by the sequential procedure described in Definition 3.1. Then

$$\lim_{T\to\infty} P[\hat{k}_T \neq k\max] = 1 - \alpha,$$

and using the rules of conditional probability,

$$\lim_{T\to\infty} P[\hat{k}_T = k\max -1 | \hat{k}_T \neq k\max] = \alpha.$$

This implies that

$$\begin{aligned}\lim_{T\to\infty} & P[\hat{k}_T = k\max -1 \cap \hat{k}_T \neq k\max] \\ &= \lim_{T\to\infty} P[\hat{k}_T = k\max -1 | \hat{k}_T \neq k\max] P[\hat{k}_T \neq k\max] \\ &= \alpha(1-\alpha).\end{aligned}$$

Now

$$\begin{aligned}\lim_{T\to\infty} & P[\hat{k}_T = k\max -2 | \hat{k}_T \neq k\max -1 \cap \hat{k}_T \neq k\max] = \alpha \\ &= \lim_{T\to\infty} \frac{P[\hat{k}_T = k\max -2 \cap \hat{k}_T \neq k\max -1 \cap \hat{k}_T \neq k\max]}{P[\hat{k}_T \neq k\max -1 \cap \hat{k}_T \neq k\max)}\end{aligned}$$

and

$$\begin{aligned}\lim_{T\to\infty} & P(\hat{k}_T \neq k\max -1 \cap \hat{k}_T \neq k\max) \\ &= \lim_{T\to\infty} P(\hat{k}_T \neq k\max -1 | \hat{k}_T \neq k\max) P(\hat{k}_T \neq k\max) \\ &= (1-\alpha)^2.\end{aligned}$$

This implies that

$$\lim_{T\to\infty} P[\hat{k}_T = k\max -2 \cap \hat{k}_T \neq k\max -1 \cap \hat{k}_T \neq k\max] = \alpha(1-\alpha)^2.$$

We can deduce, by recursion, that

$$\lim_{T\to\infty} P[\hat{k}_T = k\min \cap \hat{k}_T \neq k\min + 1 \cdots \cap \hat{k}_T \neq k\max]$$
$$= \lim_{T\to\infty} [\alpha(1-\alpha)^{k\max - k\min}]. \tag{A.6}$$

Now suppose that $k\min$ increases to infinity at a rate slower than $k\max$. From (A.6), the application of the sequential procedure implies the probability that $\hat{k}_T$ reaches $k\min$ is zero in the limit, because $k\max - k\min \to \infty$. It follows that for any given $k\min$ and $k\max$, $\hat{k}_T$ must be bounded away from $k\min$. Because $k\min$ can be any arbitrary sequence, it follows that $\hat{k}_T$ has a zero probability of increasing at a rate slower than $k\max$. Thus, with the inequality $k\min \leq k\max$, $\hat{k}_T$ must increase at the same rate as $k\max$.

Acknowledgments

The authors acknowledge grants from the Social Science and Humanities Research Council of Canada. The second author also thanks the Fonds pour la Formation de Chercheurs et l'Aide à la Recherche du Québec and the National Science Foundation for financial support.

References

Akaike, H. (1974). A new look at statistical model identification. *IEEE Transactions on Automatic Control* AC-19, 716–723.

Agiakloglou, C. and Newbold, P. (1992). Empirical evidence on Dickey–Fuller type tests. *Journal of Time Series Analysis* 13, 471–483.

Banerjee, A., Lumsdaine, R.L. and Stock, J.H. (1992). Recursive and sequential tests of the unit root and tread break hypothesis. *Journal of Business & Economic Statistics* 10, 271–287.

Berk, K. (1974). Consistent autoregressive spectral estimates. *The Annals of Statistics* 2, 489–502.

Chan, N.H. and Wei, C. (1988). Limiting distribution of least squares estimates of unstable autoregressive processes. *The Annals of Statistics*, 16, 367–401.

Dickey, D.A. and Fuller, W. (1979). Distribution of the estimators for autoregressive time series with a unit root. *Journal of the American Statistical Association* 74, 427–431.

Dickey, D.A. and Said, S.E. (1981). Testing ARMA $(p, 1, q)$ versus ARMA $(p+1, q)$. In *Proceedings of the Business and Economics Statistics Section, American Statistical Association*, pp. 318–322.

Eastwood, B.J. and Gallant, R.A. (1991). Adaptive rules for seminon-parametric estimators that achieve asymptotic normality. *Econometric Theory,* 7, 307–340.

Fuller, W. (1976). *Introduction to Statistical Time Series,* John Wiley: New York.

Hall, A. (1994). Testing for a unit root in time series with pretest data based model selection. *Journal of Business & Economic Statistics* 12, 461–470.

Hannan, E.J. and Deistler, M. (1988). *The Statistical Theory of Linear Systems,* John Wiley: New York.

Harris, R.I.D. (1992). Testing for unit roots using the augmented Dickey–Fuller test. *Economic Letters* 38, 381–386.

Lewis, R. and Reinsel, G.C. (1985). Prediction of multivariate time series by autoregressive model fitting. *Journal of Multivariate Analysis* 16, 393–411.

Nabeya, S. and Perron, P. (1994). Local asymptotic distributions related to the AR (1) model with dependent errors. *Journal of Econometrics* 62, 229–264.

Pantula, S.G. (1991). Asymptotic distributions of unit-root tests when the process is nearly stationary. *Journal of Business & Economic Statistics* 9, 63–71.

Perron, P. (1989). The great crash, the oil price shock, and the unit root hypothesis. *Econometrica* 57, 1361–1401.

Perron, P. (1990). Further Evidence of Breaking Trend Functions in Macroeconomic Time Series. unpublished manuscript, Princeton University.

Perron, P. (1992). The adequacy of asymptotic approximations in the near-integrated autoregressive model with dependent errors. *Journal of Econometrics,* in press.

Perron, P. and Vogelsang, T.J. (1992). Nonstationarity and level shifts with an application to purchasing power parity. *Journal of Business & Economic Statistics* 10, 301–320.

Press, W.H., Tuekolsky, S., Vetterling, W. and Flannery, B. (1988). *Numerical Recipies in C,* Cambridge University Press: Cambridge, U.K.

Said, S. and Dickey, D.A. (1984). Testing for unit roots in autoregressive moving-average models of unknown order. *Biometrika* 71, 599–607.

Schwartz, G. (1978). Estimating the dimension of a model. *The Annals of Statistics* 6, 461–464.

Schwert, G.W. (1989). Tests for unit roots: A Monte Carlo investigation. *Journal of Business & Economic Statistics* 7, 147–160.

Shibata, R. (1980). Asymptotic efficient selection of the order of the model for estimating parameters of a linear process. *The Annals of Statistics* 8, 147–164.

Chapter 7

The Adequacy of Asymptotic Approximations in the Near-Integrated Autoregressive Model with Dependent Errors*

Pierre Perron

Université de Montréal, C.R.D.E.,
Montréal, Québec H3C 3J7, Canada

We consider the normalized least squares estimator of the parameter in a nearly integrated first-order autoregressive model with dependent errors. In a first step we consider its asymptotic distribution as well as asymptotic expansion up to order $O_p(T^{-1})$. We derive a limiting moment generating function which enables us to calculate various distributional quantities by numerical integration. A simulation study is performed to assess the adequacy of the asymptotic distribution when the errors are correlated. We focus our attention on two leading cases: MA(1) errors and AR(1) errors. The asymptotic approximations are shown to be inadequate as the MA root gets close to -1 and as the AR root approaches either -1 or 1. Our theoretical analysis helps to explain and understand the simulation results of Schwert (1989) and DeJong *et al.* (1992) concerning the size and power of Phillips and Perron's (1988) unit root test. A companion paper, Nabeya and Perron (1994), presents alternative asymptotic frameworks in the cases where the usual asymptotic distribution fails to provide an adequate approximation to the finite-sample distribution.

*Originally Published in *Journal of Econometrics* 70, pp. 317–350.
This paper is a revised version of Sections 2, 3, and 4 of Perron (1990b). The other parts of that earlier draft are now the object of a companion paper, Nabeya and Perron (1994). Some material in the present paper were initially investigated as part of a joint project with Peter C.B. Phillips in the spring of 1986.

Keywords: Near-integrated model; functional weak convergence; simulation experiment; unit root process; autoregressive moving-average models.

1. Introduction

In an attempt to cover more general time series structures, it has become popular in econometrics to consider models which permit both the regressors and the errors to have substantial heterogeneity and dependence over time. On a theoretical level, this advance has become possible due to a new class of functional central limit theorems which provide the required asymptotic results. An integrated treatment can be found in White (1984) and Gallant and White (1988). This approach made possible the analysis of a wide class of models with substantial relaxation of the standard conditions. Examples include time series models with unit roots (e.g., Phillips, 1987a), testing for structural change in a general nonlinear framework (e.g., Andrews and Fair, 1988), and cointegration (e.g., Phillips and Ouliaris, 1990). However, little is known about the adequacy of the limiting distributions as approximations to the finite-sample distribution in such a general framework. This paper is a step in a systematic examination of this issue. We consider the leading case of a dynamic first-order autoregressive model when the errors are allowed to be dependent and provide a detailed analysis of the behavior of the associated ordinary least squares estimator. To be more precise, we consider the first-order stochastic difference equation:

$$y_t = \alpha y_{t-1} + u_t, \quad t = 1, \ldots, T, \tag{1}$$

with y_0 fixed. The least-squares estimator of α on the sequence $\{y_t\}_0^T$ is

$$\hat{\alpha} = \sum_{t=1}^{T} y_t y_{t-1} \left(\sum_{t=1}^{T} y_{t-1}^2 \right)^{-1} \tag{2}$$

The distribution of $\hat{\alpha}$ has been extensively studied, especially in the case where the errors $\{u_t\}$ are uncorrelated. Mann and Wald

(1943) and Rubin (1950) showed that $T^{1/2}(\hat{\alpha}-\alpha)(1-\alpha^2)^{-1/2}$ has a limiting N(0, 1) distribution when $|\alpha| < 1$. White (1958) showed that when $|\alpha| > 1$, the limiting distribution of $|\alpha|^T(\alpha^2-1)^{-1}(\hat{\alpha}-\alpha)$ is Cauchy provided that $y_0 = 0$ and $\{u_t\}$ is normal. White also considered the case $|\alpha| = 1$ and showed that the limiting distribution of $T(\hat{\alpha}-1)$ can be expressed in terms of the ratio of two functionals of a Wiener process (see also Phillips, 1987a). The case of the unit root ($\alpha = 1$) has attracted a great deal of attention. The asymptotic distribution of $T(\hat{\alpha}-1)$ has been tabulated by Dickey (1976) via simulation methods (see also Fuller, 1976; Dickey and Fuller, 1979) and by Evans and Savin (1981a) using numerical integration. Evans and Savin (1981b) showed how the standard limiting distributions fail to provide adequate approximations to the exact distribution when α is close to but not equal to 1.

Recently, a class of models which specifically deal with the presence of a root close to, but not necessarily equal to 1, has been studied. These specify a near-integrated process where the autoregressive parameter is defined by $\alpha = \exp(c/T)$. The constant c being a measure of the deviation from the unit root case, the model may also be described as having a root local to unity: as the sample size increases, the autoregressive parameter converges to unity. When $c < 0$, the process $\{y_t\}$ is said to be (locally) stationary, and when $c > 0$, it is said to be (locally) explosive. An expression for the limiting distribution of $T(\hat{\alpha}-\alpha)$ has been derived by Bobkoski (1983), Cavanagh (1986), Chan and Wei (1987), and Phillips (1987b).[1] In the near-integrated context, with errors that are weakly dependent, Phillips (1987b) showed that (under some conditions to be made precise later)

$$T(\hat{\alpha}-\alpha) \Rightarrow \left\{\int_0^1 J_c(r)\mathrm{d}W(r)+\delta\right\}\left\{\int_0^1 J_c(r)^2\mathrm{d}r\right\}^{-1}, \tag{3}$$

[1]This framework has been quite useful in studying various problems such as the power of tests of a unit root under local alternatives (Phillips, 1987b; Phillips and Perron, 1988; Perron, 1990a), the derivation of confidence intervals when α is near unity (Cavanagh, 1986; Stock, 1991), and the power of tests of a unit root with a continuum of observations (Perron, 1989).

with $\delta = (\sigma^2 - \sigma_u^2)/(2\sigma^2)$, $\sigma^2 = \lim_{T\to\infty} \mathrm{E}(T^{-1}S_T^2)$, $S_t = \sum_{j=1}^{t} u_j$, $\sigma_u^2 = \lim_{T\to\infty} T^{-1}\mathrm{E}(\sum_{t=1}^{T} u_t^2)$, $J_c(r) = \int_0^r \exp((r-s)c)\mathrm{d}W(r)$ with $W(r)$ the unit Wiener process.

Tabulations of the limiting distribution (3) with $\delta = 0$ have been obtained by Cavanagh (1986), Chan (1988), Nabeya and Tanaka (1990), and Perron (1989) using different procedures. These studies also assess its adequacy as an approximation to the finite-sample distribution when α is near 1. They show the approximation to be quite good in the case where $y_0 = 0$. Perron (1991a, 1991b) also considers a continuous-time approximation which performs well even in the case where $y_0 \neq 0$. These asymptotic distributions provide substantial improvements over the traditional asymptotic framework when α is in the vicinity of 1 because they are continuous with respect to the autoregressive parameter α.

However, most of the evidence about the adequacy of the approximation pertains to the case $\delta = 0$, i.e., with no correlation in the residuals.[2] The purpose of this paper is to investigate this adequacy when the errors are dependent. Section 2 discusses the limiting distribution of $T(\hat{\alpha} - \alpha)$ and extends Phillips' (1987c) $\mathrm{O}_p(T^{-1})$ expansion to the near-integrated setting. The results of Perron (1991a) are used to derive appropriate limiting moment-generating functions. These permit the calculation of the cumulative distribution functions and the moments of the asymptotic distributions. In Section 3, we present simulation experiments to compare the asymptotic results with their finite-sample counterparts. We concentrate on two leading cases, namely,

$$\text{MA(1) errors: } u_t = e_t + \theta e_{t-1}, \tag{4}$$

$$\text{AR(1) errors: } u_t = \rho u_{t-1} + e_t, \tag{5}$$

with $e_t \sim$ i.i.d. $\mathrm{N}(0, \sigma_e^2)$ random variables. The results show the asymptotic distribution to be a very poor guide to the finite-sample distribution, even for large sample sizes, when θ (in the MA case)

[2] An exception is Nabeya and Tanaka (1990) which contains some results when the errors are correlated.

or ρ (in the AR case) are close to -1. When ρ is close to 1, the approximation is not as bad, but the approach to the limiting distribution is slow. Section 4 presents results for the $O_p(1)$ asymptotic approximation in the cases where a constant or a constant and a trend are included in regression (1). Section 5 provides some concluding comments, and the appendix contains mathematical derivations.

2. The Limiting Distribution of $T(\hat{\alpha} - \alpha)$

This section considers the limiting distribution of the normalized least squares estimator $T(\hat{\alpha} - \alpha)$ in the near-integrated model with dependent errors. We also consider its asymptotic expansion up to order $O_p(T^{-1})$. As a matter of notation, we denote weak convergence in distribution by '$\Rightarrow$' and equality in distribution by '$\overset{d}{=}$'. The asymptotic analyses below can be obtained under various conditions upon the error structure. We consider those of Herrndorf (1984) involving the concept of strong mixing.

Assumption 1. (a) $E(u_t) = 0$; (b) $\sup_t E|u_t|^{\beta+\varepsilon} < \infty$ *for some* $\beta > 2$ *and* $\varepsilon > 0$; (c) $\sigma^2 = \lim_{T\to\infty} T^{-1}E(S_T^2)$ *exists and* $\sigma^2 > 0$, *where* $S_t = \sum_1^t u_j$; (d) $\{u_t\}_1^\infty$ *is strong mixing with mixing numbers* α_m *that satisfy:* $\sum_1^\infty \alpha_m^{1-2/\beta} < \infty$.

When the sequence $\{u_t\}$ is strictly stationary condition (c) is implied by (a), (b), and (d) and $\sigma^2 = 2\pi f_u(0)$, where $f_u(0)$ is the nonnormalized spectral density function of $\{u_t\}$ evaluated at frequency 0. When considering the asymptotic expansion of order $O_p(T^{-1})$ the following additional restriction will be imposed on the sequence of errors $\{u_t\}$.

Assumption 2. $\{u_t\}_1^\infty$ *is a Gaussian weakly stationary sequence.*

Consider the construction of random elements defined by $X_T(r) = T^{-1/2} \times \sigma^{-1}S_{[Tr]} = T^{-1/2}\sigma^{-1}S_{j-1}[(j-1)/T \le r < j/T$, $j = 1, \ldots, T]$. From Herrndorf (1984), the following functional central limit theorem holds under Assumption 1:

$$X_T(r) \Rightarrow W(r). \tag{6}$$

Phillips (1987b) proved (3) under the conditions of Assumption 1 using the result (6). We consider here an extension of his (1987c) result concerning the asymptotic expansion up to order $O_p(T^{-1})$ of $T(\hat{\alpha} - \alpha)$. The following theorem is proved in the appendix.

Theorem 1. *Let* $J_c(r) = \int_0^r \exp((r-s)c)\mathrm{d}W(r)$, $\delta = (\sigma^2 - \sigma_u^2)/(2\sigma^2)$, *with* $\sigma_u^2 = \lim_{T\to\infty} T^{-1}\mathrm{E}(\sum_{t=1}^T u_t^2)$ *and* σ^2 *as defined in Assumption* 1; $\gamma = y_0/(\sigma T^{1/2})$, $v^2 = 2\pi f_{u^2}(0)$, *with* $f_{u^2}(0)$ *the nonnormalized spectral density function of* $\{u_t^2 - \mathrm{E}(u_t^2)\}$ *at frequency* 0; η *is a* $\mathrm{N}(0,1)$ *variable independent of the Wiener process* $W(r)$. *Under Assumptions* 1 *and* 2:

$$T(\hat{\alpha} - \alpha) \stackrel{d}{=} H(c,\gamma)/K(c,\gamma) + O_p(T^{-1}),$$

where

$$H(c,\gamma) = \int_0^1 J_c(r)\mathrm{d}W(r) + \delta + \gamma \times \int_0^1 \exp(cr)\mathrm{d}W(r) - (v/(2\sigma^2 T^{1/2}))\eta$$

and

$$K(c,\gamma) = \int_0^1 J_c(r)^2\mathrm{d}r + 2\gamma \int_0^1 \exp(cr)J_c(r)\mathrm{d}r.$$

With the normalization $\sigma_e^2 = 1$, the following specifications of the various parameters apply. For the MA(1) case: $\sigma_u^2 = 1 + \theta^2$, $\sigma^2 = (1+\theta)^2$, hence $\delta = \theta/(1+\theta)^2$ and $\gamma = y_0/[T^{1/2}(1+\theta)]$. Also, $v^2 = 2(1 + 4\theta^2 + \theta^4)$. In the AR(1) case we have $\sigma_u^2 = (1-\rho^2)^{-1}$, $\sigma^2 = (1-\rho)^{-2}$; hence $\delta = \rho/(1+\rho)$, $\gamma = y_0(1-\rho)/T^{1/2}$, and $v^2 = 2(1+\rho^2)/(1-\rho^2)^3$. We now characterize the joint moment-generating function of $\{H(c,\gamma), K(c,\gamma)\}$, which we denote by $MGF(v,u) = \mathrm{E}[\exp\{vH(c,\gamma) + uK(c,\gamma)\}]$.

Theorem 2. *Let* $g = v/(2\sigma^2 T^{1/2})$, $d = \gamma^2(\exp(2c) - 1)/2c$, *and* $\lambda = (c^2 + 2cv - 2u)^{1/2}$. *Then the joint moment-generating function of* $\{H(c,\gamma) \cdot K(c,\gamma)\}$ *is given by*

$$MGF(v,u) = \exp(v\delta - ud + v^2 g^2/2)M_{c,\gamma}(v,u),$$

where

$$M_{c,\gamma}(v,u) = \psi_c(v,u)\exp\{-(\gamma^2/2)(v+c-\lambda) \\ \times (1-\exp(v+c+\lambda)\psi_c^2(v,u))\}$$

and

$$\psi_c(v,u) = \{2\lambda\exp(-(v+c))/[\lambda+v+c)\exp(-\lambda) \\ +(\lambda-v-c)\exp(\lambda)]\}^{1/2}.$$

Theorem 2 allows direct computation, by numerical integration, of the cumulative distribution function and the moments of the asymptotic distribution. Let the joint characteristic function of $\{H(c,\gamma), K(c,\gamma)\}$ be $\mathrm{CF}(v,u) = \mathrm{MGF}(iv,iu) = \mathrm{E}[\exp\{ivH(c,\gamma) + iuK(c,\gamma)\}]$. From Gurtand (1948), the limiting distribution function of $T(\hat{\alpha}-\alpha)$ is

$$F(z) = \frac{1}{2} - (1/\pi)\int_0^\infty \mathrm{IM}[\mathrm{CF}(v,-vz)/v]\mathrm{d}v, \qquad (7)$$

with $IM(\cdot)$ the imaginary part of the complex number. The moments of the asymptotic distribution can be obtained using Mehta and Swamy's (1978) result, such that

$$\mathrm{E}[H(c,\gamma)/K(c,\gamma)]^r \\ = \Gamma(r)^{-1}\int_0^\infty u^{r-1}\{\partial^r \mathrm{MGF}(v,-u)/\partial v^r\}_{r=0}\,du. \qquad (8)$$

These results allow computation of distributional quantities for a variety of processes. The usual asymptotic distribution (3) can be obtained by letting $g=\gamma=0$, then:

$$\mathrm{MGF}(v,u) = \exp(v\delta)\psi_c(v,u), \qquad (9)$$

where $\psi_c(v,u)$ is defined in Theorem 2. The next section makes extensive use of Theorem 2, (7), and especially (9), to calculate the asymptotic distribution of $T(\hat{\alpha}-\alpha)$ when the errors are MA(1) or AR(1) processes. The numerical integrations were performed using the subroutine QDAG of the IMSL library. The bounds of integration are $(\varepsilon, \bar{V})$, where $\bar{V}$ is set such that the square of the integrand evaluated at $\bar{V}$ is less than ε. The error criterion for the numerical integration was also set at ε. For most experiments we used $\varepsilon = 1.0\text{E}{-}07$.

3. Approximating the Exact Distribution of $T(\hat{\alpha} - \alpha)$

This section considers the adequacy of the asymptotic distribution as an approximation to the finite-sample distribution. The asymptotic values are obtained using the results of Section 2 and the finite-sample values are obtained via simulations.[3] As discussed in the introduction, we consider two leading cases where the error sequence is either an MA(1) or an AR(1) process. As we will see, these cases are sufficient to provide a rich characterization of the relationship between the finite-sample results and their asymptotic counterparts. Given that our aim is mainly oriented towards studying the effect of correlation in the errors, we consider only the case where $y_0 = 0$.

The experiment is as follows. For each of the MA and AR cases we consider three values of c, namely $c = 0$ (unit root), -5 (locally stationary), and 2 (locally explosive). For each we consider the following specifications for the errors[4]: for the MA(1) case, $\theta = -0.9$, -0.7, -0.5, -0.3, 0.5; for the AR(1) case, $\rho = -0.9$, -0.5, 0.5, 0.9, 0.95. We consider sample sizes of length $T = 25$, 50, 100, 500, 1,000, 5,000 (though we do not report results for every sample sizes in each case). The finite-sample results are obtained using 10,000 replications (5,000 when $T = 5{,}000$). The asymptotic results for the cumulative distribution function are obtained using (9) for the $O_p(1)$ asymptotic framework.[5]

---

[3]In principle, the computation of the exact finite-sample distribution is possible using Imhof's (1961) routine or a variant of it. However, given the sample sizes analyzed here (up to 5,000), such an approach becomes practically infeasible.

[4]A much wider range of experiments were performed. For the sake of brevity we report only a subset of the results. Some comments made in the text pertain to the full set of results, however. These are available upon request.

[5]Special care must be taken with the numerical integration since it involves the square root of a complex-valued quantity. The use of the principal value may not ensure the continuity of the integrand. The numerical integration must be performed over Reimann surfaces consisting here of two planes. The method is described in more detail in Perron (1989).

3.1. *The MA(1) case*

Consider first the percentage points of the distribution of $T(\hat{\alpha} - \alpha)$ in the MA(1) case presented in Table 1 for $c = 0$. With $\theta = 0.5$, the asymptotic approximation is excellent, even for small sample sizes. Further experiments (not reported) confirm this adequacy for any process with positively autocorrelated MA(1) errors.

When θ is negative the picture is rather different. Here the adequacy of the approximation depends very much upon the magnitude of θ and deteriorates quite rapidly as θ approaches -1. For different values of θ the asymptotic distribution adequately approximates the finite-sample distribution for the following sample sizes: $\theta = -0.3$, $T \geq 500$; $\theta = -0.5$, $T \geq 1{,}000$; $\theta = -0.7$, $T \geq 5{,}000$; and when $\theta = -0.9$ the asymptotic distribution is still quite far away from the exact distribution when $T = 5{,}000$. The differences are substantial. For example, the 1% point of the exact distribution when $\theta = -0.9$ and $T = 50$ corresponds approximately to the 95% point of the asymptotic distribution.

The same qualitative results hold when c is -5 or 2. Figures 1 and 2 present results for the case $\theta = -0.9$. The main difference is that the approximation is marginally better with an explosive process ($c = 2$) and marginally worse when it is stationary ($c = -5$).[6]

3.2. *The AR(1) case*

Consider now the case where the errors are AR(1). The results for $c = 0$ are presented in Table 2. The picture is quite different from the MA(1) case. For $\rho < 0$, the approximation is again inadequate and worsens as ρ approaches -1. However, for comparable values of ρ and θ, the approach of the finite-sample distribution to the asymptotic distribution is faster in the AR(1) case than it is in the MA(1) case. For $\rho = -0.5$, the asymptotic approximation is adequate for $T \geq 500$, and for $\rho = -0.9$ when $T \geq 1{,}000$.

[6]The full set of results is available upon request. Tabulated critical values for an extended range of values for θ are available in the working paper version.

Table 1. The distribution of $T(\hat{\alpha} - \hat{\alpha})$; $\alpha = \exp(c/T)$, $c = 0.0$; MA errors, $u_t = e_t + \theta e_{t-1}$.

	1.0%	2.5%	5.0%	10.0%	50.0%	90.0%	95.0%	97.5%	99.0%,	Mean	Variance
					$\theta = -0.90$						
$T = 50$	−61.57	−58.22	−55.76	−52.61	−39.95	−23.90	−19.87	−16.52	−13.64	−39.07	119.64
$T = 100$	−110.28	−106.09	−101.79	−96.30	−71.38	−37.56	−29.85	−24.68	−19.33	−69.00	482.88
$T = 500$	−427.96	−407.79	−385.31	−355.24	−192.30	−63.81	−45.28	−33.94	−24.94	−201.45	11573.26
$T = 1,000$	−733.45	−675.98	−625.27	−553.72	−234.38	−67.17	−49.53	−38.65	−29.70	−275.74	33716.06
$T = 5,000$	−1867.76	−1570.88	−1275.06	−999.41	−315.94	−92.19	−71.18	−57.97	−43.94	−447.14	158516.00
$T = \infty$	−2625.73	−2036.37	−1601.94	−1181.47	−310.56	−74.68	−53.65	−41.39	−31.46	−502.44	302105.93
O_p $(T = 50)$	−3311.05	−2217.16	−1673.02	−1199.96	−300.32	−69.54	−48.83	−36.66	−26.66	−502.44	335316.86
					$\theta = -0.70$						
$T = 50$	−49.25	−44.76	−40.64	−35.63	−16.52	−4.41	−3.05	−2.10	−1.34	−18.49	142.37
$T = 100$	−77.81	−70.13	−62.69	−53.22	−20.46	−5.23	−3.45	−2.47	−1.59	−25.32	352.78
$T = 500$	−162.58	−135.44	−115.70	−92.08	−27.12	−5.86	−3.80	−2.45	−1.46	−39.19	1347.55
$T = 1,000$	−197.33	−160.71	−133.73	−102.39	−26.75	−5.72	−3.92	−2.82	−1.91	−42.33	1882.63
$T = \infty$	−239.25	−185.36	−145.63	−107.17	−27.50	−5.90	−3.96	−2.83	−1.90	−45.05	2528.21
O_p $(T = 50)$	−256.08	−193.33	−149.26	−107.97	−26.59	−5.46	−3.55	−2.42	−1.49	−45.05	2796.20
					$\theta = -0.50$						
$T = 50$	−34.69	−30.10	−25.55	−20.45	−6.36	−0.83	−0.21	0.17	0.56	−8.77	66.38
$T = 100$	−47.17	−39.19	−32.54	−25.19	−6.86	−0.95	−0.32	0.07	0.43	−10.39	111.29
$T = 500$	−62.77	−49.76	−41.22	−31.25	−7.88	−1.00	−0.35	0.07	0.45	−12.65	192.09
$T = 1,000$	−76.00	−58.10	−45.59	−34.38	−7.97	−0.94	−0.36	0.08	0.45	−13.77	258.00
$T = \infty$	−71.60	−55.32	−43.32	−31.71	−7.63	−1.02	−0.39	0.00	0.36	−12.91	231.35
O_p $(T = 50)$	−75.77	−57.39	−44.27	−31.90	−7.38	−0.90	−0.27	0.13	0.50	−12.91	253.74

(Continued)

Table 1. (*Continued*)

	1.0%	2.5%	5.0%	10.0%	50.0%	90.0%	95.0%	97.5%	99.0%,	Mean	Variance
					$\theta = -0.30$						
$T = 50$	−22.93	−18.54	−15.17	−11.34	−2.67	0.28	0.69	0.99	1.36	−4.34	27.03
$T = 100$	−27.44	−21.43	−17.22	−12.61	−2.76	0.26	0.65	0.93	1.26	−4.77	35.98
$T = 500$	−29.43	−22.96	−18.81	−13.98	−3.08	0.23	0.63	0.93	1.19	−5.28	43.82
$T = \infty$	−31.36	−24.13	−18.79	−13.62	−2.89	0.23	0.61	0.89	1.21	−5.19	46.49
O_p $(T = 50)$	−32.95	−24.92	−19.15	−13.69	−2.79	0.28	0.66	0.96	1.29	−5.19	50.35
					$\theta = 0.50$						
$T = 50$	−6.83	−5.35	−4.06	−2.84	−0.13	1.35	1.77	2.18	2.77	−0.49	3.44
$T = 100$	−7.58	−5.67	−4.23	−2.91	−0.16	1.33	1.73	2.16	2.71	−0.54	3.79
$T = 500$	−7.19	−5.47	−4.33	−3.07	−0.22	1.30	1.71	2.13	2.63	−0.60	3.75
$T = \infty$	−7.35	−5.54	−4.21	−2.91	−0.16	1.31	1.71	2.11	2.64	−0.55	3.73
O_p $(T = 50)$	−7.77	−5.75	−4.30	−2.93	−0.14	1.32	1.74	2.15	2.69	−0.55	4.01

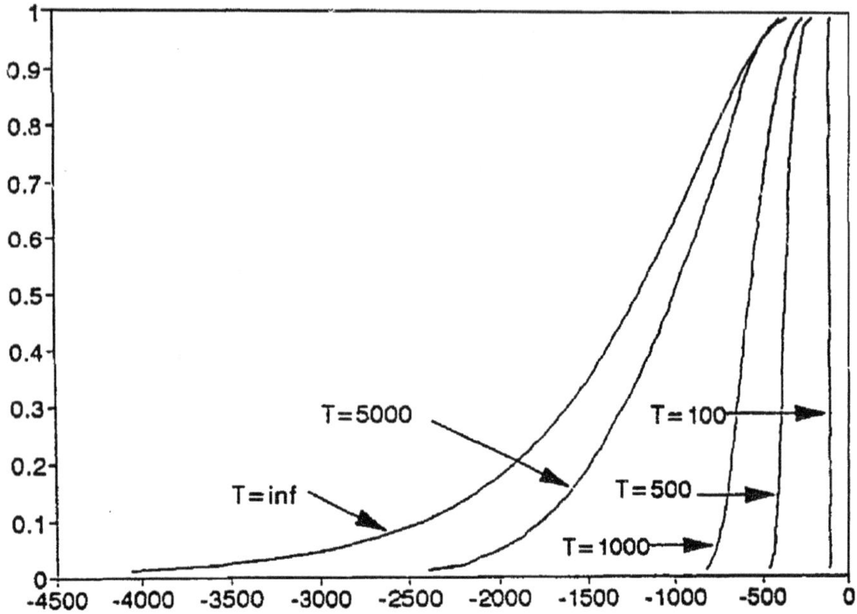

Fig. 1. The distribution of $T(\hat{\alpha}-\alpha)$; $\alpha = \exp(c/T)$, $c = -5$; MA errors, $\theta = -0.9$.

For smaller sample sizes there are important discrepancies especially when $\rho = -0.9$. However, the discrepancies are not as severe as in the MA(1) case. For example, when $T = 50$ and $\rho = -0.9$, the 1% point of the exact distribution corresponds roughly to the 30% point of the asymptotic distribution (in the MA(1) case the first percentage point of the exact distribution with $\theta = -0.9$ and $T = 50$ corresponds to the 90% point of the asymptotic distribution). Similar qualitative results hold in the stationary ($c = -5$) and explosive ($c = 2$) cases. When $c = -5$, the adequacy is marginally inferior, and when $c = 2$, it is marginally superior. These results are illustrated in Figs. 3 ($c = -5$) and 4 ($c = 2$).

The AR(1) case with positive autocorrelation offers yet a different picture. First, unlike all the cases considered so far the approach of the finite-sample distribution to its asymptotic counterpart is from a density with a larger spread to one with a smaller spread. Secondly, the differences between the finite sample and asymptotic percentage points are not substantial. For example, when $\rho = 0.9$

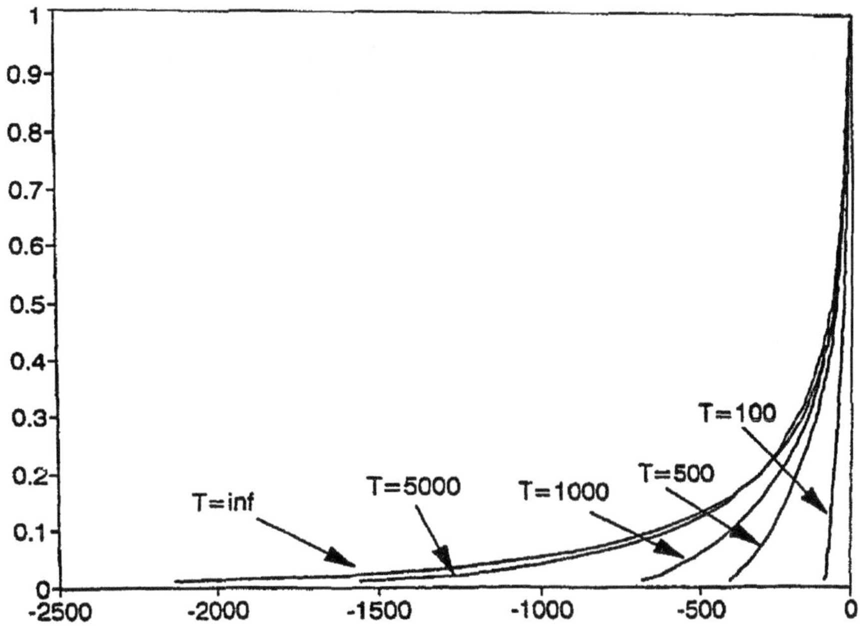

Fig. 2. The distribution of $T(\hat{\alpha}-\alpha)$; $\alpha = \exp(c/T)$, $c = 2$; MA errors, $\theta = -0.9$.

and $T = 25$, the tenth percentage point of the exact distribution corresponds roughly to the fifth percentage point of the asymptotic distribution. Nevertheless, what is interesting, and different from the MA(1) case with positive coefficient, is the fact that the approach to the asymptotic value is quite slow. The tail of the exact distribution is not well approximated until $T = 500$.

When $c = -5$ and $c = 2$, the same qualitative behavior remains as in the case where $c = 0$. The case $\rho = 0.9$ is illustrated in Figs. 5 ($c = -5$) and 6 ($c = 2$). The approximation is again marginally better with $c = 2$ and marginally worse with $c = -5$. However, some differences emerge. When $c = -5$, we notice a difference from the cases where $\rho > 0$ and $c = 0$ or 2. The right tail is better approximated by the asymptotic distribution even for quite small sample sizes (e.g., $\rho = 0.9$ and $T = 25$). Secondly, when compared to the positive MA case, there is much more movement in the percentage points as c varies.

Table 2. The distribution of $T(\hat{\alpha} - \hat{\alpha})$; $\alpha = \exp(c/T)$, $c = 0.0$; AR errors, $u_t = \rho u_{t-1} + e_t$.

	1.0%	2.5%	5.0%	10.0%	50.0%	90.0%	95.0%	97.5%	99.0%	Mean	Variance
					$\rho = -0.90$						
$T = 50$	−76.33	−69.41	−62.43	−53.52	−19.36	−4.16	−2.42	−1.46	−0.71	−24.35	359.69
$T = 100$	−122.62	−108.94	−93.53	−75.20	−23.93	−5.36	−3.34	−2.21	−1.34	−33.10	826.35
$T = 500$	−214.48	−178.94	−146.79	−112.24	−31.24	−6.58	−4.09	−2.83	−1.71	−47.75	2284.21
$T = 1{,}000$	−273.27	−216.22	−175.23	−130.44	−32.63	−6.81	−4.65	−3.42	−2.35	−53.65	3359.83
$T = \infty$	−274.69	−212.87	−167.27	−123.14	−31.71	−6.93	−4.71	−3.41	−2.35	−51.85	3329.55
O_p $(T = 50)$	−364.82	−262.57	−194.42	−133.57	−26.28	−2.21	1.35	7.27	21.63	−51.85	5662.27
					$\rho = 0.50$						
$T = 25$	−24.84	−20.50	−17.28	−13.48	−3.54	0.08	0.52	0.85	1.24	−5.31	33.31
$T = 50$	−30.92	−25.19	−20.46	−15.46	−3.86	−0.05	0.42	0.74	1.09	−6.11	47.49
$T = 100$	−36.92	−29.36	−23.45	−17.22	−3.99	−0.11	0.36	0.66	0.98	−6.74	64.12
$T = 500$	−40.14	−31.21	−25.53	−19.16	−4.42	−0.13	0.34	0.64	0.94	−7.46	79.20
$T = \infty$	−42.61	−32.84	−25.66	−18.67	−4.21	−0.14	0.30	0.60	0.91	−7.34	83.94
O_p $(T = 50)$	−45.63	−34.37	−26.56	−18.83	−4.02	−0.04	0.40	0.71	1.05	−7.34	94.12
					$\rho = 0.50$						
$T = 25$	−5.28	−3.60	−2.67	−1.73	0.27	1.76	2.31	2.82	3.46	0.10	2.55
$T = 50$	−4.47	−3.38	−2.47	−1.61	0.24	1.65	2.11	2.62	3.26	0.09	2.08
$T = 100$	−4.67	−3.42	−2.44	−1.63	0.20	1.62	2.06	2.53	3.26	0.07	2.08
$T = 500$	−4.32	−3.15	−2.47	−1.67	0.14	1.57	2.05	2.50	3.14	0.03	1.96
$T = \infty$	−4.23	−3.15	−2.35	−1.57	0.19	1.57	2.03	2.49	3.11	0.07	1.92
O_p $(T = 50)$	−4.53	−3.29	−2.41	−1.58	0.20	1.59	2.05	2.52	3.15	0.07	2.05

(*Continued*)

Table 2. (*Continued*)

	1.0%	2.5%	5.0%	10.0%	50.0%	90.0%	95.0%	97.5%	99.0%	Mean	Variance
					$\rho = 0.90$						
$T = 25$	−1.66	−0.95	−0.57	−0.28	1.25	2.83	3.44	4.06	4.86	1.27	1.67
$T = 50$	−1.04	−0.64	−0.38	−0.17	1.04	2.61	3.25	3.91	4.63	1.14	1.36
$T = 100$	−0.73	−0.45	−0.28	−0.12	0.86	2.33	3.00	3.54	4.47	1.01	1.14
$T = 500$	−0.58	−0.40	−0.26	−0.12	0.69	2.15	2.73	3.34	4.07	0.87	0.98
$T = \infty$	−0.53	−0.36	−0.23	−0.12	0.67	2.08	2.65	3.23	4.03	0.85	0.93
O_p $(T = 50)$	−0.68	−0.43	−0.26	−0.11	0.68	2.08	2.66	3.24	4.04	0.85	0.94
					$\rho = 0.95$						
$T = 25$	−1.35	−0.70	−0.40	−0.11	1.52	3.07	3.73	4.33	5.06	1.53	1.68
$T = 50$	−0.69	−0.37	−0.21	−0.05	1.33	2.89	3.56	4.24	5.02	1.42	1.42
$T = 100$	−0.43	−0.25	−0.13	−0.02	1.11	2.63	3.29	3.87	4.76	1.24	1.22
$T = 500$	−0.28	−0.17	−0.10	−0.03	0.79	2.29	2.89	3.51	4.43	0.99	1.01
$T = 1{,}000$	−0.28	−0.19	−0.11	−0.04	0.71	2.24	2.78	3.42	4.28	0.94	0.97
$T = \infty$	−0.22	−0.14	−0.08	−0.01	0.73	2.15	2.73	3.33	4.14	0.93	0.90
O_p $(T = 50)$	−0.34	−0.19	−0.10	−0.02	0.73	2.15	2.74	3.34	4.14	0.93	0.91

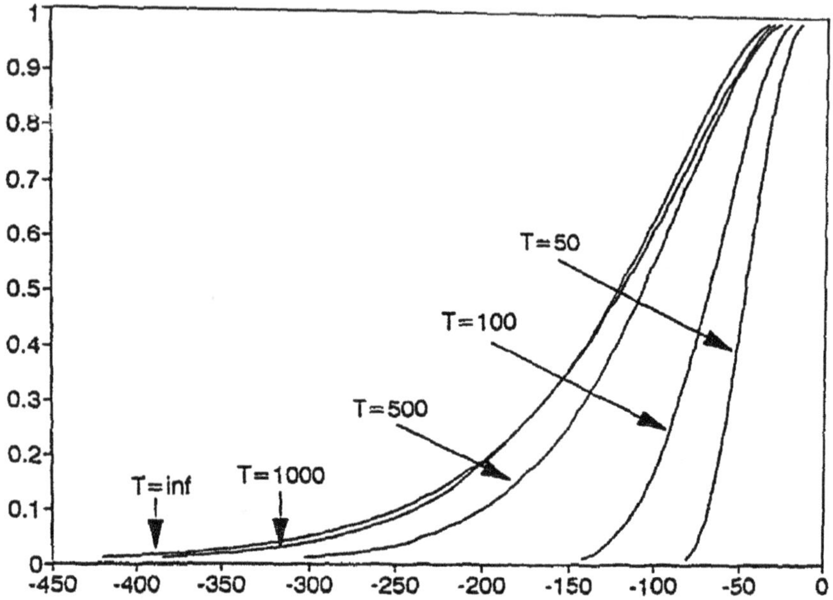

Fig. 3. The distribution of $T(\hat{\alpha}-\alpha)$; $\alpha = \exp(c/T)$, $c = -5$; AR errors, $\rho = -0.9$.

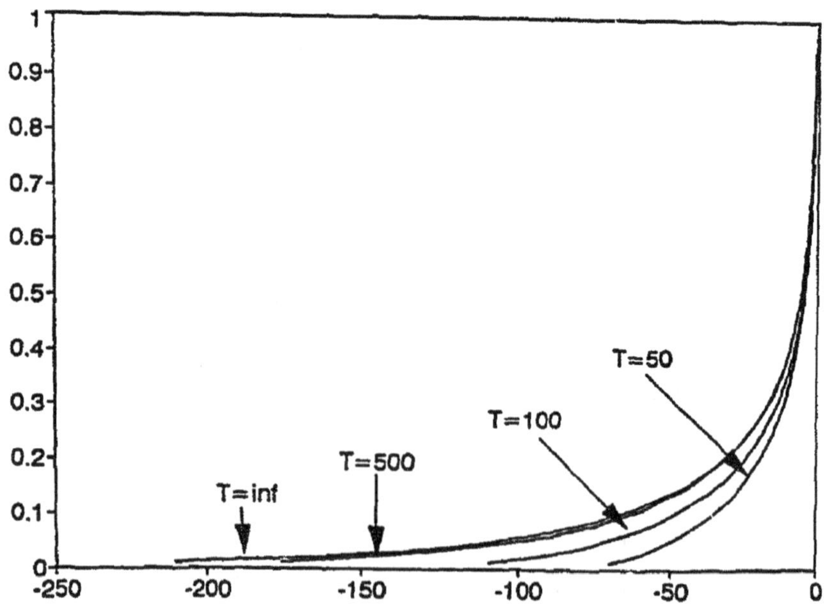

Fig. 4. The distribution of $T(\hat{\alpha} - \alpha)$; $\alpha = \exp(c/T)$, $c = 2$; AR errors, $\rho = -0.9$.

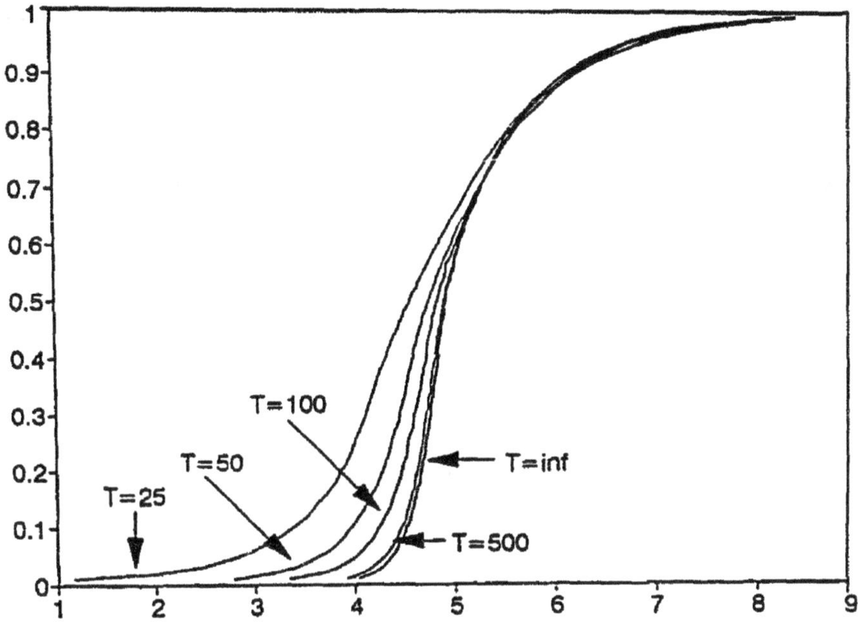

Fig. 5. The distribution of $T(\hat{\alpha} - \alpha)$; $\alpha = \exp(c/T)$, $c = -5$; AR errors, $\rho = 0.9$.

The results above suggest that the quality of the asymptotic approximation is heavily dependent on the nature and extent of the correlation in the residuals. With negative autocorrelation, the approximation becomes rapidly useless as the magnitude of this correlation increases, both in the MA(1) and AR(1) cases. In the AR(1) case with positive autocorrelation, the discrepancies are not as severe, but the approach to the asymptotic distribution remains quite slow. Only in the MA(1) case with positive autocorrelation is the approximation adequate, indeed as good as in the case without correlation.

3.3. *The moments of the distributions*

We also performed an extensive analysis of the behavior of the mean and variance of the exact and asymptotic distributions. For reason of space constraint, we report only the figures for values of c and θ or ρ presented in Tables 1 and 2. The asymptotic

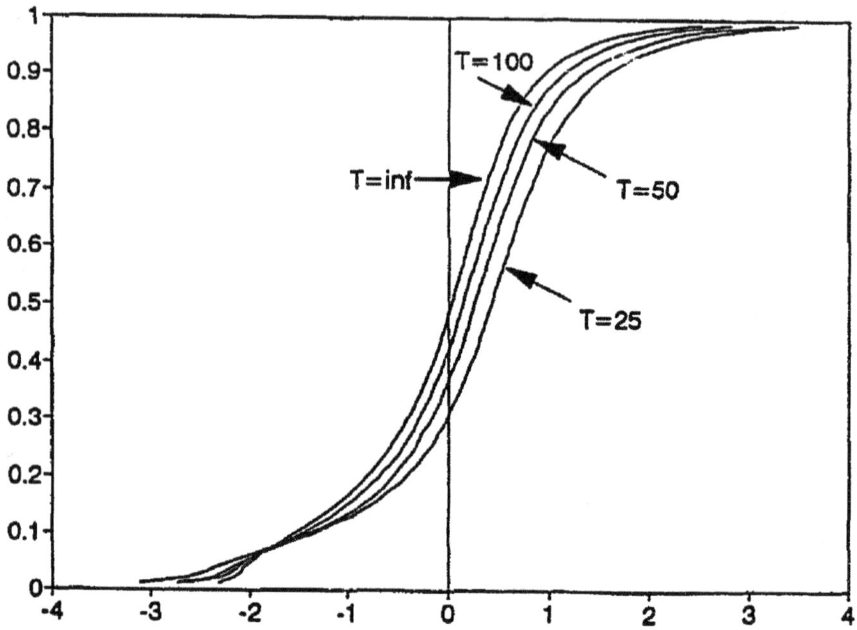

Fig. 6. The distribution of $T(\hat{\alpha}-\alpha)$; $\alpha=\exp(c/T)$, $c=2$; AR errors, $\rho=0.9$.

results were obtained using numerical integration of the function in (8).[7]

Consider first the mean of the distribution when $c = 0$. For the MA case and $\theta \geq 0$, the asymptotic value is basically equal to the exact value when $T \geq 100$. When $T < 100$, the discrepancies are minor. When $\theta < 0$, the picture is different. As θ approaches -1, it takes an increasingly larger sample size for the mean of the exact distribution to correspond to its asymptotic value. When $\theta = -0.1$, a sample of size 100 still appears enough but there are larger discrepancies with smaller sample sizes compared to the case where $\theta \geq 0$. When θ is between -0.2 and -0.5, a sample of size 500

[7]The specifications are basically the same as in the numerical integration of the distribution function, except that here the integrand does not involve complex-valued quantities, so only straightforward numerical integration routines are needed.

is needed to ensure a satisfactory approximation. When $\theta = -0.6$, a sample of size 1,000 is needed and when $\theta = -0.7$ or -0.8, the corresponding figure is $T = 5{,}000$. Finally, when $\theta = -0.9$ or -0.95 even a sample size as large as 5,000 is not sufficient to provide an adequate approximation. Of particular interest is the fact that when $\theta \leq -0.5$, the mean of the finite-sample distribution changes very rapidly as T increases. Hence, for this part of the parameter space, the asymptotic distribution provides a very bad approximation to the mean of the exact distribution when the sample size is not very large.

Consider now the AR(1) case. When $\rho \leq 0$, the picture is similar to that of the MA case, except that the discrepancies between the mean of the asymptotic and exact distributions are not as severe. When ρ is between -0.1 and -0.5, the exact mean attains its asymptotic value when T reaches somewhere between $T = 100$ and $T = 500$. When ρ is between -0.6 and -0.95, the correspondence is achieved with a sample size somewhere between $T = 500$ and $T = 1{,}000$. When $\rho > 0$, the picture is different. With ρ between 0.1 and 0.5, a sample size as small as 50 is enough to provide an adequate approximation. When $\rho = 0.7$, a sample of size 100 is needed, and with $\rho = 0.9$ or 0.95, the corresponding figure is $T = 500$. Also of interest is the fact that, when ρ is between 0.5 and 0.95, the mean is positive (unlike all the other cases considered). Note that, in all cases considered, the exact mean approaches its asymptotic counterpart in a monotonically decreasing way.[8]

With $c = -5$, a locally stationary process, the asymptotic approximation is, in general, less good than with $c = 0$ for both the MA and AR cases. That is, for a given θ or ρ and sample size T, the discrepancy between the exact mean and its asymptotic counterpart is greater. A feature that is different, however, is the fact that when θ or ρ are greater than 0.3, the approach to the asymptotic value is achieved in a monotonically increasing way. When $c = 2$, the locally explosive case, the general features are similar but the discrepancies

[8]Some slight exceptions to this rule occur for large values of T because of sampling variability induced by the simulations.

between the exact and asymptotic results are not as severe, compared to both cases where $c = 0$ or -5. As was the case with $c = 0$, the approach to the asymptotic value is monotonically increasing with T. A small difference from earlier cases is that, unless T is small and $\rho = 0.9$ or 0.95, the mean of the distribution is negative.

Tables 1 and 2 also present results for the variance of the distribution. For the case of a unit root process, $c = 0$ with MA(1) errors (Table 1), the general features are similar to the ones for the mean of the distribution. An interesting difference is that when $\theta < 0$, the approach to the asymptotic value appears to be slower. For example, when $\theta = -0.7$ and $T = 1000$, the mean of the exact distribution is quite close to the mean of the asymptotic distribution, but the variance is still quite far away. Of particular interest is the fact that, for a given θ, the exact variance approaches its asymptotic value in a monotonically increasing way. However, for a given sample size, the variance does not increase monotonically as θ approaches -1. For example, with $T = 50$ the variance of the distribution of $T(\hat{\alpha} - \alpha)$ is 66.38 ($\theta = -0.5$), 142.37 ($\theta = -0.7$), and 119.64 ($\theta = -0.9$).[9]

Much of the same features apply in the AR case (Table 2); in particular the discrepancies between the exact and asymptotic values are not as severe as for the MA(1) case. Some interesting features are, however, different from the MA(1) case. First, for a given sample size, the variance is monotonically increasing as ρ approaches -1. More importantly, when $\rho \geq 0.3$, the exact variance approaches its asymptotic value in a monotonically decreasing way unlike the case with negatively correlated MA(1) or AR(1) errors. This feature will prove of interest when considering the asymptotic expansions. The behavior of the variance when $c = -5$ or 2 is similar to that when $c = 0$. As was the case for the mean, the discrepancies are more severe with $c = -5$ and less so with $c = 2$. Apart from this, a notable difference is that in the MA(1) case with $\theta \geq 0.1$ and $c = 2$, the

[9]These features are further discussed in Nabeya and Perron (1994). In particular, the alternative asymptotic framework derived in that paper can explain them.

exact variance now decreases monotonically towards its asymptotic value.

3.4. *The $O_p(T^{-1})$ expansion*

Note first that the use of the $O_p(T^{-1})$ asymptotic expansion does not provide any improvement over the usual $O_p(1)$ asymptotic distribution when considering the mean of the distribution. Indeed both of them yield the same mean since the $O_p(T^{-1})$ expansion does not provide any location adjustment given that the extra term η has mean zero and is independent of the Wiener process $W(r)$ present in the other components.

The $O_p(T^{-1})$ asymptotic expansion provides, in contrast, an adjustment to the variance of the asymptotic distribution. However, given the independence of η and the Wiener process $W(r)$, the $O_p(T^{-1})$ expansion yields a *higher* variance than the $O_p(1)$ asymptotic distribution. As we saw, for most cases, the exact variance is *smaller* than the asymptotic variance. In these cases, the asymptotic expansion should provide *a less* accurate approximation to the exact variance than does the $O_p(1)$ asymptotic distribution. Since the asymptotic expansion does not provide an adjustment to the mean of the asymptotic distribution, it follows that, in most cases, it should provide a *poorer* approximation to the exact distribution as a whole. An instance where it could yield a better approximation to the exact variance is when the variance approaches its asymptotic value in a decreasing way. This occurs with AR(1) errors and $\rho \geq 0.3$. Even in those cases, the expansion would provide a better approximation overall only if the mean is fairly stable as T increases. In our experimental setting, this is the case only when ρ is near 0.5 and $c = 0$ or 2.

We performed many calculations related to the distribution of this asymptotic expansion. As an illustration of the qualitative results, we present in Tables 1 and 2 numerical values for $T = 50$ (other results are available upon request). Overall, the $O_p(T^{-1})$ expansion provides a worse approximation than the standard $O_p(1)$ asymptotic distribution. There are, however, several interesting features that emerge.

Consider the case of MA(1) errors. The first striking feature is that the asymptotic distribution provides an improvement in the right tail of the distribution but a worse approximation in the left tail. This asymmetric feature is interesting because it is indeed in the left tail that the asymptotic distribution is so inaccurate for negative moving average coefficients. The expansion can therefore be viewed, in a sense, as providing a useful correction where it is easy to do so, i.e., where the discrepancies between the exact and standard asymptotic distributions are least severe. Other noticeable features include the fact that the improvement is better as we move further in the right tail and worsens as the moving average coefficient varies from positive to negative.

Consider now the case of AR(1) errors. When $\rho = -0.5$, we observe the same pattern as with negative MA(1) errors; the expansion provides a worse approximation in the left tail but is surprisingly accurate in the right tail. When $\rho = -0.9$, the approximation is worse in the left tail and better for the median. However, we now observe a substantial overcorrection in the right tail such that the expansion is again a worse approximation than the standard asymptotic distribution.

As discussed above, the expansion is likely to provide a better approximation to the overall distribution in the case of AR(1) errors with positive coefficient. We now discuss this case in more detail. With $\rho = 0.5$, the asymptotic expansion provides a substantial improvement in the left tail of the distribution. For example, the first percentage point with $T = 50$ has an exact value of -4.47; the $\mathrm{O}_p(1)$ asymptotic value is -4.23 while the $\mathrm{O}_p(T^{-1})$ value is -4.53. When ρ is closer to 1, the asymptotic expansion still provides an improvement in the left tail of the distribution but not as significant. This is due to the fact that the $\mathrm{O}_p(T^{-1})$ expansion provides no adjustment to the mean of the distribution and as ρ approaches 1 the mean of the distribution decreases substantially as T increases. These observations are consistent with the fact that the $\mathrm{O}_p(T^{-1})$ expansion also provides an adequate adjustment to the variance of the distribution when $\rho = 0.5$, but that it deteriorates rapidly as ρ approaches 1. The results concerning the right tail of the distribution are quite different. Here, the $O_p(T^{-1})$ expansion provides no significant improvement over the usual $O_p(1)$ asymptotic distribution for any value of ρ considered.

The results are similar when $c = -5$, namely an improvement in the left tail of the distribution especially when $\rho = 0.5$. The approximation for the variance is better when $\rho = 0.5$, but the improvement diminishes as ρ gets closer to 1. With $c = 2$ and $\rho = 0.5$, the expansion provides a slight improvement for the 1% and 2.5% points, but a worse approximation for the 5% and 10% points. The improvements in the right tail of the distribution and in the variance are marginal. When $\rho = 0.9$ or 0.95, there is very little improvement; indeed there is little change in the $O_p(T^{-1})$ distribution as T changes.

3.5. *A possible explanation related to further research*

Some insights about the results described above can be gained by looking at the behavior of the parameter δ in the asymptotic distribution given in Theorem 1. In the MA case $\delta = \theta/(1+\theta)^2$, which becomes unbounded when θ approaches -1, and decreases to 0 as θ increases. In the AR(1) case $\delta = \rho/(1+\rho)$, which diverges to $-\infty$ as ρ approaches -1, but at a smaller rate than in the MA(1) case. When ρ approaches 1, δ approaches $\frac{1}{2}$.

These considerations lead to the following conjectures. First, the asymptotic distribution is a bad approximation when θ approaches -1 because the asymptotic distribution of $T(\hat{\alpha}-\alpha)$ with θ approaching -1 is unbounded. However, when θ approaches 1, the asymptotic distribution is still valid. In the AR(1) case, the asymptotic distribution of $T(\hat{\alpha}-\alpha)$ is again unbounded if $\rho \to -1$ as $T \to \infty$. However, the rate at which ρ may approach -1 to obtain a nondegenerate local asymptotic distribution is higher in the AR(1) case than it is in the MA(1) case. This would explain the relatively smaller discrepancies in the AR(1) case for a given equal value of θ and ρ. On the other hand, when ρ approaches $+1$ as $T \to \infty$ $T(\hat{\alpha}-\alpha)$ still has a nondegenerate asymptotic distribution but different from that given by (3). These conjectures are verified in a companion paper, Nabeya and Perron (1994), where the relevant local asymptotic distributions are derived and tabulated. It is shown that these local asymptotic distributions provide very good approximations to the finite-sample distributions. Their characterization also helps in further understanding the reasons for the failures of the usual asymptotic distribution.

4. Extensions to Cases With Fitted Intercept and Trend

In this section we consider extensions to cases where an intercept or an intercept and a trend are included in the first-order autoregression. Given the nature of the results discussed for the leading case in the previous sections, we concentrate on the quality of the standard $O_p(1)$ approximation (the asymptotic expansions and the moments of the distribution are not covered). To be more specific consider the following regressions:

$$y_t = \mu + \alpha y_{t-1} + u_t, \tag{10}$$

$$y_t = \mu + \beta t + \alpha y_{t-1} + u_t. \tag{11}$$

The least squares estimator of α in (10) is denoted $\hat{\alpha}_2$ and the one from (11) by $\hat{\alpha}_3$. We study the behavior of the statistics $T(\hat{\alpha}_i - \alpha)$, $i = 2, 3$, when the series $\{y_t\}$ is generated by (1) with $\alpha = \exp(c/T)$ and $y_0 = 0$. The cumulative distribution functions of the limiting distributions of $T(\hat{\alpha}_i - \alpha)$ can be obtained from results in Nabeya and Tanaka (1990) (see also Nabeya and Sørensen, 1993) and are stated in the following lemma.

Lemma 1. *Let* $F_i(z) = \lim_{T\to\infty} \mathrm{P}[T(\hat{\alpha}_i - \alpha) \leq z]$, $i = 2, 3$, *where* $\hat{\alpha}_2$ *and* $\hat{\alpha}_3$ *are the OLS estimates of* α *in* (10) *and* (11), *respectively. Assume that* $\{y_t\}_0^T$ *is generated by* (1) *with* $\alpha = \exp(c/T)$ *and let* δ *be defined as in Theorem* 1. *We have*

$$F_i(z) = \frac{1}{2} + (1/\pi) \int_0^\infty (1/v) IM(cf_i(v, z))dv, \quad i = 2, 3,$$

where

$$cf_i(v, z) = \exp(i(1 - 2\delta)v - c/2)[D_i(2iv, z)]^{-1/2},$$

with

$$\begin{aligned}
D_2(\lambda, z) &= c^2 \cosh(\mu)/\mu^2 + ((1 - \cosh(\mu))/\mu^4) \\
&\quad \times (2c^2\lambda - 4c\lambda(z + c) - 2\lambda^2) \\
&\quad + (\sinh(\mu)/\mu^3)(-c^3 + c^2\lambda - 2\lambda(z + c) - \lambda^2), \\
D_3(\lambda, z) &= (\cosh(\mu)/\mu^6)\{c^6 - 12c^4\lambda + 2\lambda(z + c)(-c^4 + 12c^3 - 12c^2) \\
&\quad + 48\lambda^2(c^2 - 3c + 3)\} + (1 - \cosh(\mu)/\mu^8)\{4\lambda(-c^6 + 6c^4) \\
&\quad + 16\lambda^2(c^4 - 3c^3 - 3c^2 + 18c - 18) \\
&\quad + 8\lambda(z + c)(c^5 - 3c^4 + c^4\lambda - 4\lambda^2(c^2 - 3c + 3)) \\
&\quad + 16\lambda^2(z + c)^2(-c^3 + 3c^2 - 6c + 6)\} \\
&\quad + (\sinh(\mu)/\mu^7)\{-c^7 + \lambda(c^6 + 24c^4) \\
&\quad + 4\lambda^2(-c^4 + 3c^3 - 27c^2 + 72c - 72) \\
&\quad + 2\lambda(z + c)(c^5 - 4c^4 - 12c^3 + 12c^2 \\
&\quad - c^4\lambda + 4\lambda^2(c^2 - 3c + 3)) + 16\lambda^2(z + c)^2(c^2 - 3c + 3)\},
\end{aligned}$$

where

$$\mu = (c^2 - 2\lambda(z + c))^{1/2}.$$

Lemma 1 allows us to obtain percentage points of the limiting distributions using numerical integrations.[10] Results concerning the distribution of $T(\hat{\alpha}_2 - \alpha)$ where an intercept is included are presented for the leading case $c = 0$ in Table 3 for MA(1) errors and in Table 4 for AR(1) errors. We consider only a subset of the cases discussed

[10]These were performed in double precision FORTRAN using the subroutine DCADRE of the IMSL library. Again, the bounds are $(\varepsilon, \bar{V})$, where $\bar{V}$ is such that the square of the integrand evaluated at $\bar{V}$ is less than ε. The error for the numerical integration was also set at ε. For most experiments we used $\varepsilon =$ 1.0E−07.

Table 3. The distribution of $T(\hat{\alpha} - \alpha)$; demeaned case; $\alpha = 1$; MA errors, $u_t = e_t + \theta e_{t-1}$.

	1.0%	5.0%	10.0%	50.0%	90.0%	95.0%	99.0%
			$\theta = -0.90$				
$T = 50$	−65.20	−59.72	−57.18	−47.45	−37.11	−33.90	−28.27
$T = 100$	−116.51	−108.90	−104.50	−88.19	−68.54	−62.53	−50.62
$T = 500$	−457.57	−425.06	−405.42	−306.70	−171.65	−135.29	−84.14
$T = 1{,}000$	−820.46	−741.38	−690.78	−444.78	−204.56	−160.61	−108.17
$T = 5{,}000$	−2279.55	−1784.56	−1498.46	−673.53	−288.54	−236.37	−159.01
$T = \infty$	−3978.96	−2506.52	−1978.06	−761.62	−260.23	−195.62	−119.40
			$\theta = -0.70$				
$T = 50$	−55.08	−48.22	−44.25	−29.68	−15.22	12.26	−7.65
$T = 100$	−89.32	−76.82	−69.41	−41.74	−18.74	−14.60	−9.49
$T = 500$	−205.89	−159.47	−135.26	−63.86	−22.83	−16.59	−9.08
$T = 1{,}000$	−280.21	−203.31	−168.13	−68.48	−22.86	−17.33	−11.11
$T = \infty$	−337.30	−226.71	−180.03	−69.65	−23.79	−17.87	−10.99
			$\theta = -0.50$				
$T = 50$	−42.04	−34.49	−29.95	−15.39	−5.98	−4.41	−2.28
$T = 100$	−58.50	−45.64	−38.48	−17.89	−6.48	−4.76	−2.73
$T = 500$	−85.99	−61.82	−50.14	−21.19	−7.04	−4.85	−2.33
$T = 1{,}000$	−104.93	−71.41	−57.51	−21.70	−6.88	−5.14	−2.95
$T = \infty$	−101.45	−68.44	−54.41	−21.12	−7.06	−5.20	−2.92

earlier, namely $\theta = -0.9, -0.7, -0.5$ in the MA(1) case, and $\rho = -0.9, -0.5, 0.9$ in the AR(1) case.

Consider first results pertaining to the MA(1) case in Table 3. The qualitative picture is similar to the case where no intercept is included in the regression. The quality of the asymptotic approximation deteriorates as the moving average parameter θ approaches -1 and becomes inadequate even for very large sample sizes when θ is -0.9. An interesting feature to note is that for small values of T (e.g., $T = 50, 100$) the differences in the exact critical values are small with and without an intercept included. A shift of the distribution to the left occurs as T increases, and this shift is substantial for the asymptotic distribution. The fact that the leftward movements of the distributions are greater than in the no-intercept case as T increases implies larger discrepancies between the exact and asymptotic distributions. For example, when $T = 100$ and $\theta = -0.9$,

Table 4. The distribution of $T(\hat{\alpha} - \alpha)$; demeaned case; $\alpha = 1$; AR errors, $u_t = \rho u_{t-1} + e_t$.

	1.0%	5.0%	10.0%	50.0%	90.0%	95.0%	99.0%
			$\rho = -0.90$				
$T = 50$	−83.87	−73.53	−67.38	−38.29	−14.82	−11.11	−5.89
$T = 100$	−140.38	−117.86	−103.19	−52.11	−19.45	−14.60	−8.51
$T = 500$	−279.43	−209.68	−172.48	−74.80	−25.20	−18.29	−9.90
$T = 1{,}000$	−357.98	−258.37	−204.48	−80.05	−25.75	−19.40	−12.86
$T = \infty$	−385.74	−260.19	−206.58	−79.93	−27.32	−20.51	−12.54
			$\rho = -0.50$				
$T = 50$	−39.27	−30.13	−25.00	−10.94	−3.70	−2.54	−0.94
$T = 100$	−48.86	−35.20	−28.64	−11.93	−3.92	−2.64	−1.22
$T = 500$	−56.08	−39.64	−31.68	−13.05	−4.07	−2.72	−0.97
$T = 1{,}000$	−66.18	−44.94	−35.63	−13.20	−4.10	−2.83	−1.37
$T = \infty$	−61.08	−41.12	−32.70	−12.68	−4.10	−2.89	−1.30
			$\rho = -0.90$				
$T = 25$	−4.50	−2.94	−2.26	0.01	2.43	3.18	4.76
$T = 50$	−4.03	−2.71	−2.16	−0.18	2.24	3.10	4.64
$T = 100$	−4.18	−2.85	−2.18	−0.26	2.02	2.78	4.53
$T = 500$	−4.52	−3.00	−2.26	−0.32	1.83	2.52	4.18
$T = 1{,}000$	−4.77	−3.10	−2.31	−0.31	1.71	2.46	4.09
$T = \infty$	−4.78	−3.03	−2.25	−0.29	1.71	2.44	4.00

the 1% point of the exact distribution corresponds roughly to the 99% point of the asymptotic distribution; when $T = 100$ and $\theta = -0.7$, the 5% point of the exact distribution corresponds roughly to the 50% point of the asymptotic distribution. When $\theta = -0.5$, the leftward shifts in the distribution are relatively important for both the finite-sample and asymptotic distributions. Nevertheless, the adequacy of the asymptotic approximation is worse compared to the case where no intercept is included.

The same qualitative results hold when $c = -5$ or 2. Figures 7 and 8 present results for the case $\theta = -0.9$. Compared to the case where $c = 0$, the main difference is again that the approximation is marginally better with an explosive process ($c = 2$) and marginally worse when it is stationary ($c = -5$). Compared to the case with no intercept (see Figs. 1 and 2), it is evident that for each T the

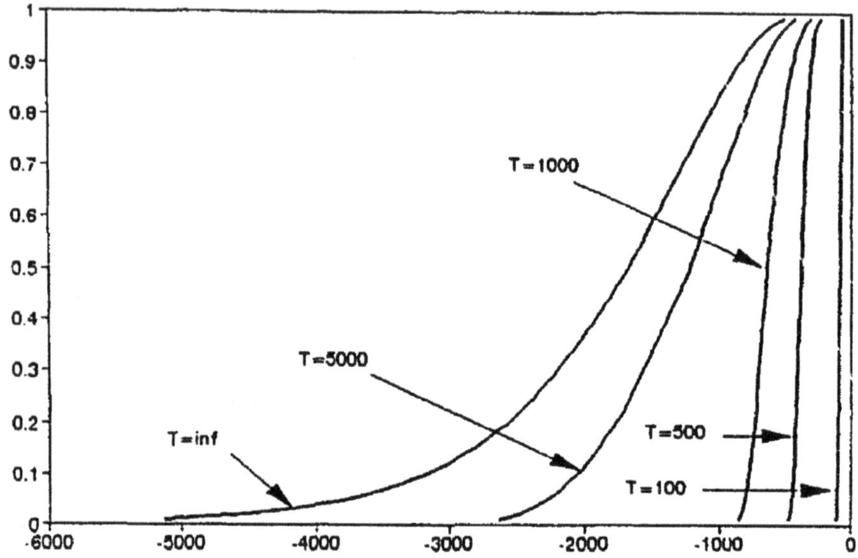

Fig. 7. The distribution of $T(\hat{\alpha}-\alpha)$; $\alpha = \exp(c/T)$, $c = -5$; MA errors, $\theta = -0.9$; demeaned case.

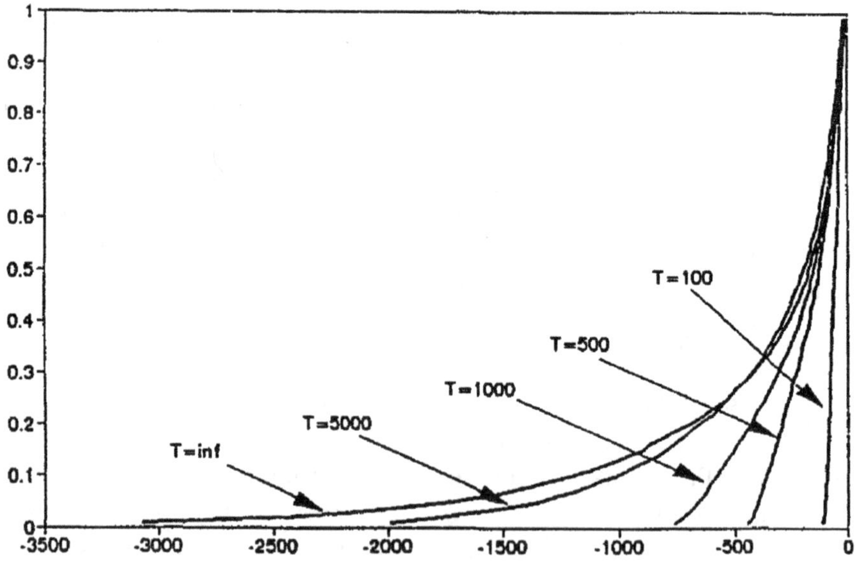

Fig. 8. The distribution of $T(\hat{\alpha}-\alpha)$; $\alpha = \exp(c/T)$, $c = 2$; MA errors, $\theta = -0.9$; demeaned case.

adequacy of the asymptotic approximation is worse when an intercept is included for both the explosive and stationary cases.

Consider now results pertaining to the case where the errors are AR(1) presented in Table 4. For negative values of ρ, the qualitative results are similar to those in the MA(1) case. The adequacy of the asymptotic approximation is worse when an intercept is included. For example, with $\rho = -0.9$ and $T = 100$, the 1% point of the exact distribution corresponds roughly to the 10% point of the asymptotic distribution when no intercept is included, whereas if one is included the 1% point of the exact distribution corresponds roughly to the 30% point of the asymptotic distribution. For $\rho = -0.5$, the asymptotic distribution could be considered an adequate approximation to the exact distribution for T greater than 100 in the no-intercept case. Now it is adequate only for T greater than 500. When $\rho = 0.9$, one first notes the leftward shift of both the exact and asymptotic distributions compared to the case where no intercept is included. Again the asymptotic distribution is a good approximation to the exact distribution when $T = 500$. It is, however, less accurate than in the no-intercept case for smaller of T (e.g., $T = 50$).

Figures 9 and 10 present results for the case $c = -5$ with $\theta = -0.9$ and 0.9. Compared to the case where $c = 0$, the main difference is again that the approximation is marginally worse. Compared to the case with no intercept (see Figs. 3 and 5), it is evident that the adequacy of the asymptotic approximation is worse when an intercept is included.

The results concerning the distribution of $T(\hat{\alpha}_3 - \alpha)$ for the case where an intercept and a trend are included are first presented for the leading case $c = 0$ in Table 5 for MA(1) errors and in Table 6 for AR(1) errors. We consider the same subset of specifications as in the intercept-only case.

The qualitative picture is again similar except that now the discrepancies between the exact and asymptotic distributions are even bigger for any sample size. Consider the following examples. First, when $\theta = -0.9$ and $T = 1{,}000$, the 1% point of the exact distribution corresponds roughly to the 90% point of the asymptotic distribution. Second, even when T is as large as 5,000,

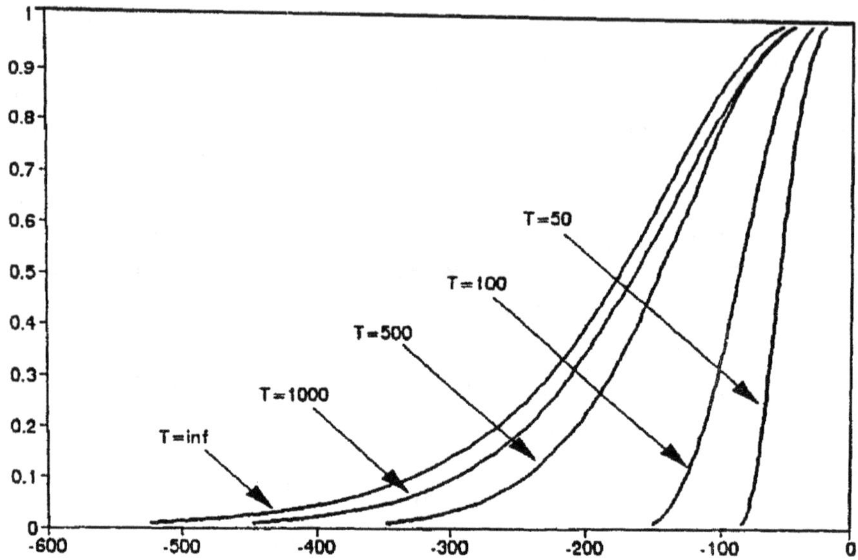

Fig. 9. The distribution of $T(\hat{\alpha} - \alpha)$; $\alpha = \exp(c/T)$, $c = -5$; AR errors, $\rho = -0.9$; demeaned case.

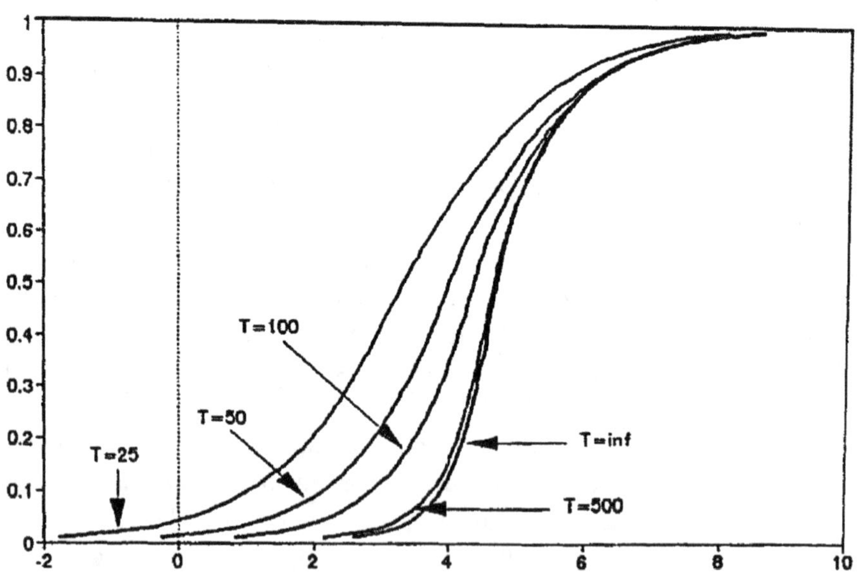

Fig. 10. The distribution of $T(\hat{\alpha} - \alpha)$; $\alpha = \exp(c/T)$, $c = -5$; AR errors, $\rho = 0.9$; demeaned case.

Table 5. The distribution of $T(\hat{\alpha} - \alpha)$; detrended case; $\alpha = 1$; MA errors, $u_t = e_t + \theta e_{t-1}$.

	1.0%	5.0%	10.0%	50.0%	90.0%	95.0%	99.0%
			$\theta = -0.90$				
$T = 50$	−69.99	−62.31	−59.65	−50.55	−41.35	−38.81	−33.44
$T = 100$	−120.76	−113.66	−110.15	−96.14	−81.92	−77.88	−70.61
$T = 500$	−481.24	−457.07	−442.54	−382.56	−298.77	−271.18	−218.21
$T = 1{,}000$	−872.31	−812.90	−780.05	−631.57	−417.01	−355.29	−263.09
$T = 5{,}000$	−2676.75	−2324.64	−2076.49	−1196.41	−601.78	−493.03	−348.43
$T = \infty$	−5377.20	−3883.05	−3252.54	−1629.49	−752.34	−600.56	−401.37
			$\theta = -0.70$				
$T = 50$	−60.49	−54.52	−51.25	−39.76	−27.64	−24.12	−18.24
$T = 100$	−99.63	−89.58	−83.76	−62.40	−40.42	−34.77	−26.12
$T = 500$	−252.88	−210.87	−188.78	−115.11	−60.87	−49.59	−33.62
$T = 1{,}000$	−347.75	−271.94	−238.93	−135.45	−61.02	−48.07	−31.97
$T = \infty$	−492.19	−355.00	−296.95	−149.05	−69.37	−55.83	−35.77
			$\theta = -0.50$				
$T = 50$	−49.92	−42.80	−39.34	−26.10	−14.52	−12.10	−7.85
$T = 100$	−72.44	−60.40	−53.90	−33.21	−17.54	−14.12	−9.84
$T = 500$	−114.36	−89.39	−76.93	−41.49	−20.00	−16.04	−10.41
$T = 1{,}000$	−137.48	−100.82	−86.42	−45.22	−19.17	−15.00	−9.63
$T = \infty$	−145.05	−106.77	−89.74	−44.99	−20.88	−16.74	−10.83

the differences are substantial, viz. the 1% point of the exact distribution corresponds roughly to the 20% point of the asymptotic distribution. The inaccuracies are not as severe when $\theta = -0.7$ or -0.5 but are nonetheless more important than in the case where no intercept or only an intercept is included in the regression.

These dramatic differences between the exact and asymptotic distributions are well captured in Figs. 11 and 12 which present results for the case $\theta = -0.9$ with $c = -5$ and 2, respectively. Comparing with Figs. 1 and 2 (no-intercept case) and Figs. 7 and 8 (intercept included), one clearly sees a gradual deterioration in the quality of the approximation as the number of included deterministic regressors increase.

Consider now results pertaining to the case of AR(1) errors presented in Table 6 for the leading case $c = 0$. For negative values of ρ, the qualitative results are similar to the those in the MA(1) case. The adequacy of the asymptotic approximation is worse when an intercept and a trend are included. For example, when $\rho = -0.9$

Table 6. The distribution of $T(\hat{\alpha} - \alpha)$; detrended case; $\alpha = 1$; AR errors, $u_t = \rho u_{t-1} + e_t$.

	1.0%	5.0%	10.0%	50.0%	90.0%	95.0%	99.0%
			$\rho = -0.90$				
$T = 50$	−89.14	−83.14	−78.70	−57.28	−32.35	−26.03	−17.65
$T = 100$	−156.14	−139.86	−130.13	−86.66	−46.90	−37.68	−24.95
$T = 500$	−362.53	−289.82	−251.28	−141.50	−68.93	−55.71	−36.29
$T = 1{,}000$	−461.75	−353.80	−303.10	−161.82	−70.29	−55.09	−35.96
$T = \infty$	−551.84	−406.08	−340.61	−171.05	−79.57	−64.12	−43.16
			$\rho = -0.50$				
$T = 50$	−50.04	−40.90	−36.30	−21.18	−10.28	−8.22	−4.98
$T = 100$	−64.15	−50.79	−43.89	−23.98	−11.47	−9.08	−5.85
$T = 500$	−77.87	−59.27	−50.21	−25.95	−12.12	−9.55	−5.88
$T = 1{,}000$	−89.14	−63.74	−54.42	−27.85	−11.51	−8.90	−5.46
$T = \infty$	−87.13	−64.20	−53.92	−27.05	−12.42	−9.88	−6.36
			$\rho = 0.90$				
$T = 25$	−8.92	−6.03	−4.94	−1.54	2.24	3.23	5.06
$T = 50$	−6.60	−4.50	−3.64	−0.95	2.38	3.48	5.39
$T = 100$	−5.72	−3.85	−3.01	−0.77	2.21	3.12	5.18
$T = 500$	−5.22	−3.50	−2.63	−0.56	1.79	2.63	4.61
$T = 1{,}000$	−5.43	−3.55	−2.69	−0.58	1.72	2.60	4.56
$T = \infty$	−5.54	−3.50	−2.62	−0.54	1.56	2.48	4.41

and $T = 50$, the 1% point of the exact distribution corresponds roughly to the 50% point of the asymptotic distribution in the no-trend case, whereas if a trend is included, it corresponds roughly to the 90% point. These discrepancies are well illustrated for the case $c = -5$ in Fig. 13 (compare to Figs. 3 and 9) which shows again the deterioration in the quality of the approximation as the number of deterministic components included in the autoregression increases. For positive values of the autoregressive coefficient (with results presented for $\rho = 0.9$), we observe, as in earlier cases, that the asymptotic distribution is a good approximation to the exact distribution for $T = 500$. However, we see that for smaller sample sizes, the quality of the approximation deteriorates as the number of included deterministic components increases, more so as T gets smaller. This feature is well illustrated in Fig. 14 which considers the case $c = -5$ and $\rho = 0.9$ to be compared to Figs. 5 and 10.

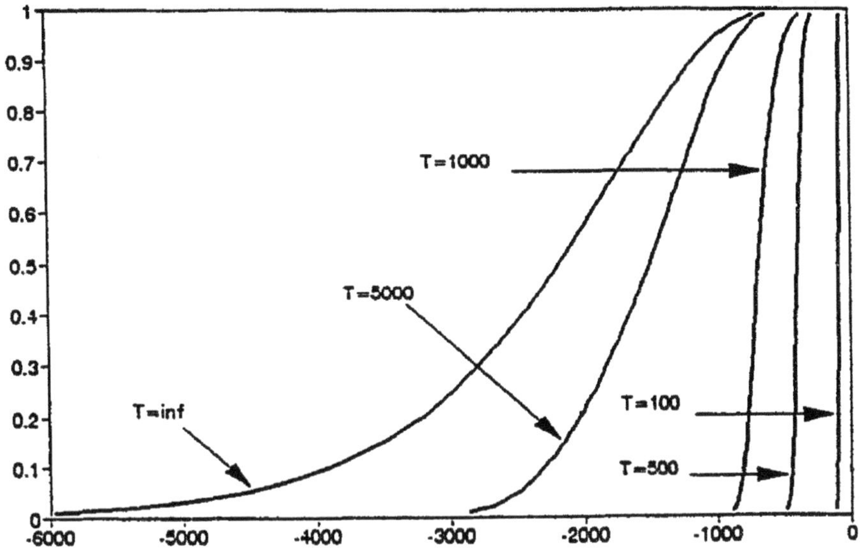

Fig. 11. The distribution of $T(\hat{\alpha} - \alpha)$; $\alpha = \exp(c/T)$, $c = -5$; MA errors, $\theta = -0.9$; detrended case.

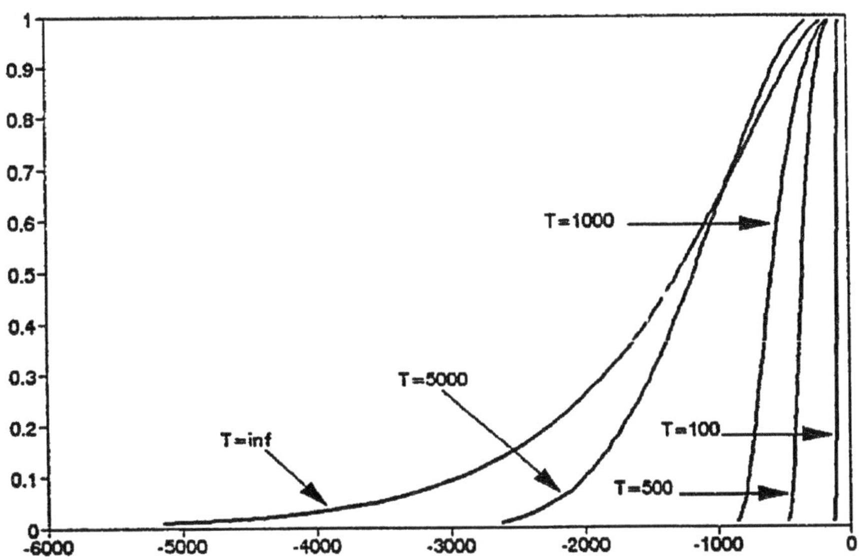

Fig. 12. The distribution of $T(\hat{\alpha} - \alpha)$; $\alpha = \exp(c/T)$, $c = 2$; MA errors, $\theta = -0.9$; detrended case.

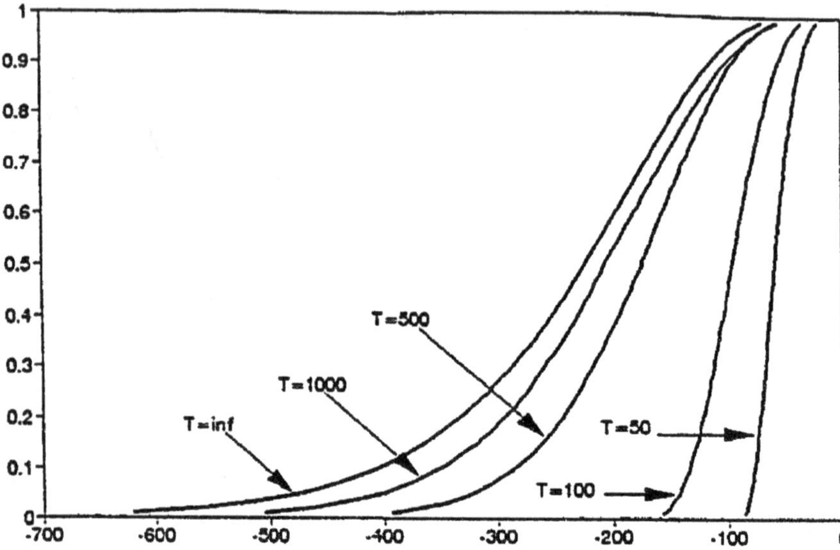

Fig. 13. The distribution of $T(\hat{\alpha} - \alpha)$; $\alpha = \exp(c/T)$, $c = -5$; AR errors, $\rho = -0.9$; detrended case.

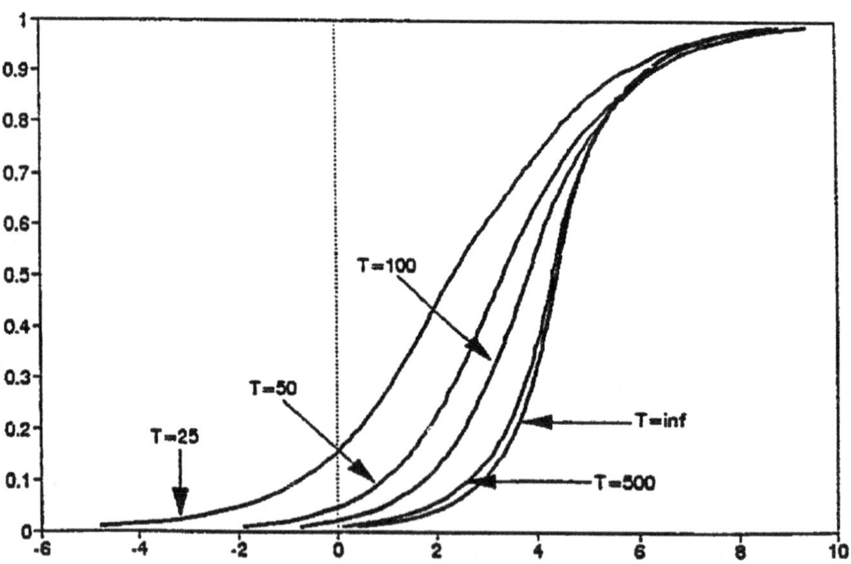

Fig. 14. The distribution of $T(\hat{\alpha} - \alpha)$; $\alpha = \exp(c/T)$, $c = -5$; AR errors, $\rho = 0.9$; detrended case.

5. Concluding Comments

We characterized and tabulated the asymptotic distribution of the normalized least squares estimator in a nearly integrated autoregressive process allowing dependence in the errors with emphasis on MA(1) and AR(1) processes. These were sufficient to provide a rich array of cases. Special attention was given to the adequacy of the standard asymptotic distribution as an approximation in finite samples. We showed that, in a substantial part of the parameter space, the approximation is seriously inadequate. An implication of our results is the inherent danger associated with an asymptotic framework that allows very general conditions with respect to the type of dependence permitted. There appears to be a tradeoff between generality in the conditions and the sample size needed for the asymptotic distribution to provide a reasonable guide to the finite-sample distribution. While this conclusion is drawn from a simple model, it should extend to more general models.

Our analysis can also be viewed as a step in a more complete analysis of the adequacy of asymptotic distributions of tests allowing general dependence and heterogeneity. Examples include the class of unit root tests proposed by Phillips (1987a) and Phillips and Perron (1988) as well as those involving the Newey–West (1987) correction in more general structural models. In practice, the statistics are used with correction factors that eliminate asymptotically the dependence of the asymptotic distributions upon nuisance parameters. Our analysis has direct implications for the behavior of such transformed statistics. Suppose that the correction factor adequately approximates the asymptotic correction necessary to eliminate the dependence upon the nuisance parameters. Our results would still imply an inadequate corrected statistic as the asymptotic distribution of the uncorrected part is far from the finite-sample distribution in an important range of the parameter space. In a sense, the finite-sample distribution would bear an adequate correspondence to the asymptotic distribution by fortuitous cancellation of approximation errors for the distribution of the original statistic and the correction factor. Such a situation is unlikely to occur as demonstrated in the case of unit root tests in the simulations of Schwert (1989) and Dejong *et al.* (1992) and the analytical results of Pantula (1991). Their results about size distortions in finite samples can be explained by the poor approximation provided by the asymptotic distribution of $T(\hat{\alpha}-1)$ which forms the basis of the transformed statistics. These issues are analyzed analytically for the Phillips–Perron (1988) test and related ones in Perron and Ng (1993) using the local asymptotic framework laid out in Nabeya and Perron (1994).

Mathematical Appendix

Proof of Theorem 1. The proof relies on Theorem 3.1 of Phillips (1987c), which shows that under the conditions of Assumptions 1 and 2 we have

$$X_T(r) \stackrel{d}{=} W(r) + O_p(T^{-1}), \tag{A.1}$$

where $\stackrel{d}{=}$ signifies equality in distribution. Using (A.1) we can prove the following lemma related to the sample moments of $\{y_t\}$. □

Lemma A.1. *Let $\{y_t\}$ be generated by (1.1) with $\alpha = \exp(c/T)$ and let the innovation sequence $\{u_t\}$ satisfy the conditions of Assumptions 1 and 2. Then:*

(a) $T^{-1/2}y_{[Tr]} \stackrel{d}{=} \sigma J_c(r) + T^{-1/2}\exp(cr)y_0 + O_p(T^{-1})$,
(b) $T^{-3/2}\sum_{t=1}^{T} y_t \stackrel{d}{=} \sigma \int_0^1 J_c(r)dr + T^{-1/2}y_0(\exp(c)-1)/c + O_p(T^{-1})$,
(c) $T^{-2}\sum_{t=1}^{T} y_t^2 \stackrel{d}{=} \sigma^2 \int_0^1 J_c(r)^2 dr + 2T^{-1/2}\sigma y_0 \int_0^1 \exp(cr)J_c(r)dr + O_p(T^{-1})$,
(d) $T^{-1}\sum_{t=1}^{T} y_{t-1}u_t \stackrel{d}{=} \sigma^2 \int_0^1 J_c(r)dW(r) + (\sigma^2 - \sigma_u^2) + T^{-1/2}\sigma y_0 \int_0^1 \exp(cr)dW(r) - T^{-1/2}(v/2\sigma^2)\eta + O_p(T^{-1})$.

Proof. The proof follows closely that of Phillips (1987c, Lemma 4.2). Using (1),

$$\begin{aligned}
T^{-1/2}y_{[Tr]} &= T^{-1/2}\sum_{j=1}^{[Tr]} \exp(([Tr]-j)c)/T)u_j \\
&\quad + T^{-1/2}\exp([Tr]c/T)y_0 \\
&= \sigma\left\{X_T(r) + c\int_0^r \exp((r-s)c)X_T(s)ds\right\} \\
&\quad + T^{-1/2}\exp([Tr]c/T)y_0.
\end{aligned}$$

Using (A.1), we deduce that

$$\begin{aligned}
T^{-1/2}y_{[Tr]} &\stackrel{d}{=} \sigma\left\{W(r) + c\int_0^r \exp((r-s)c)W(s)ds\right\} \\
&\quad + T^{-1/2}\exp(cr)y_0 + O_p(T^{-1}) \\
&= \sigma J_c(r) + T^{-1/2}\exp(cr)y_0 + O_p(T^{-1}),
\end{aligned}$$

using the fact that $J_c(r) \equiv W(r) + c\int_0^r \exp((r-s)c)W(s)ds$. The proof of parts (b) and (c) are analogous and omitted. To prove part (d) note that squaring (1), summing over t, and rearranging, we obtain

$$\begin{aligned}
T^{-1}\sum_{t=1}^{T} y_{t-1}u_t &= \frac{1}{2}\exp(-2c/T)\left\{T^{-1}y_T^2 - T^{-1}y_0^2\right. \\
&\quad \left. - T(\exp(2c/T)-1)T^{-2}\sum_{t=1}^{T} y_{t-1}^2 - T^{-1}\sum_{t=1}^{T} u_t^2\right\}.
\end{aligned}$$

Note that

$$\begin{aligned}
T^{-1}\sum_{t=1}^{T} u_t^2 &= T^{-1/2}\left[T^{-1/2}\sum_{t=1}^{T}(u_t^2 - \sigma_u^2)\right] + \sigma_u^2 \\
&\stackrel{d}{=} T^{-1/2}\zeta + \sigma_u^2 + O_p(T^{-1}),
\end{aligned}$$

where $\zeta \sim N(0, \upsilon^2)$ (see Phillips, 1987c, Lemma 4.2). Hence, using parts (a), (b), and (c),

$$\begin{aligned}
T^{-1}\sum_{t=1}^{T} y_{t-1}u_t
&\overset{d}{=} \frac{1}{2}\Bigg\{[\sigma J_c(1) + T^{-1/2}\exp(c)y_0]^2 \\
&\quad - 2c\left[\sigma^2 \int_0^1 J_c(r)^2 dr + 2T^{-1/2}\sigma y_0 \int_0^1 \exp(cr)J_c(r)dr\right] \\
&\quad - \sigma_u^2 - T^{-1/2}\zeta\Bigg\} + O_p(T^{-1}) \\
&\overset{d}{=} (\sigma^2/2)\left[J_c(1)^2 - 2c\int_0^1 J_c(r)^2 dr - 1\right] + (\sigma^2 - \sigma_u^2)/2 \\
&\quad + T^{-1/2}\left[\sigma\exp(c)y_0 J_c(1) - 2\sigma c y_0 \int_0^1 \exp(cr)J_c(r)dr - \zeta/2\right] \\
&\quad + O_p(T^{-1}).
\end{aligned}$$

The result follows by noting that $\frac{1}{2}[J_c(1)^2 - 2c\int_0^1 J_c(r)^2 dr - 1] \overset{d}{=} \int_0^1 J_c(r)dW(r)$ and that $\exp(c)J_c(1) - 2c\int_0^1 \exp(cr)J_c(r)dr \overset{d}{=} \int_0^1 \exp(cr)dW(r)$ (see Perron, 1991a) and by using the fact that $\upsilon\eta \overset{d}{=} \zeta$. Theorem 1 follows using (2) and (c–d) of Lemma A.1. □

Proof of Theorem 2. We can write $H(c,\gamma) = A(c,\gamma) + \delta - g\eta$ and $K(c,\gamma) = B(c,\gamma) - d$, where

$$A(c,\gamma) = \int_0^1 J_c(r)dW(r) + \gamma\int_0^1 \exp(cr)dW(r)$$

and

$$B(c,\gamma) = \int_0^1 J_c(r)^2 dr + 2\gamma\int_0^1 \exp(cr)J_c(r)dr + \gamma^2(\exp(2c) - 1)/2c.$$

We have

$$\begin{aligned}\mathrm{MGF}(v,u) &= \mathrm{E}[\exp\{v(A(c,\gamma)+\delta-g\eta)+u(B(c,\gamma)-d)\}]\\ &= \exp(v\delta-ud)\ \mathrm{E}[\exp(-vg\eta)]\\ &\quad\times \mathrm{E}[\exp\{vA(c,\gamma)+uB(c,\gamma)\}],\end{aligned}$$

since η is independent of the Wiener process $W(r)$, δ and d are fixed constants. With η a N(0, 1) random variable, $\mathrm{E}[\exp(-vg\eta)] = \exp(v^2g^2/2)$ and the result follows using Theorem 2 of Perron (1991a) who showed that $\mathrm{E}[\exp\{vA(c,\gamma)+uB(c,\gamma)\}] = M_{c,\gamma}(v,u)$. □

Acknowledgments

Author wish to thank him for his help during this initial stage. I also wish to thank Seiji Nabeya for useful comments. This research was supported by grants from the Social Sciences and Humanities Research Council of Canada, the Natural Sciences and Engineering Research Council of Canada and the Fonds pour la Formation de Chercheurs et l'Aide à la Recherche de Québec.

References

Andrews, D.W. and Fair, R. (1988). Inference in nonlinear econometric model with structural change. *Review of Economic Studies* 55, 615–640.

Bobkoski, M.J. (1983). Hypothesis testing in nonstationary time series, Unpublished Ph.D. dissertation (University of Wisconsin, Department of Statistics, Madison, WI).

Cavanagh, C. (1986). Roots local to unity, Manuscript, Harvard University, Department of Economics, Cambridge, MA.

Chan, N.H. (1988). The parameter inference for nearly nonstationary time series. *Journal of the American Statistical Association* 83, 857–862.

Chan, N.H. and Wei, C.Z. (1987). Asymptotic inference for nearly nonstationary AR(1) processes. *Annals of Statistics* 15, 1050–1063.

DeJong, D.N. Nankervis, J.C. Savin, N.E. and Whiteman, C.H. (1992). The power problems of unit root tests in time series with autoregressive errors. *Journal of Econometrics* 53, 323–343.

Dickey, D.A. (1976). Estimation and hypothesis testing for nonstationary time series, Unpublished Ph.D. dissertation, Iowa State University, Department of Statistics, Ames, IA.

Dickey, D.A. and Fuller, W.A. (1979). Distribution of the estimators for autoregressive time series with a unit root. *Journal of the American Statistical Association* 74, 427–431.

Evans, G.B.A. and Savin, N.E. (1981a). The calculation of the limiting distribution of the least squares estimator of the parameter in a random walk model. *Annals of Statistics* 9, 1114–1118.

Evans, G.B.A. and Savin, N.E. (1981b). Testing for unit roots: 1. *Econometrica* 49, 753–779.

Fuller, W.A. (1976). *Introduction to Statistical Time Series*, Wiley, New York, NY.

Gallant, A.R. and White, H. (1988). *A Unified Theory of Estimation and Inference for Nonlinear Dynamic Models*, Basil Blackwell, Oxford.

Gurland, J. (1948). Inversion formulae for the distribution of ratios. *Annals of Mathematical Statistics* 19, 228–237.

Herrndorf, N. (1984). A functional central limit theorem for weakly dependent sequences of random variables. *Annals of Probability* 12, 141–153.

Imhof, P.J. (1961). Computing the distribution of quadratic forms in normal variables. *Biometrika* 48, 419–426.

Mann, H.B. and Wald, A. (1943). On the statistical treatment of linear stochastic difference equations. *Econometrica* 11, 173–220.

Mehta, J.S. and Swamy, P.A.V.B. (1978). The existence of moments of some simple Bayes estimators of coefficients in a simultaneous equation model. *Journal of Econometrics* 7, 1–14.

Nabeya, S. and Perron, P. (1994). Local asymptotic distribution related to the AR(1) model with dependent errors. *Journal of Econometrics* 62, 229–264.

Nabeya, S. and Sørensen, B.E. (1993). Asymptotic distributions of the least squares estimators and test statistics in the near unit root model with non-zero initial value and local drift and trend, Manuscript Brown University, Department of Economics, Providence, RI.

Nabeya, S. and Tanaka, K. (1990). A general approach to the limiting distribution for estimators in time series regression with nonstable autoregressive errors. *Econometrica* 58, 145–163.

Newey, W. and West, K. (1987). A simple positive definite heteroskedasticity and autocorrelation consistent covariance matrix. *Econometrica* 55, 703–708.

Pantula. S.G. (1991). Asymptotic distributions of the unit root tests when the process is nearly stationary. *Journal of Business and Economic Statistics* 9, 63–71.

Perron, P. (1989). The calculation of the limiting distribution of the least squares estimator in a near-integrated model. *Econometric Theory* 5, 241–255.

Perron, P. (1990a). Tests of joint hypotheses for time series regression with a unit root, In *Advances in econometrics, 8: Cointegration, Spurious Regression, and Unit Roots* T.B. Fomby and G.F. Rhodes, eds., JAI Press, Greenwich, CT, 11, 135–159.

Perron, P. (1990b). The adequacy of limiting distributions in the AR(1) model with dependent errors, Econometric research program memo no. 349, Princeton University, Department of Economics, Princeton, NJ.

Perron, P. (1991a). A continuous-time approximation to the unstable first-order autoregressive process: The case without an intercept. *Econometrica* 59, 211–236.

Perron, P. (1991b). A continuous-time approximation to the stationary first-order autoregressive model. *Econometric Theory* 7, 236–252.

Perron, P. and Ng, S. (1993). Useful modifications to some unit root tests with dependent errors and their local asymptotic properties, Manuscript, Université de Montréal, Département de Sciences Economiques, Montréal.

Phillips, P.C.B. (1987a). Time series regression with unit roots. *Econometrica* 55, 277–302.

Phillips, P.C.B. (1987b). Towards a unified asymptotic theory for autoregression. *Biometrika* 74, 535–547.

Phillips, P.C.B. (1987c). Asymptotic expansions in nonstationary vector autoregressions. *Econometric Theory* 3, 45–68.

Phillips, P.C.B. and Ouliaris, S. (1990). Asymptotic properties of residual based tests for cointegration. *Econometrica* 58, 165–193.

Phillips, P.C.B. and Perron, P. (1988). Testing for a unit root in time series regression. *Biometrika* 75, 335–346.

Rubin, H. (1950). Consistency of maximum likelihood estimates in the explosive case, In: *Statistical Inference in Dynamic Economic Models*, T.C. Koopmans, ed., Wiley, New York, NY.

Schwert, G.W. (1989). Tests for unit roots: A Monte Carlo investigation. *Journal of Business and Economic Statistics* 7, 147–160.

Stock, J.H. (1991). Confidence intervals for the largest autoregressive root in U.S. macroeconomic time series. *Journal of Monetary Economics* 28, 435–459.

White, H. (1984). *Asymptotic Theory for Econometrician*, Academic Press, New York, NY.

White, J.S. (1958). The limiting distribution of the serial correlation coefficient in the explosive case. *Annals of Mathematical Statistics* 29, 1188–1197.

Chapter 8

Useful Modifications to Some Unit Root Tests with Dependent Errors and their Local Asymptotic Properties*

Pierre Perron and Serena Ng

Université de Montréal, C.R.D.E., Montréal, Québec H3C 3J7, Canada

Many unit root tests have distorted sizes when the root of the error process is close to the unit circle. This paper analyses the properties of the Phillips–Perron tests and some of their variants in the problematic parameter space. We use local asymptotic analyses to explain why the Phillips–Perron tests suffer from severe size distortions regardless of the choice of the spectral density estimator but that the modified statistics show dramatic improvements in size when used in conjunction with a particular formulation of an autoregressive spectral density estimator. We explain why kernel based spectral density estimators aggravate the size problem in the Phillips–Perron tests and yield no size improvement to the modified statistics. The local asymptotic power of the modified statistics are also evaluated. These modified statistics are recommended as being useful in empirical work since they are free of the size problems which have plagued many unit root tests, and they retain respectable power.

Keywords: Unit root, local asymptotic, near-integrated, twice integrated, seasonally integrated, Phillip–Perron test.

1. Introduction

Testing for the presence of unit roots and cointegration is now a common practice in applied macroeconomics. Often, one is required to

*Originally Published in *Review of Economic Studies* 63, pp. 435–463.

use statistics which appropriately account for serial correlation in the error process. Among statistics in this class, the augmented Dickey–Fuller and the Phillips–Perron tests are perhaps the most popular, as they are implemented in many statistical software packages. However, it is also by now a well-documented fact that the Phillips–Perron tests, as originally defined, suffer from severe size distortions when there are negative moving-average errors (see Phillips and Perron, 1988; Schwert, 1989, among others). Although the size of the Dickey–Fuller test is more accurate, the problem is not negligible.

The objectives of our paper are twofold. The first is to provide an understanding for the sources of size distortions in the Phillips–Perron tests and to explain how these distortions relate to the choice of the spectral density estimator. We use local asymptotic frameworks to analyze the properties of the statistics when the autoregressive (AR) or the moving average (MA) error process has a root close to the unit circle. We find that although no spectral density estimator can eliminate size distortions in these cases, kernel-based spectral density estimators tend to aggravate the size problem.

Our second objective is to compare the properties of the Phillips–Perron tests with selected statistics which can be viewed as modified Phillips–Perron tests. These modified statistics, based originally on the work of Stock (1999), are found to have exact sizes much closer to the nominal size *when used in conjunction with a particular formulation of the autoregressive spectral density estimator*. However, these attractive properties do not generalize to kernel-based spectral density estimators. Using local asymptotic analyses as developed in Nabeya and Perron (1994), we provide an explanation for these results, qualify the conditions when the modifications will alleviate size distortions, and when they will be vacuous.

This paper is organized as follows. Definitions of the statistics and estimators for the spectral density at frequency zero are given in Section 2, where the finite-sample properties of the statistics are also presented. The next three sections analyse the theoretical properties of the statistics for different choices of the spectral density estimator under three different stochastic assumptions on the error process. The local asymptotic size and power of the statistics are analysed in Section 6. Section 7 analyses selected data series and discusses the empirical relevance of the results in this paper. A conclusion completes the analysis. The main proofs are contained in a mathematical appendix, and results pertaining to the limiting behavior of the autoregressive spectral density estimator are contained in Perron and Ng (1995).

2. The Test Statistics

Consider the data generating process

$$y_t = \alpha y_{t-1} + u_t, \tag{1}$$

where $\{u_t\}$ is i.i.d. $(0, \sigma_u^2)$. White (1958) showed that the normalized least squares statistic, $T(\hat{\alpha} - 1)$, and the t-statistic for $\hat{\alpha}$, defined as $t_{\hat{\alpha}} = (\hat{\alpha} - 1)/[s_u(\Sigma_{t=1}^T y_{t-1}^2)^{-1/2}]$ with $s_u^2 = T^{-1}\Sigma_{t=1}^T \hat{u}_t^2$, have the following asymptotic distributions:

$$T(\hat{a} - 1) \Rightarrow \left(\int_0^1 W(r)dW(r)\right)\left(\int_0^1 W(r)^2 dr\right)^{-1}, \tag{2}$$

$$t_a \Rightarrow \left(\int_0^1 W(r)dW(r)\right)\left(\int_0^1 W(r)^2 dr\right)^{-1/2}, \tag{3}$$

where $W(r)$ is a standard Brownian motion on $C[0,1]$, the space of continuous functions on the interval $[0,1]$, and $\Rightarrow$ denotes weak convergence in distribution. When $\{u_t\}$ is serially correlated, Phillips (1987) showed that, under some regularity conditions, the limiting distributions of the statistics become

$$T(\hat{a} - 1) \Rightarrow \left(\int_0^1 W(r)dW(r) + \lambda\right)\left(\int_0^1 W(r)^2 dr\right)^{-1},$$

$$t_\alpha \Rightarrow (\sigma/\sigma_u)\left(\int_0^1 W(r)dW(r) + \lambda\right)\left(\int_0^1 W(r)^2 dr\right)^{-1/2},$$

where $\lambda = (\sigma^2 - \sigma_u^2)/2\sigma^2$, $\sigma_u^2 = \lim_{T\to\infty} T^{-1}E[\sum_{t=1}^T u_t^2], \sigma^2 = \lim_{T\to\infty} T^{-1}E[S_T^2]$, and $S_T = \sum_{j=1}^T u_j$. When $\{u_t\}$ is stationary, $\sigma^2 = 2\pi f_u(0)$, where $f_u(0)$ is the nonnormalized spectral density function of $\{u_t\}$ evaluated at frequency zero. The case $\sigma^2 = \sigma_u^2$

obtains with martingale difference innovations. To remove the dependence of the asymptotic distributions on the nuisance parameters σ^2 and σ_u^2, Phillips (1987) and Phillips and Perron (1988) proposed the statistics Z_α and Z_t, defined in the case of regression (1), as

$$Z_\alpha = T(\hat{a} - 1) - (s^2 - s_u^2)\left(2T^{-2}\sum_{t=1}^{T} y_{t-1}^2\right)^{-1}, \tag{4}$$

$$Z_t = (s_u/s)t_\alpha - (1/2)(s^2 - s_u^2)\left(s^2 2T^{-2} - \sum_{t=1}^{T} y_{t-1}^2\right)^{-1/2}, \tag{5}$$

where s_u^2 and s^2 are consistent estimates of σ_u^2 and σ^2 respectively. We will frequently refer to the term $(s^2 - s_u^2)(T^{-2}\sum_{t=1}^{T} y_{t-1}^2)^{-1}$ as the serial correlation correction factor. The Z_α and Z_t statistics, hereafter referred to as the *PP* tests, are often used in situations where considerations of weakly dependent errors become relevant. The asymptotic distributions of these statistics are given by (2) and (3), respectively.[1]

Stock (1999) proposed a class of statistics which exploits the feature that a series converges with different rates of normalization under the null and the alternative hypotheses. We consider two such tests, hereafter referred to as the M tests. The first statistic is:

$$MZ_\alpha = (T^{-1}y_T^2 - s^2)\left(2T^{-2}\sum_{t=1}^{T} y_t^2\right)^{-1}. \tag{6}$$

The statistic can be rewritten as

$$MZ_\alpha = Z_\alpha + (T/2)(\hat{a} - 1)^2. \tag{7}$$

For this reason, MZ_α can be seen as a modified version of Z_α. The term $(T/2)(\hat{\alpha} - 1)^2$ will subsequently be referred to as the modification factor. Under standard assumptions, convergence of $\hat{\alpha}$ to one at rate T ensures that Z_α and MZ_α are asymptotically

[1]If there are additional deterministic components (constant and trends) in regression (1), the Weiner process $W(r)$ in (2) and (3) and y_{t-1} in (4) and (5) should be replaced by their de-trended counterparts.

equivalent. The asymptotic critical values of MZ_α are therefore the same as those of Z_α, i.e., those corresponding to (2). The second statistic, *MSB*, is defined as:

$$MSB = \left(T^{-2}\sum_{t=1}^{T} y_{t-1}^2/s^2\right)^{1/2}. \tag{8}$$

Noting that the sum of squares of an $I(1)$ series is $O_p(T^2)$ but that of an $I(0)$ series is $O_p(T)$, the *MSB* statistic effectively tests the null hypothesis that the former condition is true. Under the alternative hypothesis, the statistic tends to zero. Hence, the unit root hypothesis is rejected in favor of stationarity when *MSB* is smaller than some appropriate critical value. Note that of all the tests considered, *MSB* is the only one that is bounded from below by zero. The statistic is related to Bhargava's (1986) R_1 statistic which is built upon the work of Sargan and Bhargava (1983). Critical values with y_t demeaned and detrended are provided by Stock (1999). Note that *MSB* and the *PP* tests are related as follows:

$$Z_t = MSB \cdot Z_\alpha. \tag{9}$$

This suggests a new modified *PP* test can be defined from the relation $MZ_t = MSB \cdot MZ_\alpha$, viz:

$$MZ_t = Z_t + (1/2)\left(\sum_{t=1}^{T} y_{t-1}^2/s^2\right)^{1/2}(\hat{\alpha}-1)^2. \tag{10}$$

2.1. *Estimating σ^2 and σ_u^2*

Construction of the statistics defined in the previous subsection requires estimates of σ_u^2 and/or σ^2. We consider three choices of s^2 as an estimator for σ^2. The first assumes that σ^2 is known. The second is an autoregressive spectral density estimator defined as

$$s_{\mathrm{AR}}^2 = s_{ek}^2/(1-\hat{b}(1))^2, \tag{11}$$

$s_{ek}^2 = T^{-1}\sum_{t=k+1}^{T}\hat{e}_{tk}^2, \hat{b}(1) = \sum_{j=1}^{k}\hat{b}_j$, with $\hat{b}_j$ and $\{\hat{e}_{tk}^2\}$ obtained from the autoregression:

$$\Delta y_t = b_0 y_{t-1} + \sum_{j=1}^{k} b_j \Delta y_{t-j} + e_{tk}. \tag{12}$$

When the roots of u_t are bounded away from the unit circle, consistency of the parameters in the augmented autoregression (12) has been shown by Said and Dickey (1984) to hold under the null hypothesis that $\alpha = 1$ if $k = o(T^{1/3})$. Consistency of s_{AR}^2 based upon (12) follows from Said and Dickey's results. The above formulation permits an estimator of σ^2 that converges to a strictly positive value under both the null and the alternative hypotheses. As discussed in Stock (1999), this ensures consistency of the unit root tests. An alternative autoregressive spectral density estimator can be constructed from the regression (see e.g., Berk, 1974):

$$\hat{u}_t = \sum_{j=1}^{k} b_j \hat{u}_{t-j} + e_{tk}, \tag{13}$$

where $\hat{u}_t$ are the least squares residuals from the first-order autoregression (1). Estimates of the spectral density implied by (12) and (13) are asymptotically equivalent when $\hat{\alpha}$ is a consistent estimate of $\alpha = 1$. As we will see, the advantage of defining s_{AR}^2 according to (12) is that it does not depend on $\hat{\alpha}$ (through $\hat{u}_t$), and therefore de-couples the estimation of α from the estimation of σ^2.

The third estimator of σ^2 considered is a kernel estimator based on the sample autocovariances and is defined as:

$$s_{\mathrm{WA}}^2 = T^{-1}\sum_{t=1}^{T}\hat{u}_t^2 + 2T^{-1}\sum_{k=1}^{M} w(k/M)\sum_{t=k+1}^{T}\hat{u}_t\hat{u}_{t-k}. \tag{14}$$

The window or kernel function $w(x)$ is assumed to satisfy the conditions in Andrews (1991), except that we restrict our analysis to kernels which also satisfy $w(x) = 0$ for $|x| > 1$. The parameter M therefore acts as a truncation lag, and s_{WA}^2 is consistent for σ^2

provided $M \to \infty$ and $M/T \to 0$ as $T \to \infty$. For future references, we also define

$$\psi = \int_0^1 w(x)dx. \tag{15}$$

Throughout this paper, we make use of the following estimator of σ_u^2:

$$s_u^2 = T^{-1} \sum_{t=1}^{T} \hat{u}_t^2. \tag{16}$$

2.2. *Finite sample properties of the statistics*

One problem with the *PP* tests is that they suffer from noticeable size distortions when the root of the error process is close to the unit circle. This can be seen from the results presented in Table 1, which are based on 1000 replications of the DGP $y_t = y_{t-1} + u_t$, with $(1 - \rho L)u_t = (1 + \theta L)e_t$. The regression is $y_t = \mu + \alpha y_{t-1} + v_t$. The size of the *PP* tests is evaluated at various sample sizes and the results for $T = 200$ and 500 are selected for discussion. In column one, we report results based on the Bartlett window using an automatic selection of the bandwidth as discussed in Andrews (1991). Column two presents results for the autoregressive spectral density estimator formulated according to (12).[2] Results with σ^2 assumed known are reported in column three. For the sake of comparison, the size of the t_ρ statistic proposed by Dickey and Fuller (1979) and extended by Said and Dickey (1984) is given in the last column.[3] Critical values are taken as the left tail 5 percentage point of the distribution given in Fuller (1976).

The simulations in Table 1 show that the *PP* tests are always too liberal when ρ is negative. However, for positive and large AR(1) errors, the tests are conservative when constructed with s_{WA}^2, but too liberal when constructed with s_{AR}^2 or the true value σ^2. For moving average noise functions, the *PP* tests are too liberal with all

[2] The truncation lag is 6 for $T = 200$ and 8 for $T = 500$.

[3] The truncation lag is based on the deterministic rule $4(T/100)^{1/4}$.

Table 1. Exact size of the Phillips–Perron tests; 5% nominal size.

	$Z(\alpha)$			$Z(t)$			
Estimate of σ^2	s^2_{WA}	s^2_{AR}	σ^2	s^2_{WA}	s^2_{AR}	σ^2	ADF(t)
			(a) $T = 200$				
i.i.d. errors	0.044	0.079	0.039	0.030	0.058	0.026	0.045
MA(1)−0.800	0.983	0.736	0.687	0.981	0.903	0.922	0.356
−0.500	0.419	0.114	0.082	0.387	0.162	0.139	0.055
−0.200	0.101	0.072	0.038	0.070	0.046	0.027	0.035
0.200	0.033	0.073	0.048	0.020	0.052	0.030	0.038
0.500	0.024	0.060	0.041	0.009	0.034	0.025	0.037
0.800	0.026	0.059	0.045	0.014	0.032	0.025	0.041
AR(1)−0.800	0.689	0.236	0.220	0.661	0.418	0.414	0.048
−0.500	0.211	0.110	0.071	0.207	0.130	0.104	0.060
−0.200	0.077	0.077	0.059	0.083	0.091	0.073	0.051
0.200	0.035	0.092	0.061	0.042	0.101	0.066	0.066
0.500	0.015	0.098	0.063	0.023	0.093	0.061	0.050
0.800	0.004	0.075	0.088	0.015	0.074	0.079	0.048
			(b) $T = 500$				
i.i.d. errors	0.051	0.073	0.055	0.048	0.066	0.054	0.042
MA(1)−0.800	0.971	0.631	0.623	0.967	0.858	0.869	0.283
−0.500	0.345	0.085	0.078	0.332	0.135	0.119	0.064
−0.200	0.104	0.063	0.063	0.098	0.073	0.070	0.057
0.200	0.036	0.062	0.047	0.039	0.055	0.052	0.047
0.500	0.038	0.054	0.052	0.035	0.047	0.044	0.059
0.800	0.038	0.058	0.062	0.047	0.062	0.059	0.046
AR(1)−0.800	0.653	0.170	0.172	0.638	0.276	0.274	0.046
−0.500	0.195	0.073	0.069	0.184	0.082	0.081	0.059
−0.200	0.083	0.050	0.060	0.080	0.055	0.062	0.045
0.200	0.041	0.057	0.059	0.045	0.057	0.062	0.058
0.500	0.021	0.064	0.057	0.033	0.056	0.055	0.045
0.800	0.013	0.065	0.080	0.024	0.053	0.067	0.052

estimators of σ^2 when θ is negative. Size distortions are noticeable even when θ is around -0.5, with the unit root hypothesis always being rejected as θ approaches -1. The size of t_ρ in the AR cases are very accurate. This is not surprising given that the autoregressive noise functions considered are of finite order. However, t_ρ also has substantial size distortions in the negative MA case, though the problem is less severe than with the *PP* tests.

Since t_ρ does not require an estimate of the spectral density function, inadequacy of the spectral density estimator could potentially explain why size distortions are larger with the *PP* tests. The effect of the choice of the kernel on the size of the *PP* tests was analysed in Kim and Schmidt (1990) via simulations. These authors experimented with the Bohman, the Bartlett, and the Parzen windows and found the choice of the kernel did not make a significant difference as far as size distortions were concerned. Our own simulations also found the Quadratic window, with and without pre-whitening, incapable of resolving the size problem.[4] Table 1 also suggests that replacing s^2_{WA} by s^2_{AR} will not improve the size of the *PP* tests. Indeed, size problems persist even when σ^2 is assumed known.

The properties of the M tests are presented in Table 2. When we use s^2_{WA} with a Bartlett window,[5] size distortions in the M tests are significant, just as in the *PP* tests. When we use s^2_{AR} to construct the statistics, the size distortions in both negative MA and AR cases are dramatically smaller than those of the *PP* tests, and the exact sizes of the M tests are quite close to the nominal size of 5%. The M tests based upon s^2_{AR} (but not s^2_{WA}) continue to have much better properties than the *PP* tests even at larger sample sizes. This suggests an intricate interplay between the choice of s^2 and the properties of the M tests.

In the next three sections, the dependence of the statistics on the choice of the spectral density estimator will be analysed. A result that will emerge repeatedly is that the standard asymptotic distributions, i.e., (2) and (3), are reasonably good guides to the finite-sample distributions of the M tests but are poor guides to the *PP* tests when the root of the error process is close to the unit circle. Since the *PP* tests are built upon $T(\hat{\alpha} - 1)$, the size problem with the *PP* tests follows naturally from the analyses of Nabeya and Perron

---

[4]The quadratic kernel was reported to have good properties by Andrews (1991), and Andrews and Monahan (1992) showed that the properties can be further improved with pre-whitening.

[5]Hyslop (1991) found the same results using the Parzen window.

Table 2. Exact size of the modified tests; 5% nominal size.

	$MZ(\alpha)$			MSB			$MZ(t)$		
Estimate of σ^2	s^2_{WA}	s^2_{AR}	σ^2	s^2_{WA}	s^2_{AR}	σ^2	s^2_{WA}	s^2_{AR}	σ^2
(a) $T = 200$									
i.i.d. errors	0.037	0.072	0.034	0.055	0.093	0.045	0.023	0.049	0.018
MA(1)−0.800	0.981	0.089	0.000	0.984	0.103	0.000	0.961	0.063	0.000
−0.500	0.383	0.052	0.005	0.425	0.063	0.018	0.302	0.038	0.005
−0.200	0.088	0.062	0.027	0.119	0.079	0.039	0.044	0.037	0.011
0.200	0.032	0.072	0.042	0.042	0.086	0.061	0.014	0.048	0.026
0.500	0.022	0.057	0.040	0.037	0.076	0.053	0.007	0.032	0.024
0.800	0.023	0.056	0.045	0.043	0.072	0.057	0.014	0.031	0.024
AR(1)−0.800	0.629	0.034	0.008	0.667	0.041	0.008	0.604	0.040	0.017
−0.500	0.163	0.068	0.028	0.190	0.084	0.038	0.159	0.069	0.033
−0.200	0 058	0.061	0.038	0.069	0.076	0.046	0.066	0.073	0.056
0.200	0.019	0.073	0.045	0.032	0.090	0.057	0.037	0.089	0.057
0.500	0.007	0.085	0.054	0.013	0.107	0.065	0.022	0.092	0.059
0.800	0.003	0.065	0.074	0.005	0.087	0.095	0.015	0.072	0.079
(b) $T = 500$									
i.i.d. errors	0.037	0.061	0.043	0.052	0.069	0.054	0.044	0.060	0.048
MA(1)−0.800	0.961	0.010	0.000	0.970	0.012	0.000	0.950	0.010	0.005
−0.500	0.304	0.042	0.021	0.336	0.049	0032	0.285	0.060	0.040
−0.200	0.072	0.055	0.047	0.100	0.063	0.055	0.088	0.063	0.053
0.200	0.025	0.052	0.034	0.034	0.063	0.047	0.039	0.051	0.047
0.500	0.028	0.045	0.040	0.038	0.059	0.057	0.034	0.046	0.043
0.800	0.024	0.048	0.044	0.038	0.058	0.062	0.046	0.059	0.057
AR(1)−0.800	0.624	0.027	0.030	0.653	0.036	0.041	0.594	0.038	0.041
−0.500	0.169	0.040	0.035	0.192	0.055	0.045	0.150	0.048	0.047
−0.200	0.059	0.029	0.037	0.083	0.042	0.048	0.073	0.049	0.052
0.200	0.027	0.042	0.046	0.042	0.066	0.064	0.044	0.053	0.058
0.500	0.013	0.048	0.039	0.025	0.057	0.064	0.032	0.055	0.055
0.800	0.009	0.046	0.063	0.009	0.060	0.081	0.023	0.053	0.064

(1994) and Perron (1996), where it was shown that (2) is a poor approximation for $T(\hat{\alpha} - 1)$ when θ is close to -1 or $|\rho|$ close to 1.

We extend previous work in the literature by deriving the local asymptotic distributions for the *PP* and the *M* tests for various estimators of the spectral density function. This allows us to explain why the size distortions are much larger with s^2_{WA} than with s^2_{AR}, and why the size of the *M* tests are much better than those of the

PP tests. Our local asymptotic analysis is based on the following framework:

$$y_t = (1 + c/T)\, y_{t-1} + u_t. \tag{17}$$

The assumption that $y_0 = e_0 = 0$ is maintained throughout. The series $\{y_t\}$ has an autoregressive root local to unity with noncentrality parameter c. A unit root obtains when $c = 0$. We shall restrict attention to cases where $\{u_t\}$ is either a pure AR or a pure MA process with roots local to the boundary of -1 and/or 1. Our derivations will be based on a regression without a constant or a trend, but it is straightforward to generalize the results to encompass additional deterministic terms. In most cases, one can substitute $W(r)$ by a demeaned or a detrended Wiener process without altering any of the analytics.

3. The Nearly-White-Noise, Nearly-Integrated Case

In this section, we analyse the properties of the tests for the case where a large negative moving-average component is present. Following the framework used by Nabeya and Perron (1994), we specify the data-generating process as:

$$y_t = (1 + c/T)y_{t-1} + u_t, \tag{18}$$

$$u_t = e_t + \theta_T e_{t-1}, \tag{19}$$

$$\theta_T = -1 + \delta/\sqrt{T}. \tag{20}$$

Here, y_t is an ARMA(1, 1) with an autoregressive root local to unity and a moving average coefficient local to -1. While the roots cancel and $\{y_t\}$ is a white noise process in the limit, it is nearly integrated in finite samples. Nabeya and Perron (1994) showed that the rate of $\sqrt{T}$ at which θ_T approaches -1 ensures a nondegenerate limit for the least-squares estimator $\hat{\alpha}$, with

$$(\hat{\alpha} - 1) \Rightarrow -\left(1 + \delta^2 \int_0^1 J_c(r)^2 dr\right)^{-1}, \tag{21}$$

where $J_c(r) = \int_0^r \exp((r-s)c)dW(s)$. Since the limit of $(\hat{\alpha}-1)$ has a negative support, $T(\hat{\alpha}-1)$ is unbounded and diverges to $-\infty$. Note that $\hat{\alpha}$ is not a consistent estimate of α in this local asymptotic framework. Furthermore,

$$s_u^2 \Rightarrow \sigma_e^2\left(2-\left(1+\delta^2\int_0^1 J_c(r)^2dr\right)^{-1}\right). \tag{22}$$

Hence, s_u^2 is also not a consistent estimate of $\sigma_u^2 = 2\sigma_e^2$. Since s_u^2 is based on the estimated residuals, the inconsistency of $\hat{\alpha}$ directly affects the properties of s_u^2. Straightforward calculations show that t_α also diverges to $-\infty$. The following theorems, proved in the Appendix, characterize the asymptotic distributions of Z_α, Z_t, MZ_α and MSB for the different estimators of σ^2. We start with the case where σ^2 is assumed known, followed by the autoregressive spectral density estimator.

Theorem 3.1. *Let $\{y_t\}$ be defined by (18)–(20). Let the estimator of σ^2 be $\sigma_T^2 = \sigma_e^2\delta^2/T$ which is assumed known when constructing the statistics. Let $e_\infty = \lim_{T\to\infty} e_T/\sigma_e$. Then as $T\to\infty$:*

(a) $T^{-1}Z_\alpha \Rightarrow -(1/2)(1+\delta^2\int_0^1 J_c(r)^2dr)^{-2}$;
(b) $T^{-1}Z_t \Rightarrow (1+\delta^2\int_0^1 J_c(r)^2dr)^{-3/2}/(2\delta)$;
(c) $MZ_\alpha \Rightarrow -((e_\infty+\delta J_c(1))^2-\delta^2)(2(1+\delta^2\int_0^1 J_c(r)^2dr))^{-1}$;
(d) $MSB \Rightarrow (1+\delta^2\int_0^1 J_c(r)^2dr)^{1/2}/\delta$.

Theorem 3.2. *Let $\{y_t\}$ be defined by (18)–(20). Let the estimator of σ^2 be the autoregressive spectral density estimator s_{AR}^2 based on (11). Furthermore, assume that $k/T\to 0$ and $k\to\infty$ as $T\to\infty$. The asymptotic distributions of Z_α, Z_t, MZ_α, and MSB are the same as those given in Theorem 3.1.*

Remark 1. The spectral density at frequency zero is zero in the limit for the present model, a situation ruled out by assumption in Berk (1974). We cannot appeal to his results or those of Said and Dickey (1984) to prove consistency of s_{AR}^2. However, we show in Perron and Ng (1995) that $\hat{b}(1)$ based upon (12) diverges to infinity, and that s_{ek}^2 is bounded. Thus, s_{AR}^2 tends to zero, the asymptotic value of σ_T^2.

The equivalence of the results in Theorems 3.1 and 3.2 follows from the consistency of s^2_{AR} for σ^2.

The spectral density at frequency zero plays no role in the *PP* tests in the limit because it converges to 0 whereas s^2_u is $O_p(1)$. Since a normalization factor of T^{-2} for the sample moments of y^2_{t-1} is too strong for this nearly-integrated nearly-white noise process, the normalized serial correlation correction factor is $O_p(T)$, the same as the rate at which $T(\hat{\alpha}-1)$ diverges. In consequence, the *PP* tests diverge at rate T. We note that upon simplification, the limiting distribution of $T^{-1}Z_\alpha$ is the same as that of $-(1/2)(\hat{\alpha}-1)^2$, the modification factor.

Although the normalized least squares estimator, the correction for serial correlation, and the modification factor in MZ_α all explode at rate T, the distribution of MZ_α is bounded. This is because the explosive terms in the *PP* tests are offset by the modification factor asymptotically. As this last term is absent from the *PP* tests, their asymptotic equivalence with the *M* tests breaks down under the assumptions of Theorems 3.1 and 3.2.

Size distortions associated with MZ_α are due to the discrepancies between (3) and the distribution implied by (6) evaluated at the null hypothesis of $c=0$. The dramatically smaller size distortions reported in Table 2 for MZ_α suggest that while the approximation provided by (3) for MZ_α is not perfect, it is a substantial improvement compared to Z_α.

Theorem 3.3. *Let $\{y_t\}$ be defined by (18)–(20) and let σ^2 be estimated by S^2_{WA} defined by (13). Let ψ be defined by (15) and assume that $M\to\infty$ and $M/T\to 0$ as $T\to\infty$. Then:*

(a) $(MT)^{-1}Z_\alpha \Rightarrow -\psi\delta^2(\int_0^1 J_c(r)^2dr)(1+\delta^2\int_0^1 J_c(r)^2dr)^{-3}$;

(b) $(MT)^{-1/2}Z_t \Rightarrow -(\psi(\delta^2/2)\int_0^1 J_c(r)^2dr)^{1/2}(1+\delta^2\int_0^1 J_c(r)^2 dr)^{-3/2}$;

(c) $(MT)^{-1}MZ_\alpha \Rightarrow -\psi\delta^2(\int_0^1 J_c(r)^2dr)(1+\delta^2\int_0^1 J_c(r)^2dr)^{-3}$;

(d) $(MT)^{1/2}MSB \Rightarrow (1+\delta^2\int_0^1 J_c(r)^2dr)^{3/2}(2\psi\delta^2\int_0^1 J_c(r)dr)^{-1/2}$.

Remark 2. The *PP* tests in Theorem 3.3 based upon s^2_{WA}explode to $-\infty$ at a rate of MT, faster than the rate T when σ^2 is presumed known or estimated by s^2_{AR}. This is consistent with the finding of Pantula (1991), who used a different local asymptotic framework to show that when σ^2 is estimated by the Bartlett window, Z_α and Z_t are unbounded, with the rate of divergence to $-\infty$ depending on the rate at which θ approaches -1 as well as the rate of increase of the truncation lag. With Bartlett weights of $1-k/(M+1)$, $\psi = 1/2$. Our results show that different choices of the weighting function will only affect the magnitude of ψ but not the rate at which the statistics explode. This also confirms the result of Kim and Schmidt (1990) that size distortions cannot be eliminated by an alternative choice of the kernel.

An intuitive explanation of these results is as follows. The quantity s^2_{WA} is based on a weighted sum of sample autocovariances which inherit inconsistency from $\hat{\alpha}$.[6] Divergence arises from the fact that each sample autocovariance is $O_p(T)$, and the explosive terms cumulate as the M lags are summed up. For this reason, the serial correlation correction factor in the *PP* tests is $O_p(MT)$ and dictates the rate of divergence of the statistics. As well, the modification factor in MZ_α, $(T/2)(\hat{\alpha}-1)^2$, is only $O_p(T)$, and is not strong enough to create an offsetting effect. Thus, MZ_α also diverges to negative infinity and has the same asymptotic distribution as Z_α.

The rate of divergence of the *PP* tests depends on whether s^2_{AR} or s^2_{WA} is used. For the former, divergence is at rate T and for the latter it is MT. For this reason, size distortions reported in Table 1 are larger for s^2_{WA} than for s^2_{AR}.

Remark 3. Phillips and Ouliaris (1990) showed that the *PP* tests will be inconsistent against stationary alternatives under the standard asymptotic framework (i.e., θ fixed as T increases) if residuals under the null hypothesis, $\Delta y_t = u_t$, are used to construct s^2_{WA} and s^2_u. Theorem 3.3 shows that the use of the estimated residuals, $\{\hat{u}_t\}$, will lead to increasing size distortions in the present local asymptotic

[6] An implication is that pre-whitening procedures which depend on the least-squares residuals will also be affected by inconsistency of $\hat{\alpha}$.

framework. Thus, estimators of the form (14) are inadequate whether one uses the estimated residuals or the residuals under the null.

In view of the inconsistency of $\hat{\alpha}$, the formulation of the autoregressive spectral density estimator based on Δy_t is not asymptotically equivalent to that based on $\hat{u}_t$. Specifically, (12) does not have a first-order dependence on $\hat{\alpha}$ as compared to a regression based on $\hat{u}_t$. Simulations show that if s_B^2 based upon (13) is used, the statistics will behave much like those reported in Table 1 for the Bartlett window.[7] Thus, among those estimators for σ^2 considered, the one that is adequate in both the standard and the local asymptotic framework is that based on the autoregression of the form (12) using first differences of $\{y_t\}$ instead of the estimated autoregression residuals.

Although our discussion has focused on the properties of MZ_α, the same intuition applies to *MSB* and MZ_t. For example, under Theorems 3.1 and 3.2, we have $T^{-2}\sum_{t=1}^{T} y_{t-1}^2$ tending to ∞, but s_{AR}^2 and the true value of σ^2 both converge to zero at the same rate. The resulting distribution for *MSB* is therefore neither degenerate nor explosive. It can also be seen that *MSB* tends to zero in Theorem 3.3 because of the unusually slow rate of convergence of $\sum_{t=1}^{T} y_t^2$. Since the test rejects the null hypothesis of a unit root when the statistic is small, the probability that *MSB* rejects the null hypothesis converges to one.

4. The Nearly-Twice Integrated Case

This section studies the behavior of the statistics in the presence of autoregressive errors with a large positive coefficient. The data-generating process is:

$$y_t = (1 + c/T)y_{t-1} + u_t \tag{23}$$

$$u_t = (1 + \phi/T)u_{t-1} + e_t, \quad \phi < 0, \tag{24}$$

[7]Intuitively, the reason is that the moment matrix of regressors in (13) does not converge to the population moments, and estimates from (13) cannot be used to obtain a consistent estimate of σ^2.

where $e_t \sim$ i.i.d. $(0, \sigma_e^2)$. As $T \to \infty$, $y_t = 2\,y_{t-1} - y_{t-2} + e_t$, a process with two unit roots, hence the terminology 'nearly-twice integrated.' Nabeya and Perron (1994) showed that

$$T(\hat{\alpha}-1) \Rightarrow Q_c(J_\phi(1))^2 \left(2\int_0^1 Q_c(J_\phi(r))^2 dr\right)^{-1}, \tag{25}$$

where $Q_c(J_\phi(r)) = \int_0^r \exp((r-v)c)J_\phi(v)dv$ and $J_\phi(v) = \int_0^v \exp((v-s)\phi)dW(s)$. Note that unlike the previous model, $\hat{\alpha}$ is consistent, in the sense that $\hat{\alpha} - (1 + c/T) \to 0$. It can be verified that

$$\begin{aligned} T^{-1}s_u^2 &= T^{-2}\Sigma\hat{u}_t^2 \Rightarrow \lambda\sigma_e^2, \\ \lambda &= \int_0^1 J_\phi(r)^2 dr - \left(Q_c(J_\phi(1))^2 - 2c\int_0^1 Q_c(J_\phi(r))^2 dr\right)^2 \\ &\quad \times \left(4\int_0^1 Q_c(J_\phi(r))^2 dr\right)^{-1}. \end{aligned} \tag{26}$$

A consequence of this is that t_α diverges to ∞, even though the normalized least-squares estimator does not. The following theorems give the limits of the statistics for various ways of estimating σ^2.

Theorem 4.1. *Let $\{y_t\}$ be a stochastic process given by (23)–(24). Let the estimator of σ^2 be $T^{-2}\sigma_T^2 = \sigma_e^2/\phi^2$ which is assumed known in constructing the statistics. Then as $T \to \infty$, we have:*

(a) $Z_\alpha \Rightarrow (Q_c(J_\phi(1))^2 - 1/\phi^2)(2\int_0^1 Q_c(J_\phi(r))^2 dr)^{-1}$;
(b) $Z_t \Rightarrow (\phi/2)(Q_c(J_\phi(1))^2 - 1/\phi^2)(\int_0^1 Q_c(J_\phi(r))^2 dr)^{-1/2}$;
(c) $MZ_\alpha \Rightarrow (Q_c(J_\phi(1))^2 - 1/\phi^2)^2(2\int_0^1 Q_c(J_\phi(r))^2 dr)^{-1}$;
(d) $MSB \Rightarrow (\phi^2\int_0^1 Q_c(J_\phi(r))^2 dr)^{1/2}$.

Theorem 4.2. *Let $\{y_t\}$ be defined by (23) and (24). Let the estimator of σ^2 be the autoregressive spectral density estimator s_{AR}^2 based on (12) with the truncation lag, k, chosen such that $k^3/T \to 0$ and $k \to \infty$ as $T \to \infty$. Let η represent the limiting distribution of $T(\hat{b}(1) - b(1))$. We have*

(a) $Z_\alpha \Rightarrow (Q_c(J_\phi(1))^2 - 1/(c+\phi+\eta)^2)(2\int_0^1 Q_c(J_\phi(r))^2 dr)^{-1}$;

(b) $Z_t \Rightarrow (1/2)(c+\phi+\eta)(Q_c(J_\phi(1))^2 - 1/(c+\phi+\eta)^2)(\int_0^1 Q_c(J_\phi(r))^2 dr)^{-1/2}$;
(c) $MZ_\alpha \Rightarrow (Q_c(J_\phi(1))^2 - 1(c+\phi+\eta)^2)(2\int_0^1 Q_c(J_\phi(r))^2 dr)^{-1}$;
(d) $MSB \Rightarrow ((c+\phi+\eta)^2 \int_0^1 Q_c(J_\phi(r))^2 dr)^{1/2}$.

Remark 4. Said and Dickey's proof of consistency of the coefficients $\hat{b}_i$ in (12) requires that $1 - b(1)$ be bounded away from zero, a condition which is not satisfied in the limit since the autoregressive coefficient for $\{u_t\}$ converges to 1. Indeed, $\sigma_T^2 \to \infty$ for the data-generating process in question. However, in this limiting case, Δy_t is an integrated process, and the coefficients on Δy_{t-i} are order T consistent (see Park and Phillips, 1988 for the case of a finite-order autoregression). Hence, s_{AR}^2 also tends to ∞ and is equivalent to σ_T^2 asymptotically.

Even though s_{AR}^2 and σ_T^2 both tend to ∞, the limiting distributions of the statistics are different. This is because $T^{-2}s_{\text{AR}}^2 \to \sigma_e^2/(c+\phi+\eta)^2$ as shown and defined in Perron and Ng (1995), whereas the true spectral density at frequency zero satisfies $T^{-2}\sigma_T^2 = \sigma_e^2/\phi^2$. Thus, the limiting distributions of the statistics based on s_{AR}^2 contain the variable η even under the null hypothesis that $c = 0$.

The modification factor in MZ_α can be written as $(2T)^{-1}[T(\hat{\alpha} - 1)]^2$ which converges to 0 in view of (25). Furthermore, the serial correlation correction factor converges to 0 upon normalization by $T^{-2}\Sigma_{t=1}^T y_{t-1}^2$ because the latter diverges to ∞. Thus, given the consistency of $\hat{\alpha}$, the modifications to the PP tests are vacuous, and Z_α and MZ_α are asymptotically equivalent.

Theorem 4.3. *Let $\{y_t\}$ be defined by (23) and (24) and let σ^2 be estimated by s_{WA}^2 as defined by (14) with truncation lag M. Let λ be defined by (26), $Q_c(J_\phi(r))$ by (25) and ψ by (15). If $M \to \infty$ and $M/T \to 0$ as $T \to \infty$, we have:*

(a) $Z_\alpha \Rightarrow (Q_c(J_\phi(1))^2(2\int_0^1 Q_c(J_\phi(r))^2 dr)^{-1}$;
(b) $(T/M)^{-1/2} Z_t \Rightarrow (1/2)(Q_c(J_\phi(1))^2)(2\psi\lambda \int_0^1 Q_c(J_\phi(r))^2 dr)^{-1/2}$;
(c) $MZ_\alpha \Rightarrow Q_c(J_\phi(1))^2(2\int_0^1 Q_c(J_\phi(r))^2 dr)^{-1}$;
(d) $(T/M)^{-1/2} MSB \Rightarrow (\int_0^1 Q_c(J_\phi(r))^2 dr)^{1/2}(2\psi\lambda)^{-1/2}$.

Remark 5. Theorem 4.3 shows that Z_α and MZ_α have the same limiting distribution as $T(\hat{\alpha}-1)$ given by (25). Even though $(s^2_{\mathrm{WA}} - s^2_u)$ diverges to infinity at rate MT, $T^{-2}\Sigma_{t=1}^{T} y_{t-1}^2$ diverges at an even faster rate of T^2. The normalized serial correlation correction factor therefore also has no role in the asymptotic distributions of the *PP* and *M* tests. The fast rate of convergence of $\sum_{t=1}^{T} y_{t-1}^2$ also explains why Theorem 4.3 requires only that $M/T \to 0$ and $M \to \infty$ as $T \to \infty$. This condition is more flexible than in the standard asymptotic case where M needs to be $o(T^{1/4})$ (see Phillips, 1987).

Remark 6. Since most of the statistics are bounded under the parameterization of this section, the discrepancies between the approximate and the exact distributions are also smaller than the ones reported for the negative moving-average model. The size distortions for Z_α and MZ_α observed in Tables 1 and 2 when ρ is close to 1 is primarily due to the fact that the local asymptotic distribution of $T(\hat{\alpha}-1)$ is nonnegative, but the critical values based on (2) are negative. This discrepancy is offset in part by the presence of ϕ and η, which shifts the local asymptotic distributions to the left of $T(\hat{\alpha}-1)$. For certain parameterizations, tests based on critical values from (2) will reject the null hypothesis of one unit root in favor of stationarity even when there are nearly two unit roots.

5. The Nearly-Integrated Seasonal Model

This section studies the behavior of the statistics in the presence of autoregressive errors with a large negative coefficient. Consider the data-generating process:

$$y_t = (1 + c/T)y_{t-1} + u_t, \tag{27}$$

$$u_t = -(1 + \phi/T)u_{t-1} + e_t, \quad \phi < 0. \tag{28}$$

This model can be written as $y_t = [(1+c/T) - (1+\phi/T)]y_{t-1} + (1 + (c+\phi)/T)y_{t-2} + e_t$. As $T \to \infty$, the process becomes a seasonal model of period two with a root on the unit circle. That is, $y_t = y_{t-2} + e_t$.

As shown in Nabeya and Perron (1994),

$$\hat{\alpha} \Rightarrow 1 - \left(2\int_0^1 B(r)^2 dr\right)\left(\int_0^1 (C(r)^2 + B(r)^2)dr\right)^{-1}, \qquad (29)$$

where $A(r) = (\phi - c)[Qc(J_{\phi,1}(r)) - Q_c(J_{\phi,2}(r))] + 2J_{c,1}(r)$, $B(r) = J_{\phi,1}(r) - J_{\phi,2}(r), C(r) = A(r) - B(r), J_{c,1}(s) = \int_0^s \exp((s-\upsilon)c)\, dW_1(\upsilon), J_{\phi,i}(s) = \int_0^s \exp((s-\upsilon)\phi)dW_i(\upsilon), Q_c(J_{\phi,i}(r)) = \int_0^r \exp((r-s)c)J_{\phi,i}(s)ds$ for $i = 1, 2$, $W_1(r)$ and $W_2(r)$ being independent Wiener processes. The autoregressive coefficient in (27) approaches -1 at rate T. This is necessary to obtain a nondegenerate limit of $\hat{\alpha}$. Note that this is faster than the rate of $\sqrt{T}$ used in the MA(1) case, a difference that will be useful in explaining the relative size distortions in the two models (see Remark 8). As in the moving-average case of Section 2, $\hat{\alpha}$ is not a consistent estimate of α, and $T(\hat{\alpha} - 1)$ diverges to $-\infty$ as T increases. However, like the twice-integrated model, the error process has an autoregressive root on the unit circle, and the limiting variance of $\{u_t\}$ is infinite. Specifically,

$$T^{-1}s_u^2 \Rightarrow \lambda_1 \sigma_e^2, \qquad (30)$$

where $\lambda_1 = (1/2)(\int_0^1 B(r)^2 dr \int_0^1 C(r)^2 dr)(\int_0^1 C(r)^2 + B(r)^2 dr)^{-1}$. It is easy to show that t_α diverges to $-\infty$ at rate $\sqrt{T}$. The following two theorems characterize the distributions of the various statistics when σ^2 is presumed known and when estimated by s_{AR}^2.

Theorem 5.1. *Let $\{y_t\}$ be a stochastic process generated by (27)–(28). Let the estimator of σ^2 be $\sigma_T^2 = \sigma_e^2/(2+\phi/T)^2$ which is assumed known in constructing the statistics. As $T \to \infty$, we have:*

(a) $T^{-1}Z_\alpha \Rightarrow -2(\int_0^1 B(r)^2 dr)^2(\int_0^1 C(r)^2 + B(r)^2 dr)^{-2}$;
(b) $T^{-1}Z_t \Rightarrow -\sqrt{2}(\int_0^1 B(r)^2 dr)^2(\int_0^1 C(r)^2 + B(r)^2 dr)^{-3/2}$;
(c) $MZ_\alpha \Rightarrow ((A(1)^2/2) - 1)(\int_0^1 C(r)^2 + B(r)^2 dr)^{-1}$;
(d) $MSB \Rightarrow (\frac{1}{2}\int_0^1 C(r)^2 + B(r)^2 dr)^{1/2}$.

Theorem 5.2. *Let $\{y_t\}$ be defined by (27) and (28). Let the estimator of σ^2 be the autoregressive density estimator s_{AR}^2 defined by (11) with the truncation lag k chosen to be such that $k^3/T \to 0$ and $k \to \infty$ as $T \to \infty$. The statistics Z_α, Z_t, MZ_α, and MSB have* *the same limiting distributions as given in Theorem* 5.1 *for the case where* σ^2 *is known.*

Remark 7. We show in Perron and Ng (1995) that s^2_{AR} converges to $\sigma^2_e/4$, the same limit as that of the true value $\sigma^2_T = \sigma^2_e/(2+\phi/T)^2$. The asymptotic equivalence of the results in the two theorems then follows.

While Z_α is driven by $(T/2)(\hat{\alpha}-1)^2$ in the limit, this explosive term is being offset by the modification factor in MZ_α. In both cases of negative residual serial correlation, the problem with the *PP* tests arises because $\hat{\alpha}$ is not consistent for α, and the asymptotic equivalence between MZ_α and Z_α breaks down.

Theorem 5.3. *Let* $\{y_t\}$ *be defined by* (27) *and* (28) *and* ψ *by* (15). *Let* σ^2 *be estimated by* s^2_{WA} *as defined by* (14) *with truncation lag* M. *Let* $\lambda_2 = ((\int_0^1 B(r)^2dr)/(\int_0^1 C(r)^2 + B(r)^2dr))^2(\int_0^1 C(r)^2 - B(r)^2dr)$. *If* $M \to \infty$ *and* $M/T \to 0$, *then as* $T \to \infty$,

(a) $(MT)^{-1}Z_\alpha \Rightarrow (-4\psi\lambda_2)(\int_0^1 C(r)^2 + B(r)^2dr)^{-1}$;
(b) $(MT)^{-1/2}Z_t \Rightarrow -(2\psi\lambda_2)^{1/2}(\int_0^1 C(r)^2 + B(r)^2)^{-1/2}$;
(c) $(MT)^{-1}MZ_\alpha \Rightarrow (-4\psi\lambda_2)(\int_0^1 C(r)^2 + B(r)^2dr)^{-1}$;
(d) $(MT)^{1/2}MSB \Rightarrow (\int_0^1 C(r)^2 + B(r)^2)^{1/2}(8\psi\lambda_2)^{-1/2}$.

Remark 8. Theorem 5.3 shows that the statistics are divergent for any choice of the truncation lag. This is because $\hat{\alpha}$ is not a consistent estimate of α, and the sample autocovariances constructed on the basis of $\hat{u}_t$ are also inconsistent estimates of the true autocovariances (see Remark 2).

Size distortions are expected to be smaller in this model with negative autocorrelation than in the earlier model with negative moving average errors. This is because for an autoregressive coefficient ρ and a moving-average coefficient θ of the same size, the implied value of ϕ is larger than the implied value of δ for a given T. This is due to the fact that our local framework specifies a different rate of approach to the boundary -1, namely $\sqrt{T}$ in the MA case and T in the AR case. In this sense, a ρ of -0.8 is further away from the boundary than a

θ of -0.8. The consequences of these different rates of normalization are reflected in the simulations presented earlier.

6. Size and Power

In this section, we assess the properties of the modified tests based on the autoregressive spectral density estimator. We first consider the local asymptotic size of the tests by computing the p-value of the statistics in Theorems 3.2, 4.2 and 5.2 with $c = 0$ using critical values from the standard asymptotic distributions.[8] These results are presented in the top panel of Figs. 1–3. We then tabulate the 5% critical values from 10,000 replications of the respective asymptotic distributions and construct size-adjusted power functions for values of c between -15 and 5. For $T = 200$, this corresponds to values of α between 0.925 and 1.025, and for $T = 500$ between 0.97 and 1.01. The integrals in the asymptotic distributions are approximated by partial sums of i.i.d.$N(0, 1)$ random variables with 1,000 steps. These experiments are performed for a set of noncentrality parameters (δ in the first model, and ϕ in models two and three).

For the nearly-integrated, nearly-white-noise model, we consider values of δ between 1 and 20 (for $T = 500$, θ varies between -0.96 and -0.11). The results are presented in Fig. 1. The top panel is the asymptotic size of the statistics. For small values of δ, tests based on the standard asymptotic critical values under-reject the unit root hypothesis, but the local asymptotic size becomes more accurate as δ increases. The local asymptotic size is similar to the exact size reported in Table 2 especially when σ^2 is assumed known. The local asymptotic size is still close to the exact size when σ^2 is estimated by s^2_{AR} except when T is small and θ is very close to -1. Although the exact size is slightly liberal in those cases, it is actually closer to the nominal size of the test. Thus, the finite-sample bias of s^2_{AR} is such that it reduces size distortions. The bottom three panels

[8]For MZ_α and MZ_t, the lower five percentage points are -8.1 and -1.95 as given in Fuller (1976). For MSB, the case without a constant was not provided by Stock (1999). Approximating the distribution of $(\int_0^1 W(r)^2 dr)^{1/2}$ gives critical values for the lower and upper five percentage points of 0.23 and 1.28, respectively.

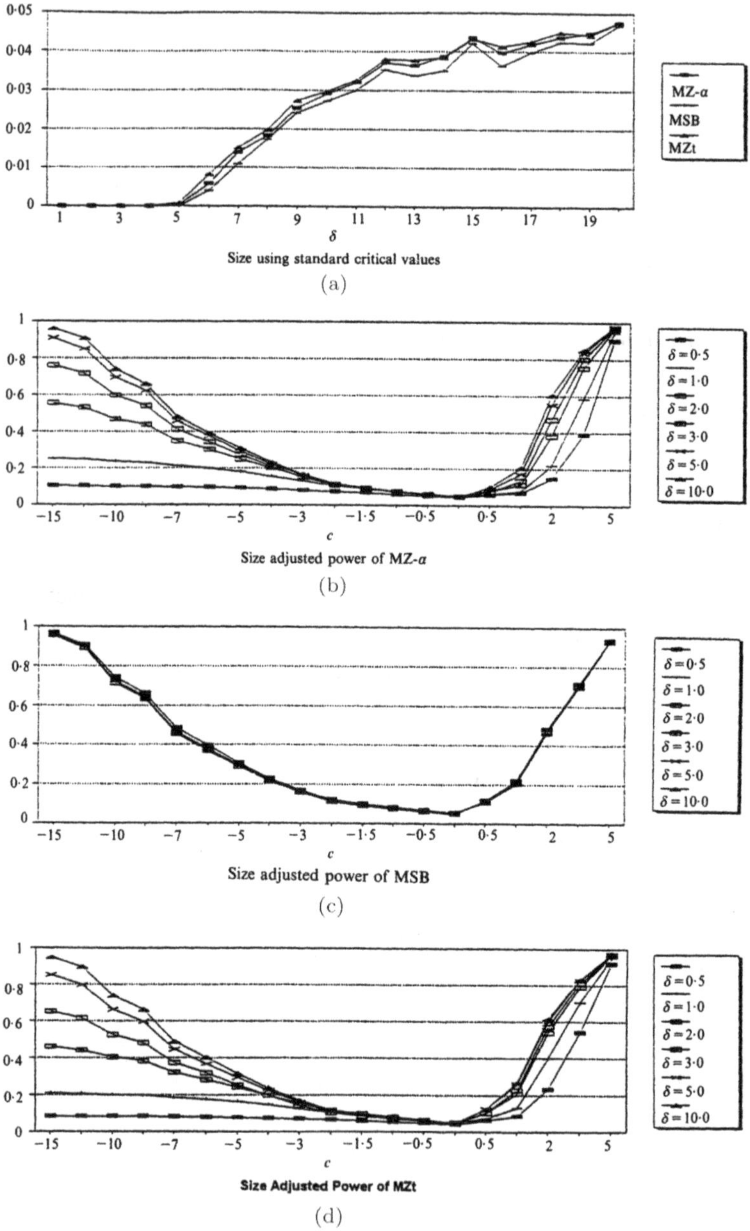

Fig. 1. Nearly-integrated, nearly-white noise model.

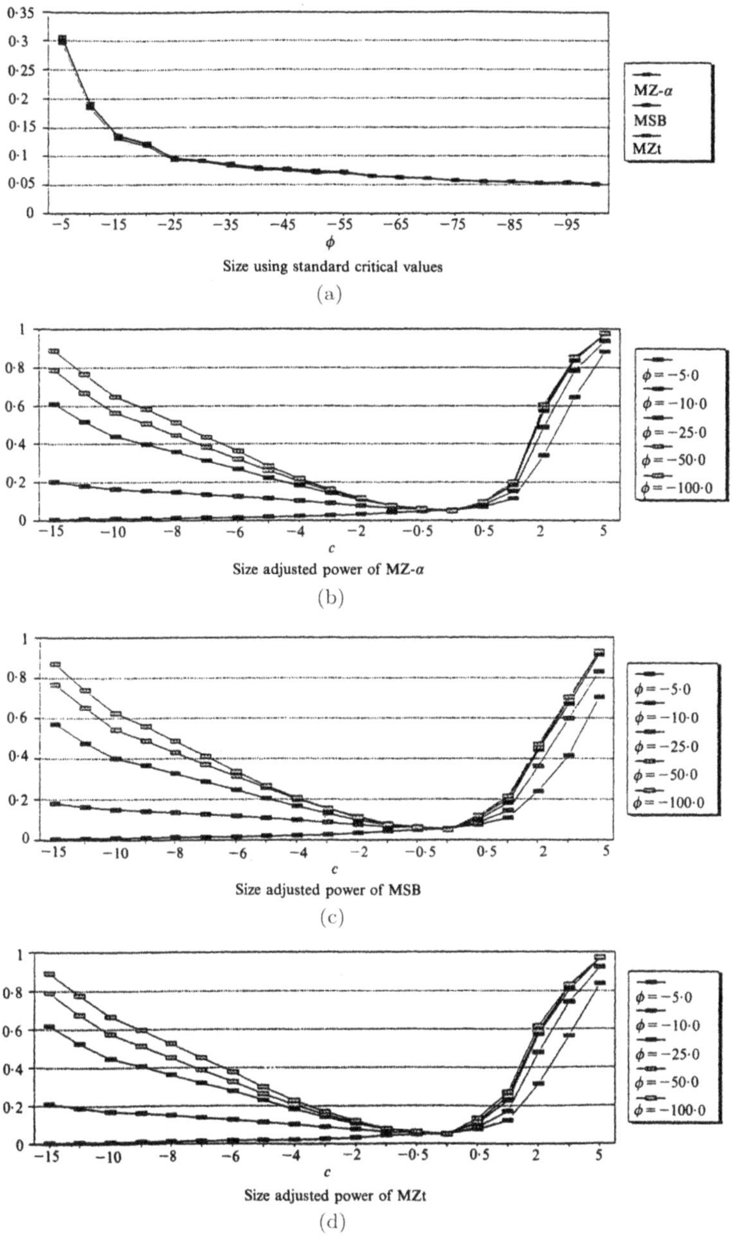

Fig. 2. Nearly-twice-integrated model.

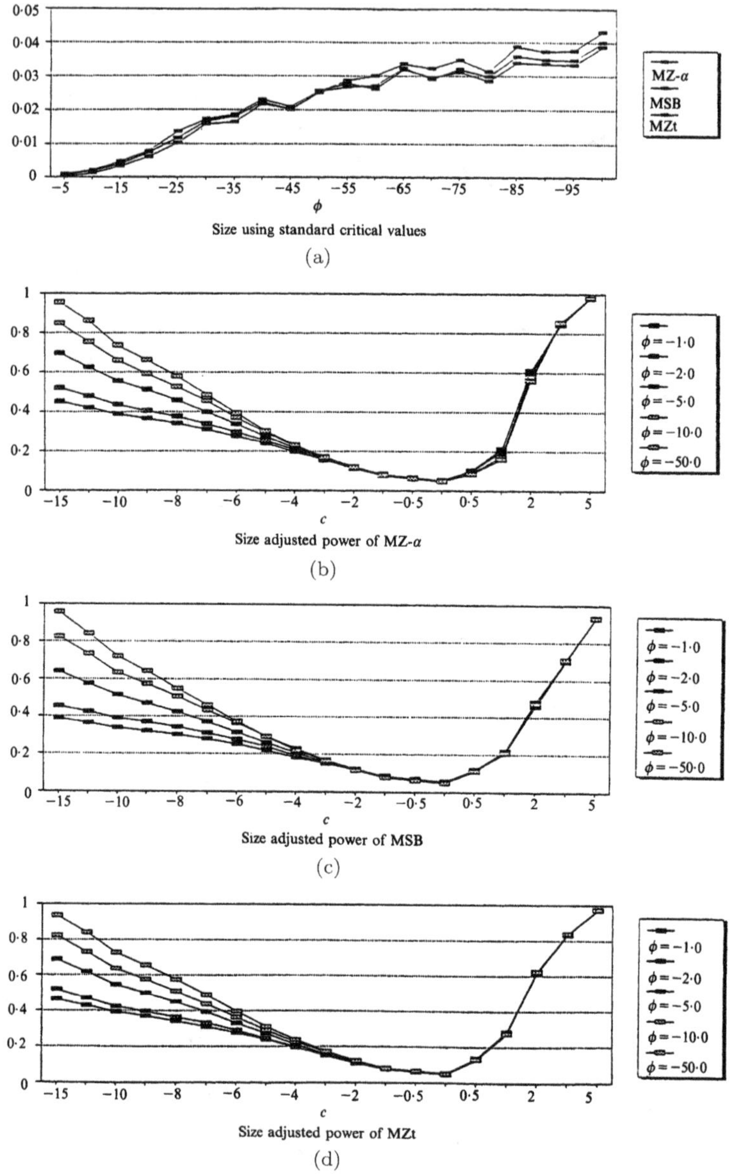

Fig. 3. Nearly-seasonally-integrated model.

are the size adjusted local asymptotic power for MZ_α, MSB and MZ_t, respectively. As expected, local asymptotic power is higher against explosive alternatives (with $c > 0$) than against stationary alternatives ($c < 0$). For values of δ close to zero, MZ_α and MZ_t apparently has no local asymptotic power. This can be explained by the fact that the distributions of these statistics are independent of c when δ approaches zero. However, for MSB the power function is independent of the value of δ.

For the nearly-twice-integrated model, ϕ is varied between -5 and -100, corresponding to values of ρ between 0.975 and 0.50 for $T = 200$. We see from the top panel of Fig. 2 that the tests over-reject the null hypothesis for small values of ϕ but the size of the tests improves as ϕ increases. The size implied by the local asymptotic results is larger than the finite sample size reported in Table 2, but once again, the exact size is closer to the nominal size for the parameters and sample sizes considered. As for local asymptotic power, all three tests have low power when ϕ is small but it increases monotonically as ϕ increases.

For the nearly-seasonally-integrated model, ϕ is also varied between -5 and -100 implying values of ρ between -0.975 and -0.50 for $T = 200$. The results are shown in Fig. 3. The tests based on the standard asymptotic critical values tend to under reject the unit root hypothesis unless ϕ is large. This result parallels the finite sample size reported in Table 2. Unlike the negative MA case, MZ_α and MZ_t do not suffer from power loss at small values of ϕ. All three M tests have comparable power. As in the nearly-twice-integrated model, there is a strong dependence of power on the values of ϕ.

Our local asymptotic simulations suggest that except when values of the noncentrality parameters are extremely small, the local asymptotic power of the tests are good. More importantly, the sizes of the tests based on standard critical values are usually within reasonable range of the nominal size. This is a useful result since it suggests that standard asymptotic critical values can be used with little loss of accuracy.

The finite sample power of the statistics using the standard critical values are reported in Table 3 for $T = 200$ and $T = 500$. The

Table 3. Power of the modified statistics using the autoregressive spectral density estimator (5% one-tailed tests).

	MZ(α)			*MSB*			*MZ*(t)			*ADF*(t)		
α	0.95	0.90	0.85	0.95	0.90	0.85	0.95	0.90	0.85	0.95	0.90	0.85
					(a) $T = 200$							
i.i.d. errors	0.521	0.857	0.950	0.563	0.888	0.962	0.403	0.746	0.901	0.258	0.690	0.925
MA(1)−0.800	0.554	0.898	0.973	0.592	0.922	0.980	0.442	0.820	0.945	0.964	1.000	1.000
−0.500	0.420	0.791	0.929	0.460	0.821	0.951	0.302	0.654	0.853	0.315	0.815	0.977
−0.200	0.521	0.838	0.949	0.573	0.871	0.962	0.394	0.747	0.894	0.280	0.677	0.932
0.200	0.529	0.851	0.960	0.585	0.885	0.972	0.391	0.761	0.906	0.238	0.688	0.904
0.500	0.488	0.833	0.939	0.550	0.872	0.953	0.365	0.710	0.901	0.207	0.623	0.883
0.800	0.420	0.756	0.906	0.463	0.794	0.923	0.296	0.629	0.816	0.147	0.480	0.769
AR(1)−0.800	0.333	0.622	0.780	0.379	0.671	0.815	0.222	0.485	0.656	0.266	0.705	0.928
−0.500	0.447	0.809	0.925	0.499	0.847	0.952	0.332	0.698	0.855	0.250	0.699	0.936
−0.200	0.478	0.839	0.960	0.518	0.867	0.973	0.377	0.729	0.903	0.267	0.680	0.914
0.200	0.508	0.833	0.943	0.554	0.867	0.962	0.413	0.742	0.887	0.264	0.627	0.877
0.500	0.496	0.784	0.938	0.551	0.825	0.952	0.384	0.682	0.857	0.232	0.562	0.848
0.800	0.445	0.713	0.829	0.499	0.754	0.860	0.366	0.599	0.747	0.198	0.411	0.603

(*Continued*)

Table 3. (*Continued*)

	$MZ(\alpha)$			MSB			$MZ(t)$			$ADF(t)$		
α	0.95	0.90	0.85	0.95	0.90	0.85	0.95	0.90	0.85	0.95	0.90	0.85
(b) $T = 500$												
i.i.d. errors	0.893	0.985	0.994	0.916	0.988	0.997	0.814	0.969	0.987	0.901	1.000	1.000
MA(1)−0.800	0.521	0.779	0.923	0.559	0.809	0.937	0.391	0.716	0.890	1.000	1.000	1.000
−0.500	0.842	0 957	0.982	0.877	0.967	0.988	0.775	0.912	0.959	0.953	1 000	1.000
−0.200	0.885	0.980	0.997	0.916	0.988	0.999	0.810	0.961	0.988	0.900	1.000	1.000
0.200	0.884	0.988	0.998	0.916	0.991	1.000	0.899	0.970	0.986	0.898	1.000	1.000
0.500	0.893	0.983	0.996	0.919	0.988	0.998	0.812	0.959	0.981	0.896	0.999	1.000
0.800	0.867	0.982	0.997	0.893	0.987	0.997	0.774	0.963	0.983	0.936	1.000	1.000
AR(1)−0.800	0.944	1.000	1.000	0.966	1.000	1.000	0.836	1.000	1.000	0.895	1.000	1.000
−0.500	0.970	1.000	1.000	0.983	1.000	1.000	0.918	1.000	1.000	0.915	1.000	1.000
−0.200	0.981	1.000	1.000	0.987	1.000	1.000	0.930	1.000	1.000	0.920	1.000	1.000
0.200	0.964	1.000	1.000	0.980	1.000	1.000	0.929	1.000	1.000	0.899	1.000	1.000
0.500	0.971	1.000	1.000	0.980	1.000	1.000	0.904	1.000	1.000	0.861	1.000	1.000
0.800	0.925	1.000	1.000	0.954	1.000	1.000	0.829	1.000	1.000	0.777	1.000	1.000

power of the augmented Dickey–Fuller statistic, t_ρ, is also reported. Since the power functions are unadjusted for size distortions, the power reported for t_ρ in the negative MA case is misleadingly high and should be interpreted with caution.

There are several features of note. First, MZ_α tends to be more powerful than MZ_t, consistent with a finding of Phillips and Ouliaris (1990) and Nabeya and Tanaka (1990) that the t-statistic has a lower asymptotic power curve than the normalized least-squares estimator. Second, the power of MZ_α and MSB matches that of t_ρ in finite samples except in the negative AR case, where the power discrepancies can be traced to the fact the M tests are undersized. Indeed, in the positive AR(1) case when the exact size is very close to the nominal 5% (see Table 2), the power of the M tests exceeds that of t_ρ (see, in particular, the cases $\rho = 0.5$ and 0.9). Third, comparing the results for $T = 200$ and $T = 500$ in Table 3, the rate at which the power of the M tests increases is model dependent. The M tests have lowest power when there is a large negative AR component in the noise function and when the sample size is small, but power increases rapidly as T increases. Although the power of these tests is much higher in small samples in the negative MA case, it increases more slowly with T. These results are due to the fact that the distinction between the null and the alternative hypotheses sharpens more rapidly in the AR models as T increases (recall the rate of approach to -1 is T in the AR case but is $\sqrt{T}$ in the MA case).

7. Empirical Issues

Our theoretical analysis shows, and simulations confirm, that the modified statistics have more robust properties than the unmodified unit root tests. The M tests have much better size and higher power than t_ρ in the negative MA case; comparable size and higher power in the positive AR case; and lower power in the negative AR case due to size distortions on the conservative side. Since negative AR errors are rarely found in economic data, one can generally expect size and power advantages in the M tests over the PP tests and t_ρ.

The real strength of the M tests is in the case of negative MA errors when most unit root tests have size distortions. This is an important consideration in assessing the empirical importance of the M tests because negative MA errors are not uncommon in macroeconomic time series. Such errors can arise for a number of reasons, among which is the presence of additive outliers. Franses and Haldrup (1994) showed that the presence of systematic outliers in an $I(1)$ series will shift the limiting distribution of unit-root tests to the left, and therefore imply an over-rejection of the unit-root hypothesis. Using data on the U.S.-Finland real exchange rate, which has an apparent outlier in 1918 and possibly six others, the authors find that using dummy variables to remove the outliers before constructing t_ρ will lead to stronger evidence of nonstationarity. However, one can also use unit-root tests that are robust to the presence of negative MA errors. Vogelsang (1999) exploits this property of the M tests and rejects a unit root in the real exchange rate data using MZ_α and MZ_t. This shows that the modified statistics will have power. Use of the modified tests in this context has the additional advantage of not having to identify the outliers.

To illustrate when the modifications in the M tests are important, we now consider unit-root tests on two series: real GDP and inflation.[9] The values for Z_α and Z_t for real GDP are -5.94 and -1.62 respectively, and the kernel-based normalized spectral density at frequency zero $(s^2_{\mathrm{WA}}/s^2_u)$ based on a 4-lag Bartlett window is 1.66. The values for MZ_α, MZ, and MSB are -9.69, -2.98 and 0.307, respectively.[10] The normalized autoregressive spectral density at frequency zero $(s^2_{\mathrm{AR}}/s^2_u)$ estimated with $k = 6$ is 2.58. Both spectral density estimates are far from zero, suggesting that the series is neither nearly stationary nor nearly-seasonally integrated. The ARMA(1,1) estimates are 1.002 and 0.25, respectively. As the

[9]We use quarterly log (GDPQ) for output, and the first difference of monthly log (PUNEW) for inflation. The data are taken from CITIBASE. The estimation period is 54:1 to 93:3. The inflation series is demeaned prior to the construction of the statistics, while the GNP series is detrended by a linear time trend.

[10]The 5% critical values for MSB with a constant is 0.191, and 0.164 when there is a constant and a trend.

MA coefficient is far away from the problematic parameter space, the modification factors in the M tests should have vacuous effects. Indeed, both Z_α and MZ_α fail to reject the unit root hypothesis. The t_ρ statistic (12 lags) yields a similar conclusion with a value of -1.23.

For the inflation series, Z_α and Z_t are -184.2 and -10.73 with 4 lags in s^2_{WA}, and are -322.25 and -13.49 with 12 lags. The values for s^2_{AR}/s^2_u are 0.99 and 1.91 respectively, and are larger as the truncation lag increases further. By contrast, MZ_α, MZ_t and MSB are -13.9, -2.59 and 0.18 with $k=6$, and -8.14, -1.96 and 0.24 with $k=12$. The values for s^2_{AR}/s^2_u are 0.10 and 0.05 respectively, much smaller than the kernel-based estimates. The ARMA(1, 1) estimates are 0.98 and -0.73, respectively.[11] Although the autoregressive coefficient is very close to one and one would not expect the unit root hypothesis to be rejected, the large negative MA coefficient suggests the size of the PP tests might be distorted. The results of the unit root tests are therefore as predicted by our analysis: the larger the truncation lag in s^2_{WA}, the faster the PP tests diverge. The autoregressive spectral density estimator is accurate even when σ^2 is small, and along with the modification factor, produces a test statistic that is well behaved. The M tests therefore suggest the presence of a unit root. Use of Z_α would have rejected the null hypothesis, while use of t_ρ would have yielded ambiguous results (the value with 8 lags is -2.53, close to a rejection at the 10% level).

A negative MA component also appears in important macroeconomic variables such as the unemployment rate and monthly consumption. The M tests analysed in this paper can be seen as practical remedies to potentially serious inference problems. In Ng and Perron (1995), we show that the presence of negative MA components can also affect cointegration analysis. The modified tests are shown in that context to play a useful role in determining the

[11]Perron (1994) suggests that a negative moving average component should be present in the inflation series if the monetary authorities react to offset inflationary/disinflationary pressures that are inconsistent with a price level target path. This makes inflation strongly mean-reverting.

variable on which a cointegrating regression should be normalized, and the consequence in terms of the accuracy of the coefficient estimates could be dramatic.

We close this section with some comments on the implementation of the M statistics when the data are trending over time or when the mean of the series is unknown. In the latter case, one simply treats y_t in (6) and (8) as being de-meaned, and adds a constant to the autoregression (12) when constructing the spectral density estimator. In the case of trending data, the y_t that appears in (6) and (8) is de-trended. However, we still include only a constant in (12). The reason for this is that the estimate of the spectral density function at frequency zero is still consistent under the null hypothesis and its limit remains bounded above zero under the alternative of trend-stationarity. Including a time trend in (12) would yield similar asymptotic results but simulations (not reported) indicate that the size of the resulting M tests is less accurate. As well, the truncation lag has been fixed in the simulations. A data-dependent rule for selecting k and a more effective treatment of the deterministic terms in s^2_{AR} are under investigation by the authors.

8. Conclusions

When the root of the error process is close to the unit circle, many commonly used unit root tests have size distortions. However, simple modifications which have negligible effects in a standard asymptotic framework can lead to tests with substantially more accurate sizes. The proviso is that the modifications be used in conjunction with an estimate of the spectral density at frequency zero that is consistent in both the standard asymptotic framework and in the local framework used here. Kernel-based spectral density estimators using estimated residuals do not satisfy these criteria and tend to aggravate the size problem. On the other hand, the autoregressive spectral density estimator formulated on the basis of an augmented autoregression with first differences of the data serves this purpose. When appropriately implemented, the modified statistics have robust properties and are useful tests for a unit root.

The statistics will also be useful in cointegration analysis where serial correlation in the noise function is often encountered.

It is important to put into perspective the properties of the modified statistics vis-à-vis the general issue of distinguishing between unit roots and stationary processes. As discussed in Campbell and Perron (1991) these two types of processes are observationally equivalent in the sense that for any stationary process, there will exist a unit root process which approximates it arbitrarily well and vice-versa. To be concrete about the implication of this result, consider the case of MA(1) errors with a negative coefficient θ. The near observational equivalence implies that when using unit root tests with asymptotic critical values, there will exist values of θ in the range $(-1, x)$ for some $-1 < x < 0$, say, such that liberal size distortions will surface. The value of x will depend on the sample size and the test used, but it will always approach -1 as the sample size increases; i.e., the range over which size distortions occur will diminish. The problem, as shown in previous simulations including some presented in this paper, is that the rate at which the range shrinks can be very slow.

In practice, the problem is that for conventional tests (e.g., the PP tests or t_ρ) and sample sizes commonly encountered, this value of x where size distortions start to be important is too far away from -1 (e.g., somewhere around -0.4, when $T = 100$). This has been the cause of some concern because this range includes values (e.g., between -0.8 and -0.4) which are of practical relevance (e.g., Schwert, 1987) and for which we would rather not classify unit-root processes with such moving-average coefficients as stationary ones.

The class of modified statistics discussed in this paper can be viewed as tests with a much smaller range of size distortions (e.g., for θ between -1 and -0.9) for any given common sample size. This can be useful in practice because classifying unit root processes with values of θ in this range as stationary is likely to be of less concern. It is important to note that this improvement in size is achieved while retaining reasonable power.

The above justifications for using the modified statistics are valid insofar as the aim of testing for unit roots is to classify as precisely as possible whether a process is difference or trend stationary. There are, however, instances when the objective of the analysis is otherwise and using the modified statistics may not be appropriate. Suppose the aim of unit root tests is to decide which restrictions to make in a forecasting exercise. As reported in Campbell and Perron (1991), near-stationary unit processes are better forecast using stationary models, while near-integrated stationary processes are better forecast using integrated models. To the extent that imposing a false restriction may help reduce the mean squared error in this context, it is desirable to misclassify trend stationary processes as difference stationary and vice-versa, and one would rather use the conventional Dickey–Fuller or Phillips–Perron statistics to test for unit roots. Of course, in such cases, the 'optimal' value of x is highly dependent on the overall objectives of the analysis of which unit root tests is just an important first step. On this issue, more work remains to be done.

Appendix

As a matter of notation, we shall let C_i denote (not necessarily the same) constants throughout this appendix. For each model, we start with a series of lemmas that consider the limit of the relevant sample moments. Proofs of Lemmas 3.4, 4.2 and 5.2 are given in Perron and Ng (1995).

Proofs of results in Section 3

The following two lemmas are taken from or generalizations of results in Nabeya and Perron (1994), and the proof of Lemma 3.3 can be found in the working paper version of this paper.

Lemma 3.1. *Let* $\{y_t\}$ *be generated by* (18)–(20), *and define* $X_t = (1 + c/T)X_{t-1} + e_t, a_T = (1 - \delta/\sqrt{T})(1 - C/T), b_T = 1 - (1 - c/T)$

$(1-\delta/\sqrt{T})$ with $a_T \to 1$ and $T^{1/2}b_T \to \delta$ as $T \to \infty$. We have

(a) $y_t = a_T e_t + b_T X_t$;
(b) $\sum_1^T X_{t-1}e_t = O_p(T)$;
(c) $T^{-2}\sum_1^T X_t \Rightarrow \sigma_e^2 \int_0^1 J_c(r)^2 dr$

Lemma 3.2. *Let $\{y_t\}$ be generated according to (18)–(20) and $J_c(r)$ be defined as in (21). Let $e_\infty = \lim_{T\to\infty} e_T/\sigma_e$. Then as $T \to \infty$,*

(a) $T^{-1}\sum_t^T y_{t-1}^2 \Rightarrow \sigma_e^2 + \sigma_e^2\delta^2 \int_0^1 J_c(r)^2 dr$;
(b) $T^{-1}\sum_t^T y_{t-1}u_t \Rightarrow -\sigma_e^2$;
(c) $y_T \Rightarrow \sigma_e e_\infty + \delta\sigma_e J_c(1)$;
(d) $T^{-1}\sum_t^T u_t^2 \Rightarrow 2\sigma_e^2$.

Lemma 3.3. *Let $\{y_t\}$ be generated by (18)–(20). Let $k/T \to 0$ as $k \to \infty$ and $T \to \infty$. Then for i, $j = 1, \ldots, k$,*

(a) $$T^{-1}\sum_{t=k+1}^T y_{t-1}u_{t-1} \Rightarrow \begin{cases} \sigma_e^2 & \text{if } i = 1, \\ 0 & \text{otherwise}; \end{cases}$$

(b) $$T^{-1}\sum_{t=k+1}^T y_{t-j}u_{t-i} \Rightarrow \begin{cases} \sigma_e^2(1+\delta^2\int_0^1 J_c(r)^2 dr) & \text{if } i = j, \\ \sigma_e^2\delta^2\int_0^1 J_c(r)^2 dr & \text{if } i \neq j; \end{cases}$$

(c) $$T^{-1}\sum_{t=k+1}^T u_{t-i}u_{t-j} \Rightarrow \begin{cases} 2\sigma_e^2 & \text{if } i = j, \\ -\sigma_e^2 & \text{if } |i-j| = 1, \\ 0 & \text{otherwise}. \end{cases}$$

Throughout, it is useful to write $\sum_{t=1}^T \hat{u}_t^2 = \sum_{t=1}^T (u_t - (\hat{\alpha}-\alpha)y_{t-1})^2$, and note that $(\hat{\alpha}-\alpha) = \sum_1^T y_{t-1}u_t / \Sigma_1^T y_{t-1}^2$. Then (22) follows from Lemma 3.2.

Proof of Theorem 3.1. It is convenient to write $T^{-1}Z_\alpha = (\hat{\alpha}-1) - (1/2)(s^2 - s_u^2)/T^{-1}\sum_t^T y_{t-1}^2$. Given (19) and (20), $s^2 = \sigma_T^2 = \sigma_e^2(1+\theta_T)^2 = \sigma_e^2\delta^2/T$, the correction factor is:

$$(s^2 - s_u^2)\bigg/\left(T^{-1}\sum_t^T y_{t-1}^2\right) = (\delta^2\sigma_e^2/T)\bigg/\left(T^{-1}\sum_t^T y_{t-1}^2\right) - s_u^2\bigg/\left(T^{-1}\sum_t^T y_{t-1}^2\right). \quad \text{(A.1)}$$

It follows that,

$$T^{-1}Z_\alpha = (\hat{\alpha} - 1) + s_u^2 \Big/ \left(2T^{-1}\sum_1^T y_{t-1}^2\right) + o_p(1)$$

$$\Rightarrow -1 \Big/ \left(1 + \delta^2 \int_0^1 J_c(r)^2 dr\right)$$

$$+ \left(1 + 2\delta^2 \int_0^1 J_c(r)^2 dr\right) \Big/ \left(2\left(1 + \delta^2 \int_0^1 J_c(r) dr\right)^2\right).$$

Part (a) follows upon simplification. Part (c) follows from (a) and (b) of Lemma 3.2 and the definition of s^2. To prove part (d), note that given the definition of σ^2, $MSB = [T^{-1}\sum_t^T y_{t-1}^2/(\delta^2\sigma_e^2)]^{1/2}$, and the result follows from (a) of Lemma 3.2. Given that $Z_t = MSB \cdot Z_\alpha$, part (b) follows from parts (a) and (d). □

Proof of Theorem 3.2

Lemma 3.4. *Let* $\{y_t\}$ *be generated by* (18)–(19) *and* s_{AR}^2 *be obtained by applying OLS to* (11). *Then* $s_{\text{AR}}^2 \to 0$ *provided* $k/T \to 0$ *and* $k \to \infty$ *as* $T \to \infty$.

The proof of the Lemma is given in Perron and Ng (1995). Given that s_{AR}^2 and σ_T^2 have the same asymptotic limit, the results for Theorem 3.2 are the same as those for Theorem 3.1. □

Proof of Theorem 3.3. We start with a Lemma useful for deriving the limiting distribution of $M^{-1}s_{\text{WA}}^2$.

Lemma 3.5. *Let* $M \to \infty$ *and* $M/T \to 0$ *as* $T \to \infty$. *Then*

(a) $M^{-1}T^{-1}\sum_{k=1}^M w_k \sum_{t=k+1}^T u_t u_{t-k} \to 0$;
(b) $M^{-1}T^{-1}\sum_{k=1}^M w_k \sum_{t=k+1}^T y_{t-1} u_{t-k} \to 0$;
(c) $M^{-1}T^{-1}\sum_{k=1}^M w_k \sum_{t=k+1}^T y_{t-k-1} u_t \to 0$;
(d) $M^{-1}T^{-1}\sum_{k=1}^M w_k \sum_{t=k+1}^T y_{t-1} y_{t-k-1} \Rightarrow \sigma_e^2\psi\delta^2 \int_0^1 J_c(r)^2 dr$.

Proof. (a) Note that

$$
\begin{aligned}
T^{-1}\sum_{k=1}^{M} w_k \sum_{t=k+1}^{T} u_t u_{t-k} \\
&= T^{-1} w_1 \sum_{t=2}^{T} u_t u_{t-1} + T^{-1} \sum_{k=2}^{M} w_k \sum_{t=k+1}^{T} u_t u_{t-k}. \qquad \text{(A.2)}
\end{aligned}
$$

Consider the first term in (A.2). Using the fact that $u_t = e_t - (1 - \delta/\sqrt{T})e_{t-1}$,

$$
\begin{aligned}
w_1 T^{-1} \sum_{t=2}^{T} u_t u_{t-1} &= w_1 T^{-1} \sum_{t=2}^{T} e_t e_{t-1} - w_1(1-\delta/\sqrt{T})T^{-1} \sum_{t=2}^{T} e_{t-1}^2 \\
&\quad - w_1(1-\delta/\sqrt{T})T^{-1} \sum_{t=2}^{T} e_t e_{t-2} \\
&\quad + w_1(1-\delta/\sqrt{T})^2 T^{-1} \sum_{t=2}^{T} e_{t-1} e_{t-2}.
\end{aligned}
$$

Since $\{e_t\}$ is i.i.d., all terms vanish except the second, and the expression converges to $-w_1\sigma_e^2$ and to zero upon normalization by M^{-1}. To show that the second term in (A.2) vanishes, note that

$$
\begin{aligned}
T^{-1}\sum_{k=2}^{M} w_k \sum_{t=k+1}^{T} u_t u_{t-k} \\
&= T^{-1}\sum_{k=2}^{M} w_k \sum_{t=k+1}^{T} e_t e_{t-k} - T^{-1}(1-\delta/\sqrt{T}) \sum_{k=2}^{M} w_k \sum_{t=k+1}^{T} e_{t-1} e_{t-k} \\
&\quad - T^{-1}(1-\delta/\sqrt{T}) \sum_{k=2}^{M} w_k \sum_{t=k+1}^{T} e_t e_{t-k-1} \\
&\quad + T^{-1}(1-\delta/\sqrt{T})^2 \sum_{k=2}^{M} w_k \sum_{t=k+1}^{T} e_{t-1} e_{t-k-1}. \qquad \text{(A.3)}
\end{aligned}
$$

Now consider the first element. Define $\tilde{s}_e^2 = T^{-1}\sum_{t=1}^{T} e_t^2 + 2T^{-1}\sum_{k=1}^{M} w_k \sum_{t=k+1}^{T} e_t e_{t-k}$. We have

$$T^{-1}\sum_{k=2}^{M} w_k \sum_{t=k+1}^{T} e_t e_{t-k} = \left(\tilde{s}_e^2 - T^{-1}\sum_{t=1}^{T} e_t^2\right)\Big/ 2 - T^{-1} w_1 \sum_{t=2}^{T} e_t e_{t-1}.$$

This expression vanishes since $\tilde{s}_e^2 \to \sigma_e^2$ (provided $M/T \to 0$ as $T \to \infty$), $T^{-1}\sum_{t=1}^{T} e_t^2 \to \sigma_e^2$, and $T^{-1}\sum_{t=2}^{T} e_t e_{t-1} \to 0$ given that $\{e_t\} \sim$ i.i.d. Similar arguments apply to show that the remaining terms of (A.3) also converge to zero. To prove part (b), note that

$$\begin{aligned}
&M^{-1}T^{-1}\sum_{k=1}^{M} w_k \sum_{t=k+1}^{T} y_{t-1}u_{t-k} \\
&\quad = M^{-1}a_T w_1 T^{-1}\sum_{t=2}^{T} e_{t-1}^2 - M^{-1} w_1 a_T (1-\delta/\sqrt{T}) T^{-1}\sum_{t=2}^{T} e_{t-1}e_{t-2} \\
&\quad\quad + w_1 M^{-1} T^{1/2} b_T T^{-3/2}\left(\sum_{t=2}^{T} X_{t-1}e_{t-1} - (1-\delta/\sqrt{T})\right. \\
&\quad\quad \left.\times \sum_{t=2}^{T} X_{t-1}e_{t-2}\right) + M^{-1}T^{-1}\sum_{k=2}^{M} w_k \sum_{t=k+1}^{T} \\
&\quad\quad (a_T e_{t-1} + b_T X_{t-1})(e_{t-k} - (1-\delta/\sqrt{T})e_{t-k-1}).
\end{aligned}$$

Since $\alpha_T \to 1$ and $T^{-1}\sum_{t=2}^{T} e_{t-1}^2 \to \sigma_e^2, \alpha_T w_1 T^{-1}\sum_{t=2}^{T} e_{t-1}^2$ converges to $w_1\sigma_e^2$ and the first term vanishes upon normalization by M^{-1}. The next three terms vanish using $T^{1/2}b_T \to \delta$, $T^{-1}\Sigma e_{t-1}e_{t-2} \to \mathbf{0}$, and Lemmas 3.1 and 3.2. It remains to show that the last term vanishes. We have

$$\begin{aligned}
&M^{-1}T^{-1}\sum_{k=2}^{M} w_k \sum_{t=k+1}^{T} (a_T e_{t-1} + b_T X_{t-1})(e_{t-k} - (1-\delta/\sqrt{T})e_{t-k-1}). \\
&\quad = M^{-1}\left[a_T T^{-1}\sum_{k=2}^{M} w_k \sum_{t=k+1}^{T} e_{t-1}e_{t-k} - a_T(1-\delta/\sqrt{T})T^{-1}\right.
\end{aligned}$$

$$\times \sum_{k=2}^{M} w_k \sum_{t=k+1}^{T} e_{t-1}e_{t-k-1} + T^{1/2}b_T T^{-3/2}\left(\sum_{k=2}^{M} w_k \sum_{t=k+1}^{T} X_{t-1}e_{t-k} - (1-\delta/\sqrt{T})\sum_{k=2}^{M} w_k \sum_{t=k+1}^{T} X_{t-1}e_{t-k-1}\right)\Bigg]. \quad \text{(A.4)}$$

The first two terms of (A.4) converge to zero using arguments similar to those in part (a). Consider the third term (the behavior of the fourth is similar). Since $T^{1/2}b_T \to \delta$, we consider

$$\begin{aligned} M^{-1}T^{-3/2}\sum_{k=2}^{M} w_k \sum_{t=k+1}^{T} X_{t-1}e_{t-k} &= M^{-1}T^{-3/2}\sum_{k=2}^{M} w_k \sum_{t=k+1}^{T} \alpha^k X_{t-k-1}e_{t-k} \\ &\quad + M^{-1}T^{-3/2}\sum_{k=2}^{M} w_k \sum_{t=k+1}^{T} z_{t,k}, \end{aligned} \quad \text{(A.5)}$$

where $z_{t,k} = e_{t-k}\sum_{i=1}^{k}\alpha^{i-1}e_{t-i}$. Since $X_{t-k-1}e_{t-k} = (1/2\alpha)(X_{t-k}^2 - \alpha^2 X_{t-k-1}^2 - e_{t-k}^2)$ and $X_0 = 0$ by assumption, the first term of (A.5) simplifies to

$$(1/2\alpha)M^{-1}T^{-3/2}\sum_{k=2}^{M}\alpha^k w_k \times \left(X_{T-k}^2 - (\alpha^2-1)\sum_{t=k+1}^{T} X_{t-k-1}^2 - \sum_{t=k+1}^{T} e_{t-k}^2\right).$$

It is easy to see that the third term vanishes. For the first term, define the process on $D[0,1]$ as $X_T(s) = X_{[Ts]} = X_{j-1}, (j-1)/T \le s < j/T$ and $X_T(1) = X_T$. Now

$$\frac{1}{2\alpha}M^{-1}T^{-3/2}\sum_{k=2}^{M}\alpha^k w_k X_{T-k}^2 = \frac{1}{2\alpha}T^{-1/2}\int_0^1 \exp\left(\frac{c[Ms]}{T}\right) w\left(\frac{[Ms]}{M}\right) \times T^{-1}X_T^2\left(\frac{[T-s-1]}{T}\right) ds + o_p(1),$$

which converges to 0 since $T^{-1/2}X_T(s) \Rightarrow \sigma_e J_c(s)$. Similar arguments using the facts that $T(\alpha^2 - 1) \to 2c$ and $T^{-2}\sum_{t=k+1}^{T} X_{t-k-1}^2 \Rightarrow \sigma_e^2 \int_0^1 J_c(r)^2 dr$ show that the second term vanishes. Thus, the first term of (A.5) vanishes. It remains to show that the second term of (A.5) converges to zero. Since $E[z_{t,k}] = \alpha^{k-1}\sigma_e^2$, we can write the expression as

$$\begin{aligned} M^{-1}T^{-3/2}\sum_{k=2}^{M} w_k \sum_{t=k+1}^{T} (z_{t,k} - E[z_{t,k}]) \\ + M^{-1}T^{-3/2}\sum_{k=2}^{M} w_k \sum_{t=k+1}^{T} (\alpha^{k-1}\sigma_e^2) \\ = M^{-1}T^{-3/2}\sum_{k=2}^{M} w_k \sum_{t=k+1}^{T} (z_{t,k} - E[z_{t,k}]) + o(1). \end{aligned}$$

It can be shown that $\mathrm{Var}(\sum_{t=k+1}^{T} z_{t,k}) = 2\sigma_e^4\alpha^{2(k-1)}(T-k) + (T-k)\sigma_e^4(\alpha^{2k}-1)/(\alpha^2-1)$. Hence using an argument as in Newey and West (1987),

$$\begin{aligned} P\left\{\left|M^{-1}T^{-3/2}\sum_{k=2}^{M} w_k \sum_{t=k+1}^{T} (z_{t,k} - E[z_{t,k}])\right| > \varepsilon\right\} \\ \leqq \sum_{k=2}^{M} P\left\{\left|M^{-1}T^{-3/2}\sum_{t=k+1}^{T} (z_{t,k} - E[z_{t,k}])\right| > \varepsilon/CM\right\} \text{ since } w_k \leqq C \\ \leqq \sum_{k=2}^{M}\left[M^{-2}T^{-3}(T-K)(C^2M^2/\varepsilon^2)\sigma_e^4\left\{2\alpha^{2(k-1)} + \frac{\alpha^{2k}-1}{\alpha^2-1}\right\}\right] \end{aligned}$$

by the Cauchy–Schwartz inequality

$$\begin{aligned} \leqq \frac{M}{T}\frac{2\sigma_e^4}{T}\frac{(T/M)(\alpha^{2M}-1)}{T(1-\alpha^2)}C^2/\varepsilon^2 + \frac{M}{T}\frac{\sigma_e^4}{T(\alpha^2-1)} \\ \times\left(\frac{(T/M)}{T}\frac{(\alpha^{2M}-1)}{(\alpha^2-1)} - 1\right)C^2/\varepsilon^2 \to 0 \end{aligned}$$

since $T(\alpha^2-1)\to 2c$ and $(T/M)(\alpha^{2M}-1)\to 2c$ as $M/T\to 0$. This completes the proof for part (b). The proof of part (c) is analogous. To prove part (d), note that

$$\begin{aligned}
&M^{-1}T^{-1}\sum_{k=1}^{M} w_k \sum_{t=k+1}^{T} y_{t-1}y_{t-k-1}\\
&\quad= M^{-1}T^{-1}\sum_{k=1}^{M} w_k \sum_{t=k+1}^{T} \alpha_T^2 e_{t-1}e_{t-k-1}\\
&\qquad+ M^{-1}\alpha_T T^{1/2} b_T T^{-3/2}\left(\sum_{k=1}^{M} w_k \sum_{t=k+1}^{T} X_{t-1}e_{t-k-1}\right.\\
&\qquad\left.+\sum_{k=1}^{M} w_k \sum_{t=k+1}^{T} e_{t-1}X_{t-k-1}\right)\\
&\qquad+ M^{-1}Tb_T^2T^{-2}\sum_{k=1}^{M} w_k \sum_{t=k+1}^{T} X_{t-1}X_{t-k-1}.
\end{aligned}$$

The first three terms converge to zero using arguments similar to those in parts (a) and (b). We therefore concentrate on the fourth term, which we write as:

$$\begin{aligned}
&M^{-1}Tb_T^2T^{-2}\sum_{k=1}^{M} w_k \sum_{t=k+1}^{T} \alpha^k X_{t-k-1}^2\\
&\quad+M^{-1}Tb_T^2T^{-2}\sum_{k=1}^{M} w_k \sum_{t=k+1}^{T} X_{t-k-1}\left(\sum_{i=0}^{k-1}\alpha^i e_{t-i-1}\right).
\end{aligned}$$

The second term converges to zero by Lemma 3.1. Consider

$$\begin{aligned}
&Tb_T^2M^{-1}T^{-2}\sum_{k=1}^{M} w_k \sum_{t=1}^{T-k} \alpha^k X_{t-1}^2\\
&= Tb_T^2\sum_{k=1}^{M}\int_{(k-1)/M}^{k/M} w\left(\frac{[Ms]}{M}\right)\sum_{t=1}^{T-k}\int_{(t-1)/T}^{t/T}\exp\left(\frac{ck}{T}\right)T^{-1}X_T^2(r)drds
\end{aligned}$$

$$= Tb_T^2 \int_0^1 w\left(\frac{[Ms]}{M}\right) \int_0^{1-s(M/T)} \exp\left(cs\frac{M}{T}\right) T^{-1} X_T^2(r) dr ds$$

$$\Rightarrow \delta^2 \int_0^1 w(s)\sigma_e^2 \int_0^1 J_c(r)^2 dr ds = \delta^2 \psi \sigma_e^2 \int_0^1 J_c(r)^2 dr,$$

provided $M/T \to 0$. □

Lemma 3.6. *Let $\{y_t\}$ be generated by (18)–(20) and s_{WA}^2 be defined by (14), where $\hat{u}_t$ are* OLS *residuals obtained from (1). Let ψ be defined by (15) and $J_c(r)$ by (21). Then*

$$M^{-1} s_{\mathrm{WA}}^2 \Rightarrow \left(2\sigma_e^2 \delta^2 \psi \int_0^1 J_c(r)^2 dr\right)\left(1 + \delta^2 \int_0^1 J_c(r)^2 dr\right)^{-2}. \tag{A.6}$$

Proof. We first note that since $s_u^2 = O_p(1)$, the limit of $M^{-1}(s_{\mathrm{WA}}^2 - s_u^2)/2$ is the same as the limit of $M^{-1}s_{\mathrm{WA}}^2/2$. We have

$$\begin{aligned} M^{-1}(s_{\mathrm{WA}}^2 - s_u^2)/2 &= M^{-1}T^{-1}\sum_{k=1}^{M} w_k \sum_{t=k+1}^{T} \hat{u}_t \hat{u}_{t-k} \\ &= M^{-1}T^{-1}\sum_{k=1}^{M} w_k \sum_{t=k+1}^{T} \{u_t u_{t-k} - (\hat{\alpha} - \alpha) y_{t-1} u_{t-k} \\ &\quad -(\hat{\alpha} - \alpha) y_{t-k-1} u_t + (\hat{\alpha} - \alpha)^2 y_{t-1} y_{t-k-1}\}. \end{aligned} \tag{A.7}$$

Using (A.7), Lemma 3.5 and the result that $(\hat{\alpha} - \alpha) \Rightarrow -(1 + \delta^2 \int_0^1 J_c(r)^2 dr)^{-1}$ give the result stated in Lemma 3.6. Theorem 3.3 follows from Lemmas 3.1, 3.2 and 3.6 □

Proofs of results in Section 4

Parts (a) to (d) of the following Lemma is proved in Nabeya and Perron (1994), and part (e) is given in Perron and Ng (1995).

Lemma 4.1. *Let $\{y_t\}$ be a process given by* (23) *and* (24) *with $J_\phi(r)$ and $Q_c(J_\phi(r))$ as defined in* (25). *As $T \to \infty$:*

(a) $T^{-3/2}y_T \Rightarrow \sigma_e Q_c(J_\phi(1))$;
(b) $T^{-4}\sum_1^T y_t^2 \Rightarrow \sigma_e^2 \int_0^1 Q_c(J_\phi(r))^2 dr$;
(c) $T^{-3}\sum_1^T y_{t-1}u_t \Rightarrow (\sigma_e^2/2)\{Q_c(J_\phi(1))^2 - 2c\int_0^1 Q_c(J_\phi(r))^2 dr\}$;
(d) $T^{-2}\sum_1^T u_t^2 \Rightarrow \sigma_e^2 \int_0^1 (J_\phi(r)^2 dr$;
(e) $T^{-1}\sum_{t=k+1}^T \Delta y_{t-1}e_t \Rightarrow \sigma_e^2(c\int_0^1 Q_c(J_\phi(r))dW(r) + \int_0^1 J_\phi(r)\, dW(r))$ *if $k/T \to 0$ as $T \to \infty$.*

Theorem 4.1 and (26) follow from this Lemma and the definition $T^{-2}s^2 = T^{-2}\sigma_T^2 = \sigma_e^2/\phi^2$.

Proof of Theorem 4.2.

Lemma 4.2. *Let $\{y_t\}$ be a process given by* (23) *and* (24) *with $J_\phi(r)$ and $Q_c(J_\phi(r))$ as defined in* (25). *Let s_{AR}^2 be obtained by applying* OLS *to* (12) *with $k \to \infty$ and $k = o(T^{1/3})$ and let $T(\hat{b}(1) - b(1)) \to \eta$ with the random variable η defined as in Perron and Ng* (1995). *Then $T^{-2}s_{\mathrm{AR}}^2 \to \sigma_e^2/(c + \phi + \eta)^2$.*

The proof of Theorem 4.2 follows arguments analogous to those used in Theorem 4.1, with $(c + \phi + \eta)$ replacing ϕ.

Proof of Theorem 4.3. Since $T^{-1}s_u^2 \Rightarrow \lambda\sigma_e^2$, (see (26)), the limit of $M^{-1}T^{-1}s_{\mathrm{WA}}^2/2$ is the same as the limit of $M^{-1}T^{-1}(s_{\mathrm{WA}}^2 - s_u^2)/2$, which we write as

$$
\begin{aligned}
M^{-1}T^{-1}\frac{s_{\mathrm{WA}}^2 - s_u^2}{2} &= M^{-1}T^{-2}\sum_{k=1}^{M} w_k \sum_{t=k+1}^{M} \hat{u}_t\hat{u}_{t-k} \\
&= M^{-1}T^{-2}\sum_{k=1}^{M} w_k \sum_{t=k+1}^{T} u_t u_{t-k}
\end{aligned}
$$

$$-T(\hat{\alpha}-\alpha)M^{-1}T^{-3}\left(\sum_{k=1}^{M} w_k \sum_{t=k+1}^{T} y_{t-1}u_{t-k}\right.$$

$$\left.+\sum_{k=1}^{M} w_k \sum_{t=k+1}^{T} y_{t-k-1}u_t\right)+T^2(\hat{\alpha}-\alpha)^2$$

$$\times M^{-1}T^{-4}\sum_{k=1}^{M} w_k \sum_{t=k+1}^{T} y_{t-1}y_{t-k-1}. \qquad \text{(A.8)}$$

We note from (25) that $T(\hat{\alpha}-\alpha) = O_P(1)$. The next Lemma characterizes the limit of each term in (A.8).

Lemma 4.3. *Let $\{y_t\}$ be generated by (23)–(25) and let $M/T \to 0$ and $M \to \infty$ as $T \to \infty$. Let $\psi = \int_0^1 w(s)ds, Q_c(J_\phi(r))$ and $J_\phi(v)$ as defined in (25). Then*

(a) $M^{-1}T^{-2}\sum_{k=1}^{M} w_k \sum_{t=k+1}^{T} u_t u_{t-k} \Rightarrow \psi\sigma_e^2 \int_0^1 J_\phi(r)^2 dr;$
(b) $M^{-1}T^{-3}\sum_{k=1}^{M} w_k \sum_{t=k+1}^{T} y_{t-1}u_{t-k} \Rightarrow (\psi/2)\sigma_e^2(Q_c(J_\phi(1))^2 - 2c \int_0^1 Q_c(J_\phi(r))^2 dr);$
(c) $M^{-1}T^{-3}\sum_{k=1}^{M} w_k \sum_{t=k+1}^{T} y_{t-k-1}u_t \Rightarrow -(\psi/2)\sigma_e^2(Q_c(J_\phi(1))^2 - 2c \int_0^1 Q_c(J_\phi(r))^2 dr);$
(d) $M^{-1}T^{-4}\sum_{k=1}^{M} w_k \sum_{t=k+1}^{T} y_{t-1}y_{t-k-1} \Rightarrow \psi\sigma_e^2 \int_0^1 Q_c(J_\phi(r))^2 dr.$

Proof. To prove part (a), note that since $u_t = \rho u_{t-1} + e_t$, where $\rho = (1+\phi/T)$,

$$M^{-1}T^{-2}\sum_{k=1}^{M} w_k \sum_{t=k+1}^{T} u_t u_{t-k}$$

$$= M^{-1}T^{-2}\sum_{k=1}^{M} w_k \sum_{t=k+1}^{T} \rho^k u_{t-k}^2 + M^{-1}T^{-2}\sum_{k=1}^{M} w_k \sum_{t=k+1}^{T} z_{k,t} \qquad \text{(A.9)}$$

where $z_{k,t} = u_{t-k}(\sum_{t=0}^{k-1} \rho^i e_{t-i})$. Consider the first term and let $U_T(s) = u_{[Ts]} = u_{t-1}, (j-1)/T \leqq s < j/T$, and note that

$T^{-1/2}U_T(s) \Rightarrow \sigma_e J_\phi(s)$. Then

$$M^{-1}T^{-2}\sum_{k=1}^{M} w_k \sum_{t=k+1}^{T} \rho^k u_{t-k}^2$$

$$= \sum_{k=1}^{M}\int_{(k-1)/M}^{k/M} w\left(\frac{[sM]}{M}\right)\sum_{t=k+1}^{T}\int_{(t-1)/T}^{t/T} \exp\left(\phi s\frac{M}{T}\right)T^{-1}U_T(r)^2 dr ds$$

$$= \int_0^1 w\left(\frac{[sM]}{M}\right)\int_{sM/T}^{1} \exp\left(\phi s\frac{M}{T}\right)T^{-1}U_T(r)^2 dr ds$$

$$\Rightarrow \sigma_e^2\int_0^1 w(s)\int_0^1 J_\phi(r)^2 dr ds \equiv \psi\sigma_e^2\int_0^1 J_\phi(r)^2 dr,$$

provided $M/T \to 0$ and using Lemma 4.1. The second term in (A.9) converges to zero since $\sum_{t=k+1}^{T} e_t e_{t-k}$ is $O_p(\sqrt{T})$ and $M/T \to 0$ by assumption. The proofs to parts (b), (c) and (d) follow analogously using the results of Lemma 4.1.

Combining (25), Lemma 4.3 and (A.8), we have:

Lemma 4.4. *Let s^2_{WA} be defined by (14) with $M \to \infty$ and $M/T \to 0$ as $T \to \infty$. Let $\psi = \int_0^1 w(s)ds$ and λ be defined by (26). Then*

$$(MT)^{-1}s^2_{\text{WA}} \Rightarrow 2\sigma_e^2\psi\lambda.$$

The results of Theorem 4.3 follow directly from Lemmas 4.1 and 4.4. □

Proofs of results in Section 5

The following Lemma is proved in Nabeya and Perron (1994).

Lemma 5.1. *Let $\{y_t\}$ be generated by (27) and (28). Using the definitions following (29), we have, as $T \to \infty$:*

(a) $T^{-2}\sum_{t=1}^{T} y_{t-1}^2 \Rightarrow (\sigma_e^2/8)\int_0^1 (C(r)^2 + B(r)^2)dr$;
(b) $T^{-2}\sum_{t=1}^{T} y_{t-1}u_t \Rightarrow -(\sigma_e^2/4)\int_0^1 B(r)^2 dr$;
(c) $T^{-1}y_T^2 \Rightarrow (\sigma_e^2/8)A(1)^2$;
(d) $T^{-2}\sum_{t=1}^{T} u_t^2 \Rightarrow (\sigma_e^2/2)\int_0^1 B(r)^2 dr$.

To prove (30), we simply apply the above lemma to the definition of $T^{-1}s_u^2$. Theorem 5.1 uses the above lemma and the fact that $s^2 = \sigma_T^2 = \sigma_e^2/(2+\phi/T)^2 \to \sigma_e^2/4$. Since s_u^2 is $O_p(T)$ and s^2 is $O_p(1)$, $(s^2 - s_u^2)$ is dominated by s_u^2.

Proof of Theorem 5.2.

Lemma 5.2. *Let $\{y_t\}$ be generated by* (27) *and* (28). *Let s_{AR}^2 be obtained by applying* OLS *to* (12) *with $k = o(T^{1/3})$. Then $s_{\text{AR}}^2 \to \sigma_e^2/4$.*

The results of the theorem are obvious in view of Theorem 5.1.

Proof of Theorem 5.3. We begin by considering the limit of $M^{-1}T^{-1}(s_{\text{WA}}^2 - s_u^2)/2$, which we write as

$$\begin{aligned}M^{-1}T^{-1}(s_{\text{WA}}^2 - s_u^2)/2 &= M^{-1}T^{-2}\sum_{k=1}^{M} w_k \sum_{t=k+1}^{T} u_t u_{t-k}\\ &\quad - (\hat{\alpha}-\alpha)M^{-1}T^{-2}\sum_{k=1}^{M} w_k \sum_{t=k+1}^{T} y_{t-1}u_{t-k}\\ &\quad - (\hat{\alpha}-\alpha)M^{-1}T^{-2}\sum_{k=1}^{M} w_k \sum_{t=k+1}^{T} y_{t-k-1}u_t\\ &\quad + (\hat{\alpha}-\alpha)^2 M^{-1}T^{-2}\sum_{k=1}^{M} w_k \sum_{t=k+1}^{T} y_{t-1}y_{t-k-1}.\end{aligned} \tag{A.10}$$

The next Lemma characterizes the limit of each term in (A.10).

Lemma 5.3. *Let $\{y_t\}$ be generated by* (27) *and* (28) *and let $M/T \to 0$ and $M \to \infty$ as $T \to \infty$ with $\psi = \int_0^1 w(x)dx$, we have*

(a) $M^{-1}T^{-2}\sum_{k=1}^{M} w_k \sum_{t=k+1}^{T} u_t u_{t-k} \Rightarrow 0$;
(b) $M^{-1}T^{-2}\sum_{k=1}^{M} w_k \sum_{t=k+1}^{T} y_{t-1}u_{t-k} \Rightarrow 0$;
(c) $M^{-1}T^{-2}\sum_{k=1}^{M} w_k \sum_{t=k+1}^{T} y_{t-k-1}u_t \Rightarrow 0$;

(d) $M^{-1}T^{-2}\sum_{k=1}^{M} w_k \sum_{t=k+1}^{T} y_{t-1}y_{t-k-1} \Rightarrow \psi(\sigma_e^2/8)\int_0^1 (C(r)^2 - B(r)^2)dr$.

Proof. To prove part (a), note that since $u_t = \rho u_{t-1} + e_t$, where $\rho = -(1+\phi/T)$,

$$M^{-1}T^{-2}\sum_{k=1}^{M} w_k \sum_{t=k+1}^{T} u_t u_{t-k}$$

$$= M^{-1}T^{-2}\sum_{k=1}^{M} w_k \sum_{t=k+1}^{T} \rho^k u_{t-k}^2 + M^{-1}T^{-2}\sum_{k=1}^{M} w_k \sum_{t=k+1}^{T} z_{k,t} \tag{A.11}$$

where $z_{k,t} = u_{t-k}(\sum_{i=0}^{k-1}\rho^i e_{t-i})$. It is straightforward to show that the second term converges to zero. Consider the first term. We have,

$$M^{-1}T^{-2}\sum_{k=1}^{M} w_k \sum_{t=k+1}^{T} \rho^k u_{t-k}^2$$

$$= M^{-1}T^{-2}\sum_{k=1}^{M} \rho^k w_k \sum_{t=k+1}^{T} u_{t-k}^2$$

$$= M^{-1}T^{-2}\left(\sum_{j=1}^{M/2} \rho^{2j} w_{2j} \sum_{t=2j+1}^{T} u_{t-2j}^2 \right.$$

$$\left. - \sum_{j=1}^{M/2} (-\rho)^{2j-1} w_{2j-1} \sum_{t=2j}^{T} u_{t-2j+1}^2 \right)$$

$$= M^{-1}T^{-2}\left(\sum_{j=1}^{M/2} \exp\left(\frac{2j\phi}{T}\right) w_{2j} \sum_{t=2j+1}^{T} u_{t-2j}^2 \right.$$

$$\left. - \sum_{j=1}^{M/2} \exp\left(\frac{(2j-1)\phi}{T}\right) w_{2j-1} \sum_{t=2j}^{T} u_{t-2j+1}^2 \right) + o_p(1).$$

Define $w_m^*(r) = w_{2j}, 2(j-1)/M \leq r \leq 2j/M$ and let $U_T(s) = u_{[Ts]} = u_{j-1}$ for $j-1/T \leq s < j/T$. Now rewrite the first term (for sum over even terms) as:

$$\frac{1}{2}\left(\frac{M}{2}\right)^{-1}\sum_{j=1}^{M/2}\exp\left(\frac{2j\phi}{T}\right)w_{2j}\left(T^{-2}\sum_{t=2j+1}^{T}u_{t-2j}^2\right)$$

$$= \frac{1}{2}\sum_{j=1}^{M/2}\int_{2(j-1)/M}^{2j/M}\exp\left(\frac{2j\phi}{T}\right)w_{2j}ds\sum_{t=2j+1}^{T}\int_{t-1/T}^{t/T}T^{-1}U_T(r)^2dr$$

$$= \frac{1}{2}\int_0^1\exp\left(\frac{2[sM/2]\phi}{T}\right)w_M^*(s)ds\int_{(sM/2)/T}^{1}T^{-1}U_T(r)^2dr$$

$$= \frac{1}{2}\int_0^1 w(s)ds\frac{\sigma_e^2}{2}\int_0^1 B(r)^2dr = \frac{\sigma_e^2}{4}\psi\int_0^1 B(r)^2dr.$$

The second term (for sum over odd terms) can be shown to converge to the same functional, giving the result as stated in (a). Similar arguments can be used to show that the odd and even terms in parts (b) and (c) cancel. To show part (d),

$$\begin{aligned}
&M^{-1}T^{-2}\sum_{k=1}^{M}w_k\sum_{t=k+1}^{T}y_{t-1}y_{t-k-1}\\
&\quad= M^{-1}T^{-2}\sum_{k=1}^{M}w_k\sum_{t=k+1}^{T}\alpha^k y_{t-k-1}^2\\
&\qquad+ M^{-1}T^{-2}\sum_{k=1}^{M}w_k\sum_{t=k+1}^{T}y_{t-k-1}\left(\sum_{i=1}^{k}\alpha^{i-1}u_{t-i}\right).
\end{aligned} \tag{A.12}$$

Using Lemma 5.1 and provided $M/T \to \theta$ as $T \to \infty$, we have, for the first term of (A.12):

$$M^{-1}T^{-2}\sum_{k=1}^{M}w_k\sum_{t=k+1}^{T}\alpha^k y_{t-k-1}^2 \Rightarrow \psi\frac{\sigma_e^2}{8}\int_0^1(C(r)^2+B(r)^2)dr. \tag{A.13}$$

The second term of (A.12) can be written as

$$M^{-1}T^{-2}\sum_{k=1}^{M} w_k\alpha^{k-1}\sum_{t=k+1}^{T} y_{t-k-1}u_{t-k} + M^{-1}T^{-2}$$
$$\times\sum_{k=1}^{M} w_k\sum_{t=k+1}^{T} y_{t-k-1}\left(\sum_{i=1}^{k-1}\alpha^{i-1}u_{t-i}\right).$$

It is straightforward to show that the second term vanishes. For the first term,

$$M^{-1}T^{-2}\sum_{k=1}^{M} w_k\alpha^{k-1}\sum_{t=k+1}^{T} y_{t-k-1}u_{t-k} \Rightarrow -\frac{\psi\sigma_e^2}{8}\int_0^1 B(r)^2dr. \tag{A.14}$$

The result of (d) follows by combining (A.13) and (A.14). Therefore

$$M^{-1}T^{-2}(\hat{\alpha}-\alpha)^2\sum_{k=1}^{M} w_k\sum_{t=k+1}^{T} y_{t-1}y_{t-k-1}$$
$$\Rightarrow (\psi\sigma_e^2/2)\left[\left(\int_0^1 B(r)^2dr\right)\left(\int_0^1 (C(r)^2+B(r)^2)dr\right)^{-1}\right]^2$$
$$\times\left[\int_0^1 (C(r)^2+B(r)^2)dr\right]$$
$$\equiv \psi\sigma_e^2\lambda_2/2.$$

Combining (A.10), Lemma 5.3, (29) and (28), we have

Lemma 5.4. *Let s^2_{WA} be defined by (14), λ_2 be defined as in (28), let $M\to\infty$ with $M/T\to 0$ as $T\to\infty$. Then*

$$M^{-1}T^{-1}s^2_{\mathrm{WA}} \Rightarrow \sigma_e^2\psi\lambda_2.$$

The results of Theorem 5.3 follow directly from Lemmas 5.1 and 5.4.

□

Acknowledgments

The authors acknowledge grants from the Social Science and Humanities Research Council of Canada (SSHRC) and the Fonds de la Formation de Chercheurs et l'Aide à la Recherche du Québec (FCAR). The first author also thanks the National Science Foundation for financial support.

References

Andrews, D.W.K. (1991). Heteroskedastic and autocorrelation consistent covariance matrix estimation. *Econometrica* 59, 817–854.

Andrews, D.W.K. and Monahan, J. (1992). An improved heteroskedasticity and autocorrelation consistent covariance matrix estimator. *Econometrica* 60, 953–966.

Berk, K.N. (1974). Consistent autoregressive spectral estimates. *The Annals of Statistics* 2, 489–502.

Bhargava, A. (1986). On the theory of testing for unit roots in observed time series. *Review of Economic Studies* 53, 369–384.

Campbell, J.Y. and Perron, P. (1991). Pitfalls and opportunities, what macroeconomists should know about unit roots. *NBER Macroeconomics Annual* 6, 141–201.

Dickey, D.A. and Fuller, W.A. (1979). Distribution of the estimators for autoregressive time series with a unit root. *Journal of the American Statistical Association* 74, 427–431.

Franses, P.H. and Haldrup, N. (1994). The effects of additive outliers on tests of unit roots and cointegration. *Journal of Business and Economic Statistics* 12, 471–478.

Fuller, W.A. (1976). *Introduction to Statistical Time Series* (John Wiley: New York).

Hyslop, D. (1991). The performance of Phillips–Perron type unit root tests using the spectral density estimators (Manuscript, Princeton University).

Kim, K. and Schmidt, P. (1990). Some evidence on the accuracy of Phillips–Perron tests using alternative estimates of nuisance parameters. *Economic Letters* 34, 345–350.

Nabeya, S. and Perron, P. (1994). Local asymptotic distributions related to the AR(1) model with dependent errors. *Journal of Econometrics* 62, 229–264.

Nabeya, S. and Tanaka, K. (1990). Limiting powers of unit root tests in time series regressions. *Journal of Econometrics* 46, 247–71.

Newey, W.K. and West, K.D. (1987). A simple, positive semi-definite, heteroskedasticity and autocorrelation consistent covariance matrix. *Econometrica* 55, 703–708.

Ng, S. and Perron, P. (1995). Estimation and inference in nearly unbalanced, nearly cointegrated systems (Manuscript, Université de Montréal).

Pantula, S.G. (1991). Asymptotic distributions of unit root tests when the process is nearly stationary. *Journal of Business and Economic Statistics* 9, 63–71.

Park, J.Y. and Phillips, P.C.B. (1988). Regressions with integrated processes, part I. *Econometric Theory* 4, 468–498.

Perron, P. (1994). Non-stationarities and nonlinearities in Canadian inflation. In *Economic Behavior and Policy Choice under Price Stability*, Bank of Canada: Ottawa.

Perron, P. (1996). The adequacy of asymptotic approximations in the near-integrated autoregressive model with dependent errors. *Journal of Econometrics* 70, 316–350.

Perron, P. and Ng, S. (1995). Properties of the autoregressive spectral density estimator at frequency zero (Manuscript, Université de Montréal).

Phillips, P.C.B. (1987). Time series regression with unit roots. *Econometrica* 55, 277–302.

Phillips, P.C.B. and Ouliaris, S. (1990). Asymptotic properties of residual based tests for cointegration. *Econometrica* 58, 165–193.

Phillips, P.C.B. and Perron, P. (1988). Testing for a unit root in time series regression. *Biometrika* 75, 335–346.

Said, S.E. and Dickey, D.A. (1984). Testing for unit roots in autoregressive-moving average models of unknown order. *Biometrika* 71, 599–607.

Sargan, J.D. and Bhargava, A. (1983). Testing for residuals from least squares regression being generated by gaussian random walk. *Econometrica*, 51, 153–174.

Schwert, G.W. (1987). Effects of model specification on tests for unit roots in macroeconomic data. *Journal of Monetary Economics*, 20, 73–103.

Schwert, G.W. (1989). Tests for unit roots, a Monte Carlo investigation. *Journal of Business and Economic Statistics* 7, 147–160.

Stock, J.H. (1999). A Class of Tests for Integration and Cointegration. In Engle, R.F., and H. White (eds.), Cointegration, Causality and Forecasting. A Festschrift in Honour of Clive W.J. Granger. Oxford: Oxford University Press, 137–167.

Vogelsang, T.J. (1999). Two simple procedures for testing for a unit root when there are additive outliers, *Journal of Time Series Analysis*, 20, 237–252.

White, J. (1958). The limiting distribution of the serial correlation coefficient in the explosive case. *Annals of Mathematical Statistics* 29, 1188–1197.

Chapter 9

Unit Roots in the Presence of Abrupt Governmental Interventions with an Application to Brazilian Data*

Regina Celia Cati[†], Marcio G. P. Garcia[‡] and Pierre Perron[†,§,¶]

[†] *Department of Economics, Boston University, 270 Bay State Road, Boston, MA 02215, USA*
[‡] *PUC, Rio de Janeiro, RJ 22451-000, Brazil*
[§] *CRDE, Université de Montréal, Montréal, QC H3C-3J7, Canada*
[¶] *perron@bu.edu*

This paper considers econometric issues related to time-series data that have been subject to abrupt governmental interventions. The motivating example for this study is the Brazilian monthly inflation rate (1974:1–1993:6) which we use throughout for illustration. This series has been heavily influenced by the effect of the so-called shock plans implemented by various governments starting in the mid-1980s. The plans act as 'inliers' in the sense that the series is temporarily brought down to low levels before returning to its previous trend path. We analyse the effects on standard unit root tests and measures of persistence caused by the presence of these 'inliers'. We show a substantial bias in favor of concluding that the series is stationary and that shocks have temporary effects. We then construct appropriately corrected statistics which take into account the presence of the plans. These show, unlike the standard tests, that the stochastic behavior of the inflation rate was indeed unstable over this period. Simulation results are presented to support the adequacy of our corrected statistics.

Keywords: Structural change, inliers, shock plans, measures of persistence, explosive process, nonstationarity.

*Originally Published in *Journal of Applied Econometrics* 14, pp. 27–56.

1. Introduction

Nonstationarity in economic data can take various forms; for example, an autoregressive unit root or the presence of structural changes in a functional relation among a set of variables. In this paper, we discuss an alternative form of nonstationarity related to the effects of abrupt governmental interventions also referred to as 'shock plans'.

Our analysis is directly motivated by the time-series properties of the Brazilian inflation rate. This series is characterized by important increases starting in the early 1980s, turning into hyperinflation by the end of the 1980s. Yet this period of very high inflation has been marked by a few (five that are important until the early 1990s) 'shock plans' which have brought inflation to a low level for a short period of time. Intuition suggests that, in this highly volatile period with an ever-increasing trend path for inflation, standard statistical measures related to the issue of nonstationarity and the persistence of shocks would show the series to be highly persistent and nonstationary. Yet exactly the opposite occurs. Standard unit root tests suggest that inflation was stationary in that period and that shocks affected its level in a temporary manner. Indeed, standard measures suggest that inflation was 'more stationary' and less persistent in this hyperinflation period than in the 1970s when inflation was moderate.

An issue we want to analyse is whether these results are the artifact of the presence of the temporary changes created by the shock plans. To get some intuition on this issue, we can view these shock plans as creating 'inliers' whose magnitude is related to the current level of the series. Hence, if the series truly has a stochastic trend (i.e., a unit root) or even an explosive path, the magnitude of these 'inliers' are, themselves, nonstationary random variables which have a tendency to increase as inflation increases. Since these shock plans have failed, the series exhibits a tendency to return to its old (nonstationary) trend path after each episode. This is basically what contaminates the standard statistical measures, since the failures of the shock plans create a kind of spurious mean-reverting aspect to the series.

This argument is fundamentally the flip-side of the argument exposed in Perron (1989, 1990) where permanent changes in the trend function of a series with a stationary noise bias standard unit root tests and persistence measures towards accepting the unit root hypothesis and concluding that shocks have persistent effects. Here, temporary, but large, changes bias these measures in the opposite direction.

The problem is somewhat related to the analysis of Franses and Haldrup (1994) who showed how unit root tests have liberal size distortions when a series with a unit root is contaminated by additive outliers (see also Vogelsang, 'Two simple procedures for testing for a unit root when there are additive outliers', forthcoming in *Journal of Time Series Analysis*). The issue is, however, qualitatively different in two aspects. First, the occasional events occur for more than a single period, lasting usually several months. Second, and more importantly, the magnitude of the 'inliers' or shock plans is directly related to the actual level of the series and is, hence, a nonstationary random variable.

The aim of the paper is first to provide a detailed analysis of the statistical effects of such 'inliers' on standard statistical tools such as unit root tests and measures of persistence. The second goal is to provide modifications to these standard tests that directly take into account the presence of the shock plans. As we shall see, the answers obtained are dramatically different.

The structure of the paper is as follows. Section 2 describes in detail the data used and Section 3 briefly discusses the historical settings surrounding the shock plans implemented by the various Brazilian governments. The results obtained from the application of standard unit root tests and measures of persistence are presented in Section 4. The bias of the unit root tests in the presence of occasional shock plans is analysed in Section 5. Our results show that standard unit root tests are severely biased by the shock plans towards a rejection of the unit root hypothesis in favor of stationary fluctuations around a stable linear trend function. Accordingly, Section 6 considers modified versions of the tests that explicitly take into account the presence of the shock plans. Some simulations show that these modifications yield tests with correct sizes and the empirical results show a very different picture. Indeed, we no longer reject the unit root hypothesis in favor of stationary alternatives as appears intuitively plausible in a period of hyperinflation.

2. Description of the Data

The series used in this paper is the monthly Brazilian inflation rate for the period 1974:1 to 1993:6. Note that the choice of 1993:6 as the end of the sample is to avoid incorporating the Real Plan which is still in effect.[1] We use what is called the 'official inflation index'. This is actually a splice of several indices that was used by the government as the official index to all mandatory indexation schemes (for taxes, wages, etc.). This index was also widely employed by the financial markets and the central bank used it to calibrate the real interest rate. We applied two modifications to this 'official index'. First, since the price index is computed from an average of the daily prices from the beginning to the end of the month, the measured monthly inflation reflects price changes from the middle of the previous month to the middle of the current month. To obtain a better approximation of price changes from beginning to end of month, we used a geometric mean with equal weights of inflation over periods t and $t + 1$. Second, given the sudden and important changes in inflation caused by the shock plans, the usual continuity assumption that justifies the use of monthly averages breaks down. In order to mitigate the problems caused by averaging in this context, we used, for the months immediately following the plans, special price vectors computed by the government at the moment of each plan.

A graph of the inflation rate series is presented in Fig. 1. It is clear that this series is characterized in the 1980s by several sudden drops that are important in magnitude. These are the outcome of the various shock plans instituted by the government in an attempt to stop the process of high and increasing inflation.

[1] We also applied the same analysis to the Brazilian nominal interest rate, defined as the monthly compound overnight rate (basically the equivalent of the FED funds rate in the US). The two series having similar general properties, the results obtained are qualitatively similar. Hence, we do not report them.

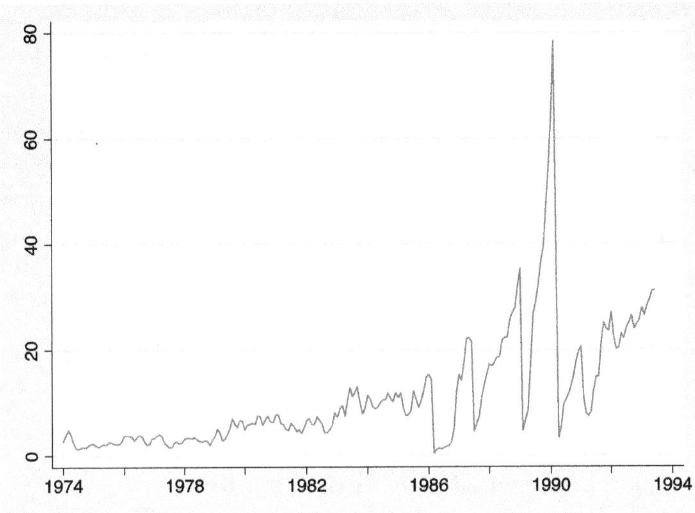

Fig. 1. Brazilian inflation rate, 1974:1 to 1993:6.

Table 1. List of shock plans in the Brazilian economy.

Name	Period	Length (months)	Decrease in inflation (%)
Cruzado	86 : 3–86:10	8	14.03 (97.7)
Bresser	87 : 7–87:9	3	16.75 (78.1)
Summer	89 : 2–89:4	3	30.64 (86.4)
Collor I	90 : 3–90:5	3	75.17 (95.8)[a]
Collor II	91 : 2–91:6	5	9.53 (46.2)

Note: [a]For this plan, the decrease is computed over the two months from 90:02 to 90:04.

Table 1 presents a summary of the various plans along with the dates we retained to define them and the magnitude of the decrease in inflation. The starting date of a plan was decided as the first month when over this month (and possibly the next one due to overlap) the decrease in inflation was at least 40% compared to its level in the preceding month. Choosing the ending date of a plan is somewhat more difficult. Our choice was guided both by historical records and by the use of dummy variables to create a real interest rate series with as few outliers as possible. More precisely, we estimated an error-correction model for the bivariate system that includes the inflation and nominal interest rates. The deterministic component included, besides the constant, various dummies for the plans. The ending dates of the plans were chosen so that the residuals from that error-correction model contained as few outliers as possible. It is important to note, however, that the results presented in this paper are not sensitive to minor variations in the choice of the ending dates for the plans.

Given the importance of the shock plans for the time-series behavior of the inflation rate series, we start with a brief historical overview.

3. A Brief History of the Shock Plans

After a short period of economic reforms in the mid-1960s the Brazilian economy grew sharply for almost a decade throughout the 1970s with a growth rate of GDP at 7.5%, on average. The yearly inflation rate was stable around 20% until the oil shock in 1973. The strategy adopted for economic development was a success because of a profitable combination of external financing and strong government support for private and public investment. This situation changed with the oil shocks in 1973 and in 1979 which were followed by an increase in the cost of external financing after the abrupt rise in interest rates at the beginning of the 1980s.

In contrast, high inflation rates and diminishing GDP growth rates were the norm starting in the 1980s (including negative rates in 1981 and, especially, in 1983). The first attempt to stabilize the economy was carried by a so-called orthodox economic team in 1982. The internal interest rate was raised above the international level; a plan for deficit reduction was proposed and a wage desindexation policy was adopted in order to restrict the internal aggregate demand. The external restriction imposed by the interest payments constraint and the lack of an international financial market that would provide financial aid inverted Brazil's former position in international trade. From 1984 on, the Brazilian trade balance was positive enough to meet international commitments. Despite the soundness of those economic decisions, the inflation rates did not fall below two digits a month. It was kept stable at around 150% per year. On the other hand, Brazilian GDP decreased by 2.0%, on average, in two years and it barely grew by 1.0%, on average, until the middle of the decade. Hence, this economic period was labeled as one of stagflation.

With the end of the dictatorship and the nomination of a new president at the beginning of 1985, the expectation was that a democracy would succeed in setting up a new economic order. However, high inflation rates and a lack of economic stability still persisted. After a long and deep recession, with high costs to the former government, the New Republic rulers decided to manage the situation without imposing more social costs. Hence, this time brought renewed discussions about the inflation rate and alternative proposals to manage it. The inertial inflation approach appeared as an alternative answer to the problem. Though the proponents of the inertial inflation approach agreed on a more orthodox diagnostic, no agreement was reached for the fight against inflation. For the inercialists, a traditional orthodox plan to stabilize the economy would imply high social costs for implementation. This alternative would require, probably, many years to bring the inflation rate down to a single-digit figure and there was reluctance to wait and accept more losses.

A monetary reform based on a general desindexation and a change of currency was the core of the Cruzado Plan in 1986, Brazil's first heterodox attempt to stabilize the economy. A price freeze was also deemed necessary to avoid extra income gain or losses during the stabilization plan. Hence, the Cruzado Plan was also followed by a general price freeze. With hindsight, it is possible to criticize the Cruzado Plan in the way they established the initial level of some key economic variables. The interest rate was set below the international level and most of the time was negative in real terms. Then a consumption bubble and price pressures from the demand side imposed pressures to put aside the Plan against inflation. In July 1986, many commodities in the supermarket disappeared and goods such as gas, gasoline, etc. were subject to rationing. However, election and political pressures delayed changes until November.

At that time, the government and its economic team could not keep the stabilization process under control, so the inflation rate again reached a two-digit figure per month (at about 14.5% in January 1987). The year following the Cruzado Plan showed high inflation rates, uncertainty and disagreement about the right economic policy to follow. At the same time, industrial production and investments started declining again.

By June, the inflation rates were out of control and the relative prices were disorganized. After a substitution of the Minister of Finance in April, a new stabilization plan was tried: the macroeconomic Consistency Plan, also called the Bresser Plan. Even though this was an attempt to correct the wrong path taken earlier it insisted on freezing prices again. As with the Cruzado Plan, the Bresser Plan was unable to solve the problem of the public deficit. A lack of control of this deficit and additional political pressures hampered any attempt to cut spending and the Finance Minister could do no more than watch further increases in inflation.

After another Finance Minister substitution in January 1988, Brazil witnessed the highest inflation rates in its history and another stabilization, called the Summer Plan, was tried a year later. It was an attempt analogous to that of Argentina. This Plan was based upon a tight monetary policy; interest rates were raised far above their historical levels and the government took the opportunity to change the feature of its internal debt (the government changed LFT (Treasury Financial Notes) by BBC (Central Bank Notes) and other longer-maturity debt instruments). Once more, the fiscal situation was not solved and the inflation rate rose again. By the end of Sarney's term, it had reached as high as 85% a month. The country was close to hyperinflation and economic chaos.

This path persisted until March 1990, when Fernando Collor de Melo became the new president. His first economic decision was a monetary reform that sequestered about 75% of all financial assets. They were converted into long-run deposits under the responsibility of the Brazilian Central Bank. The money supply fell sharply and, accordingly, so did the inflation rate, and economic activity registered a strong contraction. The plan was so efficient in the short run that, after two months, the inflation rate appeared stable at a low level. Internal debt and the interest payments decreased giving a short breathing space to the Treasury's financial operation. However, as the total debt problem was not solved and government spending was not managed, inflation went back to 20% a month in December 1990.

At the beginning of 1991, the economic team tried again to stabilize the economy through a mixture of price freezes and public spending cuts. That was the Collor II Plan which lasted less than four months. Due to a severe recession, the inflation rate stabilized around 20% a month for the rest of President Collar's term. Political problems, corruption and the increase in uncertainty are ingredients for a more general crisis that ended with the impeachment of President Collor in December 1992. He was replaced by the vice-president, Itamar Franco, who inherited an inflation rate close to 30% a month. This inflationary feature continued until July 1994 (at about 50% a month), when a new and, up to now, successful plan was introduced, the Real Plan.

4. Empirical Results with Standard Unit Root Tests

In this section we discuss empirical results obtained with the application of some standard unit root tests. By that we mean tests that do not take into account the presence of the shock plans. We start with a description of the statistics as well as a measure of persistence. The empirical results are then discussed highlighting the potential problems of this standard approach.

4.1. *The test statistics*

For our analysis we use three tests for the presence of an autoregressive unit root. The first two are by now standard tools in the analysis of univariate data, namely the Augmented Dickey–Fuller (1979) test (labeled ADF) and the Phillips–Perron (1988) test based on the normalized bias in a first-order autoregression (labeled Z_α). We also consider a new test suggested by Stock (1990) and further analysed by Perron and Ng (1996) which is a modification of the

Phillips–Perron test that is less subject to size distortions in the presence of serial correlation in the first-differences of the data (this test is labeled MZ_α). Finally, we consider a measure of the persistence of shocks based on an autoregressive spectral density estimator at frequency zero.

The class of processes considered, y_t, can be described as follows:

$$\begin{aligned} y_t &= \mu + \beta t + z_t, \\ A(L)z_t &= B(L)e_t, \end{aligned} \tag{1}$$

where $A(L) = 1 - a_1 L - a_2 L^2 - \cdots - a_p L^p$ is a pth-order autoregressive polynomial in the lag operator L (defined such that $Lx_t = x_{t-1}$). Similarly $B(L)$ is a qth-order moving-average polynomial defined by $B(L) = 1 + b_1 L + b_2 L^2 + \cdots + b_q L^q$. The errors $\{e_t\}$ are assumed to be martingale differences (e.g., uncorrelated but not necessarily homoscedastic). System (1) simply describes a process that is the sum of a deterministic time trend (a first-order polynomial in time) and a noise function modeled as an ARMA process. Of course, more general processes are possible, but for simplicity of exposition we consider this leading case of interest.

The null hypothesis is that one autoregressive root is unity, i.e., we have the factorization $A(L) = (1 - L)A^*(L)$ with all the roots of $A^*(L)$ outside the unit circle. This implies that the sum of the autoregressive coefficients is unity. The usual alternative hypothesis is that this sum is less than one (in which case z_t is stationary) but given the nature of the series analysed here we also consider the alternative hypothesis that this sum is greater than one, i.e., z_t is an explosive process.

The ADF tests of Dickey and Fuller (1979) (also extended by Said and Dickey, 1984 to the case of data having an ARMA structure) is based on the idea that a stationary and invertible ARMA process can be approximated by an autoregression. Hence, the relevant regression estimated by OLS is:

$$y_t = \eta + \gamma t + \alpha y_{t-1} + \sum_{j=1}^{k} c_j \Delta y_{t-j} + v_t. \tag{2}$$

Here, α is the sum of the autoregressive coefficients and the null hypothesis can be tested using the t-statistic for $\alpha = 1$. To select k, we use Bayesian Information Criterion (BIC). This is adequate provided the noise component z_t does not contain roots that are close to the unit circle, in particular strong negative MA components. This is indeed the case with the inflation series under investigation.[2]

The unit root test of Phillips and Perron (1988) is based on a nonparametric correction of the autoregressive estimate, $\hat{\alpha}$, in the following first-order autoregression:

$$y_t = \hat{\eta} + \hat{\gamma}t + \hat{\alpha}y_{t-1} + \hat{u}_t. \tag{3}$$

Let

$$s_u^2 = T^{-1}\sum_{t=1}^{T} \hat{u}_t^2$$

and $\tilde{y}_{t-1}$ be the residuals from a regression of y_{t-1} on a constant and a time trend. The test is defined as:

$$Z_\alpha = T(\hat{\alpha} - 1) - (s^2 - s_u^2) \Big/ \left(2T^{-2}\sum_{t=1}^{T} \tilde{y}_{t-1}^2\right), \tag{4}$$

where s^2 is a consistent estimate of the spectral density function at frequency zero of Δz_t under the null hypothesis of a unit root, denoted $h_{\Delta z}(0)$. Conventional estimators are based on a kernel method that constructs a weighted sum of the empirical autocovariances of the estimated residuals $\hat{u}_t$ (see e.g., Andrews, 1991). However, Perron and Ng ('An autoregressive spectral density estimator at frequency zero for nonstationarity tests', forthcoming in *Econometric Theory*) found this estimator to be inferior to an autoregressive spectral density estimator based on the first-differences of the data,

[2]An alternative method, following Campbell and Perron (1991) and Ng and Perron (1995), is to use a data-dependent method based on a general-to-specific recursive procedure. We also verified that the basic qualitative results are not sensitive to using this alternative strategy.

defined by

$$s^2 = s_{ek}^2/(1 - \hat{b}(1))^2 \tag{5}$$

with

$$s_{ek}^2 = T^{-1}\sum_{t=1}^{T} \hat{e}_{tk}^2, \quad \hat{b}(1) = \sum_{j=1}^{k} \hat{b}_j,$$

where $\hat{b}_j$ and $\{\hat{e}_{tk}\}$ are obtained from a kth-order augmented autoregression in Δy_t:

$$\Delta y_t = c + b_0 y_{t-1} + \sum_{j=1}^{k} b_j \Delta y_{t-j} + e_{tk}. \tag{6}$$

The consistency of s^2 for $h_{\Delta z}(0)$ under the null hypothesis of a unit root follows from the results of Said and Dickey (1984) and Berk (1974). Again, k is selected as that value which minimizes the BIC criterion. An important point to note is that s^2 is bounded above by zero even under the alternative of a stationary noise function z_t. This ensures the consistency of the modified statistic which we now describe.

Stock (1990) proposed a class of statistics which exploits the feature that a series converges with different rates of normalization under the null and alternative hypotheses. We consider one such test, referred to as MZ_α, defined by

$$MZ_\alpha = (T^{-1}\tilde{y}_T^2 - s^2) \Big/ \left(2T^{-2}\sum_{t=1}^{T} \tilde{y}_{t-1}^2\right), \tag{7}$$

where again $\tilde{y}_t$ are the residuals from a regression of y_t on a constant and a trend and s^2 is the autoregressive spectral density estimator defined by Equations (5) and (6). We can view MZ_α as a modified Phillips–Perron test. These issues are examined in detail in Perron and Ng (1996) where, in particular, it is demonstrated that this test has superior size and power properties for a wide range of data-generating processes.

A topic that has received substantial attention recently is the measure of the persistence of shocks on the level of a given series.

Here, the concept of persistence relates to the long term effect of a shock e_t in Equation (1) on the level of y_t (see e.g., Cochrane, 1988; Campbell and Mankiw, 1987). All the measures of persistence proposed are directly related to the normalized spectral density function at frequency zero of the first-differences of a series, $f_{\Delta y}(0) = h_{\Delta y}(0)/\sigma^2_{\Delta y}$ where $\sigma^2_{\Delta y}$ is the variance of the first differences of the series y_t. For example, if the series is trend-stationary, $f_{\Delta y}(0)$ is 0 and the series exhibits no persistence. For a random walk, it is one and shocks have a one-for-one effect on the long-term level of the series. When $0 < f_{\Delta y}(0) < 1$, shocks have a permanent effect but their influence is attenuated over time. If $f_{\Delta y}(0) > 1$, their effect is exacerbated over time. Hence, an estimate of $f_{\Delta y}(0)$ provides valuable information on the characteristics of a series. Here, we consider the estimate $\hat{f}_{\Delta y}(0) = \hat{h}_{\Delta y}(0)/\hat{\sigma}^2_{\Delta y}$ where

$$\hat{\sigma}^2_{\Delta y} = T^{-1}\sum_{t=1}^{T}(\Delta y_t - \overline{\Delta y})^2,$$

the sample variance of Δy_t and where $\hat{h}_{\Delta y}(0)$ is an autoregressive spectral density estimate at frequency 0 defined by

$$\hat{h}_{\Delta y}(0) = s^2_{ek}/(1 - \hat{d}(1))^2 \tag{8}$$

with

$$s^2_{ek} = T^{-1}\sum_{t=1}^{T}\hat{e}^2_{tk}, \quad \hat{d}(1) = \sum_{j=1}^{k}\hat{d}_j,$$

where $\hat{d}_j$ and $\{\hat{e}_{tk}\}$ are obtained from the following kth-order augmented autoregression in Δy_t:

$$\Delta y_t = c + \sum_{j=1}^{k} d_j \Delta y_{t-j} + e_{tk}. \tag{9}$$

Note that Equation (9) differs from (6) in that the lagged level y_{t-1} is not included. This ensures consistency under both the null and stationary alternative hypotheses and a more efficient estimator under the null hypothesis of a unit root. The truncation lag is again selected using the BIC criterion.

4.2. *Empirical results*

We applied the tests discussed above to the Brazilian monthly inflation rate series for the period 1974:1–1993:6. The results are presented in Tables 2 and 3. The strategy adopted was to conduct the tests for the full sample and various subsamples with and without shock plans. Consider first the results for the full sample. All three unit root tests concur for an overwhelming rejection of the null hypothesis of a unit root in favor of stationary fluctuations. All statistics are significant at the 1% level (the critical values, from Fuller, 1976, are –29.5 for Z_α and MZ_α and –3.96 for ADF). We

Table 2. Empirical results for unit root tests.

					ADF		
Sample	Z_α	k	MZ_α	k	t_α	$\hat{\alpha}$	k
74:1–93:6	-41.55^a	2	-37.57^a	2	-6.61^a	0.75	1
74:1–79:12	-9.12	2	-7.55	2	-2.43	0.79	2
74:1–84:12	-22.92^c	4	-19.17^d	4	-2.63	0.84	4
74:1–89:8	-45.20^a	1	-38.25^a	1	-5.09^a	0.72	1
80:1–89:8	-31.22^a	1	-26.67^b	1	-3.75^b	0.75	0
80:1–93:6	-30.65^a	1	-27.87^b	1	-5.65^a	0.73	1
85:1–93:6	-19.04^d	1	-17.38	1	-4.48^a	0.73	1

Note: The superscripts a, b, c and d denote significance in favor of stationary alternatives at the 1%, 2.5%, 5% and 10% levels, respectively.

Table 3. Empirical results for $\hat{f}_{\Delta y}(0)$.

	k										
Sample	1	2	3	4	5	6	7	8	9	10	*kbic* (k)
74:1–93:6	1.58	0.97	0.86	0.65	0.59	0.51	0.41	0.35	0.25	0.20	0.97 (2)
74:1–79:12	1.74	0.70	0.59	0.37	0.48	0.37	0.22	0.29	0.29	0.27	0.70 (2)
74:1–84:12	1.08	0.43	0.57	0.33	0.31	0.29	0.34	0.29	0.25	0.21	0.33 (4)
74:1–89:8	1.09	0.93	0.67	0.42	0.34						1.09 (1)
80:1–89:8	1.09	0.94	0.68	0.43	0.35						1.09 (1)
80:1–93:6	1.58	0.97	0.87	0.66	0.60	0.52	0.42	0.36	0.25	0.21	0.97 (2)
85:1–93:6	1.61	1.00	0.88	0.68	0.62	0.53	0.43	0.37	0.26	0.22	1.00 (2)

note, for further comparisons, that the measure of persistence given by the estimate of the spectral density function at the origin of the first-differences of the data is 0.97 with the order k chosen by BIC, a value substantially above 0 which contrasts with the unit root tests. Note, however, that this measure of persistence is sensitive to the order k selected. Indeed, we see a substantial decrease in $\hat{f}_{\Delta y}(0)$ as k increases.

Consider now the results for various subsamples, starting with those that do not contain shock plans. For 1974:1 to 1979:12, all tests agree on a nonrejection of the unit root at any conventional significance level. For 1974:1 to 1984:12, the results are mixed; the ADF test does not allow for a rejection while the Z_α and MZ_α tests do so at the 5% and 10% levels, respectively. Overall, the results suggest that the period prior to the shock plans and the high inflation is characterized by stochastic nonstationarity and persistence of shocks. This is confirmed by the estimate $\hat{f}_{\Delta y}(0)$ which is 0.70 for the period 1974:1 to 1979:12, again well above 0 (for the period 1974:1 to 1984:12 it is 0.33).

We now turn to the results concerning the subsamples that contain shock plans. The results for the subperiods 1980:1 to 1993:6 and 1985:1 to 1993:6 are very similar to those for the full sample. Most tests agree for a rejection of the unit root, the rejections being stronger using the sample 1980:1 to 1993:6. It is of interest to note that the estimates of the measure of persistence $\hat{f}_{\Delta y}(0)$ are 0.97 and 1.00 (almost the same as that for the full sample). These values suggest substantial persistence of shocks contrary to the unit root tests. In particular, it is important to note that $\hat{f}_{\Delta y}(0)$ is higher when a strong rejection of the unit root occurs (i.e., when including the plans) than when a rejection is not possible (i.e., not including subperiods with plans). These results offer a conflicting picture of the properties of the data.

Finally, we also consider the subsamples 1974:1 to 1989:8 and 1980:1 to 1989:8 to assess whether the results are due to the presence of the very large increase in the inflation rate during the period surrounding the Collor I plan. Hence, for these two subsamples, only the first three plans are present. The results show the same pattern as with the full sample, namely a strong rejection of the unit root and a value of $\hat{f}_{\Delta y}(0)$ at 1.09 which suggests, on the contrary, high persistence. As with the full sample, this estimate decreases substantially as the order k increases. Hence, the results are robust to excluding the most dramatic period associated with the Collor I plan.

The results of this section suggest the following perplexing conclusion. The inflation rate is characterized by stochastic nonstationarity and persistence of shocks prior to the emergence of very high levels of inflation and the institution of the various shocks plans. The opposite holds for the period of high inflation with occasional shocks plans. For that period, fluctuations in inflation appear as stationary deviations around a stable linear trend function and shocks accordingly have effects that dissipate quickly (given the low value of the sum of the autoregressive coefficients).

These results are perplexing because they are contrary to what intuition would suggest. Indeed, one would expect nonstationary (or erratic) behavior to occur especially in a period of uncontrolled growth in inflation and a failed attempt at stabilizing its level. Yet standard tests suggest the opposite.

Our argument is that the results are simply artifacts created by the occasional presence of short but important shock plans. Indeed, the plans act in such a way that the level of the series is brought temporarily to a low level. Since the plans in the period considered have all failed quickly, inflation has returned to its old trend path. This is a manifestation of a mean-reverting behavior that also characterizes a stationary series. Since the decreases and subsequent increases are so important they are likely to contaminate the statistical tests used.

The question of interest is then whether inflation, in periods when shock plans are not into effect, is characterized by a trend path that is unstable (stochastically nonstationary with a unit root) or is even of an explosive nature. To answer this question, the tests used so far must be modified to isolate the effect of the shock plans. These modifications are the object of the following sections. Before presenting them, we first turn to the issue of the possible bias on the unit root tests and the measure of persistence caused by the shock plans.

5. Biases Caused by the Presence of 'Inliers'

In this section we present simple simulation experiments that aim at quantifying the bias on the size of the unit root tests and the mean of the persistence measure created by the presence of shock plans (or 'inliers') that are short-lived but important in magnitude. The results will show how shock plans can create spurious mean-reverting behavior that would lead an investigator to conclude that the time series is stationary over the whole sample when using unit root tests.

5.1. *Description of the experiments*

The data are first generated according to the following simple random walk with drift interrupted by occasional 'inliers' or shock plans:

$$\begin{aligned} y_t &= y_0 + \mu t + S_t \quad t = 1, \ldots, T \quad \text{and} \quad t \notin \{t_{i,j}\}, \\ y_t &= a \quad \text{for} \quad t \in \{t_{i,j}\} \quad (j = 1, \ldots, p;\ i = 1, \ldots, n_j), \end{aligned} \tag{10}$$

where

$$S_t = \sum_{j=1}^{t} e_j.$$

Here, $t_{i,j}$ refers to the time index of the ith observation of plan j. There are p shock plans and each contains n_j observations. The series is a random walk with drift μ except when a plan is in effect, in which case the level of the series drops to a value a. To complete the specifications, the errors $\{e_t\}$ are independent $N(0,\ 1)$ random variables, the initial condition is $y_0 = a$ (so that the plans, in effect, bring the level of inflation to its initial value).[3]

[3]We also considered a slight modification of this data-generating process when the plans bring the level of inflation to half its value the month before the plan. The results, reported in a previous version of the paper, are qualitatively similar and, hence, for brevity we omit them.

We used the following specific values for the parameters. First $a = 4$ which can be viewed as an initial level of 4% for the inflation rate. There are $p = 3$ plans irrespective of the sample size and each plan contains $n_j = 6$ ($j = 1, 2, 3$) observations corresponding to plans lasting 6 months. A key parameter is the drift μ which specifies how fast the deterministic trend component increases. We consider four values ranging from mild to rapid growth: $\mu = 0.1$, 0.2, 0.4 and 0.8. The specification of this trend component is important because it basically dictates the magnitude of the decrease occurring with a shock plan. The faster the rate of growth, the larger the decrease and the likely importance of the spurious effect on the unit root tests. We consider three different sample sizes, $T = 150$, 250 and 500. Associated with each of these are the starting dates of the plans. These are $\{40, 70, 120\}$ for $T = 150$, $\{150, 170, 220\}$ for $T = 250$, and $\{250, 350, 450\}$ for $T = 500$. It is important to note that as the sample size increases the number of plans remains the same but the decreases caused by the plans are more important since they occur when the level of the series is higher.

Given the possibility that the noise component for Brazilian inflation is explosive, we also considered experiments with such an explosive process interrupted by shock plans. The set-up is exactly the same as described above except that the process describing the behavior of the series when shock plans are not in effect is given by:

$$\begin{aligned} y_t &= y_0 + \mu t + Z_t, \\ Z_t &= \alpha Z_{t-1} + e_t, \end{aligned} \tag{11}$$

with $Z_0 = 1$. In our experiments, we considered $\alpha = 1.01$ and $\alpha = 1.02$. All the other parameter configurations are exactly as for the unit root case.

While these data-generating processes are simple they are rich enough to obtain an overview of the bias on unit root tests caused by temporary shock plans or 'inliers'.

5.2. *Description of the results*

We used 1000 replications for each specification to compute the exact size of the unit root tests. The nominal size of the test is 5% and the

Table 4. Exact size of unit root tests in the presence of shock plans (5% nominal size).

	$T = 150$			$T = 250$			$T = 500$		
	Size	Mean	S.E.	Size	Mean	S.E.	Size	Mean	S.E.
(a) Z_α									
$\mu = 0.1$	0.59	−23.21	8.03	0.78	−34.66	13.82	0.92	−58.93	22.70
$\mu = 0.2$	0.80	−26.75	6.68	0.97	−45.17	8.93	1.00	−82.40	11.96
$\mu = 0.4$	0.98	−30.39	3.06	1.00	−51.46	3.08	1.00	−93.63	3.90
$\mu = 0.8$	1.00	−31.62	1.27	1.00	−53.12	1.34	1.00	−96.81	1.46
(b) MZ_α									
$\mu = 0.1$	0.51	−21.35	7.34	0.76	−32.10	12.80	0.92	−55.25	21.01
$\mu = 0.2$	0.75	−24.74	6.15	0.97	−41.85	8.08	1.00	−76.70	10.75
$\mu = 0.4$	0.97	−28.19	2.75	1.00	−47.54	2.70	1.00	−86.68	3.41
$\mu = 0.8$	1.00	−29.34	1.13	1.00	−49.03	1.16	1.00	−89.47	1.27
(c) ADF									
$\mu = 0.1$	0.48	−3.25	0.67	0.76	−4.03	1.00	0.91	−5.38	1.44
$\mu = 0.2$	0.69	−3.50	0.53	0.96	−4.67	0.73	1.00	−7.04	1.46
$\mu = 0.4$	0.94	−3.76	0.21	1.00	−5.00	0.48	1.00	−9.01	0.94
$\mu = 0.8$	1.00	−3.84	0.08	1.00	−5.03	0.28	1.00	−9.55	0.13

critical values are taken from Fuller (1976) (–21.8 for Z_α and MZ_α and –3.41 for ADF). We also report the mean and standard deviation of the statistics. The program was coded in Gauss using the RNDNS routine to generate the random numbers. To allow proper comparison of differences across cases, we used the same starting seed for each entry, arbitrarily set at 12345. The results for the unit root case are presented in Table 4.

We first note that in all cases the tests are severely oversized, so much as to be useless to provide a characterization of the non-stationary nature of the series. Consider the case of Z_α presented in panel (a) of Table 4. With a sample size $T = 150$ and a small drift $\mu = 0.1$, the test would incorrectly reject the unit root in favor of stationary deviations around a linear trend in 59% of the cases. This false rate of rejection increases as the drift μ and the sample size T increases, and quickly reaches 100% (for example, when $\mu = 0.4$ and $T = 250$, which roughly characterizes the Brazilian inflation series).

The same qualitative results hold for the tests MZ_α (panel (b)) and ADF (panel (c)). The rates of rejections are only marginally lower compared to those with Z_α. It is interesting to note that the mean of the statistics seems to approach some limiting value as μ increases, keeping a fixed sample size. This limiting value is well below the respective 5% critical value. The concentration also increases given that the standard error decreases. This implies a limiting rate of rejections of 100% as μ increases keeping T fixed but without the tests diverging to minus infinity. On the other hand, when μ is kept fixed and T increases the means of the statistics grow more negative (perhaps diverging to minus infinity) but the standard errors also increase.

The results presented here clearly show that short but abrupt shock plans can bias the tests statistics against the unit root hypothesis in favor of stationary fluctuations around a stable linear trend function. This is an undesirable feature since the time span covered by the plans are very short compared to the whole sample (18 'months' in samples of 150 to 500 'months').

Consider now the behavior of the persistence measure $\hat{f}_{\Delta y}(0)$. To better assess potential biases, we slightly generalized the data-generating process described by Equation (10) to include serial correlation in the errors to permit some variations in the true values of the persistence of shocks. Hence, we now have

$$S_t = \sum_{j=1}^{t} v_t,$$

with v_t generated by the first-order autoregression $v_t = \rho v_{t-1} + e_t$. With $e_t \sim$ i.i.d. N(0, 1), $f_{\Delta y}(0) = (1 - \rho^2)/(1 - \rho)^2$. We used three values of ρ, namely 0.0, 0.5 and –0.5 with corresponding values of $f_{\Delta y}(0)$ given by 1.0, 3.0 and 1/3. The results are presented in Table 5.[4] These are quite striking. For large T and/or large μ, the mean of $\hat{f}_{\Delta y}(0)$ is close to 1.0 when k is less than 6 (the length of the

[4]To conserve space, only the means and not the standard deviations are presented. The latter carry little useful information. They are generally quite small. They decrease with an increase in T or μ and increase with an increase in the order k.

Table 5. Mean of $\hat{f}_{\Delta y}(0)$ with shock plans.

	k										
	1	2	3	4	5	6	7	8	9	10	*kbic*
					(a) $\rho = 0.0$						
$T = 150$											
$\mu = 0.1$	0.99	0.98	0.98	0.97	0.96	0.46	0.45	0.44	0.44	0.43	0.73
$\mu = 0.2$	1.00	1.00	1.00	1.01	1.01	0.40	0.39	0.39	0.39	0.39	0.53
$\mu = 0.4$	1.01	1.01	1.02	1.02	1.03	0.36	0.36	0.36	0.36	0.36	0.36
$\mu = 0.8$	1.01	1.01	1.02	1.03	1.03	0.35	0.35	0.35	0.36	0.36	0.35
$T = 250$											
$\mu = 0.1$	1.00	0.99	0.99	0.99	0.99	0.43	0.43	0.43	0.43	0.43	0.51
$\mu = 0.2$	1.00	1.00	1.01	1.01	1.01	0.36	0.36	0.37	0.37	0.37	0.37
$\mu = 0.4$	1.00	1.01	1.01	1.02	1.02	0.34	0.35	0.35	0.35	0.35	0.34
$\mu = 0.8$	1.00	1.01	1.01	1.02	1.02	0.34	0.34	0.35	0.35	0.35	0.34
$T = 500$											
$\mu = 0.1$	1.00	1.00	1.00	1.00	1.00	0.39	0.39	0.39	0.39	0.39	0.40
$\mu = 0.2$	1.00	1.00	1.01	1.01	1.01	0.35	0.35	0.35	0.35	0.35	0.35
$\mu = 0.4$	1.00	1.00	1.01	1.01	1.01	0.34	0.34	0.34	0.34	0.34	0.34
$\mu = 0.8$	1.00	1.00	1.01	1.01	1.01	0.34	0.34	0.34	0.34	0.34	0.34
					(b) $\rho = 0.5$						
$T = 150$											
$\mu = 0.1$	1.19	1.25	1.26	1.24	1.23	0.56	0.63	0.64	0.63	0.63	0.82
$\mu = 0.2$	1.15	1.20	1.21	1.20	1.19	0.52	0.56	0.57	0.57	0.57	0.70
$\mu = 0.4$	1.06	1.08	1.09	1.10	1.10	0.41	0.43	0.43	0.43	0.44	0.45
$\mu = 0.8$	1.01	1.02	1.03	1.04	1.04	0.35	0.36	0.36	0.37	0.37	0.35
$T = 250$											
$\mu = 0.1$	1.18	1.24	1.25	1.26	1.25	0.56	0.63	0.65	0.65	0.65	0.64
$\mu = 0.2$	1.12	1.15	1.17	1.17	1.17	0.48	0.52	0.53	0.54	0.54	0.53
$\mu = 0.4$	1.02	1.04	1.04	1.05	1.05	0.37	0.38	0.38	0.38	0.38	0.37
$\mu = 0.8$	1.01	1.01	1.02	1.02	1.02	0.34	0.35	0.35	0.35	0.35	0.34
$T = 500$											
$\mu = 0.1$	1.16	1.22	1.24	1.25	1.25	0.54	0.61	0.63	0.63	0.64	0.57
$\mu = 0.2$	1.06	1.08	1.09	1.09	1.10	0.40	0.43	0.44	0.44	0.44	0.41
$\mu = 0.4$	1.01	1.01	1.02	1.02	1.02	0.35	0.35	0.35	0.35	0.35	0.35
$\mu = 0.8$	1.00	1.01	1.01	1.01	1.01	0.34	0.34	0.34	0.34	0.34	0.34
					(c) $\rho = -0.5$						
$T = 150$											
$\mu = 0.1$	0.73	0.79	0.77	0.77	0.77	0.37	0.31	0.31	0.31	0.31	0.57
$\mu = 0.2$	0.90	0.95	0.93	0.95	0.94	0.36	0.33	0.34	0.34	0.34	0.47
$\mu = 0.4$	0.98	1.00	1.00	1.01	1.01	0.35	0.34	0.35	0.35	0.35	0.35
$\mu = 0.8$	1.00	1.01	1.01	1.02	1.03	0.35	0.35	0.35	0.35	0.36	0.35
$T = 250$											
$\mu = 0.1$	0.83	0.88	0.87	0.88	0.87	0.36	0.32	0.33	0.32	0.33	0.39
$\mu = 0.2$	0.96	0.98	0.98	0.99	0.99	0.35	0.33	0.34	0.34	0.34	0.35
$\mu = 0.4$	0.99	1.00	1.00	1.01	1.01	0.34	0.34	0.34	0.34	0.35	0.34
$\mu = 0.8$	1.00	1.01	1.01	1.01	1.02	0.34	0.34	0.34	0.35	0.35	0.34
$T = 500$											
$\mu = 0.1$	0.90	0.94	0.93	0.94	0.94	0.36	0.32	0.33	0.33	0.33	0.35
$\mu = 0.2$	0.98	0.99	0.99	0.99	1.00	0.34	0.33	0.34	0.34	0.34	0.34
$\mu = 0.4$	1.00	1.00	1.00	1.00	1.01	0.34	0.34	0.34	0.34	0.34	0.34
$\mu = 0.8$	1.00	1.00	1.01	1.01	1.01	0.34	0.34	0.34	0.34	0.34	0.34

plans used in the data-generating process) and close to 0.34 when k is greater than or equal to 6, irrespective of the true value of $f_{\Delta y}(0)$. For small values of T and/or μ the means are somewhat higher when $\rho = 0.5$ and somewhat lower when $\rho = -0.5$. The mean of the estimates when k is selected by BIC correspond to that obtained with large k's when T and/or μ is large and to some value in between 1.0 and 0.34 when T and μ are small. Some theoretical explanations for this peculiar bias are provided in the next subsection.

These simulation results go a long way to explain the empirical results discussed earlier for the inflation rate. We saw, indeed, that with any subsample that includes shock plans, the estimate $\hat{f}_{\Delta y}(0)$ was close to 1.0 when k is small but that it decreased substantially with an increase in k.

Table 6 presents the results when the data are generated by an explosive process. Here, we present the probability of rejecting the null hypothesis of a unit root in favor of stationary fluctuations (along with the mean and standard error of the statistics).[5] The results show again that, even with an explosive noise component, the presence of shock plans induces a strong bias in spuriously concluding that the process is trend-stationary. This bias increases as μ increases (in which case the shock plans are more important) but, unlike in the unit root case, decreases as T increases. This false rejection in favor of stationary fluctuations also decreases as α increases.

5.3. *Some theoretical explanations*

To provide explanations for the simulation results, we consider data generated by

$$y_t = y_0 + \mu t + S_t \quad t \neq T_{Bj} + 1, \ldots, T_{Bj} + n,$$
$$y_t = y_0 \quad t = T_{Bj} + 1, \ldots, T_{Bj} + n$$

[5]The rejection rates against explosive alternatives are as follows. For $\alpha = 1.01$, they are less than 1% for $T = 150$ and 250. At $T = 500$, they range from 0.03 for $\mu = 0.1$ to 0.19 for $\mu = 0.8$. When $\alpha = 1.02$, they range from 0.03 to 0.09 for $T = 150$, and from 0.02 to 0.20 for $T = 250$. With $T = 500$, they reach 100% for any value of μ.

Table 6. Probability of rejecting in favor of stationarity when the noise function is explosive.

	$T = 150$			$T = 250$			$T = 500$		
	Size	Mean	S.E.	Size	Mean	S.E.	Size	Mean	S.E.
				(a) Z_α					
(i) $\alpha = 1.01$									
$\mu = 0.1$	0.68	−24.12	7.68	0.82	−35.27	12.35	0.39	−21.74	12.46
$\mu = 0.2$	0.78	−26.12	6.92	0.83	−38.69	13.94	0.58	−24.93	18.45
$\mu = 0.4$	0.96	−29.72	4.30	0.91	−45.24	12.85	0.64	−31.06	24.87
$\mu = 0.8$	1.00	−31.59	1.32	1.00	−51.74	3.39	0.71	−41.86	30.14
(ii) $\alpha = 1.02$									
$\mu = 0.1$	0.25	−17.66	6.80	0.09	−14.99	6.74	0.00	2.67	0.42
$\mu = 0.2$	0.41	−18.96	8.34	0.18	−16.32	10.17	0.00	2.66	0.82
$\mu = 0.4$	0.71	−22.18	9.60	0.33	−18.92	13.79	0.00	2.62	1.52
$\mu = 0.8$	0.90	−27.12	7.12	0.60	−24.08	16.60	0.00	2.49	2.66
				(b) MZ_α					
(i) $\alpha = 1.01$									
$\mu = 0.1$	0.60	−22.22	7.07	0.80	−32.58	11.55	0.30	−20.28	11.79
$\mu = 0.2$	0.73	−24.11	6.42	0.81	−35.67	13.12	0.49	−23.23	17.36
$\mu = 0.4$	0.94	−27.54	3.98	0.90	−41.67	12.16	0.64	−28.91	23.36
$\mu = 0.8$	1.00	−29.30	1.17	1.00	−47.74	3.12	0.71	−38.93	28.23
(ii) $\alpha = 1.02$									
$\mu = 0.1$	0.17	−16.15	6.37	0.07	−13.73	6.34	0.00	2.89	0.40
$\mu = 0.2$	0.30	−17.37	7.91	0.17	−14.94	9.59	0.00	2.88	0.79
$\mu = 0.4$	0.60	−20.38	9.18	0.27	−17.32	12.97	0.00	2.84	1.45
$\mu = 0.8$	0.87	−25.08	6.78	0.54	−22.10	15.59	0.00	2.71	2.51
				(c) ADF					
(i) $\alpha = 1.01$									
$\mu = 0.1$	0.53	−3.30	0.66	0.78	−3.96	0.88	0.16	−2.83	0.95
$\mu = 0.2$	0.69	−3.45	0.57	0.80	−4.14	1.03	0.32	−2.98	1.48
$\mu = 0.4$	0.91	−3.71	0.34	0.90	−4.55	1.01	0.53	−3.35	2.10
$\mu = 0.8$	1.00	−3.83	0.09	1.00	−4.96	0.30	0.70	−4.04	2.57
(ii) $\alpha = 1.02$									
$\mu = 0.1$	0.11	−2.68	0.74	0.05	−2.27	0.65	0.00	0.70	0.10
$\mu = 0.2$	0.21	−2.75	0.90	0.12	−2.31	1.02	0.00	0.70	0.18
$\mu = 0.4$	0.44	−2.97	1.10	0.21	−2.44	1.36	0.00	0.70	0.24
$\mu = 0.8$	0.79	−3.44	0.77	0.39	−2.80	1.62	0.00	0.68	0.39

for $j = 1, \ldots, p$ and where

$$S_t = \sum_{j=1}^{t} u_t$$

with u_t some 'mixing' process with $\sigma^2 = \lim_{T\to\infty} T^{-1}E(S_T^2)$. Hence, the series is trending with an integrated noise function. There are p shock plans occurring at times T_{Bj} that last n periods. We posit that T_{Bj} is a fixed proportion of the total sample such that $T_{Bj}/T = \lambda_j$. We simplify by considering the detrended version defined by:

$$\begin{aligned} y_t^{\mathrm{d}} &= S_t \quad t \neq T_{Bj}+1, \ldots, T_{Bj}+n, \\ y_t^{\mathrm{d}} &= -\mu T_{Bj} \quad t = T_{Bj}+1, \ldots, T_{Bj}+n. \end{aligned} \tag{12}$$

Representation (12) helps to make clear the relationship between additive outliers and shock plans. As a special case, with $n = 1$, we have a unit root process with additive outliers that increase as the sample size increases. Hence, this differentiates our set-up from the one analysed in Franses and Haldrup (1994) and Vogelsang (forthcoming).

We first consider the limit of the autoregressive coefficient in the following AR(1) process estimated by least-squares:

$$y_t^{\mathrm{d}} = \hat{\alpha} y_{t-1}^{\mathrm{d}} + \hat{u}_t.$$

Standard derivations show the following limit for α:

$$\hat{\alpha} \Rightarrow \frac{\sigma^2 \int_0^1 W(r)^2 dr + \mu^2 \left(\sum_{j=1}^{n} \lambda_j^2\right)(n-1)}{\sigma^2 \int_0^1 W(r)^2 dr + \mu^2 \left(\sum_{j=1}^{n} \lambda_j^2\right) n} < 1.$$

Hence, the presence of a plan biases $\hat{\alpha}$ below one even asymptotically. The extent of the bias is greatest for plans of short duration. If the plans are long-lasting the bias decreases. This is expected since long-lasting plans can be perceived as permanent changes in levels which, as shown in Perron (1990), bias the autoregressive coefficient towards one whether the noise function is stationary or integrated. This asymptotic bias helps to explain the strong rejections of the unit root when shock plans are present. The result is also qualitatively different from the standard additive outlier case analysed in Franses and Haldrup (1994) where the limit of $\hat{\alpha}$ is still one but nuisance parameters are introduced that produce distorted tests.

Consider now the limiting behavior of the autocorrelation function of the first-differences of y_t^{d} given by:

$$\begin{aligned}
\Delta y_t^{\mathrm{d}} &= u_t & t &\neq T_{Bj}+1,\ldots,T_{Bj}+n+1\\
&= -\mu T_{Bj} - S_{T_{Bj}} & t &= T_{Bj}+1\\
&= 0 & t &= T_{Bj}+2,\ldots,T_{Bj}+n\\
&= S_{T_{Bj}+p+1} + \mu T_{Bj} & t &= T_{Bj}+n+1
\end{aligned}$$

for $j = 1,\ldots,p$. It is easy to show that the following limits hold:

$$T^{-2}\sum_{t=q+1}^{T}(\Delta y_{t-q}^{\mathrm{d}})^2 \to 2\mu^2\left(\sum_{j=1}^{n}\lambda_j^2\right) \tag{13}$$

and

$$\begin{aligned}
T^{-2}\sum_{t=q+1}^{T}\Delta y_t^{\mathrm{d}}\Delta y_{t-q}^{\mathrm{d}} &\to 0 & q &\neq n,\\
T^{-2}\sum_{t=q+1}^{T}\Delta y_t^{\mathrm{d}}\Delta y_{t-q}^{\mathrm{d}} &\to -\mu^2\left(\sum_{j=1}^{n}\lambda_j^2\right) & q &= n.
\end{aligned} \tag{14}$$

Hence,

$$\hat{\alpha}_q = T^{-2}\sum_{t=q+1}^{T}\Delta y_t^{\mathrm{d}}\Delta y_{t-q}^{\mathrm{d}}\Bigg/T^{-2}\sum_{t=q+1}^{T}(\Delta y_{t-q}^{\mathrm{d}})^2,$$

the correlation coefficient of order q, converges to 0 unless the order q is the same as the length of the plan n, in which case it converges to –1/2. This somewhat generalizes the result of Franses and Haldrup (1994). They showed that additive outliers induce a negative MA component in the level of the series. It is easy to see that if the magnitude of the outlier increases their result implies that the MA component converges to –1. This corresponds to the case where $n = 1$ here since a correlation coefficient of –1/2 corresponds to that of an noninvertible MA process. Our result shows that, more generally, a large plan will induce a noninvertible MA(n) process.

Consider finally, the limit behavior of the persistence measure applied to y_t^{d} defined by Equation (8) using the autoregression

$$\Delta y_t^{\mathrm{d}} = \sum_{j=1}^{k} d_j \Delta y_{t-j}^{\mathrm{d}} + e_t.$$

We can use Equations (13) and (14) to show that if $k < n$, the OLS estimates $\hat{d}_j$ ($j = 1, \ldots, k$) all converge to 0. Also, $T^{-1}s_{ek}^2$ and

$$T^{-1}\hat{\sigma}_{\Delta y^{\mathrm{d}}}^2 = T^{-2}\sum_{t=1}^{T}(\Delta y_t^{\mathrm{d}})^2,$$

both converge to

$$2\mu^2\left(\sum_{j=1}^{n}\lambda_j^2\right)$$

and, hence $\hat{f}_{\Delta y^{\mathrm{d}}}(0)$ converges to 1. This explains the bias reported in the simulations showing that the limit of the persistence measure is 1 irrespective of the correlation structure of the errors if k is chosen less than the length of the plan.

When k is chosen at least as large as the length of the plan, things are different and depend on the specific values of k and n. It is difficult to obtain a general closed-form solution valid for any pairs (k, n) but it is relatively easy to compute the limit of $\hat{f}_{\Delta y^{\mathrm{d}}}(0)$ numerically. As an example, consider the case where $\mathrm{int}(n/k) = 1$ which corresponds to the simulation experiments reported in Table 5 when k is between 6 and 10 (since $n = 6$). In this case, it is easy to show that $\hat{d}_j \to 0$ for $j \neq n$ and $\hat{d}_n \to -0.5$. Hence, $\hat{d}(1) \to -0.5$. Also,

$$T^{-1}s_{ek}^2 \to \left(\frac{3}{2}\right)\mu^2\left(\sum_{j=1}^{n}\lambda_j^2\right), \quad T^{-1}\hat{\sigma}_{\Delta y^{\mathrm{d}}}^2 \to 2\mu^2\left(\sum_{j=1}^{n}\lambda_j^2\right).$$

Hence, $\hat{f}_{\Delta y^{\mathrm{d}}}(0) \to 1/3$. This corresponds, indeed, very closely to the values obtained by simulations for the various data-generating processes (see Table 5). It is important to reiterate that this limiting

Table 7. Theoretical limit of $\hat{f}_{\Delta y}(0)$.

	n									
k	1	2	3	4	5	6	7	8	9	10
1	0.33	1.00	1.00	1.00	1.00	1.00	1.00	1.00	1.00	1.00
2	0.17	0.33	1.00	1.00	1.00	1.00	1.00	1.00	1.00	1.00
3	0.10	0.33	0.33	1.00	1.00	1.00	1.00	1.00	1.00	1.00
4	0.07	0.22	0.33	0.33	1.00	1.00	1.00	1.00	1.00	1.00
5	0.05	0.22	0.33	0.33	0.33	1.00	1.00	1.00	1.00	1.00
6	0.04	0.12	0.22	0.33	0.33	0.33	1.00	1.00	1.00	1.00
7	0.03	0.12	0.22	0.33	0.33	0.33	0.33	1.00	1.00	1.00
8	0.02	0.10	0.22	0.22	0.33	0.33	0.33	0.33	1.00	1.00
9	0.02	0.10	0.18	0.22	0.33	0.33	0.33	0.33	0.33	1.00
10	0.02	0.07	0.18	0.22	0.22	0.33	0.33	0.33	0.33	0.33
11	0.01	0.07	0.18	0.22	0.22	0.33	0.33	0.33	0.33	0.33
12	0.01	0.06	0.10	0.18	0.22	0.22	0.33	0.33	0.33	0.33
13	0.01	0.06	0.10	0.18	0.22	0.22	0.33	0.33	0.33	0.33
14	0.01	0.04	0.10	0.18	0.22	0.22	0.22	0.33	0.33	0.33
15	0.01	0.04	0.09	0.18	0.18	0.22	0.22	0.33	0.33	0.33

value is valid irrespective of the true persistence of shocks, hence it is difficult to speak about biases *per se*.

Table 7 presents the limiting value of $\hat{f}_{\Delta y^{\mathrm{d}}}(0)$ for pairs $(k,\ n)$ with k and n ranging from 1 to 15 and 1 to 10, respectively. These show that the limit decreases as k increases for a given n. Also, for a given n, the limit depends only on the value of $\mathrm{int}(k/n)$. Since, the inflation rate is affected by plans that last between 3 and 8 periods, our theoretical results help explain why the persistence measures reported in Table 3 are decreasing as k increases.

The simulation and theoretical results in this section explain the rejections reported in the previous section for the Brazilian inflation rate. Indeed, our experiments clearly show that shock plans induce a strong bias in unit root tests in concluding for stationarity whether the true noise component has a unit or explosive root. On the other hand, the measure of persistence $\hat{f}_{\Delta y}(0)$ is biased towards 1.0 (thereby suggesting persistence) when a small-order k is selected. To verify the claim that the noise component of the inflation rate is not stationary, it remains to devise unit root tests that are immune to the presence of the shock plans. This is the object of the next section.

6. Corrected Versions of Unit Root Tests

We now present modifications to the unit root tests that take into account the presence of the shock plans. The strategy is similar to that used in Perron (1989, 1990) in the case of permanent changes in level or slope of the trend function. The idea is to take the shock plans from the noise function to the trend function. More precisely, it is the movements in and out of the periods called 'plans' that are isolated. This is not a statement about the deterministic nature of the timing and magnitude of the plans. Rather, it is to be viewed as a device to isolate their effect so that the tests can meaningfully assess the stochastic properties of the series when shock plans are not into effect. Since the timing of the plans is well documented and relates to governmental interventions, we treat the dates of their occurrence as known rather than as random variables to be estimated. In proposing alternative versions of the unit root tests, one must keep in mind that, in practical situations, the data-generating process may not exactly correspond to, say, the one described by process (10) used in the simulations. Some flexibility is warranted. To that effect, we present various versions that can capture more realistic situations.

It is useful to first define some notations. Let $da(j)_t$ denote a dummy variable taking value 1 when the time index t corresponds to the first month when plan j takes effect, and 0 otherwise. Similarly, let $db(j)_t$ be a dummy variable taking value 1 when the time index t corresponds to the first month *after* the end of plan j, and 0 otherwise. Finally, let $D(j)_t$ be a dummy variable taking value 1 when the time index t corresponds to one of the months when plan j is in effect, and 0 otherwise.

The first modification to the ADF test is simply to include these dummies for each plan in autoregression (2). Hence, the relevant regression is:

$$y_t = \eta + \gamma t + \sum_{j=1}^{p}(k_j da(j)_t + \lambda_j db(j)_t$$

$$+ \phi_j D(j)_t) + \alpha y_{t-1} + \sum_{i=1}^{k} c_i \Delta y_{t-i} + v_t. \tag{15}$$

The test statistic, denoted ADF(C_A), is again constructed as the t-statistic for testing that α, the sum of the autoregressive coefficients, is unity. The number of lagged first-differences of the data, k, is again selected using the BIC criterion.

It is useful at this point to discuss the role played by the various dummies. First, note that $da(j)_t$ and $db(j)_t$ are used to allow removing the influence of the plans under the null hypothesis of a unit root. On the other hand, $D(j)_t$ is used to remove the influence of the plans under the alternative hypothesis of stationarity. This can be seen by noting that, with a unit root or an explosive process, $da(j)_t$ acts as a one-time blip that becomes a permanent decrease in level (the beginning of the plan). The dummy $db(j)_t$ also acts as a one-time blip that becomes permanent thereby allowing an increase in level that marks the end of the plan. When the series is stationary, $D(j)_t$ acts as a temporary level shift that marks the occurrence of the plan.

This first strategy for a modified test is akin to the 'innovational outlier' modeling device which implies that the movements in and out of plans depends on the correlation structure of the noise component. An alternative strategy is to adopt an 'additive outlier' modeling device where the movements in and out of plans are supposed not to be influenced by the noise component of the series. To that effect, we can posit a data-generating process of the form

$$y_t = a + bt + \sum_{j=1}^{p} \tau_j D(j)_t + z_t,$$

where z_t is defined by Equation (1). Using an autoregressive approximation, the regression[6] is:

$$y_t = \eta + \gamma t + \sum_{i=0}^{k+1} \sum_{j=1}^{p} \phi_{j,i} D(j)_{t-i} + \alpha y_{t-1} + \sum_{i=1}^{k} c_i \Delta y_{t-i} + v_t. \quad (16)$$

---

[6]In principle, one could use a two-step strategy by first eliminating the deterministic components and performing the unit root test on the detrended series. This, however, leads to substantial problems of inference when the deterministic components include step dummy variables (see Perron and Vogelsang, 1992).

The relevant statistic, denoted ADF(C_B), is again the t-statistic for testing that $\alpha = 1$. The modifications to the test Z_α are similar. For version A (akin to the 'innovational outlier' strategy), it involves using the following first-order autoregression:

$$y_t = \eta + \gamma t + \sum_{j=1}^{p} (k_j da(j)_t + \lambda_j db(j)_t + \phi_j D(j)_t) + \alpha y_{t-1} + v_t. \tag{17}$$

Denote the OLS estimate of α by $\breve{\alpha}$ and the sample variance of the residuals, $\breve{v}_t$, by

$$\breve{s}_u^2 = T^{-1} \sum_{t=1}^{T} \breve{v}_t^2.$$

Also, denote by $\breve{y}_{t-1}$ the residuals from the following regression:

$$y_{t-1} = \eta + \gamma t + \sum_{j=1}^{p} (k_j da(j)_t + \lambda_j db(j)_t + \phi_j D(j)_t) + \xi_t. \tag{18}$$

The first modified version of the Z_α test can now be described as follows:

$$Z_\alpha(C_A) = T(\breve{\alpha} - 1) - (\breve{s}^2 - \breve{s}_u^2) \Big/ \left(2T^{-2} \sum_{t=1}^{T} \breve{y}_{t-1}^2 \right),$$

where $\breve{s}^2$ is an estimate of the spectral density at frequency zero of the residuals v_t described below. The modified test corresponding to the 'additive outlier' strategy, denoted $Z_\alpha(C_B)$, is similar except that regression (17) is replaced by the regression

$$y_t = \eta + \gamma t + \sum_{i=0}^{1} \sum_{j=1}^{p} \phi_{j,i} D(j)_{t-i} + \alpha y_{t-1} + v_t \tag{19}$$

and regression (18) is replaced by the regression

$$y_{t-1} = \eta + \gamma t + \sum_{i=0}^{1} \sum_{j=1}^{p} \phi_{j,i} D(j)_{t-i} + \xi_t. \tag{20}$$

Since the test MZ_α is constructed directly from detrended series, only the 'additive outlier' strategy is entertained to construct a modified statistic given by:

$$MZ_\alpha(C) = (T^{-1}\ddot{y}_T^2 - \breve{s}^2) \Big/ \left(2T^{-2}\sum_{t=1}^{T}\ddot{y}_{t-1}^2\right),$$

where $\ddot{y}_t$ are the residuals from the following regression:

$$y_t = \eta + \gamma t + \sum_{j=1}^{p}\phi_j D(j)_t + \ddot{y}_t.$$

The modification to the spectral density estimator of the residuals v_t is different. Here, we consider only the deterministic components that are relevant under the null hypothesis, namely the constant and the dummies $da(j)_t$ and $db(j)_t$. Denoting this estimator by $\breve{s}^2$, it is defined by Equation (5) but with the autoregression (6) replaced by:

$$\Delta y_t = \eta + \sum_{i=0}^{k}\sum_{j=1}^{p}(k_{j,i}da(j)_{t-i} + \lambda_{j,i}db(j)_{t-i})$$
$$+ b_0 y_{t-1} + \sum_{i=1}^{k} b_i \Delta y_{t-i} + e_{tk}. \tag{21}$$

Finally, the modified estimator of the persistence measure is given by $\breve{f}_{\Delta y}(0) = \breve{h}_{\Delta y}(0)/\breve{\sigma}_{\Delta y}^2$ where

$$\breve{\sigma}_{\Delta y}^2 = T^{-1}\sum_{t=1}^{T}(\widetilde{\Delta y_t})^2,$$

the sample variance of Δy_t corrected for the plans, i.e., the sum of squared residuals from the following regression:

$$\Delta y_t = \eta + \sum_{j=1}^{p}(k_j da(j)_t + \lambda_j db(j)_t) + e_t.$$

The quantity $\breve{h}_{\Delta y}(0)$ is the autoregressive spectral density estimate at frequency 0 defined by $\breve{h}_{\Delta y}(0) = s_{ek}^2/(1 - d(1))^2$, with

$$\breve{s}_{ek}^2 = T^{-1}\sum_{t=1}^{T} \breve{e}_{tk}^2, \quad \breve{d}(1) = \sum_{j=1}^{k} \breve{d}_j,$$

where $\breve{d}_j$ and $\{\breve{e}_{tk}\}$ are obtained from the following kth-order augmented autoregression in Δy_t:

$$\Delta y_t = \eta + \sum_{i=0}^{k}\sum_{j=1}^{p}(k_{j,i}da(j)_{t-i} + \lambda_{j,i}db(j)_{t-i}) + \sum_{i=1}^{k} b_i\Delta y_{t-i} + e_{tk}.$$

6.1. *Simulation results*

The modifications described above leave the asymptotic distributions of the unit root tests unchanged under the null hypothesis (compared to the unmodified statistics applied to series in the class described by Equation (1)), provided the plans are treated as fixed in length as the sample size increases. However, for the asymptotic distributions to provide satisfactory approximations to the finite sample distributions, the shock plans must be of relatively short duration. In this section we present simulations whose aim is to verify whether the usual asymptotic distribution provides a satisfactory approximation and if the modifications are effective in making the test immune to the presence of the shock plans. To do this, we examine the exact size of the modified tests using the same experiments as in Section 5 (exactly the same generated series are used). We also performed additional experiments where the processes are generated under the alternative hypothesis of trend-stationarity interrupted by shock plans. In this case, the data are generated using the same specifications except that the process describing the behavior of the series when no plans are in effect is given by

$$y_t = a + \mu t + \alpha y_{t-1} + e_t, \tag{22}$$

instead of the random walk with drift described by Equation (10). To examine power against stationary fluctuations, we specify $\alpha = 0.8$ and

0.9. We also investigated power against explosive alternatives using the process described by Equation (11) with $\alpha = 1.01$ and $\alpha = 1.02$.

The results for size and power are presented in Tables 8 and 9. The first feature of interest is that the exact sizes of the tests are, in most cases, very close to the nominal 5% size.[7] Hence, the

Table 8. Exact size of unit root tests corrected for shock plans (5% nominal size).

	$T = 150$		$T = 250$		$T = 500$	
	Left	Right	Left	Right	Left	Right
			(a) $Z_\alpha(C_A)$			
$\mu = 0.1$	0.060	0.083	0.045	0.066	0.047	0.053
$\mu = 0.2$	0.057	0.089	0.044	0.064	0.048	0.052
$\mu = 0.4$	0.060	0.083	0.046	0.062	0.048	0.050
$\mu = 0.8$	0.084	0.067	0.057	0.051	0.055	0.044
			(b) $Z_\alpha(C_B)$			
$\mu = 0.1$	0.056	0.080	0.043	0.061	0.045	0.050
$\mu = 0.2$	0.053	0.081	0.044	0.060	0.046	0.050
$\mu = 0.4$	0.058	0.079	0.047	0.055	0.046	0.048
$\mu = 0.8$	0.082	0.060	0.060	0.049	0.049	0.046
			(c) $MZ_\alpha(C)$			
$\mu = 0.1$	0.027	0.126	0.024	0.102	0.033	0.090
$\mu = 0.2$	0.023	0.131	0.024	0.101	0.033	0.091
$\mu = 0.4$	0.022	0.123	0.024	0.098	0.034	0.088
$\mu = 0.8$	0.037	0.093	0.033	0.082	0.040	0.076
			(d) ADF(C_A)			
$\mu = 0.1$	0.088	0.073	0.067	0.066	0.060	0.055
$\mu = 0.2$	0.086	0.073	0.067	0.066	0.060	0.055
$\mu = 0.4$	0.084	0.074	0.066	0.066	0.060	0.055
$\mu = 0.8$	0.079	0.075	0.064	0.066	0.059	0.055
			(e) ADF(C_B)			
$\mu = 0.1$	0.094	0.055	0.065	0.051	0.055	0.043
$\mu = 0.2$	0.098	0.055	0.066	0.046	0.058	0.043
$\mu = 0.4$	0.107	0.052	0.072	0.046	0.059	0.043
$\mu = 0.8$	0.154	0.043	0.101	0.043	0.065	0.040

[7]One exception is the $MZ_\alpha(C)$ which is conservative in the left tail and liberal in the right tail for small-sample sizes but the discrepancies disappear with larger samples.

Table 9. Power of the unit root tests corrected for shock plans.

	$T = 150$				$T = 250$				$T = 500$			
$\alpha =$	0.8	0.9	1.01	1.02	0.8	0.9	1.01	1.02	0.8	0.9	1.01	1.02
					(a) $Z_\alpha(C_A)$							
$\mu = 0.1$	0.90	0.40	0.12	0.74	1.00	0.84	0.48	0.97	1.00	1.00	0.97	1.00
$\mu = 0.2$	0.90	0.39	0.13	0.74	1.00	0.84	0.47	0.97	1.00	1.00	0.97	1.00
$\mu = 0.4$	0.91	0.41	0.12	0.74	1.00	0.85	0.47	0.97	1.00	1.00	0.97	1.00
$\mu = 0.8$	0.93	0.47	0.10	0.72	1.00	0.88	0.46	0.97	1.00	1.00	0.97	1.00
					(b) $Z_\alpha(C_B)$							
$\mu = 0.1$	0.91	0.39	0.11	0.73	1.00	0.84	0.46	0.97	1.00	1.00	0.97	1.00
$\mu = 0.2$	0.91	0.39	0.11	0.73	1.00	0.84	0.46	0.97	1.00	1.00	0.97	1.00
$\mu = 0.4$	0.92	0.41	0.11	0.73	1.00	0.86	0.46	0.97	1.00	1.00	0.97	1.00
$\mu = 0.8$	0.94	0.47	0.09	0.72	1.00	0.89	0.44	0.97	1.00	1.00	0.97	1.00
					(c) $MZ_\alpha(C)$							
$\mu = 0.1$	0.79	0.26	0.17	0.78	1.00	0.77	0.56	0.97	1.00	1.00	0.97	1.00
$\mu = 0.2$	0.79	0.26	0.17	0.78	1.00	0.76	0.56	0.97	1.00	1.00	0.97	1.00
$\mu = 0.4$	0.79	0.27	0.16	0.77	1.00	0.77	0.56	0.97	1.00	1.00	0.97	1.00
$\mu = 0.8$	0.83	0.31	0.14	0.76	1.00	0.83	0.54	0.97	1.00	1.00	0.97	1.00
					(d) ADF(C_A)							
$\mu = 0.1$	0.93	0.43	0.12	0.72	1.00	0.82	0.46	0.97	1.00	1.00	0.97	1.00
$\mu = 0.2$	0.93	0.43	0.12	0.72	1.00	0.82	0.46	0.97	1.00	1.00	0.97	1.00
$\mu = 0.4$	0.93	0.43	0.12	0.72	1.00	0.82	0.46	0.97	1.00	1.00	0.97	1.00
$\mu = 0.8$	0.91	0.40	0.12	0.72	1.00	0.81	0.46	0.97	1.00	1.00	0.97	1.00
					(e) ADF(C_B)							
$\mu = 0.1$	0.95	0.46	0.08	0.70	1.00	0.86	0.43	0.97	1.00	1.00	0.97	1.00
$\mu = 0.2$	0.95	0.47	0.08	0.70	1.00	0.86	0.43	0.97	1.00	1.00	0.97	1.00
$\mu = 0.4$	0.96	0.52	0.08	0.69	1.00	0.88	0.43	0.96	1.00	1.00	0.97	1.00
$\mu = 0.8$	0.97	0.63	0.06	0.66	1.00	0.91	0.40	0.96	1.00	1.00	0.97	1.00

modifications are successful in providing tests that are immune to the presence of shock plans and the usual asymptotic distribution provides a good approximation to the finite sample distribution. The second feature to note is that the tests still have reasonable power. Furthermore, the power function appears little influenced by different rates of growth and it increases rapidly as the sample size increases. We, therefore, conclude that the modifications are adequate. Comparing the different tests, the corrected versions of ADF are more powerful than those of Z_α and MZ_α with trend-stationary alternatives. When the alternative is that of an explosive process, the reverse relation holds. The differences are, however, small.

Table 10 presents the mean of $\breve{f}_{\Delta y}(0)$ the persistence measure corrected for plans, for each case. We see that the means are very close to the true values, especially when the order k is chosen using the BIC criterion.

6.2. *Empirical results*

We applied the modified unit root tests to the Brazilian inflation rate series using the dates for the plans as specified in Table 1. The results are presented in Table 11 for the various subsamples that incorporate plans. For any subsample that includes the five plans, the results point to a strong rejection of the unit root but this time in favor of an explosive alternative. This is confirmed by the estimates of the persistence measure presented in Table 12, which show a rather erratic behavior as k varies. For example, with the full sample (74:1–93:6) the estimate is 1.67 for $k = 2$ (chosen by BIC) but 68.44 with $k = 4$. This should not be too surprising since, if the process is indeed explosive, the persistence measure is not well defined.

Consider now the two subsamples that only include the first three plans and exclude the period surrounding the Collor I plan, namely 1974:1 to 1989:8 and 1980:1 to 1989:8. For the latter, there is no evidence against a unit root (against either alternatives) and the persistence measure is fairly stable around 1.0 for different values of k. For 1974:1 to 1989:8, the conclusions are mixed; $Z_\alpha(C_A)$, ADF(C_A) and ADF(C_B) suggest an explosive process (though less strongly than with the full sample), and the other tests and the persistence measures suggest a unit root process.

Hence, there is strong evidence that the shock plans are responsible for the spurious finding of stationarity using standard tests and that once these are taken into account the evidence strongly supports an explosive path (interrupted by shock plans) when the whole sample is analysed. However, the explosive behavior appears due to the presence of the very high inflation period surrounding the period of the Collor I plan. As one referee pointed out, a look at the graph of the inflation series might suggest the possibility that this

Table 10. Mean of $\breve{f}_{\Delta y}(0)$ corrected for shock plans.

	k	1	2	3	4	5	6	7	8	9	10	*kbic*
						(a) $\rho = 0.0$						
$T = 150$	$\mu = 0.1$	0.97	0.93	0.90	0.86	0.83	0.79	0.77	0.74	0.73	0.71	0.97
	$\mu = 0.2$	0.98	0.94	0.91	0.86	0.82	0.79	0.76	0.74	0.73	0.71	0.98
	$\mu = 0.4$	1.00	0.97	0.94	0.88	0.81	0.78	0.75	0.73	0.72	0.70	1.00
	$\mu = 0.8$	1.07	1.08	1.05	0.93	0.78	0.74	0.72	0.70	0.68	0.67	1.07
$T = 250$	$\mu = 0.1$	0.99	0.96	0.94	0.92	0.90	0.88	0.87	0.85	0.84	0.83	0.99
	$\mu = 0.2$	0.99	0.96	0.94	0.92	0.90	0.88	0.87	0.85	0.84	0.83	0.99
	$\mu = 0.4$	1.00	0.98	0.96	0.93	0.89	0.87	0.86	0.85	0.83	0.82	1.00
	$\mu = 0.8$	1.04	1.04	1.02	0.96	0.87	0.85	0.84	0.82	0.81	0.80	1.04
$T = 500$	$\mu = 0.1$	0.99	0.98	0.97	0.96	0.95	0.94	0.93	0.92	0.92	0.91	0.99
	$\mu = 0.2$	0.99	0.98	0.97	0.96	0.94	0.94	0.93	0.92	0.92	0.91	0.99
	$\mu = 0.4$	1.00	0.99	0.98	0.96	0.94	0.93	0.93	0.92	0.92	0.91	1.00
	$\mu = 0.8$	1.02	1.02	1.01	0.97	0.93	0.92	0.91	0.91	0.90	0.89	1.02
						(b) $\rho = 0.5$						
$T = 150$	$\mu = 0.1$	2.87	2.75	2.67	2.54	2.46	2.36	2.32	2.28	2.26	2.50	2.87
	$\mu = 0.2$	2.88	2.75	2.67	2.54	2.45	2.36	2.32	2.27	2.25	2.49	2.88
	$\mu = 0.4$	2.90	2.78	2.69	2.54	2.43	2.34	2.30	2.25	2.23	2.47	2.90
	$\mu = 0.8$	3.01	2.89	2.77	2.56	2.34	2.25	2.21	2.17	2.15	2.37	3.01
$T = 250$	$\mu = 0.1$	2.92	2.84	2.78	2.73	2.68	2.64	2.61	2.56	2.53	2.48	2.92
	$\mu = 0.2$	2.92	2.84	2.79	2.73	2.68	2.63	2.61	2.55	2.53	2.48	2.92
	$\mu = 0.4$	2.93	2.85	2.80	2.73	2.66	2.62	2.60	2.54	2.51	2.47	2.93
	$\mu = 0.8$	3.00	2.91	2.84	2.73	2.61	2.56	2.54	2.48	2.46	2.41	3.00
$T = 500$	$\mu = 0.1$	2.95	2.92	2.89	2.85	2.84	2.82	2.80	2.78	2.76	2.73	2.95
	$\mu = 0.2$	2.96	2.92	2.89	2.85	2.84	2.82	2.80	2.78	2.76	2.73	2.96
	$\mu = 0.4$	2.96	2.93	2.90	2.85	2.83	2.81	2.80	2.77	2.75	2.72	2.96
	$\mu = 0.8$	2.99	2.96	2.91	2.85	2.80	2.78	2.76	2.74	2.72	2.69	2.99
						(c) $\rho = -0.5$						
$T = 150$	$\mu = 0.1$	0.33	0.32	0.31	0.29	0.28	0.27	0.26	0.25	0.25	0.24	0.33
	$\mu = 0.2$	0.33	0.32	0.31	0.30	0.28	0.27	0.26	0.25	0.25	0.24	0.33
	$\mu = 0.4$	0.34	0.34	0.33	0.31	0.28	0.27	0.26	0.25	0.24	0.24	0.34
	$\mu = 0.8$	0.39	0.43	0.42	0.37	0.27	0.26	0.25	0.24	0.23	0.23	0.39
$T = 250$	$\mu = 0.1$	0.33	0.33	0.32	0.31	0.30	0.30	0.29	0.29	0.28	0.28	0.33
	$\mu = 0.2$	0.33	0.33	0.32	0.31	0.30	0.30	0.29	0.29	0.28	0.28	0.33
	$\mu = 0.4$	0.34	0.34	0.33	0.32	0.30	0.30	0.29	0.29	0.28	0.28	0.34
	$\mu = 0.8$	0.36	0.38	0.38	0.35	0.30	0.29	0.28	0.28	0.27	0.27	0.36
$T = 500$	$\mu = 0.1$	0.33	0.33	0.33	0.32	0.32	0.31	0.31	0.31	0.31	0.31	0.33
	$\mu = 0.2$	0.33	0.33	0.33	0.32	0.32	0.31	0.31	0.31	0.31	0.31	0.33
	$\mu = 0.4$	0.34	0.34	0.33	0.33	0.32	0.31	0.31	0.31	0.31	0.30	0.34
	$\mu = 0.8$	0.35	0.36	0.35	0.34	0.31	0.31	0.31	0.30	0.30	0.30	0.35

Table 11. Empirical results for unit root tests corrected for shock plans.

							ADF(C_A)			ADF(C_B)		
Sample	$Z_\alpha(C_A)$	k	$Z_\alpha(C_B)$	k	$MZ_\alpha(C)$	k	t_α	$\hat{\alpha}$	k	t_α	$\hat{\alpha}$	k
74:1–93:6	14.58^e	2	6.46^e	2	−14.03	2	4.39^e	1.14	4	5.63^e	1.22	3
74:1–89:8	2.66^e	4	−4.19	4	−5.71	4	-0.05^e	1.00	4	2.06^e	1.10	4
80:1–89:8	−10.56	4	−12.77	4	−7.76	4	−2.29	0.89	0	0.47^e	1.04	4
80:1–93:6	9.36^e	2	3.55^e	2	−10.60	2	3.54^e	1.15	4	5.10^e	1.25	3
85:1–93:6	6.53^e	2	2.88^e	2	−6.24	2	3.19^e	1.14	2	5.06^e	1.31	3

Note: The superscript e denotes significance at the 1% level in favor of explosive alternatives.

Table 12. Empirical results for $\breve{f}_{\Delta y}(0)$ corrected for shock plans.

Sample\\k	1	2	3	4	5	*kbic* (k)
74:1–93:6	0.50	1.67	12.70	68.44	6.42	1.67 (2)
74:1–89:8	1.07	0.75	0.97			1.07 (1)
80:1–89:8	0.99	0.70	0.95			0.99 (1)
80:1–93:6	0.46	1.64	18.53	14.67	2.78	0.46 (1)
85:1–93:6	0.38	1.84	43.21	2.65	0.72	0.38 (1)

period was an outlier that could be responsible for additional biases. This question is addressed in the next subsection.

6.3. *Accounting for possible outliers*

The problem at hand is the following. First, the results indicate a unit root for subperiods that do not include plans. Second, using uncorrected statistics show stationarity whether one uses subsamples that include all five plans or just the first three plans. Hence, the high spike in inflation around the end of 1989 is not solely responsible for the unit root rejections. Third, for subperiods that do not include the era around the Collor I plan, the results corrected for the first three plans show again a unit root behavior. However, when considering the full sample, the corrected statistics indicate an explosive behavior, contrary to subperiods with only the first three plans. Hence, it may be possible that the noise is a unit root process but that the era around the Collor I plan is an outlier that biases the corrected tests in favor of an explosive process. Therefore, to test for outliers we work under the maintained assumption that the noise structure is a unit root process and, accordingly, we adopt the methodology suggested by Vogelsang (forthcoming) for that case.

The method is quite simple and can be described as follows. First, consider running a regression of the form

$$y_t = \mu + \beta t + \theta D(T_{\mathrm{ao}})_t + u_t \tag{23}$$

where $D(T_{\mathrm{ao}})_t$ is an additive outlier at time T_{ao}, i.e., taking value 1 if $t = T_{\mathrm{ao}}$ and 0 otherwise. Let $t_{\hat{\theta}}(T_{\mathrm{ao}})$ be the t-statistic for testing that $\theta = 0$. Since the date at which the outlier occurs is unknown, the test is based on the statistic $\tau = \sup_{T_{ao}} |t_{\hat{\theta}}(T_{\mathrm{ao}})|$. Vogelsang (forthcoming) derived the limiting distribution of τ under the assumption that the noise component u_t in Equation (23) is a unit root process and that the relative position of the date at which the outlier occurs is a fixed proportion of total sample, i.e., $T_{\mathrm{ao}}/T = \lambda$. Then $\tau \Rightarrow \sup_\lambda |H(\lambda)|$ where $H(\lambda) = W^*(\lambda)/(\int_0^1 W^*(r)^2 dr)^{1/2}$ with $W^*(r)$ the residuals from a projection of the Wiener process on the functions $\{1, r\}$. The critical values at the 1% and 5% significance levels are 3.73 and 3.31, respectively. If the test concludes in favor of a rejection, the date of the outlier is simply estimated as $\hat{T}_{\mathrm{ao}} = \arg\max_{T_{ao}} |t_{\hat{\theta}}(T_{\mathrm{ao}})|$. The corresponding observation is then dropped and the testing procedure is iterated until a nonrejection occurs.

Applying this procedure to the inflation rate series showed all the observations from 1989:8 to 1990:4 (inclusive) to be categorized as outliers. This is the period of very rapid growth in inflation just prior to the implementation of the Collor I plan.

Having identified the outliers, the unit root test proceeds by adding one-time dummy variables for each outlier and as many lagged values as there are lags of the inflation rate in the relevant regressions (understanding, of course, that we do not duplicate identical dummy regressors). The outcome of the unit root tests and persistence measure so-corrected for outliers are presented in Tables 13 and 14. Consider first the unit root test correcting for outliers but not correcting for the four remaining plans (Table 13, panel A). The conclusion is basically the same as before, namely a strong rejection in favor of a stationary process. Surprisingly, the

Table 13. Empirical results for unit root tests treating 1989:08–1990:04 as an outlier.

(A) Not corrected for the presence of plans

					ADF		
Sample	Z_α	k	MZ_α	k	t_α	$\hat{\alpha}$	k
74:1–93:6	-53.01^a	1	-45.92^a	1	-5.62^a	0.75	1
80:1–93:6	-40.43^a	1	-34.59^a	1	-4.98^a	0.73	1
85:1–93:6	-26.33^b	1	-22.41^c	1	-4.03^a	0.71	1

Note: The superscripts a, b and c denote significance in favor of stationary alternatives at the 1%, 2.5% and 5% levels, respectively.

(B) Corrected for the presence of plans

							ADF(C_A)			ADF(C_B)		
Sample	$Z_\alpha(C_A)$	k	$Z_\alpha(C_B)$	k	$MZ_\alpha(C)$	k	t_α	$\hat{\alpha}$	k	t_α	$\hat{\alpha}$	k
74:1–93:6	-21.54^d	1	-27.47^b	1	-19.65^d	1	-2.81^g	0.91	1	-0.48^e	0.98	3
80:1–93:6	-21.14^d	1	-25.84^b	1	-16.47	1	-2.98^g	0.88	1	-0.83^e	0.95	3
85:1–93:6	-10.15	1	-13.60	1	-7.75	1	-2.28^f	0.89	1	-0.18^e	0.99	3

Note: The superscripts b and d denote significance in favor of stationary alternatives at the 2.5% and 10% levels, respectively; and the superscripts e, f, and g denote significance in favor of explosive alternatives at the 1%, 5% and 10% levels, respectively.

Table 14. Empirical results for $\hat{f}_{\Delta y}(0)$ treating 1989:08 to 1990:04 as an outlier.

Sample	k								
	1	2	3	4	5	6	7	8	*kbic* (k)
(a) Not corrected for the presence of plans									
74:1–93:6	3.90	2.39	2.12	1.62	1.46	1.26	1.02	0.87	2.39 (2)
80:1–93:6	3.96	2.44	2.17	1.65	1.50	1.29	1.05	0.89	2.44 (2)
85:1–93:6	4.15	2.57	2.26	1.74	1.59	1.37	1.10	0.95	2.57 (2)

Sample	k					
	1	2	3	4	5	*kbic* (k)
(b) Corrected for the presence of plans						
74:1–93:6	1.10	0.59	1.59	0.83	0.55	1.10 (1)
80:1–93:6	1.02	0.53	1.55	0.81	0.53	1.02 (1)
85:1–93:6	0.91	0.41	1.08	0.51	0.25	0.91 (1)

persistence measure indicate a greater degree of nonstationarity with estimates close to 2.5 using any subsample (Table 14, panel A). If one corrects for the presence of both the outliers and the plans, the unit root tests give mixed results (Table 13, panel B); a borderline rejection in favor of an explosive process with the ADF versions and a borderline rejection in favor of a stationary process with the Z_α versions. The estimates of the persistence measures, (Table 14, Panel B) give a value close to 1.0 irrespective of the subsample used. We view these results as not casting strong enough evidence to reject the unit root hypothesis, consistent with the results from the other subsamples.

7. Conclusions

This paper has considered issues related to tests for a unit root and a measure of the persistence of shocks when a time series of data is contaminated by large level shifts that are of short duration. These temporary events are labeled as 'inliers' or 'shock plans' following our applications to the Brazilian inflation rate. We first showed that standard unit root tests are severely biased in favor of rejecting the unit root against stationary fluctuations when shock plans are present (whether the noise component be characterized by a unit or explosive root). On the other hand, the persistence measure considered shows a strong bias towards one when the order k selected is small.

Hence, a practical recommendation is to complement the application of standard unit root tests with the calculation of measures of persistence. An outcome where the unit root tests reject in favor of stationary fluctuations and the measure of persistence is well above 0 can be a sign that the series is contaminated by 'inliers' or 'shock plans'. To avoid the bias present when applying standard unit root tests, our study proposed corrected versions of three unit root tests. These corrected versions are shown to be adequate in terms of size and power.

The application of our corrected tests showed that the noise function of the Brazilian inflation rate is an explosive process if the period around the end of 1989 is not treated as an outlier. If the latter is treated as an outlier (as the tests suggest), the noise component is better characterized as a unit root process.

The macroeconomic interpretation of our results is a support of the inflation inertia hypothesis which essentially states that shocks to inflation are highly persistent (see, among others, Arida and Lara-Resende, 1985; Bacha, 1988; Bresser Pereira and Nakano, 1986; Lopes, 1984; Modiano, 1988; Novaes, 1991; Pastore, 1994; Simonsen, 1988). This behavior of the inflation process is mainly explained by the widespread indexation to lagged inflation (backward-looking indexation) and to a highly passive monetary policy that easily accommodates inflationary pressures while aiming at keeping unemployment low.

Note finally that, while the methodology developed in this paper is directly motivated by and applied to the Brazilian inflation rate, the tools developed will be of direct application to a wide variety of cases where a series is affected by temporary but important events (for example, wars, strikes, etc.).

Acknowledgments

We thank the financial support from the Programme d'Analyses et de Recherches Économiques Appliquées au Développement International (PARADI) financed by l'Agence Canadienne de Développement International (ACDI) and administered by the Centre de Recherche et Développement en Economique (CRDE), Université de Montréal. The first author acknowledges financial support for parts of this research from the Department of Economics at the Universidade de São Paulo as well as the Centre de Recherche et Développement en Économique (Université de Montréal) for a postdoctoral fellowship. The third author also acknowledges financial support from the Social Sciences and Humanities Research Council of Canada (SSHRC), the Natural Sciences and Engineering Council of Canada (NSERC), and the Fonds de la Formation de Chercheurs et l'Aide á la Recherche du Québec (FCAR). Contract grant sponsors: ACDI; Universidade de São Paulo; Université de Montréal; SSHRC; NSERC; FCAR.

References

Andrews, D.W.K. (1991). Heteroskedastic and autocorrelation consistent covariance matrix estimation. *Econometrica* 59, 817–854.

Arida, P. and Lara-Resende, A. (1985). Inertial inflation and monetary reform in Brazil. In *Inflation and Indexation: Argentina, Brazil and Israel,* J. Williamson (ed.), MIT Press, Cambridge, MA, pp. 27–45.

Bacha, E.L. (1988). Moeda, inércia e conflito: Reflexões sobre políticas de estabilização no Brasil. *Pesquisa e Planejamento Económico* 18, 1–16.

Berk, K.N. (1974). Consistent autoregressive spectral estimates. *The Annals of Statistics* 2, 489–502.

Bresser Pereira, L. and Y. Nakano (1986). Inertial inflation and heterodox shocks in Brazil. In *Inertial Inflation, Theories of Inflation and the Cruzado Plan,* J. M. Rego (ed.), Editora Paz e Terra, Rio de Janeiro.

Campbell, J. Y. and N. G. Mankiw (1987). Are output fluctuations transitory? *Quarterly Journal of Economics* 102, 857–880.

Campbell, J. Y. and P. Perron (1991). Pitfalls and opportunities: what macroeconomists should know about unit roots. In *NBER Macroeconomics Annual,* (eds.) O. J. Blanchard and S. Fisher (eds.), MIT Press, Cambridge, MA, 141–201.

Cochrane, J.H. (1988). How big is the random walk in GNP? *Journal of Political Economy* 96, 893–920.

Dickey, D.A. and W.A. Fuller (1979). Distribution of the estimators for autoregressive time series with a unit root. *Journal of the American Statistical Association* 74, 427–431.

Franses, P.H. and N. Haldrup (1994). The effect of additive outliers on tests for unit roots and cointegration. *Journal of Business and Economic Statistics* 12, 471–478.

Fuller, W.A. (1976). *Introduction to Statistical Time Series,* Wiley, New York.

Lopes, F.L. (1984). Inflação inertial, hiperinflação e desinflação: Notas e conjecturas. *Revista Economia Politica* 5, 135–151.

Modiano, E.M. (1988). The Cruzado first attempt: the Brazilian stabilization program of 1986. In *Inflation Stabilization,* M. Bruno, R. Dornbush, S. Fisher and G. D. Tella (eds.), MIT Press, Cambridge, MA.

Ng, S. and P. Perron (1995). Unit root tests in ARMA models with data dependent methods for the selection of the truncation lag. *Journal of the American Statistical Association* 90, 268–281.

Novaes, A.D. (1991). Um teste de hipotese da inflação inercial no Brasil. *Pesquisa e Planejamento Econdmico* 21, 377–396.

Pastore, A.C. (1994). Déficit público, a sustentabilidade do crescimento das dívidas interna e externa, senhoriagem e inflação: Uma análise do regime monetário Brasileiro. *Revista de Econometria* 14, 177–234.

Perron, P. (1989). The great crash, the oil price shock and the unit root hypothesis. *Econometrica* 57, 1361–1401.

Perron, P. (1990). Testing for a unit root in a time series with a changing mean. *Journal of Business and Economic Statistics* 8, 153–162.

Perron, P. and S. Ng (1996). Useful modifications to unit root tests with dependent errors and their local asymptotic properties. *Review of Economic Studies* 63, 435–463.

Perron, P. and T.J. Vogelsang (1992). Testing for a unit root in a time series with a changing mean: Corrections and extensions. *Journal of Business and Economic Statistics* 10, 467–470.

Phillips, P.C.B. and P. Perron (1988). Testing for a unit root in time series regression. *Biometrika* 75, 335–346.

Said, S.E. and D.A. Dickey (1984). Testing for unit roots in autoregressive-moving average models of unknown order. *Biometrika* 71, 599–607.

Simonsen, M.H. (1988). Price stabilization and income policies: Theory and the Brazilian case study. In *Inflation Stabilization*, M. Bruno, R. Dornbush, S. Fisher and G. D. Tella (eds.), MIT Press, Cambridge, MA.

Stock, J.H. (1990). A class of tests for integration and cointegration. Mimeo, Kennedy School of Government, Harvard University.

Chapter 10

Lag Length Selection and the Construction of Unit Root Tests with Good Size and Power*

Serena Ng[†,‡] and Pierre Perron[§,¶]

[†] *Department of Economics, Johns Hopkins University, Baltimore, MD 21218, USA*

[‡] *Serena.Ng@jhu.edu*

[§] *Department of Economics, Boston University, 270 Bay State Rd., Boston, MA 02215, USA*

[¶] *perron@bu.edu*

It is widely known that when there are errors with a moving-average root close to −1, a high order augmented autoregression is necessary for unit root tests to have good size, but that information criteria such as the *AIC* and the *BIC* tend to select a truncation lag (k) that is very small. We consider a class of Modified Information Criteria (*MIC*) with a penalty factor that is sample dependent. It takes into account the fact that the bias in the sum of the autoregressive coefficients is highly dependent on k and adapts to the type of deterministic components present. We use a local asymptotic framework in which the moving-average root is local to −1 to document how the *MIC* performs better in selecting appropriate values of k. In Monte-Carlo experiments, the *MIC* is found to yield huge size improvements to the DF^{GLS} and the feasible point optimal P_T test developed in Elliott *et al.* (1996). We also extend the M tests developed in Perron and Ng (1996) to allow for *GLS* detrending of the data. The

*Originally Published in *Econometrica* 69, pp. 1519–1554.

MIC along with *GLS* detrended data yield a set of tests with desirable size and power properties.

Keywords: Integrated processes; truncation lag; *GLS* detrending; information criteria.

1. Introduction

Many unit root tests have been developed for testing the null hypothesis of a unit root against the alternative of stationarity. While the presence or absence of a unit root has important implications, many remain skeptical about the conclusions drawn from such tests. This concern is justifiable, as these tests generally suffer from two problems. First, many tests have low power when the root of the autoregressive polynomial is close to but less than unity (e.g., DeJong *et al.*, 1992). Second, the majority of the tests suffer from severe size distortions when the moving-average polynomial of the first differenced series has a large negative root (e.g., Schwert, 1989; Perron and Ng, 1996). Although less severe, the problem also arises when there is a large negative autoregressive root in the residuals. The consequence is over-rejections of the unit root hypothesis. While few economic time series are found to have negative serial correlation of the autoregressive type, many do exhibit a large negative moving-average root. The inflation rate is an example, and omitted outliers have also been shown to induce a negative moving-average root in the error process (e.g., Franses and Haldrup, 1994). It is therefore desirable to have powerful unit root tests that are robust to size distortions.

The implementation of unit root tests often necessitates the selection of an autoregressive truncation lag, say, k. This is required in the autoregression used to form the Said-Dickey-Fuller test. It is also required in constructing an autoregressive estimate of the spectral density at frequency zero. However, simulations have repeatedly shown a strong association between k and the severity of size distortions and/or the extent of power loss (e.g., Ng and Perron, 1995; Lopez, 1997). To date, there exists no operational procedure for selecting k that yields robust and satisfactory results. The problem is that while a small k is adequate for finite order autoregressive errors and *ARMA* processes with small moving-average components, a large k is generally necessary for noise functions with a moving-average root that is large. For the latter class of models, selection rules such as the Schwarz or Bayesian Information Criterion (*BIC*) and the Akaike Information Criterion (*AIC*) tend to select values of k that are generally too small for unit root tests to have good sizes.

This paper sets out to resolve two problems. One is to enhance the power of several tests that have been shown to have small size distortions. More precisely, in Perron and Ng (1996), we analyzed a class of modified tests, originally proposed by Stock (1990), and showed that these tests are far more robust to size distortions than other unit root tests in the literature when the residuals have negative serial correlation. Exploiting the findings in Elliott *et al.* (1996) (hereafter ERS) and Dufour and King (1991) that local *GLS* detrending of the data yields substantial power gains, we apply the idea of *GLS* detrending to the modified tests and show that nonnegligible size and power gains can be made when used in conjunction with an autoregressive spectral density estimator at frequency zero provided the truncation lag is appropriately selected.

Our second task is to provide an improved procedure for choosing the truncation lag. We argue that the penalty of k assigned to overfitting in information criteria such as the *AIC* under-estimates the cost of a low order model when the unit root process has a negative moving-average root and, hence, tends to select a lag length that is too small. We suggest a class of modified information criteria that takes better account of the cost of underfitting. The modified *AIC* is shown to lead to substantial size improvements over standard information criteria in all the unit root tests considered. Taking the two steps together, *GLS* detrending along with the selection rule for k provide procedures which allow for the construction of unit root tests with much improved size and power. Additionally, using *GLS* detrended data to estimate the spectral density at frequency zero is found to have favorable size and power implications.

The paper proceeds as follows. After setting up the preliminaries in Section 2, Section 3 considers the class of M tests with GLS detrending. Section 4 explains why information criteria will underestimate k and Section 5 proposes an alternative penalty factor in the context of integrated data. Section 6 presents the properties of the modified information criteria using a local asymptotic framework and shows why the AIC version is preferable. Simulation results are presented in Section 7, and the procedures are illustrated using inflation rates for the G7 countries in Section 8. Section 9 offers brief concluding remarks. Mathematical derivations are given in the Appendix.

2. Preliminaries

We assume that we have $T+1$ observations generated by (for $t = 0, \ldots, T$):

$$y_t = d_t + u_t, \quad u_t = \alpha u_{t-l} + v_t, \tag{1}$$

where $E(u_0^2) < \infty$ (even as $T \to \infty$), $v_t = \delta(L)e_t = \sum_{j=0}^{\infty} \delta_j e_{t-j}$ with $\sum_{j=0}^{\infty} j|\delta_j| < \infty$ and $\{e_t\} \sim$ i.i.d. $(0, \sigma_e^2)$. The nonnormalized spectral density at frequency zero of v_t is given by $\sigma^2 = \sigma_e^2 \delta(1)^2$. Also, $T^{-1/2} \sum_{t=1}^{[sT]} v_t \Rightarrow \sigma W(s)$, where $\Rightarrow$ denotes weak convergence in distribution and $W(s)$ is a Wiener process defined on $C[0,1]$. In (1), $d_t = \psi' z_t$, where z_t is a set of deterministic components. We consider $d_t = \sum_{i=0}^{p} \psi_i t^i$, with special focus on $p = 0, 1$, though the analysis remains valid for more general cases.[1] Of interest is testing the null hypothesis $\alpha = 1$ against $\alpha < 1$. The DF test due to Dickey and Fuller (1979) and Said and Dickey (1984) is the t statistic for $\beta_0 = 0$ in the autoregression:

$$\Delta y_t = d_t + \beta_0 y_{t-1} + \sum_{j=1}^{k} \beta_j \Delta y_{t-j} + e_{tk}. \tag{2}$$

[1]Most of the results presented here, in particular Theorem 1, hold with the less restrictive assumption that $\{e_t\}$ is a martingale difference sequence with respect to the sigma algebra $\mathscr{F}_t = \{e_s;\, s \le t\}$ with $\sigma_e^2 = \lim_{T\to\infty} T^{-1} \sum_{t=1}^{T} E(e_t^2|\mathscr{F}_{t-1})$.

In Perron and Ng (1996), we analyzed the properties of three tests: MZ_α, MZ_t, and *MSB*, collectively referred to as the M tests. These are defined as (for the no deterministic term case):

$$MZ_\alpha = (T^{-1}y_T^2 - s_{AR}^2)\left(2T^{-2}\sum_{t=1}^{T} y_{t-1}^2\right)^{-1}, \tag{3}$$

$$MSB = \left(T^{-2}\sum_{t=1}^{T} y_{t-1}^2/s_{AR}^2\right)^{1/2}, \tag{4}$$

and $MZ_t = MZ_a \times MSB$. All three tests are based on s_{AR}^2, an autoregressive estimate of the spectral density at frequency zero of v_t. Let $\hat{\beta}(1) = \sum_{i=1}^{k}\hat{\beta}_i$, $\hat{\sigma}_k^2 = (T-k)^{-1}\sum_{t=k+1}^{T}\hat{e}_{tk}^2$, with $\hat{\beta}_i$ and $\{\hat{e}_{tk}\}$ obtained from (2) estimated by *OLS*. Then,

$$s_{AR}^2 = \hat{\sigma}_k^2/(1-\hat{\beta}(1))^2. \tag{5}$$

The M tests for $p = 0$ and 1 can be obtained upon replacing y_{t-1} and y_T by the residuals obtained from least squares detrending. The MZ_α and MZ_t tests can be viewed as modified versions of the Phillips (1987) and Phillips-Perron (1988) Z_α and Z_t tests, referred to as the Z tests below (see Perron and Ng, 1998). The Z tests suffer from severe size distortions when v_t has a negative moving-average root. When the root is close to minus one (e.g., -0.8), the rejection rate can be as high as 100% (see Schwert, 1989). Perron and Ng (1996) showed that the M tests have dramatically smaller size distortions than most (if not all) unit root tests in the literature in cases of negative moving-average errors *if the autoregressive spectral density estimator defined above is used in conjunction with a suitably chosen* k. The three M tests have similar properties and we use MZ_α for illustration.

3. The Tests Under *GLS* Detrending

To improve power, ERS proposed local to unity *GLS* detrending of the data. For any series $\{x_t\}_{t=0}^{T}$, define $(x_0^{\bar{\alpha}}, x_t^{\bar{\alpha}}) \equiv (x_0, (1-\bar{\alpha}L)x_t)$, $t = 1, \ldots, T$, for some chosen $\bar{\alpha} = 1 + \bar{c}/T$. The *GLS* detrended series is defined as

$$\tilde{y}_t \equiv y_t - \hat{\psi}' z_t, \tag{6}$$

where $\hat{\psi}$ minimizes $S(\bar{\alpha}, \psi) = (y^{\bar{\alpha}} - \psi' z^{\bar{\alpha}})'(y^{\bar{\alpha}} - \psi' z^{\bar{\alpha}})$. If v_t is i.i.d. normal, the point optimal test of the null hypothesis $\alpha = 1$ against the alternative hypothesis $\alpha = \bar{\alpha}$ is the likelihood ratio statistic, $L = S(\bar{\alpha}) - S(1)$, where $S(\bar{\alpha}) = \min_{\psi} S(\bar{\alpha}, \psi)$. ERS considered a feasible point optimal test that takes into account that v_t may be serially correlated. The statistic is

$$P_T = [S(\bar{\alpha}) - \bar{\alpha} S(1)]/s^2_{AR}, \tag{7}$$

where s^2_{AR} is defined as in (5). The value of $\bar{c}$ is chosen such that the asymptotic local power function of the test is tangent to the power envelope at 50% power. For $p = 0$ this is -7.0 and for $p = 1$, it is -13.5. The P_T test did not perform well in simulations when k was selected by the *BIC*. Large size distortions were recorded. ERS then recommended the DF^{GLS} statistic (whose local asymptotic power function also lies close to the Gaussian local power envelope) as the t-statistic for testing $\beta_0 = 0$ from the following regression estimated by *OLS*:

$$\Delta \tilde{y}_t = \beta_0 \tilde{y}_{t-1} + \sum_{j=1}^{k} \beta_j \Delta \tilde{y}_{t-j} + e_{tk}. \tag{8}$$

3.1. *The M^{GLS} tests*

While the power gains of the DF^{GLS} from using *GLS* detrended data are impressive, simulations also show that the test exhibits strong size distortions when v_t is an $MA(1)$ with a negative coefficient. For $T = 100$, ERS report rejection rates of 30 and 50% for $p = 0$ and 1, respectively, when the nominal rate is 5%. An explanation is provided in Section 4. Size distortions, however, are less of an issue with the M tests in theory as we showed in Perron and Ng (1996). In practice, it does require us to have a way to find the appropriate k. Suppose for the moment that this lag length selection problem is solved. Since the power gains from the DF^{GLS} over the DF come from the use of *GLS* detrended data, it is natural to consider the M

tests under *GLS* detrending. We now state the asymptotic properties of the M^{GLS} tests.

Theorem 1. *Let y_t be generated as in (1) with $\alpha = 1 + c/T$ and consider data obtained from local GLS detrending at $\bar{\alpha} = 1 + \bar{c}/T$. Also, let s^2_{AR} be a consistent estimate of σ^2. If $p = 0$, $MZ_\alpha^{GLS} \Rightarrow 0.5(J_c(1)^2 - 1)(\int_0^1 J_c(r)^2 dr)^{-1}$ and $MSB^{GLS} \Rightarrow (\int_0^1 J_c(r)^2 dr)^{1/2}$. If $p = 1$, $MZ_\alpha^{GLS} \Rightarrow 0.5(V_{c,\bar{c}}(1)^2 - 1)(\int_0^1 V_{c,\bar{c}}(r)^2 dr)^{-1}$ and $MSB^{GLS} \Rightarrow (\int_0^1 V_{c,\bar{c}}(r)^2 dr)^{1/2}$, where $J_c(r)$ is an Ornstein–Uhlenbeck process defined by $dJ_c(r) = cJ_c(r)dr + dW(r)$ with $J_c(0) = 0$, and $V_{c,\bar{c}}(r) = J_c(r) - r[\lambda J_c(1) + 3(1-\lambda)\int_0^1 sJ_c(s)ds]$, $\lambda = (1-\bar{c})/(1-\bar{c}+\bar{c}^2/3)$. Results for MZ_t^{GLS} can be obtained in view of the fact that $MZ_t^{GLS} = MZ_\alpha^{GLS} \times MSB^{GLS}$.*

The proof is based on standard results for sample moments of *GLS* detrended data and therefore omitted. The MZ_α^{GLS}, MZ_t^{GLS}, and MSB^{GLS} have similar size and power properties and will be referred to as the M^{GLS} tests. We have computed, via simulations, the local asymptotic power functions of the M tests with *OLS* and *GLS* detrending. These showed the M^{GLS} tests to be uniformly more powerful than the M tests and also to have local asymptotic power functions that are indistinguishable from those of the DF^{GLS} and the Gaussian asymptotic local power envelope. A graph is therefore not presented. The important point is that the M^{GLS} tests, like the DF^{GLS}, achieve local asymptotic power that is approximately the same as the feasible point-optimal test. We also consider two modified feasible point optimal tests and derive their limiting distributions:

$$p = 0: \quad MP_T^{GLS} = \left[\bar{c}^2 T^{-2} \sum_{t=1}^{T} \tilde{y}_{t-1}^2 - \bar{c}T^{-1}\tilde{y}_T^2\right] \bigg/ s^2_{AR}$$

$$\Rightarrow \bar{c}^2 \int_0^1 J_c(r)^2 dr - \bar{c}J_c(1)^2,$$

$$p = 1: \quad MP_T^{GLS} = \left[\bar{c}^2 T^{-2} \sum_{t=1}^{T} \tilde{y}_{t-1}^2 + (1-\bar{c})T^{-1}\tilde{y}_T^2\right] \bigg/ s^2_{AR}$$

$$\Rightarrow \bar{c}^2 \int_0^1 V_{c,\bar{c}}(r)^2 dr + (1-\bar{c})V_{c,\bar{c}}(1)^2. \tag{9}$$

Table 1. Asymptotic critical values[a].

Percentile	MZ_α^{GLS}	MZ_t^{GLS}, ADF^{GLS}	MSB^{GLS}	P_T^{GLS}, MP_T^{GLS}
		Case: $p = 0$, $\bar{c} = -7.0$		
0.01	−13.8	−2.58	0.174	1.78
0.05	−8.1	−1.98	0.233	3.17
0.10	−5.7	−1.62	0.275	4.45
		Case: $p = 1$, $\bar{c} = -13.5$		
0.01	−23.8	−3.42	0.143	4.03
0.05	−17.3	−2.91	0.168	5.48
0.10	−14.2	−2.62	0.185	6.67

Note: [a]For MZ_α^{GLS}, MZ_t^{GLS}, and ADF^{GLS} with $p = 0$, the critical values are from Fuller (1976). The other entries were obtained using 20,000 simulations and 5,000 steps to approximate the Wiener processes.

These tests are based on the same motivation that leads to the definition of the M tests in Stock (1990), namely, to provide functionals of sample moments that have the same asymptotic distributions as well known unit root tests. The MP_T is important because its limiting distribution coincides with that of the feasible point optimal test considered in ERS.

Critical values for the limiting distributions of the M^{GLS} tests were obtained by numerical simulations that approximate the Wiener process $W(r)$ by partial sums of $N(0, 1)$ random variables with 5,000 steps and 20,000 simulations. These are summarized in Table 1.

3.2. *The construction of the autoregressive spectral density estimator and the $\overline{M}^{GLS}$*

All that is required for Theorem 1 and the results of ERS to hold is that s_{AR}^2 converges to σ^2 under the null hypothesis of a unit root, and for the tests to be consistent, that Ts_{AR}^2 diverges under the alternative of stationarity. The attractiveness of GLS detrending is that it estimates the deterministic function with more precision and leads to reduced bias in $\hat{\beta}_i$. Since this should lead to a more precise estimate of $\hat{\beta}(1)$, we also consider estimating s_{AR}^2 using the autoregression (8) based on GLS detrended data. An additional advantage is that this estimator is invariant to the slope of the trend function.[2] Experimentation with different values of $\bar{c}$ when constructing the autoregressive spectral density gave similar results. Hence, $\bar{c} = -7.0$ and -13.5 will continue to be used. The tests so constructed are denoted $\overline{Z}^{GLS}$ and $\overline{M}^{GLS}$. These are to be distinguished from the Z^{GLS} and M^{GLS} statistics that use least squares detrended data to estimate s^2_{AR}.

4. The Size Issue, the Selection of *k*, and Information Criteria

To see the influence of the lag order k, we consider the finite sample size of the tests $\overline{M}Z_\alpha^{GLS}$, DF^{GLS}, and $\overline{Z}_\alpha^{GLS}$. The setup, which will be used throughout, is to consider data generated by (1) with $\alpha = 1$ and v_t either an $MA(1)$, i.e., $v_t = e_t + \theta e_{t-1}$, or an $AR(1)$, i.e., $v_t = \rho v_{t-1} + e_t$, with $e_t \sim$i.i.d. $N(0, 1)$. The errors are generated from the rndn() function in Gauss with 999 as the seed, and 5,000 replications are used. We use the nominal size of 5% as the benchmark.

Reported in Tables 2 and 3 are the sizes of the tests for $p = 0, 1$ at selected values of θ and ρ. We report results for $T = 100$ and 250 to highlight the fact that the size issue in the negative moving average case is not a small sample problem.

Several features of the results for MA errors are of note. First, for a θ of the same absolute value, a negative θ always requires a larger lag to obtain a more accurate size. Second, for a positive θ, the size of the tests are significantly better when k is even. Third, the larger the number of deterministic terms, the more distant are the exact from the nominal sizes. Fourth, there is always a value of k that will yield an $\overline{M}Z_\alpha^{GLS}$ with a reasonably accurate size when $\theta = -0.80$. For example, for $p = 0$ when $T = 100$ and 250, exact sizes of 0.077 and 0.055, respectively, are achievable with lags of 9 and 8.

[2]When $p = 1$, Perron and Ng (1996), Stock (1990), and ERS estimate Equation (2) with only a constant by ordinary least-squares. This still ensures a consistent estimate under the null hypothesis but the omission of the trend term implies a noninvariance with respect to the slope of the trend function in finite samples under the alternative hypothesis. This noninvariance is such that the tests lose power as the slope increases.

Table 2. Exact size of the tests at selected values of k: $p = 0$.

MA Case	T	θ	$k = 0$	1	2	3	4	5	6	7	8	9	10
$\overline{Z}_{\alpha}^{GLS}$	100	−0.8	0.994	0.904	0.808	0.740	0.698	0.672	0.654	0.644	0.636	0.633	0.627
		−0.5	0.618	0.322	0.224	0.193	0.176	0.181	0.182	0.192	0.197	0.207	0.216
		0.0	0.072	0.082	0.089	0.104	0.112	0.125	0.137	0.152	0.164	0.179	0.190
		0.5	0.010	0.143	0.056	0.116	0.100	0.129	0.135	0.153	0.158	0.179	0.187
		0.8	0.006	0.172	0.036	0.147	0.071	0.155	0.108	0.166	0.138	0.188	0.169
	250	−0.8	0.987	0.861	0.725	0.641	0.592	0.559	0.539	0.524	0.515	0.507	0.504
		−0.5	0.553	0.258	0.166	0.124	0.112	0.110	0.106	0.112	0.111	0.115	0.116
		0.0	0.060	0.062	0.063	0.066	0.069	0.075	0.077	0.081	0.087	0.087	0.092
		0.5	0.008	0.111	0.047	0.080	0.065	0.077	0.076	0.083	0.084	0.091	0.095
		0.8	0.004	0.145	0.030	0.108	0.049	0.099	0.062	0.096	0.074	0.100	0.085
$\overline{M}Z_{\alpha}^{GLS}$	100	−0.8	0.990	0.828	0.543	0.319	0.187	0.129	0.095	0.086	0.079	0.077	0.078
		−0.5	0.595	0.241	0.126	0.097	0.084	0.092	0.097	0.106	0.120	0.132	0.141
		0.0	0.065	0.070	0.082	0.095	0.104	0.119	0.130	0.145	0.158	0.172	0.185
		0.5	0.008	0.139	0.054	0.113	0.097	0.126	0.132	0.150	0.156	0.176	0.186
		0.8	0.004	0.168	0.035	0.143	0.068	0.153	0.106	0.164	0.136	0.186	0.168
	250	−0.8	0.987	0.828	0.597	0.407	0.260	0.166	0.107	0.075	0.055	0.044	0.035
		−0.5	0.545	0.227	0.121	0.081	0.067	0.063	0.062	0.066	0.070	0.071	0.074
		0.0	0.058	0.059	0.061	0.064	0.066	0.072	0.076	0.079	0.084	0.084	0.088
		0.5	0.008	0.110	0.045	0.078	0.064	0.076	0.075	0.082	0.083	0.090	0.095
		0.8	0.004	0.145	0.029	0.108	0.048	0.099	0.061	0.095	0.074	0.100	0.084
DF^{GLS}	100	−0.8	0.991	0.843	0.624	0.455	0.322	0.243	0.187	0.157	0.129	0.114	0.104
		−0.5	0.605	0.263	0.147	0.108	0.089	0.087	0.081	0.079	0.076	0.078	0.075
		0.0	0.073	0.070	0.072	0.074	0.072	0.072	0.070	0.070	0.071	0.074	0.073
		0.5	0.010	0.124	0.046	0.081	0.063	0.075	0.074	0.077	0.078	0.075	0.071
		0.8	0.005	0.150	0.030	0.107	0.043	0.089	0.054	0.082	0.059	0.080	0.064

(*Continued*)

Table 2. (*Continued*)

MA Case	T	θ	$k=0$	1	2	3	4	5	6	7	8	9	10
	250	−0.8	0.985	0.823	0.613	0.449	0.332	0.247	0.194	0.155	0.128	0.109	0.099
		−0.5	0.535	0.227	0.123	0.084	0.068	0.064	0.062	0.062	0.061	0.058	0.060
		0.0	0.057	0.054	0.054	0.053	0.056	0.055	0.058	0.058	0.056	0.055	0.054
		0.5	0.009	0.103	0.043	0.070	0.055	0.061	0.057	0.058	0.057	0.060	0.060
		0.8	0.004	0.133	0.027	0.090	0.039	0.074	0.046	0.068	0.049	0.065	0.051
AR Case	T	ρ	$k=0$	1	2	3	4	5	6	7	8	9	10
$\overline{Z}_{\alpha}^{GLS}$	100	−0.8	0.802	0.286	0.286	0.295	0.300	0.308	0.314	0.322	0.328	0.340	0.346
		0.8	0.000	0.102	0.115	0.135	0.143	0.158	0.172	0.191	0.201	0.217	0.226
	250	−0.8	0.737	0.170	0.171	0.171	0.173	0.176	0.179	0.182	0.185	0.184	0.183
		0.8	0.000	0.069	0.071	0.077	0.082	0.084	0.090	0.094	0.093	0.099	0.100
$\overline{M}Z_{\alpha}^{GLS}$	100	−0.8	0.784	0.027	0.035	0.050	0.065	0.081	0.089	0.113	0.120	0.140	0.149
		0.8	0.000	0.100	0.114	0.134	0.142	0.157	0.170	0.190	0.199	0.216	0.225
	250	−0.8	0.730	0.043	0.041	0.047	0.048	0.050	0.058	0.062	0.064	0.071	0.071
		0.8	0.000	0.069	0.071	0.077	0.082	0.084	0.090	0.094	0.092	0.099	0.100
DF^{GLS}	100	−0.8	0.791	0.073	0.074	0.077	0.079	0.080	0.076	0.081	0.075	0.083	0.078
		0.8	0.000	0.066	0.070	0.069	0.068	0.069	0.071	0.072	0.071	0.077	0.079
	250	−0.8	0.718	0.057	0.054	0.056	0.054	0.055	0.057	0.057	0.057	0.059	0.058
		0.8	0.000	0.059	0.061	0.063	0.062	0.060	0.060	0.060	0.058	0.058	0.054

Table 3. Exact size of the tests at selected values of k: $p = 1$.

MA Case	T	θ	$k = 0$	1	2	3	4	5	6	7	8	9	10
$\overline{Z}_{\alpha}^{GLS}$	100	−0.8	1.000	0.994	0.965	0.940	0.918	0.906	0.893	0.888	0.882	0.882	0.883
		−0.5	0.810	0.470	0.340	0.299	0.290	0.309	0.323	0.358	0.376	0.413	0.450
		0.0	0.053	0.076	0.100	0.136	0.169	0.222	0.259	0.313	0.346	0.400	0.430
		0.5	0.001	0.169	0.059	0.163	0.155	0.226	0.253	0.303	0.343	0.397	0.432
		0.8	0.001	0.219	0.023	0.229	0.091	0.284	0.190	0.350	0.297	0.422	0.389
	250	−0.8	1.000	0.992	0.960	0.927	0.899	0.875	0.859	0.848	0.839	0.833	0.830
		−0.5	0.801	0.404	0.253	0.202	0.181	0.174	0.176	0.179	0.188	0.199	0.209
		0.0	0.047	0.057	0.064	0.077	0.081	0.094	0.104	0.113	0.129	0.142	0.154
		0.5	0.002	0.127	0.037	0.092	0.071	0.097	0.095	0.120	0.122	0.143	0.153
		0.8	0.001	0.176	0.014	0.129	0.038	0.131	0.068	0.139	0.099	0.162	0.136
$\overline{M}Z_{\alpha}^{GLS}$	100	−0.8	1.000	0.925	0.592	0.345	0.220	0.185	0.159	0.163	0.163	0.185	0.183
		−0.5	0.750	0.252	0.104	0.093	0.099	0.132	0.159	0.204	0.234	0.285	0.317
		0.0	0.032	0.055	0.080	0.115	0.152	0.206	0.243	0.300	0.332	0.390	0.419
		0.5	0.001	0.161	0.054	0.155	0.148	0.219	0.246	0.299	0.338	0.392	0.427
		0.8	0.001	0.211	0.021	0.224	0.087	0.277	0.186	0.346	0.294	0.417	0.386
	250	−0.8	1.000	0.978	0.800	0.514	0.276	0.147	0.087	0.058	0.048	0.040	0.036
		−0.5	0.786	0.311	0.126	0.074	0.057	0.059	0.065	0.069	0.082	0.100	0.106
		0.0	0.040	0.049	0.057	0.070	0.077	0.088	0.099	0.108	0.124	0.136	0.149
		0.5	0.002	0.124	0.034	0.090	0.069	0.095	0.094	0.117	0.121	0.142	0.150
		0.8	0.001	0.174	0.014	0.128	0.036	0.130	0.066	0.137	0.098	0.160	0.135
DF^{GLS}	100	−0.8	1.000	0.977	0.815	0.600	0.415	0.293	0.203	0.167	0.125	0.116	0.098
		−0.5	0.838	0.375	0.176	0.112	0.087	0.084	0.074	0.070	0.068	0.069	0.065
		0.0	0.070	0.072	0.067	0.072	0.069	0.073	0.072	0.071	0.069	0.068	0.061
		0.5	0.003	0.160	0.045	0.091	0.064	0.073	0.062	0.066	0.061	0.067	0.062
		0.8	0.002	0.209	0.021	0.127	0.036	0.099	0.046	0.085	0.053	0.076	0.061

(*Continued*)

Table 3. (*Continued*)

MA Case	T	θ	$k=0$	1	2	3	4	5	6	7	8	9	10
	250	−0.8	1.000	0.984	0.875	0.685	0.504	0.362	0.259	0.194	0.150	0.120	0.099
		−0.5	0.809	0.352	0.161	0.101	0.072	0.061	0.059	0.053	0.053	0.056	0.052
		0.0	0.053	0.055	0.055	0.056	0.056	0.054	0.054	0.054	0.055	0.055	0.054
		0.5	0.002	0.126	0.034	0.071	0.048	0.057	0.053	0.054	0.052	0.056	0.050
		0.8	0.001	0.172	0.015	0.098	0.024	0.075	0.032	0.063	0.038	0.058	0.042
AR Case	T	ρ	$k=0$	1	2	3	4	5	6	7	8	9	10
$\overline{Z}_\alpha^{GLS}$	100	−0.8	0.960	0.497	0.509	0.526	0.549	0.574	0.593	0.616	0.640	0.669	0.690
		0.8	0.000	0.123	0.162	0.213	0.252	0.312	0.351	0.410	0.441	0.489	0.514
	250	−0.8	0.956	0.319	0.327	0.330	0.335	0.345	0.352	0.360	0.369	0.375	0.379
		0.8	0.000	0.070	0.077	0.089	0.099	0.119	0.129	0.150	0.159	0.176	0.183
$\overline{M}Z_\alpha^{GLS}$	100	−0.8	0.941	0.001	0.008	0.030	0.058	0.104	0.140	0.191	0.233	0.290	0.329
		0.8	0.000	0.121	0.160	0.211	0.249	0.309	0.349	0.408	0.439	0.487	0.513
	250	−0.8	0.952	0.011	0.015	0.020	0.028	0.037	0.050	0.062	0.076	0.092	0.106
		0.8	0.000	0.070	0.076	0.089	0.099	0.119	0.128	0.150	0.159	0.175	0.183
DF^{GLS}	100	−0.8	0.967	0.069	0.067	0.067	0.060	0.068	0.059	0.065	0.063	0.065	0.065
		0.8	0.000	0.072	0.067	0.070	0.071	0.069	0.069	0.074	0.066	0.072	0.071
	250	−0.8	0.959	0.052	0.051	0.052	0.050	0.049	0.050	0.050	0.051	0.052	0.053
		0.8	0.000	0.057	0.055	0.057	0.054	0.055	0.058	0.056	0.055	0.059	0.058

In contrast, there does not exist a k that will result in an acceptable size for $\overline{Z}_{\alpha}^{GLS}$, and increasing the sample size will not reduce the size distortions adequately. For the sample sizes reported, the exact sizes for $\overline{Z}_{\alpha}^{GLS}$ are well above 0.5 at $\theta = -0.8$, and size distortions are nonnegligible even when $\theta = -0.5$. Fifth, the DF^{GLS} also exhibits a dependence on k. At $T = 100$, the size is 0.104 at $k = 10$, but is 0.624 when $k = 2$.

For AR errors, size discrepancies between $\overline{Z}_{\alpha}^{GLS}$ and $\overline{M}_{\alpha}^{GLS}$ also exist when ρ is large and negative, albeit not as dramatic as in the MA case. The results reveal that the $\overline{M}Z_{\alpha}^{GLS}$ is inferior to the DF^{GLS} when $p = 1$, the sample size is small, and $|\rho|$ is large. For large negative AR coefficients, the $\overline{M}Z_{\alpha}^{GLS}$ has a more accurate size at k greater than 1. For large positive AR coefficients, the $\overline{M}Z_{\alpha}^{GLS}$ is oversized for small T, though the size improves as T increases.

Clearly, the choice of k is crucial especially in the case of negative moving average errors. Various practical solutions have been used against this problem. In ERS, the BIC is used to select k but they set the lower bound to be 3, because even larger size distortions would have resulted if zero was the lower bound. An alternative method is the sequential t test for the significance of the last lag considered in Ng and Perron (1995). It has the ability to yield higher k's than the BIC when there are negative moving-average errors and reduce size distortions. But, the sequential test tends to overparameterize in other cases. This, as does ERS's implementation of the BIC, leads to less efficient estimates and subsequently to power losses. Neither approach is fully satisfactory.

The AIC and the BIC belong to the class of information based rules (hereafter, IC) where the chosen value of k is $k_{ic} = \arg\min_{k \in [0,\ldots,k_{\max}]} IC(k)$ where[3]

$$IC(k) = \ln(\hat{\sigma}_k^2) + (k + p + 1)C_T/(T - k_{\max}), \qquad (10)$$

$\hat{\sigma}_k^2 = (T - k_{\max})^{-1} \sum_{t=k_{\max}+1}^{T} \hat{e}_{tk}^2$, $C_T/T \to 0$, as $T \to \infty$, and $C_T > 0$. The various criteria differ in C_T, the weight applied to overfitting,

[3]See Ng and Perron (2000) for an analysis pertaining to the adequacy of the particular formulation of the information criterion stated.

but all use k as the penalty to overfitting. We argue below that, with integrated data, this penalty may be a poor approximation to the cost of underfitting.

5. An Alternative Penalty for Integrated Data

To motivate the main idea, we consider the derivation of the *AIC* for data generated by a finite order $AR(k_0)$ with normal errors and a unit root (i.e., the DGP is (2) with $k = k_0$, $\beta_0 = 0$, and $e_{tk} = e_t \sim$ i.i.d. $N(0,1)$). The family of parametric models fitted is given by (8). For notation, we let $\beta^0(k) = (0, \beta_1, \ldots, \beta_k)'$, $\hat{\beta}(k) = (\hat{\beta}_0, \hat{\beta}_1, \ldots, \hat{\beta}_k)'$, $\beta^0_{-0}(k) = (\beta_1, \ldots, \beta_k)'$, and $\hat{\beta}_{-0}(k) = (\hat{\beta}_1, \ldots, \hat{\beta}_k)'$. In what follows, we adopt the treatment of Gourieroux and Monfort (1995, pp. 307–309). The goal is to select an autoregressive order between 0 and some upper bound $k_{\max}$. We assume this upper bound satisfies $k_{\max}/T \to 0$ as $T \to \infty$. Let $f(\Delta y|\beta^0(k))$ be the likelihood function of the data $(\Delta y_{k_{\max}+1}, \ldots, \Delta y_T)$ conditional on the initial observations $(y_0, \ldots, y_{k_{\max}})$. This ensures that each competing model is evaluated with the same number of effective observations, namely $T - k_{\max}$. The Kullback distance between the true probability distribution and the estimated parametric model is $Q = E_0[\ln(f(\Delta y|\beta^0(k))) - \ln(f(\Delta y|\hat{\beta}(k)))]$ with sample analog:

$$\tilde{Q} = (T - k_{\max})^{-1} \sum_{t=k_{\max}+1}^{T} \ln(f(\Delta y_t|\beta^0(k))) - (T - k_{\max})^{-1} \sum_{t=k_{\max}+1}^{T} \ln(f(\Delta y_t|\hat{\beta}(k))).$$

As discussed in Gourieroux and Monfort (1995), Akaike's suggestion was to find a Q^* such that $\lim_{T\to\infty} E[T(Q - Q^*)] = 0$ so that Q^* is unbiased for Q to order T^{-1}. Let $X_t = (\tilde{y}_{t-1}, Z_t)$ with $Z_t = (\Delta\tilde{y}_{t-1}, \ldots, \Delta\tilde{y}_{t-k})$ and

$$\Phi_T(k) = (1/\hat{\sigma}_k^2)(\hat{\beta}(k) - \beta^0(k))' \sum_{t=k_{\max}+1}^{T} X_t X_t'(\hat{\beta}(k) - \beta^0(k)),$$

where $\hat{\sigma}_k^2 = (T - k_{\max})^{-1} \sum_{t=k_{\max}+1}^{T} \hat{e}_{tk}^2$. Using Taylor series expansions, we have $TQ = \Phi_T(k)/2 + o_p(1)$ and $T\tilde{Q} = -\Phi_T(k)/2 + o_p(1)$. Since $T(Q - \tilde{Q}) = \Phi_T(k) + o_p(1)$, $\lim_{T\to\infty} E[T(Q - Q^*)] = 0$ if $Q^* = \tilde{Q} + \Phi_T(k)$ and the remainder term is uniformly integrable. Now consider the term $\Phi_T(k)$ in the context of integrated data. Given the asymptotic block diagonality of the matrix $D_T \sum_{t=k_{\max}+1}^{T} X_t X_t' D_T$ with $D_T = \text{diag}((T - k_{\max})^{-1}, (T - k_{\max})^{-1/2}, \ldots, (T - k_{\max})^{-1/2})$, we have

$$\begin{aligned}\Phi_T(k) &= \left(\frac{1}{\hat{\sigma}_k^2}\right) \hat{\beta}_0^2 \sum_{t=k_{\max}+1}^{T} \tilde{y}_{t-1}^2 \\ &\quad + \left(\frac{1}{\hat{\sigma}_k^2}\right) (\hat{\beta}_{-0}(k) - \beta_{-0}^0(k))' \sum_{t=k_{\max}+1}^{T} Z_t Z_t' (\hat{\beta}_{-0}(k) - \beta_{-0}^0(k)) \\ &\quad + o_p(1) \\ &= (1/\hat{\sigma}_k^2)\hat{\beta}_0^2 \sum_{t=k_{\max}+1}^{T} \tilde{y}_{t-1}^2 + \chi_k^2 + o_p(1) \end{aligned} \tag{11}$$

where χ_k^2 is a Chi-square random variable with k degrees of freedom that is asymptotically independent of the first term. Hence, assuming the remainder term is uniformly integrable, a Q^* that will satisfy $\lim_{T\to\infty} E[T(Q - Q^*)] = 0$ is

$$\begin{aligned} Q^* &= (T - k_{\max})^{-1} \sum_{t=k_{\max}+1}^{T} \ln(f(\Delta y_t | \beta^0(k))) \\ &\quad - (T - k_{\max})^{-1} \sum_{t=k_{\max}+1}^{T} \ln(f(\Delta y_t | \hat{\beta}(k))) \\ &\quad + \left[(1/\hat{\sigma}_k^2)\hat{\beta}_0^2 \sum_{t=k_{\max}+1}^{T} \tilde{y}_{t-1}^2 + k \right]. \end{aligned}$$

Since the first term is common to all models and the second term is proportional to $-((T - k_{\max})/2)\ln(\hat{\sigma}_k^2)$ under normality, minimizing

Q^* is equivalent to minimizing

$$MAIC(k) = \ln(\hat{\sigma}_k^2) + \frac{2(\tau_T(k) + k)}{T - k_{\max}}, \tag{12}$$

where $\tau_T(k) = (\hat{\sigma}_k^2)^{-1}\hat{\beta}_0^2 \sum_{t=k_{\max}+1}^{T} \tilde{y}_{t-1}^2$ and $\hat{\sigma}_k^2 = (T - k_{\max})^{-1} \sum_{t=k_{\max}+1}^{T} \hat{e}_{tK}^2$. Note that the same result holds in the general case where the data are generated by (1). The important step is the relation given by (11), which can be shown to hold in the general case, in the sense that the remainder term is $o_p(1)$ uniformly in k for $0 < k \leq K_T$, where $K_T = o(T^{1/3})$. This follows using the results of Berk (1974) and Said and Dickey (1984) provided $k \to \infty$, $k^3/T \to 0$, and $T^{-1/2} \sum_{j=k+1}^{\infty} |\beta_j| \to 0$.

As a natural generalization of the $MAIC(k)$, we propose a class of *Modified Information Criteria (MIC)* that selects k as $k_{mic} = \arg\min_k MIC(k)$ where

$$MIC(k) = \ln(\hat{\sigma}_k^2) + \frac{C_T(\tau_T(k) + k)}{T - k_{\max}}, \tag{13}$$

with $C_T > 0$ and $C_T/T \to 0$ as $T \to \infty$. The *MBIC* obtains with $C_T = \ln(T - k_{\max})$.

There are two important elements that distinguish (13) from standard information criteria. The first is the imposition of the null hypothesis $\beta_0 = 0$. This idea of imposing the null hypothesis in model selection when the ultimate objective is hypothesis testing appears new and may have implications beyond the unit root issue.[4] This is an interesting avenue of research, but is beyond the scope of this paper. The second element is that we retain a term in the penalty function that is stochastic. Now, as $T \to \infty$, $(1/\hat{\sigma}_k^2)\hat{\beta}_0^2 \sum_{t=k_{\max}+1}^{T} \tilde{y}_{t-1}^2 \Rightarrow (\sigma^2/\sigma_e^2)[\widetilde{W}(1)^2 - 1]^2/[2\int_0^1 \widetilde{W}(r)^2 dr] \equiv \tau(\widetilde{W})$, say, where $\widetilde{W}$ is a detrended Wiener process, whose exact form depends on the deterministic components and the method of detrending. In principle, we could use the mean of $\tau(\widetilde{W})$, which is independent of k, instead of $\tau_T(k)$, to construct Q^* and the objective function would then reduce to the standard *AIC*. A central theme

[4]We thank a referee for bringing this feature to our attention.

of our argument for retaining $\tau_T(k)$ is that, unless both T and k are very large, it varies substantially with k, especially when a negative *MA* component is present. In a sense, the imposition of the null hypothesis, $\beta_0 = 0$, allows us to avoid using asymptotic expected values to approximate the penalty factor. We can therefore hope that $\tau_T(k)$ will better capture the relevant cost of selecting different orders in finite samples since it depends not only on k, but also on the nature of the deterministic components and the detrending procedure. Hence, the penalty term will also depend on these factors.

To get an idea of the dependence of $\tau_T(k)$ on k, Fig. 1 graphs the average over 5,000 replications of $\tau_T(k)$ against k for $p = 0, 1$ using the DGP $\Delta y_t = e_t + \theta e_{t-1}$ with $e_t \sim$ i.i.d. $N(0,1)$. For positive θ, $\tau_T(k)$ is indeed fairly invariant to k (as is the case for *AR* processes, not reported). But, there is a clear inverse and nonlinear relation between $\tau_T(k)$ and k for negative values of θ. This relation exists even when θ is -0.5 for both $p = 0$ and $p = 1$. The basic reason for the dependence of $\tau_T(k)$ on k is that the bias in the sum of the estimated autoregressive coefficients (i.e., $\hat{\beta}_0$) is very high for small values of k. This suggests that $\tau_T(k)$ will play a role in determining the appropriate k when there are negative moving-average errors. To make this claim precise, the next subsection uses a local asymptotic framework to analyze the properties of $\tau_T(k)$ and the *MIC*.

6. Local Asymptotic Properties

As discussed above, $\tau_T(k)$ will likely influence the choice of k when there is a large negative moving-average root, a situation for which standard unit root tests exhibit severe size distortions. To understand the issue of lag length selection in this setting, a useful device is an asymptotic framework in which the moving average coefficient is parameterized as local to -1.

$$\begin{aligned} y_t - \alpha_T y_{t-1} &= e_t + \theta_T e_{t-1}, \\ \alpha_T = 1 + c/T, &\quad \theta_T = -1 + \delta/\sqrt{T}, \end{aligned} \tag{14}$$

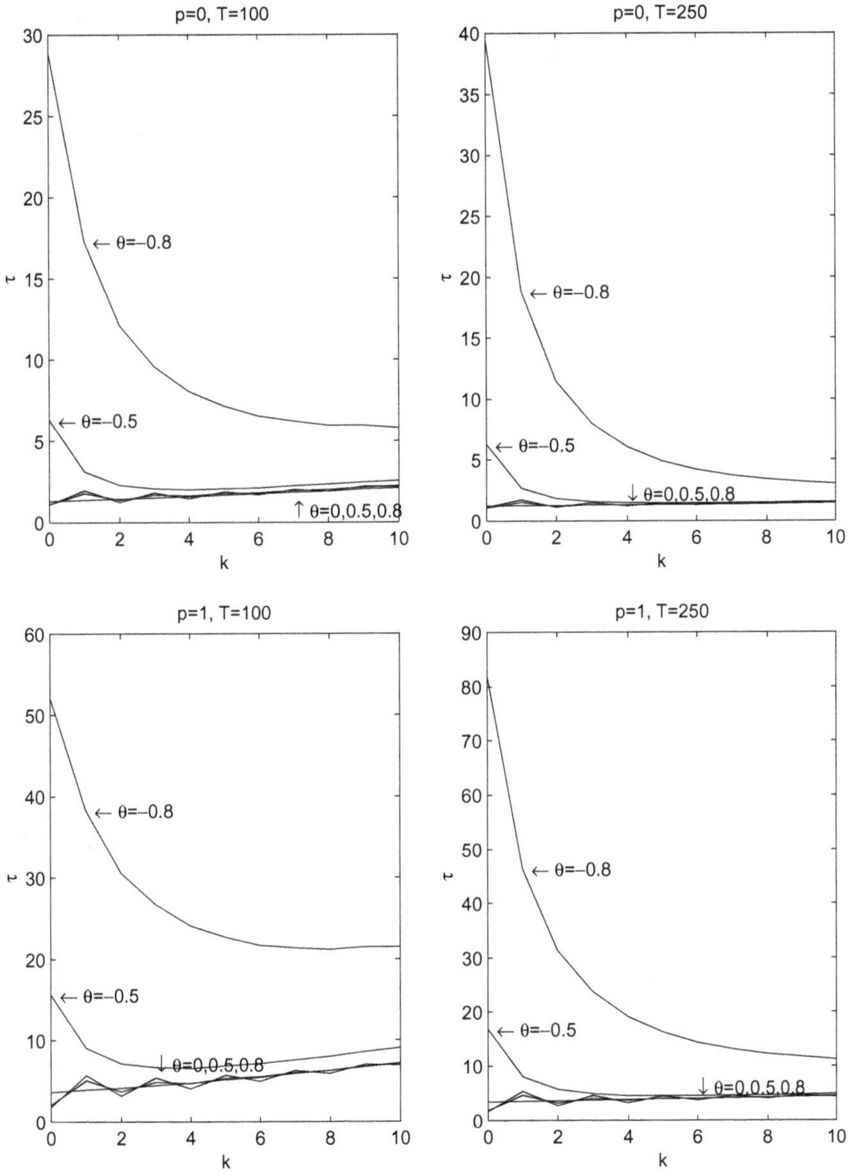

Fig. 1. $\tau_T(\mathrm{k})$.

$e_t \sim$ i.i.d. $(0, \sigma_e^2)$, $y_0 = e_0 = 0$, $\delta > 0$, and $c < 0$. As T increases to infinity, $\alpha_T \to 1, \theta_T \to -1$, and y_t becomes a white noise process. However, y_t is a *nearly integrated nearly white noise process* in finite samples. Nabeya and Perron (1994) used this asymptotic framework to derive approximations to the finite sample properties of the least-squares estimator in a first-order autoregressive model with negative moving-average errors and showed these approximations not only to be superior to those obtained from standard asymptotic distributions but also to be quite accurate even for small sample sizes. Perron and Ng (1996) used the same setup to analyze the size and power of various unit root tests based upon augmented autoregressions. Our aim here is to use the same device to first obtain useful qualitative features about the relative properties of the *IC* and *MIC*, and second, to guide us to an appropriate choice of the penalty weight C_T.

6.1. *Local asymptotic properties of* $\tau_T(k)$

The following Theorem (proved in the appendix) summarizes the general properties of $\tau_T(k)$ in this local asymptotic framework.

Theorem 2. *Let* y_t *be generated by* (14), $Z_1 = (1 + \delta^2 \int_0^1 J_c(r)^2 dr)/[\delta^2 \int_0^1 J_c(r)^2 dr]^2$ *when* $p = 0$, *and* $Z_1 = (1 + \delta^2 \int_0^1 V_{c,\bar{c}}(r)^2 dr)/[\delta^2 \int_0^1 V_{c,\bar{c}}(r)^2 dr]^2$ *when* $p = 1$. *Let* $k \to \infty$, $K_{1T} = o(T^{1/2})$, $T^{1/2}/K_{2T} = o_p(1)$, *and* $K_{3T} = o(T)$. *Then,* (i) *uniformly in* $0 < k \leq K_{1T}$: $(k^2/T)\tau_T(k) \Rightarrow Z_1$; (ii) *if* $k/T^{1/2} \to \kappa$, $\tau_T(k) \Rightarrow \vartheta(\kappa)^2 Z_2$, *where* $E(Z_2) = E(Z_1)$ *and* $\vartheta(\kappa) = 2\delta \exp(-2\kappa\delta)/(1 - \exp(-2\kappa\delta))$; *and* (iii) *uniformly in* $K_{2T} \leq k \leq K_{3T}$, $\tau_T(k) \Rightarrow Z_3$, *where* Z_3 *is* $O_p(1)$.

Theorem 2 shows that the local limit of $\tau_T(k)$ is unbounded unless k increases at least at rate $T^{1/2}$. Since $\tau_T(0)$ diverges at rate T (see Nabeya and Perron (1994)), and $\tau_T(k)$ diverges at rate T/k^2 if $k^2/T \to 0$, the cost to overfitting is not linear in k as is assumed for the *IC*, and Fig. 1 bears this out. Theorem 2 also highlights the fact that $\tau_T(k)$ will depend, via Z_1, on the nature of the deterministic components and the method of detrending.

Given Theorem 2 and the fact that $\hat{\sigma}_k^2$ is consistent for σ_e^2 irrespective of the rate of increase of k (shown in the appendix), it can be shown that, in this local asymptotic framework, the limiting distribution of the DF test diverges to $-\infty$ unless k grows at least at rate $T^{1/2}$.[5] Hence, the unit root will always be rejected if standard critical values are used. This explains why the DF test tends to be oversized when k is small. In Perron and Ng (1996, 1998), we showed that $k^2 s_{AR}^2 = O_p(1)$ but we also need Ts_{AR}^2 to be $O_p(1)$ for the M tests to be bounded under the null hypothesis in this local framework. But $k^2 s_{AR}^2$ and Ts_{AR}^2 can both be bounded only if $k/T^{1/2} \to \kappa$ for some $0 < \kappa < \infty$. In the next section, we will consider model selection procedures with this requirement in mind.

6.2. *Local asymptotic properties of the MIC and IC*

In the standard asymptotic framework, with data generated by (1), $\tau_T(k)$ is invariant to k for T and k large. It follows that k_{mic} grows at a logarithmic rate from Hannan and Deistler (1988). We now consider the properties of k chosen by the MIC and the IC in the local asymptotic framework with the additional technical requirement that $C_T^2/T \to 0$. Using Theorem 2 and an expansion for $\ln(\hat{\sigma}_k^2)$ derived in the appendix, we have the following expansions for the MIC and IC.

Lemma 1. *Let y_t be generated by* (14) *and K_{1T}, K_{2T}, and K_{3T} as defined in Theorem* 2; *then* (i) *uniformly in $0 < k \le K_{1T}$:*

$$MIC(k) = \ln(\sigma_e^2) + \frac{1}{k} + \frac{C_T}{k^2} Z_1 + C_T \frac{k}{T} + O_p(T^{-1/2}); \tag{15}$$

(ii) *for $k/T^{1/2} \to \kappa$, with $\vartheta(\kappa)$ defined in Theorem* 2 :

$$MIC(k) = \ln(\sigma_e^2) + \frac{\vartheta(\kappa)}{\sqrt{T}} + \frac{C_T}{T}\vartheta(\kappa)^2 Z_2 + C_T \frac{k}{\sqrt{T}} + O_p(T^{-1/2}); \tag{16}$$

[5] This is a more general result than Pantula (1991) who used a different local asymptotic framework for modeling the negative moving average component in unit root tests. Pantula's result applies only when $k = O_p(T^{1/4})$.

(iii) *uniformly in* $K_{2T} \leq k \leq K_{3T}$:

$$MIC(k) = \ln(\sigma_e^2) + \frac{2\delta}{\sqrt{T}\exp(2k\delta/\sqrt{T})} + \frac{C_T}{T}Z_3 + C_T\frac{k}{T} + O_p(T^{-1/2}). \qquad (17)$$

The expansions for the $IC(k)$ *are the same except that the terms involving the* Z*'s are absent.*

The objects of ultimate interest are the values k_{ic} and k_{mic} chosen by the IC and the MIC, respectively, over all possible values of k.

Theorem 3. *Let* y_t *be generated by* (14) *and consider the rate of increase of* k *as* $T \to \infty$. *Define*:

$$F_{MIC(1)}(\kappa) = \frac{-4\delta^2\exp(-2\kappa\delta)}{\sqrt{T}[1-\exp(-2\kappa\delta)]^2} + C_T\left(\frac{-16\delta^3}{T}\frac{\exp(-2\kappa\delta)^2 Z_2}{[1-\exp(-2\kappa\delta)]^3} + \frac{1}{\sqrt{T}}\right);$$

$$F_{MIC(2)}(\kappa) = -\frac{\sqrt{C_T/T}}{\kappa^2} + \sqrt{C_T/T} - \frac{2C_T^2}{\kappa^3 T}Z_1;$$

$$F_{IC}(\kappa) = \frac{-4\delta^2\exp(-2\kappa\delta)}{[1-\exp(-2\kappa\delta)]^2} + C_T.$$

When C_T *is fixed,* $k_{mic}/(\kappa_{mic(1)}T^{1/2}) \to 1$ *where* $\kappa_{mic(1)}$ *solves* $F_{MIC(1)}(\kappa) = 0$, *and* $(k_{ic}/\kappa_{ic}T^{1/2}) \to 1$, *where* κ_{ic} *solves* $F_{IC}(\kappa) = 0$. *When* $C_T \to \infty, k_{ic}/\sqrt{T/C_T} \to 1$. *When* $C_T \to \infty$ *and* $C_T^3/T \to d(\geq 0)$, $k_{mic}/(\kappa_{mic(2)}\sqrt{T/C_T}) \to 1$, *where* $\kappa_{mic(2)}$ *solves* $F_{MIC(2)}(\kappa) = 0$. *If* $C_T \to \infty$ *and* $C_T^3/T \to \infty$, $k_{mic}/(2Z_1T)^{1/3} \to 1$.

Provided C_T is constant (as is the case with Akaike's value of 2), both the IC and the MIC will choose k proportional to $T^{1/2}$ asymptotically, much faster than the logarithmic rate attained under the standard asymptotic framework. The rate of $\sqrt{T}$ is optimal for unit root tests because this is the rate that will render the class of

DF as well as the *M* tests nondivergent under the null hypothesis in this local asymptotic framework. However, when C_T increases with T, this optimal rate is not achieved.

The proportionality factor for the rate of increase of k_{ic} when C_T increases with T is one and highlights a major shortcoming of the *IC* in such cases, namely a complete independence to δ, the parameter of the local asymptotic framework. For the k_{mic} when C_T increases with T, the proportionality factor depends on the parameters of the model via Z_1. This effect, however, still vanishes as T increases. This lack of flexibility does not arise when C_T is fixed since κ_{mic} and κ_{ic} always maintain a dependence on δ. Although k_{ic} increases at the appropriate rate of $\sqrt{T}$, it too has drawbacks because κ_{ic} is invariant to T and it is the same whether $p = 0$ or $p = 1$. In contrast, κ_{mic} adapts to the sample size and the trend function via a second order factor involving Z_1. Thus, the *MIC* with C_T fixed maintains a proportionality factor that is the most flexible with respect to T, the regression model, as well as parameters of the DGP.

If the weight C_T on the penalty factor is too large, that is $C_T^3/T \to \infty$, the *MIC* chooses a rate that is independent of C_T, namely $T^{1/3}$, while the *IC* continues to select a k that is proportional to $\sqrt{T/C_T}$. This, however, is a result of theoretical rather than practical interest since almost all information criteria that have been suggested are such that $C_T^3/T \to 0$.

To illustrate these properties, we consider the *MAIC* and *AIC* with $C_T = 2$ and the *MBIC* and *BIC* with $C_T = \ln(T)$. We denote the proportionality factors by κ_{maic}, κ_{aic}, κ_{mbic}, and κ_{bic}, respectively. Numerical values of κ are reported in Table 4 for selected values of δ and T under the null hypothesis of a unit root with $c = 0$. In the case of the *MIC* these values are random variables because of the presence of Z_1. The numerical calculations reported use its expected value.[6] Table 4 shows first that the κ_{mic}'s are drastically

[6]To get the expected value we use the following crude approximation. For $p = 0$, $E[Z_1] \approx (E[1 + \delta^2 \int_0^1 W(r)^2 dr])/\delta^4 E([\int_0^1 W(r)^2 dr]^2) = (4 + 2\delta^2)/\delta^4$. For $p = 1$, $E[Z_1] \approx E[1 + \delta^2 \int_0^1 V_{0,\bar{c}}(r)^2 dr])/\delta^4 E(\int_0^1 V_{0,\bar{c}}(r)^2 dr)^2$. For $\bar{c} = -13.5$, tedious but straightforward algebra gives $E[Z_1] \approx (1 + \delta^2(0.102475)/(1 + \delta^4(0.102475)^2)$.

Table 4. Values of the proportionality factor κ.

$T\backslash\delta$		$p=0$						$p=1$					
		0.50	1.00	1.50	2.00	5.00	8.00	0.50	1.00	1.50	2.00	5.00	8.00
100	*bic*	0.99	0.96	0.93	0.89	0.67	0.54	0.99	0.96	0.93	0.89	0.67	0.54
	aic	0.69	0.65	0.61	0.57	0.39	0.30	0.69	0.65	0.61	0.57	0.39	0.30
	mbic	3.59	1.73	1.24	1.04	0.68	0.54	6.34	2.89	1.87	0.40	0.71	0.54
	maic	1.70	0.88	0.68	0.59	0.39	0.30	2.96	1.37	0.90	0.70	0.39	0.30
150	*bic*	0.99	0.96	0.93	0.90	0.69	0.55	0.99	0.96	0.93	0.90	0.69	0.55
	aic	0.69	0.65	0.61	0.57	0.39	0.30	0.69	0.65	0.61	0.57	0.39	0.30
	mbic	3.58	1.73	1.24	1.04	0.69	0.55	6.41	2.92	1.88	1.42	0.72	0.56
	maic	1.63	0.85	0.67	0.59	0.39	0.30	2.87	1.32	0.88	0.69	0.39	0.30
250	*bic*	0.99	0.97	0.94	0.90	0.70	0.57	0.99	0.67	0.94	0.90	0.70	0.57
	aic	0.69	0.65	0.61	0.57	0.39	0.30	0.69	0.65	0.61	0.57	0.39	0.30
	mbic	3.55	1.71	1.24	1.04	0.71	0.57	6.46	2.93	1.89	1.42	0.73	0.57
	maic	1.55	0.82	0.66	0.58	0.39	0.30	2.75	1.27	0.85	0.67	0.39	0.30
500	*bic*	0.99	0.97	0.94	0.91	0.72	0.58	0.99	0.97	0.94	0.91	0.72	0.58
	aic	0.69	0.65	0.61	0.57	0.39	0.30	0.69	0.65	0.61	0.57	0.39	0.30
	mbic	3.47	1.68	1.23	1.04	0.72	0.58	6.46	2.92	1.89	1.42	0.75	0.59
	maic	1.44	0.79	0.65	0.58	0.39	0.30	2.60	1.20	0.81	0.65	0.39	0.30
5000	*bic*	0.99	0.98	0.96	0.93	0.76	0.63	0.99	0.98	0.96	0.93	0.76	0.63
	aic	0.69	0.65	0.61	0.57	0.39	0.30	0.69	0.65	0.61	0.57	0.39	0.30
	mbic	3.06	1.52	1.16	1.02	0.76	0.63	6.12	2.74	1.78	1.36	0.78	0.63
	maic	1.14	0.71	0.62	0.57	0.39	0.30	2.12	1.00	0.72	0.60	0.39	0.30
50000	*bic*	0.99	0.98	0.96	0.94	0.79	0.66	0.99	0.98	0.96	0.94	0.79	0.66
	aic	0.69	0.65	0.61	0.57	0.39	0.30	0.69	0.65	0.61	0.57	0.39	0.30
	mbic	2.56	1.34	1.08	0.99	0.79	0.67	5.43	2.42	1.59	1.25	0.80	0.66
	maic	0.92	0.67	0.61	0.57	0.39	0.30	1.69	0.85	0.66	0.58	0.39	0.30

larger than the κ_{ic}'s when δ is small. Second, the κ_{ic}'s vary much less with δ than the κ_{mic}'s. Third, the κ_{mic}'s are larger when $p = 1$ than when $p = 0$, but the κ_{ic}'s are invariant to the order of the deterministic terms in the regression.

Values of k selected by the IC and the MIC are in Table 5. These are constructed using the expected value of the MIC given by (for the

Table 5. k chosen by minimizing the asymptotic approximation of the MIC and IC (18).

		$p=0$						$p=1$					
$T\backslash\delta$		0.50	1.00	1.50	2.00	5.00	8.00	0.50	1.00	1.50	2.00	5.00	8.00
100	*bic*	4	4	4	4	3	2	4	4	4	4	3	2
	aic	6	6	6	5	3	3	6	6	6	5	3	3
	mbic	16	8	5	4	3	2	29	13	8	6	3	2
	maic	17	8	6	5	3	3	29	13	9	7	3	3
150	*bic*	5	5	5	4	3	3	5	5	5	4	3	3
	aic	8	8	7	7	4	3	8	8	7	7	4	3
	mbic	19	9	6	5	3	3	35	15	10	7	3	3
	maic	20	10	8	7	4	3	35	16	10	8	4	3
250	*bic*	6	6	6	6	4	3	6	6	6	6	4	3
	aic	10	10	9	9	6	4	10	10	9	9	6	4
	mbic	23	11	8	7	4	3	43	19	12	9	4	3
	maic	24	13	10	9	6	4	43	20	13	10	6	4
500	*bic*	8	8	8	8	6	5	8	8	8	8	6	5
	aic	15	14	13	12	8	6	15	14	13	12	8	6
	mbic	31	15	11	9	6	5	57	26	16	12	6	5
	maic	32	17	14	13	8	6	58	27	18	14	8	6
5000	*bic*	24	23	23	22	18	15	24	23	23	22	18	15
	aic	49	46	43	40	27	21	49	46	43	40	27	21
	mbic	74	37	28	24	18	15	148	66	43	33	19	15
	maic	80	50	44	40	27	21	150	71	50	42	27	21
50000	*bic*	67	66	65	63	53	45	67	66	65	63	53	45
	aic	154	147	137	128	88	68	154	147	137	128	88	68
	mbic	174	91	74	67	54	45	369	164	108	85	54	45
	maic	207	151	138	128	88	68	379	190	147	130	88	68

$IC(k)$ the term with Z_1 is absent)

$$E[MIC(k)] = \ln(\sigma_e^2) + \frac{(1-\theta_T^2)(-\theta_T)^{2(k+1)}}{1-(-\theta_T)^{2(k+1)}} + C_T\left(\left[\frac{(1-\theta_T^2)(-\theta_T)^{2k}}{1-(-\theta_T)^{2(k+1)}}\right]^2 E(Z_1) + \frac{k}{T}\right). \quad (18)$$

This expression is valid uniformly in $0 < k \leq K_{3T}$ for $K_{3T} = o(T)$ and is obtained using Equation (A.4) and Lemma A.5. When δ is small and especially when $p = 1$, the k_{mic} are several folds larger than the k_{ic}. As well, differences between k_{aic} and k_{maic} are nontrivial even for sample sizes deemed unrealistically large. Table 5 also provides some guidance for the k's that would likely be selected in practice. When $\theta = -0.9$, for example, the local asymptotic results for $\delta = 1$ and $T = 100$ suggest that $k_{bic} = 4$ for both $p = 0$ and 1, but that the k_{maic} are 8 and 13, respectively. The local asymptotic results therefore support the earlier claim that the *IC* lacks the flexibility to handle DGP's with a large negative moving average root and/or alternative deterministic components. The issue of practical interest is whether in finite samples and for realistic values of the moving average parameter, such large discrepancies remain and what are the implications for unit root tests. The rest of the analysis presents evidence to this effect.

6.3. *The finite sample properties of k_{mic}*

To evaluate the finite sample properties of the *MIC vis-à-vis* the *IC*, we performed the following simulation experiment. For a given DGP, we construct the $\overline{MZ}_{\alpha}^{GLS}$ and DF^{GLS} tests at each $k \in [0, 10]$, and record the exact sizes. We then find the so-called optimal values of k^*, denoted k^*_{mz} and k^*_{dfgls}, as the first k with a size closest to within 0.03 of the nominal size of 0.05. If no such k exists, k^* is the k for which the exact size of the test has the smallest absolute deviation from the nominal size of 5%. We then obtain k_{bic} as the median value selected by the *BIC* over the range 0 and 10. A similar procedure is used to obtain k_{mbic}, k_{aic}, and k_{maic}. The setup for the simulations is as outlined in Section 4.

In Tables 6 ($p = 0$) and 7 ($p = 1$), we report the values for k^*_{mz}, k^*_{dfgls}, k_{ic}, and k_{mic} for selected parameters for $C_T = 2$ and $C_T = \ln(T - k_{\max})$. While the k_{ic}'s are in the same range as the k^*'s for many cases, there are important differences between the two when $\theta < 0$. The k_{mic} is much closer to k^*_{mz} and k^*_{dfgls} in negative *MA* models than the k_{ic}. For example, when $\theta = -0.8$, k^*_{mz} and k^*_{dfgls}

Table 6. Selected k for $p = 0$.

	MA Case: $v_t = (1 + \theta L)e_t$											
	$T = 100$						$T = 250$					
θ	k^*_{mz}	k^*_{dfgls}	k_{bic}	k_{aic}	k_{mbic}	k_{maic}	k^*_{mz}	k^*_{dfgls}	k_{bic}	k_{aic}	k_{mbic}	k_{maic}
−0.8	8	8	2	4	5	6	7	10	4	6	5	7
−0.7	6	10	2	3	3	4	6	7	3	5	4	5
−0.6	5	7	1	2	2	3	5	6	2	4	3	4
−0.5	3	7	1	2	2	2	4	4	2	3	2	3
−0.4	2	4	1	2	1	2	2	2	1	2	2	2
−0.3	2	3	1	1	1	1	2	2	1	2	1	2
−0.2	1	2	0	1	1	1	1	1	1	1	1	1
−0.1	1	1	0	0	0	0	1	1	0	1	0	1
0.0	0	0	0	0	0	0	0	0	0	0	0	0
0.1	0	0	0	0	0	0	0	0	0	1	0	1
0.2	0	0	0	1	0	1	1	1	1	1	1	1
0.3	2	2	1	1	1	1	2	1	1	2	1	2
0.4	2	2	1	2	1	2	2	2	1	2	2	2
0.5	2	2	1	2	2	2	2	2	2	3	2	3
0.6	2	2	2	3	2	3	2	2	2	4	2	4
0.7	2	2	2	4	2	4	2	2	3	5	3	5
0.8	2	2	3	5	3	5	2	2	4	7	4	7

(*Continued*)

Table 6. (*Continued*)

	AR Case: $(1-\rho L)v_t = e_t$											
	$T=100$						$T=250$					
ρ	k^*_{mz}	k^*_{dfgls}	k_{bic}	k_{aic}	k_{mbic}	k_{maic}	k^*_{mz}	k^*_{dfgls}	k_{bic}	k_{aic}	k_{mbic}	k_{maic}
−0.8	1	1	1	1	1	1	1	1	1	1	1	1
−0.7	1	1	1	1	1	1	1	1	1	1	1	1
−0.6	1	1	1	1	1	1	1	1	1	1	1	1
−0.5	1	1	1	1	1	1	1	1	1	1	1	1
−0.4	1	1	1	1	1	1	1	1	1	1	1	1
−0.3	1	1	1	1	1	1	1	1	1	1	1	1
−0.2	1	1	0	1	1	1	1	1	1	1	1	1
−0.1	1	1	0	0	0	0	1	1	0	1	0	1
0.0	0	0	0	0	0	0	0	0	0	0	0	0
0.1	0	0	0	0	0	0	0	0	0	1	0	1
0.2	1	0	0	1	0	1	1	1	1	1	1	1
0.3	1	1	1	1	1	1	1	1	1	1	1	1
0.4	0	1	1	1	1	1	1	1	1	1	1	1
0.5	1	1	1	1	1	1	1	1	1	1	1	1
0.6	1	1	1	1	1	1	1	1	1	1	1	1
0.7	1	1	1	1	1	1	1	1	1	1	1	1
0.8	1	1	1	1	1	1	1	1	1	1	1	1

Note: The entries are the median values of the selected k using 5,000 replications.

Table 7. Selected k for $p = 1$.

	MA Case: $v_t = (1 + \theta L)e_t$											
	$T = 100$						$T = 250$					
θ	k^*_{mz}	k^*_{dfgls}	k_{bic}	k_{aic}	k_{mbic}	k_{maic}	k^*_{mz}	k^*_{dfgls}	k_{bic}	k_{aic}	k_{mbic}	k_{maic}
−0.8	6	10	0	2	6	6	7	10	3	5	7	8
−0.7	4	8	1	2	4	5	5	8	3	4	5	6
−0.6	4	6	1	2	3	4	4	6	2	3	4	4
−0.5	3	6	1	2	2	3	3	4	2	3	3	3
−0.4	2	3	1	1	2	2	2	3	1	2	2	2
−0.3	2	2	0	1	1	1	2	2	1	1	1	2
−0.2	1	4	0	1	1	1	1	1	1	1	1	1
−0.1	0	1	0	0	0	1	1	1	0	1	0	1
0.0	0	0	0	0	0	0	0	0	0	0	0	0
0.1	1	0	0	1	0	0	1	0	0	1	0	1
0.2	1	2	0	1	0	1	1	1	1	1	0	1
0.3	0	2	1	1	0	1	2	2	1	2	1	2
0.4	2	2	1	2	1	2	2	2	1	2	2	2
0.5	2	2	1	2	2	2	2	2	2	3	2	3
0.6	2	2	2	3	2	2	2	2	3	4	2	4
0.7	2	2	2	4	2	4	2	2	3	5	3	5
0.8	2	2	3	5	2	4	4	4	4	7	4	6

(*Continued*)

Table 7. (*Continued*)

	AR Case: $(1-\rho L)v_t = e_t$											
	$T = 100$						$T = 250$					
ρ	k^*_{mz}	k^*_{dfgls}	k_{bic}	k_{aic}	k_{mbic}	k_{maic}	k^*_{mz}	k^*_{dfgls}	k_{bic}	k_{aic}	k_{mbic}	k_{maic}
−0.8	3	1	1	1	1	1	4	1	1	1	1	1
−0.7	3	1	1	1	1	1	1	1	1	1	1	1
−0.6	2	1	1	1	1	1	1	1	1	1	1	1
−0.5	1	1	1	1	1	1	1	1	1	1	1	1
−0.4	1	1	1	1	1	1	1	1	1	1	1	1
−0.3	1	1	1	1	1	1	1	1	1	1	1	1
−0.2	1	1	0	1	1	1	1	1	1	1	1	1
−0.1	0	1	0	0	0	0	0	1	0	1	0	1
0.0	0	0	0	0	0	0	0	0	0	0	0	0
0.1	1	0	0	1	0	0	1	0	0	1	0	1
0.2	1	1	1	1	0	1	1	1	1	1	0	1
0.3	1	1	1	1	0	1	1	1	1	1	1	1
0.4	1	1	1	1	1	1	1	1	1	1	1	1
0.5	1	1	1	1	1	1	1	1	1	1	1	1
0.6	1	1	1	1	1	1	1	1	1	1	1	1
0.7	1	1	1	1	1	1	1	1	1	1	1	1
0.8	1	1	1	1	1	1	1	1	1	1	1	1

Note: The entries are the median values of the selected k using 5,000 replications.

are 10 at $T = 250$ and $p = 1$; the *BIC* chooses 3 over half the time whereas k_{maic} has 8 as the median value. The k_{aic} is closer to k^*_{mz} and k^*_{dfgls} than the k_{bic}, but still not as close as the k_{maic}. For the *AR* case, the k_{maic} indeed selects the appropriate order of one, but when $p = 1$, this sometimes differs from k^*_{mz}. This does not reflect a weakness of the *MIC* but rather that the $\overline{M}^{GLS}$ tests are undersized when the autoregressive coefficient is large and negative. The results show that k_{maic} corresponds closely to the k^*_{dfgls} in the AR case.

7. Size and Power of the Tests in Finite Samples

In this section, we need to establish two things: first, that the *MIC* chooses values of k that are appropriate in the sense of minimizing size distortions under the null hypothesis and not overparameterizing under the alternative; second, that the M^{GLS} and $\overline{M}^{GLS}$ statistics have power comparable to the DF^{GLS} and dominate the M tests. Whenever an estimate of σ^2 is required, s^2_{AR} based upon the regression (8) is used. Because the *MAIC* dominates all other criteria from both theoretical and numerical perspectives we only consider the M^{GLS} tests constructed using the *MAIC*. In the simulations, the lower bound is always zero to reduce the chance of overparameterizing when a large k is not necessary. The upper bound is $k_{\max} = \text{int}(12(T/100)^{1/4})$.[7] The results are tabulated for $T = 100, 150$, and 250. The 10 statistics being considered are detailed in the notes to Table 8 and Table 10.

For $p = 0$, the results for size are reported in Table 8. The M and M^{GLS} tests still have inflated sizes at $T = 100$ when $\theta = -0.8$. However, use of *GLS* detrended data to construct s^2_{AR} produces substantial size improvements. The DF^{GLS} based on the *BIC* yields an exact size around 0.4 at $\theta = -0.8$, but the *MAIC* improves the size to 0.107. The *MAIC* also yields a substantial size improvement over $DF^{GLS}(t)$, which has a size of 0.35 when $\theta = -0.8$. The results also

[7]In practice, the upper bound should be set such that the constraint does not bind. In simulations, the upper bound occasionally binds when $T = 100$ but not for larger sample sizes (e.g., $T = 150$).

Table 8. Size of the tests; $p = 0$.

MA Case: $v_t = e_t + \theta e_{t-1}$, $e_t \sim N(0,1)$											
T	θ	M_{mic}	M_{mic}^{GLS}	$\overline{M}_{mic}^{GLS}$	DF_{bic}	DF_{bic}^{GLS}	DF_t^{GLS}	DF_{mic}^{GLS}	P_{Tbic}	$\overline{P}_{Tmic}^{GLS}$	$\overline{M}P_{Tmic}^{GLS}$
100	−0.8	0.058	0.041	0.036	0.394	0.411	0.356	0.107	0.429	0.034	0.036
	−0.5	0.028	0.013	0.056	0.082	0.104	0.095	0.073	0.109	0.048	0.051
	0.0	0.018	0.005	0.043	0.072	0.077	0.072	0.050	0.094	0.036	0.040
	0.5	0.027	0.013	0.062	0.081	0.082	0.080	0.048	0.118	0.048	0.054
	0.8	0.042	0.018	0.067	0.095	0.097	0.091	0.039	0.145	0.054	0.062
150	−0.8	0.034	0.018	0.026	0.407	0.390	0.304	0.095	0.420	0.025	0.027
	−0.5	0.025	0.009	0.054	0.085	0.093	0.086	0.066	0.091	0.047	0.049
	0.0	0.017	0.005	0.049	0.062	0.075	0.070	0.051	0.085	0.040	0.046
	0.5	0.028	0.011	0.054	0.066	0.072	0.071	0.046	0.090	0.047	0.050
	0.8	0.040	0.013	0.060	0.079	0.077	0.070	0.034	0.108	0.047	0.052
250	−0.8	0.016	0.006	0.022	0.370	0.323	0.236	0.088	0.380	0.021	0.021
	−0.5	0.036	0.012	0.058	0.075	0.089	0.079	0.063	0.086	0.049	0.054
	0.0	0.021	0.006	0.047	0.059	0.062	0.062	0.045	0.069	0.039	0.042
	0.5	0.034	0.009	0.056	0.059	0.068	0.069	0.045	0.078	0.047	0.051
	0.8	0.044	0.014	0.059	0.066	0.062	0.058	0.042	0.081	0.050	0.053

(*Continued*)

Table 8. (*Continued*)

						AR Case: $v_t = \rho v_{t-1} + e_t, e_t \sim N(0,1)$					
T	ρ	M_{mic}	M^{GLS}_{mic}	$\overline{M}^{GLS}_{mic}$	DF_{bic}	DF^{GLS}_{bic}	DF^{GLS}_{t}	DF^{GLS}_{mic}	P_{Tbic}	$\overline{P}^{GLS}_{Tmic}$	$\overline{M}P^{GLS}_{Tmic}$
100	−0.8	0.002	0.000	0.015	0.067	0.082	0.076	0.049	0.063	0.013	0.014
	−0.5	0.010	0.003	0.037	0.067	0.076	0.071	0.046	0.087	0.028	0.033
	0.5	0.029	0.012	0.064	0.082	0.087	0.082	0.053	0.108	0.051	0.057
	0.8	0.047	0.025	0.090	0.091	0.081	0.077	0.053	0.125	0.068	0.077
150	−0.8	0.003	0.000	0.021	0.062	0.069	0.067	0.048	0.051	0.019	0.020
	−0.5	0.013	0.004	0.039	0.060	0.074	0.069	0.045	0.072	0.031	0.035
	0.5	0.030	0.010	0.055	0.063	0.069	0.066	0.047	0.081	0.046	0.051
	0.8	0.043	0.017	0.070	0.067	0.071	0.069	0.047	0.092	0.055	0.063
250	−0.8	0.007	0.001	0.029	0.056	0.057	0.054	0.040	0.049	0.026	0.028
	−0.5	0.022	0.006	0.046	0.052	0.057	0.058	0.047	0.062	0.040	0.042
	0.5	0.028	0.008	0.052	0.056	0.061	0.061	0.045	0.070	0.043	0.046
	0.8	0.037	0.010	0.060	0.060	0.061	0.063	0.051	0.073	0.051	0.056

Notes: The results reported for the M_{mic} tests are those corresponding to MZ_α. Since the numerical results for MZ_α, MZ_t, and MSB are similar, we use the label M in the tables. Three variants of the M tests are considered. The M_{mic} is the original MZ_a, the M^{GLS}_{mic} uses GLS detrended data to construct the statistic but uses least squares detrending in the estimation of s^2_{AR}, and $\overline{M}^{GLS}_{mic}$ uses GLS detrending at $\bar{c} = -7.0$ when constructing s^2_{AR}. When the BIC is used, $k_{\max} = 6$ and $k_{\min} = 3$ as in ERS. For the MIC, $k_{max} = \text{int}(12(T/100)^{1/4})$, $k_{min} = 0$, and $C_T = 2$, giving the $MAIC$. The DF^{GLS}_t is based on the 10% sequential t test for the significance of the last lag.

confirm that the P_T test constructed with k chosen by BIC suffers from substantial size distortions. However, the $\overline{P}_T$ and $\overline{MP}_T^{GLS}$ show remarkably little size distortions when the $MAIC$ is used to select k.

To assess the relative power of the tests, we tabulate the rejection rates at $\bar{\alpha} = 1 + \bar{c}/T$. For $p = 0$ and $\bar{c} = -7.0$, the values of $\bar{\alpha}$ are 0.93, 0.953, and 0.972 for $T = 100, 150$, and 250, respectively.[8] The size-adjusted power of the tests are reported in Table 9. Although the tests should reject the null hypothesis 50% of the time in large samples, none of the tests quite achieve the 50% (asymptotic) target for sample sizes less than 200. Thus, in finite samples, the need to account for serial correlation reduces the power of the tests across the board. However, the GLS based tests have more power than tests that do not implement GLS detrending. Furthermore, the power increase from $T = 100$ to $T = 150$ is quite substantial, especially when we take into account the fact that we are evaluating at a closer alternative at $T = 150$ than at $T = 100$.

For $p = 1$, the results for size are reported in Table 10. The size distortions are somewhat larger when there is a linear trend in the regression, but there are clearly gains from using GLS detrended data to construct s_{AR}^2 and selecting k with the $MAIC$. The resulting $\overline{M}^{GLS}$ tests have size close to the nominal size for all parameter values, while the M and M^{GLS} tests continue to have size close to 0.15. As well, the $MAIC$ yields exact sizes for the DF^{GLS} and the P_T tests that are far more accurate than those obtained using the sequential t test and the BIC.

The size-adjusted power of the tests for $p = 1$ are reported in Table 11. Power is again evaluated at $\bar{\alpha} = 1 + \bar{c}/T$. For $\bar{c} = -13.5$, $\bar{\alpha}$ is 0.865 for $T = 100$, 0.910 for $T = 150$, and 0.946 for $T = 250$. Compared to the size-adjusted power reported in Table 9, all GLS based tests have lower power when $p = 1$. However, there are nonnegligible power gains from GLS detrending.

[8]We also performed extensive power simulations with other values of α and T. The results are qualitatively similar and do not affect the conclusions concerning the relative performance of the different tests.

Table 9. Size-adjusted power of the tests; $p = 0$.

MA Case: $v_t = e_t + \theta e_{t-1}$, $e_t \sim N(0,1)$											
T	θ	M_{mic}	M_{mic}^{GLS}	$\overline{M}_{mic}^{GLS}$	DF_{bic}	DF_{bic}^{GLS}	DF_t^{GLS}	DF_{mic}^{GLS}	P_{Tbic}	$\overline{P}_{Tmic}^{GLS}$	$\overline{M}P_{Tmic}^{GLS}$
100	−0.8	0.193	0.237	0.311	0.241	0.419	0.412	0.385	0.313	0.326	0.311
	−0.5	0.253	0.351	0.344	0.191	0.448	0.429	0.385	0.414	0.362	0.355
	0.0	0.315	0.452	0.458	0.131	0.432	0.436	0.454	0.418	0.471	0.442
	0.5	0.259	0.332	0.324	0.122	0.414	0.402	0.370	0.402	0.351	0.347
	0.8	0.180	0.257	0.264	0.134	0.366	0.335	0.343	0.348	0.296	0.281
150	−0.8	0.180	0.252	0.296	0.262	0.448	0.417	0.415	0.381	0.307	0.292
	−0.5	0.267	0.387	0.384	0.195	0.450	0.430	0.393	0.421	0.398	0.396
	0.0	0.339	0.463	0.460	0.152	0.422	0.405	0.454	0.413	0.467	0.453
	0.5	0.263	0.382	0.380	0.160	0.441	0.418	0.405	0.429	0.393	0.398
	0.8	0.233	0.346	0.363	0.147	0.382	0.362	0.404	0.387	0.377	0.374
250	−0.8	0.190	0.319	0.334	0.287	0.441	0.420	0.437	0.419	0.335	0.327
	−0.5	0.264	0.404	0.396	0.201	0.477	0.455	0.428	0.458	0.425	0.409
	0.0	0.338	0.481	0.477	0.152	0.464	0.446	0.483	0.455	0.485	0.478
	0.5	0.258	0.411	0.411	0.169	0.439	0.424	0.431	0.441	0.422	0.413
	0.8	0.246	0.385	0.395	0.143	0.432	0.413	0.419	0.434	0.406	0.405

(Continued)

Table 9. (*Continued*)

AR Case: $v_t = \rho v_{t-1} + e_t, e_t \sim N(0,1)$											
T	ρ	M_{mic}	M_{mic}^{GLS}	$\overline{M}_{mic}^{GLS}$	DF_{bic}	DF_{bic}^{GLS}	DF_t^{GLS}	DF_{mic}^{GLS}	P_{Tbic}	$\overline{P}_{Tmic}^{GLS}$	$\overline{MP}_{Tmic}^{GLS}$
100	−0.8	0.246	0.349	0.350	0.154	0.433	0.429	0.404	0.384	0.362	0.342
	−0.5	0.330	0.437	0.432	0.145	0.428	0.420	0.455	0.405	0.452	0.437
	0.5	0.279	0.365	0.368	0.115	0.338	0.330	0.385	0.379	0.385	0.374
	0.8	0.196	0.251	0.241	0.084	0.297	0.292	0.298	0.261	0.276	0.269
150	−0.8	0.275	0.398	0.416	0.161	0.477	0.437	0.432	0.442	0.419	0.409
	−0.5	0.332	0.453	0.454	0.161	0.404	0.390	0.468	0.423	0.463	0.458
	0.5	0.307	0.419	0.411	0.140	0.417	0.410	0.446	0.407	0.436	0.430
	0.8	0.232	0.333	0.330	0.103	0.341	0.324	0.362	0.325	0.349	0.353
250	−0.8	0.325	0.479	0.472	0.157	0.485	0.480	0.493	0.468	0.477	0.471
	−0.5	0.300	0.452	0.447	0.179	0.493	0.477	0.466	0.469	0.459	0.460
	0.5	0.336	0.443	0.436	0.158	0.447	0.434	0.456	0.428	0.468	0.456
	0.8	0.279	0.375	0.366	0.131	0.386	0.382	0.381	0.381	0.387	0.383

Note: Power is evaluated at $\bar{\alpha} = 1 + \bar{c}/T, \bar{c} = -7.0$. These are 0.93, 0.953, and 0.972 for $T = 100$, $T = 150$, and $T = 250$, respectively.

Table 10. Size of the tests; $p = 1$.

MA Case: $v_t = e_t + \theta e_{t-1}$, $e_t \sim N(0,1)$											
T	θ	M_{mic}	M_{mic}^{GLS}	$\overline{M}_{mic}^{GLS}$	DF_{bic}	DF_{bic}^{GLS}	DF_t^{GLS}	DF_{mic}^{GLS}	P_{Tbic}	$\overline{P}_{Tmic}^{GLS}$	$\overline{M}P_{Tmic}^{GLS}$
100	−0.8	0.140	0.123	0.059	0.509	0.567	0.528	0.123	0.562	0.061	0.061
	−0.5	0.037	0.025	0.035	0.092	0.107	0.111	0.069	0.145	0.036	0.036
	0.0	0.011	0.005	0.018	0.075	0.074	0.078	0.039	0.161	0.017	0.018
	0.5	0.032	0.022	0.048	0.084	0.085	0.094	0.038	0.190	0.046	0.047
	0.8	0.048	0.033	0.056	0.096	0.105	0.109	0.017	0.250	0.053	0.056
150	−0.8	0.074	0.061	0.027	0.581	0.586	0.477	0.082	0.570	0.028	0.028
	−0.5	0.032	0.017	0.028	0.087	0.103	0.104	0.049	0.114	0.028	0.029
	0.0	0.017	0.008	0.023	0.066	0.067	0.072	0.038	0.111	0.023	0.024
	0.5	0.025	0.016	0.032	0.068	0.084	0.085	0.029	0.142	0.032	0.034
	0.8	0.058	0.035	0.058	0.080	0.088	0.085	0.018	0.181	0.055	0.057
250	−0.8	0.025	0.016	0.012	0.566	0.531	0.381	0.064	0.554	0.012	0.012
	−0.5	0.031	0.016	0.033	0.094	0.105	0.099	0.048	0.103	0.034	0.035
	0.0	0.023	0.011	0.029	0.056	0.059	0.065	0.038	0.083	0.030	0.031
	0.5	0.032	0.012	0.037	0.062	0.066	0.070	0.028	0.097	0.036	0.037
	0.8	0.059	0.032	0.059	0.062	0.067	0.060	0.026	0.119	0.059	0.060

(*Continued*)

Table 10. (*Continued*)

AR Case: $v_t = \rho v_{t-1} + e_t, e_t \sim N(0,1)$											
T	ρ	M_{mic}	M_{mic}^{GLS}	$\overline{M}_{mic}^{GLS}$	DF_{bic}	DF_{bic}^{GLS}	DF_t^{GLS}	DF_{mic}^{GLS}	P_{Tbic}	$\overline{P}_{Tmic}^{GLS}$	$\overline{M}P_{Tmic}^{GLS}$
100	−0.8	0.000	0.000	0.000	0.072	0.079	0.084	0.034	0.067	0.001	0.000
	−0.5	0.006	0.003	0.012	0.071	0.074	0.077	0.037	0.122	0.013	0.013
	0.5	0.033	0.021	0.042	0.076	0.083	0.088	0.038	0.178	0.040	0.043
	0.8	0.081	0.056	0.093	0.081	0.086	0.091	0.043	0.210	0.080	0.085
150	−0.8	0.001	0.000	0.003	0.061	0.062	0.069	0.028	0.043	0.001	0.004
	−0.5	0.017	0.006	0.024	0.059	0.070	0.075	0.038	0.090	0.014	0.023
	0.5	0.037	0.018	0.044	0.056	0.070	0.074	0.039	0.127	0.041	0.043
	0.8	0.045	0.022	0.056	0.063	0.066	0.073	0.043	0.142	0.056	0.055
250	−0.8	0.001	0.000	0.003	0.057	0.054	0.058	0.028	0.009	0.004	0.004
	−0.5	0.017	0.006	0.024	0.046	0.047	0.053	0.038	0.025	0.023	0.023
	0.5	0.037	0.018	0.044	0.059	0.063	0.068	0.039	0.046	0.042	0.043
	0.8	0.045	0.022	0.056	0.054	0.056	0.061	0.043	0.054	0.052	0.055

Note: The results reported for the M_{mic} tests are those corresponding to MZ_α. Since the numerical results for MZ_α, MZ_t, and MSB are similar, we use the label M in the tables. Three variants of the M tests are considered. The M_{mic} is the original MZ_α, the M_{mic}^{GLS} uses GLS detrended data to construct the statistic but uses least squares detrending in the estimation of s_{AR}^2, and $\overline{M}_{mic}^{GLS}$ uses GLS detrending at $\bar{c} = -13.5$ when constructing s_{AR}^2. When the BIC is used, $k_{\max} = 6$ and $k_{\min} = 3$ as in ERS. For the MIC, $k_{\max} = \text{int}(12(T/100)^{1/4})$, $k_{\min} = 0$, and $C_T = 2$, giving the $MAIC$. The DF_t^{GLS} is based on the 10% t test for the significance of the last lag.

Table 11. Size-adjusted power of the tests; $p = 1$.

		MA Case: $v_t = e_t + \theta e_{t-1}$, $e_t \sim N(0,1)$									
T	θ	M_{mic}	M_{mic}^{GLS}	$\overline{M}_{mic}^{GLS}$	DF_{bic}	DF_{bic}^{GLS}	DF_t^{GLS}	DF_{mic}^{GLS}	P_{Tbic}	$\overline{P}_{Tmic}^{GLS}$	$\overline{M}P_{Tmic}^{GLS}$
100	−0.8	0.296	0.266	0.348	0.283	0.398	0.414	0.357	0.355	0.350	0.348
	−0.5	0.320	0.246	0.300	0.258	0.421	0.420	0.324	0.324	0.311	0.305
	0.0	0.450	0.266	0.439	0.175	0.362	0.341	0.432	0.488	0.448	0.442
	0.5	0.196	0.147	0.207	0.161	0.316	0.310	0.259	0.320	0.224	0.218
	0.8	0.149	0.166	0.154	0.169	0.282	0.269	0.254	0.260	0.163	0.155
150	−0.8	0.236	0.269	0.316	0.309	0.444	0.416	0.354	0.326	0.316	0.314
	−0.5	0.283	0.348	0.325	0.291	0.434	0.415	0.358	0.384	0.333	0.330
	0.0	0.407	0.470	0.457	0.200	0.384	0.352	0.462	0.369	0.466	0.465
	0.5	0.312	0.347	0.348	0.212	0.365	0.365	0.375	0.354	0.344	0.348
	0.8	0.171	0.209	0.217	0.201	0.317	0.310	0.309	0.302	0.227	0.225
250	−0.8	0.247	0.285	0.296	0.371	0.460	0.430	0.375	0.406	0.300	0.297
	−0.5	0.308	0.370	0.366	0.278	0.459	0.441	0.392	0.422	0.374	0.371
	0.0	0.380	0.458	0.451	0.250	0.437	0.418	0.447	0.432	0.455	0.456
	0.5	0.311	0.374	0.365	0.258	0.426	0.406	0.388	0.441	0.367	0.363
	0.8	0.225	0.268	0.271	0.233	0.364	0.354	0.310	0.388	0.280	0.282

(*Continued*)

Table 11. (*Continued*)

T	ρ	M_{mic}	M_{mic}^{GLS}	$\overline{M}_{mic}^{GLS}$	DF_{bic}	DF_{bic}^{GLS}	DF_t^{GLS}	DF_{mic}^{GLS}	P_{Tbic}	$\overline{P}_{Tmic}^{GLS}$	$\overline{M}P_{Tmic}^{GLS}$
AR Case: $v_t = \rho v_{t-1} + e_t, e_t \sim N(0,1)$											
100	−0.8	0.257	0.308	0.296	0.204	0.380	0.372	0.385	0.256	0.301	0.299
	−0.5	0.332	0.396	0.382	0.205	0.367	0.366	0.431	0.303	0.398	0.393
	0.5	0.235	0.265	0.270	0.133	0.262	0.264	0.286	0.247	0.279	0.280
	0.8	0.114	0.127	0.129	0.098	0.167	0.165	0.206	0.144	0.150	0.144
150	−0.8	0.311	0.363	0.349	0.240	0.435	0.409	0.409	0.345	0.351	0.347
	−0.5	0.388	0.433	0.423	0.240	0.404	0.376	0.437	0.385	0.438	0.429
	0.5	0.296	0.338	0.345	0.207	0.338	0.313	0.366	0.333	0.358	0.355
	0.8	0.189	0.211	0.218	0.133	0.239	0.228	0.251	0.211	0.228	0.224
250	−0.8	0.379	0.452	0.441	0.245	0.472	0.456	0.468	0.464	0.446	0.442
	−0.5	0.372	0.453	0.451	0.300	0.509	0.484	0.445	0.445	0.452	0.457
	0.5	0.331	0.387	0.385	0.233	0.391	0.372	0.392	0.401	0.395	0.388
	0.8	0.259	0.280	0.288	0.189	0.314	0.292	0.289	0.312	0.285	0.284

Note: Power is evaluated at $\bar{\alpha} = 1 + c/T$, $\bar{c} = -13.5$. There are 0.865, 0.910, and 0.946 for $T = 100$, $T = 150$, and $T = 250$, respectively.

A feature of interest is that the *MIC* allows the $\overline{M}^{GLS}$ tests to have an exact size that is closer to 5% than fixing k at k^*_{mz} assuming a known DGP. For example, in the *MA* case with $\theta = -0.8$ and $p = 1$, the size is 0.059 using the *MIC* and 0.159 using a value of k fixed at $k^*_{mz} = 6$. When $T = 150$, the corresponding values are 0.027 with the *MAIC* and 0.093 at $k^*_{mz} = 8$ (not reported). Hence, the *MAIC* chooses a k that adapts to the particular sample analyzed. Size improvements obtained using the *MIC* compared to fixing k at k^* are even larger when the number of deterministic components increases (see Perron and Rodríguez, 1998 for the case of a trend with a change in slope).

The size reported for $\overline{M}^{GLS}$ is remarkable, given that when $\theta < 0$, the Z tests based on kernel estimates of σ^2 reject with probability close to one under the null hypothesis. The *DF* tests have higher size-unadjusted power in the case of negative *MA* errors. Such high probabilities of rejections are misleading, however, as they reflect size distortions. Although the DF^{GLS}_{bic} has higher size-adjusted power than $\overline{M}^{GLS}_{mic}$ for some parameter values, this result must also be put into perspective. Consider $\theta = -0.8$, $T = 100$, and $p = 1$. While the size-adjusted power of the DF^{GLS}_{bic} is higher than the $\overline{M}^{GLS}_{mic}$ by 5 basis points, its size is more distorted by 45 basis points. This is a substantial increase in the probability of a Type I error. The DF^{GLS}_{mic} fares better, but there is still a size advantage in favor of the $\overline{M}^{GLS}_{mic}$.

Overall, the $\overline{M}^{GLS}_{mic}$ tests have better sizes, while the DF^{GLS}_{mic} has better power. A drawback of the $\overline{M}^{GLS}_{mic}$ test is that it is substantially undersized for pure AR processes with a coefficient close to -1. Although economic time series rarely display such a property, this nonetheless translates into lower (size unadjusted) power. As discussed in Section 6.2, this is not due to a weakness of the *MIC* in selecting k, but rather is an inherent property of the $\overline{M}^{GLS}$ tests. The trade-offs between the DF^{GLS}_{mic} and the $\overline{M}^{GLS}_{mic}$ are to be determined by the practitioner.

8. Empirical Illustrations

To illustrate the ease of application of our procedures and the differences in inference that can result, we consider inflation series from the G7 countries. Data for the *GDP* deflator are taken from the OECD International Sector Database. We constructed annualized quarterly inflation rate series covering the period 1960:2 to 1997:2 ($T+1=149$ observations) as (400 times) the logged differences of successive quarters. All estimation results are presented in Table 12. As a preliminary step, we first estimated the simple $ARMA(1,1)$ model for each series. The results show autoregressive roots that are indeed close to one (ranging from 0.92 for Germany to 0.98 for France). Also, the *MA* coefficients are negative for all series, ranging from −0.27 for Italy to −0.72 for France. This suggests that such series are likely affected by the types of problems we discussed.

The first step to constructing the tests is to generate *GLS* detrended series as defined by (6) with $\bar{c}=-7$ as the noncentrality

Table 12. Empirical results for inflation series from the G7 countries.

		USA	Canada	UK	Japan	Italy	France	Germany
	$AR(\alpha)$:	0.95	0.95	0.93	0.87	0.92	0.98	0.92
	$MA(\theta)$:	−0.32	−0.51	−0.55	−0.42	−0.27	−0.72	−0.54
BIC	k	2	1	1	1	0	3	3
	$\overline{Z}_\alpha^{GLS}$	−4.04	-9.63^b	-26.81^a	-34.12^a	-7.39^c	-5.70^c	-26.58^a
	$\overline{MZ}_\alpha^{GLS}$	−3.71	-8.13^b	-18.05^a	-23.95^a	-7.17^c	−3.49	-16.57^a
	DF^{GLS}	−1.37	-2.07^b	-3.31^a	-3.85^a	-1.95^c	−1.36	-3.00^a
	$\overline{P}_T^{GLS}$	7.39	3.36^c	1.40^a	1.01^a	4.06^c	7.92	1.96^b
	$\overline{MP}_T^{GLS}$	6.60	3.12^b	1.41^a	1.02^a	3.42^c	7.01	1.98^b
MAIC	k	2	2	8	12	5	7	5
	$\overline{Z}_\alpha^{GLS}$	−4.04	-6.98^c	-14.75^a	-13.35^a	−2.71	−3.27	-18.83^a
	$\overline{MZ}_\alpha^{GLS}$	−3.71	−5.48	-6.00^c	−3.17	−2.49	−1.07	-8.82^b
	DF^{GLS}	−1.37	-1.69^c	1.77^c	−1.33	−1.08	−0.77	-2.21^b
	$\overline{P}_T^{GLS}$	7.39	4.92	4.13^c	7.64	11.68	25.85	3.42^c
	$\overline{MP}_T^{GLS}$	6.60	4.58	4.17^c	7.73	9.83	22.89	3.44^c

Note: *a*, *b*, and *c* denotes a statistic significant at the 1%, 5%, and 10% level, respectively.

parameter (since inflation series are nontrending; otherwise use $\bar{c} = -13.5$). The next step is to estimate by OLS the autoregression (8) for all values of k ranging from 0 to some maximal order, say $k_{\max}$. We use $k_{\max} = \text{int}(12(T/100)^{1/4})$ but other values are valid. In the model selection stage, each autoregression uses the same number of effective observations, namely $T - k_{\max}$. The selected k_{maic} is obtained as the value of k that minimizes (12). Given k_{maic}, the DF_{maic}^{GLS} is obtained as the t-statistic on $\beta_0 = 0$ from re-estimation of (8) with $T - k_{maic}$ effective observations. The same regression is used to construct s_{AR}^2 according to (5). Then, the $\overline{M}^{GLS}$ tests are constructed from (3) and (4), the $\overline{P}_T^{GLS}$ test from (7), and the $\overline{MP}_T^{GLS}$ test from (9), all with GLS detrended series $\tilde{y}_t$. Using the BIC, the steps are the same except that the term $\tau_T(k)$ is omitted from (12).

The results show interesting differences between using the $MAIC$ or the BIC to select the lag length. First, for all countries, except the US, the selected lag length is higher with $MAIC$ than with BIC. Second, in all cases (except the US), the evidence in favor of stationarity is weaker with the $MAIC$ than with the BIC. For example, with the BIC, the $\overline{MZ}_{\alpha,bic}^{GLS}$ and DF_{bic}^{GLS} tests suggest a rejection at the 1% significance level for Japan and Germany, at the 5% level for Canada, and at the 10% level for Italy. When using the $MAIC$, there are only rejections at the 5% level for Germany (showing that the tests still have power) and at the 10% level for the UK. For Canada, the DF_{maic}^{GLS} is significant at the 10% level but barely. For the US and France, there is no difference in outcome. Of interest also is the fact that the $\overline{Z}_{\alpha,maic}^{GLS}$ still shows strong rejections at the 1% level for Japan, the UK, and Germany. This accords with the simulation results that the $\overline{Z}_{\alpha}^{GLS}$ test has high size distortions even when constructed using the $MAIC$.

9. Conclusions

The analysis of ERS shows that detrending data by GLS yields power gains for unit root tests. We find that GLS detrending also allows for a more precise autoregressive spectral density estimate and ensures that it is invariant to the parameters of the trend function. We also show that the *BIC* or *AIC* are not sufficiently flexible for unit root tests. We suggest a modification, the *MAIC*, which is evidently more robust when there are negative moving-average errors. Use of the *MAIC* in conjunction with *GLS* detrended data results in a class of $\overline{M}^{GLS}$ tests that have good size and power. The key distinction between the *MAIC* and standard information criteria is that the former takes account of the fact that the bias in the estimate of the sum of the autoregressive coefficients is highly dependent on k. In this paper, we show that the *MAIC* is useful for the DF^{GLS} test and the construction of the autoregressive estimate of the spectral density at frequency zero. We also expect the *MAIC*, or the more general *MIC*, to be useful in a broader range of applications because macroeconomic data are known to be highly persistent, and in such cases, the bias in the estimate of the sum of the autoregressive coefficients should depend on the order of the autoregression. However, it should also be noted that while the suggested information criterion is useful in the context of unit root tests as we have shown, it may not be appropriate when selecting the lag order in other contexts. The general usefulness of the *MIC* also awaits further investigation.

Appendix

Proof of Theorem 2

In this appendix, for simplicity, we consider the derivations in the case where the DGP is (14) and no deterministic component is present in the regression (8), in which case y_t is used instead of $\tilde{y}_t$. All the results carry over to the more general case with least-squares or *GLS* detrending. With *GLS* detrending, the stated results remain exactly the same when $p = 0$ and are as stated in the text with $V_{c,\bar{c}}(r)$ replacing $J_c(r)$ when $p = 1$. As a matter of notation, we let $\Rightarrow$ denote weak convergence in distribution and $\rightarrow$ convergence in probability. Also, to alleviate notational burden, we let $N = T - k_{max}$. Note that we can write the DGP as

$$\Delta y_t = -\sum_{i=1}^{k}(-\theta_T)^i \Delta y_{t-i} + \frac{c}{T}\sum_{i=0}^{k}(-\theta_T)^i y_{t-i-1}$$
$$+ e_t - (-\theta_T)^{k+1} e_{t-k-1}.$$

Let

$$z_{t-j} = \sum_{i=0}^{k-j}(-\theta_T)^i \Delta y_{t-i-j} = \frac{c}{T}\sum_{i=0}^{k-j}(-\theta_T)^i y_{t-i-j-1}$$
$$+ e_{t-j} - (-\theta_T)^{k+1-j} e_{t-k-1};$$

the DGP can also be written as

$$\Delta y_t = \frac{c}{T} y_{t-1} - (-\theta_T) z_{t-1} + \frac{c}{T}\sum_{i=1}^{k}(-\theta_T)^i y_{t-i-1}$$
$$+ e_t - (-\theta_T)^{k+1} e_{t-k-1} = \beta_0 y_{t-1} + \sum_{i=1}^{k} \beta_i z_{t-i} + e_{tk} \quad \text{(A.1)}$$

With $\beta_0 = c/T$, $\beta_1 = \theta_T$, $\beta_i = 0$ for $i > 1$, and

$$e_{tk} = \frac{c}{T}\sum_{i=1}^{k}(-\theta_T)^i y_{t-i-1} + e_t - (-\theta_T)^{k+1} e_{t-k-1}. \quad \text{(A.2)}$$

Since $\{z_{t-i}, i = 1, \ldots, k\}$ is a linear function of $\{\Delta y_{t-i}, i = 1, \ldots, k\}$, the *OLS* estimate of β_0 and the sum of squared residuals from the regression

$$\Delta y_t = \beta_0 y_{t-1} + \sum_{i=1}^{k} \beta_i \Delta y_{t-i} + e_{tk} \quad \text{(A.3)}$$

have the same properties as the corresponding quantities from estimating (A.1). We start with a Lemma that will be used extensively.

Lemma A.1. *Let* $J_c(r) = \int_0^r \exp((r-s)c)dW(s)$, $K_{3T} = o(T)$, *and* $\delta_{ij} = 1$ *if* $i = j$ *and 0 otherwise with* $i, j = 1, \ldots, k$. *Then,* *uniformly for* $i, j \leq k_{\max} \leq K_{3T} = o(T)$: (a) $N^{-1}\sum_{t=k_{\max}+1}^{T} y_{t-1}^2 = \sigma_e^2(1+\delta^2\int_0^1 J_c(r)^2 dr) + O_p(T^{-1/2})$; (b) $N^{-1}\sum_{t=k_{\max}+1}^{T} y_{t-1}z_{t-j} = \delta_{1j}\sigma_e^2 + O_p(T^{-1/2})$; (c) $N^{-1}\sum_{t=k_{\max}+1}^{T} z_{t-i}z_{t-j} = \sigma_e^2(\delta_{ij} + (-\theta_T)^{2(k+1)-i-j}) + O_p(T^{-1/2})$; (d) $N^{-1}\sum_{t=k_{\max}+1}^{T} \Delta y_t y_{t-1} = -\sigma_e^2 + O_p(T^{-1/2})$; (e) $N^{-1}\sum_{t=k_{\max}+1}^{T} \Delta y_t z_{t-j} = -\delta_{1j}\sigma_e^2 + O_p(T^{-1/2})$.

Proof. From Nabeya and Perron (1994), we can write $y_t = a_T e_t + b_T X_t$ where $X_t = (1+c/T)X_{t-1} + e_t$, $a_T = (-\theta_T)(1+c/T)$, $b_T = 1 - (1+c/T)(-\theta_T)$, $a_T \to 1, T^{1/2}b_T \to \delta$ as $T \to \infty$. Also, $\sum_{t=1}^{T} X_{t-1}e_t = O_p(T)$. For part (a), we have:

$$
\begin{aligned}
N^{-1}&\sum_{t=k_{\max}+1}^{T} y_{t-1}^2 \\
&= N^{-1}\sum_{t=k_{\max}+1}^{T} (a_T^2 e_{t-1}^2 + b_T^2 X_{t-1}^2 + 2a_T b_T X_{t-1}e_{t-1}) \\
&= \sigma_e^2\left(1+\delta^2\int_0^1 J_c(r)^2 dr\right) + O_p(T^{-1/2}).
\end{aligned}
$$

For part (b), we have

$$
\begin{aligned}
N^{-1}&\sum_{t=k_{\max}+1}^{T} y_{t-1}z_{t-j} \\
&= N^{-1}\sum_{t=k_{\max}+1}^{T} (a_T e_{t-1} + b_T X_{t-1}) \\
&\quad\times \left(\frac{c}{T}\sum_{i=0}^{k-j}(-\theta_T)^i y_{t-i-j-1} - (-\theta_T)^{k+1-j}e_{t-k-1} + e_{t-j}\right).
\end{aligned}
$$

Since $X_t = \sum_{j=1}^{t}\exp((t-j)c/T)e_j$, $X_{t-1} = X_{t-k-1} + \sum_{j=1}^{k}\exp((t-j)c/T)e_j$. We have $\sum_{t=k_{\max}+1}^{T} e_t e_s = O_p(T^{1/2})$ for $t \neq s, T^{1/2}b_T = O_p(1)$, and $\sum_{t=k_{\max}+1}^{T} X_{t-1}e_t = O_p(T)$, hence, $N^{-1}\sum_{t=k_{\max}+1}^{T} y_{t-1}z_{t-1} - \sigma_e^2 = O_p(T^{-1/2})$ and $N^{-1}\sum_{t=k_{\max}+1}^{T} y_{t-1}z_{t-j} = O_p(T^{-1/2})$

for $j > 1$ uniformly in $0 < i,\ j \leq k_{\max} \leq K_{3T}$. For part (c),

$$\begin{aligned}
N^{-1} \sum_{t=k_{\max}+1}^{T} z_{t-i} z_{t-j} &= N^{-1} \sum_{t=k_{\max}+1}^{T} [e_{t-i} - e_{t-k-1}(-\theta_T)^{k+1-i}] \\
&\quad \times [e_{t-j} - e_{t-k-1}(-\theta_T)^{k+1-j}] + O_p(T^{-1/2}) \\
&= \sigma_e^2(\delta_{ij} + (-\theta_T)^{2(k+1)-i-j}) + O_p(T^{-1/2}).
\end{aligned}$$

Part (d) follows using the fact that

$$\begin{aligned}
N^{-1} \sum_{t=k_{\max}+1}^{T} \Delta y_t y_{t-1} &= cN^{-2} \sum_{t=k_{\max}+1}^{T} y_{t-1}^2 + N^{-1} \\
&\quad \times \sum_{t=k_{\max}+1}^{T} (a_T e_{t-1} + b_T X_{t-1})(e_t + \theta_T e_{t-1}),
\end{aligned}$$

and part (e) using the fact that

$$\begin{aligned}
N^{-1} \sum_{t=k_{\max}+1}^{T} \Delta y_t z_{t-i} &= N^{-1} \sum_{t=k_{\max}+1}^{T} \left(\frac{c}{T} y_{t-1} + e_t - (-\theta_T) e_{t-1}\right) \\
&\quad \times \left(\frac{c}{T} \sum_{j=0}^{k-i} (-\theta_T)^j y_{t-i-j-1} + e_{t-i} - e_{t-k-1}(-\theta_T)^{k+1-i}\right).
\end{aligned}$$

The proof of Theorem 2 follows directly from the following Lemma concerned with the limit of $\hat{\beta}_0$, the *OLS* estimate of β_0 from the regression (A.1), which is equivalent to the *OLS* estimate of β_0 from the autoregression (8). □

Lemma A.2. *Consider the OLS estimate of β_0 from the autoregression (8). Let $x = \delta^2 \int_0^1 J_c(r)^2 dr$ when $p = 0$ and $x = \delta^2 \int_0^1 V_{c,\bar{c}}(r)^2 dr$ when $p = 1$. Then, (i) uniformly in $0 < k \leq K_{1T}$: $k\hat{\beta}_0 \Rightarrow -x^{-1}$; (ii) if $k/T^{1/2} \to \kappa$, $T^{1/2}\hat{\beta}_0 \Rightarrow A$, where A is a random variable with $E(A) = -\vartheta(\kappa)E(x^{-1})$, $\vartheta(\kappa) = 2\delta\exp(-2\kappa\delta)/(1-\exp(-2\kappa\delta))$; (iii) uniformly in $K_{2T} \leq k \leq K_{3T}$, $T^{1/2}\hat{\beta}_0 = O_p(1)$.*

(i) *Case* $0 < k \leq K_{1T}$. Let $X_t' = (y_{t-1}, Z_t')$, $Z_t = (z_{t-1}, \ldots, z_{t-k})'$. Define the following two $(k+1)$ by $(k+1)$ matrices:

$$\widehat{R}_T = N^{-1} \sum_{t=k_{\max}+1}^{T} X_t X_t' = N^{-1} \begin{bmatrix} \sum_{t=k_{\max}+1}^{T} y_{t-1}^2 & \sum_{t=k_{\max}+1}^{T} y_{t-1} Z_t' \\ \sum_{t=k_{\max}+1}^{T} Z_t y_{t-1} & \sum_{t=k_{\max}+1}^{T} Z_t Z_t' \end{bmatrix},$$

$$\widetilde{R} = \sigma_e^2 \begin{bmatrix} 1+x & e' \\ e & I_k + \iota\iota' \end{bmatrix},$$

where $x = \delta^2 \int_0^1 J_c(r)^2 dr$, $e' = (1, 0_{1\times k})$ and ι is a $k \times 1$ vector of ones. The inverse of $\widetilde{R}$ is

$$\widetilde{R}^{-1} = \sigma_e^{-2} \frac{1}{1+(k+1)x} \times \begin{bmatrix} k+1 & -k & 1 & 1 & & 1 \\ -k & k(1+x) & -(1+x) & -(1+x) & & -(1+x) \\ 1 & -(1+x) & k(1+x) & -x & & -x \\ \cdot & \cdot & \cdot & \cdot & \cdot & \cdot \\ 1 & -(1+x) & & -x & k(1+x) & -x \\ 1 & -(1+x) & -x & -x & -x & k(1+x) \end{bmatrix}.$$

In what follows, the structure of $\widetilde{R}^{-1}$ will be exploited. Let

$$\widehat{V}_T = \left(N^{-1} \sum_{t=k_{\max}+1}^{T} \Delta y_t y_{t-1}, N^{-1} \sum_{t=k_{\max}+1}^{T} \Delta y_t z_t' \right)',$$

$$\widetilde{V} = (-1, -1, 0_{k-1}')'.$$

Note that $\hat{\theta} = (\hat{\beta}_0, \hat{\beta}_1, \ldots, \hat{\beta}_k)' = \widehat{R}_T^{-1}\widehat{V}_T$.

Lemma A.3. (a) $\max_{0<k\le K_{1T}} \|\widehat{R}_T - \widetilde{R}\| = O_p(K_{1T}/\sqrt{T})$, (b) $\max_{0<k\le K_{1T}} \|\widehat{R}_T^{-1} - \widetilde{R}^{-1}\| = O_p(K_{1T}/\sqrt{T})$.

Proof. Using Lemma A.1, we have $\widehat{R}_{ij} - \widetilde{R}_{ij} = C_{ij}$ where $C_{ij} = O_p(T^{-1/2})$ uniformly in $0 < k \le K_{1T}$ for i, $j = 1, \ldots, k+1$ and part (a) follows. To prove part (b), note that the lower $k \times k$ submatrix of $\widetilde{R}$ is, say, $\widetilde{R}_k = (I_k + \iota\iota')$. Then $(I_k + \iota\iota')^{-1} = I_k - (1 + \iota'\iota)^{-1}\iota\iota'$ has eigenvalues of 1 and $(1 + \iota'\iota)^{-1}$. Hence $\max_{0<k\le K_{1T}} \|\widetilde{R}_k^{-1}\| = O_p(1)$. Using arguments as in Perron and Ng (1998), the $(k+1) \times (k+1)$ matrix $\widetilde{R}$ also has eigenvalues of the same order as $\widetilde{R}_k$. Hence $\max_{0<k\le K_{1T}} \|\widetilde{R}^{-1}\| = O_p(1)$. Let $Q_T = \|\widehat{R}_T^{-1} - \widetilde{R}^{-1}\|$. Then $\max_{0<k<K_{1T}} Q_T \le \max_{0<k<K_{1T}} \|\widehat{R}_T^{-1}\| \|\widehat{R}_T - \widetilde{R}\| \|\widetilde{R}^{-1}\| = O_p(K_{1T}/\sqrt{T})$ by arguments analogous to Berk (1974). □

Lemma A.4. (a) $\max_{0<k\le K_{1T}} \|\widehat{V}_T - \widetilde{V}\| = O_p(\sqrt{K_{1T}/T})$; (b) $\max_{0<k\le K_{1T}} \|\widehat{V}_T\| = O_p(1)$.

Proof. Part (a) follows directly from Lemma A.1 (d,e). Since $\widetilde{V} = (-1, -1, 0_{1\times k-1})'$, $\max_{0<k\le K_{1T}} \|\widetilde{V}\| = O_p(1)$ and $\max_{0<k\le K_{1T}}$ $\|\widehat{V}_T\| = O_p(1)$ in view of part (a) of the Lemma.

Let $e' = (1, 0_{1\times k})$. We have

$$\hat{\beta}_0 = e'\widehat{R}_T^{-1}\widehat{V}_T = e'(\widehat{R}_T^{-1} - \widetilde{R}^{-1})(\widehat{V}_T - \widetilde{V}) + e'(\widehat{R}_T^{-1} - \widetilde{R}^{-1})\widetilde{V}$$
$$+ e'\widetilde{R}^{-1}(\widehat{V}_t - \widetilde{V}) + e'\widetilde{R}^{-1}\widetilde{V} = A + B + C + D.$$

For a matrix M, denote the i, jth element of M^{-1} by m^{ij}. We now consider the limit of each of the four terms. For A,

$$\begin{aligned}
&\max_{0<k\leq K_{1T}} \|ke'(\widehat{R}_T^{-1} - \widetilde{R})(\widehat{V}_T - \widetilde{V})\|^2 \\
&\quad \leq K_{1T}^2 \sum_{i=1}^{k} \|\widehat{r}^{1i} - \widetilde{r}^{1i}\|^2 \|\widehat{V}_{i,T} - \widetilde{V}_i\|^2 \\
&\quad \leq c\frac{K_{1T}^2}{T} \sum_{i=1}^{k} \|\widehat{r}^{1i} - \widetilde{r}^{1i}\|^2 \leq c\left(\frac{K_{1T}^2}{T}\right)^2 \to 0
\end{aligned}$$

since $\sum_{i=1}^{k} \|\widehat{r}^{1i} - \widetilde{r}^{1i}\|^2 \leq \|\widehat{R}_T^{-1} - \widetilde{R}^{-1}\|^2 = O_p(k^2/T) \leq O_p(K_{1T}^2/T)$. For the term B, we need $\max_{0<k\leq K_{1T}} \|ke'(\widehat{R}_T^{-1} - \widetilde{R}^{-1})\widetilde{V}\|$. Recall that $\widehat{R}_T = \widetilde{R} + C$, where $C_{ij} = O_p(T^{-1/2}), i, j = 1, \ldots, k+1$. Then $\widehat{R}_T^{-1} = \widetilde{R}^{-1} + \widetilde{R}^{-1}C\widehat{R}^{-1} + O_p(T^{-1})$. We have (ignoring, without loss, the $O_p(T^{-1})$ term):

$$\begin{aligned}
&\|e'(\widehat{R}_T^{-1} - \widetilde{R}^{-1})\widetilde{V}\| \\
&\quad = \|e'\widetilde{R}^{-1}C\widetilde{R}^{-1}\widetilde{V}\| = \left\| \sum_{j=1}^{k+1} (\widetilde{r}^{1j} + \widetilde{r}^{2j}) \sum_{i=1}^{k+1} \widetilde{r}^{1i} C_{ij} \right\| \\
&\quad \leq \sum_{j=1}^{k+1} \|\widetilde{r}^{1j} + \widetilde{r}^{2j}\| \sum_{i=1}^{k+1} \|\widetilde{r}^{1i}\| \|C_{ij}\| \\
&\quad \leq c^* \sum_{j=1}^{k} \|\widetilde{r}^{1j} + \widetilde{r}^{2j}\| \sum_{i=1}^{k+1} \|\widetilde{r}^{1i}\|,
\end{aligned}$$

where $c^* = \max_{i,j} C_{ij}$. But the sum of the norms of the elements of the first row of $\widetilde{R}^{-1}$ is $[2k + (k-1)][1 + (k+1)x]^{-1} = O_p(1)$, and the sum of the norms of the sum of the first two elements of each row of $\widetilde{R}^{-1}$ is given by $(1 + kx + (k-1)x)/(1 + (k+1)x) = O_p(1)$. Since $c^* = O_p(T^{-1/2})$, $\max_{0<k\leq K_{1T}} \|ke'(\widehat{R}_T^{-1} - \widetilde{R}^{-1})\widetilde{V}\| = O_p(K_{1T}/T^{1/2})$.

For the term C, first note that

$$(\widetilde{V}_T - \widetilde{V})' = N^{-1}\left[\left(\sum_{t=k_{\max}+1}^{T} \Delta y_t y_{t-1} + 1\right), \left(\sum_{t=k_{\max}+1}^{T} \Delta y_t z_{t-1} + 1\right), \sum_{t=k_{\max}+1}^{T} \Delta y_t z_{t-2}, \ldots, \sum_{t=k_{\max}+1}^{T} \Delta y_t z_{t-k}\right].$$

Thus,

$$\begin{aligned}
&\max_{0<k\leq K_{1T}} \|ke'\widetilde{R}^{-1}(\widehat{V}_T - \widetilde{V})\| \\
&\quad \leq \frac{K_{1T}}{1+(K_{1T}+1)x}\left[\left\|(K_{1T}+1)\left(N^{-1}\sum_{t=k_{\max}+1}^{T} \Delta y_t y_{t-1} + 1\right)\right\|\right. \\
&\qquad - \left\|K_{1T}\left(N^{-1}\sum_{t=k_{\max}+1}^{T} \Delta y_t z_{t-1} + 1\right)\right\| \\
&\qquad \left. + N^{-1}\sum_{j=2}^{k}\left\|\sum_{t=k_{\max}+1}^{T} \Delta y_t z_{t-j}\right\|\right] \leq (cK_{1T}/N^{1/2}) \to 0.
\end{aligned}$$

For D,

$$\max_{0<k\leq K_{1T}} (ke'\widetilde{R}^{-1}\widetilde{V}) = \frac{-K_{1T}}{1+(K_{1T}+1)x} \to \frac{-1}{x} \equiv \frac{-1}{\delta^2 \int_0^1 J_c(r)^2 dr}.$$

Combining the results, we have $k\hat{\beta}_0 = o_p(1) + o_p(1) + o_p(1) + O_p(1) = O_p(1)$. Thus, $k\hat{\beta}_0 \Rightarrow (\delta^2 \int_0^1 J_c(r)^2 dr)^{-1}$ uniformly in $0 < k \leq K_{1T}$.

(ii) *Case* $k/T^{1/2} \to \kappa$:

For the case where $k/T^{1/2} \to \kappa$, we first note that the relevant matrix $\widetilde{R}$ is given by

$$\sigma_e^2 \begin{bmatrix} 1+x & e' \\ e & I_k + \mu\mu' \end{bmatrix}$$

where $\mu' = [(-\theta_T)^k, \ldots, (-\theta_T)]$. The limiting vector $\widetilde{V}$ remains unchanged. Now we have $\sqrt{T}\hat{\beta}_0 = \sqrt{T}e'\widetilde{R}^{-1}\widetilde{V} + \sqrt{T}W$, where $W = A + B + C$. It is easy to see that $E(\sqrt{T}W) = 0$ in large samples. So, we need to evaluate $\sqrt{T}e'\widetilde{R}^{-1}\widetilde{V}$, (minus) the sum of the first two elements of $\widetilde{R}^{-1}$. By partitioned matrix inversion, these are given by

$$\begin{aligned}
&\widetilde{R}^{-1}_{11} : [(1+x) - e'(I_k + \mu\mu')^{-1}e]^{-1},\\
&\widetilde{R}^{-1}_{12} : -[(1+x) - e'(I_k + \mu\mu')^{-1}e]^{-1}e'(I_k + \mu\mu')^{-1}e.
\end{aligned}$$

Since $(I_k + \mu\mu')^{-1} = I_k - \mu\mu'/\lambda$, $\lambda = 1 + \mu'\mu$, $e'(I_k + \mu\mu')^{-1}e = e'e - (e'\mu\mu'e/\lambda) = 1 - [(-\theta_T)^{2^k}/\lambda]$. Hence, $1 - e'(I_k + \mu\mu')^{-1}e = (-\theta_T)^{2k}/\lambda$ and substituting for $\lambda = (1 - (-\theta_T)^{2(k+1)})/(1 - (-\theta_T)^2)$,

$$\begin{aligned}
\sqrt{T}e'\widetilde{R}^{-1}\widetilde{V} &= \frac{-(-\theta)^{2k}(1-\theta_T)^2}{[1-(-\theta)^{2(k+1)}]x + (-\theta)^{2k}(1-\theta_T)^2}\\
&= \frac{-(-\theta)^{2k}(1-\theta_T)^2}{[1-(-\theta)^{2(k+1)}]x} + O_p(T^{-1/2}), \qquad \text{(A.4)}
\end{aligned}$$

which converges to $-\vartheta(\kappa)/x$ setting $k = \kappa T^{1/2}$ and taking the limit as $T \to \infty$.

(iii) *Case* $K_{2T} \le k \le K_{3T}$.

Note that $(-\theta_T)^k \to 0$ if $k/\sqrt{T} \to \infty$. Let

$$\widetilde{R} = \sigma_e^2 \begin{bmatrix} 1+x & e' \\ e & I_k \end{bmatrix}$$

and note that

$$\widetilde{R}^{-1} = \sigma_e^2 \begin{bmatrix} x^{-1} & -x^{-1} & 0_{1\times k-1} \\ -x^{-1} & (1+x)x^{-1} & 0 \\ 0_{k-1\times 1} & 0 & I_{k-1} \end{bmatrix}.$$

Thus, $\|\widetilde{R}\| = O_p(1)$ and $\|\widetilde{R}^{-1}\| = O_p(1)$. From Lemma A.l (a,b,c), $\max_{K_{2T}\le k\le K_{3T}} \|\widehat{R}_T - \widetilde{R}\| = O_p(K_{3T}/T^{1/2})$. Also, $\max_{K_{2T}\le k\le K_{3T}} \|\widehat{R}_T\| \le \max_{K_{2T}\le k\le K_{3T}} \|\widehat{R}_T - \widetilde{R}\| + \max_{K_{2T}\le k\le K_{3T}} \|\widetilde{R}\| = O_p(K_{3T}/T^{1/2})$ since $\max_{k_{2T}\le k\le K_{3T}} \|\widetilde{R}\| = O_p(1)$ and $K_{3T}/T^{1/2} \to \infty$. Since $\max_{K_{2T}\le k\le K_{3T}} \|\widehat{R}_T^{-1}\| \le 1/\max_{K_{2T}\le k\le K_{3T}} \|\widehat{R}_T\|$, $\max_{K_{2T}\le k\le K_{3T}} \|\widehat{R}_T^{-1}\| \le O_p(T^{1/2}/K_{3T})$. It follows that

$$\begin{aligned}\max_{K_{2T}\le k\le K_{3T}} \|\widehat{R}_T^{-1} - \widetilde{R}^{-1}\| &\le \max_{K_{2T}\le k\le K_{3T}} \|\widehat{R}_T^{-1}\|\|\widehat{R}_T - \widetilde{R}\|\|\widetilde{R}^{-1}\| \\ &\le O_p(T^{1/2}/K_{3T})O_p(K_{3T}/T^{1/2})O_p(1) = O_p(1).\end{aligned}$$

We also have $\max_{K_{2T}\le k\le K_{3T}} \|\widehat{V}_T - \widetilde{V}\| = O_p(K_{3T}^{1/2}/T^{1/2})$ and $\|\widetilde{V}\| = O_p(1)$. Thus,

$$\begin{aligned}T^{1/2}\hat{\beta}_0 &= T^{1/2}e'\widehat{R}_T^{-1}\widehat{V}_T \\ &= T^{1/2}e'(\widehat{R}_T^{-1} - \widehat{R}^{-1})(\widehat{V}_T - \widetilde{V}) + T^{1/2}e'(\widehat{R}_T^{-1} - \widetilde{R}^{-1})\widehat{V} \\ &\quad + T^{1/2}e'\widetilde{R}^{-1}(\widehat{V}_T - \widetilde{V}) + T^{1/2}e'\widetilde{R}^{-1}\widetilde{V} \\ &= A + B + C + D.\end{aligned}$$

For A, we have

$$\begin{aligned}&\max_{K_{2T}\le k\le K_{3T}} \|T^{1/2}e'(\widehat{R}_T^{-1} - \widetilde{R}^{-1})(\widehat{V}_T - \widetilde{V})\|^2 \\ &\quad\le \max_{K_{2T}\le k\le K_{3T}} T\sum_{i=1}^{k} \|\hat{r}^{1i} - \tilde{r}^{1i}\|^2\|\widehat{V}_{i,T} - \widetilde{V}_i\|^2 \\ &\quad\le \max_{K_{2T}\le k\le K_{3T}} c\sum_{i=1}^{k} \|\hat{r}^{1i} - \tilde{r}^{1i}\|^2 \\ &\quad\le \max_{K_{2T}\le k\le K_{3T}} c\|\widehat{R}_T^{-1} - \widetilde{R}^{-1}\|^2 = O_p(1).\end{aligned}$$

For B, $\|T^{1/2}e'(\widehat{R}_T^{-1} - \widetilde{R}^{-1})\widetilde{V}\| = \|T^{1/2}e'\widetilde{R}^{-1}C\widetilde{R}^{-1}\widetilde{V}\| + O_p(T^{-1/2})$ for some $\widetilde{R}$ such that $\widehat{R}_T = \widetilde{R} + C$, where the matrix C has elements that are $O_p(T^{-1/2})$. Now

$$e'\widetilde{R}^{-1} = [x^{-1}, -x^{-1}, 0, 0, 0, \dots, 0],$$
$$\widetilde{R}^{-1}\widetilde{V} = [0, -1, 0, 0, \dots, 0]'.$$

Therefore $C\widetilde{R}^{-1}\widetilde{V} = -C_{.2}$ where $C_{.2}$ is the second column of the matrix C. Then $e'\widetilde{R}^{-1}C\widetilde{R}^{-1}\widetilde{V} = x^{-1}(-C_{12} + C_{22})$. Since C_{12} and $C_{22} = O_p(T^{-1/2})$, $\max_{K_{2T}\le k\le K_{3T}} \|T^{1/2}e'\widetilde{R}^{-1}C\widetilde{R}^{-1}\widetilde{V}\| = O_p(1)$. For the third term C,

$$\begin{aligned}
&\max_{K_{2T}\le k\le K_{3T}} \|T^{1/2}e'\widetilde{R}^{-1}(\widehat{V}_T - \widetilde{V})\| \\
&= \max_{K_{2T}\le k\le K_{3T}} \|T^{1/2}[x^{-1}, -x^{-1}, \dots, 0](\widehat{V}_T - \widetilde{V})\| \\
&= \max_{K_{2T}\le k\le K_{3T}} \left\| \frac{T^{1/2}}{x}\left[(\widehat{V}_{1,T} - \widetilde{V}_1) - (\widehat{V}_{2,T} - \widetilde{V}_2)\right]\right\| \\
&\le \frac{T^{1/2}}{x}\frac{c}{T^{1/2}} = O_p(1).
\end{aligned}$$

For D,

$$T^{1/2}e'\widetilde{R}^{-1}\widetilde{V} = T^{1/2}\left[-x^{-1} + x^{-1}\right] = 0.$$

Combining the results, we have $T^{1/2}\hat{\beta}_0 = O_p(1)$ uniformly in $K_{2T} \le k \le K_{3T}$. □

Proof of Lemma 1

Lemma A.5. *Consider the estimated autoregression* (8) *when the DGP is given by* (14). *Let* $\hat{\sigma}_k^2 = N^{-1}\sum_{t=1}^{T}\hat{e}_{tk}^2$; *then uniformly over* $0 < k < K_{3T}$,

$$\ln(\hat{\sigma}_k^2) = \ln(\sigma_e^2) + \frac{(-\theta_T)^{2(k+1)}}{1-(-\theta_T)^{2(k+1)}}(1-\theta_T^2) + O_p(T^{-1/2}).$$

Proof. Let $E_k = (e_{k_{\max}+1k}, \dots, e_{Tk})'$, where $e_{tk} = cT^{-1}\sum_{i=1}^{k}(-\theta_T)^i y_{t-i-1} + e_t - (-\theta_T)^{k+1}e_{t-k-1}$. Let $X = (X_1, \dots, X_T)'$. Then

$$\hat{\sigma}_k^2 = N^{-1}E_k'E_k - N^{-1}E_k'X(X'X)^{-1}X'E_k.$$

By direct calculations,

$$N^{-1}E_k'E_k = N^{-1}\sum_{t=k_{\max}+1}^{T} e_t^2 + (-\theta_T)^{2(k+1)}N^{-1} \times \sum_{t=k_{\max}+1}^{T} e_{t-k-1}^2 + O_p(T^{-1/2}) \tag{A.5}$$

$$= \sigma_e^2(1+(-\theta_T)^{2(k+1)}) + O_p(T^{-1/2}). \tag{A.6}$$

Consider $N^{-1}E_k'X = (N^{-1}\sum_{t=k_{\max}+1}^{T} y_{t-1}e_{tk}, N^{-1}\sum_{t=k_{\max}+1}^{T} Z_{t-j}' e_{tk})$. Using results established in the proof of Theorem 2,

$$\begin{aligned} N^{-1}&\sum_{t=k_{\max}+1}^{T} y_{t-1}e_{tk} \\ &= N^{-1}\sum_{t=k_{\max}+1}^{T}(a_Te_{t-1}+b_TX_{t-1}) \\ &\quad\times\left(e_t-(-\theta_T)^{k+1}e_{t-k-1}+\frac{c}{T}\sum_{i=1}^{k}(-\theta_T)^i y_{t-i-1}\right) \\ &= O_p(T^{-1/2}) \end{aligned}$$

and

$$\begin{aligned} N^{-1}\sum_{t=k_{\max}+1}^{T} z_{t-j}e_{tk} &= N^{-1}\sum_{t=k_{\max}+1}^{T}(e_{t-j}-(-\theta_T)^{k+1-j}e_{t-k-1}) \\ &\quad\times(e_t-(-\theta_T)^{k+1}e_{t-k-1})+O_p(T^{-1/2}) \\ &= (-\theta_T)^{2(k+1)-j}N^{-1}\sum_{t=k_{\max}+1}^{T} e_{t-k-1}^2 + O_p(T^{-1/2}), \\ &= \sigma_e^2(-\theta_T)^{2(k+1)-j} + O_p(T^{-1/2}). \end{aligned}$$

Therefore,

$$\begin{aligned} N^{-1}E_k'X &= \sigma_e^2[0, \theta_T^{2(k+1)-1}, \ldots, (-\theta_T)^{2(k+1)-k}] + O_p(T^{-1/2}), \\ &= \sigma_e^2(\theta_T)^{k+1}[0, \mu'] + O_p(T^{-1/2}), \end{aligned}$$

where $\mu' = [(-\theta_T)^k, \ldots, (-\theta_T)]$. From Lemma A.l,

$$\begin{aligned} N^{-1}(X'X) &= \sigma_e^2 \begin{bmatrix} 1+x & e' \\ e & I_k + \mu\mu' \end{bmatrix} + O_p(T^{-1/2}) \\ &\equiv R + O_p(T^{-1/2}). \end{aligned} \tag{A.7}$$

In view of the properties of $N^{-1}E_k'X$,

$$N^{-1}E_k'X(X'X)^{-1}X'E_k = (-\theta_T)^{2(k+1)}\mu'R^{-1}(k)\mu + O_p(T^{-1/2}),$$

where $R^{-1}(k)$ is the lower right $k \times k$ submatrix of R^{-1}. By partitioned matrix inversion,

$$\begin{aligned} R^{-1}(k) &= [I_k + \mu\mu' - e'e/(1+x)]^{-1}, \\ &= [J_k + \mu\mu']^{-1}, \end{aligned}$$

where $J_k = I_k - ee'/(1+x)$. For a $k \times k$ matrix A and $k \times 1$ vectors u and v, $(A+uv')^{-1} = A^{-1} - A^{-1}u(1+v'A^{-1}u)^{-1}v'A^{-1}$. Hence, for $\lambda = 1 + \mu' J_k^{-1}\mu$,

$$\begin{aligned} R^{-1}(k) &= J_k^{-1} - J_k^{-1}\mu\mu' J_k^{-1}/\lambda, \\ \mu' R^{-1}(k)\mu &= \mu' J_k^{-1}\mu - (\mu' J_k^{-1}\mu)^2/\lambda, \\ &= \frac{\mu' J_k^{-1}\mu}{1 + \mu' J_k^{-1}\mu}. \end{aligned}$$

Since $J_k = I_k - ee'/(1+x)$, $J_k^{-1} = I_k + ee'/x$, and

$$\mu' J_k^{-1}\mu = \mu'\mu + \mu' ee'\mu/x = \mu'\mu + (-\theta_T)^{2k}/x.$$

We also have $\mu'\mu = \sum_{i=1}^k (-\theta_T)^{2i}$, hence

$$\mu'\mu = \frac{(-\theta_T)^2 - (-\theta_T)^{2(k+1)}}{1 - (-\theta_T)^2}, \quad 1 + \mu'\mu = \frac{1 - (-\theta_T)^{2(k+1)}}{1 - (-\theta_T)^2}.$$

After some simplifications,

$$\frac{\mu' J_k^{-1}\mu}{1+\mu' J_k^{-1}\mu} = \frac{(-\theta_T)^2[1-(-\theta_T)^{2k}] + (-\theta_T)^{2k}[1-(-\theta_T)^2]/x}{[1-(-\theta_T)^{2(k+1)}] + (-\theta_T)^{2k}[1-(-\theta_T)^2]/x},$$

$$= (-\theta_T)^2\left[\frac{1-(-\theta_T)^{2k}}{1-(-\theta_T)^{2(k+1)}}\right] + O_p(T^{-1/2}).$$

Hence,

$$N^{-1}E_k' X(X'X)^{-1}X'E_k = (-\theta)^{2(k+1)}(-\theta_T)^2\left[\frac{1-(-\theta_T)^{2k}}{1-(-\theta_T)^{2(k+1)}}\right] + O_p(T^{-1/2}).$$

It follows that

$$\hat\sigma_k^2 = \sigma_e^2\left(1+(-\theta_T)^{2(k+1)} - (-\theta_T)^{2(k+2)}\left[\frac{1-(-\theta_T)^{2k}}{1-(-\theta_T)^{2(k+1)}}\right]\right) + O_p(T^{-1/2}),$$

$$\ln(\hat\sigma_k^2) = \ln(\sigma_e^2) + \left[\frac{(-\theta)^{2(k+1)}(1-(-\theta_T)^2)}{1-(-\theta_T)^{2(k+1)}}\right] + O_p(T^{-1/2}).$$

Lemma 1 is a consequence of Theorem 2 and Lemma A.5. □

Proof of Theorem 3

The strategy of the proof is to first determine the rate of increase of k through a global minimization of the objective function. Once this rate is found, the proportionality factor is deduced. In the following, we use the notation $k \propto f(T)$ for the case where $k/f(T) \to c$ for some finite constant c strictly above 0. We also let $d \geq 0$ be a finite constant not necessarily the same throughout.

For both the *IC* and *MIC*, it is readily seen that the objective function is $O_p(C_T/T^{1/2})$ when $k/T^{1/2} \to \kappa(>0)$, and that it is $O_p(C_T k/T)$ when $k/T^{1/2} \to \infty$. Hence, irrespective of the choice of C_T, having $k/T^{1/2} \to \kappa$ will always, in large samples, yield a smaller value of the objective function than with $k/T^{1/2} \to \infty$. Hence, we only need to compare the cases $k/T^{1/2} \to \kappa(>0)$ and $k/T^{1/2} \to 0$.

We start with the analysis for the *IC*. When $k/T^{1/2} \rightarrow 0$, there are two cases of interest. The first is when $k^2C_T/T \rightarrow d$ in which case the objective function is $O_p(1/k)$. The minimal value is obtained if k is chosen as the maximal value permissible, which means $k \propto \sqrt{T/C_T}$. The minimized objective function is then $O_p(\sqrt{C_T/T})$. The second case is when $T/k^2C_T \rightarrow d$ and the objective function is $O_p(kC_T/T)$, which is minimized by taking the minimal permissible rate of increase for k, again $k \propto \sqrt{T/C_T}$, and the minimized objective function is then $O_p(\sqrt{C_T}/\sqrt{T})$. So the case with $k/T^{1/2} \rightarrow \kappa(>0)$ yields an objective function of order $O_p(C_T/T^{1/2})$ and the case $k/T^{1/2} \rightarrow 0$ an objective function of order $O_p(\sqrt{C_T}/\sqrt{T})$. Hence, $k \propto \sqrt{T/C_T}$. To derive the constant of proportionality, note that if C_T is fixed, $k = \kappa\sqrt{T}$ and the objective function is (ignoring constants)

$$\frac{2\delta}{\sqrt{T}}\vartheta(\kappa) + \frac{C_T}{\sqrt{T}}\kappa.$$

Taking derivatives and equating to zero yields the stated solution. When $C_T \rightarrow \infty$ as $T \rightarrow \infty$, the objective function (ignoring constants) is

$$\frac{1}{\kappa\sqrt{T/C_T}} + \kappa\frac{C_T\sqrt{T/C_T}}{T}$$

and the value of κ that minimizes this expression is 1. Consider now the *MIC*(k). The first case is where

$$\frac{C_T}{k} \rightarrow d \quad \text{and} \quad \frac{C_T k^2}{T} \rightarrow d. \tag{A.8}$$

Then the objective function when $k/T^{1/2} \rightarrow 0$ is $O_p(k^{-1})$. Hence, the highest possible rate is chosen and is given by $k \propto \sqrt{T/C_T}$ and the minimized objective function is of order $O_p(\sqrt{C_T/T})$. Since the objective function when $k/T^{1/2} \rightarrow \kappa(>0)$ is $O_p(C_T/T^{1/2})$, we have $k \propto T^{1/2}$ if C_T is fixed and $k \propto \sqrt{T/C_T}$ if $C_T \rightarrow \infty$. However, in view of the requirement that $C_T/k \rightarrow d$ this implies the need for $C_T^3/T \rightarrow d$. To obtain the constant of proportionality when C_T is fixed, we substitute $k = \kappa\sqrt{T}$ in (16) and equate its first derivative to zero and solve for κ. When C_T is increasing, we substitute $k = \kappa\sqrt{T/C_T}$ in (15), and again equate the first derivative to zero and solve for κ.

Consider now the case where $C_T^3/T \to \infty$. There are two cases to cover when either of the conditions in (A.8) is not satisfied. Suppose first that $k^3/T \to d$, in which case $k/C_T \to d$. The objective function in the case $k/T^{1/2} \to 0$ is $O_p(C_T/k^2)$ and the highest rate for k is chosen, namely $k \propto T^{1/3}$. The minimized objective function is then of order $O_p(C_T/T^{2/3})$, which is of a smaller order than the case with $k/T^{1/2} \to \kappa(> 0)$. Hence, $k \propto T^{1/3}$ is the resulting rate. If $T/k^3 \to d$, in which case $C_T k^2/T \to \infty$, the objective function is of order $O_p(kC_T/T)$ and the smallest rate for k is optimal. So again, $k \propto T^{1/3}$ and the minimized objective function is also of order $O_p(C_T/T^{2/3})$. To obtain the factor of proportionality, let $k = \kappa T^{1/3}$. The objective function is then (ignoring constants)

$$\frac{1}{\kappa T^{1/3}} + \frac{C_T Z_1}{\kappa^2 T^{2/3}} + \frac{C_T \kappa}{T^{2/3}}.$$

When $C_T^3/T \to \infty$, the first term is negligible compared to the last two. Hence, minimizing the last two terms with respect to κ yields $\kappa = (2Z_1)^{1/3}$.

Acknowledgments

We thank three anonymous referees and Alain Monfort, the co-editor, for comments on earlier drafts. The authors thank the Social Sciences and Humanities Research Council of Canada (SSHRC), the Natural Sciences and Engineering Research Council of Canada (NSERC), and the Fonds de la Formation de Chercheurs et l'Aide à la Recherche du Québec (FCAR) for financial support.

References

Berk, K.N. (1974). Consistent autoregressive spectral estimates. *The Annals of Statistics* 2, 489–502.

DeJong, D.N., Nankervis, J.C., Savin, N.E. and Whiteman C.H. (1992). The power problem of unit root tests in time series with autoregressive errors. *Journal of Econometrics* 53, 323–343.

Dickey, D.A. and Fuller, W.A. (1979). Distribution of the estimators for autoregressive time series with a unit root. *Journal of the American Statistical Association* 74, 427–431.

Dufour, J.M. and King, M. (1991). Optimal invariant tests for the autocorrelation coefficient in linear regressions with stationary and nonstationary errors *Journal of Econometrics* 47, 115–143.

Elliott, G., Rothenberg, T.J. and Stock, J.H. (1996). Efficient tests for an autoregressive unit root. *Econometrica* 64, 813–836.

Franses, P.H. and Haldrup, N. (1994). The effects of additive outliers on tests of unit roots and cointegration. *Journal of Business and Economic Statistics* 12, 471–478.

Fuller, W.A. (1976). *Introduction to Statistical Time Series* John Wiley, New York.

Gourieroux, C. and Monfort, A. (1995). *Statistical and Econometric Models* Vol. 1. Cambridge University Press, Cambridge.

Hannan, E. J. and Deistler, M. (1988). *The Statistical Theory of Linear Systems* John Wiley, New York.

Lopez, J. H. (1997). The power of the ADF test. *Economics Letters* 57, 5–10.

Nabeya, S. and Perron, P. (1994). Local asymptotic distribution related to the AR(1) model with dependent errors. *Journal of Econometrics* 62, 229–264.

Ng, S. and Perron, P. (1995). Unit root tests in ARMA models with data dependent methods for the selection of the truncation lag. *Journal of the American Statistical Association* 90, 268–281.

Ng, S. and Perron, P. (2000). A note on the selection of time series models. Manuscript, Department of Economics, Boston College.

Pantula, S.G. (1991). Asymptotic distributions of unit root tests when the process is nearly stationary. *Journal of Business and Economic Statistics* 9, 63–71.

Perron, P. and Ng, S. (1996). Useful modifications to unit root tests with dependent errors and their local asymptotic properties. *Review of Economic Studies* 63, 435–465.

Perron, P. and Ng, S. (1998). An autoregressive spectral density estimator at frequency zero for nonstationarity tests. *Econometric Theory* 14, 560–603.

Perron, P. and Rodríguez, G. (1998). GLS detrending, efficient unit root tests and structural change. Manuscript, Boston University.

Phillips, P.C.B. (1987). Time series regression with unit roots. *Econometrica* 55, 277–302.

Phillips, P.C.B. and Perron, P. (1988). Testing for a unit root in time series regression. *Biometrika* 75, 335–346.

Said, S.E. and Dickey, D.A. (1984). Testing for unit roots in autoregressive-moving average models of unknown order. *Biometrika* 71, 599–607.

Schwert, G.W. (1989). Tests for unit roots: A monte carlo investigation. *Journal of Business and Economic Statistics* 7, 147–160.

Stock, J.H. (1990). A class of tests for integration and cointegration. Manuscript, Harvard University.

Chapter 11

A Simple Modification to Improve the Finite Sample Properties of Ng and Perron's Unit Root Tests*

Pierre Perron[†,§] and Zhongjun Qu[‡,¶]

[†]*Department of Economics, Boston University, 270 Bay State Rd., Boston, MA 02215, USA*
[‡]*Department of Economics, University of Illinois at Urbana-Champaign, 1206 S. Sixth Street, Champaign, IL 61820, USA*
[§]*perron@bu.edu*
[¶]*zqu@uiuc.edu*

The tests introduced by Ng and Perron [Ng, S., Perron, P. (2001). Lag length selection and the construction of unit root tests with good size and power. *Econometrica* 69, 1519–1554] have the drawback that for nonlocal alternatives the power can be very small. The aim of this note is to point out an easy solution to this power reversal problem, which in addition leads to tests having an exact size even closer, in most cases, to nominal size. It involves using OLS instead of GLS detrended data when constructing the modified information criterion.

Keywords: Integrated processes; truncation lag; GLS detrending; information criteria.

Ng and Perron (2001), henceforth NP, introduced a class of unit root tests that have a local asymptotic power function close to the Gaussian local power envelope (see Elliott *et al.*, 1996, henceforth ERS). Also, with the use of a modified information criterion, the

*Originally Published in *Economics Letters* 94, pp. 12–19.

tests have exact size close to nominal size even in the presence of a large negative moving-average component. This is achieved without sacrificing power for local alternatives. Hence, they have become often used and are now available as options in several popular statistical software packages (for a review that highlights the importance of these tests, see Haldrup and Jansson, 2006). Consider a univariate time series $\{y_t;\ t = 1, \ldots, T\}$ generated by

$$\begin{aligned} y_t &= z_t'\gamma + v_t, \\ v_t &= \alpha v_{t-1} + u_t, \end{aligned} \tag{1}$$

where u_t is a linear process of the form $u_t = \Sigma_{j=0}^{\infty} c_j e_{t-j}$ with $\Sigma_{j=0}^{\infty} j|c_j| < \infty$ and $e_t \sim$ i.i.d. $(0,\ \sigma_e^2)$. Also, $\sigma_e^2 = \lim_{T\to\infty} T^{-1}E(\Sigma_{t=1}^{T} u_t)^2$ is the 'long run variance' or (2π) times the spectral density function at frequency zero of u_t. The initial condition is such that $y_1 = O_p(1)$. The vector z_t is a set of deterministic components, usually $z_t = (1, t, \ldots, t_p)'$ and in most applications $p = 0$ (nontrending data) or $p = 1$ (trending data). The null hypothesis is that $\alpha = 1$ and the alternative hypothesis is $|\alpha| < 1$.

The tests are easy to apply and can be described succinctly as follows. We first have tests collectively referred to as the M^{GLS} tests. The three members of this class, which have very similar properties, are the following:

$$\mathrm{MZ}_{\alpha}^{\mathrm{GLS}} = \frac{T^{-1}\tilde{y}_T^2 - s_{\mathrm{AR}}^2}{2T^{-2}\sum_{t=1}^{T}\tilde{y}_{t-1}^2}, \quad \mathrm{MSB}^{\mathrm{GLS}} = \left(\frac{T^{-2}\sum_{t=1}^{T}\tilde{y}_{t-1}^2}{s_{\mathrm{AR}}^2}\right)^{1/2},$$

and $\mathrm{MZ}_t^{\mathrm{GLS}} = \mathrm{MSB}^{\mathrm{GLS}} * \mathrm{MZ}_{\alpha}^{\mathrm{GLS}}$, where $\tilde{y}_t = y_t - z_t'\hat{\gamma}^{\mathrm{GLS}}$, with $\hat{\gamma}^{\mathrm{GLS}}$ the so-called GLS estimate of γ obtained from the least-squares regression of $y_t^{\bar{\alpha}}$ on $z_t^{\bar{\alpha}}$, where the quasi-differenced series are $y_t^{\bar{\alpha}} = y_t - \bar{\alpha}y_t^{\bar{\alpha}}$ $(t = 2, \ldots, T)$ and $y_1^{\bar{\alpha}} = y_1(z_t^{\bar{\alpha}}$ is defined similarly). Following ERS, $\bar{\alpha} = 1 + \bar{c}/T$, with $\bar{c} = -7$ for $p = 0$ and $\bar{c} = -13.5$ for $p = 1$. Also of interest is the feasible point optimal test of ERS, $P_T = [S(\bar{\alpha}) - \bar{\alpha}S(1)]/s_{\mathrm{AR}}^2$, where $S(\alpha) = \inf_{\gamma} \Sigma_{t=1}^{T}(y_t^{\alpha} - \gamma z_t^{\alpha})^2$, and a modified version suggested by NP

$$\begin{aligned}\mathrm{MP}_T^{\mathrm{GLS}} &= \left[\bar{c}^2 T^{-2}\sum_{t=1}^{T}\tilde{y}_{t-1}^2 - \bar{c}T^{-1}\tilde{y}_T^2\right]\Big/ s_{\mathrm{AR}}^2 \quad \text{for } p=0;\\ &= \left[\bar{c}^2 T^{-2}\sum_{t=1}^{T}\tilde{y}_{t-1}^2 + (1-\bar{c})T^{-1}\tilde{y}_T^2\right]\Big/ s_{\mathrm{AR}}^2 \quad \text{for } p=1.\end{aligned}$$

In all these tests, s_{AR}^2 is an autoregressive spectral density estimate of σ^2 defined by $s_{\mathrm{AR}}^2 = [\hat{\sigma}_{ek}^2/(1-\Sigma_{i=1}^k \hat{b}_i)^2]$ where $\hat{\sigma}_{ek}^2 = T^{-1}\Sigma_{t=k+1}^T \hat{e}_{tk}^2$ with $\hat{b}_i$ and $\hat{e}_{tk}$ obtained from the following OLS regression

$$\Delta\tilde{y}_t = \hat{b}_0\tilde{y}_{t-1} + \sum_{i=1}^{k}\hat{b}_i\Delta\tilde{y}_{t-i} + \hat{e}_{tk}. \tag{2}$$

Note that the $\mathrm{ADF}^{\mathrm{GLS}}$ test of ERS is the t-statistic associated with $\hat{b}_0$ in this last regression.

A crucial component to obtain tests with good size in finite samples is a new method to select the autoregressive order k based on a Modified Akaike Information Criterion (MAIC) which chooses a value $k_{\mathrm{MAIC}} = \arg\min_{k\in[0,k_{\max}]}\mathrm{MAIC}(k)$ where

$$\mathrm{MAIC}(k) = \ln(\hat{\sigma}_k^2) + 2(\tau_T(k)+k)/(T-k_{\max}) \tag{3}$$

with $\tau_T(k) = (\hat{\sigma}_k^2)^{-1}\hat{b}_0^2\Sigma_{t=k_{\max}+1}^T\tilde{y}_{t-1}^2$ and $\hat{\sigma}_k^2 = (T - k_{\max})^{-1}\Sigma_{t=k_{\max}+1}^T\hat{e}_{tk}^2$. The upper bound is usually set to $k_{\max} = \mathrm{int}(12(T/100)^{1/4})$, where int() denotes the integer part, but other values are possible. NP showed that when all tests discussed above are constructed using the MAIC, the exact size is close to nominal size even in the presence of a large negative moving-average component, though some can be conservative in some cases. Also, for local alternatives the power of the tests is close to the Gaussian local asymptotic power envelope.

A drawback of these tests is that for nonlocal alternatives the power can be very small. In fact, for a given sample size T, the power can decrease as α gets further away from 1. This power reversal was pointed out by Seo (2005), though it has been known to the authors for quite some time. An alternative specification considered by NP, which uses OLS detrended data instead of GLS detrended data in the autoregression (2) to construct the spectral density estimate, does not suffer much from this problem but it leads to tests that are too conservative in common sample sizes and, hence, was not recommended. The power problem is quite important. Indeed, if the underlying process is i.i.d. the test has very low power (see below). Hence, when dealing with a random walk and performing the test both in levels and first-differences, one is likely to conclude that the series is integrated of order 2.

The aim of this note is to point out that there is an easy solution to this power reversal problem, which in addition leads to tests having an exact size even closer to nominal size for all specifications considered in NP. The idea is to use a hybrid of the two specifications considered by NP. The autoregressive spectral density estimate is still constructed using GLS detrended data as specified above, but the selection of the autoregressive order k uses the MAIC constructed with OLS detrended data, i.e., replacing $\tilde{y}_t$ by $\hat{y}_t$, the residuals from a regression of y_t on z_t and using the autoregression[1]

$$\Delta\hat{y}_t = \breve{b}_0\hat{y}_{t-1} + \sum_{i=1}^{k}\breve{b}_i\Delta\hat{y}_{t-i} + \breve{e}_{tk}. \tag{4}$$

So our modification involves a two step procedure. First, use Equation (4) to construct $\mathrm{MAIC}(k) = \ln(\breve{\sigma}_k^2)+2(\tau_T(k)+k)/(T-k_{\max})$ with $\tau_T(k) = (\breve{\sigma}_k^2)^{-1}\breve{b}_0^2\,\Sigma_{t=k_{\max}+1}^{T}\hat{y}_{t-1}^2$ and $\breve{\sigma}_k^2 = (T-k_{\max})^{-1}\Sigma_{t=k_{\max}+1}^{T}\breve{e}_{tk}^2$. Once the order is selected as the minimizer of $\mathrm{MAIC}(k)$, say k_{MAIC}, estimate Equation (2) with k_{MAIC} lags to construct s_{AR}^2 or the $\mathrm{ADF}^{\mathrm{GLS}}$ test.

Table 1 presents the size of the tests $\mathrm{MZ}_\alpha^{\mathrm{GLS}}$ (the properties of $\mathrm{MSB}^{\mathrm{GLS}}$ and $\mathrm{MZ}_t^{\mathrm{GLS}}$ are similar), $\mathrm{ADF}^{\mathrm{GLS}}$, P_T and $\mathrm{MP}_T^{\mathrm{GLS}}$. The data are generated by Equation (1) with $\alpha = 1$, $v_1 = 0$, and u_t is either an

[1]One can also use a regression in levels with the deterministic components added as regressors, i.e., $\Delta y_t = c + \delta t + \breve{b}_0 y_{t-1} + \Sigma_{i=1}^{k}\breve{b}_i\Delta y_{t-1} + \hat{e}_{tk}$. The results are virtually the same.

Table 1. Exact size of the tests; $T = 100$.

	$p = 0$				$p = 1$			
θ	$\text{MZ}_\alpha^{\text{GLS}}$	ADF^{GLS}	P_T	MP_T^{GLS}	$\text{MZ}_\alpha^{\text{GLS}}$	ADF^{GLS}	P_T	MP_T^{GLS}
MA case: $u_t = e_t + \theta e_{t-1}$; $e_t \sim$ i.i.d. $N(0, 1)$								
–0.8	0.067	0.143	0.061	0.065	0.122	0.211	0.122	0.124
($T = 150$)	(0.034)	(0.103)	(0.031)	(0.033)	(0.055)	(0.130)	(0.055)	(0.056)
–0.5	0.063	0.082	0.052	0.059	0.046	0.083	0.046	0.047
0.0	0.048	0.055	0.038	0.044	0.021	0.046	0.020	0.021
0.5	0.058	0.048	0.046	0.052	0.037	0.032	0.035	0.036
0.8	0.064	0.038	0.052	0.058	0.046	0.018	0.043	0.045
ρ	$\text{MZ}_\alpha^{\text{GLS}}$	ADF^{GLS}	P_T^{GLS}	MP_T^{GLS}	$\text{MZ}_\alpha^{\text{GLS}}$	ADF^{GLS}	P_T^{GLS}	MP_T^{GLS}
AR case: $u_t = \rho u_{t-1} + e_t$; $e_t \sim$ i.i.d. $N(0, 1)$								
–0.8	0.015	0.051	0.013	0.015	0.001	0.039	0.001	0.001
–0.5	0.037	0.051	0.031	0.034	0.014	0.042	0.014	0.015
0.5	0.066	0.056	0.052	0.060	0.049	0.043	0.044	0.047
0.8	0.088	0.055	0.067	0.078	0.090	0.047	0.076	0.081

MA(1) process of the form $u_t = e_t + \theta e_{t-1}$ or an AR(1) of the form $u_t = \rho u_{t-1} + e_t$, with $e_t \sim$ i.i.d. $N(0, 1)$, and in both cases $u_1 = e_1$. The sample size is $T = 100$ and 50,000 replications are used. We consider the values $\theta, \rho = -0.8$, –0.5, 0.0, 0.5, 0.8 (for the MA(1) case with $\theta = -0.8$ we also report the results with $T = 150$). The nominal size of the tests is 5%. Comparing the results with those in NP, this version of the tests has size closer to 5% in almost all cases. When $p = 0$, the size of the $\text{MZ}_\alpha^{\text{GLS}}$, P_T and MP_T^{GLS} tests is close to 5% for all cases, except with a large negative autoregressive coefficient, in which case they are conservative as in NP. The ADF^{GLS} test has, as expected, an exact size close to 5% with autoregressive errors but is substantially liberal in the case of MA(1) errors with $\theta = -0.8$. These size distortions remain even with $T = 150$ and larger sample sizes not reported (consistent with the theoretical results in NP). In the case of trending data with $p = 1$, the same picture emerges except that the $\text{MZ}_\alpha^{\text{GLS}}$, P_T and MP_T^{GLS} tests have some liberal size distortions with $T = 100$, which, however, disappear with $T = 150$ or larger.

Consider now the power of the tests, which we simulated for two cases. First, with $u_t \sim$ i.i.d. $N(0,\ 1)$ and second with MA(1) errors with $\theta = -0.8$ as specified above. The value of α is varied between 1 and 0 (and the graphs report results in terms of the value of $1 - \alpha$). Three versions of the tests are considered: (a) the original procedure recommended by NP, labeled GLS–GLS; (b) our modification labeled OLS–GLS; and (c) tests constructed using OLS detrended data throughout (to assess the extent of the power gain of using GLS detrended data). The sample size is $T = 100$ and again 50,000 replications are used. We present results for the ADF^{GLS} test (Figs. 1 for $p = 0$ and 2 for $p = 1$) and the $\text{MZ}_\alpha^{\text{GLS}}$ test (Fig. 3 for $p = 0$ and 4 for $p = 1$). The left panel reports the power while the right panel reports the autoregressive order selected. Note that the lag selected is the same for the OLS–GLS and OLS–OLS procedures.

The results show that the original procedure recommended by NP and our modification indeed have higher power than tests based on OLS detrended data when α is 'local' to 1. But when α decreases the original NP version has power that decreases rapidly, especially with a negative moving-average component in which case the power is very low. This reversal does not occur with our modification (though in the i.i.d. case there is some slight decrease) and the power for values of α away from 1 is close to what can be achieved using OLS detrended data throughout (which also can exhibit slight power decreases).

The reason for the difference in power when using OLS or GLS detrended data to construct the MAIC can be seen by looking at the right panels, which report the autoregressive orders selected. When the MAIC is constructed with GLS detrended data, the value of k selected increases as α moves away from the null value 1. Such is not the case when the MAIC is constructed with OLS detrended data. The smaller lags translate into higher power. Hence, using OLS detrended data to construct the MAIC allows the selection of the appropriate order under the null hypothesis to achieve an exact size close to nominal size and is also appropriate in selecting a small lag under the alternative to achieve higher power.

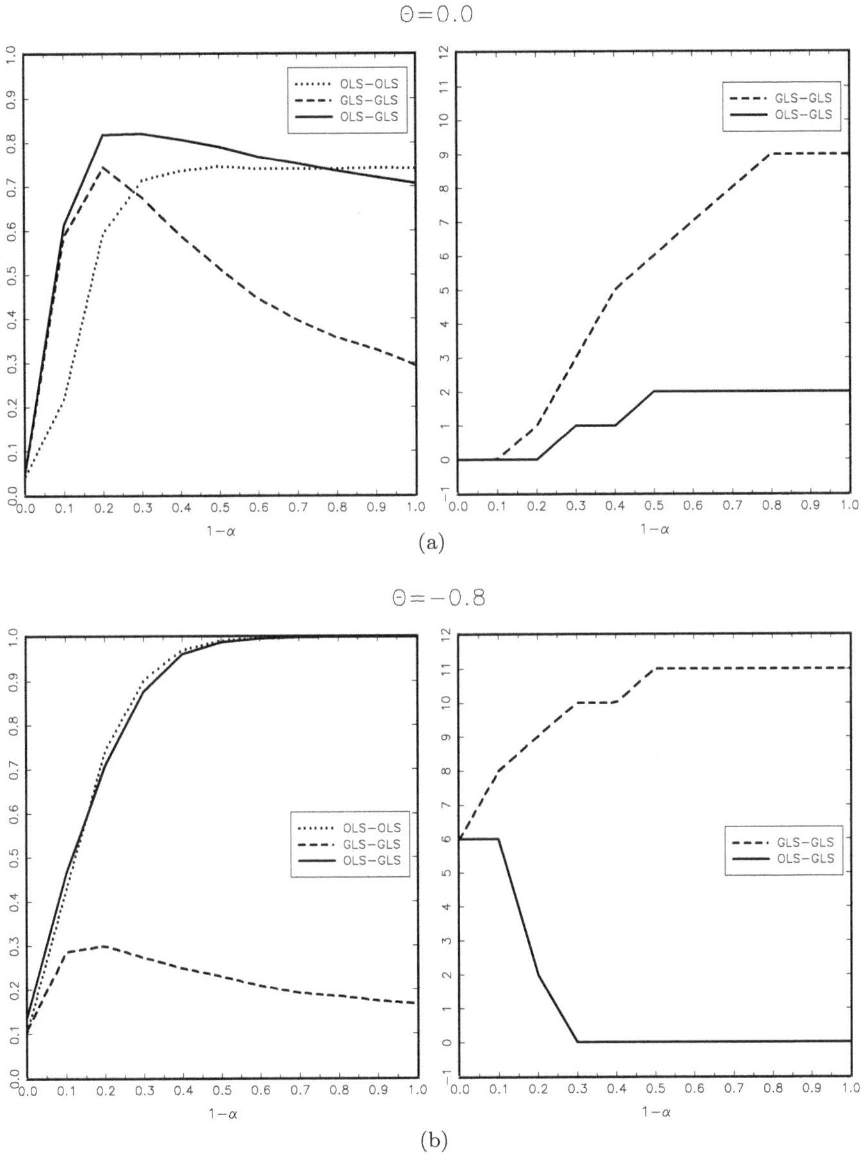

Fig. 1. Power of the ADF test (a) and median of the lag order selected (b) ($T = 100$, MA(1) errors, constant only).

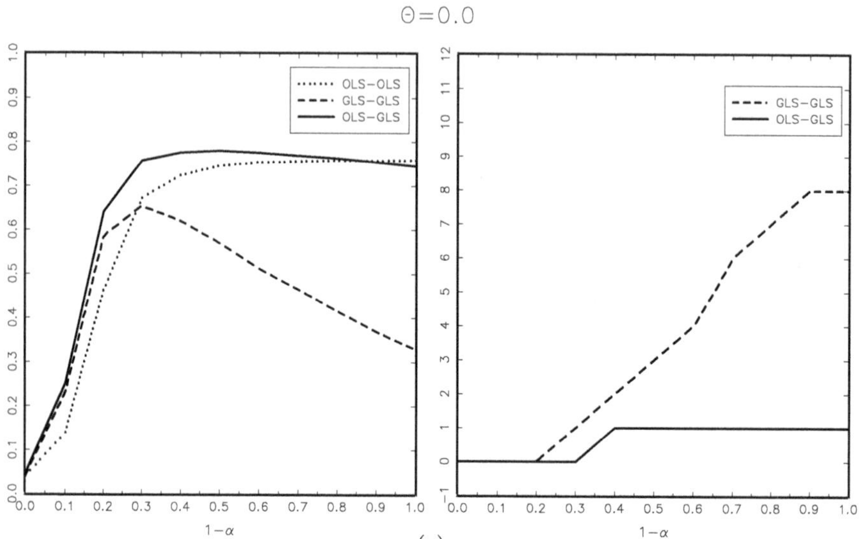

(a)

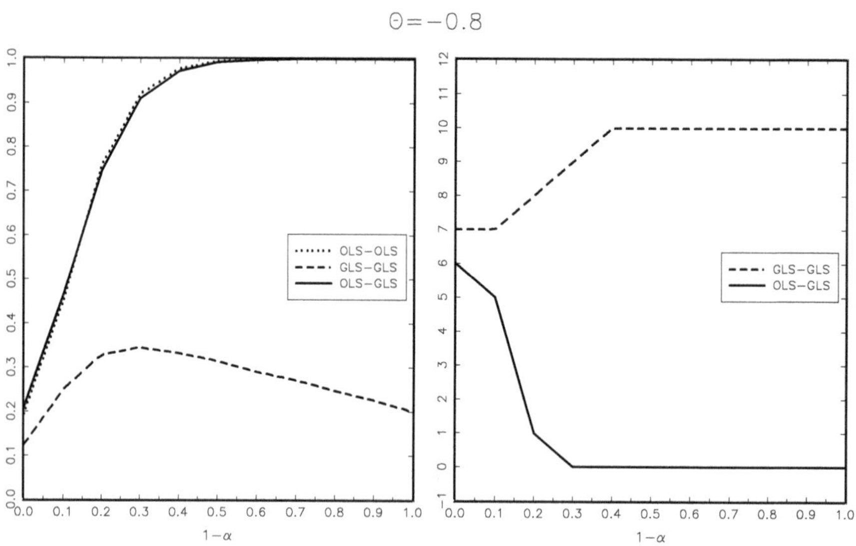

(b)

Fig. 2. Power of the ADF test (a) and median of the lag order selected (b) ($T = 100$, MA(1) errors, constant and linear trend).

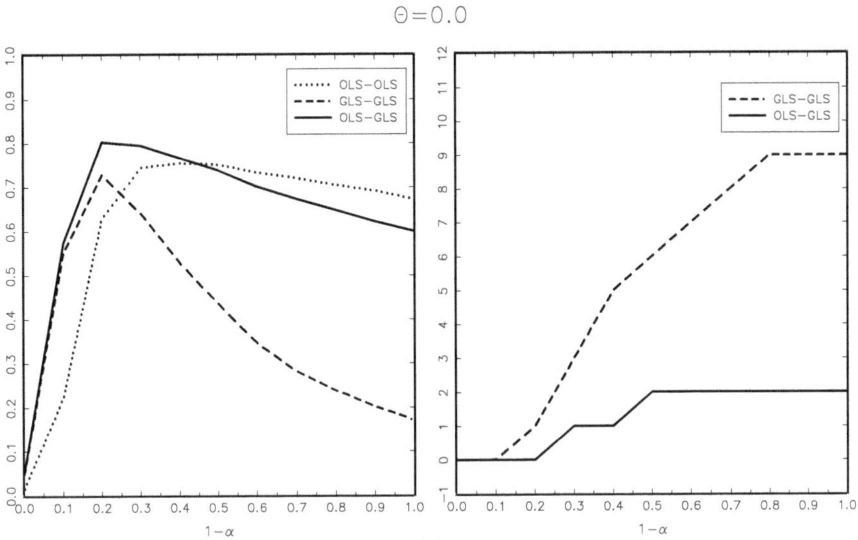

(a)

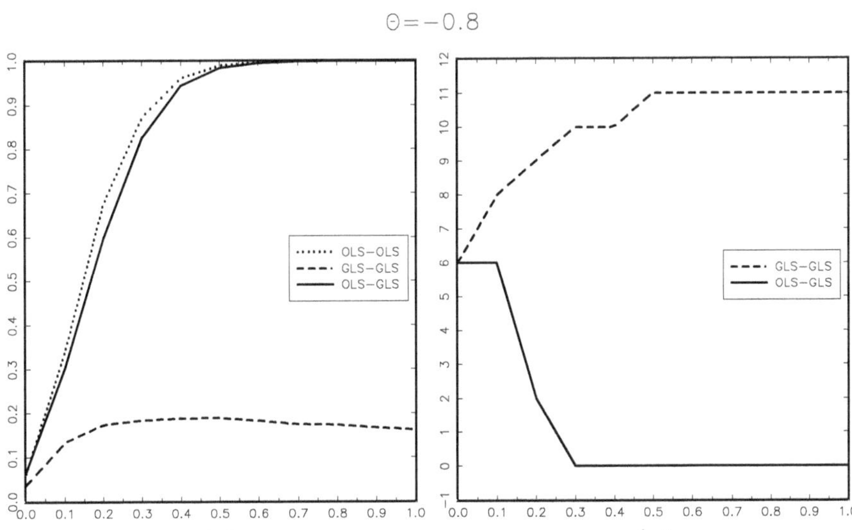

(b)

Fig. 3. Power of the MZ_α test (a) and median of the lag order selected (b) ($T = 100$, MA(1) errors, constant only).

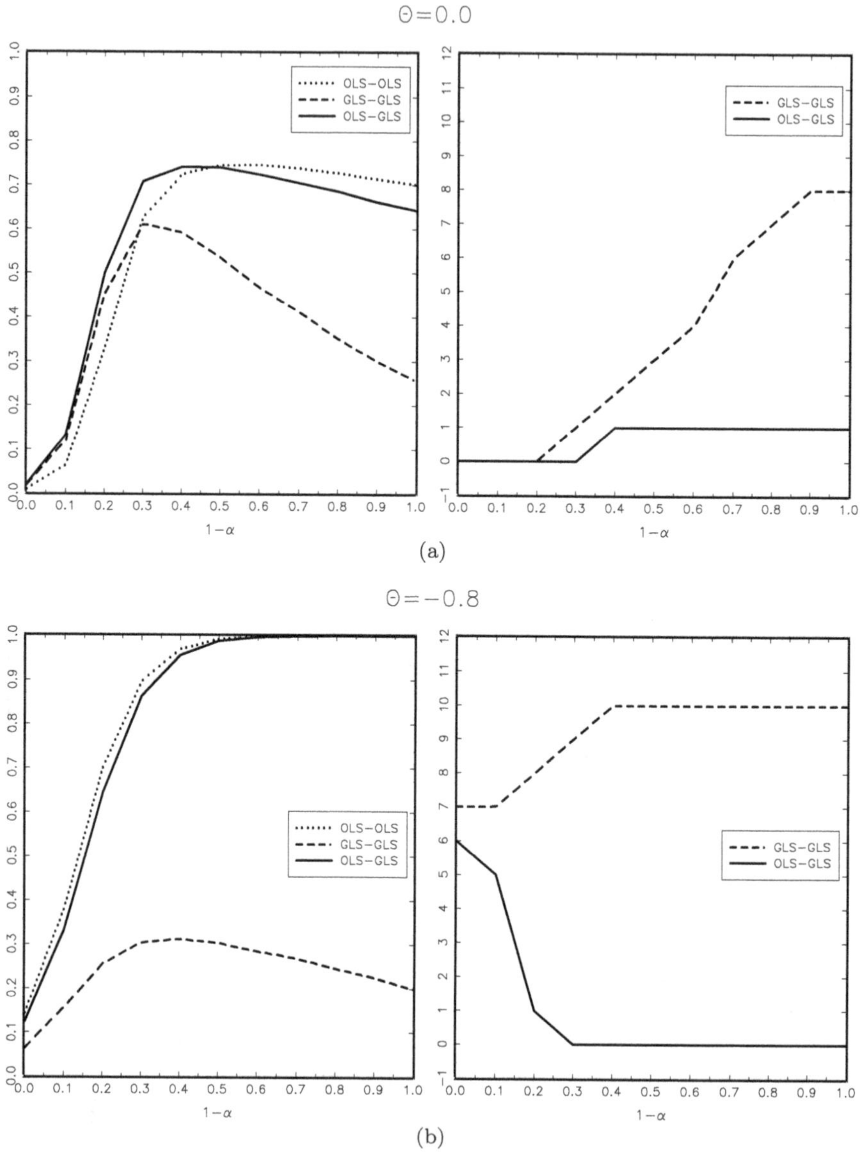

Fig. 4. Power of the MZ_α test (a) and median of the lag order selected (b) ($T = 100$, MA(1) errors, constant and linear trend).

Using OLS detrended data is preferable to using GLS detrended data not only when using the MAIC but also when using any information criterion such as the AIC. The reason is the following. When α is not local to one, the estimate of the coefficients of the trend function is biased when using GLS detrended data. This implies that the series $\tilde{y}_t$ do not have mean zero or may be trending. The autoregression (2) is then mis-specified and the estimate $\hat{b}_0$ is biased and the bias varies with the lag order. This bias induces serial correlation in the errors and, as a result, additional lags are needed to compensate for that. With the MAIC the effect is more pronounced since the additional penalty term involves $\hat{b}_0$ which is sensitive to the lag order irrespective of the underlying process. With OLS detrended data, the series $\hat{y}_t$ has basically mean zero so that this problem does not occur.

In summary, our modification is easy to implement, leads to tests with finite sample size even closer to nominal size and a power function that is not severely reduced for nonlocal alternatives. The $\text{MZ}_\alpha^{\text{GLS}}$, P_T and MP_T^{GLS} tests have similar properties, their exact size is close to nominal size in the case of MA processes but can be quite conservative with autoregressive errors with a negative coefficient. On the other hand, the ADF^{GLS} has liberal size distortions in the case of negative moving-average errors but have size close to nominal size and slightly higher power with autoregressive errors. Since errors with a negative moving average component are more likely than errors with a negative autoregressive coefficient, the $\text{MZ}_\alpha^{\text{GLS}}$, P_T and MP_T^{GLS} tests should be preferable to the ADF^{GLS} test.

References

Elliott, G., Rothenberg, T.J. and Stock, J.H. (1996). Efficient tests for an autoregressive unit root. *Econometrica* 64, 813–836.

Haldrup, N. and Jansson, M. (2006). Improving size and power in unit root testing. In *Palgrave Handbook of Econometrics.* Vol. 1: Econometric Theory, Patterson, K., Mills, T.C. (eds.), Palgrave Macmillan, pp. 252–277.

Ng, S. and Perron, P. (2001). Lag length selection and the construction of unit root tests with good size and power. *Econometrica* 69, 1519–1554.

Seo, M.H. (2005). Improving unit root tests by a generalization of the autoregressive spectral density estimator at frequency zero. Unpublished Manuscript, Department of Economics, London School of Economics.

Part B

Unit Roots with Trend Breaks

Chapter 12

The Great Crash, the Oil Price Shock, and the Unit Root Hypothesis*

Pierre Perron

Department of Economics, Princeton University, Princeton, NJ 08544, USA; Centre de Recherche et Développement en Economique, Université de Montréal, C.P. 6128, Succ. A, Montréal, Canada, H3C-3J7

We consider the null hypothesis that a time series has a unit root with possibly nonzero drift against the alternative that the process is 'trend-stationary'. The interest is that we allow under both the null and alternative hypotheses for the presence of a one-time change in the level or in the slope of the trend function. We show how standard tests of the unit root hypothesis against trend stationary alternatives cannot reject the unit root hypothesis if the true data generating mechanism is that of stationary fluctuations around a trend function which contains a one-time break. This holds even asymptotically. We derive test statistics which allow us to distinguish the two hypotheses when a break is present. Their limiting distribution is established and selected percentage points are tabulated. We apply these tests to the Nelson–Plosser data set and to the postwar quarterly real GNP series. In the former, the break is due to the 1929 crash and takes the form of a sudden change in the level of the series. For 11 out of the 14 series analyzed by Nelson and Plosser we can reject at a high confidence level the unit root hypothesis. In the case of the postwar quarterly real GNP series, the break in the trend function occurs at the time of the oil price shock (1973) and takes the form of a change in the slope. Here again we can reject the null hypothesis of a unit root. If one is ready to postulate that the 1929 crash and the slowdown in growth after 1973 are not realizations of an underlying time-invariant

*Originally Published in *Econometrica*, 57, pp. 1361–1401 and in Erratum; *Econometrica*, 61, pp. 248–249.

stochastic process but can be modeled as exogenous, then the conclusion is that most macroeconomic time series are not characterized by the presence of a unit root. Fluctuations are indeed stationary around a deterministic trend function. The only 'shocks' which have had persistent effects are the 1929 crash and the 1973 oil price shock.

Keywords: Hypothesis testing; intervention analysis; structural change; stochastic trends; deterministic trends; functional weak convergence; Wiener process; macroeconomic time series.

1. Introduction

The unit root hypothesis has recently attracted a considerable amount of work in both the economics and statistics literature. Indeed, the view that most economic time series are characterized by a stochastic rather than deterministic nonstationarity has become prevalent. The seminal study of Nelson and Plosser (1982) which found that most macroeconomic variables have a univariate time series structure with a unit root has catalyzed a burgeoning research program with both empirical and theoretical dimensions.

Nelson and Plosser's study was followed by a series of empirical analyses which basically confirmed their findings. Some (Stulz and Wasserfallen (1985) and Wasserfallen (1986), among others) applied a similar Dickey–Fuller (1979) statistical methodology to other economic series. On the statistical front, there emerged an interest in developing alternative approaches to test the unit root hypothesis. Examples include: the class of tests proposed by Phillips and Perron (1988) and the methodology suggested by Campbell and Mankiw (1987, 1988) and Cochrane (1988) using an estimate of the spectral density at frequency zero. Empirical applications of these methodologies generally reaffirmed the conclusion that most macroeconomic time series have a unit root (e.g., Perron, 1988).

These studies had many effects on economic theorizing. They seem to confirm previous analyses which had advanced the unit root hypothesis for particular economic series, e.g., consumption (Hall, 1978), velocity of money (Gould and Nelson, 1974), and stock prices (Samuelson, 1973). They also launched a series of theoretical investigations with implications consistent with a unit root, e.g., Blanchard and Summers (1986) for employment. Furthermore, a considerable stock of statistical tools was developed for more general models with integrated variables; these include the cointegration framework (Engle and Granger, 1987) and multivariate systems (Stock and Watson, 1988; Phillips and Durlauf, 1986).

As far as macroeconomic theories are concerned, the most important implication of the unit root revolution, is that under this hypothesis random shocks have a permanent effect on the system. Fluctuations are not transitory. This implication, as forcefully argued by Nelson and Plosser, has profound consequences for business cycle theories. It runs counter to the prevailing view that business cycles are transitory fluctuations around a more or less stable trend path. It is therefore of importance to assess carefully the reliability of the unit root hypothesis as an empirical fact.

The aim of this paper appears startling, given the results in the above mentioned literature. Our conclusion is that most macroeconomic time series are not characterized by the presence of a unit root and that fluctuations are indeed transitory. Only two events (shocks) have had a permanent effect on the various macroeconomic variables: the Great Crash of 1929 and the oil price shock of 1973.

Of course, to reach such a conclusion, a particular postulate must be introduced which differentiates our approach from the previous ones. This postulate is that the Great Crash and the oil price shock were not a realization of the underlying data-generating mechanism of the various series. In this sense, we consider these shocks as exogenous. The exogeneity assumption is not a statement about a descriptive model for the time series representation of the variables. It is used here as a device to remove the influence of these shocks from the noise function. A more detailed discussion of these issues and their implications can be found in Section 6.

These two shocks are rather different in nature. On one hand, the Great Crash created a dramatic drop in the mean of most aggregate variables. On the other hand, the 1973 oil price shock was followed by a change in the slope of the trend for most aggregates, i.e., a slowdown in growth. In this light, our aim is to show that most macroeconomic variables are 'trend-stationary' if one allows a single change in the intercept of the trend function after 1929 and a single change in the slope of the trend function after 1973.

Our approach is in the spirit of the 'intervention analysis' suggested by Box and Tiao (1975). According to their methodology, 'aberrant' or 'outlying' events can be separated from the noise function and be modeled as changes or 'interventions' in the deterministic part of the general time series model. Using such a strategy makes it 'possible to distinguish between what can and what cannot be explained by the noise' (Box and Tiao, 1975, p. 72). These 'interventions' are assumed to occur at a known date. The same strategy is used in the present analysis in that we consider the time of the changes in the trend function as fixed rather than as a random variable to be estimated.

To make our point as unambiguous as possible, we use the same data set as Nelson and Plosser, as well as the real GNP series analyzed by Campbell and Mankiw. The data set used by Nelson and Plosser contains 14 macroeconomic variables sampled annually. All series end in 1970 and contain only one break, the 1929 Great Crash. We shall not analyze the unemployment rate series since there is a general agreement that it is stationary. The real GNP series is postwar quarterly from 1947:I to 1986:III and so contains only one break as well, the 1973 oil shock. Furthermore, to make our analysis as similar as possible to previous ones, the statistical methodology applied here is an extension of the Dickey–Fuller methodology (as used by Nelson and Plosser) to test for the presence of a single unit root in a univariate time series.

The plan of the paper is as follows. Section 2 motivates the ensuing analysis and presents the alternative models considered. Section 3 shows that usual tests will not be able to reject the unit root hypothesis if in fact the deterministic trend of the series has a single break (either in the intercept or the slope). In Section 4, we develop formal statistical tests of the null hypothesis of a unit root which can distinguish the unit root hypothesis from that of a stationary series around a trend which has a single break. The asymptotic distributions under the null hypothesis are derived and tabulated. Empirical results from applying these procedures are presented in Section 5. Section 6 contains a discussion of some issues raised by our analysis and suggestions for future research. All theorems are proved in Appendix A.

2. Motivation

The null hypothesis considered is that a given series $\{y_t\}_0^T$ (of which a sample of size $T+1$ is available) is a realization of a time series process characterized by the presence of a unit root and possibly a nonzero drift. However, the approach is generalized to allow a one-time change in the structure occurring at a time $T_B (1 < T_B < T)$. Three different models are considered under the null hypothesis: one that permits an exogenous change in the level of the series (a 'crash'), one that permits an exogenous change in the rate of growth, and one that allows both changes. These hypotheses are parameterized as follows:

Null hypotheses:

Model (A) $y_t = \mu + dD(TB)_t + y_{t-1} + e_t,$

Model (B) $y_t = \mu_1 + y_{t-1} + (\mu_2 - \mu_1)DU_t + e_t,$

Model (C) $y_t = \mu_1 + y_{t-1} + dD(TB)_t + (\mu_2 - \mu_1)DU_t + e_t,$ where

$$D(TB)_t = 1 \quad \text{if } t = T_B + 1, \quad 0 \text{ otherwise;}$$
$$DU_t = 1 \quad \text{if } t > T_B, \quad 0 \text{ otherwise; and}$$
$$A(L)e_t = B(L)v_t,$$

$v_t \sim$ i.i.d. $(0, \sigma^2)$, with $A(L)$ and $B(L)$ pth and qth order polynomials, respectively, in the lag operator L.

The innovation series $\{e_t\}$ is taken to be of the ARMA(p, q) type with the orders p and q possibly unknown. This postulate allows the series $\{y_t\}$ to represent quite general processes. More general conditions are possible and will be used in subsequent theoretical derivations.

Instead of considering the alternative hypothesis that y_t is a stationary series around a deterministic linear trend with time invariant parameters, we shall analyze the following three possible alternative models:

Alternative hypotheses:

Model (A) $y_t = \mu_1 + \beta t + (\mu_2 - \mu_1)DU_t + e_t,$

Model (B) $y_t = \mu + \beta_1 t + (\beta_2 - \beta_1)DT_t^* + e_t,$

Model (C) $y_t = \mu_1 + \beta_1 t + (\mu_2 - \mu_1)DU_t + (\beta_2 - \beta_1)DT_t + e_t$

where

$$DT_t^* = t - T_B, \quad \text{and} \quad DT_t = t \quad \text{if } t > T_B \quad \text{and 0 otherwise.}$$

Here, T_B refers to the time of break, i.e., the period at which the change in the parameters of the trend function occurs. Model (A) describes what we shall refer to as the crash model. The null hypothesis of a unit root is characterized by a dummy variable which takes the value one at the time of break. Under the alternative hypothesis of a 'trend-stationary' system, Model (A) allows for a one-time change in the intercept of the trend function. For the empirical cases we have in mind, T_B is the year 1929 and $\mu_2 < \mu_1$. Model (B) is referred to as the 'changing growth' model. Under the alternative hypothesis, a change in the slope of the trend function without any sudden change in the level at the time of the break is allowed. Under the null hypothesis, the model specifies that the drift parameter μ changes from μ_1 to μ_2 at time T_B. In the empirical examples presented in Section 5, T_B is the first quarter of 1973 and $\beta_2 < \beta_1$, reflecting a slowdown in growth following the oil shock. Model (C) allows for both effects to take place simultaneously, i.e., a sudden change in the level followed by a different growth path.

To motivate the use of these three models as possible alternatives to the unit root with drift hypothesis, we present in this section some descriptive analyses for three series: 'nominal wages' (1900–1970), 'quarterly real GNP' (1947:I–1986:III) and 'common stock prices' (1871–1970).

Figure 1 shows a plot of the logarithm of the nominal wage series. A feature of this graph is the marked decrease between 1929 and 1930. Apart from this change, the trend appears fairly stable (same slope) over the entire period. The solid line is the estimated trend line from a regression on a constant, a trend and a dummy variable taking a value of 0 prior and at 1929 and value 1 afterwards. Table 1 presents the results from estimating (by OLS) a regression of the Dickey–Fuller type, i.e.,:

$$y_t = \tilde{\mu} + \tilde{\beta} t + \tilde{\alpha} y_{t-1} + \sum_{i=1}^{k} \tilde{c}_i \Delta y_{t-i} + \tilde{e}_t. \tag{1}$$

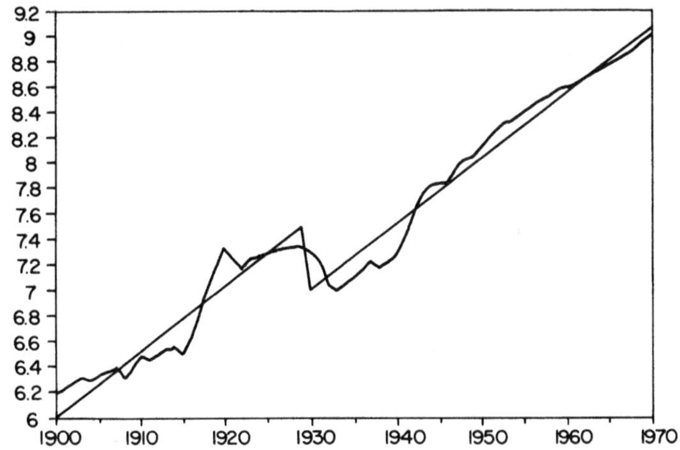

Fig. 1. Logarithm of 'nominal wages'.

Note: The broken straight line is a fitted trend (by OLS) of the form $\tilde{y}_t = \tilde{\mu} + \tilde{\gamma} DU_t + \tilde{\beta} t$ where $DU_t = 0$ if $t \leq 1929$ and $DU_t = 1$ if $t > 1929$.

Table 1. Regression analysis for the wages, quarterly GNP, and common stock price series.

Regression: $y_t = \tilde{\mu} + \tilde{\beta} t + \tilde{\alpha} y_{t-1} + \sum_{i=1}^{k} \tilde{c}_t \Delta y_{t-i} + \tilde{e}_t$								
Series/Period	k	$\tilde{\mu}$	$t_{\tilde{\mu}}$	$\tilde{\beta}$	$t_{\tilde{\beta}}$	$\tilde{\alpha}$	$t_{\tilde{\alpha}}$	$S(\tilde{e})$
(a) Wages								
1900–1970[a]	2	0.566	2.30	0.004	2.30	0.910	−2.09	0.060
1900–1929	7	4.299	2.84	0.037	2.73	0.304	−2.82	0.0803
1930–1970	8	1.632	3.60	0.012	2.64	0.735	−3.19	0.0269
(b) Common stock prices								
1871–1970[a]	2	0.481	2.02	0.003	2.37	0.913	−2.05	0.158
1871–1929	3	0.3468	2.13	0.0063	2.70	0.732	−2.29	0.1209
1930–1970	4	−0.5312	−1.64	0.0166	1.96	0.788	−1.89	0.1376
(c) Quarterly real GNP								
1947:I–1986:III	2	0.386	2.90	0.0004	2.71	0.946	−2.85	0.010
1947:I–1973:I	2	0.637	3.04	0.0008	2.99	0.910	−3.02	0.0099
1973:II–1986:III	1	0.883	2.23	0.0008	2.27	0.878	−2.23	0.0102

Note: [a]Results taken from Nelson and Plosser (1982, Table 5).

The first row presents the full sample regression. The coefficient on the lag dependent variable is 0.910 with a t statistic for the hypothesis that $\alpha = 1$ of -2.09. Using the critical values tabulated by Dickey and Fuller, we cannot reject the null hypothesis of a unit root. When the sample is split in two (pre-1929 and post-1929), the estimated value of α decreases dramatically: 0.304 for the pre-1929 sample and 0.735 for the post-1929 sample. However, due to the small samples available, the t statistics are not large enough (in absolute value) to reject the hypothesis that $\alpha = 1$, even at the 10% level.

Two features are worth emphasizing from this example: (a) the full sample estimate of α is markedly superior to any of the split sample estimates and relatively close to one. It appears that the 1929 crash is responsible for the near unit root value of α; and (b) the split sample regressions are not powerful enough to reject the hypothesis that $\alpha = 1$ even though the estimates are well below one. It would be useful, in this light, to have a more powerful procedure based on the full sample that would allow the 1929 break to be exogenous.

Figure 2 graphs the postwar quarterly real GNP series. Here, the series behave according to Model (B) where there is no sharp change in the level of the series at the 1973:I break point but rather a change in the slope. The solid line is a fitted trend where a dummy variable is included in the regression, taking the value 0 prior and at 1973:I and the value $(t - 105)$ after 1973:I (1973:I being the 105th observation in the sample). Table 1 compares regressions of the form (1) with full and split samples. Again, the estimate of α is lower in both subsamples than with the full sample (given the quarterly nature of the series, the difference is important). The same features discussed above appear to hold when there is a change in the slope of the trend function.

As a final example, consider the common stock price series graphed in Fig. 3. The break point is again in 1929 but in this case there appears to be both a sudden change in the level of the series in 1929 and a higher growth rate after. The solid line is the estimated trend with two dummy variables added, an intercept dummy (0 prior and at 1929, 1 after 1929) and a slope dummy (0 prior and at 1929

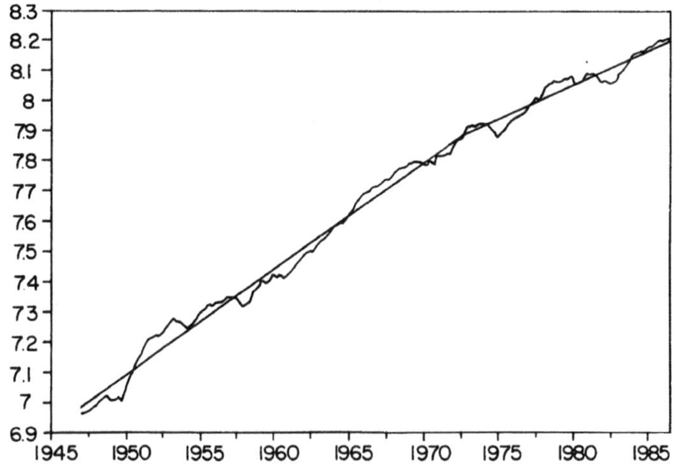

Fig. 2. Logarithm of 'postwar quarterly real GNP'.
Note: The broken straight line is a fitted trend (by OLS) of the form: $\tilde{y}_t = \tilde{\mu} + \tilde{\beta}_t + \tilde{\gamma} DT_t^*$ where $DT_t^* = 0$ if $t \leq$ 1973:I and $DT_t^* = t - T_B$ if $t >$ 1973:I $= T_B$.

and t after 1929). The estimated values of α (in regression (1)) with the full sample are 0.913 but are only 0.732 using the pre-1929 sample and 0.788 using the post-1929. Here again, the t statistics are not large enough, however, to reject the unit root hypothesis at even the 10% level using any of the subsamples.[1]

Table 2 presents the autocorrelation function of the 'detrended series' for the full set of variables analyzed by Nelson and Plosser, along with the postwar quarterly real GNP series. All series are detrended according to Model (A) (with a constant, a trend, and an intercept dummy) except for the postwar Quarterly Real GNP

[1]Dickey–Fuller tests for the presence of a unit root using split samples are presented in Appendix B for all the series considered. The results are presented for values of k ranging from 1 to 12. These results show that (i) the conclusions drawn are not sensitive to the value of k chosen; (ii) for some series it is possible to reject the unit root hypothesis, especially when considering the post-1929 subsample. Furthermore, the statistical significance of the lagged first-differences (not reported) suggest that a large value of k may be needed. For example, the t statistics on the eighth lagged first-difference is often statistically significant. A similar pattern will occur in the full sample tests reported in subsequent sections.

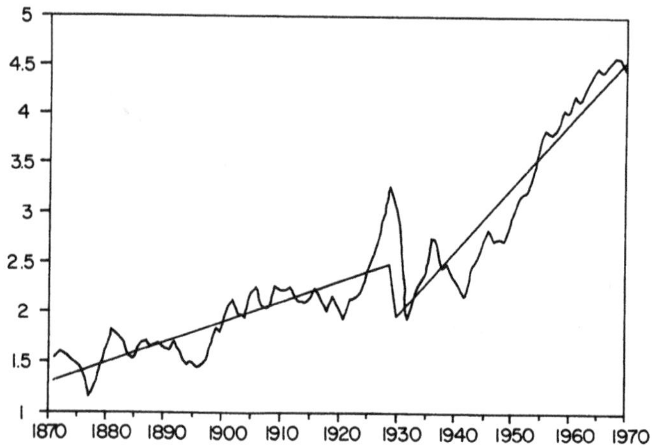

Fig. 3. Logarithm of 'common stock prices'.
Note: The broken straight line is a fitted trend (by OLS) of the form $\tilde{y}_t = \tilde{\mu} + \tilde{\gamma}_1 DU_t + \tilde{\beta}_t + \tilde{\gamma}_2 DT_t$ where $DU_t = DT_t = 0$ if $t \leq 1929$ and $DU_t = 1$, $DT_t = t$ if $t > 1929$.

Series (with a slope dummy instead of the intercept dummy, Model (B)) and the real wage and common stock price series (with both a slope and intercept dummy, Model (C)). Unlike the 'standard' detrended series (see Table 4 of Nelson–Plosser), the autocorrelations decay quite rapidly for all variables except for the consumer prices and velocity series. This behavior of the autocorrelation function is certainly not the one usually associated with either a random walk or a detrended random walk. Indeed, the 'detrended' series appear stationary.

The results of this section motivate the analysis presented in the following sections. We first investigate the effects of the two types of changes in the trend function that we consider on the statistical properties of autoregressive estimates of the type found in regression (1) (both in finite samples and asymptotically). We find that such changes create a spurious unit root that may not vanish, even asymptotically. To overcome the problem of the low power associated with testing for a unit root using split samples, formal test statistics, which permit the presence of either or both an intercept and a slope shift, are developed in Section 4.

Table 2. Sample autocorrelations of the 'detrended' series.

Series	Period	T	Variance	r_1	r_2	r_3	r_4	r_5	r_6
Real GNP	A 1909–1970	62	0.010	0.77	0.45	0.23	0.11	0.05	0.04
Nominal GNP	A 1909–1970	62	0.023	0.68	0.31	0.12	0.08	0.11	0.12
Real per capita GNP	A 1909–1970	62	0.012	0.81	0.54	0.33	0.20	0.13	0.09
Industrial production	A 1860–1970	111	0.017	0.71	0.44	0.32	0.17	0.08	0.12
Employment	A 1890–1970	81	0.005	0.82	0.59	0.43	0.30	0.20	0.15
GNP deflator	A 1889–1970	82	0.015	0.82	0.63	0.45	0.31	0.17	0.06
Consumer prices	A 1860–1970	111	0.066	0.96	0.89	0.80	0.71	0.63	0.54
Wages	A 1900–1970	71	0.016	0.76	0.47	0.26	0.12	0.03	−0.03
Real wages	C 1900–1970	71	0.003	0.74	0.40	0.12	−0.12	−0.27	−0.33
Money stock	A 1889–1970	82	0.023	0.87	0.69	0.52	0.38	0.25	0.11
Velocity	A 1869–1970	102	0.036	0.90	0.79	0.70	0.62	0.57	0.52
Interest rate	A 1900–1970	71	0.587	0.77	0.58	0.38	0.25	0.15	0.11
Common stock prices	C 1871–1970	100	0.066	0.80	0.53	0.36	0.20	0.10	0.08
Quarterly GNP	B 47:I 86:III	159	0.001	0.94	0.83	0.70	0.57	0.45	0.35

Note: A, B, and C denote the detrending procedure corresponding to the given model under the alternative hypothesis.

3. The Effect of a Shift in the Trend Function on Tests for a Unit Root

To assess the effects of the presence of a shift in the level of the series or a shift in the slope (at a single point of time) on tests for the presence of a unit root, we first present a small Monte Carlo experiment. Consider first the 'crash hypothesis' (Model (A)). We generated 10,000 replications of a series $\{y_t\}$ of length 100 defined by

$$y_t = \mu_1 + (\mu_2 - \mu_1)DU_t + \beta t + e_t \quad (t = 1, \ldots, T), \tag{2}$$

where $DU_t = 1$ if $t > T_B$, and 0 otherwise.

For simplicity, $\mu_1 = 0$, $\beta = 1$, $T_B = 50$, $T = 100$ and the innovations e_t are i.i.d. $N(0, 1)$. For the 'changing growth' hypothesis, a similar setup is considered except that y_t is generated by

$$y_t = \mu + \beta_1 t + (\beta_2 - \beta_1) DT_t^* + e_t \quad (t = 1, \ldots, T), \tag{3}$$

where $DT_t^* = t - T_B$ if $t > T_B$, and 0 otherwise.

Again, $\mu = 0$, $\beta_1 = 1$, $T_B = 50$, $T = 100$, and $e_t \sim$ i.i.d. $N(0,1)$. For each replication, we computed the autoregressive coefficient $\tilde{\alpha}$ in the following regression, using ordinary least squares:

$$y_t = \tilde{\mu} + \tilde{\beta} t + \tilde{\alpha} y_{t-1} + \tilde{e}_t. \tag{4}$$

Figure 4 graphs the cumulative distribution function of $\tilde{\alpha}$ when the data generating process (D.G.P.) is given by (2) for various values of μ_2. This experiment reveals that as the magnitude of the crash increases (μ_2 decreases), the c.d.f of $\tilde{\alpha}$ becomes more concentrated at a value ever closer to 1. The corresponding mean and variance of the sample of $\tilde{\alpha}$ generated are shown in Table 3. Figure 5 graphs the c.d.f. of $\tilde{\alpha}$ when the D.G.P. is given by (3) for various values of β_2. As β_2 diverges from β_1, again, the c.d.f. becomes more concentrated

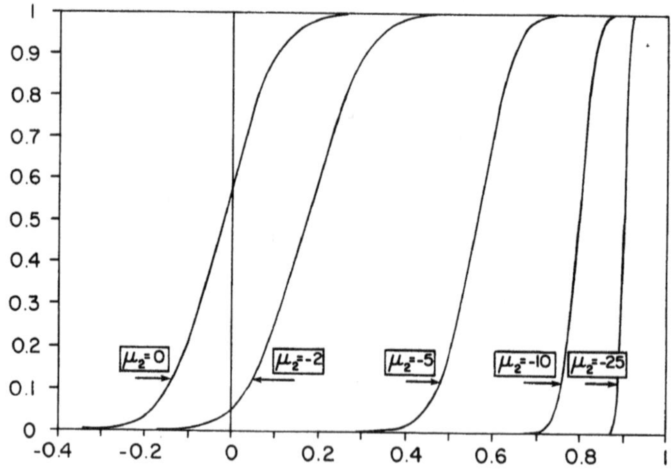

Fig. 4. C.D.F. of $\tilde{\alpha}$ under the 'crash' model.

Note: $\tilde{\alpha}$ is the estimated autoregressive parameter in regression (4). The data-generating mechanism is given by Equation (2) with $\mu_1 = 0$, $\beta = 1.0$ and $\{e_t\}$ i.i.d. $N(0, 1)$, $T = 100$ and $T_B = 50$.

Table 3. Mean and variance of $\tilde{\alpha}$.

	$\mu_2 = 0$	$\mu_2 = -2$	$\mu_2 = -5$	$\mu_2 = -10$	$\mu_2 = -25$
(a) Crash Simulations, $\mu_1 = 0$, $\beta = 1$					
Mean	−0.019	0.172	0.558	0.795	0.899
Variance	0.00986	0.01090	0.00471	0.00089	0.00009
	$\beta_2 = 1.0$	$\beta_2 = 0.9$	$\beta_2 = 0.7$	$\beta_2 = 0.4$	$\beta_2 = 0.0$
(b) Breaking Trend Simulations, $\beta_1 = 1$, $\mu = 0$					
Mean	−0.019	0.334	0.825	0.949	0.981
Variance	0.00986	0.00938	0.00094	0.00009	0.00001

Note: See notes to Fig. 4 for case (a) and Fig. 5 for case (b).

and closer to one. The computed mean and variance of $\tilde{\alpha}$ presented in Table 3 confirms this behavior.[2]

What emerges from this experiment is that if the magnitude of the shift is significant, one could hardly reject the unit root hypothesis even if the series is that of a trend (albeit with a break) with i.i.d. disturbances. In particular, one would conclude that the shocks have permanent effects. Here, the shocks clearly have no permanent effects, only the one-time shift in the trend function is permanent.

To analyze the effect of an increase in the sample size on the distribution of $\tilde{\alpha}$ with a shift of a given magnitude, we derive the asymptotic limit of $\tilde{\alpha}$. To this end, we again consider processes generated by Models (A), (B), or (C) under the alternative hypotheses,

[2]Note that when the error structure is i.i.d., $\tilde{\alpha}$ is free of nuisance parameters and hence can be used as a formal test statistic on the same ground as the t statistic. However, we also performed a similar experiment with the t statistic on $\tilde{\alpha}(\alpha = 1)$ in regression (4) as well as in a regression with additional lags of first-differences as regressors. The results obtained show the same behavior. If anything, the t statistic with extra lags of first-differences as regressors shows a still greater bias toward nonrejection of the null hypothesis of a unit root. These results are available upon request. We prefer to report our result in terms of the behavior of the estimator $\tilde{\alpha}$ instead of its t statistic because it makes clear that what causes the nonrejection is not due solely to the behavior of the variance estimator. What is of importance is that $\tilde{\alpha}$ is biased towards unity.

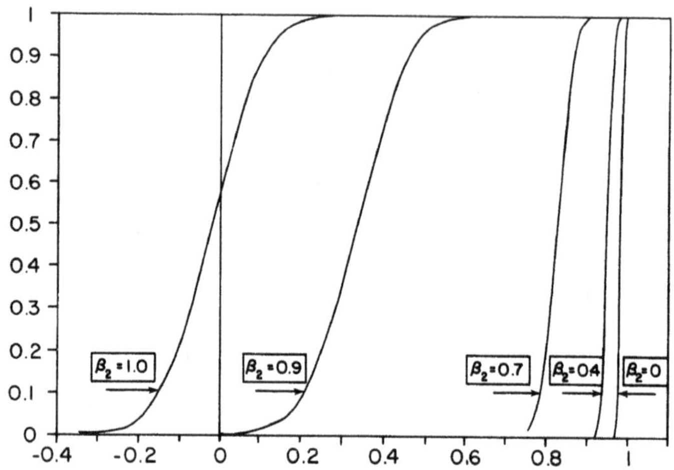

Fig. 5. C.D.F. of $\tilde{\alpha}$ under the 'breaking trend' model.
Note: $\tilde{\alpha}$ is the estimated autoregressive parameter in regression (4). The data-generating mechanism is given by Equation (3) with $\mu = 0$, $\beta_1 = 1.0$, $\{e_t\}$ i.i.d. $N(0, 1)$, $T = 100$, $T_B = 50$.

but we enlarge the framework by allowing general conditions on the error structure $\{e_t\}$. Many such sets of conditions are possible and would allow us to carry out the asymptotic theory. For simplicity, we use the 'mixing-type' conditions of Phillips (1987) and Phillips and Perron (1988). These are stated in Assumption 1.

Assumption 1. *(a)* $E(e_t) = 0$ *all* t; *(b)* $\sup_t E|e_t|^{\beta+\varepsilon} < \infty$ *for some* $\beta > 2$ *and* $\varepsilon > 0$; *(c)* $\sigma^2 = \lim_{T\to\infty} T^{-1}E(S_T^2)$ *exists and* $\sigma^2 > 0$, *where* $S_T = \sum_1^T e_t$; *(d)*$\{e_t\}_1^\infty$ *is strong mixing with mixing numbers* α_m *that satisfy*: $\sum_1^\infty \alpha_m^{1-2/\beta} < \infty$.

These conditions are general enough to permit the series $\{e_t\}$ to be generated by a finite order ARMA(p, q) process with Gaussian innovations. To carry out the asymptotic analysis, we shall require that both the prebreak and postbreak samples increase at the same rate as the total number of observations, T, increases. To this effect, we assume, for simplicity, that $T_B = \lambda T$ for all T. We refer to λ as the 'break fraction'. The asymptotic limits are taken as T increases to infinity in a sequence that ensures an integer value of T_B for a given λ. This type of increasing sequence is assumed throughout the paper. The results proved in Appendix A are presented in the following theorem.

Theorem 1. *Let $\{y_t\}_0^T$ be a sample of size $T+1$ generated under one of the alternative hypotheses with the innovations $\{e_t\}$ satisfying Assumption 1. Let '$\rightarrow$' denote convergence in probability. Furthermore, let $T_B = \lambda T$, for all T and $0 < \lambda < 1$; then as $T \rightarrow \infty$:*

(a) *The 'crash hypothesis': Under Model* (A)

$$\tilde{\alpha} \rightarrow \{[\mu_1 - \mu_2]^2 A + \gamma_1\}\{[\mu_1 - \mu_2]^2 A + \sigma_e^2\}^{-1}$$

where

$$A = [\lambda - 4\lambda^2 + 6\lambda^3 - 3\lambda^4], \quad \gamma_1 = \lim_{T\rightarrow\infty} T^{-1}\sum_1^T E(e_t e_{t-1}),$$

and

$$\sigma_e^2 = \lim_{T\rightarrow\infty} T^{-1}\sum_1^T E(e_t^2).$$

(b) *The 'breaking trend hypothesis': Under either Model* (B) *or* (C)

(i) $\tilde{\alpha} \rightarrow 1$,

(ii) $T(\tilde{\alpha} - 1) \rightarrow \{3(-1 + 4\lambda - 5\lambda^2 + 2\lambda^3)\} \cdot\{2(-3 + 4\lambda - 3\lambda^2 + 3\lambda^3 - \lambda^4)\}^{-1}$.

Part (a) of Theorem 1 shows that under the crash hypothesis, the limit of $\tilde{\alpha}$ depends on the relative magnitude of $[\mu_1 - \mu_2]^2$ A and σ_e^2. In particular, this limit gets closer to one as $[\mu_1 - \mu_2]^2$ increases. Another feature is that the limit of $\tilde{\alpha}$ is always greater than the true first-order autoregressive coefficient of the stationary part of the series, γ_1/σ_e^2. However, since $\tilde{\alpha}$ does not converge to 1, the usual statistics for testing that $\alpha = 1$, such as $T(\tilde{\alpha} - 1)$ or the t statistic on $\tilde{\alpha}$, would eventually reject the null hypothesis of a unit root. Nevertheless, added to the generally poor power properties of tests for a unit root is the consideration that the limit of $\tilde{\alpha}$ is inflated above its true value. These conditions are such that it could be difficult to reject the unit root hypothesis in finite samples.

There is another interpretation to the results under the crash hypothesis. As stated in model (A), the change in the intercept of the trend function is given by $(\mu_2 - \mu_1)$, a fixed value. This implies that in the asymptotic derivations we are considering a shift which decreases relative to the level of the series as the sample size increases. It may be more appropriate to specify the change in the intercept as a magnitude relative to the level of the series at the time of the break. Since at this period the level of the series is proportional to T_B, we can specify $(\mu_2 - \mu_1)$ as a proportion of T_B, say, $(\mu_2 - \mu_1) = \gamma T_B$. In this case the 'crash' is proportional to the level of the series. Since $T_B \to \infty$ as $T \to \infty$, it is clear, from part (a), that under this interpretation, $\tilde{\alpha} \to 1$.

Such ambiguity does not occur under the 'breaking trend hypothesis' (Models (B) or (C)) as is shown by part (b) of Theorem 1. Here, the limit of $\tilde{\alpha}$ is 1 irrespective of the behavior of the intercept and the limit of $T(\tilde{\alpha} - 1)$ is invariant to the relative magnitude of the shift (β_2 versus β_1). The expression in part (b, ii) is a function of λ. However, it varies from 0 to 1/2 for values of λ in the range (0, 1). Since the one-sided 5% asymptotic critical value of $T(\tilde{\alpha} - 1)$ is -21.8 under the null hypothesis of a unit root (Fuller (1976)), Theorem 2 implies that the unit root hypothesis could not be rejected, even asymptotically.[3]

These results could be extended to more general test statistics, such as the t statistics. Nevertheless, the picture is clear. Tests of the unit root hypothesis are not consistent against 'trend stationary' alternatives when the trend function contains a shift in the slope. Although they are not inconsistent against a shift in the intercept of the trend function (if the change is fixed as T increases), their power is likely to be substantially reduced due to the fact that the limit of the autoregressive coefficient is inflated above its true value. When

[3] After the first draft of this paper was written, we became aware of a result similar to part (b, i) of Theorem 1 proved by Rappoport and Reichlin (1989). In fact, in the case of deterministic trends with multiple shifts in slope, they prove the following more general result: 'If the true model contains $K + 1$ segments, then any fitted model involving K or less segments will, asymptotically, yield a larger sum of squared residuals than [a difference stationary] model' (p. 9).

interpreting the 'crash' as proportional to the level of the series as T increases, $\tilde{\alpha}$ unambiguously converges to one and implies a considerable loss in power. There is therefore a need to develop alternative statistical procedures that could distinguish a process with a unit root from a process stationary around a breaking trend function.

4. Alternative Statistical Procedures

In this section, we extend the Dickey–Fuller testing strategy to ensure a consistent testing procedure against shifting trend functions. We shall present several ways to do so, all of which are asymptotically equivalent, and discuss the main differences between each.

Consider first detrending the raw series $\{y_t\}$ according to either model (A), (B), or (C). Let $\{\tilde{y}_t^i\}$, $i = A, B, C$ be the residuals from a regression of y_t on (1) $i = A$: a constant, a time trend, and DU_t; (2) $i = B$: a constant, a time trend, and DT_t^*; (3) $i = C$: a constant, a time trend, DU_t, and DT_t. Furthermore, let $\tilde{\alpha}^i$ be the least squares estimator of α in the following regression:

$$\tilde{y}_t^i = \tilde{\alpha}^i \tilde{y}_{t-1}^i + \tilde{e}_t \quad (i = A, B, C;\ t = 1, 2, \ldots, T). \tag{5}$$

Up to this point the extensions from the no break model are straightforward enough. However, matters are not so simple concerning the distribution of the statistics of interest, namely the normalized bias, $T(\tilde{\alpha}^i - 1)$, and the t statistic on $\tilde{\alpha}^i$, $t_{\tilde{\alpha}}i$ $(i = A, B, C)$. Needless to say, the only manageable analytical distribution theory is asymptotic in nature. But two additional features are also present over the usual procedure: (a) extra regressors and (b) the split sample nature of these extra regressors. To this effect, we derive the asymptotic distribution of $T(\tilde{\alpha}^i - 1)$ and $t_{\tilde{\alpha}}i$ under the null hypothesis of a unit root. As in Section 3, we require that the break point T_B increases at the same rate as the total sample size T. Again, for simplicity, it is assumed that $T_B = \lambda T$ with both T and T_B integer valued.

The method of proof is similar to that of Phillips (1987) and Phillips and Perron (1988). We use weak convergence results that hold for normalized functions of the sum of the innovations when the latter are assumed to satisfy Assumption 1. The limiting distributions obtained under this general setting are then specialized to the i.i.d. case. The asymptotic distributions in the i.i.d. case are evaluated using simulations, and critical values are tabulated. We then show how the results can be extended to innovations $\{e_t\}$ that follow the general ARMA(p, q) process in the same way that the Dickey–Fuller regressions are modified by adding extra lags of first-differences of the data as regressors.

The main results concerning the asymptotic distributions of the normalized bias estimators and the t statistics of the autoregressive coefficient under the null hypothesis of a unit root are presented in the next theorem.

Theorem 2. *Let $\{y_t\}$ be generated under the null hypothesis of model i ($i = A, B, C$) with the innovation sequence $\{e_t\}$ satisfying Assumption* 1. *Let $\Rightarrow$ denote weak convergence in distribution and $\lambda = T_B/T$ for all T. Then, as $T \to \infty$:*

(a) $T(\tilde{\alpha}^i - 1) \Rightarrow H_i/K_i$,
(b) $t_{\tilde{\alpha}i} \Rightarrow (\sigma/\sigma_e)H_i/(g_iK_i)^{1/2}$,

where

$$H_A = g_A D_1 - D_5\psi_1 - D_6\psi_2; \quad K_A = g_A D_2 - D_4\psi_2 - D_3\psi_1;$$
$$H_B = g_B D_1 + D_5\psi_3 + D_8\psi_4; \quad K_B = g_B D_2 + D_7\psi_4 + D_3\psi_3;$$
$$H_C = g_C D_9 + D_{13}\psi_5 - D_{14}\psi_6; \; K_C = g_C D_{10} - D_{12}\psi_6 + D_{11}\psi_5;$$

with

$$\psi_1 = 6D_4 + 12D_3; \quad \psi_2 = 6D_3 + (1-\lambda)^{-1}\lambda^{-1}D_4;$$
$$\psi_3 = (1+2\lambda)(1-\lambda)^{-1}D_7 - (1+3\lambda)D_3;$$
$$\psi_4 = (1+2\lambda)(1-\lambda)^{-1}D_3 - (1-\lambda)^{-3}D_7;$$
$$\psi_5 = D_{12} - D_{11}; \quad \psi_6 = \psi_5 + (1-\lambda)^2 D_{12}/\lambda^3;$$

and

$$D_1 = \left(\frac{1}{2}\right)(w(1)^2 - \sigma_e^2/\sigma^2) - w(1)\int_0^1 w(r)dr;$$

$$D_2 = \int_0^1 w(r)^2 dr - \left[\int_0^1 w(r)dr\right]^2;$$

$$D_3 = \int_0^1 rw(r)dr - \left(\frac{1}{2}\right)\int_0^1 w(r)dr;$$

$$D_4 = \int_0^\lambda w(r)dr - \lambda\int_0^1 w(r)dr;$$

$$D_5 = w(1)/2 - \int_0^1 w(r)dr;$$

$$D_6 = w(\lambda) - \lambda w(1);$$

$$D_7 = \int_\lambda^1 rw(r)dr - \lambda\int_\lambda^1 w(r)dr - ((1-\lambda)^2/2)\int_0^1 w(r)dr;$$

$$D_8 = ((1-\lambda^2)/2)w(1) - \int_\lambda^1 w(r)dr;$$

$$D_9 = \int_0^1 w(r)^2 dr - \lambda^{-1}\left(\int_0^\lambda w(r)dr\right)^2 - (1-\lambda)^{-1}\left(\int_\lambda^1 w(r)dr\right)^2;$$

$$D_{10} = (w(1)^2 - \sigma_e^2/\sigma^2)/2 - \lambda^{-1}w(\lambda)\int_0^\lambda w(r)dr$$
$$- (w(1) - w(\lambda))(1-\lambda)^{-1}\int_\lambda^1 w(r)dr;$$

$$D_{11} = \int_0^1 rw(r)dr - \left(\frac{1}{2}\right)(1+\lambda)\int_0^1 w(r)dr + \left(\frac{1}{2}\right)\int_0^\lambda w(r)dr;$$

$$D_{12} = \int_0^\lambda rw(r)dr - (\lambda/2)\int_0^\lambda w(r)dr;$$

$$D_{13} = (1-\lambda)w(1)/2 + w(\lambda)/2 - \int_0^1 w(r)dr;$$

$$D_{14} = \lambda w(\lambda)/2 - \int_0^\lambda w(r)dr;$$

$$g_A = 1 - 3(1-\lambda)\lambda; \quad g_B = 3\lambda^3; \quad g_C = 12(1-\lambda)^2;$$

and where $w(r)$ *is the unit Wiener process defined on* $C[0,1]$, $\sigma^2 = \lim_{T\to\infty} E[T^{-1}S_T^2]$, $S_T = \sum_1^T e_t$, *and* $\sigma_e^2 = \lim_{T\to\infty} E[T^{-1}\sum_1^T e_t^2]$.

Theorem 2 provides a representation for the limiting distribution of the normalized least squares estimators and their t statistics in terms of functionals of Wiener processes. These limiting distributions are functions of the parameter λ, the ratio of the prebreak sample size to total sample size. It is easy to verify that when λ is either 0 or 1, the limiting distributions are identical over all models and are given by:

$$T(\tilde{\alpha}-1) \Rightarrow H/K \quad \text{and} \quad t_{\tilde{\alpha}} \Rightarrow (\sigma/\sigma_e)H/K^{1/2}$$

where

$$\begin{aligned} H = {} & \left(\frac{1}{2}\right)(w(1)^2 - \sigma_e^2/\sigma^2) + 12\left[\int_0^1 rw(r)dr - \left(\frac{1}{2}\right)\int_0^1 w(r)dr\right] \\ & \cdot \left[\int_0^1 w(r)dr - \left(\frac{1}{2}\right)w(1)\right] - w(1)\int_0^1 w(r)dr, \\ K = {} & \int_0^1 w(r)^2 dr - 12\left(\int_0^1 rw(r)dr\right)^2 \\ & + 12\int_0^1 w(r)dr\int_0^1 rw(r)dr - 4\left(\int_0^1 w(r)dr\right)^2. \end{aligned}$$

These latter asymptotic distributions correspond to those derived by Phillips and Perron (1988) in the case where no dummy variables are included.

The expressions for the limiting distributions in Theorem 2 depend on additional nuisance parameters, apart from λ, namely σ^2 and σ_e^2. As in Phillips (1987) and Phillips and Perron (1988), σ_e^2 is the variance of the innovations and σ^2 is, in the case of weakly stationary innovations, equal to $2\pi f(0)$ where $f(0)$ is the spectral density of $\{e_t\}$ evaluated at frequency zero. When the innovation sequence $\{e_t\}$ is independent and identically distributed, $\sigma^2 = \sigma_e^2$ and, in that case, the limiting distributions are invariant with respect to nuisance parameters, except λ.

Therefore, when $\sigma^2 = \sigma_e^2$, percentage points of the limiting distributions can be tabulated for given values of λ. Tables 4–9, present selected percentage points that will allow us to carry hypothesis

Table 4. Percentage points of the asymptotic distribution of $T(\tilde{\alpha}-1)$ in Model A; time of break relative to total sample size: λ.

$\lambda =$	0.1	0.2	0.3	0.4	0.5	0.6	0.7	0.8	0.9
1%	−34.17	−35.85	−35.07	−34.44	−34.07	−35.83	−35.59	−34.86	−34.65
2.5%	−28.93	−30.35	−29.92	−29.26	−29.00	−29.80	−29.61	−29.40	−29.35
5%	−25.04	−26.00	−25.90	−25.40	−25.25	−25.56	−25.99	−25.82	−25.40
10%	−21.45	−22.16	−21.93	−21.61	−21.55	−21.79	−22.33	−22.10	−21.48
90%	−4.57	−5.19	−5.13	−4.28	−3.85	−4.36	−5.15	−5.32	−4.62
95%	−3.40	−3.90	−3.80	−2.83	−2.38	−2.92	−3.86	−3.87	−3.27
97.5%	−2.35	−2.92	−2.85	−1.69	−1.42	−1.89	−2.78	−2.84	−2.13
99%	−1.28	−1.70	−1.60	−0.61	−0.40	−0.78	−1.58	−1.78	−1.39

Table 5. Percentage points of the asymptotic distribution of $t_{\tilde{\alpha}}$ in Model A; time of break relative to total sample size: λ.

$\lambda =$	0.1	0.2	0.3	0.4	0.5	0.6	0.7	0.8	0.9
1%	−4.30	−4.39	−4.39	−4.34	−4.32	−4.45	−4.42	−4.33	−4.27
2.5%	−3.93	−4.08	−4.03	−4.01	−4.01	−4.09	−4.07	−3.99	−3.97
5%	−3.68	−3.77	−3.76	−3.72	−3.76	−3.76	−3.80	−3.75	−3.69
10%	−3.40	−3.47	−3.46	−3.44	−3.46	−3.47	−3.51	−3.46	−3.38
90%	−1.38	−1.45	−1.43	−1.26	−1.17	−1.28	−1.42	−1.46	−1.37
95%	−1.09	−1.14	−1.13	−0.88	−0.79	−0.92	−1.10	−1.13	−1.04
97.5%	−0.78	−0.90	−0.83	−0.55	−0.49	−0.60	−0.82	−0.89	−0.74
99%	−0.46	−0.54	−0.51	−0.21	−0.15	−0.26	−0.50	−0.57	−0.47

testing. The critical values are obtained via simulation methods. We briefly describe the steps involved. First, we generate a sample of size 1,000 of i.i.d. $N(0,\ 1)$ random deviates, $\{e_t\}$. We then construct sample moments of the data which converge weakly to the various functionals of the Wiener process involved in the representation of the asymptotic distributions. For example, as $T \to \infty$, $T^{-1/2}\sum_1^T e_t \Rightarrow w(1)$, $T^{-1/2}\sum_1^{T_B} e_t \Rightarrow w(\lambda)$, $T^{-3/2}\sum_1^T \sum_{j=1}^t e_j \Rightarrow \int_0^1 w(r)\,dr$, $\sum_{t=1}^T (\sum_{j=1}^{t-1} e_j)e_t \Rightarrow (\frac{1}{2})(w(1)^2 - 1)$, etc. With a sample size of 1,000 and i.i.d. $N(0,\ 1)$ variates, we can expect the approximation to be quite accurate. Once the various functionals are evaluated, we construct the expressions in Theorem 2 and obtain one realization of the limiting distributions of the statistics $T(\tilde{\alpha}^i - 1)$, $t_{\tilde{\alpha}^i}$ $(i = A, B, C)$.

Table 6. Percentage points of the asymptotic distribution of $T(\tilde{\alpha}-1)$ in Model B; time of break relative to total sample size: λ.

$\lambda =$	0.1	0.2	0.3	0.4	0.5	0.6	0.7	0.8	0.9
1%	−34.34	−37.16	−38.07	−39.21	−39.77	−40.08	−38.70	−36.18	−34.69
2.5%	−28.74	−31.97	−32.78	−33.42	−33.60	−33.21	−32.31	−31.45	−29.42
5%	−25.00	−27.16	−28.61	−29.23	−29.65	−29.51	−28.68	−27.24	−25.25
10%	−21.26	−23.10	−24.20	−25.04	−25.40	−25.15	−24.30	−23.01	−21.24
90%	−4.27	−5.09	−5.92	−6.62	−6.96	−6.71	−6.08	−5.26	−4.45
95%	−3.12	−3.85	−4.50	−5.06	−5.31	−5.15	−4.59	−3.82	−3.16
97.5%	−2.20	−2.69	−3.30	−3.90	−4.14	−3.94	−3.36	−2.72	−2.21
99%	−1.11	−1.58	−2.19	−2.50	−3.01	−2.54	−2.20	−1.50	−1.24

Table 7. Percentage points of the asymptotic distribution of $t_{\tilde{\alpha}}$ in Model B; time of break relative to total sample size: λ.

$\lambda =$	0.1	0.2	0.3	0.4	0.5	0.6	0.7	0.8	0.9
1%	−4.27	−4.41	−4.51	−4.55	−4.56	−4.57	−4.51	−4.38	−4.26
2.5%	−3.94	−4.08	−4.17	−4.20	−4.26	−4.20	−4.13	−4.07	−3.96
5%	−3.65	−3.80	−3.87	−3.94	−3.96	−3.95	−3.85	−3.82	−3.68
10%	−3.36	−3.49	−3.58	−3.66	−3.68	−3.66	−3.57	−3.50	−3.35
90%	−1.35	−1.48	−1.59	−1.69	−1.74	−1.71	−1.61	−1.49	−1.34
95%	−1.04	−1.18	−1.27	−1.37	−1.40	−1.36	−1.28	−1.16	−1.04
97.5%	−0.78	−0.87	−0.97	−1.11	−1.18	−1.11	−0.97	−0.87	−0.77
99%	−0.40	−0.52	−0.69	−0.75	−0.82	−0.78	−0.67	−0.54	−0.43

We replicate this procedure 5,000 times and obtain the critical values from the sorted vector of replicated statistics. This procedure is performed for each statistic with nine values of the parameter λ, the ratio of prebreak sample size to total sample size.[4]

Several features are worth mentioning with respect to these critical values. First, as expected, for a given size of the test, the critical values are larger (in absolute value) for each model than the standard Dickey–Fuller critical values when considering the left tail. One would therefore expect a loss in power. Secondly, although the critical values are not significantly influenced by the value of the

[4]For some evidence on the adequacy of this method to obtain critical values for limiting distributions involving functions of Wiener processes, see Chan (1988).

Table 8. Percentage points of the asymptotic distribution of $T(\tilde{\alpha}-1)$ in Model C; time of break relative to total sample size: λ.

$\lambda =$	0.1	0.2	0.3	0.4	0.5	0.6	0.7	0.8	0.9
1%	−36.17	−39.97	−42.98	−45.52	−44.07	−44.75	−43.02	−41.48	−36.58
2.5%	−30.65	−34.92	−36.48	−37.12	−37.56	−37.72	−37.50	−35.16	−31.82
5%	−26.63	−29.95	−32.47	−33.22	−33.79	−33.19	−33.11	−30.70	−27.16
10%	−22.68	−25.50	−27.90	−29.39	−29.41	−29.04	−28.14	−25.79	−22.62
90%	−4.74	−5.85	−7.35	−8.43	−8.84	−8.55	−7.41	−6.17	−4.89
95%	−3.41	−4.34	−5.50	−6.67	−7.19	−6.79	−5.66	−4.52	−3.52
97.5%	−2.51	−3.19	−4.14	−5.37	−5.82	−5.47	−4.33	−3.35	−2.49
99%	−1.31	−2.14	−2.82	−3.96	−4.39	−4.24	−2.80	−2.02	−1.28

Table 9. Percentage points of the asymptotic distribution of $t_{\tilde{\alpha}}$ in Model C; time of break relative to total sample size: λ.

$\lambda =$	0.1	0.2	0.3	0.4	0.5	0.6	0.7	0.8	0.9
1%	−4.38	−4.65	−4.78	−4.81	−4.90	−4.88	−4.75	−4.70	−4.41
2.5%	−4.01	−4.32	−4.46	−4.48	−4.53	−4.49	−4.44	−4.31	−4.10
5%	−3.75	−3.99	−4.17	−4.22	−4.24	−4.24	−4.18	−4.04	−3.80
10%	−3.45	−3.66	−3.87	−3.95	−3.96	−3.95	−3.86	−3.69	−3.46
90%	−1.44	−1.60	−1.78	−1.91	−1.96	−1.93	−1.81	−1.63	−1.44
95%	−1.11	−1.27	−1.46	−1.62	−1.69	−1.63	−1.47	−1.29	−1.12
97.5%	−0.82	−0.98	−1.15	−1.35	−1.43	−1.37	−1.17	−1.04	−0.80
99%	−0.45	−0.67	−0.81	−1.04	−1.07	−1.08	−0.79	−0.64	−0.50

parameter λ, the maximum (in absolute value) occurs around the value $\lambda = 0.5$, i.e., for a break at mid-sample.[5] In the left tail, critical values of the statistics are smallest (in absolute value) when λ is close to 0 or 1. This is to be expected since, as previously mentioned, the critical values are identical to those of Dickey and Fuller when $\lambda = 0, 1$.

Some critical values are worth noticing. Consider the t statistics. Under Model (A), the 'crash hypothesis', the 5% critical value has a minimum (over values of λ) of −3.80. Under Models (B) and (C), the

[5]The simulated critical values suggest that the limiting distributions are symmetric around $\lambda = 0.5$. This feature seems intuitively plausible. We have not, however, proved that such is the case.

corresponding figures are -3.96 and -4.24 respectively. The critical values under the various models are therefore noticeably smaller than the standard Dickey–Fuller critical value of -3.41 (see Dickey, 1976; and Fuller, 1976).

These sets of results can be used to perform hypothesis testing. One simply picks the critical value corresponding to the sample value of λ at the chosen significance level. Since we only provide critical values for a selected grid of λ's, the procedure suggested is to choose the critical value corresponding to the value of λ nearest to its sample value, i.e., T_B/T. Given that the differences in the critical values over adjacent values for λ in the tables are not substantially different, this procedure should not cause misleading inferences.

4.1. *Extensions to more general error processes: Case* (1)

Using regressions (5) ($i = A, B, C$) and the critical values in Tables 4–9 is valid only in the case where the innovation sequence $\{e_t\}$ is uncorrelated. When there is additional correlation, as one would expect, an extension is necessary. Two approaches are possible. One is to follow the approach suggested by Phillips (1987) and Phillips and Perron (1988). This involves finding a set of transformed statistics that would converge weakly to the limiting distributions expressed in Theorem 2 with $\sigma^2 = \sigma_e^2$. The other approach is that suggested by Dickey and Fuller (1979) and Said and Dickey (1984).

Consider first the extension to Phillips' (1987) procedure. It is useful first to write the limiting distributions of Theorem 2 in a different, more compact form. To do so, we adopt the framework suggested by Ouliaris *et al.* (1988). Define $w_i(r)$ ($i = A, B, C$) to be the stochastic process on [0, 1] such that $w_i(r)$ is the projection residual of a Wiener process $w(r)$ on the subspace generated by the following set of functions: (1) $i = A$:1, r, $du(r)$, where $du(r) = 0$ if $r \le \lambda$ and $du(r) = 1$ if $r > \lambda$; (2) $i = B$: 1, r, $dt^*(r)$, where $dt^*(r) = 0$ if $r \le \lambda$ and $dt^*(r) = r - \lambda$ if $r > \lambda$; (3) $i = C$: 1, r, $du(r)$, $dt(r)$, where $dt(r) = 0$ if $r \le \lambda$ and $dt(r) = r$ if $r > \lambda$. Adopting this notation, an alternative representation of the limiting distributions in Theorem 2 is given by:

$$T(\tilde{\alpha}^i - 1) \Rightarrow \left(\int_0^1 w_i(r)dw(r) + \delta\right)\left(\int_0^1 w_i(r)^2 dr\right)^{-1} \quad (i = A, B, C)$$

and

$$t_{\tilde{\alpha}}i \Rightarrow (\sigma/\sigma_e)\left(\int_0^1 w_i(r)dw(r) + \delta\right)\left(\int_0^1 w_i(r)^2 dr\right)^{-1/2} \quad (i = A, B, C)$$

where $\delta = (\sigma^2 - \sigma_e^2)/(2\sigma^2)$.

Now, define $\tilde{\sigma}_i^2$ and $\tilde{\sigma}_{ei}^2$ to be, respectively, any consistent estimator of σ^2 and σ_e^2 based on the estimated residuals from regression (5)($i = A, B, C$).[6] Also define S_i^2 ($i = A, B, C$) to be the residual sum of squares from the regression y_{t-1} on (1) $i = A$: 1, t, DU_t; (2) $i = B$: 1, t, DT_t^*; and (3) $i = C$: 1, t, DU_t, DT_t. We then define the transformed statistics as:

$$Z(\tilde{\alpha}^i) = T(\tilde{\alpha}^i - 1) - T^2(\tilde{\sigma}_i^2 - \tilde{\sigma}_{ei}^2)/2S_i^2 \quad (i = A, B, C), \tag{6}$$

$$Z(t_{\tilde{\alpha}}i) = (\tilde{\sigma}_{ei}/\tilde{\sigma}_i)\, t_{\tilde{\alpha}}i - T\left(\tilde{\sigma}_i^2 - \tilde{\sigma}_{ei}^2\right)/2\tilde{\sigma}_i S_i \quad (i = A, B, C). \tag{7}$$

Following Ouliaris *et al.* (1988), it is straightforward to show that:

$$Z(\tilde{\alpha}^i) \Rightarrow \left(\int_0^1 w_i(r)dw(r)\right)\left(\int_0^1 w_i(r)^2 dr\right)^{-1} \quad (i = A, B, C) \tag{8}$$

and

$$Z(t_{\tilde{\alpha}}i) \Rightarrow \left(\int_0^1 w_i(r)dw(r)\right)\left(\int_0^1 w_i(r)^2 dr\right)^{-1/2} \quad (i = A, B, C). \tag{9}$$

The limiting distributions in (8) and (9) are those whose critical values are presented in Tables 4–9 derived using the representation given by Theorem 2.

The other approach adopts the procedures suggested by Dickey and Fuller (1979, 1981) and Said and Dickey (1984) which add extra lags of the first differences of the data as regressors in Equation (5).

---

[6] See Phillips (1987), Phillips and Perron (1988), and Perron (1988) for details.

This extended framework is characterized by the following regression (again estimated by OLS):

$$\tilde{y}_t^i = \tilde{\alpha}^i \tilde{y}_{t-1}^i + \sum_{j=1}^{k} \tilde{c}_j \Delta \tilde{y}_{t-j}^i + \tilde{e}_t \quad (i = A, B, C), \tag{10}$$

where

$$\Delta \tilde{y}_t^i = \tilde{y}_t^i - \tilde{y}_{t-1}^i.$$

In the above representation, $\tilde{\alpha}$ is the OLS estimator of α, the sum of the autoregressive coefficients and the test is again that $\alpha = 1$. The parameter k specifies the number of extra regressors added. In a simple AR(p) process, $k = p$. In a more general ARMA(p, q) process with p and q unknown, k must increase at a controlled rate with the sample size. Arguments similar to those developed by Said and Dickey can be used to show that the limiting distributions of the statistics $t_{\tilde{\alpha}^i}$ $(i = A, B, C)$ are the same when the innovation sequence is an ARMA(p, q) process and regression (10) is used, as they are when the errors are i.i.d. and regression (5) is used. However, slightly more restrictive assumptions are needed with respect to the innovation sequence $\{e_t\}$ and the truncation parameter k for this asymptotic equivalence to hold. They are detailed in the following Assumption (see Said and Dickey, 1984).

Assumption 2. (a) $A(L)e_t = B(L)v_t$; (b) v_t *is a sequence of i.i.d.* $(0, \sigma^2)$ *random variables with finite* $(4+\delta)$*th moment for some* $\delta > 0$; (c) $k \to \infty$ *and* $T^{-1}k^3 \to 0$ *as* $T \to \infty$.

4.2. *Extensions to more general error processes: Case* (2)

A possible drawback of the methods suggested above is that they imply that the change in the trend function occurs instantaneously. Given, for instance, that the Great Depression was not an instantaneous event but lasted several years, one may wish to allow for such a transition period during changes in the trend function. One way to model this is to suppose that the economy reacts gradually to a shock to the trend function.[7] Consider, for instance, Model (A) where a crash occurs. A plausible specification of the trend function, say η_t^A, is given by:

$$\eta_t^A = \mu_1 + \beta t + \gamma\psi(L)DU_t \tag{11}$$

where $\psi(L)$ is a stationary and invertible polynomial in L with $\psi(0) = 1$ and $\gamma = \mu_2 - \mu_1$. The long run change in the trend function is given by $\gamma\psi(1)$ while the immediate impact is simply γ. A similar framework holds for models (B) and (C).

One way to incorporate such a gradual change in the trend function is to suppose that the economy responds to a shock to the trend function the same way as it reacts to any other shock, i.e., to impose $\psi(L) = B(L)^{-1}A(L)$ (see Section 2). In the literature on outliers specification the framework suggested here is analogous to the so-called 'innovational outlier' model whereas the framework considered in Section 4.1 is analogous to the 'additive outlier' model (see, e.g., Tsay, 1986). We can then implement tests for the presence of a unit root in a framework that directly extends the Dickey–Fuller strategy by adding dummy variables in regression (1). The following regressions, corresponding to Models (A), (B), and (C) are constructed by nesting the corresponding models under the null and alternative hypotheses:

$$y_t = \hat{\mu}^A + \hat{\theta}^A DU_t + \hat{\beta}^A t + \hat{d}^A D(TB)_t + \hat{\alpha}^A y_{t-1} + \sum_{i=1}^{k} \hat{c}_i \Delta y_{t-i} + \hat{e}_t, \tag{12}$$

$$y_t = \hat{\mu}^B + \hat{\theta}^B DU_t + \hat{\beta}^B t + \hat{\gamma}^B DT_t^* + \hat{\alpha}^B y_{t-1} + \sum_{i=1}^{k} \hat{c}_i \Delta y_{t-i} + \hat{e}_t, \tag{13}$$

[7] Again, this treatment is analogous to the methodology proposed by Box and Tiao (1975) concerning intervention analyses.

$$y_t = \hat{\mu}^C + \hat{\theta}^C DU_t + \hat{\beta}^C t + \hat{\gamma}^C DT_t + \hat{d}^C D(TB)_t + \hat{\alpha}^C y_{t-1} + \sum_{i=1}^{k} \hat{c}_i \Delta y_{t-i} + \hat{e}_t. \tag{14}$$

The null hypothesis of a unit root imposes the following restrictions on the true parameters of each model: Model (A), the 'crash hypothesis': $\alpha^A = 1$, $\beta^A = 0$, $\theta^A = 0$; Model (B), the 'breaking slope with no crash': $\alpha^B = 1$, $\gamma^B = 0$, $\beta^B = 0$; and Model (C), where both effects are allowed: $\alpha^C = 1$, $\gamma^C = 0$, $\beta^C = 0$. Under the alternative hypothesis of a 'trend stationary' process, we expect $\alpha^A, \alpha^B, \alpha^C < 1$; β^A, β^B, $\beta^C \neq 0$; θ^A, θ^C, γ^B, $\gamma^C \neq 0$. Finally, under the alternative hypothesis, d^A, d^C, and θ^B should be close to zero while under the null hypothesis they are expected to be significantly different from zero.

The asymptotic distribution of the t statistics $t_{\tilde{\alpha}}A$ and $t_{\tilde{\alpha}}C$ in (12) and (14), are the same, respectively, as the asymptotic distribution of $t_{\tilde{\alpha}}A$ and $t_{\tilde{\alpha}}C$ in (10). However such a correspondence does not hold for the t statistic $t_{\tilde{\alpha}}B$ in (13). Apart from the one-time dummy variable $d(TB)_t$, regressions (13) and (14) are equivalent; hence the asymptotic distribution of $t_{\tilde{\alpha}}B$ is identical to the asymptotic distribution of $t_{\tilde{\alpha}}C$. This implies that in the above framework, it is not possible to test for a unit root under the maintained hypothesis that the trend function has a change in slope with the two segments joined at the time of the change. Consequently, tests for the presence of a unit root in Model (B) will have less power using regression (13) than using (10) where the asymptotic critical values of the t statistic on α are smaller (in absolute value). However, it is still possible to test for a unit root with constant drift against a trend stationary process as in Model (B) and use the critical values of Table 5. One simply runs the regression:

$$y_t = \hat{\mu}^B + \hat{\beta}^B t + \hat{\gamma}^B DT_t^* + \hat{\alpha}^B y_{t-1} + \sum_{i=1}^{k} \hat{c}_i \Delta y_{t-i} + \hat{e}_t. \tag{15}$$

The asymptotic distribution of the t statistic $t_{\tilde{\alpha}}B$ in (15) is the same as the asymptotic distribution of the t statistic $t_{\tilde{\alpha}}B$ in (10). Given that the regressor DU_t is absent from (15), this case, however, implies that the change in drift is not permitted under the null hypothesis.

Finally, note that it is possible to apply Phillips' nonparametric procedure using regressions (12)–(14) without the lagged first-differenced regressors and applying the corrections given by (6) and (7). However, such a procedure has the unattractive feature of imposing only a one-period adjustment to the change in the trend function. In the notation of (11), it imposes $\psi(L) = 1 - \psi_1 L$ where ψ_1 is the coefficient on the first lag in the polynomial $B(L)^{-1}A(L)$.

The procedures outlined in this section permit testing for the presence of a unit root in a quite general time series process which allows a one-time break in the mean of the series or its rate of growth (or both). In the next section, we apply these procedures in the specific context of breaks at the time of the 1929 crash and the 1973 oil shock.

5. Empirical Applications

We apply the test statistics derived in the previous section to the data set used by Nelson and Plosser and to the postwar quarterly real GNP series. The data set considered by Nelson and Plosser consists of 14 major macroeconomic series sampled at an annual frequency. We omit the analysis of the unemployment rate series given that it is generally perceived as being stationary. The sample varies for each series with a starting date between 1869 and 1909. However, each series ends in 1970. Given that we entertain the hypothesis that only the 1929 Great Crash and the 1973 oil price shock caused a major change in trend function, each series in this data set contains only one break. It is therefore possible to apply the tests described in the previous section. Similarly, the quarterly postwar real GNP series contains a single break as the sample goes from 1947:I to 1986:III. Following Nelson and Plosser, we consider the logarithm of each series except for the interest rate for which we use the level.

Of the 13 series in the Nelson–Plosser data set that we analyze, preliminary investigations showed that 11 were potentially well-characterized by a trend function with a constant slope but with a major change in their level occurring right after the year 1929. For these series, the maintained hypothesis is, therefore, that of Model (A) and given that the Great Crash did not occur instantaneously but lasted several years, we apply regression (12) to carry out our testing procedure. The two series that were not modeled as such are the 'real wages' and 'common stock price' series. For these series, it appeared that not only a change in the level occurred after 1929 but there was also an increase in the slope of the trend function after this date. For these reasons, the maintained hypothesis is that of Model (C), and we use regression (14) to implement our tests.

The postwar quarterly real GNP series offers yet a different picture. The 1973 oil price shock did not cause a significant drop in the level of the series. However, after that date, the slope of the trend function has sensibly decreased. This phenomenon is consistent with the much discussed slowdown in the growth rate of real GNP since the mid-seventies; see, for example, the recent symposium in the *Journal of Economic Perspectives* (1988). For these reasons, the maintained hypothesis is that of Model (B). Given the inherent difficulty in testing for a unit root allowing lagged effects for the change in the trend function, we apply regression (10) ($i = B$) to carry the testing procedure. Modeling the change in the trend function following the 1973 oil price shock as instantaneous, at least appears more plausible than the change that occurred during the Great Depression.

Table 10 presents the corresponding estimated regressions for each series along with the t statistic on the parameters for the following respective hypotheses: $\mu = 0$, $\beta = 0$, $\theta = 0$, $\gamma = 0$, $d = 0$, and $\alpha = 1$. Recall that under the hypothesis of a unit root process $\mu \neq 0$ (in general), $\beta = 0$, $\theta = 0$ (except in regression Models (C)), $\gamma = 0$, $d \neq 0$, and $\alpha = 1$. Under the alternative hypothesis of stationary fluctuations around a deterministic breaking trend function: $\mu \neq 0$, $\theta \neq 0$, $\beta \neq 0$, $\gamma \neq 0$ (in general), $d = 0$, and $\alpha < 1$.

The value of k chosen is determined by a test on the significance of the estimated coefficients $\hat{c}_i$. We actually used a fairly liberal procedure choosing a value of k equal to say k^* if the t statistic on $\hat{c}_i$ was greater than 1.60 in absolute value and the t statistic on $\hat{c}_l$ for $l > k^*$ was less than 1.60 (with a maximum value for k of 8, except for the postwar quarterly real GNP series where we used a maximum of 12). This liberal procedure is justified in the sense that including too many extra regressors of lagged first-differences does not affect the size of the test but only decreases its power. Including too few lags may have a substantial effect on the size of the test.

Consider first the series for which we applied Model (A). To evaluate the significance of the t statistic on $\hat{\alpha}$, we use the critical value presented in Table 5 with a value of λ closest to the ratio of prebreak sample size to total sample size. Of the 11 series in that group, the unit root hypothesis cannot be rejected even at the 10% level for three of them: 'consumer prices', 'interest rate' and 'velocity'. However, we can reject the null hypothesis of a unit root at that 2.5% level or better for all other eight series. We can reject it at the 1% level for the following series: 'real GNP', 'nominal GNP', 'industrial production', 'employment', and 'wages', and at the 2.5% level for the series 'real per capita GNP', 'GNP deflator', and 'money stock'. In some cases the coefficient $\hat{\alpha}$, which is an estimate of the sum of the autoregressive coefficients, is dramatically different from one. For example, it is 0.282 for 'real GNP' and 0.322 for 'industrial production'.

Given that the unit root hypothesis can be rejected for the eight series mentioned above, we can assess the significance of the other coefficients using the fact that the asymptotic distribution of their t statistic is standardized normal. In all cases, the estimated coefficients on the constant ($\hat{\mu}$), the postbreak dummy ($\hat{\theta}$), and the trend ($\hat{\beta}$) are significant at least at the 5% level. All series showed a trend function with a positive slope and a significant decrease in level just after 1929. For these eight series the coefficient on the break dummy ($\hat{d}$) is not significant. These results strongly suggest that, except for the 'consumer price', 'velocity', and 'interest rate' series, the underlying process is one of stationary fluctuations around a deterministic trend function.

Table 10. Tests for a unit root.

(a) Regression (12), Model A; $y_t = \hat{\mu} + \hat{\theta}DU_t + \hat{\beta}t + \hat{d}D(TB)_t + \hat{\alpha}y_{t-1} + \sum_{i=1}^{k} \hat{c}_i \Delta y_{t-i} + \hat{e}_t$

$T_b = 1929$	T	λ	k	$\hat{\mu}$	$t_{\hat{\mu}}$	$\hat{\theta}$	$t_{\hat{\theta}}$	$\hat{\beta}$	$t_{\hat{\beta}}$	$\hat{d}$	$t_{\hat{d}}$	$\hat{\alpha}$	$t_{\hat{\alpha}}$	$S(\hat{e})$
Real GNP	62	0.33	8	3.441	5.07	−0.189	−4.28	0.0267	5.05	−0.018	−0.30	0.282	-5.03^{a}	0.0509
Nominal GNP	62	0.33	8	5.692	5.44	−0.360	−4.77	0.0359	5.44	0.100	1.09	0.471	-5.42^{a}	0.0694
Real per capita GNP	62	0.33	7	3.325	4.11	−0.102	−2.76	0.0111	4.00	−0.070	−1.09	0.531	-4.09^{b}	0.0555
Industrial production	111	0.63	8	0.120	4.37	−0.298	−4.58	0.0323	5.42	−0.095	−0.99	0.322	-5.47^{a}	0.0875
Employment	81	0.49	7	3.402	4.54	−0.046	−2.65	0.0057	4.26	−0.025	−0.77	0.667	-4.51^{a}	0.0295
GNP deflator	82	0.49	5	0.669	4.09	−0.098	−3.16	0.0070	4.01	0.026	0.53	0.776	-4.04^{b}	0.0438
Consumer prices	111	0.63	2	0.065	1.12	−0.004	−0.21	0.0005	1.75	−0.036	−0.79	0.978	−1.28	0.0445
Wages	71	0.41	7	2.38	5.45	−0.190	−4.32	0.0197	5.37	0.085	1.36	0.619	-5.41^{a}	0.0532
Money stock	82	0.49	6	0.301	4.72	−0.071	−2.59	0.0121	4.18	0.033	0.68	0.812	-4.29^{b}	0.0440
Velocity	102	0.59	0	0.050	0.932	−0.005	−0.20	−0.0002	−0.35	−0.136	−2.01	0.941	−1.66	0.0663
Interest rate	71	0.41	2	−0.018	−0.088	−0.343	−2.06	0.0105	2.64	0.197	0.64	0.976	−0.45	0.2787

(b) Regression (14), Model C; $y_t = \hat{\mu} + \hat{\theta}DU_t + \hat{\beta}t + \hat{\gamma}DT_t + \hat{d}D(TB)_t + \hat{\alpha}y_{t-1} + \sum_{i=1}^{k} \hat{c}_i \Delta y_{t-i} + \hat{e}_t$

$T_b = 1929$	T	λ	k	$\hat{\mu}$	$t_{\hat{\mu}}$	$\hat{\theta}$	$t_{\hat{\theta}}$	$\hat{\beta}$	$t_{\hat{\beta}}$	$\hat{\gamma}$	$t_{\hat{\gamma}}$	$\hat{d}$	$t_{\hat{d}}$	$\hat{\alpha}$	$t_{\hat{\alpha}}$	$S(\hat{e})$
Common stock prices	100	0.59	1	0.353	4.09	−1.051	−4.29	0.0070	4.43	0.0139	3.98	0.128	0.76	0.718	-4.87^{b}	0.1402
Real wages	71	0.41	8	2.115	4.33	−0.190	−3.71	0.0107	3.79	0.0066	3.33	0.031	0.78	0.298	-4.28^{c}	0.0330

(c) Regression (10), Model B; $y_t = \tilde{\mu} + \tilde{\beta}t + \tilde{\gamma}DT_t^* + \tilde{y}_t$; $\tilde{y}_t = \tilde{\alpha}\tilde{y}_{t-1} + \sum_{i=1}^{k} \tilde{c}_i \Delta y_{t-i} + \tilde{e}_t$

$T_b = 1973{:}\text{I}$	T	λ	k	$\tilde{\mu}$	$t_{\tilde{\mu}}$	$\tilde{\beta}$	$t_{\tilde{\beta}}$	$\tilde{\gamma}$	$t_{\tilde{\gamma}}$	$\tilde{\alpha}$	$t_{\tilde{\alpha}}$	$S(\tilde{e})$
Quarterly real GNP	159	0.66	10	6.977	1160.51	0.0087	97.73	−0.0031	−12.06	0.86	-3.98^{c}	0.0097

Note: a, b, and c denote statistical significance at the 1%, 2.5%, and 5% level respectively.

Consider now panel (b) of Table 10 which presents the results for the 'common stock price', and 'real wages' series estimated under Model (C). We can reject the null hypothesis of a unit root at the 2.5% level for 'common stock prices' and at the 5% level for the 'real wages' series. In both cases, the constant ($\hat{\mu}$), the postbreak constant dummy ($\hat{\theta}$), the trend ($\hat{\beta}$), and the postbreak slope dummy ($\hat{\gamma}$) are highly significant, while the break dummy ($\hat{d}$) is not. The coefficients $\hat{\alpha}$ for the 'real wages' series is very low at 0.298 while for the 'common stock price' it is at 0.718 showing substantial mean reversion effects. This finding about the 'common stock price' series is particularly striking given the vast amount of theoretical and empirical studies supporting the random walk hypothesis in this situation.

Finally, panel (c) of Table 10 presents the results for the postwar quarterly real GNP series using regression (10) corresponding to Model (B). In this case, the null hypothesis that $\alpha = 1$ can be rejected at the 5% level with an estimated coefficient $\tilde{\alpha}$ equal to 0.86. This result is especially significant given the usual poor power properties of tests for a unit root against stationary alternatives when using a data set with a small span sampled frequently (see, e.g., Perron (1991) and Shiller and Perron (1985)). The other estimated coefficients in panel (c) confirm the relevance of the 'trend stationary' model versus the 'unit root' model. The coefficient on the postbreak slope dummy coefficient ($\tilde{\gamma}$) is highly significant. The estimated regression is therefore consistent with an underlying process characterized by stationary fluctuations around a deterministic trend function with a decrease in the slope after 1973.[8]

Table B.3 in Appendix B presents the estimated value for the sum of the autoregressive coefficient, α, and its t statistic for the null hypothesis $\alpha = 1$, for all values of the truncation lag parameter k

[8]Basically, the same estimates of the sum of the autoregressive coefficient α and its t statistic were obtained using regressions (13) and (14), for all values of k, showing some robustness for the results presented. We also applied Phillips' nonparametric procedure to the detrended series (Equations (6) and (7), $i = B$). These test statistics did not allow the rejection of the null hypothesis of a unit root.

between 1 and 12. In general, the results are quite robust to which value of k is selected.[9]

The results presented in this section are quite striking. The unit root hypothesis can be rejected for all but three series. To obtain these results, only a rather weak postulate needed to be imposed, namely the presence of a one-time change in the trend function. We claim that this is a weak postulate for the following reasons. As shown by Nelson and Plosser and Campbell and Mankiw, all the series analyzed have a unit root if the trend function is not allowed to change. This view implies that the 1929 crash is simply one big outlier in the innovation sequence. On the other hand, it also implies that the post-1973 growth slowdown is a succession of smaller innovations and that the mean of the innovations is different for the pre-1973 and the post-1973 period. These alternative interpretations are, we think, less appealing than the hypothesis of a break in the trend function, especially given that we allow such a break under both the null and alternative hypotheses.

Given that the 'consumer price', 'velocity', and 'interest rate' series appear to be characterized by the presence of a unit root, it seems worthwhile to see if this feature is stable in the pre and post 1929 samples. The subsample Dickey–Fuller type regressions are presented in Appendix B for values of k between 1 and 12. Consider first the 'consumer price index'. With $k = 2$, the estimated value of the sum of the autoregressive coefficients, $\tilde{\alpha}$ is 0.965 for the pre-1929 sample with a t statistic of -1.28. We therefore cannot reject the null hypothesis of a unit root for this subperiod. However, for the post-1929 sample, the picture is rather different; when $k = 7$, $\tilde{\alpha}$ has a value of 0.704 with a t statistic of -4.56 significant at the 1% level. The nonrejection of the unit root hypothesis using the full sample is due to the pre-1929 sample. After 1929, the unit root is no longer present.

[9]One notable exception is the quarterly real GNP series where the t statistic is significant at the 5% level with $k = 2$ or $k = 10$. It is significant at the 2.5% level with $k = 11$, with a value of -4.32. We choose to report the result of $k = 10$ because the 10th lagged first-difference was highly significant (t statistic of 2.29) while the 11th and 12th lags were not.

The case for the 'velocity' series has a special feature. It has been well documented that the US velocity series declined steadily until 1946, remained at a fairly constant level until 1970, before increasing (see, for example, Poole (1988) and Gould and Nelson (1974)). For this reason, we discuss applications of the standard Dickey–Fuller procedure for the following three samples: 1869–1929, 1930–1945, 1946–1970. We cannot reject the unit root hypothesis with the pre-1929 sample even though $\tilde{\alpha}$ is only 0.865 ($k = 0$). However, the picture is dramatically different using post-1929 samples. For the period 1930–1945, with $k = 1$, $\tilde{\alpha}$ is estimated at -0.011 with a t statistic (for testing $\alpha = 1$) of -3.44 which is significant at the 5% level. Though this result should be taken with caution due to the small number of observations, it is suggestive of a quite different behavior. The period 1946–1970 affords a modest amount of additional information and yields still more dramatic results: with $k = 4$, $\tilde{\alpha}$ is equal to 0.00 with a t statistic (for $\alpha = 1$) of -5.86, significant at the 1% level using the Dickey–Fuller critical values. It therefore appears that the 'velocity' and 'consumer price' series yield similar results: the presence of a unit root before 1929 but not after.

The results for the interest rate series indicate that, in this case, the unit root hypothesis cannot be rejected at usual significance levels for both subsamples (even though α is estimated at 0.540 ($k = 3$) with the pre-1929 sample). Indeed, given our previous results, we can conclude that only the 'interest rate' series is characterized by the presence of a unit root after 1929. All the other series are better construed as stationary fluctuations around a deterministic trend function for this period.

6. Discussion and Concluding Comments

When testing for the presence of a unit root in a time series of data against the hypothesis of stationary fluctuations around a deterministic trend function, the use of a long span of data has definite advantages. It allows tests with larger power compared to using a smaller span, in most cases even if the latter allows more observations (see Shiller and Perron (1985) and Perron (1991)). The drawback, however, is that a data set with a large span has more chance to include a major event which one would rather consider as an outlier or as exogenous given its relative importance. The arguments in this paper rest on the postulate that two such events have occurred in the 20th century: the 1929 Great Crash and the slowdown in growth after the oil shock of 1973. We therefore considered, as a relevant alternative, a trend function with a change in the intercept in 1929 and a change in the slope after 1973.

Let us discuss, in more detail, what are the relevant issues in drawing particular conclusions about the nature of economic fluctuations from our results. It is particularly important to put our results into perspective and also highlight what has not been shown.

The first important issue to point out is that we have not provided a formal unconditional statistical model of the time series properties of the various aggregates. A rejection of the null hypothesis of a unit root conditional on the possibility of shifts in the underlying trend function at known dates does not imply that the various series can be modeled as stationary fluctuations around a completely deterministic breaking trend function. As a matter of general principle, a rejection of the null hypothesis does not imply acceptance of a particular alternative hypothesis. However, since the tests were designed to have power against a specific class of alternative hypotheses, it is useful to look among close members of that class to propose an interesting statistical model for the various aggregates. Only with such a model is it possible to provide forecasts with appropriate standard errors.

We certainly do not entertain the view that the trend function including its changes are deterministic. This would imply that one would be able to forecast with certainty future changes. This is indeed quite unappealing. What we have in mind in specifying our class of maintained hypotheses can be parameterized as follows:

$$y_t = \eta_t + Z_t, \quad \eta_t = \mu_t + \beta_t t, \tag{16}$$

where $A(L)Z_t = B(L)e_t; e_t \sim$ i.i.d.$(0, \sigma^2)$; $\mu_t = \mu_{t-1} + V(L)v_t$, and $\beta_t = \beta_{t-1} + W(L)w_t$. Here, the Z_t's are (not necessarily stationary) deviations from the trend function η_t. The intercept and the slope of the trend functions, μ_t and β_t, are themselves random variables modeled as integrated processes with $W(L)$, $V(L)$ stationary and invertible polynomials. However, the important distinction is that the timing of the occurrence of the shocks v_t and w_t are rare relative to the sequence of innovations $\{e_t\}$, for example, poisson processes with arrival rates specified such that their occurrences are rare relative to the frequency of the realizations in the sequence $\{e_t\}$. The intuitive idea behind this type of modeling is that the coefficients of the trend function are determined by long-term economic fundamentals (e.g., the structure of the economic organization, population growth, etc.) and that these fundamentals are rarely changed. In our examples, v_t is nonzero in 1929 (the great depression) and w_t is nonzero in 1973 (the oil price shock).

In this sense, our exogeneity assumption about the changes in the trend function is a device that allows taking these shocks out of the noise function into the trend function without specific modeling of the stochastic nature of the behavior of μ_t, and β_t. It is in this sense that our approach does not provide an unconditional representation of the time series properties of the various variables.

Estimation of models of the form (16) by specifying a probability distribution for the error sequences $\{e_t, w_t, v_t\}$ is clearly an important avenue of future research. Interesting recent advances on this topic have been provided by Hamilton (1987) and Lam (1988) where the slope of the trend function is allowed to take two different values and the changes are modeled as a binomial process. However, no methods are currently available to test whether Z_t is integrated or not in this framework.[10] Problems in estimation of models of the form (16) are further compounded by the fact that, according to our view, only one nonzero realization of both v_t and w_t would be present in the data set typically available for the series of interest.

In the above framework, the purpose of this paper is to test whether Z_t is an integrated process or not, i.e., to test whether the shocks $\{e_t\}$ have persistent effects that do not vanish over a long horizon. Our approach is to remove from the noise function two events that occurred at two dates where we believe positive occurrences of the shocks $\{v_t, w_t\}$ happened and to model them as

[10]Both authors studied the behavior of the postwar quarterly real GNP with possible shifts in the slope of the trend function. Hamilton (1987) imposes Z_t to be an integrated process while Lam (1988) leaves Z_t unconstrained.

part of the trend function. The fact that we model these changes as exogenous implies that our results are conditional. That is, conditional upon the presence of a change in the trend function in 1929 and 1973, the fluctuations are transitory (i.e., Z_t is stationary).

An important direction for future research is to make this conditional result into an unconditional statement. This could, in principle, be achieved by a direct test for structural change in the trend function. In a sense, our procedure allows such a test, but conditional on a change occurring at a fixed known date. Hence, problems of pretesting and 'data mining' could be raised regarding the role of looking, *ex-post*, at the data on the choice of the date.[11] Accordingly, what is needed is a test for structural changes in the trend function occurring at unknown dates. The problem, however, is that care must be applied to ensure that the test has an adequate size under both the unit root and trend-stationary hypotheses. No such test is currently available in the literature. We hope to report, in the near future, developments in this area and applications in this context.

However, an important issue of observational equivalence could not even be settled by such a formal test for structural change. Consider, for instance, the following limiting case in the crash model.[12] A trend-stationary model with a break and where the errors have zero variance is observationally equivalent to a unit root model with drift where the errors have a high probability of being zero but are occasionally nonzero and finite. In general, when the variance of the errors is nonzero, the two models will be nearly observationally equivalent with the disturbances in the unit root model having fat tails. We are able to make a distinction in our empirical result through the mixing conditions (see Assumptions 1 and 2) which prohibit fat tailed disturbances. Any formal test for structural change would, presumably, also have to impose some mixing condition prohibiting fat-tailed disturbances, thereby not resolving this issue of near-observational equivalence.

---

[11]See, for example, Christiano (1992).

[12]This issue and its following illustration were raised by a referee.

In fact, any test for the presence of a unit root against trend-stationary alternatives is subject to another type of observational equivalence, as recently argued by Cochrane (1991) and Blough (1988). Indeed, in finite samples, any trend-stationary process is nearly observationally equivalent to a unit root process with a strong mean-reversion component, i.e., where the errors have a moving-average component with a root near minus one.[13] The fact that we reject the unit root hypothesis excluding the event of 1929 suggests that if there is a unit root at all the correlation structure of the innovation sequence must exhibit substantial mean reversion.

To sum up: trend stationary processes with a break are nearly observationally equivalent to unit root processes with strong mean-reversion and a fat-tailed distribution for the error sequence. Whichever view one adopts cannot be decided by data alone. Nevertheless, the picture under any of these views is basically the same: shocks had little, if any, persistence effect over a long horizon. Only those associated with the Great Depression and the oil price shock significantly altered the long run behavior of the series.

While choosing one view over the other is a matter of convenience for interpreting the data, it has profound implications for a multitude of statistical procedures. Indeed, under the unit root view one must ensure the validity of the procedures under fat-tailed disturbances, and at the moment very few are appropriate in a time series context. Hence for all practical purposes, it may be more advantageous to adopt the trend-stationary view with breaks and detrend our series accordingly prior to analyzing the remaining noise.

[13]This observational equivalence problem only disappears asymptotically. In other words, in finite samples, any test for a unit root with ARMA errors should have zero power. Formally, the critical values should be determined such that the test has a given fixed size over all possible values of the nuisance parameters (here, the additional correlation in the errors). Given the near observational equivalence, any such test would have zero power by definition. The unit root tests can be rationalized by arguing that we are willing to have the wrong size over some of the parameter space exactly because for all practical purposes it does not matter whether we label a series as trend stationary or difference stationary with a strong mean-reversion component.

Appendix A

The conditions imposed by Assumption 1 permit us to use a functional weak convergence result due to Herrndorf (1984). Let $S_t = \sum_{j=1}^{t} e_J(S_0 = 0)$ and define the following variable lying in the space $D[0, 1]$:

$$\begin{aligned} X_T(r) &= \sigma^{-1}T^{-1/2}S_{[Tr]} \\ &= \sigma^{-1}T^{-1/2}S_{j-1}, \quad (j-1)/T \leq r < j/T \quad (j = 1, \ldots, T) \\ X_T(1) &= \sigma^{-1}T^{-1/2}S_T. \end{aligned}$$

Herrndorf's theorem states that under the conditions of Assumption 1, $X_T(r) \Rightarrow w(r)$ where $\Rightarrow$ denotes weak convergence to the associated probability measure and $w(r)$ is the unit Wiener process defined on $C[0, 1]$, the space of all continuous functions on the interval $[0, 1]$. Following Phillips (1987) and Phillips and Perron (1988), it is easy to derive the following lemma related to functions of S_t. These results will be used in proving both Theorems 1 and 2.

Lemma A.1: *Let* $S_t = \sum_{j=1}^{t} e_j(S_0 = 0)$ *and assume that the innovation sequence* $\{e_t\}$ *satisfies the conditions of Assumption* 1. *Furthermore, let* $T_B = \lambda T$ *for all* T; *then as* $T \to \infty$:

(a) $T^{-3/2}\sum_{1}^{T_B} S_j \Rightarrow \sigma \int_0^{\lambda} w(r)dr;$

(b) $T^{-2}\sum_{1}^{T_B} S_j^2 \Rightarrow \sigma^2 \int_0^{\lambda} w(r)^2dr;$

(c) $T^{-5/2}\sum_{1}^{T_B} jS_j \Rightarrow \sigma \int_0^{\lambda} rw(r)dr;$

(d) $T^{-3/2}\sum_{1}^{T_B} je_j \Rightarrow \sigma \left(\lambda w(\lambda) - \int_0^{\lambda} w(r)dr\right);$

(e) $T^{-1}\sum_{1}^{T} S_{j-1}e_j \Rightarrow (\sigma^2/2)(w(1)^2 - \sigma_e^2/\sigma^2).$

The results in (a)–(d) are simple extensions of those in Phillips (1987) and Phillips and Perron (1988); part (e) is proved in Phillips (1987). The previous results can be recovered by simply letting $\lambda = 1$ in which case $T_B = T$. For example, to prove part (a):

$$T^{-3/2}\sum_{1}^{T_B} S_j = \sigma \sum_{1}^{T_B} \int_{i-1/T}^{i/T} T^{-1/2}\sigma^{-1} S_{[Tr]} dr$$

$$= \sigma \int_0^{\lambda} X_T(r) dr \Rightarrow \sigma \int_0^{\lambda} w(r) dr$$

using Herrndorf's weak convergence result and the continuous mapping theorem (see, e.g., Billingsley (1968)). The proofs of (b), (c), and (d) are entirely analogous.

Proof of Theorem 1: We consider first the most general model (C) in which both β (the slope) and μ (the intercept) are allowed to change, i.e.:

$$\begin{aligned} y_t &= \mu_1 + \beta_1 t + e_t \quad (0 \leq t \leq T_B). \\ y_t &= \mu_2 + (\beta_2 - \beta_1) T_B + \beta_2 t, \beta_2 t + e_t \quad (T_B < t \leq T). \end{aligned} \tag{A.1}$$

The following lemma provides a convenient representation of the sample moments of $\{y_t\}$.

Lemma A.2: *Assume that $\{y_t\}_0^T$ is generated according to (A.1) with the innovation sequence $\{e_t\}$ satisfying Assumption 1; then,*

(a)

$$\begin{aligned} \sum_{1}^{T} t y_{t-1} &= (1/6)[\beta_1(3\lambda - \lambda^3) + \beta_2(2 - 3\lambda + \lambda^3)]T^3 \\ &\quad + \left(\frac{1}{2}\right)\left[\mu_1\lambda^2 + \mu_2(1-\lambda^2) + \lambda(1-\lambda)(\beta_1 - \beta_2)\right] T^2 \\ &\quad + [(3/2)\mu_1\lambda + (3/2)(1-\lambda)\mu_2 - \mu_2 - (1/3)\beta_1\lambda \\ &\quad - (1/3)\beta_2(1-\lambda)]T + \left[T^{-3/2}\sum_{1}^{T} t e_{t-1}\right] T^{3/2} + o_p(T), \end{aligned}$$

(b)

$$\sum_1^T y_{t-1} = \left(\frac{1}{2}\right)[\beta_1(\lambda^2 - 2\lambda) + \beta_2(1 + 2\lambda - 3\lambda^2)]T^2$$

$$+ \left[\lambda\mu_1 + (1-\lambda)\mu_2 + (3/2)\beta_1\lambda - \left(\frac{1}{2}\right)\beta_2(1+3\lambda)\right]T$$

$$+ \left[T^{-1/2}\sum_1^T e_{t-1}\right]T^{1/2} - (\mu_2 - \mu_1),$$

(c)

$$\sum_1^T y_{t-1}^2 = (1/3)\left[\beta_1^2\lambda^2(3-2\lambda) + \beta_2^2(1-\lambda)^3 + 3\beta_1\beta_2\lambda(1-\lambda)^2\right]T^3$$

$$+ \left[\mu_1\beta_1\lambda^2 + \mu_2\beta_2(1-\lambda^2) - \left(\frac{1}{2}\right)(\beta_1\lambda + \beta_2(1-\lambda)^2)\right]T^2$$

$$+2\left[\lambda(\beta_1 - \beta_2)T^{-1/2}\sum_{T_B+1}^T e_t + \beta_1 T^{-3/2}\right.$$

$$\left.\times\sum_1^{T_B} te_t - \beta_2 T^{-3/2}\sum_{T_B+1}^T te_t\right]T^{3/2} + \left[\lambda\mu_1^2 + (1/6)\beta_1^2\lambda^2\right.$$

$$+\mu_1\beta_1\lambda + (1-\lambda)\mu_2^2 + (1/6)\beta_2^2(1-\lambda)$$

$$\left.+\mu_2\beta_2(1-\lambda) - 2\mu_2\beta_2 + T^{-1}\sum_1^T e_{t-1}^2\right]T + o_p(T),$$

(d)

$$\sum_1^T y_t y_{t-1} = (1/3)[\beta_1^2\lambda^2(3-2\lambda) + \beta_2^2(1-\lambda)^3$$

$$+ 3\beta_1\beta_2\lambda(1-\lambda)^2]T^3 + [\mu_1\beta_1\lambda^2 + \mu_2\beta_2(1-\lambda^2)]T^2$$

$$+2\left[\beta_1 T^{-3/2}\sum_{1}^{T_B} te_{t-1} + \beta_2 T^{-3/2}\sum_{T_B+1}^{T} te_{t-1}\right.$$

$$\left.+\lambda(\beta_1-\beta_2)T^{-1/2}\sum_{T_B+1}^{T} e_t\right] T^{3/2}$$

$$+\left[\mu_1^2\lambda+\mu_2^2(1-\lambda)-(1/3)(\lambda\beta_1^2+(1-\lambda)\beta_2^2)\right.$$

$$\left.+\lambda(\beta_2^2-\beta_1\beta_2)+\lambda\beta_1(\mu_2+\mu_1)-2\lambda\mu_2\beta_2\right]T+o_p(T).$$

It is straightforward but tedious to show that $\tilde{\alpha}$ is given by $\tilde{\alpha} = A/D$ where

$$\begin{aligned} A = {} & (T^2/2+T/2)\sum ty_{t-1}\sum y_t \\ & -(T^3/3+T^2/2+T/6)\sum y_{t-1}\sum y_t - T\sum ty_{t-1}\sum ty_t \\ & +(T^2/2+T/2)\sum y_{t-1}\sum ty_t + (T^4/12-T^2/12)\sum y_t y_{t-1} \end{aligned} \tag{A.2}$$

and

$$\begin{aligned} D = {} & (T^4/12-T^2/12)\sum y_{t-1}^2 - T\left(\sum ty_{t-1}\right)^2 \\ & +(T^2+T)\sum ty_{t-1}\sum y_{t-1} - (T^3/3+T^2/2+T/6)\left(\sum y_{t-1}\right)^2 \end{aligned} \tag{A.3}$$

where all summations run from 1 to T. To prove part (b) we derive the limits of $T^{-7}A$ and $T^{-7}D$ using Lemma A.2:

$$\begin{aligned} T^{-7}D = {} & (1/12)T^{-3}\sum y_{t-1}^2 - T^{-6}\left(\sum ty_{t-1}\right)^2 + T^{-3}\sum ty_{t-1} \\ & \cdot T^{-2}\sum y_{t-1} - (1/3)T^{-4}\left(\sum y_{t-1}\right)^2 + o_p(1) \end{aligned}$$

$$\begin{aligned}
&= (1/36)[\beta_1^2\lambda^3 + \beta_2^2(1-3\lambda+2\lambda^3) + 3\beta_2\beta_1\lambda(1-\lambda^2)] \\
&\quad - (1/36)[\beta_1\lambda(3-\lambda^2) + \beta_2(2-3\lambda+\lambda^3)]^2 \\
&\quad + (1/12)[\beta_1\lambda(3-\lambda^2) + \beta_2(2-3\lambda+\lambda^3)][\beta_1\lambda(2-\lambda) \\
&\quad + \beta_2(1-\lambda)^2] - (1/12)[\beta_1\lambda(2-\lambda) + \beta_2(1-\lambda)^2]^2 + o_p(1).
\end{aligned}$$

Simple algebra yields:

$$T^{-7}D \to (1/36)(-3\lambda^2 + 4\lambda^3 - 3\lambda^4 + 3\lambda^5 - \lambda^6)(\beta_1 - \beta_2)^2. \quad \text{(A.4)}$$

Similarly,

$$\begin{aligned}
T^{-7}A &= T^{-3}\sum ty_{t-1}\cdot T^{-2}\sum y_{t-1} - (1/3)T^{-4}\left(\sum y_{t-1}\right)^2 \\
&\quad - T^{-6}\left(\sum ty_{t-1}\right)^2 + (1/12)T^{-3}\sum y_t y_{t-1} + o_p(1) \\
&= (1/12)[\beta_1\lambda(3-\lambda^2) + \beta_2(2-3\lambda+\lambda^3)]\cdot[\beta_1\lambda(2-\lambda) \\
&\quad + \beta_2(1-\lambda)^2] - (1/12)[\beta_1\lambda(2-\lambda) + \beta_2(1-\lambda)^2]^2 \\
&\quad -(1/36)[\beta_1\lambda(3-\lambda^2) + \beta_2(2-3\lambda+\lambda^3)]^2 \\
&\quad + (1/36)[\beta_1^2\lambda^3 + \beta_2^2(1-3\lambda+2\lambda^3) + 3\beta_2\beta_1\lambda(1-\lambda^2)] + o_p(1)
\end{aligned}$$

and

$$T^{-7}A \to (1/36)(-3\lambda^2 + 4\lambda^3 - 3\lambda^4 + 3\lambda^5 - \lambda^6)(\beta_1 - \beta_2)^2.$$

Therefore, $\tilde{\alpha} = T^{-7}A/T^{-7}D \to 1$ proving part (b, i). To prove part (b, ii), note that $T(\tilde{\alpha} - 1) = T^{-6}(A - D)/T^{-7}D$. Simple manipulation yields:

$$\begin{aligned}
T^{-6}(A-D) &= (1/12)T^{-2}\sum y_t y_{t-1} - (1/12)T^{-2}\sum y_{t-1}^2 \\
&\quad + T^{-5}\sum ty_{t-1}\sum y_{t-1} + (1/6)T^{-3}y_T\sum y_{t-1}
\end{aligned}$$

$$
\begin{aligned}
&- \left(\frac{1}{2}\right) T^{-4} y_T \sum t y_{t-1} - (1/12) T^{-4} \left(\sum y_{t-1}\right)^2 + o_p(1) \\
&= (1/24)[\beta_1^2\lambda^2 + \beta_2^2(1-\lambda)^2 + 2\beta_1\beta_2\lambda(1-\lambda)] \\
&\quad + (1/12)[\beta_1\lambda(3-\lambda^2) + \beta_2(2-3\lambda+\lambda^3)][\beta_1\lambda(2-\lambda) \\
&\quad + \beta_2(1-\lambda)^2] + (1/12)[(\beta_1-\beta_2)\lambda + \beta_2]\cdot[\beta_1\lambda(2-\lambda) \\
&\quad + \beta_2(1-\lambda)^2] - (1/12)[(\beta_1-\beta_2)\lambda + \beta_2][\beta_1\lambda(3-\lambda^2) \\
&\quad + \beta_2(2-3\lambda+\lambda^3)] - (1/12)[\beta_1\lambda(2-\lambda) \\
&\quad + \beta_2(1-\lambda)^2]^2 + o_p(1)
\end{aligned}
$$

and

$$
T^{-6}(A-D) \to (1/24)(\beta_1-\beta_2)^2[-\lambda^2 + 4\lambda^3 - 5\lambda^4 + 2\lambda^5].
$$

Therefore,

$$
\begin{aligned}
T(\tilde{\alpha}-1) \to\ & (3/2)[-1+4\lambda-5\lambda^2+2\lambda^3] \\
& \cdot[-3+4\lambda-3\lambda^2+3\lambda^3-\lambda^4]^{-1}
\end{aligned}
$$

proving part (b, ii).

Proof of Part (a): The proof of part (a) is straightforward but tedious. The following arguments present the main steps. First, the expressions for the moments of $\{y_t\}$ in Lemma A.2 can be written as follows:

$$
\sum_1^T t y_{t-1} = a_1T^3 + b_1T^2 + c_1T^{3/2} + d_1T + o_p(T),
$$

$$
\sum_1^T y_{t-1} = a_2T^2 + b_2T + c_2T^{1/2} + d_2,
$$

$$
\sum_1^T y_{t-1}^2 = a_3T^3 + b_3T^2 + c_3T^{3/2} + d_3T + o_p(T),
$$

$$
\sum_1^T y_t y_{t-1} = a_4T^3 + b_4T^2 + c_4T^{3/2} + d_4T + o_p(T).
$$

The coefficients a_i, b_i, c_i, and d_i, $(i = 1, \ldots, 4)$ are given by the corresponding expressions in Lemma A.2 where the condition $\beta_1 = \beta_2 = \beta$ is imposed. From (A.3), the denominator of $\tilde{\alpha}$ is given by:

$$\begin{aligned}
T^{-5}D = {} & [a_3/12 - a_1^2 + a_1a_2 - a_2^2/3]T^2 \\
& + [b_3/12 - 2a_1b_1 + a_1b_2 + b_1a_2 + a_1a_2 - (2/3)a_2b_2 - a_2^2/2]T \\
& + [c_3/12 - 2a_1c_1 + a_1c_2 + c_1a_2 - (2/3)a_2c_2]T^{1/2} \\
& + [d_3/12 - a_3/12 - b_1^2 - 2a_1d_1 + a_1d_2 + b_1b_2 + d_1a_2 \\
& + a_1b_2 + b_1a_2 - b_2^2/3 - (2/3)a_2d_2 - a_2b_2 - a_2^2/6] + o_p(1).
\end{aligned}$$

Now, tedious algebra shows that the $0(T^2)$, $0(T)$, and $0(T^{1/2})$ coefficients all cancel out and the sum of the $0(1)$ coefficients yields the following result:

$$T^{-5}D \to (1/12)\sigma_e^2 + (1/12)[\mu_1 - \mu_2]^2[\lambda - 4\lambda^2 + 6\lambda^3 - 3\lambda^4].$$

To analyze the limit of $T^{-5}A$, the numerator of $\tilde{\alpha}$, we need the expansions for Σy_t, and $\Sigma t y_t$. These are given by:

$$\begin{aligned}
\sum_1^T t y_t = {} & a_1T^3 + (b_1 + \beta/2)T^2 + c_1T^{3/2} + (1/3)(d_1 \\
& + (5/6)\beta + \mu_2)T + o_p(T), \\
\sum_1^T y_t = {} & a_2T^2 + (b_2 + \beta)T + c_2T^{1/2} + d_2.
\end{aligned}$$

Then we can write, using (A.2):

$$\begin{aligned}
T^{-5}A = {} & [a_1a_2 - a_2^2/3 - a_1^2 + a_4/12]T^2 \\
& + [a_1a_2 + (b_2 + \beta)(a_1/2 - a_2/3) + (b_1 + \beta/2)(a_2/2 - a_1) \\
& - a_2^2/2 + b_4/12 + a_1b_2/2 - a_1b_1 - a_2b_2/3 + b_1a_2/2]T \\
& + [a_1c_2 + c_1a_2 - (2/3)a_2c_2 - 2a_1c_1 + c_4/12]T^{1/2}
\end{aligned}$$

$$+ [(b_2 + \beta)(b_1/2 + a_1/2 - b_2/3 - a_2/2)$$
$$+ (1/3)(d_1 + (5/6)\beta + \mu_2)(a_2/2 - a_1)$$
$$+ (b_1 + \beta/2)(b_2/2 + a_2/2 - b_1) + d_1a_2/2 + b_1a_2/2 - a_2d_2/3$$
$$- a_2b_2/2 - a_2^2/6 - d_1a_1 + a_1d_2/2 + d_4/12 + a_4/12].$$

Again, tedious algebra shows that the $0(T^2)$, $0(T)$, and $0(T^{1/2})$ terms all cancel out and that the sum of the coefficients remaining yields the following result:

$$T^{-5}A \to (1/12)\gamma_1 + (1/12)[\mu_1 - \mu_2]^2[\lambda - 4\lambda^2 + 6\lambda^3 - 3\lambda^4].$$

This proves part (a).

Proof of Theorem 2: We first note some invariance properties of the estimators $\tilde{\alpha}^i(i = A, B, C)$: $\tilde{\alpha}^A$ is invariant with respect to μ and d; $\tilde{\alpha}^B$ is invariant to μ_1 and μ_2; and $\tilde{\alpha}^C$ is invariant to d, μ_1 and μ_2. Hence, without loss of generality, we can study the limiting distribution of $T(\tilde{\alpha}^i - 1)$ and $t_{\tilde{\alpha}^i}$ under the null hypothesis that the sequence $\{y_t\}$ is generated according to:

$$y_t = y_{t-1} + e_t \quad (t = 1, \dots, T) \tag{A.5}$$

with the innovation sequence satisfying Assumption 1. It is also straightforward to show that under (A.5), $\tilde{\alpha}^i(i = A, B, C)$ are asymptotically equivalent to the least-squares estimators $\hat{\alpha}^i(i = A, B, C)$ in the following regressions:

$$y_t = \hat{\mu}^A + \hat{\theta}^A DU_t + \hat{\beta}^A t + \hat{\alpha}^A y_{t-1} + \hat{e}_t^A, \tag{A.6}$$
$$y_t = \hat{\mu}^B + \hat{\beta}^B t + \hat{\gamma}^B DT_t^* + \hat{\alpha}^B y_{t-1} + \hat{e}_t^B, \tag{A.7}$$
$$y_t = \hat{\mu}^C + \hat{\theta}^C DU_t + \hat{\beta}^C t + \hat{\gamma}^C DT_t + \hat{\alpha}^C y_{t-1} + \hat{e}_t^C, \tag{A.8}$$

where $DU_t = 1$ and $DT_t = DT_t^* = 0$ if $t \le T_B$ and $DU_t = 0$, $DT_t^* = t - T_B$, $DT_t = t$ if $t > T_B$. Since we are also concerned with regressions of the type (A.6)–(A.8) later in the text, we shall derive the limiting distributions concerning $\tilde{\alpha}^i$ $(i = A, B, C)$ using the representation of $\hat{\alpha}^i$ $(i = A, B, C)$. Note, however, that $\hat{\alpha}^A$ in (A.6) and $\hat{\alpha}^C$ in (A.9) are not invariant to the value of the parameter d under the null hypothesis. To achieve invariance, one must introduce a dummy variable $D(TB)_t$ taking value 1 at $t = T_B + 1$ and 0 elsewhere, as is done in regressions (12) and (14). Under the null hypothesis, $\hat{\alpha}^B$ is not invariant to a drift taking two distinct values: μ_1 and μ_2. To achieve invariance, the variable DU_t must be introduced in (A.7). However, this affects the limiting distribution of $\hat{\alpha}^B$ which becomes equivalent to that of $\hat{\alpha}^C$. Hence, one cannot analyze directly the case of a joint segmented trend function in a one-step type regression. We now turn to the proof of the theorem deriving the limiting distribution of $T(\hat{\alpha}^i - 1)$ and $t_{\hat{\alpha}^i}$ $(i = A, B, C)$ in (A.6)–(A.8).

The following Lemma provides weak convergence results for the sample moments of the data and will be used extensively. Its proof is a simple extension of Lemma A.1 and follows the methods of Phillips (1987) and Phillips and Perron (1988).

Lemma A.3: *Let $\{y_t\}_0^T$ be a stochastic process generated according to* (A.5) *with the innovation sequence satisfying Assumption* 1. *Furthermore, let $T_B = \lambda T$; then as $T \to \infty$:*

(a) $T^{-3/2} \sum_1^{T_B} y_t \Rightarrow \sigma \int_0^\lambda w(r)dr$,
(b) $T^{-5/2} \sum_1^{T_B} ty_t \Rightarrow \sigma \int_0^\lambda rw(r)dr$,
(c) $T^{-2} \sum_1^{T_B} y_t^2 \Rightarrow \sigma^2 \int_0^\lambda w(r)^2 dr$,
(d) $T^{-1} \sum_1^{T} y_{t-1} e_t \Rightarrow (\sigma^2/2)(w(1)^2 - \sigma_e^2/\sigma^2)$.

Parts (a), (b), and (c) are simple generalizations of results in Phillips (1987) and Phillips and Perron (1988) where $T_B = T$ and hence $\lambda = 1$; part (d) is proved in Phillips (1987).

Using the property that a regression of the form $y = X_1\hat{\beta}_1 + X_2\hat{\beta}_2 + \hat{e}$ yields a numerically identical estimator $\hat{\beta}_2$ as obtained in a regression of the form $y^* = X_2^*\hat{\beta}_2 + \hat{e}$ where y^* and X_2^* are projections of y and X_2, respectively, on the space spanned by the vectors in X_1, (A.6), (A.7), and (A.8), can be written as:

$$Y_{it} = \hat{\beta}^i X_{1it} + \hat{\psi}_i X_{2it} + \hat{\alpha}^i y_{it-1}^* + \hat{e}_{it} \quad (t = 1, \ldots, T; i = A, B, C), \tag{A.9}$$

where

$$\hat{\psi}_A = \hat{\theta}^A; \hat{\psi}_B = \hat{\gamma}^B; \hat{\psi}_C = \hat{\gamma}^C,$$

$$Y_{At} = Y_{Bt} = y_t - \overline{Y},$$

$$y^*_{At-1} = y^*_{Bt-1} = y_{t-1} - \overline{Y}_{-1},$$

$$X_{1At} = X_{1Bt} = t - \bar{t},$$

$$X_{1Ct} = t - c_1 \quad \text{if} \quad t \leq T_B \quad \text{and} \quad = t - c_2 \quad \text{otherwise},$$

$$X_{2At} = DU_t - \lambda,$$

$$X_{2Bt} = -\bar{t}^* \quad \text{if} \quad t \leq T_B \quad \text{and} \quad = t - T_B - \bar{t}^* \quad \text{otherwise},$$

$$X_{2Ct} = t - c_1 \quad \text{if} \quad t \leq T_B \quad \text{and} \quad = 0 \quad \text{otherwise},$$

$$Y_{Ct} = y_t - A \quad \text{if} \quad t \leq T_B \quad \text{and} \quad = y_t - B \quad \text{otherwise},$$

$$y^*_{Ct-1} = y_{t-1} - A' \quad \text{if} \quad t \leq T_B \quad \text{and} \quad = y_{t-1} - B' \quad \text{otherwise},$$

and

$$\bar{t} = T^{-1}\sum_1^T t = (T+1)/2,$$

$$\bar{t}^* = T^{-1}\sum_1^{T-T_B} t = T(1-\lambda)^2/2 + (1-\lambda)/2,$$

$$c_1 = \bar{t} - T(1-\lambda)/2, c_2 = \bar{t} + T\lambda/2,$$

$$A = \overline{Y} + \lambda^{-1}T^{-1}\sum_1^{T_B}(y_t - \overline{Y}),$$

$$B = \overline{Y} - (1-\lambda)^{-1}T^{-1}\sum_1^{T_B}(y_t - \overline{Y}),$$

$$A' = \overline{Y}_{-1} + \lambda^{-1}T^{-1}\sum_1^{T_B}(y_{t-1} - \overline{Y}_{-1}),$$

$$B' = \overline{Y}_{-1} - (1-\lambda)^{-1}T^{-1}\sum_{1}^{T_B}(y_{t-1} - \overline{Y}_{-1}),$$

$$\overline{Y} = T^{-1}\sum_{1}^{T} y_t,\ \overline{Y}_{-1} = T^{-1}\sum_{1}^{T} y_{t-1}.$$

Now, let $Y_i' = (y_{i,0}^*, \ldots, y_{i,T-1}^*)$, $E' = (e_1, \ldots, e_T), Z_i' = [X_{1i}, X_{2i}]$, $X_{1i} = (X_{1t,1}, \ldots, X_{1i,T})$, $X_{2i} = (X_{2i,1}, \ldots, X_{2i,T})$; then, under the null hypothesis,

$$\begin{aligned}\hat{\alpha}^i - 1 &= (Y_i' Y_i')^{-1}Y_i'E - (Y_i' Y_i)^{-1}Y_i' Z_i \\ &\quad \times [Z_i'(I - P_{Y_i})Z_i]^{-1}Z_i'[I - P_{Y_i}]E \qquad \text{(A.10)}\end{aligned}$$

where $P_{Y_i} = Y_i(Y_i' Y_i)^{-1}Y_i'$. Furthermore, define the following terms:

$$Z_i' Z_i = \begin{bmatrix} a_i & b_i \\ b_i & c_i \end{bmatrix},\quad Z_i' Y_i = \begin{bmatrix} H_i \\ J_i \end{bmatrix},\quad Z_i' E = \begin{bmatrix} K_i \\ L_i \end{bmatrix}.$$

Straightforward manipulation yields the following representation for $\hat{\alpha}^i - 1$:

$$\begin{aligned}\hat{\alpha}^i - 1 &= [(a_i c_i - b_i^2)Y_i' E - c_i H_i K_i + b_i J_i K_i - a_i J_i L_i + b_i H_i L_i] \\ &\quad /[(a_i c_i - b_i^2)Y_i' Y_i - a_i J_i^2 - c_i H_i^2 + 2b_i H_i J_i] \\ &\equiv E_i/F_i. \qquad \text{(A.11)}\end{aligned}$$

The quantities involved are defined as follows with their respective limits as $T \to \infty$, obtained using Lemma A.3:

$$a_A = a_B = T^3/12 - T/12,\ T^{-3}a_A = T^{-3}a_B \to 1/12,$$

$$b_A = -(1-\lambda)\lambda T^2/2,\ T^{-2}b_A \to -(1-\lambda)\lambda/2,$$

$$c_A = (1-\lambda)\lambda T,\ T^{-1}c_A \to (1-\lambda)\lambda,$$

$$Y_A' Y_A = Y_B' Y_B = \sum_{1}^{T}(y_{t-1} - \overline{Y}_{-1})^2,$$

$$T^{-2}Y_A' Y_A = T^{-2}Y_B' Y_B \Rightarrow \sigma^2\left[\int_0^1 w(r)^2 dr - \left(\int_0^1 w(r)dr\right)^2\right],$$

$$Y_A' E = Y_B' E = \sum_1^T (y_{t-1} - \overline{Y}_{-1}) e_t,$$

$$T^{-1} Y_A' E = T^{-1} Y_B' E \Rightarrow (\sigma^2/2)(w(1)^2 - \sigma_e^2/\sigma^2) - \sigma^2 w(1) \int_0^1 w(r)dr,$$

$$H_A = H_B = \sum_1^T t(y_{t-1} - \overline{Y}_{-1}),$$

$$T^{-5/2} H_A = T^{-5/2} H_B \Rightarrow \sigma \int_0^1 rw(r)dr - (\sigma/2) \int_0^1 w(r)dr,$$

$$J_A = \sum_1^{T_B} (y_{t-1} - \overline{Y}_{-1}),\ T^{-3/2} J_A$$

$$\Rightarrow \sigma \int_0^\lambda w(r)dr - \sigma\lambda \int_0^1 w(r)dr,$$

$$K_A = K_B = \sum_1^T tu_t - \bar{t} \sum_1^T u_t, T^{-3/2} K_A = T^{-3/2} K_B$$

$$\Rightarrow (\sigma/2) w(1) - \sigma \int_0^1 w(r)dr,$$

$$L_A = \sum_1^{T_B} (e_t - \overline{E}), T^{-1/2} L_A \Rightarrow \sigma w(\lambda) - \sigma\lambda w(1),$$

$$b_B = \sum_{T_{B+1}}^T (t - \bar{t})(t - T_B),\ T^{-3} b_B \to (1-\lambda)^2 (1 + 2\lambda)/12,$$

$$c_B = T_B \bar{t}^{*2} + \sum_1^{T-T_B} (t - \bar{t}^*)^2,\ T^{-3} c_B \to (1-\lambda)^3 (1 + 3\lambda)/12,$$

$$J_B = \sum_{T_B+1}^T (t - T_B)(y_{t-1} - \overline{Y}_{-1}),$$

$$T^{-5/2} J_B \Rightarrow \sigma \int_\lambda^1 rw(r)dr - \lambda\sigma \int_\lambda^1 w(r)dr - ((1-\lambda)^2/2)\sigma \int_0^1 w(r)dr,$$

$$L_B = \sum_{T_B+1}^{T} (t - T_B)(e_t - \overline{E}), T^{-3/2} L_B$$

$$\Rightarrow ((1 - \lambda^2)/2)\sigma w(1) - \sigma \int_{\lambda}^{1} w(r)dr,$$

$$a_C = \sum_{1}^{T_B} (t - c_1)^2 + \sum_{T_B+1}^{T} (t - c_2)^2, T^{-3} a_C$$

$$\to (1 - \lambda)^2/12 + \lambda^3/12,$$

$$b_C = \sum_{1}^{T_B} (t - c_1)^2, \; T^{-3} b_C \to \lambda^3/12,$$

$$c_C = \sum_{1}^{T_B} (t - c_1)^2, \; T^{-3} c_C \to \lambda^3/12,$$

$$Y_C' Y_C = \sum_{1}^{T_B} (y_{t-1} - A')^2 + \sum_{T_B+1}^{T} (y_{t-1} - B')^2,$$

$$T^{-2} Y_C' Y_C \Rightarrow \sigma^2 \left[\int_0^1 w(r)^2 dr - \lambda^{-1} \left(\int_0^{\lambda} w(r)dr \right)^2 \right.$$

$$\left. - (1 - \lambda)^{-1} \left(\int_{\lambda}^{1} w(r)dr \right)^2 \right],$$

$$Y_C' E = \sum_{1}^{T_B} (y_{t-1} - A')e_t + \sum_{T_B+1}^{T} (y_{t-1} - B')e_t,$$

$$T^{-1} Y_C' E \Rightarrow (\sigma^2/2)(w(1)^2 - \sigma_e^2/\sigma^2) - \lambda^{-1}\sigma^2 w(\lambda) \int_0^{\lambda} w(r)dr$$

$$-\sigma^2 (w(1) - w(\lambda))(1 - \lambda)^{-1} \int_{\lambda}^{1} w(r)dr,$$

$$H_C = \sum_{1}^{T_B} (t - c_1)(y_{t-1} - A') + \sum_{T_B+1}^{T} (t - c_2)(y_{t-1} - B'),$$

$$T^{-5/2}H_C \Rightarrow \sigma\left[\int_0^1 rw(r)dr - \left(\frac{1}{2}\right)(1+\lambda) \times \int_0^1 w(r)dr + \left(\frac{1}{2}\right)\int_0^\lambda w(r)dr\right],$$

$$J_C = \sum_1^{T_B}(t-c_1)(y_{t-1}-A'),\ T^{-5/2}J_C \Rightarrow \sigma\left[\int_0^\lambda rw(r)dr - (\lambda/2)\int_0^\lambda w(r)dr\right],$$

$$K_C = \sum_1^{T_B}(t-c_1)e_t + \sum_{T_B+1}^{T}(t-c_2)e_t,$$

$$T^{-3/2}K_C \Rightarrow \sigma\left[\left(\frac{1}{2}\right)(1-\lambda)w(1) + \left(\frac{1}{2}\right)w(\lambda) - \int_0^1 w(r)dr\right],$$

$$L_C = \sum_1^{T_B}(t-c_1)e_t,\ T^{-3/2}L_C \Rightarrow \sigma\left[(\lambda/2)w(\lambda) - \int_0^\lambda w(r)dr\right].$$

Now, using (A.11), we can write the statistics as follows:

$$T(\hat{\alpha}^A - 1) = T^{-5}E_A/T^{-6}F_A, \tag{A.12}$$

$$t_{\hat{\alpha}}A = T^{-5}E_A/[\hat{S}_A^2 \cdot T^{-6}F_A \cdot T^{-4}(a_Ac_A - b_A^2)]^{1/2}, \tag{A.13}$$

$$T(\hat{\alpha}^i - 1) = T^{-7}E_i/T^{-8}F_i, \quad i = B, C, \tag{A.14}$$

$$t_{\hat{\alpha}}^i = T^{-7}E_i/[\hat{S}_i^2 \cdot T^{-8}F_i \cdot T^{-6}(a_ic_i - b_i^2)]^{1/2}, i = B, C, \tag{A.15}$$

where $\hat{S}_i^2 = T^{-1}\sum_{i=1}^{T}\hat{e}_t^i$.

The results of Theorem 2 follow taking the limits of the expressions in (A.12)–(A.15) as $T \to \infty$, using the weak convergence results of the relevant moments (given above) and the fact that $\hat{S}_i^2 (i = A,\ B,\ C)$ converges in probability to σ_e^2 as $T \to \infty$.

Appendix B

Extended set of results for tests of a unit root using split and full samples (See Tables B.1–B.3).

Table B.1. Tests for a unit root on pre-1929 samples.

Regression: $y_t = \tilde{\mu} + \tilde{\beta}t + \tilde{\alpha}y_{t-1} + \sum_{i=1}^{k} \tilde{d}_i \Delta y_{t-i} + \tilde{e}_t$

		$k=1$	$k=2$	$k=3$	$k=4$	$k=5$	$k=6$	$k=7$	$k=8$	$k=9$	$k=10$	$k=11$	$k=12$
Real GNP	$\tilde{\alpha}$	0.44	0.42	0.43	0.32								
	$t_{\tilde{\alpha}}$	−2.33	−1.83	−1.43	−1.27								
Nominal GNP	$\tilde{\alpha}$	0.60	0.52	0.45	0.44	0.57							
	$t_{\tilde{\alpha}}$	−2.14	−2.04	−1.80	−1.12	−0.89							
Real per capita GNP	$\tilde{\alpha}$	0.39	0.37	0.37	0.23								
	$t_{\tilde{\alpha}}$	−2.44	−1.91	−1.51	−1.36								
Industrial production	$\tilde{\alpha}$	0.69	0.72	0.65	0.64	0.69	0.73	0.68	0.71				
	$t_{\tilde{\alpha}}$	−3.14	−2.57	−3.00	−2.83	−2.18	−1.78	−1.93	−1.43				
Employment	$\tilde{\alpha}$	0.76	0.80	0.80	0.78	0.82	0.74	0.61	0.56				
	$t_{\tilde{\alpha}}$	−2.16	−1.65	−1.72	−1.91	−1.43	−2.05	−2.85	−2.63				
GNP deflator	$\tilde{\alpha}$	0.84	0.80	0.78	0.77	0.79	0.75	0.74	0.71				
	$t_{\tilde{\alpha}}$	−2.44	−2.79	−2.66	−2.54	−2.10	−2.19	−2.05	−2.05				
Consumer prices	$\tilde{\alpha}$	0.95	0.97	0.96	0.96	0.96	0.97	0.96	0.96				
	$t_{\tilde{\alpha}}$	−1.89	−1.28	−1.37	−1.50	−1.47	−1.29	−1.37	−1.41				
Wages	$\tilde{\alpha}$	0.76	0.73	0.68	0.66	0.63	0.54	0.30	0.43				
	$t_{\tilde{\alpha}}$	−2.32	−2.22	−2.35	−2.09	−1.90	−2.08	−2.82	−1.67				
Real wages	$\tilde{\alpha}$	0.44	0.50	0.49	0.43	0.24	0.28	0.09	0.68				
	$t_{\tilde{\alpha}}$	−2.83	−2.23	−2.05	−1.89	−2.21	−1.71	−1.84	−0.63				
Money stock	$\tilde{\alpha}$	0.75	0.71	0.55	0.61	0.55	0.46	0.25	−0.04				
	$t_{\tilde{\alpha}}$	−2.88	−2.89	−4.20	−2.71	−2.53	−2.54	−3.01	−3.37				
Velocity	$\tilde{\alpha}$	0.89	0.90	0.91	0.90	0.93	0.91	0.89	0.90				
	$t_{\tilde{\alpha}}$	−1.52	−1.33	−1.22	−1.17	−0.82	−0.93	−1.12	−1.08				
Interest rate	$\tilde{\alpha}$	0.73	0.71	0.54	0.54								
	$t_{\tilde{\alpha}}$	−1.73	−1.53	−2.06	−1.59								
Common stock prices	$\tilde{\alpha}$	0.81	0.86	0.73	0.75	0.76	0.75	0.64	0.68	0.61			
	$t_{\tilde{\alpha}}$	−1.95	−1.27	−2.29	−1.83	−1.60	−1.55	−2.09	−1.75	−2.06			
Quarterly real GNP[a]	$\tilde{\alpha}$	0.93	0.91	0.92	0.93	0.93	0.92	0.91	0.90	0.90	0.88	0.86	0.89
	$t_{\tilde{\alpha}}$	−2.51	−3.02	−2.52	−2.23	−2.18	−2.41	−2.60	−2.83	−2.68	−3.04	−3.40	−2.62

Note: [a]The sample for Quarterly Real GNP is 47:I–73:I.

Table B.2. Tests for a unit root on post-1929 samples.

		Regression: $y_t = \tilde{\mu} + \tilde{\beta}t + \tilde{\alpha}y_{t-1} + \sum_{i=1}^{k} \tilde{d}_i \Delta y_{t-i} + \tilde{e}_t$											
		$k=1$	$k=2$	$k=3$	$k=4$	$k=5$	$k=6$	$k=7$	$k=8$	$k=9$	$k=10$	$k=11$	$k=12$
Real GNP	$\tilde{\alpha}$	0.72	0.75	0.75	0.78	0.74	0.66	0.56	0.33	0.22	0.18	0.30	0.09
	$t_{\tilde{\alpha}}$	−3.48	−3.00	−2.57	−1.96	−2.16	−2.65	−3.03	−5.32	−3.97	−3.19	−2.05	−2.34
Nominal GNP	$\tilde{\alpha}$	0.76	0.87	0.87	0.89	0.86	0.79	0.72	0.63	0.56	0.51	0.41	0.40
	$t_{\tilde{\alpha}}$	−3.37	−1.87	−1.60	−1.35	−1.66	−2.40	−2.93	−4.10	−3.39	−3.20	−2.91	−2.24
Real per capita GNP	$\tilde{\alpha}$	0.78	0.79	0.79	0.82	0.78	0.71	0.63	0.43	0.34	0.32	0.46	0.30
	$t_{\tilde{\alpha}}$	−3.05	−2.91	−2.55	−1.96	−2.19	−2.64	−3.03	−5.36	−3.91	−3.08	−1.89	−2.13
Industrial production	$\tilde{\alpha}$	0.68	0.69	0.66	0.63	0.71	0.64	0.53	0.23	0.03	−0.11	−0.03	−0.24
	$t_{\tilde{\alpha}}$	−2.84	−2.75	−2.65	−2.52	−1.76	−2.13	−2.55	−5.75	−4.38	−3.61	−2.40	−2.33
Employment	$\tilde{\alpha}$	0.78	0.80	0.80	0.83	0.80	0.72	0.67	0.47	0.55	0.42	0.24	0.10
	$t_{\tilde{\alpha}}$	−3.14	−2.66	−2.34	−1.80	−2.02	−2.64	−2.70	−5.61	−2.94	−3.24	−3.30	−2.79
GNP deflator	$\tilde{\alpha}$	0.85	0.92	0.91	0.87	0.87	0.83	0.72	0.76	0.75	0.81	0.75	0.68
	$t_{\tilde{\alpha}}$	−2.60	−1.32	−1.68	−2.29	−2.25	−2.64	−4.47	−3.08	−2.57	−1.73	−2.14	−2.48
Consumer prices	$\tilde{\alpha}$	0.84	0.91	0.90	0.88	0.85	0.81	0.70	0.68	0.62	0.72	0.67	0.58
	$t_{\tilde{\alpha}}$	−2.92	−1.64	−1.76	−2.10	−2.50	−2.72	−4.55	−3.45	−3.21	−1.98	−2.04	−2.39
Wages	$\tilde{\alpha}$	0.78	0.88	0.89	0.88	0.85	0.81	0.77	0.76	0.78	0.75	0.66	0.60
	$t_{\tilde{\alpha}}$	−3.37	−1.69	−1.41	−1.46	−1.77	−2.12	−2.27	−3.19	−2.02	−2.48	−3.06	−2.81
Real wages	$\tilde{\alpha}$	0.72	0.71	0.70	0.75	0.68	0.54	0.58	0.36	0.40	0.23	0.14	−0.35
	$t_{\tilde{\alpha}}$	−3.33	−3.10	−2.55	−1.84	−2.14	−2.79	−2.05	−3.31	−2.14	−2.34	−1.89	−2.11
Money stock	$\tilde{\alpha}$	0.89	0.93	0.92	0.91	0.89	0.87	0.86	0.79	0.67	0.63	0.59	0.39
	$t_{\tilde{\alpha}}$	−3.01	−1.85	−2.03	−2.55	−2.50	−2.59	−2.57	−4.57	−5.93	−3.61	−2.83	−4.05
Velocity	$\tilde{\alpha}$	0.61	0.67	0.63	0.71	0.69	0.66	0.60	0.52	0.49	0.63	0.50	0.36
	$t_{\tilde{\alpha}}$	−3.82	−2.70	−2.80	−1.97	−1.95	−1.91	−2.11	−2.33	−2.16	−1.43	−1.82	−2.54
Interest rate	$\tilde{\alpha}$	1.07	0.98	1.01	1.03	1.04	1.06	0.97					
	$t_{\tilde{\alpha}}$	1.11	−0.28	0.14	0.37	0.51	0.60	−0.25					
Common stock prices	$\tilde{\alpha}$	0.65	0.84	0.79	0.79	0.75	0.64	0.58	0.64	0.62	0.66	0.58	0.52
	$t_{\tilde{\alpha}}$	−3.48	−1.58	−1.99	−1.89	−2.13	−3.07	−2.96	−1.99	−2.00	−1.63	−1.64	−1.56
Quarterly real GNP[a]	$\tilde{\alpha}$	0.88	0.84	0.85	0.84	0.82	0.80	0.77	0.80	0.73	0.65	0.52	0.52
	$t_{\tilde{\alpha}}$	−2.23	−2.74	−2.39	−2.44	−2.48	−2.63	−2.80	−2.19	−2.78	−3.30	−4.03	−3.17

Note: [a]The sample for Quarterly Real GNP is 73:II–86:III.

Table B.3. Tests for a unit root on the full samples.

	(1)[a]		$k=1$	$k=2$	$k=3$	$k=4$	$k=5$	$k=6$	$k=7$	$k=8$	$k=9$	$k=10$	$k=11$	$k=12$
Real GNP	(12)	$\tilde{\alpha}$	0.71	0.68	0.66	0.63	0.62	0.55	0.43	0.28	0.19	0.19	0.15	0.13
		$t_{\tilde{\alpha}}$	−4.04	−4.06	−3.86	−3.73	−3.48	−3.87	−4.81	−5.03	−4.89	−4.14	−4.16	−4.20
Nominal GNP	(12)	$\tilde{\alpha}$	0.70	0.69	0.69	0.70	0.66	0.60	0.56	0.47	0.46	0.41	0.23	0.34
		$t_{\tilde{\alpha}}$	−4.43	−4.17	−3.87	−3.58	−4.01	−4.70	−4.88	−5.42	−5.18	−5.38	−7.86	−5.98
Real per capita GNP	(12)	$\tilde{\alpha}$	0.76	0.73	0.72	0.70	0.70	0.64	0.53	0.43	0.39	0.42	0.37	0.34
		$t_{\tilde{\alpha}}$	−3.62	−3.58	−3.37	−3.21	−2.94	−3.27	−4.09	−4.08	−3.89	−3.14	−3.23	−3.17
Industrial production	(12)	$\tilde{\alpha}$	0.67	0.65	0.59	0.56	0.61	0.57	0.48	0.32	0.40	−0.37	−0.29	−0.32
		$t_{\tilde{\alpha}}$	−4.63	−4.46	−4.84	−4.69	−3.83	−3.98	−4.65	−5.47	−4.15	−4.05	−4.45	−4.08
Employment	(12)	$\tilde{\alpha}$	0.78	0.80	0.77	0.76	0.78	0.73	0.67	0.60	0.64	0.62	0.59	0.58
		$t_{\tilde{\alpha}}$	−3.77	−3.29	−3.72	−3.78	−3.12	−3.94	−4.51	−4.76	−3.59	−3.58	−3.59	−3.49
GNP deflator	(12)	$\tilde{\alpha}$	0.84	0.82	0.81	0.78	0.78	0.75	0.74	0.71	0.70	0.69	0.67	0.68
		$t_{\tilde{\alpha}}$	−3.79	−3.99	−3.89	−4.16	−4.04	−4.20	−4.10	−4.32	−4.28	−4.34	−4.35	−4.00
Consumer prices	(12)	$\tilde{\alpha}$	0.97	0.98	0.97	0.97	0.97	0.97	0.97	0.96	0.96	0.96	0.97	0.97
		$t_{\tilde{\alpha}}$	−1.87	−1.28	−1.68	−2.19	−2.06	−1.89	−2.01	−2.01	−1.90	−1.93	−1.49	−1.44
Wages	(12)	$\tilde{\alpha}$	0.77	0.76	0.74	0.73	0.71	0.68	0.62	0.62	0.64	0.63	0.60	0.67
		$t_{\tilde{\alpha}}$	−4.31	−4.15	−4.25	−4.05	−4.21	−4.73	−5.41	−4.91	−4.62	−4.63	−4.69	−3.64
Real wages	(14)	$\tilde{\alpha}$	0.68	0.63	0.57	0.52	0.49	0.47	0.38	0.30	0.29	0.28	0.27	0.29
		$t_{\tilde{\alpha}}$	−3.87	−3.97	−4.11	−4.06	−3.89	−3.62	−4.02	−4.28	−4.11	−3.82	−3.64	−3.37
Money stock	(12)	$\tilde{\alpha}$	0.87	0.87	0.85	0.86	0.84	0.81	0.78	0.76	0.74	0.74	0.72	0.77
		$t_{\tilde{\alpha}}$	−3.94	−3.61	−3.86	−3.66	−3.82	−4.29	−4.69	−4.56	−4.36	−4.29	−4.32	−3.18
Velocity	(12)	$\tilde{\alpha}$	0.93	0.94	0.95	0.96	0.97	0.97	0.96	0.95	0.96	0.96	0.94	0.98
		$t_{\tilde{\alpha}}$	−1.82	−1.53	−1.41	−1.11	−0.74	−0.79	−0.94	−1.09	−0.91	−0.89	−1.22	−0.49
Interest rate	(12)	$\tilde{\alpha}$	1.01	0.98	0.97	0.98	0.98	0.99	1.004	1.01	1.01	1.03	0.99	0.97
		$t_{\tilde{\alpha}}$	0.12	−0.45	−0.53	−0.34	−0.31	−0.24	0.07	0.22	0.12	0.46	−0.08	−0.40
Common stock prices	(14)	$\tilde{\alpha}$	0.72	0.73	0.72	0.74	0.76	0.76	0.75	0.75	0.73	0.67	0.62	0.60
		$t_{\tilde{\alpha}}$	−4.87	−4.39	−4.43	−3.91	−3.53	−3.52	−3.51	−3.41	−3.52	−4.19	−4.50	−4.35
Quarterly real GNP	(10)	$\tilde{\alpha}$	0.92	0.90	0.91	0.90	0.90	0.90	0.89	0.89	0.88	0.86	0.84	0.87
	$i=B$	$t_{\tilde{\alpha}}$	−3.33	−3.97	−3.47	−3.18	−3.16	−3.39	−3.51	−3.42	−3.55	−3.98	−4.32	−3.28

Note: [a]The number in column (1) refers to the regression model used to test for a unit root using the full samples. See the discussion in the text.

Acknowledgments

The author wish to thank Brian Campbell, Larry Christiano, Jean-Marie Dufour, Clive Granger, Whitney Newey, Hashem Pesaran, the referees, and the editor for useful comments. Christian Dea and Nicholas Marceau provided useful research assistance. This research was supported by the Social Sciences and Humanities Research Council of Canada, the Natural Sciences and Engineering Council of Canada, and Québec's F.C.A.R. grants. The first draft of this paper was written while the author was Assistant Professor at the Université de Montréal.

References

Billingsley, P. (1968). *Convergence of Probability Measures.* John Wiley: New York.

Blanchard, O.J. and Summers, L.H. (1986). Hysteresis and the European Unemployment Problem. In *NBER Macroeconomics Annual 1986*, by Stanley Fisher. (ed.), M.I.T. Press: Cambridge, pp. 15–77.

Blough, S.R. (1988). On the impossibility of testing for unit roots and cointegration in finite samples. Working Paper No. 211, John Hopkins University.

Box, G.E.P. and Tiao, G.C. (1975). Intervention analysis with applications to economic and environmental problems. *Journal of the American Statistical Association* 70, 70–79.

Campbell, J.Y. and Mankiw, N.G. (1987). Permanent and transitory components in macroeconomic fluctuations. *American Economic Review, Papers and Proceedings* 77, 111–117.

Campbell, J.Y. and Mankiw, N.G. (1988). Are output fluctuations transitory? *Quarterly Journal of Economics* 102, 857–880.

Chan, N.H. (1988). The parameter inference for nearly nonstationary time series. *Journal of the American Statistical Association* 83, 857–862.

Christiano, L. J. (1992). Searching for a break in GNP. *Journal of Business & Economic Statistics*, 10, 237–250.

Cochrane, J.H. (1988). How big is the random walk in GNP. *Journal of Political Economy* 96, 893–920.

Cochrane, J.H. (1991). A critique of the application of unit root tests. *Journal of Economic Dynamics and Control*, 15(2), 275–284.

Dickey, D.A. (1976). Estimation and hypothesis testing for nonstationary time series. Unpublished Ph.D. dissertation, Iowa State University, Ames.

Dickey, D.A. and Fuller, W.A. (1979). Distribution of the estimators for autoregressive time series with a unit root. *Journal of the American Statistical Association* 74, 427–431.

Dickey, D.A. and Fuller, W.A. (1981). Likelihood ratio statistics for autoregressive time series with a unit root. *Econometrica* 49, 1057–1072.

Engle, R.F. and Granger, C.W.J. (1987). Co-integration and error correction: Representation, estimation and testing. *Econometrica* 55, 251–276.

Fuller, W.A. (1976). *Introduction to Statistical Time Series.* John Wiley and Sons: New York.

Gould, J.P. and Nelson, C.R. (1974). The stochastic structure of the velocity of money. *American Economic Review* 64, 405–418.

Hall, R.E. (1978). Stochastic implication of the life cycle — permanent income hypothesis: Theory and evidence. *Journal of Political Economy* 86, 971–987.

Hamilton, J.D. (1987). A new approach to the economic analysis of nonstationary time series and the business cycles. *Econometrica* 57, 357–384.

Herrndorf, N. (1984). A functional central limit theorem for weakly dependent sequences of random variables. *Annals of Probability* 12, 141–153.

Lam, P.S. (1988). The generalized Hamilton model: Estimation and comparison with other models of economic time series. Mimeo, The Ohio State University.

Nelson, C.R. and Plosser, C.I. (1982). Trends and random walks in macroeconomic time series. *Journal of Monetary Economics* 10, 139–162.

Ouliaris, S. Park, J.Y. and Phillips, P.C.B. (1988). Testing for a unit root in the presence of a maintained trend. In *Advances in Econometrics and Modelling,* by Baldev Raj. (ed.), Klewer Academic Publishers: Dordrecht, pp. 7–28.

Perron, P. (1991). Test consistency with varying sampling frequency. *Econometric Theory* 7, 341–368.

Perron, P. (1988). Trends and random walks in macroeconomic time series: Further evidence from a new approach. *Journal of Economic Dynamics and Control* 12, 297–332.

Phillips, P.C.B. (1987). Time series regression with unit roots. *Econometrica,* 55, 277–302.

Phillips, P.C.B. and Durlauf, S.N. (1986). Multiple time series regression with integrated processes. *Review of Economic Studies* 53, 473–496.

Phillips, P.C.B. and Perron, P. (1988). Testing for a unit root in time series regression. *Biometrika* 75, 335–346.

Poole, W. (1988). Monetary policy lessons of recent inflation and disinflation. *Journal of Economic Perspectives* 2, 73–100.

Rappoport, P. and Reichlin, L. (1989). Segmented trends and nonstationary time series. *The Economic Journal,* 99, 168–177.

Said, S.E. and Dickey, D.A. (1984). Testing for unit roots in autoregressive — moving average models of unknown order. *Biometrika* 71, 599–608.

Samuelson, P.A. (1973). Proof that properly discounted present values of assets vibrate randomly. *The Bell Journal of Economics and Management Science* 4, 369–374.

Shiller, R.J. and Perron, P. (1985). Testing the random walk hypothesis: Power versus frequency of observation. *Economics Letters* 18, 381–386.

Stock, J.H. and Watson, M.W. (1988). Testing for common trends. *Journal of the American Statistical Association* 83, 1097–1107.

Stulz, R.M. and Wasserfallen, W. (1985). Macroeconomic Time Series, Business Cycles and Macroeconomic Policies. In *Understanding Monetary Regimes*, Karl Brunner, Allan H.Metzler, (eds.), Carnegie-Rochester Conference Series on Public Policy, 22, 9–54.

Tsay, R.S. (1986). Time series model specification in the presence of outliers. *Journal of the American Statistical Association* 81, 132–141.

Wasserfallen, W. (1986). Non-stationarities in macro-economic time series — further evidence and implications. *Canadian Journal of Economics* 19, 498–510.

Erratum

In the article by P. Perron "The Great Crash, the Oil Price Shock and the Unit Root Hypothesis" (*Econometrica* 57 (1989), 1361–1401; henceforth referred to as P89), tests for a unit root allowing for the possible presence of a one-time change in the intercept and/or slope of a series were proposed. He considered two general classes of models: (a) the "additive outlier (AO) model", appropriate when the change is sudden, and (b) the "innovational outlier (IO) model", appropriate when the change is gradual. We point out that the asymptotic distributions of the statistics associated with the AO model are different from those stated in P89 and also that they depend on the correlation structure of the data when a change in intercept is involved, even if the appropriate order of the autoregression is used. A simple modification is available which yields statistics having the same asymptotic distributions (invariant to nuisance parameters) as stated earlier. None of the modifications discussed here affect the empirical results in any way. All proofs of the results stated are in Perron and Vogelsang (1993). The notation is the same as in P89, taking into account the following misprints on pp. 1373–1374:

$$H_C = g_C D_{10} + D_{13}\psi_5 - D_{14}\psi_6,\ K_C = g_C D_9 - D_{12}\psi_6 + D_{11}\psi_5,$$
$$\psi_6 = \psi_5 + (1-\lambda)^3 D_{12}/\lambda^3,\ g_B = \lambda^3/3,\ \text{and } g_C = (1-\lambda)^3/12.$$

Also on p. 1396, $T^{-3}a_C \to (1-\lambda)^3/12 + \lambda^3/12$.

Let $\{\tilde{y}_t^i; i = A, B, C\}$ be the residuals from a regression of y_t on $\{1, t, DU_t\}$ $(i = A)$, $\{1, t, DT_t^*\}$ $(i = B)$, $\{1, t, DU_t, DT_t\}$ $(i = C)$. The tests based on the AO model are $T(\tilde{\alpha}^i - 1)$ and the t statistic for the null hypothesis $\alpha = 1, t_{\tilde{\alpha}^i}$, in the second step regression:

$$\tilde{y}_t^i = \tilde{\alpha}^i \tilde{y}_{t-1}^i + \tilde{u}_t^i \quad (i = A, B, C; t = 2, \ldots, T). \tag{1}$$

The limiting result expressed in Theorem 2 of P89 are incorrect. The proof included in the appendix of P89 is, however, correct since it established the limiting distributions of similar statistics in regressions (12), (14), and (15) associated with the IO model. The asymptotic results stated in P89 for this latter case remain valid. The error occurred in making the incorrect presumption that the limiting distributions were the same in both cases. The limiting distributions of $T(\tilde{\alpha}^i - 1)$ and $t_{\tilde{\alpha}^i}$ are instead:

$$T(\tilde{\alpha}^i - 1) \Rightarrow [H_i/g_i + L_i M_i][K_i/g_i]^{-1}, \tag{2}$$

$$t_{\tilde{\alpha}^i} \Rightarrow [H_i/g_i + L_i M_i][(\sigma_e^2/\sigma^2 + L_{i^2})K_i/g_i]^{-1/2}, \tag{3}$$

for $i = A, C$, where $L_A = D_4/[(1-\lambda)\lambda] + \psi_1/(2g_A)$, $M_A = w(\lambda) - \lambda^{-1}\int_0^\lambda w(r)dr - \lambda\psi_1/(2g_A)$, $L_C = D_4/[(1-\lambda)\lambda] + 6D_{12}/\lambda^2 - (1-\lambda)\psi_5/(2g_C)$, $M_C = w(\lambda) - \lambda^{-1}\int_0^\lambda w(r)dr - 6D_{12}/\lambda^2$. Let, for $r > \lambda$, $w_B(r) = w(r) - \int_0^1 w(r)dr + (\psi_3 + \psi_4)r/g_B - \psi_3/(2g_B) - (1+\lambda^2)\psi_4/(2g_B)$. For Model B, we have:

$$T(\tilde{\alpha}^B - 1) \Rightarrow \left[H_B + \psi_4 \int_\lambda^1 w_B(r)dr\right] K_B^{-1}, \tag{4}$$

$$t_{\tilde{\alpha}^B} \Rightarrow (\sigma/\sigma_e)\left[H_B + \psi_A \int_\lambda^1 w_B(r)dr\right][g_B K_B]^{-1/2}. \tag{5}$$

Consider first results pertaining to Models A and C. Using regression (10) of P89 does not eliminate the dependency of the t statistic on nuisance parameters even when the noise function is an autoregressive process of known order. Perron and Vogelsang (1993) show that a revised procedure that avoids this problem is to estimate:

$$\tilde{y}_t^i = \sum_{j=0}^{k} \hat{\omega}_j D(TB)_{t-j} + \hat{\alpha}^i \tilde{y}_{t-1}^i + \sum_{j=1}^{k} \hat{c}_j \Delta\tilde{y}_{t-j}^i + \hat{u}_t \quad (i = A, C). \tag{6}$$

The asymptotic distributions of the t statistics for $\alpha = 1$ in (6) are those stated in Theorem 2 of P89 with $\sigma = \sigma_e$. Consider now the

Table 1. Percentage points of the asymptotic distribution of $t_{\tilde{\alpha}}$; Model B.

λ	1.0%	2.5%	5.0%	10.0%	90.0%	95.0%	97.5%	99.0%
0.1	−4.15	−3.81	−3.52	−3.23	−1.19	−0.85	−0.55	−0.22
0.2	−4.34	−4.01	−3.72	−3.41	−1.35	−1.01	−0.70	−0.33
0.3	−4.41	−4.14	−3.85	−3.54	−1.52	−1.19	−0.87	−0.53
0.4	−4.48	−4.15	−3.91	−3.61	−1.69	−1.35	−1.06	−0.68
0.5	−4.49	−4.17	−3.93	−3.65	−1.80	−1.47	−1.21	−0.85
0.6	−4.50	−4.18	−3.94	−3.65	−1.85	−1.56	−1.29	−0.96
0.7	−4.49	−4.13	−3.89	−3.60	−1.84	−1.57	−1.35	−1.06
0.8	−4.41	−4.09	−3.83	−3.55	−1.76	−1.53	−1.33	−1.11
0.9	−4.29	−3.95	−3.72	−3.42	−1.67	−1.45	−1.26	−1.06

Table 2. Percentage points of the asymptotic distribution of $T(\tilde{\alpha} - 1)$; Model B.

λ	1.0%	2.5%	5.0%	10.0%	90.0%	95.0%	97.5%	99.0%
0.1	−34.27	−29.24	−24.90	−21.00	−3.61	−2.40	−1.48	−0.59
0.2	−37.02	−31.57	−27.45	−23.09	−4.44	−3.13	−2.04	−0.92
0.3	−37.89	−33.10	−29.10	−24.55	−5.41	−3.92	−2.70	−1.55
0.4	−38.95	−33.48	−29.60	−25.33	−6.39	−4.69	−3.47	−2.09
0.5	−38.80	−33.52	−29.94	−25.57	−6.99	−5.34	−3.99	−2.70
0.6	−38.42	−33.84	−29.74	−25.57	−7.19	−5.52	−4.34	−3.01
0.7	−38.02	−33.10	−28.97	−24.69	−6.77	−5.16	−4.09	−3.12
0.8	−37.26	−32.05	−27.97	−23.79	−5.96	−4.68	−3.79	−2.87
0.9	−35.19	−30.27	−26.07	−21.94	−5.16	−4.10	−3.37	−2.57

case of Model B. The limiting distribution of the t statistic, obtained using (10) in P89 (or the $Z(t_{\tilde{\alpha}^B})$ transform in (7)), is given by (5) with $\sigma_e = \sigma$ when the noise component is an ARMA process. Similarly, the limiting distribution of the $Z(\tilde{\alpha}^B)$ transform in (6) of P89 is given by (4) with $\sigma_e = \sigma$. These limiting distributions are different from those stated in P89. A new set of critical values is presented in Tables 1 and 2. These still allow rejection of the unit root at the 5% level for the only series to which this test was applied (post-war real GNP).

This erratum was prepared by Pierre Perron and Timothy J. Vogelsang.

Reference

Perron, P. and Vogelsang, T. J. (1993). A note on the asymptotic distributions of unit root tests in the additive outlier models with breaks. *Revista de Econometria* 13, 181–201.

Chapter 13

Testing for a Unit Root in a Time Series with a Changing Mean*

Pierre Perron

Department of Economics, Princeton University
Princeton, NJ 08544, USA
Centre de Recherche et Développement en Economique
Montréal, Québec H3C 3J7, Canada

This study considers testing for a unit root in a time series characterized by a structural change in its mean level. My approach follows the 'intervention analysis' of Box and Tiao (1975) in the sense that I consider the change as being exogenous and as occurring at a known date. Standard unit-root tests are shown to be biased toward nonrejection of the hypothesis of a unit root when the full sample is used. Since tests using split sample regressions usually have low power, I design test statistics that allow the presence of a change in the mean of the series under both the null and alternative hypotheses. The limiting distribution of the statistics is derived and tabulated under the null hypothesis of a unit root. My analysis is illustrated by considering the behavior of various univariate time series for which the unit-root hypothesis has been advanced in the literature. This study complements that of Perron (1989), which considered time series with trends.

Keywords: Functional weak convergence; hypothesis testing; intervention analysis; nonstationarity; structural change.

*Originally Published in *Journal of Business and Economic Statistics*, 8, pp. 153–162.

1. Introduction

Testing for the presence of a unit root has become a problem of great concern to economists. Theoretical advances by, among others, Dickey and Fuller (1979, 1981), Fuller *et al.* (1981), Said and Dickey (1984), Phillips (1987), and Phillips and Perron (1988) have permitted the development and applications of formal tests of this hypothesis. Useful reviews and applications of these procedures can be found in the work of Dickey *et al.* (1986) and Perron (1988). The unit-root hypothesis in a time series of data has indeed far-reaching implications with respect to economic theory and the interpretation of empirical evidence. Since the seminal study of Nelson and Plosser (1982), the view that most macroeconomic time series are best construed as exhibiting some kind of stochastic nonstationarity has become prevalent. It is argued as well that the total variability of a series over time is explained in greater part by variations in permanent shocks than by variations in transitory components.

Sometimes, however, a quick glance at the graph of a time series reveals the presence of a sudden change in the mean level of the series at a given time period. This change may appear so big and sudden compared to the variability exhibited over the rest of the sample period that one may wish to isolate its effect and consider this particular period as an 'outlier event' or as exogenous. Following the 'intervention analysis' of Box and Tiao (1975), one may wish to remove a particular sudden change from the noise function and introduce it in the deterministic part of the series. The noise function is then analyzed without this particular 'extraordinary event'.

As an example, consider the behavior of the US *ex post* real-interest rate over the period 1961:1–1986:3 constructed using the rate on three-month treasury bills deflated by the consumer price index inflation rate. Figure 1 is a graph of this series. It exhibits a marked discontinuity in its mean around the year 1980. Before 1980:3, the average real rate was close to 0 ranging from +4% to –6%. Over the period 1980:4–1986:3, however, the average real rate was close to 6%, ranging from 3% to 11%, approximately. This sudden change in mean can, therefore, be viewed as 'extraordinary' and sudden, given the general historical pattern of the series. Furthermore, it

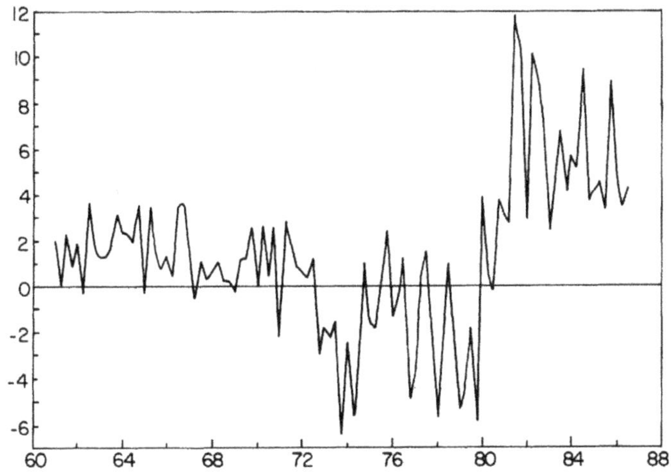

Fig. 1. *Ex post* real-interest rate (1961:1–1986:3, quarterly).

can be associated (with a slight delay) with the often-mentioned change in monetary policy initiated by the Federal Reserve Bank in October 1979. In the terminology of Box and Tiao (1975), the change in monetary policy is the 'intervention' that may have caused the sudden change in the mean of the series.

The case of the *ex post* real-interest rate has not escaped the concern of economists investigating the prevalence of stochastic nonstationarity. Recently, Walsh (1987) and Rose (1988) analyzed related series and concluded that they were characterized by the presence of a unit root. The same appears to hold with the series analyzed here. Consider a Dickey–Fuller (1979) type regression of the form

$$y_t = \mu + \alpha y_{t-1} + \sum_{i=1}^{k} c_i \Delta y_{t-i} + e_t. \tag{1}$$

Table 1 presents split-sample and full-sample Dickey–Fuller statistics for the real-interest-rate series. For the period 1961:1–1986:3 and a truncation lag parameter k equal to 3, the ordinary least squares (OLS) estimate of α is 0.872 with a t statistic for $\alpha = 1$ of -1.51; one cannot reject the null hypothesis of a unit root. If split samples are considered, however, a different picture emerges. For the

Table 1. Split-sample and full-sample Dickey–Fuller statistics.

Series/period	k	$\tilde{\mu}$	$t_{\tilde{\mu}}$	$\tilde{\alpha}$	$t_{\tilde{\alpha}}$
Real-interest rate					
1961:1–1986:3	3	0.247	0.91	0.872	−1.51
1961:1–1980:3	3	−0.028	−0.11	0.742	−1.92
1980:4–1986:3	5	7.02	2.04	−0.248	−2.19
Unemployment rate					
1948:1–1988:3	9	0.212	2.06	0.963	−2.10
1948:1–1973:4	9	0.473	2.49	0.900	−2.54
1974:1–1988:3	6	0.682	2.26	0.905	−2.33
Terms-of-trade index					
1900–1983	0	0.019	1.26	0.838	−2.64
1900–1920	4	0.126	1.31	0.784	−0.94
1921–1983	6	0.033	2.13	0.171	−4.15

Note: The regression is $y_t = \tilde{\mu} + \tilde{\alpha} y_{t-1} + \sum_{i=1}^{k} \tilde{c}_i \Delta y_{t-i} + \tilde{e}_t$.

period 1961:1–1980:3 and $k = 3$, $\hat{\alpha}$ is 0.742 with a t statistic for $\alpha = 1$ of −1.92; for the period 1980:4–1986:3 and $k = 5$, $\hat{\alpha}$ is −0.248 with a t statistic for $\alpha = 1$ of −2.19. The estimates of the sum of the autoregressive coefficients are markedly less using any split sample than the estimate using the full sample. The split samples are not large enough to permit rejection of the null hypothesis of a unit root, however.

Figure 2 presents a graph of the quarterly US unemployment series. This series has also attracted some attention, most notably in a study by Evans (1989). It can be seen that the mean of the series appears to have increased sometime in the mid-seventies. Evans considered splitting the sample into two episodes with the 1974:1 quarter as the break point. I follow the same strategy. As can be observed from the split-sample and full-sample Dickey–Fuller statistics presented in Table 1, a pattern similar to the case of the real-interest-rate series emerges. The split-sample estimates of α are below the full-sample estimate, and none of the cases allow rejection of the unit-root hypothesis. One feature that is different from the real-interest-rate case is that the change in the mean of the series is more gradual. I shall come back to the implications of this difference for my analysis.

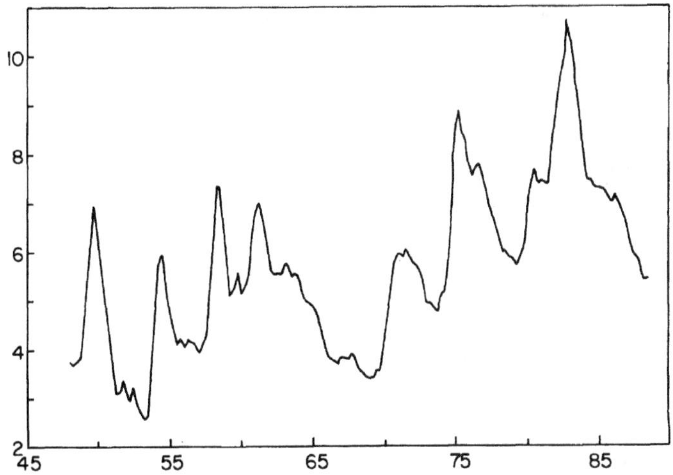

Fig. 2. US Unemployment rate (1948:1–1988:3, quarterly).

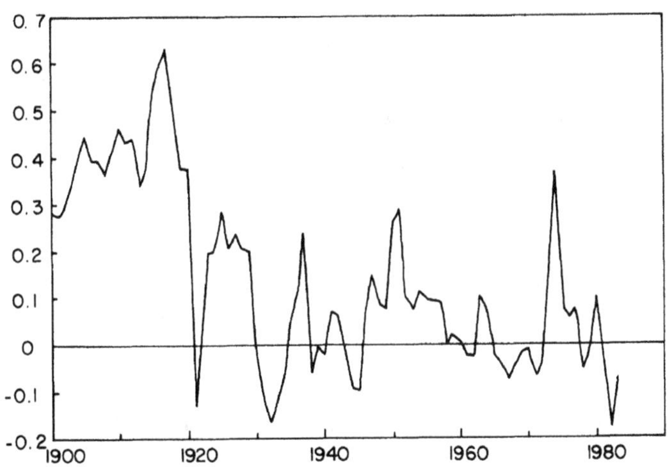

Fig. 3. Grilli–Yang commodity-price index (1900–1983, annual).

Figure 3 presents a graph of the Grilli–Yang real-commodity-price index over the period 1900–1983. This series was analyzed by Cuddington and Urzúa (1989), who investigated issues of persistence. As they pointed out, the series exhibits a marked change in mean in the year 1920. Table 1 shows that the unit-root hypothesis cannot be rejected using the full-sample estimates. It can, however, be rejected using the 1921–1983 sample at a high confidence level, but not with the 1900–1920 sample (due to a lack of power given only 21 observations).

These examples suggest the following line of investigation. First, on the supposition that the series is stationary except for this change in the mean, what are the properties of tests for a unit root as obtained, for example, from Regression (1)? Second, given that the use of split samples often implies tests with low power, can we derive test statistics for the null hypothesis of a unit root that allow the use of the full sample and permit the presence of such a change in the deterministic component? The first issue was also analyzed in a different framework by Chen and Tiao (1990). They considered random level-shift autoregressive moving average (RLARMA) models and showed how standard Box–Jenkins model-identification procedures would suggest an integrated process with the usual diagnostics indicating no misspecification. Their analysis suggests that estimated autoregressive integrated moving average models would produce forecasts with substantially higher mean squared error than an estimated RLARMA if the rate of occurrence of the shifts is small. My analysis deals with the first issue differently and puts more emphasis on the second point; that is, I concentrate on the issue of deciding whether a particular series is characterized by stationary deviations around a shifting mean function or by an integrated process in the important case in which only one shift occurs in the series.

The plan of this article is as follows: Section 2 investigates the behavior of the OLS estimate of α in Regression (1) when the data is supposed to be stationary except for a sudden change in the mean. I provide Monte Carlo evidence on the finite-sample behavior and derive the limiting distribution, which shows a nonvanishing bias. Section 3 proposes test statistics that allow such a change and derives their limiting distribution under the null hypothesis of a unit root. My method is illustrated by analyzing the series described previously.

2. Effects of a Changing Mean on Tests for a Unit Root

Under the null hypothesis, the series $\{y_t\}$ (of which a sample of size $T+1$ is available) is a realization of a time series process characterized by the presence of a unit root. The usual characterization is generalized, however, to allow a one-time change in the structure of the series occurring at a time $T_B(1 < T_B < T)$. This hypothesis can be parameterized as follows:

$$y_t = \gamma D(TB)_t + y_{t-1} + w_t, \quad t = 1, \ldots, T, \tag{2}$$

where $D(TB)_t = 1$ if $t = T_B + 1$ and 0 otherwise and $y_0 = y(0)$ is either a fixed constant or a random variable. The conditions on the sequence of innovations $\{w_t\}$ are specified such that one can use a functional central limit theorem for partial sums $\{S_t = \sum_{j=1}^{t} w_j\}$. Readers can refer to Phillips (1987) and Phillips and Perron (1988) for further details concerning these conditions. They are general enough to permit a series $\{w_t\}$ generated by any finite-order ARMA(p, q) process with Gaussian innovations.

Under the null hypothesis (2), the mean of the series $\{y_t\}$, when conditioning on the initial observation y_0, is given by y_0 up to time T_B and by $y_0 + \gamma$ afterward. Under the alternative hypothesis that the series does not contain a unit root, the model is given by

$$y_t = \mu + \gamma DU_t + e_t, \quad t = 1, \ldots, T, \tag{3}$$

where $DU_t = 0$ if $t \leq T_B$ and 1 otherwise. Again, the conditions on $\{e_t\}$ are general enough to permit an ARMA$(p+1, q)$ representation consistent with the process (2).

To assess the effects of a shift in the level of the series on tests for the presence of a unit root, I first present a small Monte Carlo experiment. Consider the behavior of the least squares estimator α in the following regression:

$$y_t = \hat{\mu} + \hat{\alpha} y_{t-1} + \hat{e}_t. \tag{4}$$

The preceding regression is one that would be used to test for a unit root $(\alpha = 1)$ if the errors were uncorrelated. Suppose, however, that the series is generated by (3) with $\mu = 0$, $T_B = 50$, $T = 100$, and $e_t \sim$ NID$(0, 1)$. We generated 10,000 replications of such a series and for each replication calculated $\hat{\alpha}$ in (4). This exercise was performed for $\gamma = 0$, 1, 2, 5, and 10. Figure 4 presents the cumulative distribution function (cdf) of $\hat{\alpha}$ in each case. The experiment reveals

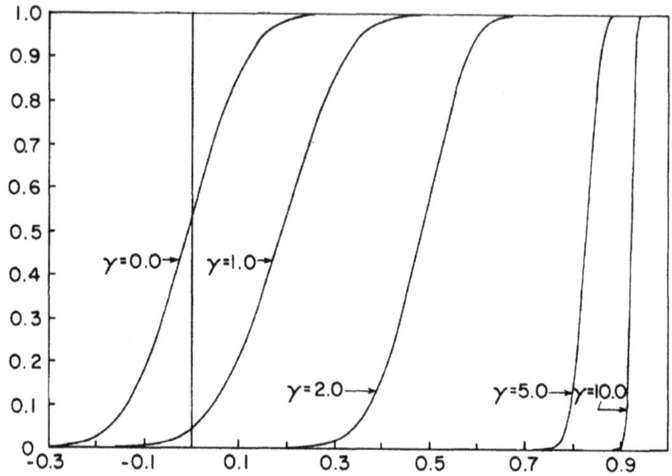

Fig. 4. Cumulative distribution function of $\hat{\alpha}$. This is the empirical cdf of $\hat{\alpha}$ in the regression $y_t = \hat{\mu} + \hat{\alpha} y_{t-1} + e_t$ when the data are generated by $y_t = \gamma DU_t + e_t$ $(t = 1, \ldots, 100)$; $DU_t = 1$ if $t > 50$ and 0 otherwise; and $e_t \sim$ i.i.d. $N(0,1)$. Simulations are based on 10,000 replications.

Table 2. Mean and variance of $\hat{\alpha}$.

γ	Mean	Variance
0.0	−0.008	0.0100
1.0	0.186	0.0131
2.0	0.478	0.0065
5.0	0.828	0.0006
10.0	0.923	0.0001

Note: The empirical mean and variance of $\hat{\alpha}$ are obtained from 10,000 replications, where $\hat{\alpha}$ is the OLS estimate in the regression $y_t = \hat{\mu} + \hat{\alpha} y_{t-1} + \hat{e}_t$. The data were generated by $y_t = \gamma DU_t + e_t$ $(t = 1, \ldots 100)$. $DU_t = 1$ if $t \geq 50$ and 0 otherwise, e_t is i.i.d. $N(0,1)$.

that as the magnitude of the change in the mean — that is, γ — increases, the cdf of $\hat{\alpha}$ becomes more concentrated at a value ever closer to 1. The corresponding means and variances of the samples of $\hat{\alpha}$ generated are shown in Table 2.

What emerges from this simulation experiment is that, if the magnitude of the change is significant, one could hardly reject the unit-root hypothesis even if the series would consist of i.i.d. disturbances around a deterministic component (albeit one with a shift in mean). In particular, one would conclude that the shocks have a permanent effect. Here, the shocks clearly have no permanent effect; only the one-time shift in the trend function is permanent. The problem is one of model misspecification.

To analyze the effect on the distribution of $\hat{\alpha}$ of an increase in the sample size with a shift of a given magnitude, I shall derive the asymptotic limit of $\hat{\alpha}$. To carry out the asymptotic analysis requires that the prebreak and postbreak samples increase at the same rate as the total number of observations, T, increases. To this effect, assume, for simplicity, that $T_B = \lambda T$ for all T. I refer to λ as the *break fraction*. The asymptotic limits are taken as T increases to infinity in a sequence that ensures an integer value for T_B for a given λ. This type of increasing sequence is assumed throughout the article.

Given these specifications and the 'mixing conditions' imposed on the sequence $\{e_t\}$, it can be shown, following Perron (1989), that, as $T \uparrow \infty$,

$$\hat{\alpha} \rightarrow [\lambda(1-\lambda)\gamma^2 + \rho_1]/[\lambda(1-\lambda)\gamma^2 + \sigma_e^2], \tag{5}$$

where $\rho_1 = \lim_{T\rightarrow\infty} T^{-1}\Sigma_1^T E(e_t e_{t-1})$, $\sigma_e^2 = \lim_{T\rightarrow\infty} T^{-1}\Sigma_1^T E(e_t^2)$, and $\rightarrow$ denotes almost sure convergence.

What the result in (5) shows is that $\hat{\alpha}$ is not a consistent estimate of the true first-order correlation coefficient of the nondeterministic part of the series $\{y_t\}$, ρ_1/σ_e^2, unless $\gamma = 0$. In particular, $\hat{\alpha}$ converges to a value greater than ρ_1/σ_e^2. This limit value approaches 1 as γ increases. Since $\hat{\alpha}$ does not converge to 1 for any fixed γ, however, the usual test statistics for testing that $\alpha = 1$, such as $T(\hat{\alpha} - 1)$ or the t statistic on $\hat{\alpha}$, would eventually reject the null hypothesis of a unit root. Nevertheless, added to the general poor power properties of tests for a unit root against stationary alternatives is now the consideration that the limit of $\hat{\alpha}$ is inflated above the true first-order correlation coefficient of the stochastic part.

The simulation experiment, along with the asymptotic result, points to the need to develop alternative statistical procedures that could distinguish a process with a unit root from a stationary series around a deterministic function with a break.

3. Tests that Allow for a Change in Mean

In this section, I extend the Dickey–Fuller testing strategy to ensure a consistent procedure when a time series is subject to a shift in its mean. I shall present several ways to do so, all of which are asymptotically equivalent, and discuss the main differences between them.

Consider first subtracting a mean from the raw series $\{y_t\}$ by allowing a change at time T_B. Let $\{\tilde{y}_t\}$ be the residuals from a regression of y_t on a constant and DU_t. Furthermore, let $\tilde{\alpha}$ be the least squares estimator of α in the following regression:

$$\tilde{y} = \tilde{\alpha}\tilde{y}_{t-1} + \tilde{e}_t, \quad t = 1, \dots, T. \tag{6}$$

Up to this point the extensions from the no-break model are straightforward enough. Matters are not so simple, however, concerning the distribution of the statistics of interest—namely, the normalized bias $T(\tilde{\alpha} - 1)$ and the t statistic for testing the null hypothesis that $\alpha = 1$, $t_{\tilde{\alpha}}$. Needless to say, the only manageable analytical distribution theory is asymptotic in nature. But two features are added over the standard Dickey–Fuller approach — (a) the presence of the extra regressor DU_t and (b) the split-sample nature of this extra regressor. To this effect, I derive the asymptotic distribution of $T(\tilde{\alpha} - 1)$ and $t_{\tilde{\alpha}}$ under the null hypothesis of a unit root. As in Section 2, I require that the break point T_B increase at the same rate as the total sample size T. Again, for simplicity, it is assumed that $T_B = \lambda T$ with both T and T_B integer valued.

The method of proof is similar to that of Phillips (1987) and Phillips and Perron (1988). I use weak convergence results that hold for normalized functions of the sum of the innovations when the latter are assumed to satisfy some appropriate 'mixing conditions'. The limiting distributions obtained under this general setting are then specialized to the i.i.d. case. The asymptotic distribution in the i.i.d. case are evaluated using simulations of functionals of Wiener processes, and critical values are tabulated. I then show how the results can be extended to innovations $\{e_t\}$ that follow the general ARMA(p, q) process.

I let $w(r)$ be a unit Wiener process defined on $C[0, 1]$, the space of all real-valued functions defined on the interval $[0, 1]$, and $\sigma^2 = \lim_{T\to\infty}[T^{-1}S_T^2]$, $S_T = \Sigma_1^T e_t$, and $\sigma_e^2 = \lim_{T\to\infty} E[T^{-1}\Sigma_1^T e_t^2]$. Denoting weak convergence in distribution by $\Rightarrow$, for $0 < \lambda < 1$,

$$T(\tilde{\alpha} - 1) \Rightarrow H/K \tag{7}$$

and

$$t_{\tilde{\alpha}} \Rightarrow (\sigma/\sigma_e)H/[\lambda(1-\lambda)K]^{1/2}, \tag{8}$$

where $H = [(1-\lambda)\lambda/2][w(1)^2 - \sigma_e^2/\sigma^2] - (1-\lambda)w(1)\int_0^\lambda w(r)dr + [w(1) - w(\lambda)][\int_0^\lambda w(r)dr - \lambda\int_0^1 w(r)dr]$ and $K = (1-\lambda)\lambda\int_0^1 w(r)^2 dr - [\int_\lambda^1 w(r)dr]^2 - (1-\lambda)[\int_0^1 w(r)dr]^2 + 2(1-\lambda)\int_0^1 w(r)dr\int_\lambda^1 w(r)dr$.

The method of proof is similar to that of Perron (1989) and is therefore omitted. Details can be found in the working-paper version of this study available on request. Equations (7) and (8) provide a representation for the limiting distribution of the normalized least squares estimator and its t statistic. These limiting distributions are functions of the parameter λ, the ratio of the prebreak sample size to total sample size. It is easy to verify that, when λ is either 0 or 1, the limiting distributions are given by $T(\tilde{\alpha} - 1) \Rightarrow H'/K'$ and $t_{\tilde{\alpha}} \Rightarrow (\sigma/\sigma_e)H'/(K')^{1/2}$, where $H' = (1/2)[w(1)^2 - \sigma_e^2/\sigma^2] - w(1)\int_0^1 w(r)dr$ and $K' = \int_0^1 w(r)^2 dr - [\int_0^1 w(r)dr]^2$.

These latter asymptotic distributions were derived by Phillips and Perron (1988) in the specific case in which no dummy variables are included. When $\lambda = 0.5$, the asymptotic distributions in (7) and (8) also correspond to the asymptotic distributions of the statistics $T(\tilde{\alpha}_{\mu d} - 1)$ and $\hat{\tau}_{\mu d}$ when $d = 2$ that were analyzed by Dickey *et al.* (1984) in the context of testing for a seasonal unit root in univariate time series.

The expressions for the limiting distributions in (7) and (8) depend on additional nuisance parameters apart from λ — namely, σ^2 and σ_e^2. As in Phillips (1987) and Phillips and Perron (1988), σ_e^2 is the variance of the innovations and σ^2 is, in the case of weakly stationary innovations, equal to $2\pi f(0)$, where $f(0)$ is the spectral density of $\{e_t\}$ evaluated at frequency 0. When the innovations $\{e_t\}$ are martingale differences, $\sigma^2 = \sigma_e^2$ and the limiting distributions are invariant with respect to nuisance parameters, except λ.

Therefore, when $\sigma^2 = \sigma_e^2$, percentage points of the limiting distributions can be tabulated for given values of λ. Furthermore, in this case, as pointed out by an associate editor, the limiting distributions are symmetric around $\lambda = 0.5$. Hence the asymptotic distributions are the same for λ and $(1-\lambda)$, and only one set of critical values needs to be tabulated. To see this, it is more useful to write the asymptotic distributions as (with $\sigma^2 = \sigma_e^2$) $T(\tilde{\alpha} - 1) \Rightarrow H^*/K^*$ and $t_{\tilde{\alpha}} \Rightarrow H^*/(\lambda(1-\lambda)K^*)^{1/2}$, where $H^* = \lambda N_1 + (1-\lambda)N_2$ and $K^* = \lambda^2 D_1 + (1-\lambda)^2 D_2$; with $N_i = \int_0^1 w_i^d(r)dw_i(r)$ and $D_i = \int_0^1 w_i^d(r)^2 dr$, $w_i^d(r) = w_i(r) - \int_0^1 w_i(r)\,dr$ is the demeaned version of the Wiener process $w_i(r)$ and $w_1(r)$ and $w_2(r)$ are independent Wiener processes. Hence the limiting distributions involve linear combinations of functionals of independent Wiener processes, the weight of the linear combination depending on the ratio λ. The asymptotic distributions are also symmetric in the general case in which $\sigma^2 \neq \sigma_e^2$ if additional conditions are imposed on the limiting variances in each subsample.

Tables 3 and 4 present selected percentage points that permit hypothesis testing. The critical values in the asymptotic case are obtained via simulation methods as in Perron (1989). I use 20,000 replications of the functionals involved in the limiting distribution (7) and (8). See that article for more details. To assess the adequacy of the asymptotic approximation, I have also included in these tables critical values of $T(\tilde{\alpha} - 1)$ and $t_{\tilde{\alpha}}$ in the finite-sample case. These were obtained using 5,000 replications of (6). In general, the asymptotic distribution is an adequate approximation to the finite-sample distribution.

Table 3. Percentage points of the distribution of $T(\tilde{\alpha} - 1)$.

	1.0%	2.5%	5.0%	10.0%	90.0%	95.0%	97.5%	99.0%
$\lambda = 0.1, 0.9$								
$T = 50$	−21.76	−18.07	−15.04	−12.09	−1.12	−0.33	0.30	0.99
$T = 100$	−21.85	−18.77	−16.14	−13.0	−1.09	−0.28	0.45	0.98
$T = 200$	−22.32	−18.80	−15.99	−12.72	−1.02	−0.29	0.35	1.04
$T = \infty$	−23.79	−19.96	−16.64	−13.36	−1.17	−0.34	0.27	1.00
$\lambda = 0.2, 0.8$								
$T = 50$	−22.95	−19.33	−16.51	−13.68	−1.44	−0.51	0.14	0.78
$T = 100$	−24.19	−20.08	−17.20	−14.33	−1.52	−0.56	0.18	0.91
$T = 200$	−24.76	−20.23	−17.05	−14.21	−1.37	−0.53	0.16	1.10
$T = \infty$	−25.03	−21.12	−18.02	−14.69	−1.55	−0.61	0.10	0.90
$\lambda = 0.3, 0.7$								
$T = 50$	−23.79	−20.10	−17.04	−14.12	−1.81	−0.79	−0.03	0.76
$T = 100$	−24.78	−20.76	−17.69	−14.84	−1.92	−0.86	−0.08	0.73
$T = 200$	−25.11	−21.01	−18.34	−15.06	−1.72	−0.78	−0.01	0.95
$T = \infty$	−25.90	−21.66	−18.55	−15.37	−1.94	−0.93	−0.14	0.71
$\lambda = 0.4, 0.6$								
$T = 50$	−24.33	−19.83	−17.0	−14.40	−2.07	−1.0	−0.32	0.80
$T = 100$	−23.98	−21.09	−18.11	−15.06	−2.07	−1.11	−0.27	0.58
$T = 200$	−24.90	−21.78	−18.50	−15.21	−2.13	−1.10	−0.25	0.57
$T = \infty$	−26.21	−22.24	−18.97	−15.71	−2.24	−1.19	−0.36	0.54
$\lambda = 0.5$								
$T = 50$	−23.45	−20.10	−17.50	−14.57	−2.18	−1.18	−0.20	0.89
$T = 100$	−25.38	−21.11	−18.41	−15.20	−2.34	−1.29	−0.37	0.54
$T = 200$	−25.10	−21.39	−18.50	−15.41	−2.27	−1.17	−0.32	0.71
$T = \infty$	−26.07	−22.06	−18.95	−15.76	−2.39	−1.35	−0.52	0.38

Notes: The entries for $\lambda = 0.5$ and $T = \infty$ are taken from Dickey *et al.* (1984, Table 6). The other entries corresponding to $T = \infty$ were obtained using 20,000 simulations based on the asymptotic distribution (7). The remaining entries, corresponding to the finite-sample cases, were obtained by directly simulating $T(\tilde{\alpha} - 1)$ from (6). In each case, 5,000 replications of a unit-root process with i.i.d. $N(0, 1)$ innovations were used.

Several features are worth mentioning with respect to these critical values. First, as expected, for a given size of the test, the critical values are larger (in absolute value) than the standard Dickey–Fuller critical values in the left tail of the distribution (see Fuller 1976). One would, therefore, expect a loss in power when

Table 4. Percentage points of the distribution of $t_{\tilde{\alpha}}$.

	1.0%	2.5%	5.0%	10.0%	90.0%	95.0%	97.5%	99.0%
$\lambda = 0.1, 0.9$								
$T = 50$	−3.90	−3.46	−3.12	−2.76	−0.51	−0.15	0.15	0.48
$T = 100$	−3.77	−3.40	−3.09	−2.78	−0.51	−0.15	0.21	0.60
$T = 200$	−3.58	−3.32	−3.06	−2.75	−0.48	−0.14	0.21	0.57
$T = \infty$	−3.67	−3.37	−3.10	−2.78	−0.55	−0.17	0.14	0.53
$\lambda = 0.2, 0.8$								
$T = 50$	−4.04	−3.65	−3.30	−2.92	−0.60	−0.24	0.06	0.36
$T = 100$	−3.86	−3.54	−3.22	−2.91	−0.64	−0.25	0.05	0.45
$T = 200$	−3.85	−3.50	−3.20	−2.89	−0.63	−0.24	0.07	0.52
$T = \infty$	−3.80	−3.49	−3.23	−2.92	−0.67	−0.28	0.04	0.44
$\lambda = 0.3, 0.7$								
$T = 50$	−4.14	−3.76	−3.39	−3.05	−0.66	−0.32	−0.01	0.34
$T = 100$	−4.05	−3.66	−3.33	−3.02	−0.73	−0.35	−0.03	0.34
$T = 200$	−3.91	−3.58	−3.34	−3.0	−0.71	−0.33	−0.00	0.39
$T = \infty$	−3.88	−3.56	−3.30	−2.99	−0.76	−0.40	−0.06	0.31
$\lambda = 0.4, 0.6$								
$T = 50$	−4.11	−3.71	−3.43	−3.08	−0.74	−0.37	−0.11	0.26
$T = 100$	−4.03	−3.68	−3.38	−3.05	−0.74	−0.42	−0.10	0.23
$T = 200$	−3.95	−3.65	−3.34	−3.02	−0.76	−0.43	−0.11	0.32
$T = \infty$	−3.92	−3.60	−3.35	−3.05	−0.81	−0.46	−0.15	0.22
$\lambda = 0.5$								
$T = 50$	−4.09	−3.72	−3.45	−3.08	−0.77	−0.41	−0.07	0.31
$T = 100$	−4.04	−3.70	−3.38	−3.08	−0.82	−0.49	−0.15	0.21
$T = 200$	−4.12	−3.61	−3.34	−3.03	−0.79	−0.45	−0.11	0.30
$T = \infty$	−3.90	−3.60	−3.34	−3.04	−0.84	−0.50	−0.20	0.15

Notes: The entries for $\lambda = 0.5$ and $T = \infty$ are taken from Dickey *et al.* (1984, Table 7). The other entries corresponding to $T = \infty$ were obtained using 20,000 simulations based on the asymptotic distribution (8). The remaining entries, corresponding to the finite-sample cases, were obtained by directly simulating $t_{\tilde{\alpha}}$ from (6). For each case, 5,000 replications of a unit-root process with i.i.d. $N(0, 1)$ innovations were used.

the alternative is a stationary process. This feature could imply that when λ is close to 0 or 1 a more powerful testing procedure could be obtained by applying the Dickey–Fuller statistics on the larger subsample. A case in point is the example of the Grilli–Yang commodity-price index discussed in Section 1. As Table 1 showed, the unit-root hypothesis is easily rejected using the 1921–1983 sample (with a p value lower than 0.01). As will be seen in Section 4, one can still reject the unit-root hypothesis using a test allowing for a changing mean but at a lower significance level (the exact p value depending on the statistic used).

Second, again in the left tail of the distribution, the critical values attain a maximum (in absolute value) around the value $\lambda = 0.5$ — that is, for a break at mid-sample. Critical values of the statistics are smallest when λ is close to 0 or 1. This is to be expected, since, as previously mentioned, the critical values are identical to those of Dickey and Fuller (1979) when $\lambda = 0, 1$.

These sets of results can be used to perform hypothesis testing. One simply picks the critical value corresponding to the sample value of λ at the chosen significance level. Since I only provide critical values for a selected grid of λ's, the procedure suggested is to choose the critical value corresponding to the value of λ nearest its sample value — that is, T_B/T. Given that the differences in the critical values over adjacent values of λ in the tables are not substantially different, this procedure should not produce misleading inferences.

3.1. *Extensions to more general error processes*

Inferences based on Regression (6), with critical values given in Tables 3 and 4, are valid only in the case in which the innovation sequence $\{e_t\}$ is uncorrelated. When there is additional correlation, as would often be expected, extensions are necessary. The first extension concerns the way in which the dynamics of the process are reflected in the change in the mean of the series. Insights into this problem can be obtained using the literature on specification of time series processes in the presence of outliers (e.g., see Tiao 1985; Tsay 1986). As stated in Equations (2) and (3), the model corresponds to the case of an additive outlier. In the case of a unit-root process, there is a single outlier at time T_B that instantaneously changes the level of the series in a permanent way. In the case of a stationary process, there is also an instantaneous change that is permanent in the level of the series. This implies that the change in the mean of the process is not affected by its dynamics.

On the other hand, the change in the mean of the series need not be instantaneous and may be affected by the dynamic specification of the noise process. In analogy with the literature on outliers in time series, such a case can be modeled by an innovational-outlier model. Under the null hypothesis, this model takes the form

$$y_t = y_{t-1} + A(L)^{-1}B(L)[v_t + \gamma D(TB)_t], \quad t = 1, \ldots, T, \tag{9}$$

where we specify an ARMA(p, q) process for e_t of the form $A(L)e_t = B(L)v_t$ with $A(L)$ and $B(L)$, pth- and qth-order polynomials in L, respectively, and $v_t \sim$ i.i.d. $(0, \sigma_v^2)$. Under the alternative hypothesis that the process does not contain a unit root, the process (3) becomes

$$y_t = \mu + A(L)^{-1}B(L)[v_t + \gamma DU_t], \quad t = 1, \ldots, T. \tag{10}$$

Equations (9) and (10) specify that the change in the mean of the series is not instantaneous but depends on the dynamic specification of the error process. The immediate impact of the change is given by γ and the longterm change is $A(1)^{-1}B(1)\gamma$.

The rest of this section specifies the appropriate testing procedure to be used in what I shall refer to as the *additive-outlier model* and the *innovational-outlier model.* In each case, however, the asymptotic distributions of the test statistics are the same and require only the critical values in Tables 3 and 4.

3.2. *The additive-outlier model*

The first approach adopts the procedure suggested by Dickey and Fuller (1979) and Said and Dickey (1984) that adds extra lags of the first differences of the data as regressors in Equation (6). This extended framework is characterized by the following regression (again estimated by OLS):

$$\tilde{y}_t = \alpha\tilde{y}_{t-1} + \sum_{j=1}^{k} c_j \Delta\tilde{y}_{t-j} + v_t, \quad t = k+1, \ldots, T, \tag{11}$$

where $\Delta\tilde{y}_t = \tilde{y}_t - \tilde{y}_{t-1}$.

In the preceding representation, α is the sum of the autoregressive coefficients and the test is again that $\alpha = 1$. The parameter k specifies the number of extra regressors added. In a simple AR(p) process, $k = p$. In a more general ARMA(p, q) process with p and q unknown, k must increase at a controlled rate with the sample size. We denote by α^* the OLS estimator of α obtained from Regression (11) and by t_{α^*} its associated t statistic for testing the null hypothesis that $\alpha = 1$. Arguments similar to those developed by Said and Dickey (1984) can be used to show that the limiting distribution of the t statistic t_{α^*} is the same when the innovation sequence is an ARMA(p, q) process and Regression (11) is used as it is when the t statistic $t_{\tilde{\alpha}}$ from Regression (6) is used and we have i.i.d. errors, provided k increases at a suitable rate with the sample size. See Said and Dickey (1984) for the exact set of conditions under which this equivalence holds.

The second approach considers extensions to the procedure suggested by Phillips (1987) and Phillips and Perron (1988). It is useful first to write the limiting distributions in (7) and (8) in a different, more compact form. To do so, I adopt the framework suggested by Ouliaris *et al.* (1989). Define $w^*(r)$ as a stochastic process on $C[0, 1]$, the space of all real-valued continuous functions on the interval $[0, 1]$, such that $w^*(r)$ is the projection residual of a Wiener process $w(r)$ on the subspace generated by the functions $\{1, du(r)\}$, where $du(r) = 1$ if $r > \lambda$ and 0 otherwise. Adopting this notation, an alternative representation of the limiting distributions in (7) and (8) is given by

$$T(\tilde{\alpha} - 1) \Rightarrow \left(\int_0^1 w * (r) dw(r) + \delta \right) \left(\int_0^1 w * (r)^2 dr \right)^{-1}$$

and

$$t_{\tilde{\alpha}} \Rightarrow (\sigma/\sigma_e) \left(\int_0^1 w * (r) dw(r) + \delta \right) \left(\int_0^1 w * (r)^2 dr \right)^{-1/2},$$

where $\delta = (\sigma^2 - \sigma_e^2)/(2\sigma^2)$.

Now define $\tilde{\sigma}^2$ and $\tilde{\sigma}_e^2$ as, respectively, any consistent estimator of σ^2 and σ_e^2 based on the estimated residuals from Regression (6). [See, among others, Perron (1988) for a discussion about the construction of such estimators.] Moreover, define S_*^2 as the residual sum of squares from the regression of y_{t-1} on a constant and DU_t. We then define the transformed statistics as

$$Z(\tilde{\alpha}) = T(\tilde{\alpha} - 1) - T^2(\tilde{\sigma}^2 - \tilde{\sigma}_e^2)/(2S_*^2) \tag{12}$$

and

$$Z(t_{\tilde{\alpha}}) = (\tilde{\sigma}_e/\tilde{\sigma})t_{\tilde{\alpha}} - T(\tilde{\sigma}^2 - \tilde{\sigma}_e^2)/(2\tilde{\sigma}S_*). \tag{13}$$

Following Ouliaris *et al.* (1989), it is straightforward to show that

$$Z(\tilde{\alpha}) \Rightarrow \left(\int_0^1 w^*(r)\, dw(r)\right)\left(\int_0^1 w^*(r)^2\, dr\right)^{-1} \tag{14}$$

and

$$Z(t_{\tilde{\alpha}}) \Rightarrow \left(\int_0^1 w^*(r)\, dw(r)\right)\left(\int_0^1 w^*(r)^2\, dr\right)^{-1/2} \tag{15}$$

The limiting distributions in (14) and (15) are those whose critical values are presented in Tables 3 and 4, derived using the representations given by (7) and (8).

3.3. *The innovational-outlier model*

Testing the null hypothesis of a unit root in the innovational-outlier model specified by Equations (9) and (10) can be achieved by considering the following regression, estimated by OLS:

$$y_t = \mu + \gamma DU_t + dD(TB)_t + \alpha y_{t-1} + \sum_{j=1}^{k} c_j \Delta y_{t-j} + v_t. \tag{16}$$

We denote by $\hat{\alpha}$ the OLS estimator of α in (16) and by $t_{\hat{\alpha}}$ its associated t statistic for testing $\alpha = 1$. Equation (16) is similar to the two-steps regression procedure given by (6) and the prior detrending. It involves only a one-step regression by estimating the trend function and the dynamics of the process simultaneously, however. Such a specification implies that the change in the mean of the series does not occur instantaneously, and its effect on the level of y_t depends on the dynamics of the process. From standard arguments used in proving (7) and (8) and the results of Said and Dickey (1984), it is clear that the asymptotic distribution of $t_{\hat{\alpha}}$ in (16) is the same as the asymptotic distribution of t_{α^*} in (11); hence the critical values from Table 4 can again be used for inferences.

4. Applications and Discussions

I applied my testing procedure to the three series discussed in Section 1. Results are presented in Tables 5, 6, and 7 for the statistics t_{α^*}, $t_{\hat{\alpha}}$, $Z(\tilde{\alpha})$, and $Z(t_{\tilde{\alpha}})$ applied to the real-interest-rate series, the unemployment series, and the terms-of-trade-index series. For the statistics $Z(\tilde{\alpha})$ and $Z(t_{\tilde{\alpha}})$, I need consistent estimators of σ^2 and σ_e^2. I used $\tilde{\sigma}_e^2 = T^{-1}\sum_1^T \tilde{e}_t^2$, where $\tilde{e}_t^2$ are the residuals from Regression (6). To construct a consistent estimator of σ^2, I follow a procedure similar to that of Newey and West (1987), using a

Table 5. Full-sample unit-root tests with changing mean: Additive-outlier method.

Series	T	$\tilde{\mu}$	$t_{\tilde{\mu}}$	$\tilde{\gamma}$	$t_{\tilde{\gamma}}$	k	α^*	t_{α^*}
Interest rate	103	0.079	0.03	5.56	9.45	3	0.632	−2.68
Unemployment rate	163	4.77	40.23	2.47	12.54	6	0.875	−3.87*
Terms-of-trade index	84	0.412	16.59	−0.37	−12.73	1	0.498	−5.25*

Note: The regressions are $y_t = \tilde{\mu} + \tilde{\gamma} DU_t + \tilde{y}_t$ and $\tilde{y}_t = \alpha^* \tilde{y}_{t-1} + \sum_{i=1}^{k} c_i^* \Delta\tilde{y}_{t-i} + e_t^*$.
*A statistic significant at the 5% level.

Table 6. Full-sample unit-root tests with changing mean: Additive-outlier method, using Phillips–Perron statistics.

Series	Statistic	*l* = 1	6	12	18	24
Interest rate	$Z(\tilde{\alpha})$	−69.32*	−90.09*	−114.12*	−128.55*	−141.97*
	$Z(t_{\tilde{\alpha}})$	−7.35*	−7.87*	−8.51*	−8.90*	−9.25*
Unemployment rate	$Z(\tilde{\alpha})$	−18.83*	−21.82*	−16.76	−12.50	−11.08
	$Z(t_{\tilde{\alpha}})$	−3.05	−3.28	−2.87	−2.47	−2.32
Terms-of-trade index	$Z(\tilde{\alpha})$	−35.57*	−27.41*	−21.26*	−19.69*	−16.93
	$Z(t_{\tilde{\alpha}})$	−4.64*	−4.23*	−3.92*	−3.86*	−3.78*

Note: *A statistic significant at the 5% level.

Table 7. Full-sample unit-root tests with changing mean: Innovational-outlier method.

Series	k	$\hat{\mu}$	$t_{\hat{\mu}}$	$\hat{\gamma}$	$t_{\hat{\gamma}}$	$\hat{d}$	$t_{\hat{d}}$	$\hat{\alpha}$	$t_{\hat{\alpha}}$
Interest rate	2	0.003	0.01	2.78	3.10	0.130	0.05	0.530	-3.71^*
Unemployment rate	9	0.538	3.74	0.284	3.19	0.079	0.25	0.886	-3.84^*
Terms-of-trade index	8	0.173	2.57	−0.162	−2.69	−0.424	−3.87	0.652	−2.46

Note: The regression is $y_t = \tilde{\mu} + \tilde{\gamma} DU_t + \hat{d} D(TB)_t + \hat{\alpha} y_{t-1} + \sum_{i=1}^{k} \hat{c}_i \Delta y_{t-i} + \hat{e}_t$.
*A statistic significant at the 5% level.

triangular Bartlett window. The statistic takes the form

$$\tilde{\sigma}^2 = T^{-1} \sum_{1}^{T} \tilde{e}_t^2 + 2T^{-1} \sum_{\tau=1}^{l} \omega(\tau, l) \sum_{t=\tau+1}^{T} \tilde{e}_t \tilde{e}_{t-\tau},$$

where $\omega(\tau, l) = 1 - [\tau/(1+l)]$ and l is the truncation lag parameter. Since there is no general agreement on the appropriate choice of l, I present the results for various values of this parameter. The parameter k in the construction of the t_{α^*} and $t_{\hat{\alpha}}$ statistics is chosen by a test of significance on the lagged first differences of the data in the appropriate regression.

Consider first the *ex post* real-interest-rate series. The sample is 1961:1–1986:3 and the time of break is 1980:3. Therefore, $\lambda = 79/103 \approx 0.8$, and, from Tables 3 and 4, the 5% critical value for the $Z(\tilde{\alpha})$ statistic is −18.02, and for the $Z(t_{\tilde{\alpha}})$, $t_{\hat{\alpha}}$, and t_{α^*} statistics, it is −3.23. Using the additive-outlier method, the results are mixed. One can easily reject the null hypothesis of a unit root using either the $Z(\tilde{\alpha})$ or $Z(t_{\tilde{\alpha}})$ statistics but not with the extended Dickey–Fuller t_{α^*}. The latter result is, however, quite sensitive to the value of k chosen. The t_{α^*} statistic with $k = 2$ (not reported) yields essentially the same estimate of α^* but the t statistic jumps to −3.69, which is highly significant. I have reported the results with $k = 3$ because the third lag of the first differences was significant using conventional significance level. Using the innovational-outlier method, the t statistic $t_{\hat{\alpha}}$ is −3.71, which permits rejection at the 1% level. Furthermore, the estimate of the sum of the autoregressive coefficients is very low at 0.530, suggesting a rather different picture than the one obtained when no break is allowed.

Consider now the unemployment-rate series. The sample is 1948:1–1988:3 and the time of break is 1974:1. Therefore, $\lambda = 103/163 \approx 0.6$ and the 5% critical values are, respectively, -18.97 and -3.35 for the normalized least squares estimator and its t statistic. Using the additive-outlier method, the results are again mixed. Here the extended Dickey–Fuller statistic t_{α^*} permits rejection at even the 1% level, but, for most values of the truncation lag parameter l, the $Z(\tilde{\alpha})$ and $Z(t_{\tilde{\alpha}})$ statistics do not permit rejecting the null hypothesis of a unit root. Using the innovational-outlier method, the unit-root hypothesis is easily rejected at even the 1% level. These results are to be expected, given that the smooth transition to a higher mean for the unemployment series suggests, if anything, that such a change would better be modeled by an innovational-outlier model. My results agree with those reached by Evans (1989) using a similar methodology.

Finally, consider the terms-of-trade-index series. The sample is annual from 1900–1983 and the time of break is 1920. Hence $\lambda = 21/84 = 0.25$, and the 5% critical values for $\lambda = 0.30$ are -18.55 and -3.30. For this series, the null hypothesis of a unit root is easily rejected considering the additive-outlier method with either of the three statistics $Z(\tilde{\alpha})$, $Z(t_{\tilde{\alpha}})$, and t_{α^*}. The estimated sum of the autoregressive coefficients is quite low at 0.498, suggesting a behavior quite different from a unit-root process. The unit-root hypothesis is not rejected using the innovational-outlier method, since the t statistic is estimated at -2.46. This may be due to the poor power properties of this testing procedure when a large value of the truncation lag parameter is used; here $k = 8$. It is, however, most probably due to the small-sample bias introduced by using an innovational-outlier method when the additive-outlier method appears more appropriate (by inspection of the very sharp drop in Fig. 3). Hence, contrary to Cuddington and Urzua (1989), we would favor rejecting the null hypothesis of a unit root for this terms-of-trade-index series.

These applications show how the unit-root hypothesis is sensitive to minor changes in the specification of the deterministic component of a time series. As in the previous study of Perron (1989) it appears that many macroeconomic time series are better construed as stationary fluctuations around a deterministic component that occasionally (but rarely) changes in a dramatic way. Only these one-time changes appear to have any permanent effects.

The present study provides critical values for testing for a unit root when a time series is subject to a more or less sudden change in its mean. It is important to understand the nature of the postulated behavior needed to implement the tests. In the spirit of the Box–Tiao intervention analysis, I view these changes as exogenous — that is, as due to factors other than the stochastic structure of the noise process. It is, therefore, part of the maintained hypothesis and must be permitted under both the null and alternative hypotheses. Inference is conditional on such a change.

Issues of concern may arise over the choice of the break point T_B. Usually, visual inspection is sufficient, since the method is better suited for sudden changes. More formal methods can be developed, however, using, for example, cumulative sum analysis or the behavior of the recursive autocorrelation coefficients. Moreover, MacNeill (1978) presented a methodology for dealing with changes in general polynomials. His analysis postulated uncorrelated errors in the polynomial regression and hence cannot be used to address the issue of concern in this article — namely, whether the series are characterized by stationary or nonstationary fluctuations around a possibly changing trend function. Although it appears possible to extend MacNeill's analysis to allow for serial correlation in the residuals, it is less transparent how it could be extended to allow for possible nonstationarity of the unit-root type. Future work along these lines is clearly desirable.

The charge of data mining could be raised in that the choice of T_B is selected after a look at the data and that, in this sense, the inference might be biased in favor of the alternative hypothesis. Such an argument, however, overlooks the fact that the inference is conditional on such a change and that it is imposed under both the null and alternative hypotheses. The question that can be answered using the methodology described here is really in the spirit of the Box–Tiao intervention analysis: If one chooses to overlook a particular event that occurred during a very short period, is the remaining noise in the series consistent with the hypothesis that a unit root is present?

Since the procedure is conditional on a given exogenous change, the method purposefully does not explain such a change nor does it provide a stochastic structure describing their occurrences. For the change to be plausibly taken out of the noise function, however, it is preferable to relate it to 'major' events that are known to have occurred and may have caused the structural change in the behavior of the series. In some of the cases considered in this article, these major events are the change in monetary policy by the Federal Reserve in October 1979 for the real-interest rate and the oil-price shock of 1973 and the ensuing slowdown in growth for the unemployment rate (or the change in monetary policy with respect to higher rates of inflation that occurred at this time).

Acknowledgments

I thank a referee and especially an associate editor for useful comments. Research support from the Social Sciences and Humanities Research Council of Canada and the Fonds pour la Formation de Chercheurs et l'Aide à la Recherche du Québec is acknowledged. Carlos Urzúa kindly supplied his data series.

References

Box, G.E.P. and Tiao, G.C. (1975). Intervention analysis with applications to economic and environmental problems. *Journal of the American Statistical Association* 70, 70–79.

Chen, C. and Tiao, G.C. (1990). Random level-shift time series models, ARIMA approximations, and level-shift detection. *Journal of Business and Economic Statistics* 8, 83–97.

Cuddington, J.T. and Urzúa, C.M. (1989). Trends and cycles in the net barter terms of trade: A new approach. *Economic Journal* 99, 426–442.

Dickey, D.A., Bell, W.R. and Miller, R.B. (1986). Unit roots in time series models: Tests and implications. *The American Statistician* 40, 12–26.

Dickey, D.A. and Fuller, W.A. (1979). Distribution of the estimators for autoregressive time series with a unit root. *Journal of the American Statistical Association* 74, 427–431.

Dickey, D.A. and Fuller, W.A. (1981). Likelihood ratio statistics for autoregressive time series with a unit root. *Econometrica* 49, 1057–1072.

Dickey, D.A., Hasza, D.R. and Fuller, W.A. (1984). Testing for unit roots in seasonal time series. *Journal of the American Statistical Association* 79, 355–367.

Evans, G.W. (1989). Output and unemployment dynamics in the United States: 1950–1985. *Journal of Applied Econometrics* 4, 213–238.

Fuller, W.A. (1976). *Introduction to Statistical Time Series*, John Wiley: New York.

Fuller, W.A., Hasza, D.P and Goebel, J.J. (1981). Estimation of the parameters of stochastic difference equations. *The Annals of Statistics* 9, 531–543.

MacNeill, I.B. (1978). Properties of sequences of partial sums of polynomial regression residuals with applications to tests for change of regression at unknown times. *The Annals of Statistics* 6, 422–433.

Nelson, C.R. and Plosser, C.I. (1982). Trends and random walks in macroeconomic time series. *Journal of Monetary Economics* 10, 139–162.

Newey, W.K. and West, K.D. (1987). A simple positive definite heteroskedasticity and autocorrelation consistent covariance matrix. *Econometrica* 55, 703–708.

Ouliaris, S., Park, J.Y. and Phillips, P.C.B. (1989). Testing for a Unit Root in the Presence of a Maintained Trend. In *Advances in Econometrics and Modelling*, B. Raj (ed.), Kluwer Academic Publishers: Dordrecht, Holland, pp. 7–28.

Perron, P. (1988). Trends and random walks in macroeconomic time series: Further evidence from a new approach. *Journal of Economic Dynamics and Control* 12, 297–332.

Perron, P. (1989). The great crash, the oil price shock and the unit root hypothesis. *Econometrica* 57, 1361–1401.

Phillips, P.C.B. (1987). Time series regression with unit roots. *Econometrica* 55, 277–302.

Phillips, P.C.B. and Perron, P. (1988). Testing for a unit root in time series regression. *Biometrika* 75, 335–346.

Rose, A.K. (1988). Is the real interest rate stable? *Journal of Finance* 43, 1095–1112.

Said, S.E. and Dickey, D.A. (1984). Testing for unit roots in autoregressive-moving average models of unknown order. *Biometrika* 71, 599–608.

Tiao, G.C. (1985). Autoregressive Moving Average Models, Intervention Problems and Outlier Detection in Time Series. In *Handbook of Statistics* 5, eds. E.J. Hannan, P.R. Krishnaiah and M.M. Rao, (eds.), Amsterdam: North-Holland, pp. 85–118.

Tsay, R.S. (1986). Time series model specification in the presence of outliers. *Journal of the American Statistical Association* 81, 132–141.

Walsh, C.E. (1987). Three questions concerning nominal and real interest rates. *Economic Review, Federal Reserve Bank of San Francisco* 4, 5–20.

Chapter 14

Testing for a Unit Root in a Time Series with a Changing Mean: Corrections and Extensions*

Pierre Perron[†] and Timothy J. Vogelsang[‡]

[†] *Département de Science Economique and Centre de Recherche et Développement en Economique Université de Montréal, Montréal, Québec H3C-3J7, Canada*

[‡] *Department of Economics, Princeton University Princeton, NJ 08544, USA*

Tests for a unit root allowing for the possible presence of a one-time change in the mean of the series were proposed by Perron (1990) (henceforth referred to as P90). He considered two general classes of models, (1) the *additive outlier model*, appropriate when the change is sudden, and (2) the *innovational outlier model*, appropriate when the change is gradual. In this note we discuss an error in the treatment of the asymptotic distributions of the tests associated with the additive outlier model. We point out that the asymptotic distributions of these statistics are different from those stated by P90, and also that they depend on the correlation structure of the data, even if the appropriate order of the autoregression is selected, thereby making the procedure useless in practice. We discuss in this note a very simple modification that yields statistics having the same asymptotic distributions as stated earlier and being invariant to nuisance parameters, making the additive outlier model applicable in practice.

Keywords: Outliers, functional weak convergence, hypothesis testing, intervention analysis, nonstationarity, structural change.

*Originally Published in *Journal of Business and Economic Statistics*, 10, pp. 467–470.

1. The Model and the Statistics

The additive outlier model allows for a change in the mean of a series $\{y_t\}_1^T$, say, at time T_b $(1 < T_b < T)$. Under the null hypothesis of a unit root, the model is specified as

$$y_t = \gamma D(TB)_t + y_{t-1} + e_t = y_1 + \gamma DU_t + Z_t, \quad t = 2, \ldots, T, \quad (1)$$

where $D(TB)_t = 1$ if $t = T_b + 1$ and 0 otherwise, $DU_t = 1$ if $t > T_b$ and 0 otherwise, $Z_t = Z_{t-1} + e_t (Z_0 = 0)$ with e_t a finite-order (autoregressive moving average) ARMA$(p,\ q)$ process, $A(L)e_t = B(L)w_t$ with $w_t \sim$ i.i.d. $(0, \sigma_w^2)$ with finite fourth moment. Under the alternative hypothesis,

$$y_t = \mu + \gamma DU_t + v_t, \quad (2)$$

where v_t is, similarly, a finite-order ARMA$(p + 1, q)$ process that can be written $(1 - \alpha L)A(L)v_t = B(L)w_t$. Hence model (1) is a special case of Model (2) when $\alpha = 1$. Let $\{\tilde{y}_t\}$ be the residuals from a regression of y_t on a constant and DU_t. The tests based on the additive outlier model from P90 are the normalized bias $T(\tilde{\alpha} - 1)$ and the t statistic for testing the null hypothesis $\alpha = 1$, denoted by $t_{\tilde{\alpha}}$, in the following second-step regression:

$$\tilde{y}_t = \alpha \tilde{y}_{t-1} + u_t, \quad t = 2, \ldots, T. \quad (3)$$

2. The Limiting Distributions

Let $w^*(r)$ be the projection residual of a Wiener process $w(r)$ on the subspace generated by the functions $\{1, du(r)\}$ where $du(r) = 1$ if $r > \lambda$ and 0 otherwise and λ is the ratio of prebreak sample size to total sample size, $\lambda = T_b/T$. Let α^* and t_{α^*} be the least squares estimator and its associated t statistic (for $\alpha = 1$) in the *innovational regression model*

$$y_t = \mu + \gamma DU_t + dD(TB)_t + \alpha y_{t-1} + u_t. \quad (4)$$

It was proved by Perron (1989a) that

$$T(\alpha^* - 1) \Rightarrow \left(\int_0^1 w^*(r)\, dw(r) + \delta \right) \left(\int_0^1 w^*(r)^2\, dr \right)^{-1}$$

and

$$t_{\alpha^*} \Rightarrow (\sigma/\sigma_e)\left(\int_0^1 w^*(r)\,dw(r) + \delta\right)\left(\int_0^1 w^*(r)^2\,dr\right)^{-1/2},$$

where $\delta = (\sigma^2 - \sigma_e^2)/(2\sigma^2)$, with $\sigma^2 = \lim_{T\to\infty} T^{-1}E(S_T^2)$ ($S_T = \Sigma_{t=1}^T e_t$), and $\sigma_e^2 = \lim_{T\to\infty} T^{-1}\Sigma_{t=1}^T E(e_t^2)$. [Using the notation of (7) and (8) of P90, we have $(\int_0^1 w^*(r)\,dw(r) + \delta) \equiv H/\lambda(1-\lambda)$ and $\int_0^1 w^*(r)^2\,dr) \equiv K/\lambda(1-\lambda)$]. Contrary to the assertion made by P90 [Equations (7) and (8)], these limiting results do not carry over to $T(\tilde{\alpha}-1)$ and $t_{\tilde{\alpha}}$ obtained from the additive regression model (3). The error occurred in making the incorrect presumption that the limiting distributions of the statistics were the same in both cases. Rather the limiting distributions are given by (see the Appendix)

$$T(\tilde{\alpha}-1) \Rightarrow \left[\int_0^1 w^*(r)\,dw(r) + \delta + A(\lambda)D(\lambda)\right] \times \left[\int_0^1 w^*(r)^2\,dr\right]^{-1} \tag{5}$$

and

$$t_{\tilde{\alpha}} \Rightarrow \left[\int_0^1 w^*(r)\,dw(r) + \delta + A(\lambda)D(\lambda)\right] \times \left[(\sigma_e^2/\sigma^2 + D(\lambda)^2)\int_0^1 w^*(r)^2\,dr\right]^{-1/2}, \tag{6}$$

where $D(\lambda) \equiv \lambda^{-1}\int_0^\lambda w(r)\,dr - (1-\lambda)^{-1}\int_\lambda^1 w(r)\,dr$ and $A(\lambda) \equiv w(\lambda) - \lambda^{-1}\int_0^\lambda w(r)\,dr$. The limiting distributions in (5) and (6) are free of the nuisance parameters σ and σ_e when the errors e_t are uncorrelated (in which case $\sigma_e = \sigma$ and $\delta = 0$) and percentage points could therefore be tabulated. There are two problems with this approach. First, the Phillips–Perron (1988) nonparametric correction is different and more complex than that stated in Section 3.2 of P90. Second, and more important, it can be shown that introducing lags of the data, as in (11) of P90, does not eliminate the dependency on nuisance parameters even in the case in which the noise function is an autoregressive process of known order.

There is, however, a straightforward way to modify the second-step regression (3) to avoid the previously mentioned problems. Consider the following modification:

$$\tilde{y}_t = \omega D(TB)_t + \alpha \tilde{y}_{t-1} + u_t. \tag{7}$$

Let the least squares estimator of α in (7) be $\hat{\alpha}$ and the t statistic for testing $\alpha = 1$ be $t_{\hat{\alpha}}$. Using arguments analogous to those of Perron and Vogelsang (1992, Appendix B), we have

$$T(\hat{\alpha} - 1) \Rightarrow \left(\int_0^1 w^*(r)\, dw(r) + \delta \right) \times \left(\int_0^1 w^*(r)^2\, dr \right)^{-1}, \tag{8}$$

and

$$t_{\hat{\alpha}} \Rightarrow (\sigma/\sigma_e) \left(\int_0^1 w^*(r)\, dw(r) + \delta \right) \times \left(\int_0^1 w^*(r)^2\, dr \right)^{-1/2}. \tag{9}$$

Hence, if the one-time dummy $D(TB)_t$ is included in the second-step regression, the asymptotic distributions of the statistics are the same as those stated by P90 and, in particular, the same as those associated with the innovational outlier model. The intuition for this result is that $\hat{\alpha}$ depends on $\Delta\tilde{y}_t$, which under the null hypothesis is e_t except at time $T_b + 1$ because y_t is demeaned differently in each subsample. The difference at time $T_b + 1$ does not vanish asymptotically, hence the limiting results (5) and (6). The introduction of the dummy $D(TB)_t$ effectively eliminates this observation.

Tabulated critical values in the case in which $\sigma = \sigma_e$ are in Tables 3 and 4 of P90 for both the asymptotic and the finite-sample distributions. Note, however, that the note to these tables should state that the entries for the finite-sample cases were obtained by directly simulating $T(\alpha^* - 1)$ and t_{α^*} from (4) — that is, using the innovational outlier model and not the additive outlier model. We generated similar finite-sample critical values for the distribution of $T(\hat{\alpha}-1)$ and $t_{\hat{\alpha}}$ from (7). The values obtained were very close to those in Tables 3 and 4 of P90, the differences being small enough to be attributed to sampling error. Hence (a) the asymptotic distribution is a good approximation and (b) Tables 3 and 4 of P90 can also be used for the finite-sample distribution of $T(\hat{\alpha} - 1)$ and $t_{\hat{\alpha}}$ from (7).

3. Extensions to More General Error Processes

Consider now the transformations to (7) necessary for the limiting distributions of the tests to be invariant to nuisance parameters. First, the extensions of the Phillips and Perron (1988) statistics, discussed in Section 3.2 of P90, remain valid provided that $\widetilde{\alpha}$ and $t_{\tilde{\alpha}}$ in Equations (12) and (13) of P90 are replaced by $\widehat{\alpha}$ and $t_{\hat{\alpha}}$ as defined by the preceding regression (7) (similarly, $\tilde{\sigma}^2$ and $\tilde{\sigma}_e^2$ need to be replaced by estimators based on the residuals from that regression). Second, the extension of the Said–Dickey (1984) method discussed in Section 3.2 of P90 remains valid provided that Equation (11) in that article is replaced by

$$\tilde{y}_t = \sum_{i=0}^{k} \omega_i D(TB)_{t-i} + \alpha \tilde{y}_{t-1} + \sum_{i=1}^{k} c_i \Delta \tilde{y}_{t-i} + u_t, \qquad (10)$$

where $\tilde{y}_t$ are the residuals from a projection of y_t on a constant and DU_t. The introduction of the one-time dummies $D(TB)_{t-i}$ $(i = 0, \ldots, k)$ makes the asymptotic distribution of the t statistic for $\alpha = 1$ in (10), $t_{\hat{\alpha}}$, be that stated in (9) with $\sigma = \sigma_e$ — that is, the same as in the case in which the errors are uncorrelated and Regression (7) is used. The simple introduction of the dummies $D(TB)_{t-i}(i = 0, \ldots, k)$ is sufficient to correct the problems discussed previously concerning the asymptotic distributions of the statistics in the additive outlier model. For the t statistic $t_{\hat{\alpha}}$ based on (10) and the modified Phillips–Perron $Z(t_{\hat{\alpha}})$, the appropriate asymptotic critical values are still those in Table 4 of P90 [for the $Z(\hat{\alpha})$ test the asymptotic critical values are those in Table 3 of P90].

4. Corrections of the Empirical Results

Tables 1 and 2 present the corrected versions of Tables 5 and 6 in P90. Table 1 considers the two-step procedure based on Equation (10) with lags of the data introduced to correct for serial correlation. Table 2 considers the Phillips–Perron (1988) modified statistics based on Regression (7). The qualitative results remain unchanged.

Table 1. Full-sample unit-root tests with changing mean: Additive outlier method.

Series	T	$\tilde{\mu}$	$t_{\tilde{\mu}}$	$\tilde{\gamma}$	$t_{\tilde{\gamma}}$	k	$\hat{\alpha}$	$t_{\hat{\alpha}}$
Interest rate	103	0.079	0.28	5.56	9.45	3	0.676	−2.46
Unemployment rate	163	4.77	40.23	2.47	12.54	6	0.901	−3.69*
Terms of trade Index	84	0.412	16.59	−0.37	−12.73	6	0.214	−4.47*

Note: The regressions are $y_t = \tilde{\mu} + \tilde{\gamma} DU_t + \tilde{y}_t$ and $\tilde{y}_t = \hat{\alpha}\tilde{y}_t + \Sigma_{i=1}^k \hat{c}_i \Delta\tilde{y}_{t-i} + \Sigma_{i=0}^k \hat{\theta}_i D(TB)_{t-i} + \hat{u}_t$. Asterisks denote a statistic significant at the 5% level. The breakpoints are as stated in P90 — 1980:3 for the interest rate, 1974:1 for the unemployment rate, and 1920 for the terms-of-trade index.

Table 2. Full-sample unit-root tests with changing mean: Additive outlier method, using Phillips–Perron statistics.

Series	Statistic	$l = 1$	$l = 6$	$l = 12$	$l = 18$	$l = 24$
Interest rate	$Z(\hat{\alpha})$	−69.11*	−90.54*	−113.88*	−127.50*	−140.22*
	$Z(t_{\hat{\alpha}})$	−7.37*	−7.90*	−8.52*	−8.88*	−9.21*
Unemployment rate	$Z(\hat{\alpha})$	−18.75	−23.41*	−18.56	−15.20	−13.38
	$Z(t_{\hat{\alpha}})$	−3.17	−3.52*	−3.18	−2.86	−2.72
Terms of trade index	$Z(\hat{\alpha})$	−36.41*	−31.17*	−26.34*	−24.71*	−21.41*
	$Z(t_{\hat{\alpha}})$	−4.79*	−4.55*	−4.33*	−4.26*	−4.15*

Note: *A statistic significant at the 5% level. See the notes to Table 1 for the specifications of the breakpoints.

Appendix: Mathematical Proof of (5) and (6)

First note that under the null hypothesis (1) we have $y_t = S_t + y_1$ if $t \le T_b$ and $y_t = S_t + (\gamma + y_1)$ if $t > T_b$, where $S_t = \Sigma_{j=1}^t e_j$. The detrended variables are

$$\begin{aligned} \tilde{y}_t &= y_t - T_b^{-1} \sum_{1}^{T_b} y_t = S_t - \overline{S}_a, \qquad t \le T_b, \\ \tilde{y}_t &= y_t - (T - T_b)^{-1} \sum_{T_{b+1}}^{T} y_t = S_t - \overline{S}_b, \quad t > T_b, \end{aligned} \tag{A.1}$$

where $\overline{S}_a = T_b^{-1}\Sigma_b^T 1 S_t$ and $\overline{S}_b = (T - T_b)^{-1}\Sigma_{T_{b+1}}^T S_t$. Note that, given the assumptions on the sequence $\{e_t\}$, $T^{-1/2}\overline{S}_a \Rightarrow \lambda^{-1}\int_0^\lambda w(r)\,dr$

and $T^{-1/2}\overline{S}_b \Rightarrow (1-\lambda)^{-1}\int_\lambda^1 w(r)\,dr$ (see Perron 1989b), where $\Rightarrow$ denotes weak convergence in distribution. Using (A.1),

$$\begin{aligned} \tilde{y}_t - \tilde{y}_{t-1} &= e_t, & t &\neq T_b + 1, \\ \tilde{y}_t - \tilde{y}_{t-1} &= e_t - \overline{S}_b + \overline{S}_a, & t &= T_b + 1. \end{aligned} \tag{A.2}$$

The normalized least squares estimator of α from (3) is given by $T(\tilde{\alpha}-1) = T^{-1}\Sigma_{t=2}^T \tilde{y}_{t-1}(\tilde{y}_t - \tilde{y}_{t-1})/T^{-2}\Sigma_{t=2}^T \tilde{y}_{t-1}^2$. We have, using (A.1) and (A.2),

$$\begin{aligned} \text{(a) } & T^{-2}\sum_{t=2}^{T} \tilde{y}_{t-1}^2 \\ &= T^{-2}\sum_{t=1}^{T_b}(S_t - \overline{S}_a)^2 + T^{-2}\sum_{t=T_b+1}^{T}(S_t - \overline{S}_b)^2 - T^{-2}(S_T - \overline{S}_b)^2 \\ &= T^{-2}\sum_{t=1}^{T} S_t^2 - \lambda T^{-1}\overline{S}_a^2 - (1-\lambda)T^{-1}\overline{S}_b^2 + o_p(1) \\ &\Rightarrow \sigma^2\left\{\int_0^1 w(r)^2 dr - \lambda^{-1}\left(\int_0^\lambda w(r)\,dr\right)^2 \right. \\ &\qquad \left. -(1-\lambda)^{-1}\left(\int_\lambda^1 w(r)\,dr\right)^2\right\} \\ &\equiv \sigma^2 \int_0^1 w^*(r)^2 dr, \quad \text{as required} \end{aligned} \tag{A.3}$$

and

$$\begin{aligned} \text{(b) } & T^{-1}\sum_{t=2}^{T} \tilde{y}_{t-1}(\tilde{y}_t - \tilde{y}_{t-1}) \\ &= T^{-1}\sum_{t=2}^{T_b} \tilde{y}_{t-1}(\tilde{y}_t - \tilde{y}_{t-1}) + T^{-1}\tilde{y}_{T_b}(\tilde{y}_{T_{b+1}} - \tilde{y}_{T_b}) \\ &\quad + T^{-1}\sum_{t=T_b+2}^{T} \tilde{y}_{t-1}(\tilde{y}_t - \tilde{y}_{t-1}) \end{aligned}$$

$$
\begin{aligned}
&= T^{-1}\sum_{t=2}^{T_b}(S_{t-1}-\overline{S}_a)e_t + T^{-1}(S_{T_b}-\overline{S}_a)\times(e_{T_b+1}+\overline{S}_a-\overline{S}_b)\\
&\quad + T^{-1}\sum_{t=T_b+2}^{T_b}(S_{t-1}-\overline{S}_b)e_t\\
&\Rightarrow \sigma^2\left\{\int_0^1 w(r)\,dw(r) + \delta - \lambda^{-1}w(\lambda)\int_0^\lambda w(r)\,dr\right.\\
&\quad \left. -(1-\lambda)^{-1}(w(1)-w(\lambda))\int_\lambda^1 w(r)\,dr\right\}\\
&\quad + \sigma^2\left[w(\lambda)-\lambda^{-1}\int_0^\lambda w(r)\,dr\right]\\
&\quad \times\left[\lambda^{-1}\int_0^\lambda w(r)\,dr - (1-\lambda)^{-1}\int_\lambda^1 w(r)\,dr\right]\\
&\equiv \sigma^2\left(\int_0^1 w^*(r)\,dw(r) + \delta + A(\lambda)D(\lambda)\right), \quad \text{as required.}
\end{aligned}
\tag{A.4}
$$

This proves (5). To prove (6), note that $t_{\tilde{\alpha}} = T^{-1}\Sigma_{t=2}^T\tilde{y}_{t-1}(\tilde{y}_t - \tilde{y}_{t-1})/[\tilde{\sigma}^2T^{-2}\Sigma_{t=2}^T\tilde{y}_{t-1}^2]^{1/2}$, where $\tilde{\sigma}^2 = T^{-1}\Sigma_{t=2}^T\tilde{u}_t^2$, with $\tilde{u}_t$ the estimated residuals from (3). Given (A.3) and (A.4), we only need the limit of $\tilde{\sigma}^2$, which is obtained as follows:

$$
\begin{aligned}
\tilde{\sigma}^2 &= T^{-1}\sum_{t=2}^T(\tilde{y}_t-\tilde{\alpha}\tilde{y}_{t-1})^2\\
&= T^{-1}\sum_{t=2}^T(\tilde{y}_t-\tilde{y}_{t-1})^2 + o_P(1)\\
&= T^{-1}\sum_{t=1}^T e_t^2 + T^{-1}(\overline{S}_a-\overline{S}_b)^2 + o_P(1)\\
&\Rightarrow \sigma_e^2 + \sigma^2\left[\lambda^{-1}\int_0^\lambda w(r)dr - (1-\lambda)^{-1}\int_\lambda^1 w(r)dr\right]^2\\
&\equiv \sigma_e^2 + \sigma^2 D(\lambda)^2.
\end{aligned}
$$

Acknowledgments

Financial support is acknowledged from the Fonds pour la Formation de Chercheurs et l'Aide à la Recherche du Québec (Perron) and the International Finance Section at Princeton University (Vogelsang).

References

Perron, P. (1989a). Testing for a unit root in a time series with a changing mean. Working Paper 3489, Université de Montréal, Centre de Recherche et Développement en Economique.

Perron, P. (1989b). The great crash, the oil price shock and the unit root hypothesis. *Econometrica* 57, 1361–1401.

Perron, P. (1990). Testing for a unit root in a time series with a changing mean. *Journal of Business & Economic Statistics* 8, 153–162.

Perron, P. and Vogelsang, T.J. (1992). Nonstationarity and level shifts with an application to purchasing power parity. *Journal of Business & Economic Statistics* 10, 301–320.

Phillips, P.C.B. and Perron, P. (1988). Testing for a unit root in time series regression. *Biometrika* 75, 335–346.

Said, S.E. and Dickey, D.A. (1984). Testing for unit roots in autoregressive-moving average models of unknown order. *Biometrika* 71, 599–608.

Chapter 15

Nonstationarity and Level Shifts with an Application to Purchasing Power Parity*

Pierre Perron† and Timothy J. Vogelsang‡

†*Department of Economics, Princeton University, Princeton, NJ 08544, USA; Centre de Recherche et Développement en Economique, Université de Montréal, Montréal, Québec, H3C-3J7, Canada*

‡*Department of Economics, Princeton University, Princeton, NJ 08544, USA*

This study considers testing for a unit root in a time series characterized by a structural change in its mean. The analysis is in the spirit of Perron (1990a), who showed that the existence of such a shift in a stationary time series biases the usual tests for a unit root toward nonrejection. The approach is, however, different given that we suppose the date of the change to be unknown. The statistic of interest is then the minimal t statistic over all possible breakpoints in regressions similar to those proposed by Perron (1990a). Other related statistics are also discussed. We derive and tabulate the asymptotic distributions of interest. Most of the emphasis, however, is given to the tabulation of finite-sample critical values using simulation experiments. Particular attention is given to the effect, on the finite-sample critical values, of various procedures to select the appropriate order of the estimated autoregressions. We apply the tests to analyze the issue of purchasing power parity between the United States and the United Kingdom and also between the United States and

*Originally Published in *Journal of Business and Economic Statistics*, 10, pp. 301–320.

Finland, whose real exchange rates are characterized by apparent shifts in level when using particular price indexes.

Keywords: Deterministic trends; hypothesis testing; simulation experiment; stochastic trends; structural change.

1. Introduction

Testing for the presence of a unit root in a time series of data has become a topic of great concern to economists. Since the seminal work of Nelson and Plosser (1982), the view that most macroeconomic variables are best construed as exhibiting some kind of stochastic nonstationarity has become prevalent. The unit-root hypothesis has far-reaching implications with respect to both economic theory and the interpretation of empirical evidence. Indeed, under the unit-root hypothesis, random shocks have a permanent effect on the system; that is, fluctuations are not transitory.

To illustrate these implications, consider issues raised by the purchasing power parity (PPP) hypothesis. A mild version of this hypothesis asserts that, given two countries, movements in the nominal exchange rate and in their respective price indexes adjust over time to leave a constant relative purchasing power between the two countries (e.g., see Dornbusch, 1988). In statistical terms, this hypothesis implies that fluctuations in the real exchange rate between any two given countries are stationary — that is, that the shocks to this relation have temporary effects. On the other hand, a sufficient condition for a violation of PPP is that the real exchange rate is characterized by the presence of a unit root. In this case, shocks have permanent effects, and there is no tendency to return to a stable value. This implication has been used by many authors to test the PPP hypothesis (e.g., Corbae and Ouliaris, 1988; Enders, 1988, 1989; Taylor and McMahon, 1988). In this framework, failure to reject the unit-root hypothesis is taken as evidence against PPP.

In a recent study, Dornbusch and Vogelsang (1991) provided an empirical analysis of this issue using a long span of annual data. The analysis is performed with the real exchange rate between the US and many countries using, when possible, two price indexes, a consumer price index (CPI) and a gross national product (GNP) (or gross domestic product [GDP]) deflator. Unlike many studies, the unit root for the real exchange rate is usually rejected. This rejection can be explained by the higher power of the tests when a long span of data is used (see Perron, 1991a). In some instances, however, the unit root is rejected with one price index but not the other.

To illustrate the latter problem, consider first Figs. 1 and 2, which present a graph of the real exchange rate between the US and the UK. Figure 1 is the real exchange rate based on the CPI, and Fig. 2 is a similar series based on the GNP deflator. The sources of the data are detailed in Appendix A. A standard Dickey–Fuller (1979) test applied to the GNP-deflator-based series rejects the unit-root hypothesis at a high significance level (t statistic of –3.23 with eight lags of first-differences of the data added). The same test applied to the CPI-based measure, however, implies a nonrejection (t statistics of –1.59 with four additional lags). As can be seen from the graphs, the CPI-based measure appears to exhibit a permanent change in level at some point between 1940 and 1950. Such is not the case with the GNP-deflator-based series. Figures 3 and 4 present a similar example using the real exchange rate between the US and Finland.

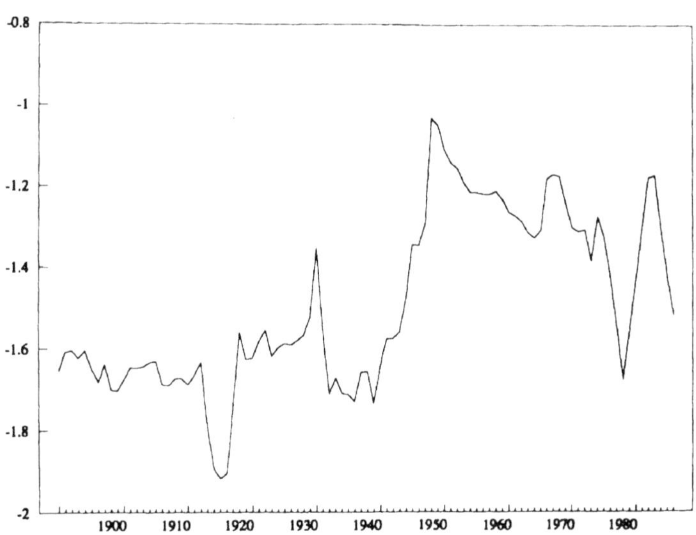

Fig. 1. Logarithm of the US/UK real exchange rate based on the consumer price indexes (CPI); 1892–1988, annual.

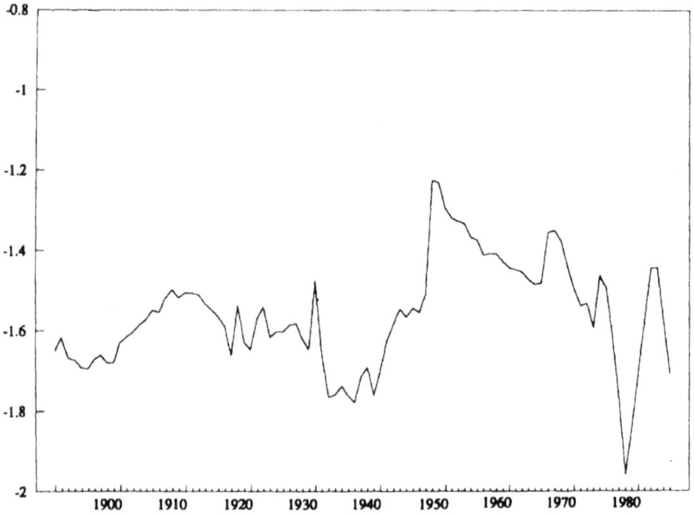

Fig. 2. Logarithm of the US/UK real exchange rate based on the GNP (GDP) deflators as the price indexes; 1892–1988, annual.

Here, the opposite result holds. The unit root is easily rejected using the CPI-based measure (t statistic of -5.74 with one lag) but not with the GDP-based measure (t statistic of -1.62 with two lags). As seen from the graphs, the GDP-based measure appears to exhibit a permanent decrease in level after 1940.

In the examples just depicted, the nonrejection of the unit-root hypothesis is associated with an apparent permanent change in the level of the series. Such issues were recently analyzed by Perron (1990a). In that article it was shown that, if a series is characterized by stationary fluctuations with a one-time permanent change in level, the usual tests for a unit root are biased toward nonrejection. Perron (1990a) also proposed a class of statistics to test the unit-root hypothesis allowing for the possible presence of a one-time change in mean. The distribution theory underlying the derivation of the critical values assumed that the date of the change in mean was exogenous (or uncorrelated with the data). In the examples discussed previously, this assumption appears inappropriate. It is difficult to single out any major exogenous event that could have caused the change in mean for one particular version of the real exchange rate measure while not for the other. It appears more plausible to view the

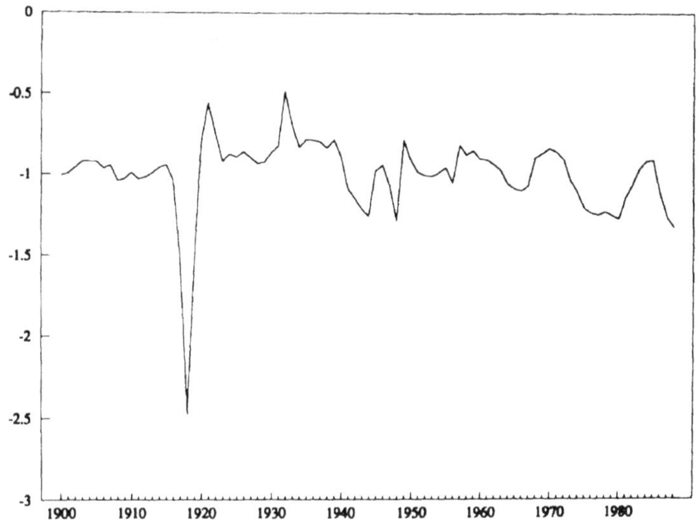

Fig. 3. Logarithm of the US/Finland real exchange rate based on the consumer price indexes (CPI); 1900–1987, annual.

possible change in mean as occurring at an unknown date, in which case the distribution theory underlying the test is different. There is also some validity to the argument that the breakpoint ought to be viewed always as an unknown parameter irrespective of the fact that one may be able to identify an exogenous event associated with the change (e.g., see Christiano, 1988). Nevertheless, the distribution theory obtained assuming a known breakpoint may often be a good approximation to the appropriate distribution if no attempts were made to select the breakpoint by systematically analyzing the data prior to this choice (see Perron, 1990b).

The aim of this article is to provide a class of statistics to test for the unit-root hypothesis allowing for a possible change in the level of the series occurring at an unknown date. The models considered are similar to those of Perron (1990a) — namely, an *additive outlier* (AO) *model* for a sudden change and an *innovational outlier* (IO) *model* appropriate for a gradual change. The class of statistics considered is similar to those of Banerjee *et al.* (1992), Zivot and Andrews (1992), and Perron (1990b), who considered trending variables. They are based on the minimal value, over all possible breakpoints, of

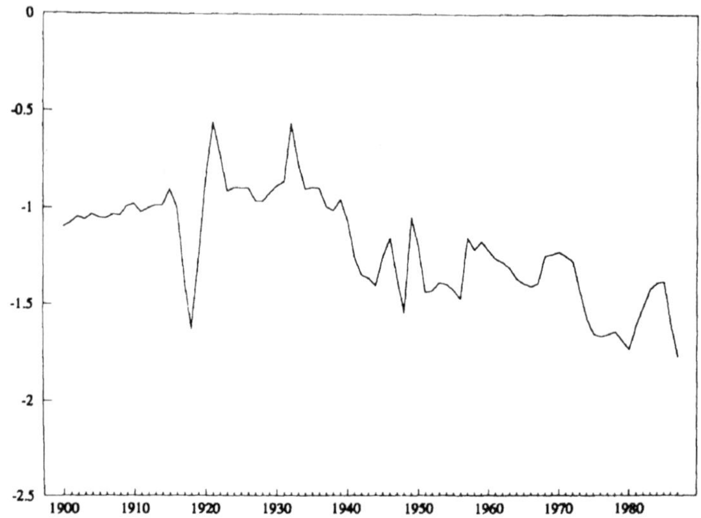

Fig. 4. Logarithm of the US/Finland real exchange rate based on the GNP (GDP) deflators as the price indexes; 1900–1987, annual.

the t statistic on the sum of the autoregressive coefficients in the appropriate augmented autoregression. We also consider a class of statistics based on the same t statistic but now evaluated at the breakpoint that yields the lowest (or highest) t statistic on the coefficient associated with the change in mean. This permits imposing the mild *a priori* imposition of a one-sided change in mean and allows tests with possibly higher power. Note that our approach is directly oriented at the issue of testing for a unit root and not testing for structural change *per se*. We view the possible presence of a structural change as part of the maintained hypothesis, and the parameters associated with it can accordingly be viewed as nuisance parameters for the inference that is performed. This contrasts with the recent literature pertaining to tests for structural change with unknown break dates in dynamic models with or without trending variables (e.g., see Andrews, 1990; Chu, 1989; Chu and White, 1990; Hansen, 1990, 1992; Perron, 1991b).

Though we derive the asymptotic distribution of the statistics, we put more emphasis on the finite-sample distributions obtained using simulation experiments. We provide an extensive analysis of the effects of different procedures to select the order of the appropriate autoregression on the finite-sample critical values. We consider the following procedures for selecting this truncation lag parameter: (a) fixed, (b) chosen according to a significance test on the last included lag (given a prespecified maximum), (c) chosen according to an F test on additional lags (up to some prespecified maximum), and (d) chosen to minimize the relevant t statistic for testing the unit-root hypothesis. In each case, we present tabulated critical values for both the AO model and the IO model.

The plan of the article is as follows. Section 2 discusses the class of models considered and the test statistics analyzed. Section 3 presents the asymptotic distribution of the statistics under the null hypothesis of a unit-root process, still allowing for a possible change in mean. Section 4 discusses the finite-sample critical values. Section 5 presents additional experiments to assess the size and power properties of the various procedures under a variety of data-generating processes. Section 6 applies our procedures to the two real-exchange-rate examples discussed previously. Section 7 contains concluding comments. Additional material is collected in three appendixes: Appendix A contains the description and sources of the data, Appendix B derives the limiting distributions stated in the text, and Appendix C presents details on the simulation method to obtain critical values associated with these asymptotic distributions.

2. The Models and Test Statistics

We briefly discuss, in this section, the models and statistical procedures involved in testing for a unit root, allowing for the possibility of a change in the level of the series. The reader is referred to Perron (1989, 1990a) for a more detailed discussion and interpretation. Throughout this article, we assume that there is at most one change. We denote the date of break, should it occur, by T_b with $1 < T_b < T$, where T is the sample size.

2.1. *The additive outlier model*

The first model is labeled, following the literature on time series with outliers (e.g., Tiao, 1985), the additive outlier (AO) model. In this case, the change is assumed to take effect instantaneously. In particular, the effect of the change on the level of the series $\{y_t\}$, say, does not depend on the dynamics exhibited by the correlation structure of $\{y_t\}$. Under the null hypothesis of a unit root, this model can be parameterized as

$$y_t = \delta D(TB)_t + y_{t-1} + w_t, \quad t = 2, \ldots, T, \tag{1}$$

where $D(TB)_t = 1$ if $t = T_b+1$ and 0 otherwise. Here, and throughout this article, $y_1 = y(1)$ is either a fixed constant or a random variable. The sequence of errors $\{w_t\}$ is specified to be a stationary and invertible (autoregressive moving average) ARMA(p, q) process. More specifically, $A^*(L)w_t = B(L)e_t$, where e_t is i.i.d. (0, σ^2) with finite fourth moment. $A^*(L)$ and $B(L)$ are polynomials in L of order p and q, respectively, whose roots are assumed to be strictly outside the unit circle. Under Model (1), the mean of the series is $y(1)$ up to time T_b and $y(1) + \delta$ afterward. Under the alternative hypothesis, the series $\{y_t\}$ does not contain a unit root and can be characterized by the following specification:

$$y_t = c + \delta DU_t + v_t, \quad t = 2, \ldots, T, \tag{2}$$

where $DU_t = 1$ if $t > T_b$ and 0 otherwise. The sequence of errors $\{v_t\}$ is likewise a stationary and invertible ARMA$(p+1, q)$ process of the form $A(L)v_t = B(L)e_t$. The mean of the series is c up to time T_b and $c + \delta$ afterward. The null hypothesis specified by (1) is a special case of the general specification (2) when $c = y(1)$ and $A(L) = (1 - L)A^*(L)$. Hence a natural testing strategy is to first remove an estimate of the deterministic part of the series and to test if the remaining noise is characterized by the presence of a unit root.

For a fixed value of the breakpoint T_b, this leads to the following two-step procedure. First, remove the deterministic part of the series using the estimates of the regression

$$y_t = \mu + \delta DU_t + \tilde{y}_t, \quad t = 1, \ldots, T. \tag{3}$$

The test is then performed using the t statistic for $\alpha = 1$ in the regression

$$\tilde{y}_t = \sum_{i=0}^{k} \omega_i D(TB)_{t-i} + \alpha \tilde{y}_{t-1} + \sum_{i=1}^{k} c_i \Delta \tilde{y}_{t-i} + e_t,$$
$$t = k+2, \ldots, T. \tag{4}$$

Note the inclusion of the dummy variables $D(TB)_{t-i}$ $(i = 0, \ldots, k)$ in the second-step regression. They are needed to ensure that the t statistic on α in (4) has the same asymptotic distribution as in the IO model and, more important, that it be invariant to the value of the truncation lag parameter k. Note in particular that the asymptotic distribution is different if these regressors are omitted (see Perron and Vogelsang, 1991, for details). This testing strategy follows the method of Dickey and Fuller (1979) and Said and Dickey (1984) by approximating ARMA (p, q) processes by AR(k) processes. The method will be asymptotically valid provided k increases at a controlled rate as T increases (see Said and Dickey 1984). We denote the t statistic for testing $\alpha = 1$ in regression (4) as $t_{\hat{\alpha}}$ (AO, T_B, k), where T_b is the break date, k is the truncation lag parameter, and AO indicates that we are applying the two-step procedure appropriate for the AO model. The issues of interest are the specific procedures for choosing T_b and k. We return to this problem after a discussion of the IO model.

2.2. *The innovational outlier model*

Under the IO model, the change is supposed to affect the level of the series $\{y_t\}$ gradually; that is, there is a transition period. Though, in principle, the dynamic effect could take any form, a natural and simple way to model such a transition is to suppose that the economy responds to a *shock* to the trend function (here the change in mean) in the same way that it reacts to any other shocks. This assumption leads to the following specification under the null hypothesis of a unit root:

$$y_t = y_{t-1} + \psi(L)(e_t + \theta D(TB)_t), \quad t = 2, \ldots, T, \tag{5}$$

where $\psi(L) = A^*(L)^{-1}B(L)$ defines the moving average representation of the noise function with $A^*(L)$ and $B(L)$ as defined in (1). The immediate impact of the change in mean is θ, and the long-run impact is given by $\psi(1)\theta$. Under the alternative hypothesis of stationary fluctuations, the model is represented by

$$y_t = a + \phi(L)(e_t + \delta DU_t), \quad t = 2, \ldots, T, \tag{6}$$

where $\phi(L) = A(L)^{-1}B(L)$ with $A(L)$ as defined in (2). The immediate impact of the change in mean is δ, and the long-run impact is $\delta\phi(1)$. Models (5) and (6) can be nested and approximated, as in the work of Said and Dickey (1984), by the finite-order autoregressive model

$$y_t = \mu + \delta DU_t + \theta D(TB)_t + \alpha y_{t-1} + \sum_{i=1}^{k} c_i \Delta y_{t-i} + e_t, \quad t = k+2, \ldots, T. \tag{7}$$

Under the null hypothesis of a unit root, α is equal to 1 (which also implies $\delta = 0$). Hence the appropriate testing strategy is to use the t statistic for testing $\alpha = 1$ when regression (7) is estimated by ordinary least squares (OLS). We denote such a t statistic, obtained with given values of T_b and k, by $t_{\hat{\alpha}}$ (IO, T_b, k).

2.3. *Procedures to select T_b*

For both statistics $t_{\hat{\alpha}}(i, T_b, k)$ ($i =$ AO, IO), the appropriate values of T_b and k are unknown. Hence the testing procedure must take account of this fact. Consider first the modifications induced by the fact that T_b is unknown. As in the work of Banerjee *et al.* (1992), Zivot and Andrews (1992), and Perron (1990b), we analyze the behavior of the statistics $t_{\hat{\alpha}}(i, T_b^*, k) = \inf_{T_b \in (k+2,T)} t_{\hat{\alpha}}$ (i, T_b, k)($i =$AO, IO), in which different specifications about the choice of k will be analyzed. This procedure is in the tradition of tests for structural change when the date of the change is assumed unknown. Note also that such a procedure is valid whether or not there is a change under the null hypothesis; that is, we do not require that T_b be identifiable (e.g., see Davies, 1987).

We also investigate a second procedure in which T_b is chosen to minimize the t statistic for testing $\delta = 0$ in regression (3) for the AO model and in regression (7) for the IO model. Given the symmetric nature of the problem, this is also equivalent to analyzing the behavior of the statistic when T_b is chosen to maximize the t statistic for testing $\delta = 0$ in each regression. We denote the resulting statistics by $t_{\hat{\alpha}}$ $(i, T_b(\hat{\delta}), k)$ (i =AO, IO). This procedure allows the imposition of the mild *a priori* restriction of a one-sided change in the mean while still treating the date of the breakpoint as unknown. As shown in later sections, this may allow a substantial increase in power relative to the other procedure.

2.4. *Procedures to select k*

For each of the two procedures for selecting T_b, we analyze different ways of selecting the truncation lag parameter k. The first method follows the theoretical development of Said and Dickey (1984) by treating k as a fixed function of the sample size T. Given that in finite samples such a rule does not restrict the choice of k, in practice this procedure simply specifies a choice of k that is independent of the data — that is, determined *a priori*. We denote the statistic of interest under this procedure by $t_{\hat{\alpha}}$ $(i, j, \bar{k})$ (i =AO, IO; $j = T_b^*$, $T_b(\hat{\delta})$).

We also consider two procedures in which the truncation lag parameter is chosen using a data-dependent method. The first is that used by Perron (1989, 1990a). Here, for any given value of T_b, k is chosen such that the coefficient on the last included lag of the first-differences of the data is significant at some level β and the coefficient on the last included lag in higher order autoregressions is insignificant up to some *a priori* specified maximum order, say, $k$max. We use the asymptotic normality of the t statistic on these coefficients to carry out the inference. This procedure is justified in the pure autoregressive (AR) case since for an AR(p) process $c_i = 0$ for all $i > p$ and $c_i \neq 0$ for $i = p$. In this AR case, the probability that the procedure selects an order less than p is asymptotically 0 provided $k$max is greater than p. This is sufficient to ensure that the resulting statistics, denoted by $t_{\hat{\alpha}}$ $(i, j, k(t))$ (i= AO, IO; j= T_b^*,

$T_b(\hat{\delta})$), have the same asymptotic distribution as in the case in which k is fixed (see Section 3).

The other procedure investigated is similar and follows the empirical application of Said and Dickey (1984), in which again an upper bound on the order is specified. This procedure uses an F test (or likelihood ratio test) for the joint significance of the coefficients on the lagged first-differences of the data in the estimated autoregressions using a $\beta\%$ test based on the asymptotic χ_q^2 distribution in which q is the number of restrictions in the restricted model. The exact procedure is as follows: First, a maximum value of k, $k$max, is specified. For a given value of T_b, the autoregression is estimated with $k$max and ($k$max − 1) lags. The F test is used to assess whether the coefficient on the $k$maxth lag is significant, and if so, the value of k chosen is this maximum value. If not, the model is estimated with ($k$max − 2) lags. The lag ($k$max − 1) is deemed significant if either the F test for ($k$max − 2) versus ($k$max − 1) lags or the F test for ($k$max − 2) versus $k$max lags are judged significant. The procedure is repeated by lowering k until a rejection of the null hypothesis that additional lags are insignificant occurs or the lower bound $k = 0$ is attained. We denote the resulting statistics by $t_{\hat{\alpha}}\ (i, j, k(F))$ ($i =$AO, IO; $j = T_b^*$, $T_b(\hat{\delta})$).

The last data-dependent method investigated for selecting the order of the autoregression mimics the way T_b is chosen — namely, to minimize the t statistic for testing $\alpha = 1$ given some prespecified maximal value $k$max. These statistics are formally defined as $t_{\hat{\alpha}}\ (i, j, k^*) = \inf_{k \in (0, k\text{max})} t_{\hat{\alpha}}\ (i, j, k)$ ($i =$ AO, IO; $j = T_b^*$, $T_b(\hat{\delta})$). This procedure is not recommended in practice but is included since our analysis will clearly show that choosing k to maximize the change of rejection can lead to a testing procedure with highly undesirable properties.

3. The Limiting Distribution of the Statistics

Our strategy in discussing the asymptotic distribution of the statistics of interest is as follows. We first consider the case in which under the null hypothesis the noise function is uncorrelated; that is, in the

AO model (1) we specify $A^*(L) = B(L) = 1$. Similarly, in the IO model (5), $\psi(L) = 1$. In this case, both models yield the same data-generating process — namely,

$$y_t = \delta D(TD)_t + y_{t-1} + e_t, y_1 = y(1); \quad t = 2, \ldots, T, \tag{8}$$

where $e_t \sim$ i.i.d. $(0,\ \sigma^2)$ and where the initial condition is a fixed constant for simplicity. In Appendix B, we derive the limiting distribution of the statistics $t_{\hat{\alpha}}$ $(i, j, k = 0)$ $(i =$ AO, IO; $j = T_b^*$, $T_b(\hat{\delta}))$ when no lags of the first-differences of the data are included in the autoregressions; that is, $k = 0$ in (4) and (7). We discuss later how the results remain valid under more general error structures and using alternative procedures to select k.

Consider first the asymptotic distribution of the statistics when T_b is chosen to minimize the t statistic for $\alpha = 1$ in regressions (4) or (7). The first thing to note is that the same asymptotic distribution applies when considering the AO or the IO model. Note, however, that this asymptotic equivalence holds only if the one-time dummies $D(TB)_{t-i}$ $(i = 0, \ldots, k)$ are included in the second-step regression (4) in the case of the AO model. This is basically because these models are similar except for some transitional effects that do not matter asymptotically (see Perron and Vogelsang [1991] for detail). The following result is proved in Appendix B using results of Perron (1989, 1990a) and Zivot and Andrews (1992). Let $W^*(\lambda, r)$ denote the projection residual of a standard Wiener process $W(r)$ on the subspace generated by the functions $\{1, du(\lambda, r)\}$, where $du(\lambda, r) = 0$ if $r \leq \lambda$ and $du(\lambda, r) = 1$ if $r > \lambda$. We obtain the following asymptotic representation under the null hypothesis (8):

$$t_{\hat{\alpha}}(i,\ T_b^*, k = 0) \Rightarrow \inf_{\lambda \in \Lambda} \left\{ \int_0^1 W^*(\lambda, r) dW(r) \right. \\ \left. \times \left[\int_0^1 W^*(\lambda, r)^2 dr \right]^{-1/2} \right\}, \quad i = \text{AO, IO}, \tag{9}$$

where Λ is a closed subset of the interval (0, 1) and where $\Rightarrow$ denotes weak convergence in distribution. The fact that the space of values for λ is restricted to a closed subset of (0, 1) means that, under this asymptotic interpretation, some trimming ought to be applied. Banerjee *et al.* (1992) and Andrews (1990) suggested using a window $\lambda \in (0.15, 0.85)$, which implies that, in finite samples, T_b can only take values in the range $([0.15T], [0.85T])$, where $[\cdot]$ denotes the integer argument. This choice is arbitrary and, as in the work of Zivot and Andrews (1992), we consider the largest *window* possible in both the theoretical derivations and the empirical applications — namely, $(k+2,\ T-1)$. Because the behavior of some of the components of the statistic are unbounded when the endpoints are included, λ needs to be restricted to a subset of the interval (0, 1). Choosing a *rectangular window*, such as $(\varepsilon,\ 1-\varepsilon)$ for any $\varepsilon > 0$, is one way to get rid of the problem caused by the endpoints. This is not the only way, however, to downweight the influence of the endpoints. An alternative is to take the minimum of the t statistics over the full interval (0, 1) and to weight the statistic by $\lambda(1-\lambda)$. This, however, implies a different asymptotic distribution. See Deshayes and Picard (1986) for a discussion of these issues.

We believe that this restriction on the interval specified by Λ can be relaxed and that we could consider the limiting distribution defined over the full interval [0, 1]. Our proof, however, does not permit stating this explicitly because it relies on arguments by Zivot and Andrews (1992). They derived the limiting distribution of various components entering in the definition of a related t statistic and applied a continuous mapping argument to obtain an expression similar to that of (9) in a different context. The application of this continuous mapping argument is valid if all of the terms are bounded uniformly over λ. There is a need to trim the space of λ because some components are unbounded as λ approaches 0 or 1. The uniform boundedness of each component, however, is only a sufficient condition to obtain the desired result. In the case of the t statistic discussed previously, even though the individual components are unbounded as λ approaches 0 or 1, the t statistic itself remains bounded. Hence there seems to be no need to restrict the space Λ. To make this claim precise, we would need to show tightness of the finite-sample distribution as well as the convergence result derived in Appendix B. This is outside the scope of this article. These comments do not apply to the limiting distribution of the other statistics to be discussed.

To tabulate the percentage points of the asymptotic distribution represented in (9), we proceeded by simulations in a manner similar to that of Perron (1989). The method used is described in Appendix C. The resulting critical values are presented in the rows labeled $T = \infty$ in Table 1 (AO model) and Table 2 (IO model). As mentioned before, the asymptotic distribution is the same in both cases.

Consider now the asymptotic distribution of the statistics $t_{\hat{\alpha}}(i, T_b(\hat{\delta}), k)$ (i =AO, IO), where T_b is chosen to minimize the t statistic for testing $\delta = 0$, the coefficient on the change in mean in regressions (3) or (7). Unlike the previous case, the asymptotic distributions of the statistics in the AO model and the IO model are

Table 1. Percentage points of the distribution of $t_{\hat{\alpha}}(\text{AO}, T_b^*, k)$, additive outlier model, choosing T_b minimizing $t_{\hat{\alpha}}$ in (4).

	1.0%	2.5%	5.0%	10.0%	90.0%	95.0%	97.5%	99.0%
$T = 50$								
$k = 0$	−5.07	−4.73	−4.41	−4.07	−2.23	−2.02	−1.84	−1.64
$k = 2$	−4.92	−4.54	−4.21	−3.86	−2.06	−1.85	−1.66	−1.50
$k = 5$	−4.70	−4.28	−3.96	−3.63	−1.90	−1.70	−1.52	−1.36
$k = k^*$	−5.28	−5.02	−4.76	−4.45	−2.58	−2.35	−2.16	−1.95
$k = k(F)$	−5.18	−4.92	−4.64	−4.35	−2.37	−2.11	−1.94	−1.75
$k = k(t)$	−5.20	−4.95	−4.67	−4.33	−2.36	−2.13	−1.95	−1.74
$T = 100$								
$k = 0$	−5.05	−4.73	−4.41	−4.10	−2.35	−2.14	−1.95	−1.73
$k = 2$	−4.87	−4.52	−4.25	−3.96	−2.26	−2.01	−1.81	−1.57
$k = 5$	−4.67	−4.36	−4.06	−3.78	−2.13	−1.91	−1.69	−1.47
$k = k^*$	−5.42	−4.96	−4.63	−4.34	−2.58	−2.33	−2.12	−1.89
$k = k(F)$	−5.20	−4.83	−4.53	−4.22	−2.43	−2.20	−1.94	−1.73
$k = k(t)$	−5.23	−4.86	−4.51	−4.20	−2.45	−2.23	−2.00	−1.77
$T = 150$								
$k = 0$	−5.06	−4.74	−4.42	−4.12	−2.38	−2.16	−1.95	−1.64
$k = 2$	−4.90	−4.56	−4.34	−4.09	−2.29	−2.04	−1.86	−1.66
$k = 5$	−4.89	−4.49	−4.18	−3.89	−2.19	−1.97	−1.76	−1.57
$k = k^*$	−5.29	−4.91	−4.67	−4.36	−2.53	−2.30	−2.08	−1.85
$k = k(F)$	−5.18	−4.80	−4.55	−4.24	−2.42	−2.17	−1.95	−1.75
$k = k(t)$	−5.15	−4.81	−4.55	−4.23	−2.41	−2.16	−1.95	−1.75
$T = \infty$	−4.95	−4.69	−4.44	−4.19	−2.51	−2.28	−2.10	−1.85

Table 2. Percentage points of the distribution of $t_{\hat{\alpha}}(\text{IO}, T_b^*, k)$, innovational outlier model, choosing T_b minimizing $t_{\hat{\alpha}}$ in (7).

	1.0%	2.5%	5.0%	10.0%	90.0%	95.0%	97.5%	99.0%
$T = 50$								
$k = 0$	−5.07	−4.70	−4.37	−4.04	−2.19	−1.99	−1.82	−1.59
$k = 2$	−5.10	−4.72	−4.42	−4.04	−2.12	−1.89	−1.68	−1.42
$k = 5$	−5.19	−4.79	−4.42	−4.07	−2.06	−1.77	−1.48	−1.10
$k = k^*$	−5.58	−5.23	−4.93	−4.56	−2.56	−2.31	−2.09	−1.89
$k = k(F)$	−5.52	−5.06	−4.81	−4.47	−2.33	−2.04	−1.79	−1.43
$k = k(t)$	−5.51	−5.07	−4.76	−4.42	−2.34	−2.09	−1.85	−1.61
$T = 100$								
$k = 0$	−5.03	−4.71	−4.39	−4.08	−2.33	−2.12	−1.91	−1.68
$k = 2$	−5.06	−4.62	−4.39	−4.06	−2.32	−2.04	−1.78	−1.57
$k = 5$	−4.94	−4.67	−4.35	−4.11	−2.30	−1.99	−1.78	−1.38
$k = k^*$	−5.45	−5.00	−4.72	−4.41	−2.59	−2.33	−2.13	−1.86
$k = k(F)$	−5.27	−4.88	−4.60	−4.28	−2.43	−2.14	−1.90	−1.54
$k = k(t)$	−5.33	−4.88	−4.58	−4.27	−2.41	−2.17	−1.96	−1.68
$T = 150$								
$k = 0$	−5.07	−4.73	−4.41	−4.11	−2.36	−2.15	−1.93	−1.61
$k = 2$	−4.95	−4.68	−4.41	−4.13	−2.34	−2.10	−1.86	−1.65
$k = 5$	−5.11	−4.70	−4.43	−4.11	−2.30	−2.04	−1.84	−1.59
$k = k^*$	−5.36	−4.97	−4.71	−4.40	−2.56	−2.29	−2.07	−1.84
$k = k(F)$	−5.21	−4.87	−4.58	−4.24	−2.41	−2.13	−1.89	−1.63
$k = k(t)$	−5.17	−4.84	−4.58	−4.24	−2.40	−2.16	−1.88	−1.64
$T = \infty$	−4.95	−4.69	−4.44	−4.19	−2.51	−2.28	−2.10	−1.85

different. We first consider the case of the IO model. Again, let Λ be a closed subset of the interval (0, 1). It is shown in Appendix B that, under (8),

$$t_{\hat{\alpha}}(\text{IO}, T_b(\hat{\delta}), k = 0) \Rightarrow \int_0^1 W^*(\lambda^*, r)dW(r) \left\{ \int_0^1 W^*(\lambda^*, r)^2 dr \right\}^{-1/2}, \tag{10}$$

where $\lambda^* = \text{argmin}_{\lambda \in \Lambda} Z(\lambda)$ with $Z(\lambda) = H(\lambda)/K(\lambda)^{1/2}$, and

$$H(\lambda) = A[\lambda W(1) - W(\lambda)] - B(\lambda) \times [(1/2)(W(1)^2 - 1) - W(1) \int_0^1 W(r)dr] K(\lambda) = A[\lambda(1-\lambda)A - B(\lambda)^2]$$

where $A = \int_0^1 W(r)^2 dr - (\int_0^1 W(r)dr)^2$ and $B(\lambda) = \lambda \int_0^1 W(r)dr - \int_0^\lambda W(r)dr$.

Consider now the asymptotic distribution of $t_{\hat{\alpha}}(\text{AO},\ T_b(\hat{\delta}),\ k)$ for the AO model, where again T_b is chosen to minimize the t statistic for testing $\delta = 0$ in (3), denoted $t_{\hat{\delta}}$. The first thing to note is that, under the null hypothesis, $t_{\hat{\delta}}$ is unbounded as $T \to \infty$. A nondegenerate asymptotic distribution is obtained, however, if we consider the scaled version $T^{-1/2}t_{\hat{\delta}}$. Hence, in this asymptotic framework, the procedure is valid if one considers choosing T_b such that the value of $T^{-1/2}t_{\hat{\delta}}$ is minimized. Of course, in finite samples it makes no difference whether one minimizes $t_{\hat{\delta}}$ or $T^{-1/2}t_{\hat{\delta}}$. It is shown in Appendix B that, under (8),

$$t_{\hat{\alpha}}(\text{AO},\ T_b(\hat{\delta}), k=0) \Rightarrow \int_0^1 W^*(\lambda^*, r)dW(r)\left\{\int_0^1 W^*(\lambda^*, r)^2 dr\right\}^{-1/2}, \tag{11}$$

where $\lambda^* = \text{argmin}_{\lambda \in \Lambda} Q(\lambda)$ with $Q(\lambda) = B(\lambda)/[K(\lambda)/A]^{1/2}$, with A, $B(\lambda)$, and $K(\lambda)$ as defined in (10). To tabulate the critical values of the asymptotic distributions (10) and (11), we proceeded in a manner analogous to that described in Appendix C. These critical values are presented in Tables 3 and 4 in the rows labeled $T = \infty$.

3.1. *The limiting distributions in the general case*

The asymptotic distributions derived previously remain valid if additional correlation is present in the data-generating process when higher order autoregressions are applied. Consider first the case in which the noise function is an AR(p) model. It is easy to show (following Dickey and Fuller [1979]) that the asymptotic distributions are still valid provided $k \geq p$. Note that the results also hold under more general conditions. For example, some forms of heteroscedasticity such as autoregressive conditional heteroscedastic errors are allowed (see Pantula 1985). When the errors follow an ARMA process and k is a fixed function of the sample size (such that $k \to \infty$ and $k^3/T \to 0$ as $T \to \infty$), we appeal to results of Said and Dickey (1984) to argue that the limiting distributions remain unchanged.

Table 3. Percentage points of the distribution of $t_{\hat{\alpha}}(\mathrm{AO}, T_b(\hat{\delta}), k)$, additive outlier model, choosing T_b minimizing $t_{\hat{\delta}}$ in (3).

	1.0%	2.5%	5.0%	10.0%	90.0%	95.0%	97.5%	99.0%
$T = 50$								
$k = 0$	−4.24	−3.86	−3.53	−3.19	−0.87	−0.50	−0.21	0.05
$k = 2$	−4.06	−3.67	−3.35	−2.97	−0.74	−0.42	−0.14	0.12
$k = 5$	−3.74	−3.44	−3.06	−2.75	−0.67	−0.34	−0.09	0.14
$k = k^*$	−4.55	−4.20	−3.89	−3.54	−1.24	−0.90	−0.65	−0.30
$k = k(F)$	−4.47	−4.03	−3.65	−3.29	−0.82	−0.45	−0.14	0.34
$k = k(t)$	−4.47	−4.07	−3.68	−3.35	−0.88	−0.53	−0.28	0.01
$T = 100$								
$k = 0$	−4.11	−3.85	−3.55	−3.20	−0.80	−0.50	−0.19	0.24
$k = 2$	−3.99	−3.72	−3.44	−3.13	−0.76	−0.39	−0.13	0.20
$k = 5$	−4.04	−3.54	−3.29	−2.97	−0.71	−0.37	−0.12	0.19
$k = k^*$	−4.58	−4.13	−3.87	−3.53	−1.05	−0.70	−0.47	−0.19
$k = k(F)$	−4.38	−3.95	−3.64	−3.29	−0.79	−0.47	−0.11	0.24
$k = k(t)$	−4.38	−4.03	−3.66	−3.30	−0.81	−0.52	−0.20	0.26
$T = 150$								
$k = 0$	−4.10	−3.79	−3.51	−3.15	−0.81	−0.43	−0.07	0.26
$k = 2$	−4.07	−3.71	−3.43	−3.07	−0.74	−0.44	0.00	0.36
$k = 5$	−4.05	−3.62	−3.33	−3.01	−0.73	−0.34	0.04	0.39
$k = k^*$	−4.39	−4.09	−3.79	−3.41	−0.99	−0.65	−0.24	0.13
$k = k(F)$	−4.29	−3.89	−3.56	−3.22	−0.79	−0.38	−0.03	0.29
$k = k(t)$	−4.30	−3.98	−3.63	−3.26	−0.80	−0.41	−0.05	0.27
$T = \infty$	−4.18	−3.90	−3.61	−3.27	−0.79	−0.39	−0.07	0.25

Consider now the limiting distribution of the statistics in which the truncation lag parameter is chosen using a test of significance on the coefficients of the lagged first-differences of the data. The asymptotic results remain valid if the noise function is that of an AR(p) process and the upper bound $k$max is selected to be greater than p. This is basically because, asymptotically, these procedures will select an order no less than p and the tests of significance on the coefficients of the lags are asymptotically independent of the t statistic on α; see Hall (1990) for details. In the case in which the noise function is a general ARMA(p, q) process, no asymptotic results are available in the present case of estimated autoregressions. Our conjecture is that the asymptotic distributions would remain valid using an argument similar to that of Said and Dickey (1984) — namely, by requiring the upper bound ($k$max) to increase to infinity

Table 4. Percentage points of the distribution of $t_{\hat{\alpha}}(\mathrm{IO}, T_b(\hat{\delta}), k)$, innovational outlier model, choosing T_b minimizing $t_{\hat{\delta}}$ in (7).

	1.0%	2.5%	5.0%	10.0%	90.0%	95.0%	97.5%	99.0%
$T = 50$								
$k = 0$	−4.68	−4.32	−3.97	−3.63	−0.36	0.12	0.53	0.86
$k = 2$	−4.93	−4.36	−4.01	−3.62	−0.16	0.41	0.85	1.32
$k = 5$	−4.84	−4.31	−3.96	−3.63	0.17	0.81	1.35	1.80
$k = k^*$	−5.27	−4.78	−4.42	−3.99	0.26	0.89	1.37	1.80
$k = k(F)$	−5.14	−4.74	−4.31	−3.86	0.15	0.84	1.34	1.72
$k = k(t)$	−5.10	−4.70	−4.26	−3.82	−0.02	0.69	1.18	1.60
$T = 100$								
$k = 0$	−4.74	−4.39	−4.05	−3.73	−0.42	0.09	0.44	1.05
$k = 2$	−4.65	−4.31	−4.06	−3.72	−0.23	0.24	0.68	1.24
$k = 5$	−4.75	−4.37	−4.03	−3.69	−0.03	0.52	1.03	1.42
$k = k^*$	−5.03	−4.66	−4.32	−3.95	0.09	0.60	1.09	1.58
$k = k(F)$	−4.95	−4.58	−4.22	−3.84	−0.10	0.43	0.89	1.33
$k = k(t)$	−4.91	−4.49	−4.19	−3.81	−0.21	0.41	0.89	1.33
$T = 150$								
$k = 0$	−4.84	−4.41	−4.12	−3.77	−0.29	0.14	0.50	0.85
$k = 2$	−4.75	−4.45	−4.17	−3.75	−0.27	0.20	0.60	0.96
$k = 5$	−4.85	−4.45	−4.16	−3.75	−0.29	0.36	0.77	1.16
$k = k^*$	−5.05	−4.70	−4.38	−3.96	−0.13	0.49	0.88	1.22
$k = k(F)$	−4.97	−4.62	−4.27	−3.86	−0.23	0.35	0.76	1.11
$k = k(t)$	−4.97	−4.62	−4.26	−3.86	−0.24	0.37	0.73	1.01
$T = \infty$	−4.73	−4.44	−4.19	−3.86	−0.42	0.09	0.50	0.93

at a controlled rate as the sample size increases to infinity. The verification of such a conjecture is outside the scope of this article.

Things are different when the truncation lag parameter k is chosen to minimize the value of the t statistic for testing $\alpha = 1$. Here, no asymptotic results are yet available. The problem lies in the fact that the procedure for selecting k remains correlated with the t statistic on α even asymptotically. There is also a presumption that the asymptotic distributions remain affected by nuisance parameters (i.e., the particular correlation structure of the data) even asymptotically. As mentioned previously, we present results concerning this statistic to inquire about the extent to which data mining can create problems of inference, and such statistics are not recommended in practice.

4. Simulations of the Finite-Sample Critical Values

In this section, we present finite-sample critical values of the statistics $t_{\hat{\alpha}}(i, j, k)(i = \text{AO, IO}; j = T_b^*, T_b(\hat{\delta}))$ under the null hypothesis of a unit root using various procedures for selecting the truncation lag parameter k. Our aim is to assess the adequacy of the asymptotic approximation and to provide more precise percentage points on which inference can be based. For the simulations, we consider the data-generating process

$$y_t = y_{t-1} + e_t, \quad y_1 = 0. \tag{12}$$

We set $y_1 = 0$ without loss of much generality (the effect of the initial condition being minor). Similarly, we impose $\delta = 0$ in (1) and $\theta = 0$ in (5), given the statistics are asymptotically invariant to these parameters. We specify the sequence of innovations $\{e_t\}$ to be i.i.d. $N(0, 1)$ even though the finite-sample distributions of the statistics are not invariant to the correlation structure of the data. The basic justification is that the asymptotic distributions are invariant to additional correlation in the data under some regularity conditions (except perhaps when k is chosen to minimize $t_{\hat{\alpha}}$). Hence there is no loss in generality in specifying i.i.d. errors under this asymptotic interpretation. The implicit assumption is that, in finite samples, the introduction of additional lags of first-differences of the data would mostly eliminate the dependency of the distributions on the correlation structure. More important, by specifying i.i.d. errors we are better able to assess the relative impact on the finite-sample distributions of various factors such as the procedure used to select the length of the autoregression k. The effects of relaxing this and other assumptions are analyzed in Section 5.

For all statistics considered, we present results for three sample sizes — namely, $T = 50$, 100, and 150. In the case in which k is a fixed value, we give results for $k = 0$, 2, and 5. This allows us to analyze the effect of overparameterization on the finite-sample critical values. When k is chosen according to a test on the coefficients of the lagged first-differences, the size of the test is set at 10%. When choosing k using a data-dependent method, we specify the upper bound on the autoregression, $k$max, to be 5, mainly for computational reasons. The processing time becomes excessive when greater values are used. In some instances, we performed some robustness check by allowing a greater value of $k$max. The results were very similar. In practice, $k$max can be specified to be greater than 5. We also tabulated critical values for the following additional cases. When k is fixed, we considered values up to $k = 10$, the results showed no interesting differences and are not reported. In the case in which k is chosen using a test of significance, we also tabulated critical values when the size of the test is 5%. The results, available on request, are very similar to the 10%-size case. Each set of results was obtained using 2,000 replications. The program was coded with the C language, and $N(0, 1)$ random deviates were obtained from the routine RAN1 of Press *et al.* (1986). To minimize sampling variability, we used the same set of generated data for all cases that share a common value of T, the sample size.

Table 1 presents the results concerning the statistic $t_{\hat{\alpha}}(\mathrm{AO}, T_b^*, k)$ — that is, the t statistic for testing $\alpha = 1$ in the AO regression (4), where the breakpoint T_b is chosen to minimize this t statistic. Consider first the asymptotic distribution. The first thing to note is that it is considerably shifted to the left compared to the case in which the breakpoint is assumed known. Consider for instance the fifth percentage point. From Perron (1990a), the asymptotic critical value when the breakpoint is assumed to occur at mid-sample is –3.04. In contrast, when the breakpoint is assumed unknown, Table 1 shows the corresponding critical value to be –4.44. Second, the asymptotic distribution is a good approximation to the exact distribution when k is fixed at 0 (though slightly less than in the IO model to be discussed).

When k is chosen using a test of significance on the coefficients of the lagged first-differences [rows $k = k(F)$ and $k = k(t)$], the critical values are substantially smaller than those corresponding to both the fixed k case and the asymptotic distribution, especially if the sample size is small. This is due to the correlation, in finite samples, between the statistics on the coefficients of the lagged first differences and the t statistic for testing $\alpha = 1$. As discussed previously, this correlation disappears asymptotically. Our simulation results show this correlation to vanish slowly. Hence for common sample sizes care must be taken in using the asymptotic critical values in this case. When k is chosen such that the t statistic for testing $\alpha = 1$ is minimized (row $k = k^*$), the critical values are substantially smaller than all others even for large sample sizes. This illustrates the extent to which data mining can affect inference. Indeed, selecting k using such a procedure and not properly adjusting the critical values would lead to tests with inflated sizes.

Table 2 presents the critical values corresponding to the statistic $t_{\hat{\alpha}}(\mathrm{IO}, T_b^*, k)$ — that is, the t statistic for testing $\alpha = 1$ using the IO regression (7) and T_b chosen to minimize this t statistic. Many of the qualitative results are similar to the case of the AO model presented in Table 1. The asymptotic distribution is very close to the finite distribution with k fixed at 0 (especially in the left tail of the distribution), even for a sample size as small as 50. Unlike in the AO model, the critical values with k fixed are relatively stable across various values of k. Otherwise, the critical values are quite similar to those in the AO case.

We now turn to the results concerning the cases in which the time of break T_b is chosen to minimize the t statistic on the coefficient associated with the change in mean. Table 3 presents the results corresponding to the AO model, where T_b is chosen to minimize the t statistic for testing $\delta = 0$ in regression (3). Recall that this procedure allows the *a priori* imposition of a one-sided change in the mean of the series (the results also apply if one were to maximize the t statistic for testing $\delta = 0$). Hence, as stated previously, we can expect possibly higher power through higher critical value (i.e., less negative). As shown in Table 3, this is indeed the case. For instance, the asymptotic 5% critical value is –3.61 compared to –4.44 (and the 10% critical value is –3.27 compared to –4.19). Similarly, the finite-sample critical values are uniformly higher (less negative) than those for the AO model, where T_b is chosen to minimize the t statistic on α (see Table 1). Much of the same qualitative comparisons across methods hold: The critical values decrease noticeably as k increases when the sample size is small (the differences are minor when T is large), and the procedure whereby k is chosen to minimize the t statistic on α leads to the smallest (most negative) critical values.

Finally, Table 4 presents the simulation results concerning the statistics $t_{\hat{\alpha}}(\text{IO}, T_b(\hat{\delta}), k)$ using the IO model, where T_b is chosen to minimize the t statistic for testing $\delta = 0$ in (7). A feature of substantial interest is that the asymptotic critical values in the IO model are substantially smaller than in the AO model. For instance, the 5% critical value is –4.19 compared to –3.61. One would therefore expect higher power under the AO model. The other differences between the critical values in Tables 3 and 4 are similar to the differences discussed when comparing the results in Tables 1 and 2. In particular, the asymptotic distribution is, in general, a good approximation to the finite-sample distribution when k is fixed, and the critical values are not much influenced by the value of k when it is fixed. There are very few differences between the critical values using the F test or the t test on the coefficients of the lagged first-differences, and the critical values are smaller throughout (more negative) in the IO case than they are in the AO case (though here the differences are more important).

5. Finite-Sample Size and Power Simulations

In this section we perform simulation experiments to address the following issues: (a) comparison of the size and power properties of the tests using different methods for selecting the truncation lag parameter k, (b) comparison of the size and power properties of the various tests with models containing additional correlation, (c) robustness of the tests designed for the AO model when applied to a data-generating process DGP described by an IO model (and vice versa), (d) size of the tests when a break is present under the null hypothesis of a unit root, and (e) power of the tests as a function of the magnitude and timing of the break.

To address issue (a), in particular, it is useful to start with a class of DGP that does not contain a change under the null or under the alternative hypothesis. To this effect, our first set of simulation experiments considers the case in which the DGP is a member of the class of zero mean ARMA(p, q) models. More specifically, we follow Hall (1990) and consider processes of the form $\Pi_{i=1}^{3}(1-\phi(i)L)(1-\alpha L)y_t = (1+\theta L)e_t$ — that is, ARMA models in which the autoregressive order is at most 4 and the moving average order is at most 1. We select seven parameter configurations that are explicitly described in Table 5. To assess the size of the tests, we specify $\alpha = 1$ and the power is evaluated at $\alpha = 0.7$. Note, however, that α is parameterized as the largest root of the polynomial and not the sum of the autoregressive coefficients. Hence one cannot meaningfully compare power across experiments. In all cases the sample size is 100, $k\max = 5$, and 1,000 replications are used with $e_t \sim$ i.i.d. $N(0, 1)$. For the cases in which the statistics are constructed with k fixed *a priori*, we use the relevant asymptotic critical values. For the other cases, we use the finite-sample critical values as listed in the appropriate table with $T = 100$. We report results for tests of nominal size 5%.

The results are presented in Table 5 with the mnemonic $TX (X = 1, 2, 3, 4)$ used to refer to the statistic whose critical values are in Table X ($X = 1, 2, 3, 4$). Consider first experiment (1) in which the DGP is an AR(1). If $\alpha = 1$, the results merely verify that the size is close to 5% in all cases except for a slight departure for the statistics from the AO regression when k is fixed and larger than the true order 0 (because asymptotic critical values are used). The power results indicate that, for all tests, it is largest with k fixed at the true value 0. The power of the tests using a data-dependent method to select k is larger than that of the tests with k fixed for values of k other than the true value 0, however. The fact that the $k(t)$ procedure leads to tests with higher power than tests based on the $k(F)$ procedure is also of interest.

Consider now experiment (2) in which the DGP is an AR(2) with a positive root of 0.65 (hence the true value of k is 1). Here the exact size of the test is greater than the nominal size if $k = 0$ or if the procedure k^* (which selects k to minimize $t_{\hat{\alpha}}$) is used. Otherwise the size is adequate. Consider now the power results. If $k = 0$, the tests have virtually no power even if they are liberal. On the other

Table 5. Finite-sample size and power simulations (base cases with no break).

DGP	α	k	$T1$	$T2$	$T3$	$T4$
(1)						
$\phi(i) = 0$	1.0	0	0.048	0.044	0.039	0.040
$(i = 1, 2, 3)$		1	0.048	0.054	0.035	0.036
$\theta = 0.0$		2	0.030	0.050	0.023	0.034
		3	0.031	0.051	0.034	0.039
		4	0.032	0.053	0.027	0.046
		5	0.024	0.054	0.028	0.046
		k^*	0.057	0.058	0.052	0.060
		$k(F)$	0.042	0.060	0.047	0.057
		$k(t)$	0.050	0.060	0.052	0.057
	0.7	0	0.849	0.828	0.942	0.743
		1	0.722	0.694	0.832	0.589
		2	0.548	0.507	0.676	0.406
		3	0.448	0.418	0.554	0.332
		4	0.320	0.304	0.420	0.219
		5	0.275	0.261	0.315	0.203
		k^*	0.852	0.792	0.913	0.406
		$k(F)$	0.778	0.617	0.764	0.479
		$k(t)$	0.844	0.783	0.819	0.688
(2)						
$\phi(1) = 0.65$	1.0	0	0.081	0.114	0.013	0.069
$\phi(2) = 0.0$		1	0.036	0.058	0.018	0.041
$\phi(3) = 0.0$		2	0.035	0.062	0.025	0.044
$\theta = 0.0$		3	0.033	0.065	0.023	0.044
		4	0.029	0.056	0.023	0.042
		5	0.026	0.055	0.015	0.044
		k^*	0.091	0.111	0.032	0.079
		$k(F)$	0.040	0.060	0.031	0.064
		$k(t)$	0.050	0.062	0.029	0.057
	0.7	0	0.002	0.001	0.002	0.002
		1	0.344	0.320	0.468	0.260
		2	0.244	0.224	0.353	0.174
		3	0.229	0.208	0.289	0.166
		4	0.176	0.176	0.231	0.151
		5	0.153	0.162	0.190	0.133
		k^*	0.367	0.288	0.419	0.167
		$k(F)$	0.314	0.263	0.416	0.225
		$k(t)$	0.383	0.323	0.429	0.290

(*Continued*)

Table 5. (*Continued*)

DGP	α	k	$T1$	$T2$	$T3$	$T4$
(3)						
$\phi(1) = -0.65$	1.0	0	0.755	0.747	0.637	0.516
$\phi(2) = 0.0$		1	0.044	0.049	0.042	0.037
$\phi(3) = 0.0$		2	0.030	0.047	0.033	0.034
$\theta = 0.0$		3	0.033	0.050	0.033	0.038
		4	0.026	0.041	0.028	0.041
		5	0.025	0.054	0.029	0.043
		k^*	0.695	0.656	0.573	0.386
		$k(F)$	0.039	0.055	0.052	0.057
		$k(t)$	0.051	0.049	0.047	0.051
	0.7	0	1.00	1.00	1.00	1.00
		1	0.834	0.806	0.917	0.715
		2	0.669	0.622	0.767	0.521
		3	0.537	0.494	0.623	0.384
		4	0.382	0.352	0.466	0.279
		5	0.313	0.302	0.362	0.218
		k^*	1.00	1.00	1.00	1.00
		$k(F)$	0.752	0.615	0.785	0.485
		$k(t)$	0.814	0.753	0.827	0.674
(4)						
$\phi(1) = 0.65$	1.0	0	0.259	0.372	0.038	0.173
$\phi(2) = 0.8$		1	0.554	0.579	0.266	0.360
$\phi(3) = 0.0$		2	0.027	0.064	0.015	0.052
$\theta = 0.0$		3	0.032	0.061	0.018	0.057
		4	0.025	0.061	0.008	0.050
		5	0.021	0.061	0.013	0.045
		k^*	0.607	0.655	0.230	0.243
		$k(F)$	0.037	0.067	0.025	0.062
		$k(t)$	0.040	0.057	0.016	0.064
	0.7	0	0.003	0.006	0.000	0.003
		1	0.791	0.772	0.737	0.590
		2	0.117	0.104	0.126	0.089
		3	0.106	0.115	0.103	0.098
		4	0.088	0.097	0.081	0.084
		5	0.071	0.086	0.071	0.073
		k^*	0.723	0.667	0.641	0.212
		$k(F)$	0.206	0.123	0.161	0.119
		$k(t)$	0.197	0.159	0.150	0.151

(*Continued*)

Table 5. (*Continued*)

DGP	α	k	$T1$	$T2$	$T3$	$T4$
(5)						
$\phi(1) = 0.65$	1.0	0	0.125	0.183	0.026	0.099
$\phi(2) = 0.5$		1	0.061	0.081	0.042	0.060
$\phi(3) = -0.4$		2	0.064	0.092	0.041	0.078
$\theta = 0.0$		3	0.026	0.056	0.020	0.041
		4	0.025	0.060	0.019	0.049
		5	0.020	0.054	0.016	0.051
		k^*	0.147	0.188	0.060	0.112
		$k(F)$	0.048	0.073	0.045	0.071
		$k(t)$	0.060	0.071	0.040	0.068
	0.7	0	0.001	0.003	0.000	0.002
		1	0.269	0.253	0.358	0.178
		2	0.393	0.374	0.438	0.265
		3	0.195	0.183	0.223	0.136
		4	0.148	0.148	0.170	0.102
		5	0.110	0.118	0.130	0.095
		k^*	0.364	0.295	0.372	0.136
		$k(F)$	0.296	0.243	0.336	0.193
		$k(t)$	0.375	0.312	0.360	0.246
(6)						
$\theta = 0.5$	1.0	0	0.031	0.032	0.008	0.027
$\phi(i) = 0.0$		1	0.092	0.098	0.078	0.070
$(i = 1, 2, 3)$		2	0.022	0.041	0.013	0.027
		3	0.039	0.062	0.042	0.049
		4	0.027	0.051	0.021	0.044
		5	0.029	0.055	0.024	0.046
		k^*	0.077	0.081	0.062	0.069
		$k(F)$	0.051	0.061	0.049	0.060
		$k(t)$	0.052	0.060	0.043	0.059
	0.7	0	0.098	0.084	0.285	0.069
		1	0.861	0.834	0.921	0.752
		2	0.335	0.295	0.468	0.237
		3	0.452	0.424	0.550	0.329
		4	0.260	0.245	0.330	0.186
		5	0.240	0.244	0.287	0.194
		k^*	0.815	0.743	0.858	0.372
		$k(F)$	0.669	0.455	0.629	0.363
		$k(t)$	0.685	0.598	0.614	0.490

(*Continued*)

Table 5. (*Continued*)

DGP	α	k	$T1$	$T2$	$T3$	$T4$
(7)						
$\theta = -0.5$	1.0	0	0.758	0.732	0.629	0.508
$\phi(i) = 0.0$		1	0.228	0.238	0.233	0.168
$(i = 1, 2, 3)$		2	0.078	0.088	0.092	0.072
		3	0.050	0.065	0.059	0.050
		4	0.035	0.046	0.036	0.041
		5	0.032	0.059	0.036	0.043
		k^*	0.686	0.648	0.571	0.375
		$k(F)$	0.248	0.116	0.155	0.098
		$k(t)$	0.219	0.196	0.142	0.135
	0.7		1.00	1.00	1.00	1.00
		1	0.996	0.998	1.00	0.992
		2	0.915	0.904	0.955	0.822
		3	0.750	0.706	0.811	0.607
		4	0.553	0.513	0.618	0.409
		5	0.421	0.395	0.475	0.299
		k^*	1.00	1.00	1.00	0.697
		$k(F)$	0.964	0.823	0.861	0.661
		$k(t)$	0.960	0.946	0.894	0.852

Notes: DGP: $\Pi_{i=1}^{3}(1-\phi(i)L)(1-\alpha L)y_t = (1+\theta L)e_t$, $e_t \sim$ i.i.d. $N(0, 1)$; $T = 100$; 1,000 replications. T1: $t_{\hat{\alpha}}(\text{AO}, T_b^*, k)$; choosing T_b minimizing $t_{\hat{\alpha}}$ in (4). T2: $t_{\hat{\alpha}}(\text{IO}, T_b^*, k)$; choosing T_b minimizing $t_{\hat{\alpha}}$ in (7). T3: $t_{\hat{\alpha}}(\text{AO}, T_b(\hat{\delta}), k)$; choosing T_b minimizing $t_{\hat{\delta}}$ in (3). T4: $t_{\hat{\alpha}}(\text{IO}, T_b(\hat{\delta}), k)$; choosing T_b minimizing $t_{\hat{\delta}}$ in (7).

hand, the power of the tests based on the $k(t)$ or $k(F)$ procedure is comparable to the power of the tests with k fixed at 1, the most favorable case [actually, except for $T3$, the power of the test with $k(t)$ is higher]. Note that the power of the tests decreases substantially using a fixed k that is greater than the true value. Experiment (3) considers again an AR(1) but with a negative root -0.65. When $k = 0$ (underparameterization), the size of the tests is dramatically high. The same feature holds using the k^* procedure. The size of the tests with $k(t)$, $k(F)$, or k fixed *a priori* greater than 0 is adequate, however. Turning now to the power results, the tests with $k(t)$ are more powerful than those with $k(F)$ and slightly less powerful than those with k fixed *a priori* at 1. But again the power decreases rapidly as k increases.

Experiment (4) considers the case in which the DGP is an AR(3) and $k = 2$. When k is fixed below 2, the size of the tests is dramatically inflated (yet with $k = 0$, the tests have basically no power). Again, the size of the tests with $k(t)$ and $k(F)$ is adequate, whereas that of the tests with k^* is again dramatically inflated. Here, the power of the tests with $k(t)$ or $k(F)$ is greater than with k fixed at a value greater than 1. Experiment (5) considers the case of an AR(3) in which two of the roots nearly cancel and hence, though $k = 3$ is the true value, a good approximation can be expected with $k = 1$. The results are basically the same as those of experiment (2).

These experiments with AR processes lead to the following conclusions. First, the tests have best properties (size close to 5% and highest power) when k is fixed at the true value. In general, when k is selected *a priori* at any other value than the true one, the tests have undesirable properties — substantial size distortions for underparameterization and reduced power with overparameterization. The size of the tests when k is selected to minimize $t_{\hat{\alpha}}$, k^*, is roughly equal to the maximum of the size over all values of k considered. Hence it has, in general, a substantially inflated size and therefore should not be used in practice. The size of the tests with k selected with the $k(t)$ or $k(F)$ procedure is close to the nominal 5% size in all cases (provided the adjusted critical values given in Tables 1–4 are used). The power in these cases is only slightly lower than that with k fixed at the true value and greater than with k fixed at a larger value (in which case the size is adequate). Since k is not known in practice, the evidence presented by these experiments strongly argues for the use of the $k(t)$ or $k(F)$ procedures to select the order of the autoregression. Among these two procedures, $k(t)$ leads to tests with higher power in almost all cases.

Experiments (6) and (7) consider cases in which the DGP is an ARMA(1, 1). With a positive moving average parameter, similar results hold. If $k = 0$, the tests are conservative and have no power. If $k = 1$, their size is inflated. The size of the tests with $k(t)$ and $k(F)$ is close to 5%, and the power is greater than when k is fixed and greater than 1. When the moving average parameter is negative, the picture is different. With k fixed, the size is close to 5% with k

greater than 2, but the power is then low. The tests with $k(t)$ and $k(F)$ have, however, a size greater than 5% (a finding that parallels Schwert's [1989] results). Other simulations with $k$max $= 10$ showed some reduction in the size distortions.

Finally, comparing the relative performance of the different tests using, say, the procedure $k(t)$ to select the order of the autoregression, we have the following ranking. The tests issued from the AO methodology seem to be more powerful with an edge for $T1$ (choosing T_b minimizing $t_{\hat{\alpha}}$). The least powerful is that obtained using the IO framework with T_b chosen to minimize the t statistic on the dummy for the change in mean ($T4$). These rankings are, however, obtained in cases for which no change in mean is present in the data.

To address the remaining issues stated at the beginning of this section, we consider data-generating processes that explicitly incorporate a change in mean under both the null and alternative hypotheses. To this effect we tailor our next simulation experiments to models estimated using the Finland/US GDP-based real exchange rate. We first suppose that the data are generated by an AO model. For the Finland/US series with a break date at 1940, the estimation of Model (2) when $k = 1$ yields the process $y_t = -0.976 - \delta DU_t + v_t$, where $v_t = \alpha v_{t-1} + 0.305\Delta v_{t-1} + e_t(\sigma_e = 0.125)$. The estimate of α is 0.582 but to obtain useful simulations we set $\alpha = 0.8$ under the alternative and $\alpha = 1$ under the null hypothesis. The estimate of δ is –0.420 (more than three times the estimated standard deviation of the residuals). We use this value as the base case but also investigate results with larger changes. In the base case, we also impose $T = 100$ and the break date $T_b = 40$. Given the results of our first set of simulations, we now report results about the properties of the tests only when k is selected using the $k(t)$ or $k(F)$ procedure.

The results are presented in Table 6. Row (1) presents the size of the tests in the base case and shows it to be near the nominal 5% size. This illustrates that, even though the null distribution of the statistics is not invariant to δ in finite samples, values of the change in mean as large as 3.4 times the standard deviation of the errors leave the size of the tests unaffected. Though this robustness of the size holds for values of δ as large as four times the standard

Table 6. Finite-sample size and power simulations (5% nominal size), data generated by the additive outlier model (2).

DGP	k	*T1*	*T2*	*T3*	*T4*
(1) $\alpha = 1$	$k(t)$	0.048	0.046	0.035	0.075
	$k(F)$	0.040	0.052	0.036	0.075
(2) $\alpha = 1$; $\delta = -1.26$	$k(t)$	0.172	0.206	0.044	0.276
	$k(F)$	0.126	0.219	0.040	0.289
(3) $\alpha = 1$; $\delta = -2.1$	$k(t)$	0.422	0.517	0.036	0.583
	$k(F)$	0.382	0.531	0.035	0.598
(4) $\alpha = 0.8$	$k(t)$	0.636	0.572	0.756	0.747
	$k(F)$	0.564	0.506	0.682	0.670
(5) $\alpha = 0.8$; $\delta = -1.26$	$k(t)$	0.936	0.979	0.697	0.992
	$k(F)$	0.899	0.975	0.621	0.990
(6) $\alpha = 0.8$; $T_b = 25$	$k(t)$	0.634	0.586	0.740	0.732
	$k(F)$	0.570	0.501	0.653	0.641
(7) $\alpha = 0.8$; $T_b = 75$	$k(t)$	0.629	0.576	0.728	0.741
	$k(F)$	0.562	0.499	0.667	0.646

Notes: DGP: $y_t = -0.976 - \delta DU_t + v_t$; $v_t = \alpha V_{t-1} + 0.305\Delta v_{t-1} + e_t (t = 3, \ldots, T)$; $e_t \sim$ i.i.d. $N(0, \sigma_e^2)$, $\sigma_e = 0.125$, $T = 100$, $v_1 = -0.124$, $v_2 = -0.103$; base case: $T_b = 40$; $\delta = -0.420$. T1: $t_{\hat{\alpha}}(\text{AO}, T_b^*, k)$ choosing T_b minimizing $t_{\hat{\alpha}}$ in (4). T2: $t_{\hat{\alpha}}(\text{IO}, T_b^*, k)$; choosing T_b minimizing $t_{\hat{\alpha}}$ in (7). T3: $t_{\hat{\alpha}}(\text{AO}, T_b(\hat{\delta}), k)$; choosing T_b minimizing $t_{\hat{\delta}}$ in (3). T4: $t_{\hat{\alpha}}(\text{IO}, T_b(\hat{\delta}), k)$; choosing T_b minimizing $t_{\hat{\delta}}$ in (7).

deviation of the residuals, it does not for more extreme values as demonstrated by experiments 2 and 3. In experiment 2, we consider tripling δ from its base value (about 10 standard deviations) and the results show substantial size distortions (above the nominal 5% level) for all tests except $t_{\hat{\alpha}}(\text{AO}, T_b(\hat{\delta}), k)$. Experiment 3, with five times the base value for δ (about 17 standard deviations), shows this feature to be more severe as δ increases. Again, however, $t_{\hat{\alpha}}(\text{AO}, T_b(\hat{\delta}), k)$ is robust to even enormous changes in mean under the null hypothesis. These size distortions in finite samples, when a change in mean is present under the null hypothesis, may at first appear troublesome. For practical purposes, however, they do not create much problem because, based on these and other nonreported experiments, the tests show significant size distortions only for changes that are more than 10 standard deviations. Such large changes can be considered enormous and unlikely in practical instances in which the hypothesis being tested here is an issue.

Rows (4)–(7) in Table 6 consider the power of the tests. The base case is presented in row (4). The tests from the AO regressions are more powerful than those from the IO regressions. The differences are not substantial, however, and the results therefore show some robustness of the tests when applied to an inappropriately chosen model (i.e., the IO instead of the AO). As expected, the power of the tests choosing T_b minimizing $t_{\hat{\delta}}$, which impose the *a priori* restriction on the sign of the change, is greater than those choosing T_b minimizing $t_{\hat{\alpha}}$.

Row (5) considers the power of the tests when the change in mean is tripled. The power of all tests increases substantially except for the statistic $t_{\hat{\alpha}}(\text{AO}, T_b(\hat{\delta}), k)$, whose power decreases marginally. Finally, rows (6) and (7) consider the base-case model, changing the position of the break. Row (6) presents results for a break early in the sample and row (7) for a break late in the sample. The results show the power to be little affected by the dating of the break for all tests. Note that in all cases the power is greater using the $k(t)$ procedure to select the truncation lag parameter compared to the $k(F)$ procedure.

Table 7 presents similar results when the data-generating process is within the class of IO models. Again, the specific process used in the simulations is calibrated from the estimated IO model using the US/Finland GDP-based series. Under the null hypothesis, Model (5) is estimated with $k = 2$ and the results are presented in Table 7 (panel A) with the parameter estimates corresponding to those in the base case. The change in mean θ is estimated close to one standard deviation. In the simulations, we set $T = 100$ and a breakpoint at $T_b = 45$. To calibrate the process under the alternative hypothesis we use regression (7). The estimated model is described in Table 7 (panel B), where α, the sum of the autoregressive coefficients, is changed to 0.8 to obtain useful power results. The change in mean is estimated to be about 1.2 standard deviations.

Row (1) presents results pertaining to the size of the tests in the base case. As shown, the size is close to the nominal 5% value for all

Table 7. Finite-sample size and power simulations (5% nominal size).

DGP	k	*T1*	*T2*	*T3*	*T4*
A. Data: Innovational outlier model (5), null hypothesis					
(1) Base Case	$k(t)$	0.061	0.057	0.050	0.058
	$k(F)$	0.071	0.050	0.052	0.052
(2) $\theta = -0.432$	$k(t)$	0.067	0.062	0.061	0.086
	$k(F)$	0.083	0.056	0.069	0.080
(3) $\theta = -0.720$	$k(t)$	0.112	0.118	0.051	0.183
	$k(F)$	0.092	0.114	0.058	0.173
B. Data: Innovational outlier model (7), alternative hypothesis					
(4) $\alpha = 0.8$	$k(t)$	0.970	0.967	0.726	0.984
	$k(F)$	0.916	0.884	0.600	0.940
(5) $\alpha = 0.8$; $\delta = -0.45$	$k(t)$	0.977	1.00	0.816	1.00
	$k(F)$	0.980	0.998	0.740	0.999
(6) $\alpha = 0.8$; $T_b = 25$	$k(t)$	0.966	0.952	0.673	0.970
	$k(F)$	0.869	0.873	0.457	0.915
(7) $\alpha = 0.8$; $T_b = 75$	$k(t)$	0.959	0.965	0.567	0.970
	$k(F)$	0.935	0.882	0.442	0.915

Notes: For panel A, DGP: $\Delta y_t = \theta D(TB)_t + 0.180\Delta y_{t-1} - 0.355\Delta y_{t-2} + e_t (t = 4, \ldots, T)$; $e_t \sim$ i.i.d. $N(0, \sigma_e^2)$, $\sigma_e = 0.130$, $T = 100$, $y_1 = 0.333$, $y_2 = 0.340$, $y_3 = 0.352$; base case: $T_b = 45$; $\theta = -0.144$. For panel B, DGP: $y_t = -0.363 + \delta DU_t + 0.117D(TB)_t + \alpha y_{t-1} + 0.336\Delta y_{t-1} + e_t (t = 3, \ldots, T)$; $e_t \sim$ i.i.d. $N(0, \sigma_e^2)$, $\sigma_e = 0.126$, $T = 100$, $y_1 = 0.333$, $y_2 = 0.340$; base case: $T_b = 45$; $\delta = -0.15$. T1: $t_{\hat{\alpha}}(\text{AO}, T_b^*, k)$; choosing T_b minimizing $t_{\hat{\alpha}}$ in (4). T2: $t_{\hat{\alpha}}(\text{IO}, T_b^*, k)$; choosing T_b minimizing $t_{\hat{\alpha}}$ in (7). T3: $t_{\hat{\alpha}}(\text{AO}, Tb(\hat{\delta}), k)$; choosing T_b minimizing $t_{\hat{\delta}}$ in (3). T4: $t_{\hat{\alpha}}(\text{IO}, T_b(\hat{\delta}), k)$; choosing T_b minimizing $t_{\hat{\delta}}$ in (7).

tests. As in the case with data generated by an AO model, the size of the tests is above the nominal size if the change in mean is very large. This is illustrated in row (3), which shows some size distortions if the change is greater than five standard deviations. Row (2) shows that little size distortion is present if the change is about three standard deviations. Again, we do not view this feature as a practical problem for reasons discussed previously.

Row (4) presents the power of the tests in the base case. Here, the powers of the AO or IO version of the test are comparable when choosing T_b minimizing $t_{\hat{\alpha}}$, but the IO version (T4) of the test when choosing T_b minimizing $t_{\hat{\delta}}$ is substantially more powerful than its AO counterpart (T3) (although the results again show robustness to the specified model). The version of the test choosing T_b minimizing $t_{\hat{\delta}}$ appears only marginally more powerful in the IO version (compare T4 and T2) and substantially less powerful in the AO version (compare T3 and Tl).

Row (5) considers the power when the change in mean is tripled from that in the base case. The power of all tests increases, especially for $t_{\hat{\alpha}}(\text{AO}, T_b(\hat{\delta}), k)$ (T3). Rows (6) and (7) consider alternative break dates either early in the sample [row (6)] or late [row (7)]. The power of the tests is not significantly affected except for $t_{\hat{\alpha}}(\text{AO}, T_b(\hat{\delta}), k)$, whose power decreases compared to the case with a break near mid-sample. Again, in almost all cases the tests based on the procedure $k(t)$ to select the order of the autoregression are more powerful than those based on the procedure $k(F)$.

6. Application to the Purchasing Power Parity Hypothesis

In this section, we apply the proposed testing procedures to the two real exchange rates discussed in the introduction — namely, the US/UK series based on the CPI and the US/Finland series based on the GNP deflator. The motivation for using a unit-root test on the real exchange rate series as a test of the PPP hypothesis can be described as follows. First, consider the multiplicative model of the real exchange rate given by $E_tP_t^*/P_t = bw_t$ where E_t is the dollar price of the foreign currency, P_t is the US price level, P_t^* is the foreign price level, b is a constant, and w_t is a positive-disturbance term with mean 1. Given that PPP is rarely considered a theory of exchange-rate determination, we can view E_t, P_t, and P_t^* as endogenously determined. Defining $E_tP_t^*/P_t$ to be the real exchange rate, R_t, we have $R_t = bw_t$, and taking the logarithm we obtain the linear specification $r_t = r + u_t$, where r_t is the logarithm of the real exchange rate, r is a constant, and u_t is a mean zero disturbance term. The absolute version of PPP states that the real exchange rate is equal to 1 at every point in time (which would imply $r = 0$ and $u_t = 0$). Practical considerations imply a less strict interpretation. First, since only price indexes (not price levels) are observed, we can only test a relative version of PPP, which states that r must be constant but imposes no restrictions on its level. Second, given the world's inherent stochastic environment, we would expect deviations from PPP over time. If we allow only transitory shocks, u_t is then a stationary stochastic process. It is this empirical implication of PPP which is of interest. Indeed, a situation in which shocks have permanent effects is in contradiction to the version of PPP just described. It is therefore of interest to test whether shocks to the real exchange rate exhibit a long-term effect.

In practice, one must specify a general class of models in which these issues can be analyzed. For our purpose, it is useful to consider the process u_t as a member of the class of finite-order ARMA(p, q) models; that is, $A(L)u_t = B(L)e_t$, where e_t is a white-noise disturbance term. In this framework, we can test the null hypothesis that PPP does not hold by testing whether the autoregressive polynomial $A(L)$ contains a root on the unit circle. Given that the interest in testing PPP centers on the stochastic behavior of the noise function $\{u_t\}$, we may wish to specify a more flexible structure about the level of the real exchange rate series. In particular, as Figures 1–4 suggest, one may want to allow for the possibility of a structural change in mean at some unknown date. Using the methods described in this article, one can test whether the noise function $\{u_t\}$ is characterized by the presence of an autoregressive unit root.

We applied the procedures proposed in earlier sections to the US/UK CPI-based real-exchange-rate series and the US/Finland real-exchange-rate series based on the GNP deflator. We do not present results pertaining to the case in which k is unrestricted ($k = k^*$) given its poor finite-sample performance documented in Section 5. Consider first applying the AO model with the breakpoint T_b chosen to minimize the value of the t statistic on the sum of the autoregressive coefficients — that is, $t_{\hat{\alpha}}(\text{AO}, T_B^*, k)$. The results are presented in Table 8 for two methods of choosing k. In this application, both methods gave the same values; hence

Table 8. Empirical results, $t_{\hat{\alpha}}(\text{AO}, T_b^*, k)$; additive outlier model.

Series	T_b	k	$\widehat{\mu}$	$\hat{\delta}$	$\hat{\alpha}$	$t_{\hat{\alpha}}$	p value (k fixed)	p value ($k(t)$)	p value (k(F))
US/Finland	1937	1	−0.97	−0.4003	0.597	−5.06	< 0.01	0.02	0.02
(GDP)			(−31.96)	(−9.91)					
(k max = 5)									
US/UK	1943	1	−1.66	0.3529	0.692	−4.83	0.02	0.03	0.03
(CPI)			(−100.6)	(14.62)					
(k max = 5)									

Note: Regression: $y_t = \mu + \delta DU_t + \tilde{y}_t; \tilde{y} = \sum_{i=0}^{k} \omega_i D(TB)_{t-i} + \alpha \tilde{y}_{t-1} + \sum_{i=1}^{k} c_i \Delta \tilde{y}_{t-i} + e_t$.

only one regression is reported. The statistics of most interest are the estimates of α and their t statistics, as well as three sets of p values in the last three columns (reported to the nearest 1%). The first set of p values is obtained using the distribution of the t statistic in the fixed k scenario using the empirically selected value of k. This amounts, basically, to reporting the p values based on the asymptotic distribution corrected, to a first approximation, for small-sample biases. The second set of p values corresponds to the critical values of the t statistic when k and T_b are chosen so that the last lag is significant according to a 10% two-sided t test on its coefficient, and the last lag in higher order autoregressions has insignificant coefficients [$k = k(t)$ in Table 1]. The last set of p values are those corresponding to the case in which T_b and k are chosen according to a 10% joint significance test on the coefficients of the lagged first differences of the data [$k = k(F)$ in Table 1]. In both cases, we selected $k\max = 5$, and the p values are obtained using the simulations performed for $T = 100$. As the results in Table 8 show, the unit-root hypothesis can easily be rejected at the 5% level using both procedures for selecting k and T_b. Indeed, for both the US/Finland and US/UK series, both procedures yield the same values — namely, $T_b = 1937$ and $k = 1$ for US/Finland and $T_b = 1943$ and $k = 1$ for US/UK.

Table 9 presents results applying the IO model selecting T_b and k to minimize the t statistic for testing $\alpha = 1$ — that is, using the statistic $t_{\hat{\alpha}}(\text{IO}, T_b^*, k)$. The presentation of the results

Table 9. Empirical results, $t_{\hat{\alpha}}$ (IO, T_b^*, k); innovational outlier model.

Series	T_b	k	$\widehat{\mu}$	$\hat{\delta}$	$\widehat{\theta}$	$\hat{\alpha}$	$t_{\hat{\alpha}}$	p value (k fixed)	p value ($k(t)$)	p value (k(F))
US/Finland	1945	1	−0.363	−0.150	0.117	0.641	−4.43	0.05	0.07	0.07
(GDP)			(−4.35)	(−3.60)	(0.91)					
(k max = 5)										
US/UK	1944	1	−0.512	0.112	−0.072	0.690	−4.83	0.02	0.03	0.03
(CPI)			(−4.80)	(4.14)	(−1.03)					
(k max = 5)										

Note: Regression: $y_t = \mu + \delta DU_t + \theta D(TB)_t + \alpha y_{t-1} + \sum_{i=1}^{k} c_i \Delta y_{t-i} + e_t$.

Table 10. Empirical results, $t_{\hat{\alpha}}$ (AO, $T_b(\hat{\delta}), k$); additive outlier model.

			With t test on last lag				With F test on additional lags			
	T_b	$\hat{\delta}$	k	$t_{\hat{\alpha}}$	p value (fixed k)	p value ($k(t)$)	k	$t_{\hat{\alpha}}$	p value (fixed k)	p value (k(F))
US/Finland	1940	−0.4204	1	−4.78	< 0.01	< 0.01	1	−4.78	< 0.01	< 0.01
(GDP)		(−11.27)		(0.587)				(0.587)		
(k max = 5)										
US/UK	1946	0.3633	1	−4.36	< 0.01	0.01	3	−3.87	0.02	0.03
(CPI)		(15.80)		(0.702)				(0.692)		
(k max = 5)										

Note: Regression: $y_t = \mu + \delta DU_t + \overline{y}_t; \overline{y} = \sum_{i=0}^{k} \omega_i D(TB)_{t-i} + \alpha \overline{y}_{t-1} + \sum_{i=1}^{k} c_i \Delta \overline{y}_{t-i} + e_t$. t statistics are in parentheses except for the columns labeled $t_{\hat{\alpha}}$, in which the entries are the estimates of α.

follows that in Table 8. For the US/UK series, $T_b = 1944$ and $k = 1$ irrespective of the method used to select k, and the results again allow a rejection of the unit root at the 5% level. For the US/Finland series, $T_b = 1945$ (and $k = 1$) using both methods, and the p value for the null hypothesis of a unit root is 0.07. Since the noise function is stationary, we can use the fact that the asymptotic distribution of $t_{\hat{\delta}}$ is standard normal to conclude that the US/Finland series is characterized by a significant decrease in mean, whereas the US/UK series is characterized by a significant increase in mean.

Tables 10 and 11 present the results when T_b and k are selected to minimize (in the case of US/Finland) or maximize (in the case of US/UK) the t statistic on the coefficient associated with the change

Table 11. Empirical results, $t_{\hat{\alpha}}(\text{IO}, T_b(\hat{\delta}), k)$; innovational outlier model.

Series	T_b	k	$t_{\hat{\delta}}$	$\hat{\alpha}$	$t_{\hat{\alpha}}$	p value (k fixed)	p value ($k(t)$)	p value ($k(F)$)
US/Finland (GDP) (k max = 5)	1945	1	−3.60	0.641	−4.43	0.03	0.03	0.04
US/UK (CPI) (k max = 5)	1944	1	4.14	0.690	−4.83	<0.01	0.02	0.02

Note: Regression: $y_t = \mu + \delta DU_t + \theta D(TB)_t + \alpha y_{t-1} + \sum_{i=1}^{k} c_i \Delta y_{t-i} + e_t$.

in mean, $t_{\hat{\delta}}$. The application of these procedures implies the *a priori* imposition of a one-sided change in mean occurring at an unknown date. Table 10 considers the AO model — that is, using the statistic $t_{\hat{\alpha}}(\text{AO}, T_b(\hat{\delta}), k)$. For the US/Finland series, the unit root is rejected at less than the 1% level using both procedures. This shows how the mild *a priori* imposition of a one-sided change can increase the power of the test substantially. The results are similar for the US/UK series with a rejection at less than the 5% level. Note, however, that the rejection is stronger using the $k(t)$ procedure to select the order of the autoregression compared to the $k(F)$ procedure. This is consistent with the simulation results reported in Section 5. Using this statistic, the breakpoint is estimated to be 1940 for the US/Finland series and 1946 for the US/UK series. Table 11 presents similar results when using the IO model. For the US/UK series, $T_b = 1944$ and $k = 1$ irrespective of the method used, and the unit-root hypothesis is again easily rejected. For the US/Finland series, $T_b = 1945$. Here again, the unit root can be rejected at the 5% level.

The results obtained strongly suggest that both the US/Finland real exchange rate based on the CPI and the US/UK real exchange rate based on the GNP deflator are stationary series if allowance is made for the possibility of a one-time change in the mean of the series. This is contrary to the evidence obtained using a standard Dickey–Fuller test, in which case the unit root cannot be rejected. Our results show the nonrejection to be because no allowance was made for a change in mean.

7. Concluding Comments

This article has presented a class of procedures to test the null hypothesis of a unit root in a time series of data that is possibly affected by a one-time change in its mean level. We considered two classes of models, labeled the *additive outlier model* and the *innovational outlier model.* The former is best suited for series exhibiting a sudden change in mean, and the latter is best suited if the change takes place gradually. The appropriateness of these models henceforth depends on the actual series being used. Nevertheless, our simulation experiments and empirical applications showed that the results can be robust to alternative specifications.

For each of these models, we considered two types of statistics. The first considers the minimal value of the t statistic on the sum of the autoregressive coefficients over all possible breakpoints. This, in effect, allows testing the unit-root hypothesis without any a priori assumption about either the location of the breakpoint or the sign of the change in mean. In an attempt to provide statistical tests with higher power, we also analyzed a procedure whereby the breakpoint is chosen to minimize (or maximize) the t statistic on the coefficient of the change in mean, henceforth imposing the mild a priori assumption that the sign of the possible change is known (while its location remains unknown). This class of statistics allows an interesting sensitivity check on the obtained inference that may be a compromise between the more powerful case in which the breakpoint is assumed known and the least powerful case in which no structure is imposed.

Finally, for each procedure suggested, we analyzed the effects on the critical values of different assumptions about the method used for choosing the truncation lag parameter in the estimated autoregressions. In practice, we recommend using a test of significance on the coefficients of the lagged first-differences. In particular, the $k(t)$ procedure, whereby the order is selected by a test of significance of the last included lag, is quite easy to implement using standard regression output and has better size and power properties. Care, however, must be taken to correct for finite-sample biases introduced by the correlation between the final statistics of interest and the

t tests on the coefficients of these additional lags. Our tabulated critical values should prove to be useful in this respect.

Appendix A: The Data

The US/UK real-exchange-rate series based on the CPI index is annual from 1892 to 1988. The series based on the GNP deflator is also annual from 1869 to 1987. The source of the data is as follows: Nominal exchange rate series — 1869–1912 from Mitchell (1981), 1913–1949 from Board of Governors of the Federal Reserve (1943, 1976), 1950–1988 from International Monetary Fund (IMF) (1988); CPI — 1892–1969 from Mitchell (1981), 1970–1988 from IMF (1988); GNP deflator — 1868–1975 from Friedman and Schwartz (1982), 1976–1987 from IMF (1988). The US/Finland real-exchange rate series based on the CPI index is annual from 1900 to 1988, whereas that based on the GDP deflator is from 1900 to 1987. The source of the data is as follows: Nominal exchange rate series — 1900–1988 from the Bank of Finland; CPI — 1900–1985 from the Bank of Finland, 1986–1988 from the IMF (1988); GDP deflator — 1900–1985 from the Bank of Finland, 1986–1987 from IMF (1988). The sources of the US data are for the GNP deflator — 1869–1975 from Friedman and Schwartz (1982), 1976–1988 from IMF (1988); for the CPI — 1860–1970 from the US Bureau of the Census (1976), 1971–1988 from IMF (1988).

The choice of deflators (CPI or GNP deflator) was based primarily on data availability. Although a producer price index (PPI) is preferable due to fixed weights and a larger portion of traded goods, PPI series are not readily available for historical time series. Over such long periods as those investigated here, however, even indexes such as PPI have drawbacks due to changes in their weights from splicing and changes in the representative basket.

Appendix B: Mathematical Derivations

In the proofs that follow, we only show weak convergence of the finite-sample distribution without proving tightness. The proof is made complete using arguments similar to those of Zivot and Andrews

(1992) showing that the statistics can be expressed as continuous mappings from $D[0, 1]$ to R^1 provided we take the infimum with respect to λ over a closed subset of the interval (0, 1).

Given that the statistics of interest are constructed applying a particular regression for all values of T_b, it will, in general, be the case that the postulated value of the break in that regression does not correspond to the true value of the break in the data-generating process (8). Hence we cannot appeal to a finite-sample invariance with respect to δ. As the following proofs show, however, such an invariance property holds asymptotically. To make the argument precise, let the true value of the break be denoted by T_b'. Under the null hypothesis, the data-generating process is

$$y_t = \delta D(TB')_t + y_{t-1} + e_t = \delta D(TB)_t + y_{t-1} + e_t^*, \tag{B.1}$$

where $D(TB')_t = 1$ if $t = T_b' + 1$ and 0 otherwise, $D(TB)_t = 1$ if $t = T_b + 1$ and 0 otherwise, and $e_t^* = e_t + \delta(D(TB')_t - D(TB)_t)$. Let $S_t = \sum_{j=1}^{t} e_j$ $(t = 1, \ldots, T)$ with $S_1 = e_1 = 0$ and $DU_t' = 1$ if $t > T_b'$ and 0 otherwise. We can express y_t as

$$y_t = y(1) + \delta DU_t' + S_t. \tag{B.2}$$

Proof of (9). We prove (9) for the AO model only. The result for the IO model follows, with appropriate modifications using results of Perron (1990a) and Zivot and Andrews (1992), and details for this case are omitted. Consider first applying regression (3) for a given value T_b. We assume $T_b \leq T_b'$ without loss of generality, given that the asymptotic distribution is invariant to T_b'. Throughout, we denote $\lambda = T_b/T$ and $\lambda' = T_b'/T$. Simple algebra yields

$$\begin{aligned}
\tilde{y}_t &= y_t - \overline{Y}_a = S_t - \overline{S}_a && \text{if } t \leq T_b \\
\tilde{y}_t &= y_t - \overline{Y}_b = S_t - \overline{S}_b - \delta(1-\lambda')/(1-\lambda) && \text{if } T_b < t \leq T_b' \\
\tilde{y}_t &= y_t - \overline{Y}_b = S_t - \overline{S}_b + \delta - \delta(1-\lambda')/(1-\lambda) && \text{if } T_b' < t \leq T,
\end{aligned} \tag{B.3}$$

with $\overline{Y}_a = T_b^{-1}\sum_1^{T_b} y_t = \lambda^{-1}T^{-1}\sum_1^{T_b} y_t, \overline{Y}_b = (T - T_b)^{-1} \times \sum_{T_b+1}^{T} y_t = (1-\lambda)^{-1}T^{-1}\sum_{T_b+1}^{T} y_t, \overline{S}_a = T_b^{-1}\sum_1^{T_b} S_t = \lambda^{-1}T^{-1}$

$\sum_1^{T_b} S_t$, and $\overline{S}_b = (T - T_b)^{-1} \sum_{T_{b+1}}^T S_t = (1-\lambda)^{-1} T^{-1} \sum_{T_{b+1}}^T S_t$. Denote the residuals from a projection of $\tilde{y}$ on $D(TB)_t$ by $\tilde{y}_t^*$, and similarly the residuals from a projection of $\tilde{y}_{t-1}$ on $D(TB)_t$ by $\tilde{y}_{t-1}^{**}$. We have

$$\begin{aligned}
\tilde{y}_t^* &= S_t - \overline{S}_a && \text{if } t \le T_b \\
\tilde{y}_t^* &= 0 && \text{if } t = T_b + 1 \\
\tilde{y}_t^* &= S_t - \overline{S}_b - \delta(1-\lambda')/(1-\lambda) && \text{if } T_b + 1 < t \le T_b' \\
\tilde{y}_t^* &= S_t - \overline{S}_b + \delta - \delta(1-\lambda')/(1-\lambda) && \text{if } T_b' < t \le T.
\end{aligned} \tag{B.4}$$

Similarly,

$$\begin{aligned}
\tilde{y}_{t-1}^{**} &= S_{t-1} - \overline{S}_a && \text{if } t \le T_b \\
\tilde{y}_{t-1}^{**} &= 0 && \text{if } t = T_b + 1 \\
\tilde{y}_{t-1}^{**} &= S_{t-1} - \overline{S}_b - \delta(1-\lambda')/(1-\lambda) && \text{if } T_b + 1 < t \le T_b' + 1 \\
\tilde{y}_{t-1}^{**} &= S_{t-1} - \overline{S}_b + \delta - \delta(1-\lambda')/(1-\lambda) && \text{if } T_b' + 1 < t \le T.
\end{aligned} \tag{B.5}$$

Using (B.4) and (B.5), we deduce that

$$\begin{aligned}
\tilde{y}_1^* - \tilde{y}_{t-1}^{**} &= e_t && \text{if } t \ne T_b + 1 \text{ and } t \ne T_b' + 1 \\
&= 0 && \text{if } t = T_b + 1 \\
&= e_{T_{b+1}'} + \delta && \text{if } t = T_b' + 1.
\end{aligned} \tag{B.6}$$

The t statistic for testing $\alpha = 1$ in Regression (4) if $k = 0$ is given by

$$\begin{aligned}
t_{\hat{\alpha}} &= (\hat{\alpha} - 1) \left[\sum_2^T (\tilde{y}_{t-1}^{**})^2 \right]^{1/2} \Big/ \hat{s} \\
&= T^{-1} \sum_2^T \tilde{y}_{t-1}^{**} (\tilde{y}_t^* - \tilde{y}_{t-1}^{**}) \Big/ \left[\hat{s}^2 T^{-2} \sum_2^T (\tilde{y}_{t-1}^{**})^2 \right]^{1/2},
\end{aligned} \tag{B.7}$$

where $\hat{s}^2 = T^{-1} \sum_2^T \hat{u}_t^2$ with $\hat{u}_t = \tilde{y}_t - \hat{\omega} D(TB)_t - \hat{\alpha} \tilde{y}_{t-1}$. Consider first the numerator in (B.7). Simple manipulations yield

$$T^{-1}\sum_{2}^{T}\tilde{y}_{t-1}^{**}(\tilde{y}_t^{*}-\tilde{y}_{t-1}^{**})=T^{-1}\sum_{2}^{T}S_{t-1}e_t-\left(T^{-1/2}\sum_{1}^{T_b}e_t\right)$$

$$\times\lambda^{-1}T^{-3/2}\sum_{1}^{T_b}S_t-\left(T^{-1/2}\sum_{T_b-1}^{T}e_t\right)$$

$$\times(1-\lambda)^{-1}T^{-3/2}\sum_{T_b+1}^{T}S_t+o_p(1).$$

Using lemma A.3 of Perron (1989), we deduce that

$$T^{-1}\sum_{2}^{T}\tilde{y}_{t-1}^{**}(\tilde{y}_t^{*}-\tilde{y}_{t-1}^{**})\Rightarrow\sigma^2\left\{\frac{1}{2}(W(1)^2-1)-\lambda^{-1}W(\lambda)\right.$$

$$\times\int_0^\lambda W(r)dr-(1-\lambda)^{-1}[W(1)$$

$$\left.-W(\lambda)]\left[\int_\lambda^1 W(r)dr\right]\right\}$$

$$\equiv\sigma^2\int_0^1 W^*(\lambda,r)dW(r).\tag{B.8}$$

Consider now the denominator in (B.7). We have, using (B.5) and some manipulations,

$$T^{-2}\sum_{2}^{T}(\tilde{y}_{t-1}^{**})^2=T^{-2}\sum_{2}^{T_b}(S_{t-1}-\overline{S}_a)^2$$

$$+T^{-2}\sum_{T_{b+2}}^{T_{b'+1}}(S_{t-1}-\overline{S}_b-\delta(1-\lambda')/(1-\lambda))^2$$

$$+T^{-2}\sum_{T_{b'+2}}^{T}(S_{t-1}-\overline{S}_b+\delta-\delta(1-\lambda')/(1-\lambda))^2$$

$$
\begin{aligned}
&= T^{-2}\sum_{2}^{T} S_{t-1}^{2} - \lambda^{-1}T^{-3}\left(\sum_{1}^{Tb} S_t\right)^2 - (1-\lambda)^{-1} \\
&\quad \times T^{-3}\left(\sum_{Tb+1}^{T} S_{t-1}\right)^2 + o_p(1) \Rightarrow \sigma^2\left\{\int_0^1 W(r)^2 dr\right. \\
&\quad \left. - \lambda^{-1}\left(\int_0^{\lambda} W(r)dr\right)^2 - (1-\lambda)^{-1}\left(\int_{\lambda}^{1} W(r)dr\right)^2\right\} \\
&\equiv \sigma^2 \int_0^1 W^*(\lambda, r)^2 dr. \qquad \text{(B.9)}
\end{aligned}
$$

Tedious but straightforward manipulations show that $\hat{s}^2 \to \sigma^2$ in probability. The result in (9) is then proved using (B.7)–(B.9).

Proof of (10). To prove (10), we need only show that the limiting distribution of $t_{\hat{\delta}}$ is $Z(\lambda)$ when $t_{\hat{\delta}}$ is the t statistic for $\delta = 0$ in regression (7) if $k = 0$ under the null hypothesis that y_t is generated by (8). As in the work of Perron (1989, 1990a), the presence of the regressor $D(TB)_t$ in (7) does not change the asymptotic distribution of the t statistic. To simplify, we therefore omit this regressor and express the t statistic as

$$
t_{\hat{\delta}} = \{[Z'M_X Z]^{-1}[Z'M_X E]\}_{11}\{\hat{s}^2[(Z'M_X Z)^{-1}]_{11}\}^{-1/2} + o_p(1) \qquad \text{(B.10)}
$$

where $Z = [DU_t y_{t-1}][(T-1)\times 2]$, $X' = [1, 1, \ldots, 1][1\times(T-1)]$, $E = [e_2^*, e_3^*, \ldots, e_T^*]$ $[1\times(T-1)]$, $M_x = I - X(X'X)^{-}X'$, and $\hat{s}^2 = T^{-1}\sum_2^T \hat{u}_t^2$ with $\hat{u}_t$ the regression residuals from applying OLS to (7) and e_t^* defined in (B.l). Note that $\hat{s}^2 \to \sigma^2$ (in probability) (see Perron 1990a). Straightforward algebra yields

$$
[(Z'M_X Z)^{-1}]_{11} = \left[\sum_2^T y_{t-1}^2 - T^{-1}\left(\sum_2^T y_{t-1}\right)^2\right]\Big/ J, \qquad \text{(B.11)}
$$

where $J = \lambda(1-\lambda)T[\sum_2^T y_{t-1}^2 - T^{-1}(\sum_2^T y_{t-1})^2] - [\lambda\sum_2^T y_{t-1} - \sum_2^{T_b} y_{t-1}]^2$. Note that, using lemma A.3 of Perron (1989),

$$T^{-3}J \Rightarrow \sigma^2\lambda(1-\lambda)\left[\int_0^1 W(r)^2dr - \left(\int_0^1 W(r)dr\right)^2\right]$$

$$-\sigma^2\left[\lambda\int_0^1 W(r)dr - \int_0^\lambda W(r)dr\right]^2 = \sigma^2\lambda(1-\lambda)A$$

$$-\sigma^2 B(\lambda)^2 \equiv \sigma^2 K(\lambda)/A. \tag{B.12}$$

Moreover, $T[(Z'M_XZ)^{-1}]_{11} \Rightarrow \sigma^2 A/[K(\lambda)/A] = \sigma^2A^2/K(\lambda)$, using (B.11) and (B.12). Finally, simple algebra, using in particular (B.l), shows that

$$T^{1/2}\{[Z'M_xZ]^{-1}[Z'M_xE]\}_{11}$$

$$= (1/T^{-3}J)\left\{\left[T^{-2}\sum_2^T y_{t-1}^2 - T^{-3}\left(\sum_2^T y_{t-1}\right)^2\right]\right.$$

$$\times\left[T^{-1/2}\sum_{T_b+1}^T e_t^* - (1-\lambda)T^{-1/2}\sum_2^T e_t^*\right]$$

$$-\left[\lambda T^{-3/2}\sum_2^T y_{t-1} - T^{-3/2}\sum_2^{T_b} y_{t-1}\right]$$

$$\left.\times\left[T^{-1}\sum_2^T y_{t-1}e_t^* - T^{-3/2}\sum_2^T y_{t-1}T^{-1/2}\sum_2^T e_t^*\right]\right\}.$$

It is straightforward to see that the asymptotic distributions of $T^{-1/2}\sum_{T_b+1}^T e_t^*, T^{-1/2}\sum_2^T e_t^*$, and $T^{-1}\sum_2^T y_{t-1}e_t^*$ are identical to those of $T^{-1/2}\sum_{T_{b+1}}^T e_t, T^{-1/2}\sum_2^T e_t$, and $T^{-1}\sum_2^T y_{t-1}e_t$. Using this fact, (B.12), and lemma A.3 of Perron (1989), we have

$$T^{1/2}\{[Z'M_XZ]^{-1}[Z'M_XE]\}_{11}$$
$$\Rightarrow \sigma^2(A/K(\lambda))\left\{A[\lambda W(1) - W(\lambda)] - B(\lambda)\left[\frac{1}{2}(W(1)^2 - 1)\right.\right.$$
$$\left.\left. - W(1)\int_0^1 (W(r)dr)\right]\right\} \equiv \sigma^2 AH(\lambda)/K(\lambda). \tag{B.13}$$

Using (B.10)–(B.13) we deduce, as required, that $t_{\hat{\delta}} \Rightarrow [\sigma^2 AH(\lambda)/K(\lambda)]/[\sigma^4 A^2/K(\lambda)]^{1/2} = H(\lambda)/K(\lambda)^{1/2}$.

Proof of (11). To prove (11), we need only show that the limiting distribution of $T^{-1/2}t_{\hat{\delta}}$ is $Q(\lambda)$ when $t_{\hat{\delta}}$ is the t statistic for $\delta = 0$ in regression (3) and y_t is generated by (8). The t statistic is not invariant, in finite samples, with respect to the value of δ. The following proof, however, shows that its asymptotic distribution is invariant. First note that

$$t_{\hat{\delta}} = \hat{\delta}/[\hat{s}^2(X'X)^{-1}_{22}]^{1/2}, \tag{B.14}$$

where $(X'X)^{-1}_{22} = T/[T_b(T - T_b)] = [T\lambda(1-\lambda)]^{-1}$; $\hat{s}^2 = T^{-1}\sum_{t=1}^{T}\hat{u}_t^2 = T^{-1}\sum_{t=1}^{T}(y_t - \hat{\mu} - \hat{\delta}DU_t)^2$ where $\hat{\mu} = (\lambda T)^{-1}\sum_{t=1}^{T_b} y_t$; and

$$\hat{\delta} = [(1-\lambda)T]^{-1}\sum_{t=T_b+1}^{T} y_t - [\lambda T]^{-1}\sum_{t=1}^{T_b} y_t. \tag{B.15}$$

Using (B.2), we can write (after some rearrangements)

$$\hat{\delta} - \delta = [(1-\lambda)T]^{-1}\sum_{T_b+1}^{T} S_t - [\lambda T]^{-1}\sum_{1}^{T_b} S_t - \delta(\lambda' - \lambda)/(1-\lambda). \tag{B.16}$$

Using lemma A.3 of Perron (1989), we deduce, from (B.16), that

$$\begin{aligned}
T^{-1/2}(\hat{\delta}-\delta) &\Rightarrow \sigma\left\{(1-\lambda)^{-1}\int_{\lambda}^{1} W(r)dr - \lambda^{-1}\int_{0}^{\lambda} W(r)dr\right\} \\
&= \sigma[(1-\lambda)\lambda]^{-1}\left\{\lambda\int_{0}^{1} W(r)dr - \int_{0}^{\lambda} W(r)dr\right\} \\
&\equiv \sigma[(1-\lambda)\lambda]^{-1}B(\lambda). \qquad\qquad (B.17)
\end{aligned}$$

Note that (B.17) implies that $T^{-1/2}(\hat{\delta}-\delta)$ has the same asymptotic distribution as $T^{-1/2}\hat{\delta}$, since $T^{-1/2}\delta \to 0$ as $T \to \infty$. Using similar arguments,

$$\begin{aligned}
\hat{s}^2 &= T^{-1}\sum_{t=1}^{T} y_t^2 - [\lambda T^2]^{-1}\left(\sum_{t=1}^{T_b} y_t\right)^2 - [(1-\lambda)T^2]^{-1}\left(\sum_{t=T_b+1}^{T} y_t\right)^2 \\
&= T^{-1}\sum_{t=1}^{T} S_t^2 - [\lambda T^2]^{-1}\left(\sum_{t=1}^{T_b} S_t\right)^2 - [(1-\lambda)T^2]^{-1}\left(\sum_{t=T_b+1}^{T} S_t\right)^2 \\
&\quad + o_p(T);
\end{aligned}$$

and

$$\begin{aligned}
T^{-1}\hat{s}^2 &\Rightarrow \sigma^2\left\{\int_{0}^{1} W(r)^2dr - \lambda^{-1}\left(\int_{0}^{\lambda} W(r)dr\right)^2 \right. \\
&\qquad \left. - (1-\lambda)^{-1}\left(\int_{\lambda}^{1} W(r)^2dr\right)^2\right\} \\
&= \sigma^2[(1-\lambda)\lambda]^{-1}\left\{(1-\lambda)\lambda\int_{0}^{1} W(r)^2dr - (1-\lambda)\right. \\
&\qquad \left. \times\left(\int_{0}^{\lambda} W(r)dr\right)^2 - \lambda\left(\int_{\lambda}^{1} W(r)dr\right)^2\right\} \\
&= \sigma^2[(1-\lambda)\lambda]^{-1}K(\lambda)/A, \text{ after rearrangements.} \qquad (B.18)
\end{aligned}$$

Using (B.14), (B.17), and (B.18), we obtain

$$\begin{aligned} T^{-1/2}t_{\hat{\delta}} &= T^{-1/2}\hat{\delta}/[T^{-1}\hat{s}^2[\lambda(1-\lambda)]^{-1}]^{1/2} \\ &\Rightarrow \sigma[(1-\lambda)\lambda]^{-1}B(\lambda)/[\sigma^2[(1-\lambda)\lambda]^{-2}K(\lambda)/A]^{1/2} \\ &\equiv B(\lambda)/[K(\lambda)/A]^{1/2} \end{aligned}$$

as required.

Appendix C: Simulation of the Asymptotic Distributions

In this appendix, we briefly describe the method used to simulate the asymptotic distributions stated in the text. We give details for the limiting random variable described by (9). The method used to tabulate the percentage points associated with the limiting distributions stated in (10) and (11) is analogous. First note that we can write the components of (9) as

$$\begin{aligned} \int_0^1 W^*(\lambda, r)dW(r) = \frac{1}{2}(W(1)^2 - 1) - \lambda^{-1}W(\lambda)\int_0^\lambda W(r)dr \\ - (1-\lambda)^{-1}(W(1) - W(\lambda))\int_\lambda^1 W(r)dr, \end{aligned} \tag{C.1}$$

and

$$\begin{aligned} \int_0^1 W^*(\lambda, r)^2 dr = \int_0^1 W(r)^2 dr - \lambda^{-1}\left(\int_0^\lambda W(r)dr\right)^2 \\ - (1-\lambda)^{-1}\left(\int_\lambda^1 W(r)dr\right)^2 \end{aligned} \tag{C.2}$$

(see Perron 1990a). We used expressions (C.1) and (C.2) to obtain critical values via simulation methods. First, we generate a sample of size 1,000 of i.i.d. $N(0,\ 1)$ random deviates, $\{e_t\}$. We then construct sample moments of the data that converge weakly to the various functionals of the Wiener process involved in the representation of the asymptotic distribution. For example, as $n \to \infty$,

$n^{-1/2}\sum_1^n e_t \Rightarrow W(1)$, $n^{-1/2}\sum_1^{[\lambda n]} e_t \Rightarrow W(\lambda)$, $n^{-3/2}\sum_1^n \sum_{j=1}^t e_j \Rightarrow \int_0^1 W(r)dr$, $n^{-1}\sum_1^n(\sum_{j=1}^{t-1} e_j)e_t \Rightarrow \frac{1}{2}(W(1)^2 - 1)$, and so forth. With a sample size of $n = 1,000$ and i.i.d. $N(0,\ 1)$ variates, we can expect the approximation to be quite accurate. Once the various functionals are evaluated, we construct the expressions in (C.1) and (C.2) and obtain one realization of the limiting random variable $\int_0^1 W^*(\lambda, r)dW(r)[\int_0^1 W^*(\lambda, r)^2 dr]^{-1/2} \equiv C(\lambda)$, say, for a given value of λ. For a given set of simulated errors $\{e_t\}_{t=1}^n$, we repeat this procedure for all values of $\lambda = j/1,000$; $j = 2, \ldots, 999$. The statistic of interest is then the minimum of $C(\lambda)$ over all values of λ. The procedure is repeated 10,000 times, and the critical values are obtained from the sorted array of the minimal values of $C(\lambda)$ obtained at each replication.

Acknowledgments

We thank anonymous referees and an associate editor for their comments. Financial Support is acknowledged from the Social Sciences and Humanities Research Council of Canada, the Natural Sciences and Engineering Council of Canada, and the Fonds pour la Formation de Chercheurs et l'Aide à la Recherche du Québec (Perron) and the International Finance Section at Princeton University (Vogelsang).

References

Andrews, D.W.K. (1990). Tests for parameter instability and structural change with unknown change point. Discussion Paper 943, Yale University, Cowles Foundation.

Banerjee, A. Lumsdaine, R.L. and Stock, J.H. (1992). Recursive and sequential tests of the unit-root and trend–break hypotheses: Theory and international evidence. *Journal of Business & Economic Statistics* 10, 271–287.

Board of Governors of the Federal Reserve (1943). *Banking and Monetary Statistics, 1914–1940*, National Capital Press: Washington, DC.

Board of Governors of the Federal Reserve (1976). *Banking and Monetary Statistics, 1941–1970*, National Capital Press: Washington, DC.

Christiano, L.J. (1988). Searching for breaks in GNP. Working Paper 2695, National Bureau of Economic Research, Cambridge, MA.

Chu, C.-S.J. (1989). New tests for parameter constancy in stationary and non-stationary regression models. Mimeo, Department of Economics: University of California San Diego.

Chu, C.-S.J. and White, H. (1990). Testing for structural change in some simple time series models. Unpublished manuscript, Department of Economics: University of California at San Diego.

Corbae, D. and Ouliaris, S. (1988). Cointegration and Tests of PPP. *Review of Economics and Statistics* 4, 508–511.

Davies, R.B. (1987). Hypothesis testing when a nuisance parameter is present only under the alternative. *Biometrika* 74, 33–43.

Deshayes, J. and Picard, D. (1986). Off-line Statistical Analysis of Change-point Models Using Non-parametric and Likelihood Methods. In *Detection of Abrupt Changes in Signals and Dynamical Systems*, M. Basseville and A. Benveniste, (eds.), Springer-Verlag: New York, pp. 103–168.

Dickey, D.A. and Fuller, W.A. (1979). Distribution of the estimators for autoregressive time series with a unit root. *Journal of the American Statistical Association* 74, 427–431.

Dornbusch, R. (1988). *Exchange Rates and Inflation*, MIT Press: Cambridge, MA.

Dornbusch, R. and Vogelsang, T. (1991). Real Exchange Rates and Purchasing Power Parity. In *Trade Theory and Economic Reform: North, South and East, Essays in Honor of Bela Balassa*, J. de Melo and A. Sapir, (eds.), Basil Blackwell: Cambridge, MA, pp. 3–24.

Enders, W. (1988). ARIMA and cointegration-tests of purchasing power parity. *Review of Economics and Statistics* 4, 504–508.

Enders, W. (1989). Unit roots and the real exchange rate before World War I: The case of Britain and the USA. *Journal of International Money and Finance* 8, 59–73.

Friedman, M. and Schwartz, A.J. (1982). *Monetary Trends in the United States and the United Kingdom: Their Relation to Income, Prices and Interest Rates, 1867–1975*, The University of Chicago Press: Chicago.

Hall, A. (1990). Testing for a unit root in time series with pretest data based model selection. Mimeo, Department of Economics: North Carolina State University.

Hansen, B.E. (1990). Lagrange multiplier tests for parameter instability in non-linear models. Mimeo, Department of Economics: University of Rochester.

Hansen, B.E. (1992). Tests for parameter instability in regressions with I(1) processes. *Journal of Business & Economic Statistics* 10, 321–335.

International Monetary Fund (1988). *International Financial Statistics: Yearbook*, Washington, DC, Author.

Mitchell, B.R. (1981). *European Historical Statistics, 1750–1975*, Columbia University Press: New York.

Nelson, C.R. and Plosser, C.I. (1982). Trends and random walks in macroeconomic time series. *Journal of Monetary Economics* 10, 139–162.

Pantula, S.G. (1985). Estimation of Autoregressive Models with ARCH Errors. In *Proceedings of the Business and Economic Statistics Section, American Statistical Association*, pp. 490–495.

Perron, P. (1989). The great crash, the oil price shock and the unit root hypothesis. *Econometrica* 57, 1361–1401.

Perron, P. (1990a). Testing for a unit root in a time series with a changing mean. *Journal of Business & Economic Statistics* 8, 153–162.

Perron, P. (1990b). Further evidence on breaking trend functions in macroeconomic variables. Memorandum 350, Princeton University, Econometric Research Program.

Perron, P. (1991a). Test consistency with varying sampling frequency. *Econometric Theory* 7, 341–368.

Perron, P. (1991b). A test for changes in a polynomial trend function for a dynamic time series. Mimeo, Department of Economics: Princeton University.

Perron, P. and Vogelsang, T.J. (1991). Testing for a unit root in a time series with a changing mean: Corrections and extensions. Mimeo, Department of Economics: Princeton University.

Press, W.H. Flanery, B.P., Teukolsky, S.A. and Vetterling, W.T. (1986). *Numerical Recipes: The Art of Scientific Computing*, Cambridge University Press: Cambridge, U.K.

Said, S.E. and Dickey, D.A. (1984). Testing for unit roots in autoregressive-moving average of unknown order. *Biometrika* 599–608.

Schwert, C.W. (1989). Testing for unit roots: A Mone Carlo investigation. *Journal of Business & Economic Statistics* 147–160.

Taylor, M. and McMahon, P. (1988). Long-run PPP in the 1920s. *European Economic Review* 32, 179–197.

Tiao, G.C. (1985). Autoregressive moving average models, intervention problems and outlier detection in time series. In *Handbook of Statistics 5*, E. J. Hannan, P. R. Krishnaiah, and M. M. Rao, (eds.), North-Holland: Amsterdam, pp. 85–118.

U.S. Bureau of the Census (1976). *The Statistical History of the United States, From Colonial Times to the Present*, Basic Books: New York.

Zivot, E. and Andrews, D.W.K. (1992). Further evidence on the great crash, the oil-price shock, and the unit-root hypothesis. *Journal of Business & Economic Statistics* 10, 251–270.

Chapter 16

Further Evidence on Breaking Trend Functions in Macroeconomic Variables*

Pierre Perron

Université de Montréal, C.R.D.E.,
Montréal, Québec, Canada H3C 3J7

This study first reexamines the findings of Perron (1989) regarding the claim that most macroeconomic time series are best construed as stationary fluctuations around a deterministic trend function if allowance is made for the possibility of a shift in the intercept of the trend function in 1929 (a crash) and a shift in slope in 1973 (a slowdown in growth). Unlike that previous study, the date of possible change is not fixed *a priori* but is considered as unknown. We consider various methods to select the break points and the asymptotic and finite sample distributions of the corresponding statistics. A detailed discussion about the choice of the truncation lag parameter in the autoregression and of its effect on the critical values is also included. Most of the rejections reported in Perron (1989) are confirmed using this approach. Secondly, this paper investigates an international data set of post-war quarterly real GNP (or GDP) series for the G-7 countries. Our results are compared and contrasted to those of Banerjee *et al.* (1992) and Zivot and Andrews (1992). In contrast to the theoretical results contained in these papers, we derive the limiting distribution of the sequential test without trimming.

Keywords: Hypothesis testing, structural change; stochastic trends; deterministic trends; simulation experiment; unit root

*Originally Published in *Journal of Econometrics*, 80, pp. 355–385.

1. Introduction

In a previous paper, Perron (1989), we argued that many macroeconomic time series could be represented as stationary fluctuations around a deterministic trend function if allowance is made for a possible change in its intercept in 1929 (a crash) and in its slope in 1973 (a slowdown in growth). The test statistics were constructed by adding dummy variables for different intercepts and slopes, extending the standard Dickey–Fuller procedure. The asymptotic distribution theory underlying the critical values obtained under the different models assumed that the dating of the break points was known *a priori*, or more precisely, that the dates chosen were uncorrelated with the data.

This postulate has been criticized, most notably by Christiano (1992) who argued that the choice of these dates had to be viewed, to a large extent, as being correlated with the data. This is an important problem because both the finite sample and asymptotic distributions of the statistics depend upon the extent of the correlation between the choice of the break points and the data. There is a sense, as argued before, in which the choice of these dates can be regarded as independent of the data. First, the dates used in the previous study were chosen *ex-ante* and not modified *ex-post*. Secondly, these dates are related to exogenous events for which economic theory would suggest the effects that actually happened; e.g., the stock market crash of 1929 with the ensuing dismantle of the economic organization and the exogenous sudden change in oil prices with the resulting alteration of international economic coordination and policies.

In the sense described above the choice of the dates can be viewed as uncorrelated with the data. There is, however, a validity to the argument that it is only *ex-post* (after looking at the data) that we can say that the changes that followed these exogenous events actually occurred as predicted by the theory. Furthermore, many other exogenous events did not have the major impact that some theories would have predicted. In this sense, the choice of the break points must be viewed as being correlated, at least to some extent, with the data. To what extent is a difficult and practically impossible question to answer. At the very least the choices were not perfectly correlated with the data as no attempts were systematically made to maximize the chances that the unit root be rejected nor to find where, according to some test criteria, were the most likely dates of change.

While we still believe that the assumption about the exogeneity of the choice of the break points is a good first approximation to the true extent of the correlation with the data, it is useful to investigate how robust the results are to different postulates. The aim of this paper is to take the extreme view where the choice of the break points is effectively made to be perfectly correlated with the data. This case is instructive to study because if one can still reject the unit-root hypothesis under such a scenario it must be the case that it would be rejected under a less stringent assumption.

We proceed as follows for the practical implementation. Again, as in the previous analysis, only one possible break point is allowed for any single series. This break point is first chosen such that the t-statistic for testing the null hypothesis of a unit root is smallest among all possible break points. We also consider choosing the break point that corresponds to a minimal t-statistic on the parameter of the change in the trend function. This allows the mild *a priori* imposition of a one-sided change which permits substantial gains in power. We also investigate various issues regarding the choice of the autoregressive truncation lag and its effect on the finite sample critical values.

Our paper is closely related to and complements those of Banerjee *et al.* (1992) and Zivot and Andrews (1992) in that similar procedures and series are analyzed. We extend their analysis in several directions. On a methodological level, we consider the asymptotic distribution of the sequential test based on the minimal value of the unit-root tests over possible break points. We show the results of Zivot and Andrews (1992) to be valid without any trimming at the end points. The proof, which is of interest in itself, is based on projection arguments and introduces a method that can be applied to a variety of frameworks. Concerning the empirical results, our analysis is more extensive and shows that alternative procedures can lead to conclusions that are less favorable to the unit root than suggested in these two studies. We pay particular attention to the importance of the selection of the truncation lag on the outcome of the tests.

The paper is organized as follows. Section 2 reviews the statistics involved. Section 3 discusses their asymptotic distribution under the null hypothesis of a unit root and Section 4 their finite sample distribution using simulation methods. Section 5 contains simulation experiments providing information about their size and power under various data-generating processes. Section 6 presents the empirical results for the Nelson and Plosser (1982) data set and Section 7 analyzes an international data set of post-war quarterly real GNP series. Section 8 offers concluding comments and a mathematical appendix contains the derivation of the limiting distributions.

2. The Models and Statistics

In this section, we briefly review the statistical procedures used to test for a unit root allowing for the presence of a change in the trend function occurring at most once. The reader is referred to Perron (1989) for details on the models. Throughout, T_{b} denotes the time at which the change in the trend function occurs. The first model allows only a change in the intercept under both the null and alternative hypotheses. Furthermore, this change is assumed to occur gradually and in a way that depends on the correlation structure of the noise function. This was termed the 'innovational outlier model', and the unit-root test is performed using the t-statistic for testing $\alpha = 1$ in the following regression:

$$y_t = \mu + \theta \mathrm{D}U_t + \beta t + \delta \mathrm{D}(T_{\mathrm{b}})_t + \alpha y_{t-1} + \sum_{i=1}^{k} c_i \Delta y_{t-i} + e_t, \quad (1)$$

where $\mathrm{D}U_t = 1(t > T_{\mathrm{b}})$ and $\mathrm{D}(T_{\mathrm{b}})_t = 1(t = T_{\mathrm{b}} + 1)$ with $1(\cdot)$ the indicator function. Regression (1), like the others that follow, is estimated by OLS and is in the spirit of the Dickey–Fuller (1979) and Said–Dickey (1984) methodology whereby autoregressive moving average processes are approximated by autoregressive processes.

Under the second model, both a change in the intercept and the slope are allowed at time T_b. The test is performed using the t-statistic for the null hypothesis that $\alpha = 1$ in the regression:

$$y_t = \mu + \theta \mathrm{DU}_t + \beta t + \gamma \mathrm{DT}_t + \delta \mathrm{D}(T_b)_t + \alpha y_{t-1} + \sum_{i=1}^{k} c_i \Delta y_{t-i} + e_t, \tag{2}$$

with $\mathrm{DT}_t = 1(t > T_b)t$. Under the third model, a change in the slope is allowed but both segments of the trend function are joined at the time of break. Here the change is presumed to occur rapidly and corresponds to the 'additive outlier model' in the terminology of Perron (1989). We use the following two-step procedure. First, the series is detrended using the following regression where $\mathrm{DT}_t^* = 1(t > T_b)(t - T_b)$:

$$y_t = \mu + \beta t + \gamma \mathrm{DT}_t^* + \tilde{y}_t. \tag{3a}$$

The test is then performed using the t-statistic for $\alpha = 1$ in the regression:

$$\tilde{y}_t = \alpha \tilde{y}_{t-1} + \sum_{i=1}^{k} c_i \Delta \tilde{y}_{t-i} + e_t. \tag{3b}$$

We denote by $t_{\hat{\alpha}}(i, T_b, k)(i = 1, 2, 3)$, the t-statistic for testing $\alpha = 1$ under model i with a break date T_b and truncation lag parameter k (using regressions (1), (2) and (3b) for $i = 1, 2$, and 3, respectively). In these regressions, T_b and k are treated as unknown. We next describe various data-dependent methods to select these values endogenously.

2.1. *Methods to choose the break date T_b*

We consider two methods to select T_b endogenously. First T_b is selected as the value which minimizes the t-statistic for testing $\alpha = 1$. We define the statistics as $t_{\alpha}^*(i) = \mathrm{Min}_{T_b \in (k+1,T)} t_{\hat{\alpha}}(i, T_b, k)(i = 1, 2, 3)$. The asymptotic distribution of $t_{\alpha}^*(1)$ and $t_{\alpha}^*(2)$ was studied by Zivot and Andrews (1992) under the condition that the range of possible values for the break point be restricted to some subset that excludes values at each end of the sample. In the next section, we show their result to remain valid even without trimming.

Secondly, T_{b} is chosen to minimize either $t_{\hat{\theta}}$, the t-statistic on the parameter associated with the change in the intercept (Model 1) or $t_{\hat{\gamma}}$, the t-statistic on the change in slope (Models 2 and 3). We denote the t-statistic on α (for a null hypothesis that $\alpha = 1$) obtained from such a procedure by $t^*_{\alpha,\theta}(1)$ for Model 1 and by $t^*_{\alpha,\gamma}(i)(i = 2, 3)$ for Models 2 and 3. More precisely, $t^*_{\alpha,\theta}(1) = t_{\hat{\alpha}}(1, T^*_{\mathrm{b}}, k)$, where T^*_{b} is such that $t_{\hat{\theta}}(T^*_{\mathrm{b}}) = \mathrm{Min}_{T_{\mathrm{b}}} \in (k+1, T) t_{\hat{\theta}}(T_{\mathrm{b}}, k)$, where again different specifications about the choice of k will be analyzed. The statistics $t^*_{\alpha,\gamma}(i)(i = 2, 3)$ are defined in an analogous fashion. This procedure allows the possibility of imposing the mild *a priori* restriction of a one-sided change, i.e., allowing the date of the change to be unknown but restricting the analysis to the cases of a 'crash' or a slowdown in growth. We also discuss the case where the break point is selected using the same procedure without any *a priori* assumption on the sign of the change. In this context the break date is selected using the maximum of the absolute value of $t_{\hat{\theta}}$ or $t_{\hat{\gamma}}$. The corresponding statistics are denoted by $t^*_{\alpha,|\theta|}(1)$ for Model 1 and $t^*_{\alpha,|\gamma|}(i)(i = 2, 3)$ for Models 2 and 3.

2.2. *Methods to select the truncation lag parameter k*

There is now evidence that using data-dependent methods to select the truncation lag parameter k leads to test statistics having better properties (stable size and higher power) than if a fixed k is chosen *a priori* (unless, of course, one happens to select that value of k which is best), see Ng and Perron (1995), Perron and Vogelsang (1992) and Hall (1994). We consider two such data-dependent methods. The first is a general to specific recursive procedure based on the t-statistic on the coefficient associated with the last lag in the estimated autoregression. More specifically, the procedure selects that value of k, say k^*, such that the coefficient on the last lag in an autoregression of order k^* is significant and that the last coefficient in an autoregression of order greater than k^* is insignificant, up to a maximum order k max. In the simulations and empirical applications below, we use a two-sided 10% test based on the asymptotic normal distribution to assess the significance of the last lags. This procedure is denoted below as 't-sig'.

Said and Dickey (1984) use yet a different method in their empirical application. It is based on testing whether additional lags are jointly significant using an 'F-test' on the estimated coefficients. First a maximum value of k, $k\max$, is specified and the autoregressions with $k\max$ and $(k\max-1)$ lags are estimated. A 10% one-tailed F-test is used to assess whether the coefficient on the $k\max$th lag is significant and if so, the value of k chosen is this maximum value. If not, the model is estimated with $(k\max-2)$ lags. The lag$(k\max-1)$ is deemed significant if either the F-test for $(k\max-2)$ versus $(k\max-1)$ lags or the F-test for $(k\max-2)$ versus $k\max$ lags are significant based on the 10% critical values of the chi-square distribution. This is repeated by lowering k until a rejection that additional lags are insignificant occurs or some lower bound is attained. In the empirical applications, the lower bound is set to $k = 1$. This procedure is denoted below as 'F-sig'.

We choose these 'general to specific' procedures rather than methods based on information criteria, such as AIC, because the latter tend to select very parsimonious models leading to tests with sometimes serious size distortions and/or power losses with data in the class of ARMA processes. Indeed, Ng and Perron (1995) show that using an information criterion leads to a selected value of k that increases to infinity, as T increases, only at the very slow rate $\log(T)$. This is consistent with many empirical results showing that using the AIC leads to very small values of k being selected (typically 0 or 1) and that often times the estimated residuals exhibit serial correlation (see Perron, 1994).

3. The Asymptotic Distribution of the Statistics

In this section, we consider the limiting distribution of the statistics. To simplify the derivations we suppose the data-generating process to be a random walk,

$$y_t = y_{t-1} + e_t, \quad (t = 1, 2, \ldots, T), \tag{4}$$

where the errors e_t are martingale differences (y_0 is some fixed value), and consider the statistics constructed with $k = 0$. Using arguments in Ng and Perron (1995), we can then state that the resulting limiting distribution remains the same when additional correlation is present and the statistics are constructed with one of the data-dependent methods to select k. This holds provided $k\max^3/T \to 0$ as $T \to \infty$. This is the same strategy as used by Zivot and Andrews (1992) and Banerjee *et al.* (1992). All statistics are asymptotically invariant to a change in intercept. Vogelsang and Perron (1994) show that they are not asymptotically invariant to a change in slope but that the asymptotic distribution corresponding to a zero change in slope is a better approximation to the finite sample distribution for values typically encountered in practice. The following Theorem concerning the asymptotic distribution of $t^*_{\hat{\alpha}}(i)(i = 1, 2, 3)$ is proved in the appendix.

Theorem 1. *Let* $\{y_t\}_0^T$ *be generated by* (4) *and denote by '*$\Rightarrow$*' weak convergence in distribution from the space* D[0, 1] *to the space* C[0, 1] *using the uniform metric on the space of functions on* [0, 1]. *Then:*

(a) *for* $i = 1, 2$:

$$\inf_{T_b \in [1,T]} t_{\hat{\alpha}}(i, T_b, k = 0) \Rightarrow \inf_{\lambda \in [0,1]} \int_0^1 W_i(r, \lambda) \mathrm{d}W(r) \bigg/ \left[\int_0^1 W_i(r, \lambda)^2 \mathrm{d}r\right]^{1/2},$$

(b) *for Model* 3:

$$\inf_{T_b \in [1,T]} t_{\hat{\alpha}}(i, T_b, k = 0) \Rightarrow \inf_{\lambda \in [0,1]} \left[\int_0^1 W_3(r, \lambda) \mathrm{d}W(r) - a \int_\lambda^1 (r - \lambda) W_0(r) \mathrm{d}r \times \int_0^1 W_3(r, \lambda)^2 \mathrm{d}r\right] \bigg/ \left[\int_0^1 W_3(r, \lambda)^2 \mathrm{d}r\right]^{1/2},$$

where $a = (\lambda^3(1-\lambda)^3/3)^{-1}$, $W_0(r)$ *and* $W_i(r,\lambda)$ *are residuals from a projection of a standard Wiener process* $W(r)$ *onto the subspace generated by the functions* $\{1,r\}(i=0), \{1,r,\mathrm{du}(r,\lambda)\}$ $(i=1), \{1,r,\mathrm{du}(r,\lambda),\mathrm{dt}^*(r,\lambda)\}(i=2)$ *and* $\{1,r,\mathrm{dt}^*(r,\lambda)\}(i=3)$, *with* $\mathrm{du}(r,\lambda)=1(r>\lambda)$ *and* $\mathrm{dt}^*(r,\lambda)=1(r>\lambda)(r-\lambda)$.

Theorem 1 differs from the results in Zivot and Andrews (1992) in two respects. First there is no need to have the hybrid metric considered in that paper. The weak convergence results hold under the uniform metric. This is achieved using arguments in Gregory and Hansen (1996) so that there is no need for a weak convergence result for the sample moments of $\mathrm{D}U_t$, or $\mathrm{D}T_t^*$. The most important and novel aspect in which our result differs is that we do not require that the possible range of values for the break point be restricted to exclude the end points. To achieve this, our proof is rather different and somewhat more involved and is based on projection arguments. The intuition is quite simple. With a break at either end points, the regressions indeed exhibit perfect multicollinearity but the coefficient on the lagged-dependent variable, α, is a linear combination of the parameter vector that is identifiable and estimable and its t-statistic is also well defined. In such cases, the regressions become equivalent to ones without dummy and the standard limiting distribution of Dickey and Fuller (1979) applies. This last result is important because it makes it unnecessary to use an arbitrary trimming near the end points, such as the 15% exclusion on both sides suggested by Banerjee *et al.* (1992). The arguments in the proof of Theorem 1 can also be applied to other context such as the cointegration tests with regime shifts considered by Gregory and Hansen (1996).

We used simulations to obtain the percentage points of the asymptotic distributions described above. These were based on 10,000 replications using partial sums of i.i.d. $\mathrm{N}(0,1)$ random variables to approximate the Wiener process and 1000 steps to compute the integrals. The critical values are presented in the rows labelled '$T=\infty$' in Table 1.

This relaxation of the need for trimming at the end points does not appear to be possible for the tests whereby the break point is

Table 1. Finite sample and asymptotic distributions.

		1.0%	2.5%	5.0%	10.0%	50.0%	90.0%	95.0%	97.5%	99.0%
(a) *Model* 1, $t^*_\alpha(1)$, *Choosing* T_b *minimizing* $t_{\hat{\alpha}}$										
$T = 60$	k(F-sig)	−5.83	−5.49	−5.21	−4.91	−3.91	−3.00	−2.70	−2.41	−1.96
	k(t-sig)	−5.92	−5.58	−5.23	−4.92	−3.91	−3.00	−2.74	−2.55	−2.25
$T = 80$	k(F-sig)	−5.77	−5.35	−5.15	−4.84	−3.87	−2.96	−2.70	−2.41	−2.12
	k(t-sig)	−5.77	−5.31	−5.09	−4.84	−3.88	−2.95	−2.73	−2.55	−2.22
$T = 100$	k(F-sig)	−5.70	−5.35	−5.09	−4.82	−3.89	−3.00	−2.74	−2.46	−2.22
	k(t-sig)	−5.70	−5.36	−5.10	−4.82	−3.87	−3.05	−2.75	−2.46	−2.22
$T = \infty$		−5.41	−5.02	−4.80	−4.58	−3.75	−2.99	−2.77	−2.56	−2.32
(b) *Model* 1, $t^*_{\alpha,\theta}(1)$, *Choosing* T_b *minimizing* $t_{\hat{\theta}}$										
$T = 60$	k(F-sig)	−5.58	−5.15	−4.88	−4.47	−3.33	−1.60	−0.84	−0.05	0.56
	k(t-sig)	−5.70	−5.21	−4.92	−4.53	−3.32	−1.79	−1.14	−0.35	0.42
$T = 80$	k(F-sig)	−5.50	−5.11	−4.85	−4.53	−3.33	−1.86	−1.06	−0.32	0.67
	k(t-sig)	−5.59	−5.09	−4.83	−4.54	−3.33	−1.92	−1.19	−0.46	0.34
$T = 100$	k(F-sig)	−5.42	−5.03	−4.80	−4.47	−3.33	−1.92	−1.33	−0.77	0.02
	k(t-sig)	−5.43	−5.05	−4.83	−4.50	−3.34	−2.02	−1.38	−0.84	−0.05
$T = \infty$		−5.15	−4.87	−4.64	−4.37	−3.39	−2.27	−1.85	−1.38	−0.70
(c) *Model* 1, $t^*_{\alpha,\lvert\theta\rvert}(1)$, *Choosing* T_b *maximizing* $\lvert t_{\hat{\theta}}\rvert$										
$T = 60$	k(F-sig)	−5.77	−5.42	−5.13	−4.80	−3.70	−1.87	−1.19	−0.39	0.24
	k(t-sig)	−5.85	−5.51	−5.18	−4.83	−3.70	−2.14	−1.34	−0.55	0.05
$T = 80$	k(F-sig)	−5.75	−5.26	−5.06	−4.77	−3.71	−2.14	−1.42	−0.79	0.11
	k(t-sig)	−5.66	−5.29	−5.04	−4.78	−3.72	−2.28	−1.67	−0.96	−0.06
$T = 100$	k(F-sig)	−5.69	−5.34	−5.03	−4.75	−3.74	−2.33	−1.80	−1.20	−0.18
	k(t-sig)	−5.68	−5.36	−5.05	−4.77	−3.71	−2.40	−1.88	−1.21	−0.34
$T = \infty$		−5.34	−5.08	−4.84	−4.59	−3.74	−2.71	−2.35	−2.01	−1.54

(*Continued*)

Table 1. (*Continued*)

		1.0%	2.5%	5.0%	10.0%	50.0%	90.0%	95.0%	97.5%	99.0%
(d) *Model* 2, $t^*_{\alpha}(2)$, *Choosing* T_b *minimizing* $t_{\hat{\alpha}}$										
$T = 70$	k(F-sig)	−6.22	−5.81	−5.52	−5.22	−4.21	−3.28	−3.00	−2.76	−2.54
	k(t-sig)	−6.32	−5.90	−5.59	−5.29	−4.24	−3.32	−3.08	−2.85	−2.67
$T = 100$	k(F-sig)	−6.07	−5.72	−5.48	−5.17	−4.17	−3.29	−3.05	−2.83	−2.58
	k(t-sig)	−6.21	−5.86	−5.55	−5.25	−4.22	−3.35	−3.13	−2.85	−2.63
$T = \infty$		−5.57	−5.30	−5.08	−4.82	−3.98	−3.25	−3.06	−2.91	−2.72
(e) *Model* 2, $t^*_{\alpha,\gamma}(2)$, *Choosing* T_b *minimizing* $t_{\hat{\gamma}}$										
$T = 70$	k(F-sig)	−5.77	−5.32	−4.95	−4.51	−2.92	−1.37	−0.93	−0.54	−0.02
	k(t-sig)	−5.77	−5.38	−4.98	−4.55	−3.04	−1.53	−1.10	−0.71	−0.27
$T = 100$	k(F-sig)	−5.50	−5.16	−4.85	−4.47	−2.91	−1.50	−1.11	−0.73	−0.30
	k(t-sig)	−5.56	−5.23	−4.91	−4.47	−2.99	−1.55	−1.19	−0.78	−0.38
$T = \infty$		−0.28	−4.95	−4.62	−4.28	−2.94	−1.64	−1.33	−0.98	−0.59
(f) *Model* 2, $t^*_{\alpha,\lvert\gamma\rvert}(2)$, *Choosing* T_b *maximizing* $\lvert t_{\hat{\gamma}}\rvert$										
$T = 70$	k(F-sig)	−6.01	−5.56	−5.25	−4.88	−3.64	−2.17	−1.82	−1.37	−0.76
	k(t-sig)	−6.07	−5.61	−5.33	−4.94	−3.72	−2.28	−1.89	−1.50	−0.85
$T = 100$	k(F-sig)	−5.72	−5.37	−5.14	−4.84	−3.54	−2.11	−1.76	−1.42	−0.89
	k(t-sig)	−5.86	−5.49	−5.19	−4.88	−3.60	−2.23	−1.87	−1.49	−0.95
$T = \infty$		−5.57	−5.20	−4.91	−4.59	−3.47	−2.15	−1.86	−1.59	−1.30
(g) *Model* 3, $t^*_{\alpha}(3)$, *Choosing* T_b *minimizing* $t_{\hat{\alpha}}$										
$T = 100$	k(F-sig)	−5.41	−4.99	−4.74	−4.44	−3.36	−2.53	−2.34	−2.21	−2.08
	k(t-sig)	−5.45	−5.11	−4.83	−4.48	−3.44	−2.60	−2.39	−2.22	−2.06
$T = 150$	k(F-sig)	−5.19	−4.85	−4.59	−4.31	−3.32	−2.47	−2.29	−2.11	−1.96
	k(t-sig)	−5.28	−4.96	−4.65	−4.38	−3.33	−2.50	−2.30	−2.13	−1.93

(*Continued*)

Table 1. (*Continued*)

		1.0%	2.5%	5.0%	10.0%	50.0%	90.0%	95.0%	97.5%	99.0%
$T = 200$	k(F-sig)	−5.19	−4.84	−4.59	−4.30	−3.30	−2.46	−2.26	−2.09	−1.96
	k(t-sig)	−5.28	−4.96	−4.65	−4.38	−3.32	−2.48	−2.27	−2.10	−1.90
$T = \infty$		−4.91	−4.62	−4.36	−4.07	−3.13	−2.32	−2.12	−1.96	−1.78
(h) *Model* 3, $t^*_{\alpha,\gamma}(3)$, *Choosing* T_b *minimizing* $t_{\hat{\gamma}}$										
$T = 100$	k(F-sig)	−5.02	−4.69	−4.40	−3.99	−2.76	−1.76	−1.46	−1.12	−0.79
	k(t-sig)	−5.26	−4.82	−4.44	−4.07	−2.83	−1.76	−1.45	−1.12	−0.83
$T = 150$	k(F-sig)	−4.89	−4.54	−4.27	−3.93	−2.74	−1.70	−1.33	−1.01	−0.64
	k(t-sig)	−5.00	−4.63	−4.36	−3.99	−2.78	−1.72	−1.40	−1.07	−0.49
$T = 200$	k(F-sig)	−4.75	−4.43	−4.13	−3.79	−2.69	−1.53	−1.23	−0.90	−0.59
	k(t-sig)	−4.77	−4.50	−4.22	−3.83	−2.72	−1.57	−1.24	−0.96	−0.56
$T = \infty$		−4.67	−4.36	−4.08	−3.77	−2.65	−1.57	−1.22	−0.90	−0.49
(i) *Model* 3, $t^*_{\alpha,\|\gamma\|}(3)$, *Choosing* T_b *maximizing* $\|t_{\hat{\gamma}}\|$										
$T = 100$	k(F-sig)	−5.29	−4.87	−4.57	−4.27	−3.15	−2.19	−1.99	−1.69	−1.40
	k(t-sig)	−5.38	−5.02	−4.67	−4.36	−3.24	−2.28	−2.04	−1.75	−1.46
$T = 150$	k(F-sig)	−5.15	−4.77	−4.49	−4.21	−3.15	−2.16	−1.89	−1.59	−1.19
	k(t-sig)	−5.23	−4.91	−4.57	−4.28	−3.18	−2.19	−1.92	−1.63	−1.30
$T = 200$	k(F-sig)	−5.02	−4.75	−4.41	−4.10	−3.07	−2.11	−1.86	−1.63	−1.29
	k(t-sig)	−5.02	−4.75	−4.41	−4.17	−3.11	−2.15	−1.91	−1.68	−1.26
$T = \infty$		−4.87	−4.58	−4.34	−4.04	−3.08	−2.14	−1.87	−1.61	−1.30

chosen with respect to the t-statistic on the coefficient of the intercept or slope change. The asymptotic distributions of $t^*_{\alpha,\theta}(1)$ and $t^*_{\alpha,|\theta|}(1)$ assuming the break point to be in some compact subset was derived in Banerjee *et al.* (1992). Similar asymptotic results are in Vogelsang and Perron (1994) for $t^*_{\alpha,\gamma}(i)$ and $t^*_{\alpha,|\gamma|}(i)(i = 2, 3)$. The critical values are reproduced in Table 1.

4. Finite Sample Critical Values

In this section we report simulation experiments to evaluate the finite sample distributions of the statistics under the null hypothesis of a unit root. We consider the leading case of a random walk where the data are generated by (4) with $y_0 = 0$ and $e_t \sim$ i.i.d. N(0,1). This allows us to assess the effects of the data dependent methods to select the truncation lag. Given the nature of the data sets analyzed in later sections, we present critical values for the following sample sizes. For Model 1, $T = 60$, 80, and 100; for Model 2, $T = 70$ and 100; and for Model 3, $T = 100$, 150, and 200. The results were obtained using 2000 replications. The program was coded using the C language and N(0,1) random deviates were obtained from the routine RAN1 of Press *et al.* (1986). For purely computational reasons, $k\max$ is set to 5. The results are presented in Table 1.

We first give some remarks on the finite sample distributions when k is fixed. The results are not reported but are available in the working paper version. In all cases, the critical values are fairly stable as k changes provided that k is held fixed when minimizing over T_b. In those cases where k is fixed, the asymptotic distribution is a good approximation to the finite sample distribution. Upon comparison with the results in Perron (1989), it is readily seen that the critical values are much lower when T_b is allowed to be data dependent than when it is considered fixed. For example, consider $t^*_\alpha(1)$ with $T = 100$ and $k = 0$, the 5% critical value is -4.93 when minimizing over T_b as opposed to -3.76 when the date of the break is considered fixed at mid-sample.

The critical values for the test constructed with k chosen according to recursive F-tests on the coefficients of the lagged first differences are presented in the rows labeled k(F-sig). For example, the 5% point with $T = 100$ is -5.09. The critical values for the test constructed with k chosen according to a t-test on the last included lag in the autoregression are presented in the rows labeled k (t-sig). The resulting values are close to the values obtained using F-sig.[1] In those cases where a data-dependent method is used to select k, the asymptotic approximation is not as good, indicating that the use of the asymptotic critical values would lead to tests that are liberal in finite samples. Comparing the distribution of the statistics $t^*_\alpha(i)(i = 1, 2, 3)$, it is interesting to note that the highest critical values (in the left tail of the distribution) occur for Model 3. This is contrary to the fixed T_b case where the highest critical values correspond to Model 1.

Consider now the critical values of the statistic $t^*_{\alpha,\theta}(1)$ and $t^*_{\alpha,\gamma}(i)(i = 2, 3)$ where T_b is chosen to minimize $t_{\hat{\theta}}$ or $t_{\hat{\gamma}}$ the t-statistic on the change in intercept or slope. The corresponding critical values are now smaller in absolute value. This is due to the *a priori* imposition of a one-sided change in the intercept of the trend function.

Consider now the statistics based upon choosing the break date maximizing the absolute value of $t_{\hat{\theta}}$ or $t_{\hat{\gamma}}, t^*_{\alpha,|\theta|}(1)$ and $t^*_{\alpha,|\gamma|}(i)(i = 2, 3)$. These statistics like $t^*_\alpha(i)$ do not impose any *a priori* condition on the sign of the change. We remark that for Models 1 and 3, the critical values in the left tail of the distribution are essentially the same between $t^*_\alpha(1)$ and $t^*_{\alpha,|\theta|}(1)$ and between $t^*_\alpha(3)$ and $t^*_{\alpha,|\gamma|}(3)$. Hence for Models 1 and 3, these two statistics are likely to have similar properties. Things are different for Model 2. The critical values in the left tail of the distribution are smaller (in absolute value) for $t^*_{\alpha,|\gamma|}(2)$ compared to $t^*_\alpha(2)$. Hence, one could expect the former to provide a more powerful test.

---

[1]The simulated critical values involving a test of significance on the lagged first differences of the data are for tests of size 10%. We chose this value on the principle that it is safer to include extra lags to achieve the correct size in finite samples (at the expense of a loss in power). However, critical values with 5% tests were also computed and are not included since they are very similar to those with the 10% tests.

5. Finite Sample Size and Power Simulations

We now discuss finite sample size and power simulations. The aim is to determine the following: how size and power are affected by the choice of k in the presence of more general error processes, and by different values of the change in intercept and slope. Finally how power varies across procedures for choosing T_b. The focus of the simulations is placed on Models 1 and 3. The data generating process (DGP) used for Model 1 is of the form

$$y_t = \theta \mathrm{D}U_t + \delta \mathrm{D}(T_b)t + \alpha y_{t-1} + \sum_{i=1}^{4} \phi(i)\Delta y_{t-i} + (1 + \psi L)e_t, \quad (5)$$

where $e_t \sim$ i.i.d. N(0,1) and $y_0 = e_0 = 0$. For Model 3, the DGP is of the form

$$y_t = \gamma \mathrm{D}T_t^* + \tilde{y}_t; \quad \tilde{y}_t = \alpha\tilde{y}_{t-1} + \sum_{i=1}^{4} \phi(i)\Delta\tilde{y}_{t-i} + (1 + \psi L)e_t. \quad (6)$$

For the size simulations, $\alpha = 1$ and for power α is set to 0.8. The sample size for all simulations is $T = 100$ and 1000 replications are used. Regressions were run for fixed $k = 0, 1, \ldots, 5$ and for k(F-sig) and k(t-sig) with $k\max = 5$. For fixed k, the 5% asymptotic critical values were used, and for k(F-sig) and k(t-sig), the appropriate 5% finite sample critical values for $T = 100$ were used. When the change in intercept or slope is nonzero, the break date is $T_b = 50$ (at mid-sample). For Model 1, we used values of δ (under the null) and θ (under the alternative) of 0, 2, 5, and 10. For Model 3, we used values of γ at 0, 0.1, 0.3, 0.5, and 1. Seven different error specifications were used: (1) $\phi(i) = 0 (i = 1, \ldots, 4)$ and $\psi = 0$; (2) $\phi(1) = 0.6$, $\phi(i) = 0$ $(i = 2, 3, 4)$ and $\psi = 0$; (3) $\phi(1) = -0.6, \phi(i) = 0$ $(i = 2, 3, 4)$ and $\psi = 0$; (4) $\phi(1) = 0.4$, $\phi(2) = 0.2$, and $\phi(3) = \phi(4) = \psi = 0$; (5) $\phi(1) = 0.3$, $\phi(2) = 0.3$, $\phi(3) = 0.24$, $\phi(4) = 0.14$, and $\psi = 0$; (6) $\phi(i) = 0$ $(i = 1, 2, 3, 4)$ and $\psi = 0.5$; (7) $\phi(i) = 0$ $(i = 1, 2, 3, 4)$ and $\psi = -0.4$. Experiment (1) has i.i.d. errors. This specification is used to isolate the effects of choosing k too large. Experiment (2) has positive correlation in the errors and is quite common in empirical data. Experiment (3) has negative correlation in the errors.

Experiments (4) and (5) have higher-order correlation and are useful in isolating the effects of picking k too small. Finally, experiments (6) and (7) have MA(1) errors.

Due to space constraints, we only include, in Table 2, the full set of results for $t_{\alpha}^{*}(3)$ and some for $t_{\alpha,\gamma}^{*}(3)$ with t-sig (the full set of results is available on request). We begin by summarizing results pertaining to the choice of k. When k is chosen less than the true order of the process, substantial size distortions often occur. In most cases the exact size is much greater than the nominal size. If k is chosen at least as big as the true order of the process, the exact size is rarely greater than the nominal size. However, power is lost if the lag structure is over parameterized. When the k(t-sig) or k(F-sig) procedure is used to pick k, the exact size is close to the nominal size in all cases except when there is a negative MA component as in experiment (7). In this case the exact size is substantially inflated above the nominal size. Power using k(t-sig) or k(F-sig) is generally quite good. It is greater than when k is larger than the true order of the process and is nearly as high as when k is set to the true order in the case of autoregressive errors. Overall, the k(t-sig) and k(F-sig) procedures have good size and power properties and clearly dominate using a fixed k. The results indicate that tests based on the k(t-sig) procedure are slightly more powerful than those based on k(F-sig).

Now consider how a change in intercept or slope affects the exact size. The tests $t_{\alpha}^{*}(1)$ and $t_{\alpha,\theta}^{*}(1)$ become oversized as δ increases. For example, consider experiment (1) for $t_{\alpha}^{*}(1)$ with k(t-sig); when $\delta = 0$ the size of the test is 0.047, when $\delta = 2$ it is 0.053, when $\delta = 5$ it is 0.096 and it rises to 0.486 when δ is as big as 10. The results in Table 2 for $t_{\alpha}^{*}(3)$ concerning models with a change in slope γ show that changes in γ do not affect the size of the tests for the range of values considered. For $t_{\alpha,\gamma}^{*}(3)$, there are slight distortions in some cases as γ increases. Additional simulations revealed that larger values of γ induce substantial size distortions. The reader is referred to Vogelsang and Perron (1994) for a more detailed analysis on this issue. It is important to note here, though, that the magnitude of δ and γ where size distortions become a problem are of the

Table 2. Finite sample size and power simulations; Model 3, $t^*_\alpha(3)$ (DGP: $y_t = \gamma DT^*_t + \tilde{y}_t$; $\tilde{y}_t = \alpha\tilde{y}_{t-1} + \sum_{i=1}^{4}\phi(i)\Delta\tilde{y}_{t-i} + (1+\psi L)e_t$, $e_t \sim$ i.i.d. N(0, 1); $T = 100$, $T_b = 50$; 2000 replications; 5% nominal size; $k\max = 5$).

	Size ($\alpha = 1.0$)					Power ($\alpha = 0.8$)				
k	$\gamma = 0$	$\gamma = 0.1$	$\gamma = 0.3$	$\gamma = 0.5$	$\gamma = 1$	$\gamma = 0$	$\gamma = 0.1$	$\gamma = 0.3$	$\gamma = 0.5$	$\gamma = 1$
(1) $\phi(i) = 0.0$ $(i = 1,\ldots,4), \psi = 0.0$										
0	0.049	0.053	0.055	0.047	0.036	0.358	0.365	0.344	0.331	0.325
1	0.044	0.049	0.048	0.042	0.037	0.287	0.299	0.283	0.277	0.277
2	0.045	0.046	0.048	0.041	0.040	0.203	0.215	0.207	0.199	0.205
3	0.038	0.039	0.042	0.040	0.047	0.160	0.177	0.169	0.165	0.167
4	0.035	0.037	0.039	0.036	0.041	0.122	0.129	0.134	0.130	0.134
5	0.035	0.035	0.039	0.038	0.039	0.110	0.123	0.125	0.116	0.117
F-sig	0.050	0.054	0.058	0.055	0.050	0.235	0.256	0.244	0.231	0.233
t-sig	0.049	0.051	0.058	0.050	0.045	0.257	0.278	0.270	0.259	0.258
$t^*_{\alpha,\gamma}(3)$	0.051	0.064	0.093	0.092	0.094	0.292	0.419	0.417	0.413	0.398
(2) $\phi(1) = 0.6$, $\psi = \phi(i) = 0.0$ $(i = 2, 3, 4)$										
0	0.000	0.000	0.000	0.000	0.001	0.000	0.000	0.000	0.000	0.000
1	0.058	0.060	0.056	0.054	0.062	0.908	0.903	0.904	0.902	0.901
2	0.046	0.048	0.049	0.049	0.055	0.753	0.758	0.761	0.753	0.756
3	0.045	0.046	0.040	0.041	0.045	0.586	0.592	0.600	0.594	0.593
4	0.037	0.040	0.034	0.038	0.044	0.405	0.426	0.424	0.417	0.417
5	0.033	0.033	0.034	0.037	0.045	0.289	0.302	0.306	0.305	0.305
F-sig	0.049	0.051	0.047	0.047	0.054	0.676	0.679	0.679	0.688	0.693
t-sig	0.049	0.047	0.038	0.046	0.049	0.760	0.773	0.774	0.778	0.785
$t^*_{\alpha,\gamma}(3)$	0.044	0.048	0.055	0.067	0.079	0.781	0.828	0.840	0.841	0.835

(*Continued*)

Table 2. (*Continued*)

	Size ($\alpha = 1.0$)					Power ($\alpha = 0.8$)				
k	$\gamma = 0$	$\gamma = 0.1$	$\gamma = 0.3$	$\gamma = 0.5$	$\gamma = 1$	$\gamma = 0$	$\gamma = 0.1$	$\gamma = 0.3$	$\gamma = 0.5$	$\gamma = 1$
(3) $\phi(1) = -0.6$, $\psi = \phi(i) = 0.0$ $(i = 2, 3, 4)$										
0	0.858	0.874	0.873	0.858	0.848	0.997	0.997	0.998	0.998	0.998
1	0.051	0.048	0.043	0.038	0.034	0.131	0.132	0.117	0.114	0.113
2	0.046	0.040	0.040	0.037	0.034	0.090	0.100	0.096	0.094	0.098
3	0.044	0.045	0.041	0.037	0.039	0.084	0.098	0.094	0.099	0.099
4	0.034	0.030	0.033	0.032	0.033	0.063	0.074	0.082	0.083	0.077
5	0.033	0.035	0.038	0.037	0.039	0.056	0.073	0.073	0.075	0.070
F-sig	0.037	0.040	0.044	0.042	0.044	0.091	0.104	0.104	0.104	0.100
t-sig	0.039	0.039	0.042	0.034	0.037	0.090	0.105	0.105	0.104	0.097
$t^*_{\alpha,\gamma}(3)$	0.046	0.062	0.082	0.076	0.073	0.113	0.192	0.180	0.178	0.167
(4) $\phi(1) = 0.4$, $\phi(2) = 0.2, \psi = \phi(3) = \phi(4) = 0.0$										
0	0.004	0.004	0.005	0.004	0.004	0.001	0.000	0.000	0.000	0.000
1	0.009	0.008	0.008	0.007	0.006	0.432	0.439	0.428	0.421	0.424
2	0.048	0.051	0.050	0.049	0.048	0.756	0.764	0.765	0.763	0.760
3	0.040	0.042	0.047	0.044	0.051	0.598	0.611	0.611	0.607	0.602
4	0.038	0.039	0.043	0.044	0.046	0.413	0.432	0.438	0.436	0.421
5	0.042	0.043	0.040	0.040	0.050	0.300	0.314	0.318	0.311	0.313
F-sig	0.040	0.048	0.049	0.047	0.050	0.582	0.593	0.600	0.591	0.593
t-sig	0.038	0.040	0.038	0.040	0.044	0.607	0.625	0.626	0.624	0.620
$t^*_{\alpha,\gamma}(3)$	0.037	0.043	0.049	0.056	0.069	0.659	0.733	0.755	0.737	0.737
(5) $\phi(1) = 0.3$, $\phi(2) = 0.3$, $\phi(3) = 0.25$, $\phi(4) = 0.14$, $\psi = 0.0$										
0	0.108	0.107	0.106	0.105	0.100	0.000	0.000	0.000	0.000	0.000
1	0.001	0.001	0.001	0.001	0.000	0.033	0.036	0.033	0.033	0.034

(*Continued*)

Table 2. (*Continued*)

	Size ($\alpha = 1.0$)					Power ($\alpha = 0.8$)				
k	$\gamma = 0$	$\gamma = 0.1$	$\gamma = 0.3$	$\gamma = 0.5$	$\gamma = 1$	$\gamma = 0$	$\gamma = 0.1$	$\gamma = 0.3$	$\gamma = 0.5$	$\gamma = 1$
2	0.002	0.002	0.001	0.001	0.000	0.566	0.568	0.569	0.571	0.578
3	0.033	0.033	0.031	0.031	0.034	0.877	0.881	0.878	0.876	0.873
4	0.071	0.073	0.078	0.077	0.080	0.904	0.898	0.900	0.897	0.899
5	0.051	0.051	0.054	0.051	0.051	0.762	0.774	0.771	0.770	0.769
F-sig	0.048	0.046	0.050	0.048	0.051	0.857	0.863	0.858	0.858	0.856
t-sig	0.038	0.037	0.039	0.037	0.037	0.855	0.859	0.864	0.860	0.859
$t^*_{\alpha,\gamma}(3)$	0.022	0.020	0.025	0.025	0.025	0.864	0.893	0.882	0.893	0.894
(6) $\psi = 0.5$, $\phi(i) = 0.0$ $(i = 1, \ldots, 4)$										
0	0.002	0.003	0.003	0.003	0.001	0.008	0.010	0.007	0.007	0.005
1	0.150	0.142	0.160	0.160	0.156	0.548	0.552	0.566	0.545	0.542
2	0.021	0.021	0.028	0.028	0.020	0.096	0.100	0.110	0.103	0.103
3	0.052	0.050	0.057	0.061	0.059	0.190	0.202	0.215	0.203	0.202
4	0.034	0.035	0.033	0.033	0.034	0.101	0.105	0.117	0.111	0.113
5	0.038	0.037	0.039	0.035	0.041	0.102	0.111	0.128	0.119	0.117
F-sig	0.053	0.055	0.062	0.065	0.061	0.206	0.213	0.222	0.210	0.209
t-sig	0.067	0.066	0.070	0.069	0.064	0.244	0.258	0.271	0.257	0.262
$t^*_{\alpha,\gamma}(3)$	0.056	0.067	0.092	0.104	0.111	0.254	0.339	0.373	0.365	0.364
(7) $\psi = -0.4$, $\phi(i) = 0.0$ $(i = 1, \ldots, 4)$										
0	0.734	0.738	0.739	0.717	0.691	0.996	0.996	0.996	0.995	0.995
1	0.202	0.210	0.206	0.189	0.174	0.751	0.763	0.727	0.720	0.715
2	0.079	0.087	0.075	0.074	0.074	0.395	0.406	0.383	0.385	0.376
3	0.044	0.048	0.050	0.046	0.046	0.252	0.263	0.254	0.251	0.245

(*Continued*)

Table 2. (*Continued*)

	Size ($\alpha = 1.0$)					Power ($\alpha = 0.8$)				
k	$\gamma = 0$	$\gamma = 0.1$	$\gamma = 0.3$	$\gamma = 0.5$	$\gamma = 1$	$\gamma = 0$	$\gamma = 0.1$	$\gamma = 0.3$	$\gamma = 0.5$	$\gamma = 1$
4	0.036	0.037	0.034	0.033	0.036	0.159	0.173	0.163	0.164	0.156
5	0.031	0.031	0.033	0.032	0.035	0.131	0.144	0.139	0.138	0.135
F-sig	0.145	0.148	0.156	0.148	0.149	0.456	0.476	0.452	0.451	0.442
t-sig	0.235	0.247	0.257	0.251	0.248	0.631	0.658	0.646	0.643	0.647
$t^*_{\alpha,\gamma}(3)$	0.174	0.236	0.287	0.272	0.273	0.587	0.690	0.673	0.665	0.659

Note: All entries refer to the statistic $t^*_\alpha(3)$ except those in the rows labeled $t^*_{\alpha,\gamma}(3)$ which correspond to this statistic constructed using the t-sig procedure.

order of 5 to 10 times the standard deviation of the errors for δ and at least 2 times the standard deviation of the errors for γ. For most macroeconomic time series (including those analyzed in later sections) intercept shifts are less than 5 standard deviations and slope changes are less than 0.7 standard deviations. Therefore, distortions caused by large changes are not a problem in practice but care should be used if a series is suspected to have a very large intercept or slope change.

We conclude by noting the effect on power of imposing the mild *a priori* condition on the sign of the change, i.e., comparing $t_{\alpha}^{*}(1)$ versus $t_{\alpha,\theta}^{*}(1)$ and $t_{3}^{*}(3)$ versus $t_{\alpha,\gamma}^{*}(3)$. It is seen that power is higher when this condition is imposed and there is indeed a nonzero change in the trend function.

6. Empirical Results for the Nelson–Plosser Data Set

Table 3 presents the empirical results for the Nelson–Plosser (1982) series for which, in Perron (1989), Model 1 was the specification of interest. A rejection of the unit root was claimed for all these series except the Consumer Price Index, Velocity and Interest Rate. Results are presented for both cases where the truncation lag is selected using the F-sig or t-sig methods and for both ways of selecting the break point. Here, $k\max$ is specified to be 10.[2] The statistics of most interest are the estimates of α and their t-statistic as well as six p-values in the last columns (reported to the nearest 1%). The first set corresponds to $t_{\alpha}^{*}(1)$ and the second to $t_{\alpha,\theta}^{*}(1)$. The first p-value in each set is based on the asymptotic distribution. It is included because it may be more robust, for example, to the presence of additional correlation than those based on finite sample

[2]The choice of $k\max$ is somewhat arbitrary. On the one hand, one would like a large value to have as unrestricted a procedure as possible. On the other hand, a large value of $k\max$ yields problems of multicollinearity in the data and also a substantial loss of power. The choice of $k\max$ was also set such that the estimated autoregressions did not show any sign of remaining correlation in the residuals as indicated by the Box–Pierce statistic. Most of the results are robust to alternative choices for $k\max$.

Table 3. Empirical results, Nelson–Plosser data; Model 1, $k\,\max = 10$.

Series	Sample	T	T_b	k	$t_{\hat{\theta}}$	$\hat{\alpha}$	$t_{\hat{\alpha}}$	$t^*_{\alpha}(1)$, p-values (asy)	(F-sig)	(t-sig)	$t^*_{\alpha,\theta}(1)$, p-values (asy)	(F-sig)	(t-sig)
Real GNP	1909–1970	62	1928	9	−5.13	0.190	−5.93	<0.01	<0.01		<0.01	<0.01	
			1928	8	−4.79	0.267	−5.50	<0.01		0.03	<0.01		0.02
Nominal GNP[a]	1909–1970	62	1928	11	−6.34	0.404	−8.16	<0.01		<0.01	<0.01		<0.01
			1928	15	−5.94	0.497	−6.21	<0.01	<0.01		<0.01	<0.01	
Real per capita GNP	1909–1970	62	1928	9	−3.73	0.313	−4.81	0.06	0.12		0.03	0.06	
			1928	7	−3.31	0.484	−4.51	0.13		0.21	0.07		0.10
Industrial production	1860–1970	111	1928	8	−5.18	0.272	−6.01	<0.01	<0.01	<0.01	<0.01	<0.01	<0.01
Employment	1890–1970	81	1928	8	−3.42	0.586	−5.14	0.02	0.05		0.01	0.02	
			1928	7	−3.11	0.650	−4.91	0.04		0.09	0.02		0.04
GNP deflator	1889–1970	82	1928	5	−3.28	0.783	−4.14	0.29	0.35	0.35			
			1919	5	−3.51	0.886	−3.24				0.58		0.54
			1919	9	−3.61	0.829	−3.87				0.27	0.28	
CPI	1860–1970	111	1939	5	2.00	0.948	−3.09	0.88	0.88	0.88			
			1919	5	−3.12	0.982	−1.16				0.98	0.96	0.96
Wages	1900–1970	71	1929	7	−4.32	0.619	−5.41	<0.01		0.02	<0.01		0.01
			1929	9	−4.10	0.635	−4.62	0.10	0.16		0.05	0.08	
Money stock	1889–1970	82	1929	7	−2.80	0.783	−4.69	0.08	0.14		0.04	0.07	
			1927	6	−2.50	0.831	−4.30	0.21		0.28			
			1928	6	−2.63	0.824	−4.28				0.12		0.15

(*Continued*)

Table 3. (*Continued*)

Series	Sample	T	T_b	k	$t_{\hat{\theta}}$	$\hat{\alpha}$	$t_{\hat{\alpha}}$	$t^*_\alpha(1)$, p-values			$t^*_{\alpha,\theta}(1)$, p-values		
								(asy)	(F-sig)	(t-sig)	(asy)	(F-sig)	(t-sig)
Velocity	1869–1970	102	1949	8	2.95	0.830	−2.81	0.95	0.94				
			1946	0	3.24	0.858	−3.29	0.81		0.81			
			1880	5	−2.74	0.928	−1.62				0.96	0.93	
			1880	0	−2.46	0.897	−2.43				0.87		0.83
Interest rate	1900–1970	71	1965	3	3.86	0.934	−1.35	>0.99	>0.99				
			1963	3	3.44	0.928	−1.35	>0.99		>0.99			
			1920	0	−4.16	1.058	1.16				>0.99		>0.99
			1918	0	−3.59	1.079	2.08				>0.99	>0.99	

Note: [a]For Nominal GNP, $k\max = 15$ (See footnote 3).

distributions. The second and third p-values correspond to the F-sig and t-sig methods, respectively. The critical values used correspond to samples of size 60, 80, or 100 whichever is closest to the actual sample size.

The empirical results show that the unit root hypothesis can be rejected at the 5% significance level or better, under either scenario about the choice of k, for Real GNP, Nominal GNP,[3] Industrial Production and Nominal Wages. For the Employment series, the finite sample p-value is 0.05 with F-sig and 0.09 with t-sig (the corresponding asymptotic p-values are 0.02 and 0.04, respectively). Hence, the unit root is also rejected for the Employment series. The Real per capita GNP and Money Stock series present a more ambiguous case. When k is chosen with the F-sig procedure, the p-value for the Real per capita GNP series is 0.12 using the finite sample distribution and 0.06 using the asymptotic distribution. The corresponding figures are 0.14 and 0.08 for the Money Stock series. These values are marginal for a rejection at the 10% level.[4]

The unit-root hypothesis cannot be rejected for the Consumer Price Index (CPI), Velocity, and Interest Rate series under any procedure. The choices of T_b and k obtained using the data-dependent methods for choosing k are different but yield the same qualitative results. The only series offering a markedly different picture from the fixed T_b case is the GNP Deflator. With k chosen according to either method the p-value is 0.35 (0.29 using the asymptotic distribution). Hence, for this series, the rejection of the unit root reported in Perron (1989) is not robust to correlation between the choice of T_b and the data.

[3]For the Nominal GNP series, $k\max$ was found binding in the sense that $k = 10$ was selected. Hence $k\max$ was increased to 15 which again was found to be binding. We did not increase $k\max$ further given the relatively few number of observations. Nevertheless, the conclusion is robust to basically any value of the truncation lag parameter k chosen.

[4]It is of interest to note, that applying the tests to the Friedman and Schwartz (1982) Real per Capita GNP (using the sample 1909–1970) allows an overwhelming rejection of the unit root under either method. See the working paper version for more details.

A comment is warranted about the choice of T_b selected according to these procedures. Except for the CPI, Velocity, and Interest Rate series (for which the unit root is not rejected), the value T_b is either 1929 (for Nominal Wage and Money Stock) or 1928 (for the other series). While 1928 does not exactly correspond to the date specified in Perron (1989), the economic interpretation remains the same. The selection of 1928 is due to the presence of the dummy variable $D(T_b)_t$ in regression (1). Hence, 1928 is often chosen because the dummy variable takes value 1 in 1929 and offers some additional fit to the 1929 crash over what the change in the intercept can do alone.

Now consider the results when T_b is chosen to minimize $t_{\hat{\theta}}$, the t-statistic on the change in intercept, i.e., imposing the one-sided restriction of a crash. When a rejection of the unit-root hypothesis occurred using $t^*_{\alpha}(1)$, it does so again here and more strongly, given that the tests have higher power. As was the case earlier, the unit root cannot be rejected for the GNP Deflator series. The results offer, however, a different picture for three series. First, for the Employment series, the unit root can be rejected at the 5% level (using any procedure) instead of 10% with the statistic $t^*_{\alpha}(1)$. More interestingly, the unit root can now be rejected at the 10% level for the Real per capita GNP and Money Stock series. For example, the p-values under the F-sig procedure are 0.06 and 0.07, respectively.

We now turn to the analysis of the Common Stock Price and Real Wage series where Model 2 is specified, i.e., allowing both a change in the intercept and the slope of the trend function. The procedures used and the presentation of the estimation results in Table 4 follow our previous analysis except that k max is now 5. Consider first the case where T_b is chosen to minimize the t-statistic on α. The date of break selected for the Common Stock Price series is 1928 (consistent with the imposition of 1929 as the break date in Perron, 1989). Both methods to choose the truncation lag yield the same model and test statistic with an asymptotic p-value of 0.02 and finite sample p-value of 0.04 for F-sig and 0.06 for t-sig. Similar results hold for the Real Wage series. The break date is 1939; the asymptotic p-value of the test is 0.03 and the finite sample ones are 0.07 with F-sig and 0.08 with t-sig.

Table 4. Empirical results, Nelson–Plosser data; Model 2, $k\max = 5$.

	Series		
	Common stock prices		Real wages
Sample	1871–1970		1900–1970
T	100		71
T_b	1928	1936	1939
k	1	3	3
$\hat{\theta}$	−1.36	−2.10	−0.25
	(−4.92)	(−4.86)	(−2.65)
$\hat{\beta}$	0.0065	0.0094	0.0086
	(4.43)	(4.93)	(5.26)
$\hat{\gamma}$	0.0141	0.0268	0.0047
	(4.61)	(4.91)	(3.38)
$\hat{x}$	0.716	0.553	0.390
$t_{\hat{x}}$	−5.50	−5.49	−5.41
$t_x^*(2)$, p-values			
(asy)	0.02		0.03
(F-sig)	0.04		0.07
(t-sig)	0.06		0.08
$t_{x,\|\gamma\|}^*(2)$, p-values			
(asy)		0.01	0.02
(F-sig)		0.02	0.04
(t-sig)		0.02	0.04
$t_{x,\gamma}^*(2)$, p-values			
(asy)		<0.01	<0.01
(F-sig)		0.01	0.02
(t-sig)		0.01	0.02

Consider now results obtained when T_b is chosen maximizing $t_{\hat{\gamma}}$ or $|t_{\hat{\gamma}}|$, the t-statistic on the coefficient of the slope change. The results are quite interesting in that the unit root is strongly rejected using either method to select the truncation lag even without the *a priori* imposition on the sign of the change in slope. The selected break date is still 1939 for the Real Wage series but now 1936 for Common Stock Price.

To compare our results with those of Zivot and Andrews (1992), note first the methodological differences involved. First, we retained the one time dummy $D(T_b)_t$ in regressions (1) and (2); we consider the F-sig procedure to select the truncation lag as well as the t-sig procedure; we consider $k\max = 5$ instead of 10 for the Real Wages and Common Stock Price series; and we also consider the case where the break date is selected using a test of significance on the coefficient of the change in slope. For the series Real GNP, Nominal GNP, Industrial Production, Nominal Wages, and Common Stock Prices our results agree with those of Zivot and Andrews (1982), namely a rejection of the unit root. Our results also show these rejections to be robust to alternative specifications for choosing the break date and the truncation lag (except for Nominal Wage using $t^*_\alpha(1)$ and F-sig). For the Employment series, our results allow a rejection at the 10% level using the finite sample critical values for the t-sig method when the break is selected minimizing the unit-root statistic (basically due to the inclusion of the one-time dummy $D(T_b)_t$ in (1)). However, our results show that a stronger rejection, at the 5% level, is possible using the F-sig method and that this rejection becomes even stronger if the mild *a priori* restriction of a one-sided change is imposed. For the Real per capita GNP and Money Stock series, the results with $t^*_\alpha(1)$ and the t-sig method are similar to those in Zivot and Andrews (1992), namely p-values of 0.21 and 0.28. Using the F-sig procedure, the p-values are substantially reduced to 0.12 and 0.14, respectively. Imposing the sign of the change *a priori* allows a rejection at the 10% level for both series using F-sig and for Real per capita GNP using t-sig. The difference for the Real Wage series is due to the different choice of k max. Our results agree for nonrejections for GNP Deflator, CPI, Velocity, and Interest Rate.

7. Results With an International Data Set for Postwar Real GNP

This Section analyzes an international data set of postwar quarterly real GNP or GDP series. The countries analyzed and the type and sampling period of the series are the following: US (GNP; 1947:1–1991:3); Canada (GDP; 1947:1–1989:1); Japan (GNP; 1957:1–1988:4); France (GDP; 1965:1–1988:3); Germany (GNP; 1960:1–1986:2); Italy (GDP; 1960:1–1985:1); and the UK (GDP; 1957:1–1986:3). The data for US are from Citibase and for Canada from the Cansim data bank. For Japan and France they are from the IFS data tape. The remaining series (UK, Germany, and Italy) are from Data Resources Inc. and are those used in Campbell and Mankiw (1989). All series are seasonally adjusted and at annual rates, except for the US and the UK which are at quarterly rates. The plots of the logarithm of most series are presented in Fig. 1. In these graphs the dashed line is the estimated trend function allowing a one-time change in slope. The date of the change varies for each series and was selected using the test $t_{\alpha}^{*}(3)$.

The results pertaining to the statistic $t_{\alpha}^{*}(3)$ are presented in Table 5. Using the asymptotic critical values, the unit root is rejected, at close to the 5% level, for all series except Italy (for Canada this rejection is not robust when using F-sig). Using the finite sample critical values, the results are not, in general, as sharp. For Japan, the unit root is strongly rejected (p-values of 0.02 and 0.03). The results are not as clear for most of the other series but some interesting cases still emerge. For the UK, using the F-sig and t-sig procedures, the p-values are 0.07 and 0.08 respectively, allowing a rejection of the unit root at the 10% level. The results for Canada, France, Germany, and the US are similar in terms of the t-statistics obtained. They range from −4.22 to −4.33 with finite sample p-values between 0.12 and 0.14 (this excludes the case of Canada with F-sig). While the unit root cannot be rejected at the 10% level, the results are not very

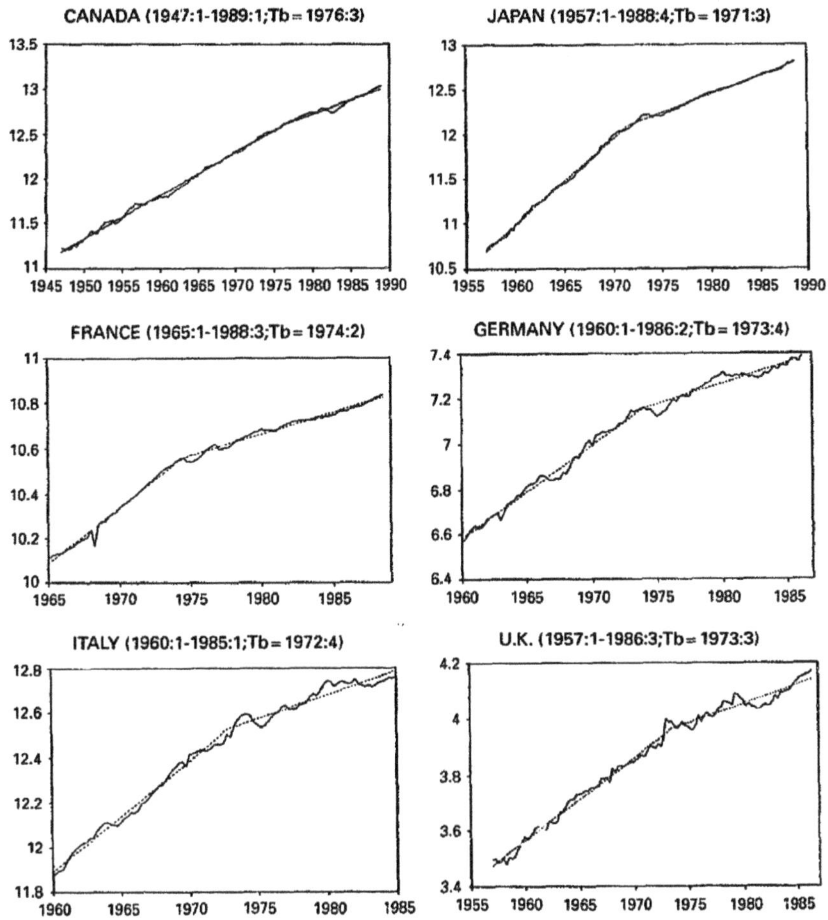

Fig. 1. Log real GNP for selected countries.

much at odds with the hypothesis that the series can be construed as stationary fluctuations around a breaking trend function. Such is not the case, however, with the GDP series from Italy. Here the p-values are large enough to cast little doubt on the unit root.

It is interesting to look at the estimated change in the slope of the trend function and the dating of the break implied by the estimation procedure. The estimated percentage decrease in the rates of growth are: US, 37%; Canada, 36%; Japan, 58%; France, 66%; Germany,

Table 5. Empirical results, Model 3, international data set.

Series	Sample	T	T_b	k	$\hat{\beta}$	$\hat{\gamma}$	$t_{\hat{\gamma}}$	$\hat{\alpha}$	$t_{\hat{\alpha}}$	$t^*_{\alpha}(3)$, p-values (asy)	(F-sig)	(t-sig)	$t^*_{\alpha,\gamma}(3)$, p-values (asy)	(F-sig)	(t-sig)
US; GNP	47:1–91:3	179	71:2	2	0.0088	−0.0024	−13.03	0.900	−4.22	0.07	0.12	0.14			
(k max = 5)			69:2	2	0.0089	−0.0024	−13.34	0.900	−4.16				0.04	0.05	0.06
Canada; GDP	47:1–89:1	169	76:3	3	0.0122	−0.0044	−21.18	0.818	−4.26	0.06		0.13			
(k max = 5)			76:3	4	0.0122	−0.0044	−21.18	0.830	−3.78	0.19	0.26				
			77:1	3	0.0121	−0.0045	−21.24	0.817	−4.25				0.03		0.06
			77:1	4	0.0121	−0.0045	−21.24	0.829	−3.77				0.10	0.13	
Japan; GNP	57:1–88:4	128	71:3	4	0.0243	−0.0142	−76.11	0.650	−5.14	<0.01	0.02		<0.01	<0.01	
(k max = 5)			71:4	4	0.0241	−0.0141	−75.19	0.659	−5.11	<0.01		0.03			
			71:3	3	0.0243	−0.0142	−76.11	0.685	−4.99				<0.01		0.02
France[a]; GDP	65:1–88:3	95	74:2	3	0.0127	−0.0080	−46.93	0.679	−4.33	0.05	0.12	0.13			
(k max = 10)			74:1	2	0.0129	−0.0081	−47.76	0.705	−4.26				0.03		0.07
			74:1	3	0.0129	−0.0081	−47.76	0.685	−4.16				0.04	0.07	
Germany; GNP	60:1–86:2	106	73:4	7	0.0104	−0.0058	−21.92	0.624	−4.33	0.05	0.12	0.13			
(k max = 10)			73:1	4	0.0106	−0.0058	−22.55	0.705	−4.00				0.06		0.11
			73:1	6	0.0106	−0.0058	−22.55	0.681	−3.82				0.09	0.14	
Italy; GDP	60:1–85:1	101	72:4	3	0.0125	−0.0071	−20.91	0.755	−3.90	0.15	0.25				
(k max = 10)			73:2	1	0.0123	−0.0071	−21.37	0.814	−3.31	0.40		0.56			
			73:3	1	0.0122	−0.0072	−21.51	0.817	−3.26				0.25	0.29	0.33
UK; GDP	57:1–86:3	118	73:3	7	0.0074	−0.0040	−18.96	0.552	−4.62	0.02	0.07	0.08			
(k max = 10)			73:2	7	0.0074	−0.0040	−18.98	0.554	−4.61				0.01	0.03	0.04

Note: [a]A dummy taking value 1 in 1968:2 was included for France to take account of the general strike in May 1968.

56%; Italy, 57%; UK, 54%. These figures are indeed quite large and suggest, besides the unit-root issue, that an important structural change has occurred. The break dates are different for each country but are all close to the year 1973, associated with the first oil price shock. They vary between 1971:2 (US) and 1976:3 (Canada). It is to be noted, however, that the method used here is not directly geared at providing a consistent estimate of the date of change. Hence, the break dates should be viewed as approximate.

As discussed, using $t^*_{\alpha,\gamma}(3)$, which select T_b based on the parameter of the change in slope, is likely to allow tests with greater power. p-values pertaining to this test are presented in the last columns of the table. Indeed, it appears more powerful. Using the t-sig procedure, the p-values for the null hypothesis of a unit root are at most 0.11 for all countries except Italy. Using the F-sig procedure, the p-values are smaller than 0.10 for US, Japan, France, and the UK; they are 0.13 and 0.14 for Canada and Germany, respectively. These results show that a simple imposition of a one-sided downward change in slope (still with an unknown break point) is enough to warrant rejection of the unit root hypothesis at close to the 10% level for all countries except Italy.

We view these results, especially given the small span of the data, as substantial evidence against the unit root. It is indeed somewhat revealing to consistently obtain such p-values given the relatively low power of unit-root tests using data over a short span (see Perron, 1991). Given that the statistical procedure used is one where an extreme assumption is made about the correlation of the choice of the break point and the data (yielding a procedure with low power compared to the case where T_{b} is assumed fixed), we view these results as consistent with the hypothesis that the series are best characterized as stationary fluctuations around a breaking trend function with a change in slope near 1973.

To compare our results with those of Banerjee *et al.* (1992) (BLS), we first note the main differences in the studies. First, the data used are slightly different in terms of both the sources and the horizon. Second, they use the one-step innovational outlier method which does not allow for a change in slope under the null hypothesis.

Third, they use a fixed value of the truncation lag set at 4 for all countries and they note that the results are robust when setting this fixed value to 8 or when using an information criterion (AIC or Schwartz) to select the order. Using these different specifications they found little evidence against the unit root for all countries except Japan.

After several investigations using both types of methods applied to both data set,[5] it turns out that the major factor responsible for the conflicting results is the method to choose the truncation lag. For example, our data-dependent methods select $k = 4$ only for Japan for which we both reject the unit root. For the other countries the implied value of the selected truncation lag is different (in no case do our methods select $k = 8$ either). We believe our methods to select the truncation lag to be better for the purpose of the unit-root tests for the following reasons documented in Ng and Perron (1994). First, fixing k to some arbitrary value can involve serious size distortions and/or power losses because the actual correlation structure of the data is not only unknown but is likely to be different across countries. However, even for data-dependent methods that implies asymptotically valid unit-root tests, there are important differences between methods based on a general to specific approach and methods based on information criteria. In the context of a model where the noise component is an ARMA process, Ng and Perron (1995) show that the latter implies a sequence of selected values for k that increases with the sample size at a logarithmic rate, a very slow rate. The finite sample implication of this result is that methods based on information criteria will tend to select very low autoregressive orders. These implied parsimonious autoregressions will often not be enough to capture important serial correlation in the data and can lead to tests with size distortions and/or power losses. These theoretical issues are consistent with the empirical results of BLS, who report values of k at 0 or 1 for all countries when using an information criterion. In no cases do our methods

---

[5]My thanks to Robin Lumsdaine for correspondence on this issue.

select such low values (except for Italy where we both agree for a nonrejection).

8. Concluding Comments

This paper documents the robustness of the results presented in Perron (1989). Unlike this previous study, we analyzed the case where the break date is explicitly correlated with the data and provided critical values to carry inference under a variety of procedures. This work is not intended as a substitute for the statistical procedures presented in that earlier paper but rather as a complement. Indeed, a case can often be made for using critical values that are based on the assumption of no correlation between the choice of the break point and the data. On the one hand, it may represent a close approximation to the actual extent of the correlation. On the other hand, each investigator may differ as to the amount of *a priori* information he or she is willing to incorporate into the analysis.

Another issue concerns the power of the tests. There appears to be a clear trade-off between power and the amount of *a priori* information one is willing to incorporate with respect to the choice of the break point. The presumption is clearly that a procedure imposing no such *a priori* information, as the ones presented in this paper, has relatively low power. In this respect, the rejection of the unit-root hypothesis, even when assuming a perfect correlation between the choice of the break point and the data, is quite strong.

Appendix. Proof of Theorem 1

To simplify cross-references, we adopt the notation of Zivot and Andrews (1992), henceforth referred to as $Z - A$. Let $S_t = \sum_{j=1}^{t} e_j (S_0 = 0)$ and $X_T(r) = \sigma^{-1}T^{-1/2}S_{[Tr]}, (j-1)/T \leq r < j/T$ (for $j = 1, \ldots, T$), where $\sigma^2 = \lim_{T\to\infty} T^{-1}E(S_T^2)$ and $[\cdot]$ denotes the integer part of the argument. Since $\{e_t\}$ is i.i.d. with finite variance, we have $X_T(r) \Rightarrow W(r)$, where $\Rightarrow$ denotes weak convergence in distribution (from the space D[0,1] to the space C[0,1] using the uniform metric on the space of functions on [0,1]) with W(r) a standard Wiener process on [0, 1]. Also, $\sigma_T^2 \equiv T^{-1}\sum_1^T e_t^2 \rightarrow_p \sigma^2$ where $\rightarrow_p$ denotes convergence in probability. Omitting the one-time dummy variable $\mathrm{D}(T_\mathrm{b})_t$ (since it is asymptotically negligible), we consider the following regressions:

$$y_t = \beta^i(\lambda) z_{tT}^i(\lambda) + \alpha^i(\lambda) y_{t-1} + e_t, \quad (t = 1, \ldots, T), \tag{A.1}$$

for models $i = 1, 2$. The vector $z_{tT}^i(\lambda)$ encompasses the deterministic components of the model and depends explicitly on λ, the break fraction, and T, the sample size. For example, $z_{t,T}^1(\lambda)' = (1, t, \mathrm{D}U_t(\lambda))$. Let $Z_T^i(\lambda, r) = \delta_T^i z_{[Tr],T}^i(\lambda)$ be a rescaled version with δ_T^i a diagonal matrix of weights. For example, $\delta_T^1 = \mathrm{diag}(1, {}^{T-1}, 1)$. We also define the limiting functions $Z^1(\lambda, r) = (1, r, \mathrm{d}u(\lambda, r))'$ where $\mathrm{d}u(\lambda, r) = 1\ (r > \lambda)$, and $Z^2(\lambda, r) = (1, r, \mathrm{d}u(\lambda, r), \mathrm{d}t^*(\lambda, r))'$ where $\mathrm{d}t^*(\lambda, r) = 1\ (r > \lambda)(r - \lambda)$. Note that, as argued in $Z - A$, we do not have $Z_T^i(\lambda, r) \Rightarrow Z^i(\lambda, r) (i = 1, 2)$ as $T \rightarrow \infty$, using the uniform metric on the space of functions on D[0,1]. The proof nevertheless remains valid without the need to introduce another metric to guarantee such convergence results. For simplicity, we henceforth drop the superscript denoting the model. Note that the following proof is valid, with trivial modifications, for a large class of deterministic components including higher-order polynomials, multiple structural changes and other types of discrete shifts.

It is convenient to first transform (A.1) as follows. Let $Pz_T(\lambda) = [Pz_{1,T}(\lambda), \ldots, Pz_{T,T}(\lambda)]$ be the linear map projecting onto the space spanned by the columns of $z_T(\lambda)' = (z_{1,T}(\lambda), \ldots, z_{T,T}(\lambda))$. By definition $Pz_T(\lambda) = z_T(\lambda)(z_T(\lambda)' z_T(\lambda))^- z_T(\lambda)'$ where $(\cdot)^-$ denotes a g-inverse. Premultiplying by $Mz_T(\lambda) \equiv (I - Pz_T(\lambda))$, (A.1) can be written, in matrix notation, as

$$Mz_T(\lambda) Y = \alpha(\lambda) Mz_T(\lambda) Y_{-1} + Mz_T(\lambda) e, \tag{A.2}$$

where $Y' = (y_1, \ldots, y_T), Y_{-1}' = (y_0, \ldots, y_{T-1})$ and $e' = (e_1, \ldots, e_T)$. The t-statistic of interest can be written as

$$\inf_{\lambda \in [0,1]} t_{\hat{\alpha}}(\lambda) = \inf_{\lambda \in [0,1]} [T^{-2} Y_{-1}' Mz_T(\lambda) Y_{-1}]^{-1/2} \times [T^{-1} Y_{-1}' Mz_T(\lambda) e]/s_T(\lambda),$$

where $s_T^2(\lambda) = T^{-1}(Y - \hat{\alpha}(\lambda)Y_{-1})'Mz_T(\lambda)(Y - \hat{\alpha}(\lambda)Y_{-1})$ with $\hat{\alpha}(\lambda)$ the OLS estimate of α in (A.2). We have

$$\begin{aligned}
T^{-2}Y_{-1}'Mz_T(\lambda)Y_{-1} &= T^{-2}\sum_{t=1}^{T} \\
&\times \left\{ y_{t-1} - z_{t,T}(\lambda)' \left[\sum_{s=1}^{T} z_{s,T}(\lambda)z_{s,T}(\lambda)'\right]^{-} \sum_{s=1}^{T} z_{s,T}(\lambda)y_{s-1} \right\}^2 \\
&= T^{-1}\sum_{t=1}^{T}\Bigg\{ T^{-1/2}S_{t-1} - Z_{t,T}(\lambda)'\delta_T \\
&\times \left[T^{-1}\sum_{s=1}^{T} \delta_T z_{s,T}(\lambda)z_{s,T}(\lambda)'\delta_T\right]^{-} T^{-1}\sum_{s=1}^{T} \delta_T z_{s,T}(\lambda)S_{s-1} \Bigg\}^2 \\
&+ o_{p\lambda}(1) = \int_0^1 \Bigg\{ \sigma X_T(r) - Z_T(\lambda, r)' \\
&\times \left[\int_0^1 Z_T(\lambda, s)Z_T(\lambda, s)'\mathrm{d}s\right]^{-} \int_0^1 Z_T(\lambda, s)\sigma X_T(s)\mathrm{d}s \Bigg\}^2 \mathrm{d}r \\
&+ o_{p\lambda}(1) = \sigma^2 \int_0^1 \{X_T(r) - Pz_T(\lambda)X_T(r)\}^2 \mathrm{d}r + o_{p\lambda}(1),
\end{aligned} \tag{A.3}$$

($o_{p\lambda}(1)$ denotes a random variable that converges in probability to 0 uniformly in λ) and:

$$\begin{aligned}
Pz_T(\lambda)X_T(r) = Z_T(\lambda, r)' &\left[\int_0^1 Z_T(\lambda, s)Z_T(\lambda, s)'\mathrm{d}s\right]^{-} \\
&\times \int_0^1 Z_T(\lambda, s)X_T(s)\mathrm{d}s.
\end{aligned}$$

Also, using developments as in $Z - A$,

$$\begin{aligned}
T^{-1}Y_{-1}'Mz_T(\lambda)e = T^{-1}\sum_{t=1}^{T} &\left\{ y_{t-1} - z_{t,T}(\lambda)' \left[\sum_{s=1}^{T} z_{s,T}(\lambda)z_{s,T}(\lambda)'\right]^{-} \right. \\
&\left. \times \sum_{s=1}^{T} z_{s,T}(\lambda)y_{s-1} \right\} e_t + o_{p\lambda}(1)
\end{aligned}$$

$$= \sigma^2 \int_0^1 X_T(r) \mathrm{d}X_T(r)$$

$$-\sigma^2 \int_0^1 P_T z(\lambda) X_T(r) \mathrm{d}X_T(r) + \mathrm{o}_{\mathrm{p}\lambda}(1). \qquad (A.4)$$

We can, therefore, express the t-statistic as a composite functional:

$$\inf_{\lambda \in [0,1]} t_{\hat{\alpha}}(\lambda) = g\left(X_T(r), \int_0^1 X_T(r)\mathrm{d}X_T(r), Pz_T(\lambda)X_T(r), \right.$$

$$\left. \int_0^1 Pz_T(\lambda)X_T(r)\mathrm{d}X_T(r), s_T(\lambda)\right) + \mathrm{o}_{\mathrm{p}\lambda}(1),$$

where

$$g = h^*\left[h\left[H_1[X_T(r), Pz_T(\lambda)X_T(r)], \quad H_2\left[\int_0^1 X_T(r)\mathrm{d}X_T(r), \right.\right.\right.$$

$$\left.\left.\left.\int_0^1 Pz_T(\lambda)X_T(r)\mathrm{d}X_T(r)\right], s_T(\lambda)\right]\right],$$

with $h^*(m) = \inf_{\lambda \in [0,1]} m(\lambda)$ for any real function $m = m(\cdot)$ on [0,1]; and for any real functions $m_1(\cdot)$, $m_2(\cdot)$, $m_3(\cdot)$ on [0,1], $h[m_1(\lambda), m_2(\lambda), m_3(\lambda)] = m_1(\lambda)^{-1/2} m_2(\lambda)/m_3(\lambda)$. The functionals H_1 and H_2 are defined by (A.3) and (A.4). The weak convergence results for each of the elements are contained in the following lemma.

Lemma A.1. *The following convergence results hold jointly:*

(a) $X_T(r) \Rightarrow W(r)$;
(b) $\int_0^1 X_T(r)\mathrm{d}X_T(r) \Rightarrow \int_0^1 W(r)\mathrm{d}W(r)$;
(c) $Pz_T(\lambda)X_T(r) \Rightarrow Pz(\lambda)W(r) \equiv Z(\lambda, r)'\left[\int_0^1 Z(\lambda, s)Z(\lambda, s)'\mathrm{d}s\right]^- \int_0^1 Z(\lambda, s)W(s)\mathrm{d}s$;
(d) $\int_0^1 Pz_T(\lambda)X_T(r)\mathrm{d}X_T(r) \Rightarrow \int_0^1 Pz(\lambda)W(r)\mathrm{d}W(r)$;
(e) $s_T^2(\lambda) = \sigma^2 + \mathrm{o}_{\mathrm{p}\lambda}(1)$.

Parts (a) and (b) are standard results, and part (e) follows using (c) and (d) and the fact that $T^{-1}\sum_1^T e_T^2 \rightarrow_p \sigma^2$. To prove part (c), we start with the following Lemma which follows from Theorem 5.5 of Billingsley (1968).

Lemma A.2. *$Pz_T(\lambda)X_T(r) \Rightarrow Pz(\lambda)W(r)$ if $X_T(r) \Rightarrow W(r)$ and for any sequence of functions $\{v_T(s)\}(0 \leq s \leq 1)$ approaching $v(s)$, we have:*

$$Pz_T(v_T(s)) \to Pz(v(s)), \tag{A.5}$$

where

$$Pz_T(v_T(s)) = Z_T(\lambda, r)' \left[\int_0^1 Z_T(\lambda, s) Z_T(\lambda, s)' \mathrm{d}s\right]^{-} \times \int_0^1 Z_T(\lambda, s) v_T(s) \mathrm{d}s,$$

and

$$Pz(v(s)) = Z(\lambda, r)' \left[\int_0^1 Z(\lambda, s) Z(\lambda, s)' \mathrm{d}s\right]^{-} \int_0^1 Z(\lambda, s) v(s) \mathrm{d}s.$$

We prove (A.5) *in two steps. First, let*

$$Pz(v_T(s)) = Z(\lambda, r)' \left[\int_0^1 Z(\lambda, s) Z(\lambda, s)' \mathrm{d}s\right]^{-} \int_0^1 Z(\lambda, s) v_T(s) \mathrm{d}s.$$

By the properties of projections in Hilbert spaces (e.g., Brockwell and Davis, 1991, p. 52):

$$Pz(v_T(s)) \to Pz(v(s)) \quad \text{if} \quad v_T(s) \to v(s). \tag{A.6}$$

Now let

$$Pz_T(v(s)) = Z_T(\lambda, r)' \left[\int_0^1 Z_T(\lambda, s) Z_T(\lambda, s)' \mathrm{d}s\right]^{-} \times \int_0^1 Z_T(\lambda, s) v(s) \mathrm{d}s.$$

We need the following Lemma stated in Parthasarathy (1977, proposition 41.19).

Lemma A.3. *Let $S_1 \subset S_2 \subset \cdots$ be an increasing sequence of subspaces in a Hilbert space $\mathscr{H}$ and let $S_\infty = \cup_i S_i$. Then $\lim_{T\to\infty} P(S_T)(x) = P(S_\infty)(x)$ for all x, where $P(S_T)(x)$ is the projection of x on the subspace S_T.*

Lemma A.3 applied to our problem implies that

$$Pz_T(\lambda)(v(s)) \to Pz(\lambda)(v(s)), \tag{A.7}$$

since we can take $\mathscr{H} = D[0,1]$ in which case $Z_T(\lambda, r) \in D[0,1]$ and $Z(\lambda, r) \in C[0,1] \subset D[0,1]$.

Next, we use the result that if for some sequence of random variables $\{X_T\}$ and $\{Y_T\}$ we have $X_T \Rightarrow X$ and $||X_T - Y_T|| \to 0$ (under some P-measure), then $Y_T \Rightarrow X$ (under the same P-measure) (e.g., Billingsley, 1968, Theorem 4.1; Parthasarathy, 1977, Corollary 51.3). Let $X = Pz(v(s))$, $X_T = Pz_T(v(s))$ and $Y_T = Pz_T(v_T(s))$. Given (A.7), we only need to show that $||Pz_T(v(s)) - Pz_T(v_T(s))|| \to 0$. This follows easily since

$$\begin{aligned}
&||Pz_T(v(s)) - Pz_T(v_T(s))||^2 \\
&\quad = ||Pz_T(v(s) - v_T(s))||^2 \\
&\quad = ||v(s) - v_T(s)||^2 - ||v(s) - v_T(s) - Pz_T(v(s) - v_T(s))||^2 \\
&\quad \leq ||v(s) - v_T(s)||^2 \to 0.
\end{aligned}$$

This completes the proof of part (c). To prove part (d), note that we have

$$\begin{aligned}
&\int_0^1 P_T z(\lambda) X_T(r) \mathrm{d}X_T(r) \\
&\qquad = \int_0^1 Z_T(\lambda, r)' \mathrm{d}X_T(r) \left[\int_0^1 Z_T(\lambda, s) Z_T(\lambda, s)' \mathrm{d}s\right]^- \\
&\qquad\quad \times \int_0^1 Z_T(\lambda, s) X_T(s) \mathrm{d}s.
\end{aligned}$$

For concreteness consider model 2 where $Z_T(\lambda, s) \equiv (Z_{1,T}(s), Z_{2,T}(\lambda, s))'$ with $Z_{1,T}(s) \equiv (1([Ts]/T)'$, $Z_{2,T}(\lambda, s) \equiv (1([Ts]/T > \lambda), 1([Ts]/T > \lambda)([Ts]/T - \lambda))'$ and $Z(\lambda, s) \equiv (Z_1(s), Z_2(s, \lambda))'$ with $Z_1(s) \equiv (1, s)'$, $Z_2(\lambda, s) \equiv (\mathrm{d}u(\lambda, s), \mathrm{d}t^*(\lambda, s))'$. Also define $Z^*_{2,T}(s) = (1, [Ts]/T - \lambda)'$ and $Z^*_2(s) = (1, (s - \lambda))'$. Using arguments as in Gregory and Hansen (1996),

$$\begin{aligned}
&\int_0^1 Z_T(\lambda,s)Z_T(\lambda,s)'\mathrm{d}s \\
&\quad= \begin{bmatrix} \int_0^1 Z_{1,T}(s)Z_{1,T}(s)'\mathrm{d}s & \int_0^1 Z_{1,T}(s)Z_{2,T}(\lambda,s)'\mathrm{d}s \\ \int_0^1 Z_{2,T}(\lambda,s)Z_{1,T}(s)'\mathrm{d}s & \int_0^1 Z_{2,T}(\lambda,s)Z_{2,T}(\lambda,s)'\mathrm{d}s \end{bmatrix} \\
&\quad= \begin{bmatrix} \int_0^1 Z_{1,T}(s)Z_{1,T}(s)'\mathrm{d}s & \int_\lambda^1 Z_{1,T}(s)Z_{2,T}^*(s)'\mathrm{d}s \\ \int_\lambda^1 Z_{2,T}^*(s)Z_{1,T}(s)'\mathrm{d}s & \int_\lambda^1 Z_{2,T}^*(s)Z_{2,T}^*(s)'\mathrm{d}s \end{bmatrix} \\
&\quad\Rightarrow \begin{bmatrix} \int_0^1 Z_1(s)Z_1(s)'\mathrm{d}s & \int_\lambda^1 Z_1(s)Z_2^*(s)'\mathrm{d}s \\ \int_\lambda^1 Z_2^*(s)Z_1(s)'\mathrm{d}s & \int_\lambda^1 Z_2^*(s)Z_2^*(s)'\mathrm{d}s \end{bmatrix} \\
&\quad= \begin{bmatrix} \int_0^1 Z_1(s)Z_1(s)'\mathrm{d}s & \int_0^1 Z_1(s)Z_2(\lambda,s)'\mathrm{d}s \\ \int_0^1 Z_2(\lambda,s)Z_1(s)'\mathrm{d}s & \int_0^1 Z_2(\lambda,s)Z_2(\lambda,s)'\mathrm{d}s \end{bmatrix} \\
&\quad= \int_0^1 Z(\lambda,s)Z(\lambda,s)'\mathrm{d}s.
\end{aligned}$$

Note that the result does not require that $Z_T(\lambda,s) \Rightarrow Z(\lambda,s)$ under the uniform metric. Similarly, we have

$$\int_0^1 Z_T(\lambda,s)X_T(s)\mathrm{d}s \Rightarrow \int_0^1 Z(\lambda,s)X(s)\mathrm{d}s.$$

Finally,

$$\begin{aligned}
&\int_0^1 Z_T(\lambda,r)'\mathrm{d}X_T(r) = T^{-1/2}\sum_1^T Z_T(\lambda,t/T)e_t \\
&\quad= \left(T^{-1/2}\sum_1^T e_t, T^{-3/2}\sum_1^T te_t, T^{-1/2}\sum_{T_b+1}^T e_t, T^{-3/2}\sum_{T_b+1}^T (t-T_b)e_t\right)
\end{aligned}$$

$$\Rightarrow \left(W(1), \int_0^1 r\mathrm{d}W(r), W(1) - W(\lambda), \int_\lambda^1 (r-\lambda)\mathrm{d}W(r)\right)$$
$$= \int_0^1 Pz(\lambda)W(r)\mathrm{d}W(r).$$

This completes the proof of part (d).

To complete the proof of the main result, we need to show continuity of the various functionals. Continuity of $h*$ and h is proved in $Z - A$.

Lemma A.4. *The functions* H_1 *and* H_2 *defined by* (A.3) *and* (A.4) *are continuous at* $(W(r),\ Pz(\lambda)W(r))$ *and* $(\int_0^1 W(r)\mathrm{d}W(r),\ \int_0^1 Pz(\lambda)W(r)\mathrm{d}W(r))$ *with* W*-probability one.*

Proof. Since H_1 and H_2 are continuous functions of their respective elements, the proof follows if each of the elements is bounded over [0,1] with W-probability one. $W(\cdot)$ is bounded with W-probability one and so is $\int_0^1 W(r)\mathrm{d}W(r)$ as discussed in $Z - A$. Using arguments similar to those in $Z - A$, $\int_0^1 Pz(\lambda)W(r)\mathrm{d}W(r)$ will be continuous if $Pz(\lambda)W(r)$ is continuous, i.e., if $\sup_{\lambda\in[0,1]} |Pz(\lambda)W(r)| < \infty$. We note that $Pz(\cdot)$ is a linear operator that maps an element on C [0,1] (the Wiener process $W(r)$ which is continuous) to a subspace defined by the functions $Z(\lambda, r)$. Continuity of $Pz(\lambda)W(r)$ follows since a linear projection map is bounded and continuous (see, e.g., Ash, 1972, p. 130 and p. 148). □

It is useful to illustrate this result by an example. Consider Model 1 where $Z(\lambda, r) = (1, r, \mathrm{d}u(\lambda, r))$. Note that

$$\int_0^1 Z(\lambda, s)Z(\lambda, s)'\mathrm{d}s = \begin{bmatrix} 1 & 1/2 & (1-\lambda) \\ 1/2 & 1/3 & (1-\lambda^2)/2 \\ (1-\lambda) & (1-\lambda^2)/2 & (1-\lambda) \end{bmatrix}.$$

If $\lambda = 0$, $\int_0^1 Z(0,s)Z(0,s)'\mathrm{d}s = A$ and if $\lambda = 1$, $\int_0^1 Z(1,s)Z(1,s)'\mathrm{d}s = B$, where

$$A = \begin{bmatrix} 1 & 1/2 & 1 \\ 1/2 & 1/3 & 1/2 \\ 1 & 1/2 & 1 \end{bmatrix} \quad \text{and} \quad B \equiv \begin{bmatrix} 1 & 1/2 & 0 \\ 1/2 & 1/3 & 0 \\ 0 & 0 & 0 \end{bmatrix}.$$

A and B are obviously singular, but a common g-inverse is given by

$$G = 12 \begin{bmatrix} 1/3 & -1/2 & 0 \\ -1/2 & 1 & 0 \\ 0 & 0 & 0 \end{bmatrix}.$$

Since the choice of g-inverse leaves a projection map unchanged, we have for $\lambda = 0, 1$:

$$Pz(\lambda)W(r) = Z^{\perp}(r)' \left[\int_0^1 Z^{\perp}(s)Z^{\perp}(s)'\mathrm{d}s \right]^{-1} \int_0^1 Z^{\perp}(s)W(s)\mathrm{d}s,$$

where $Z^{\perp}(r)' = (1, r)$, in which case the limiting distribution of $t_{\hat{\alpha}}(\lambda)(\lambda = 0, 1)$ reduces to that in the case where no dummy for structural change is included. The proof for Model 3 follows similar arguments and is therefore omitted. It uses the limiting distribution for fixed λ derived in Perron and Vogelsang (1993a, 1993b) (see also Vogelsang, 1993).

Acknowledgments

The author wishes to thank Tim Vogelsang for computational assistance beyond the call of duty. This research was supported by grants from the Social Sciences and Humanities Research Council of Canada, the Natural Sciences and Engineering Research Council of Canada and the Fonds pour la Formation de Chercheurs et l'Aide à la Recherche du Québec. John Campbell, Robert Shiller and Charles Nelson kindly provided some of the data used. I also wish to thank Don Andrews, René Garcia, Robin Lumsdaine, Serena Ng, Jim Stock and Tim Vogelsang for comments on an earlier version. This is a substantially revised version of the 1990 Princeton University Econometric Research Program Memorandum No. 350.

References

Ash, R.B. (1972). *Real Analysis and Probability.* Academic Press, New York.

Banerjee, A., Lumsdaine, R.L. and Stock, J.H. (1992). Recursive and sequential tests of the unit root and trend break hypothesis: theory and international evidence. *Journal of Business and Economic Statistics* 10, 271–287.

Billingsley, P. (1968). *Convergence of Probability Measures.* Wiley, New York.

Brockwell, P.J. and Davies, R.A. (1991). *Time Series: Theory and Methods*, 2nd edn. Springer, New York.

Campbell, J.Y. and Mankiw, N.G. (1989). International evidence on the persistence of economic fluctuations. *Journal of Monetary Economics* 23, 319–333.

Christiano, L.J. (1992). Searching for breaks in GNP. *Journal of Business and Economic Statistics* 10, 237–250.

Dickey, D.A. and Fuller, W.A. (1979). Distribution of the estimators for autoregressive time series with unit root. *Journal of the American Statistical Association* 74, 427–431.

Friedman, M. and Schwartz, A.J. (1982). *Monetary Trends in the United States and the United Kingdom: Their Relation to Income, Prices and Interest Rates.* 1867–1975, The University of Chicago Press, Chicago.

Gregory, A.W. and Hansen, B.E. (1996). Residual-based tests for cointegration in models with regime shifts. *Journal of Econometrics* 70.

Hall, A. (1994). Testing for a unit root in time series with pretest data based model selection. *Journal of Business and Economic Statistics* 12, 461–470.

Nelson, C.R. and Plosser, C.I. (1982). Trends and random walks in macroeconomic time series. *Journal of Monetary Economics* 10, 139–162.

Ng, S. and Perron, P. (1995). Unit root tests in ARMA models with data dependent methods for the selection of the truncation lag. *Journal of the American Statistical Association* 90, 268–281.

Parthasarathy, K.R. (1978). *Introduction to Probability and Measure.* Springer, New York.

Perron, P. (1989). The Great Crash, the oil price shock and the unit root hypothesis. *Econometrica* 57, 1361–1401.

Perron, P. (1991). Test consistency with varying sampling frequency. *Econometric Theory* 7, 341–368.

Perron, P. (1994). Trend, unit root and structural change in macroeconomic time series. In *Cointegration for the Applied Economists*, B.B. Rao (ed.), MacMillan Press, Basingstoke, pp. 113–146.

Perron, P. and Vogelsang, T.J. (1992). Nonstationarity and level shifts with an application to purchasing power parity. *Journal of Business and Economic Statistics* 10, 301–320.

Perron, P. and Vogelsang, T.J. (1993a). A note on the asymptotic distributions of unit root tests in the additive outlier model with breaks. *Revista de Econometria* 8, 181–202.

Perron, P. and Vogelsang, T.J. (1993b). The Great Crash, the oil price shock and the unit root hypothesis: *Erratum. Econometrica* 61, 248–249.

Press, W.H., Flanery, B.P., Teukolsky, S.A. and Vetterling, W.T. (1986). *Numerical Recipes: The Art of Scientific Computing*, Cambridge University Press, Cambridge.

Said, S.E. and Dickey, D.A. (1984). Testing for unit roots in autoregressive — moving average models of unknown order. *Biometrika* 71, 599–608.

Vogelsang, T.J. (1993). Essays on testing for nonstationarities and structural change in time series models. Unpublished Ph.d. Dissertation. Department of Economics, Princeton University, Princeton, NJ.

Vogelsang, T.J. and Perron, P. (1994). Additional tests for a unit root allowing for a break in the trend function at an unknown time. Manuscript. Department of Economics, Cornell University, Ithaca, NY.

Zivot, E. and Andrews, D.W.K. (1992). Further evidence on the Great Crash, the oil price shock and the unit root hypothesis. *Journal of Business and Economic Statistics* 10, 251–270.

Chapter 17

Additional Tests for a Unit Root Allowing for a Break in the Trend Function at an Unknown Time*

Timothy J. Vogelsang[†] and Pierre Perron

Cornell University, USA;
Boston University, USA; Centre de Recherche et Développement en Economique, Canada
[†] *tjv2@cornell.edu*

We consider unit root tests that allow a shift in trend at an unknown time. We focus on the additive outlier approach, but also give results for the innovational outlier approach. Various methods of choosing the break date are considered. New limiting distributions are derived, including the case where a shift in trend occurs under the unit root null hypothesis. Limiting distributions are invariant to mean shifts but not to slope shifts. Simulations are used to assess finite sample size and power. We focus on the effects of a break under the null and the choice of break date.

1. Introduction

During the past decade there has been an ongoing debate as to whether shocks to macroeconomic times series should be regarded as permanent or temporary. The debate was sparked by the important findings of Nelson and Plosser (1982), who concluded that most macroeconomic time series are best characterized by unit root processes, implying that shocks to these series are permanent. This view was challenged in Perron (1989, 1990), where it was shown

*Originally Published in *International Economic Review*, 39, pp. 1073–1100.

that a rejection of the unit root hypothesis is possible for many macroeconomic times series once allowance is made for a one-time shift in the trend function. Thus, many macroeconomic time series may be better characterized as having temporary shocks fluctuating around a broken deterministic trend function. Two methods of modeling the trend shifts were proposed by Perron (1989, 1990). The first, called the 'additive outlier' (AO), models the break as occurring suddenly while the second, called the 'innovational outlier' (IO), models the break as evolving more slowly over time. The choice of AO versus IO depends on the view one is taking as to the dynamics of the transition path following a break.

A key assumption of the framework proposed by Perron (1989, 1990) is that the break date of the trend function is fixed (exogenous) and chosen independently of the data. This assumption has drawn much criticism in subsequent papers, based on the argument that break dates are often chosen after looking at the data, leaving room for 'data mining'. This criticism was first pointed out by Christiano (1992). Several recent studies have proposed procedures that address the choice of break date. These include Banerjee *et al.* (1992), Perron (1997), Perron and Vogelsang (1992), and Zivot and Andrews (1992). The strategy used in all four studies was to endogenize the choice of the break date by making it data dependent. Two approaches in endogenizing the choice of the break point have been considered. Both require estimation of a Dickey–Fuller (DF) type regression at all allowable break dates. The first procedure is to choose the break date that minimizes the Dickey–Fuller t-statistic across all possible regressions. The second procedure is to choose the break date that maximizes (or minimizes) a statistic that tests the significance of one or more trend break coefficients.

Within the four studies, asymptotic results are available for many combinations of trend breaks, choice of break year, and choice of AO or IO models. For nontrending data, results are provided in Perron and Vogelsang (1992) for both the AO and IO models where the break date is chosen by minimizing the DF t-statistics and by the significance of the coefficient on a mean-break dummy variable. For trending data, Zivot and Andrews (1992) provide results for the IO models where the break date is chosen by minimizing the DF t-statistic. These include a model with an intercept shift only (crash model), a model with an intercept shift and a slope shift (crash/changing growth model), and a model with a smooth shift in slope only (changing growth model). Banerjee *et al.* (1992) give results in the IO framework for the crash model and the changing growth model where the break date is chosen both by the DF t-statistics and by the significance of a trend break dummy parameter. They do not provide results for the crash/changing growth model.

Except for Perron and Vogelsang (1992), the above studies use the assumption that no break has occurred under the unit root hypothesis when deriving the asymptotic results. It is shown in Perron and Vogelsang (1992) that for nontrending data, the unit root statistics are asymptotically invariant to a mean shift under the null hypothesis, but this invariance does not hold in finite samples. Mean and slope shifts are nuisance parameters as far as a unit root test is concerned. It is important to have results determining how mean and slope shifts affect null distributions.

This paper has several goals. First, asymptotic distributions are derived for statistics in the crash model and the crash/changing growth model parameterized in the AO framework. Statistics are considered whereby the break date is chosen by minimizing the DF t-statistic and using the significance of trend-break dummy parameters. Asymptotic results are also provided for the crash/changing growth model parameterized in the IO framework with the break date chosen via the significance of trend-break dummy parameters. These results are obtained under the assumption that no break has occurred under the null hypothesis and are new. Second, for all models asymptotic results are derived under the assumption that a break has occurred under the unit root null hypothesis. It is shown that limiting distributions of the statistics are invariant to an intercept shift, but not to a shift in slope. It is shown, however, that for slope shift magnitudes typically encountered in practice, the slope-shift asymptotics are a poor approximation to the finite sample distributions. It is only for very large slope shifts that the slope-shift asymptotics provide adequate approximations. Third, finite sample simulations are reported to illustrate how finite sample size and power is affected by shifts in the mean and slope.

The rest of the paper is organized as follows. In the next section we describe our models and statistics. In Section 3 we derive limiting null distributions. In Section 4 simulated finite sample critical values are given. Section 5 contains the results of finite sample size and power simulations. Section 6 has concluding remarks, and proofs are relegated to a mathematical Appendix.

2. The Models and Statistics

This section lays out the models and statistical procedures. The discussion is brief, and the reader is referred to Perron (1989, 1997), Zivot and Andrews (1992), and Banerjee *et al.* (1992) for further details. To be consistent with the existing literature, it is assumed that at most one break has occurred in the trend function. The date of the break (should it occur) is denoted by T_b^c with $1 < T_b^c < T$, where T is the sample size. (The superscript 'c' denotes the 'correct' break date). Since the break date is assumed to be unknown, regressions are estimated using break dates, T_b, that are different from the correct break date. The models are labeled as follows: Model 1 allows a shift in the intercept, Model 2 allows both a shift in intercept and slope, and Model 3 allows a 'smooth' shift in the slope by requiring the end points of the two segments of the broken trend to be joined.

2.1. *The additive outlier model*

The AO model applies to cases where the break is assumed to occur instantly and is not affected by the dynamics of the series. They are given by,

$$y_t = \mu + \beta t + \theta DU_t^c + z_t, \qquad \text{(Model 1)} \qquad (1.1)$$

$$y_t = \mu + \beta t + \theta DU_t^c + \gamma DT_t^c + z_t, \qquad \text{(Model 2)} \qquad (1.2)$$

$$y_t = \mu + \beta t + \gamma DT_t^c + z_t, \qquad \text{(Model 3)} \qquad (1.3)$$

where $DU_t^c = 1(t > T_b^c)$, $DT_t^c = 1(t > T_b^c)(t - T_b^c)$ and $1(\cdot)$ is the indicator function. The error z_t is specified to be an ARMA$(p+1, q)$ process defined as $A(L)z_t = B(L)e_t$, where e_t is i.i.d. $(0, \sigma^2)$ with finite fourth moment. $A(L)$ and $B(L)$ are polynomials in L of order $p+1$ and q, respectively, where $A(L)$ can be factored as $A(L) = (1 - \alpha L)A^*(L)$, and $A^*(L)$ is a pth order polynomial in L. It is assumed that $A^*(L)$ and $B(L)$ have all roots outside the unit circle. The initial condition, y_0, is assumed to be a fixed constant. Under the null hypothesis of a unit root $\alpha = 1$, $\mu = y_0$ and $z_t = z_{t-1} + \psi^*(L)e_t$, where $\psi^*(L) \equiv A^*(L)^{-1}B(L)$. Under the alternative hypothesis $|\alpha| < 1$, and z_t is the stationary process $z_t = \psi(L)e_t$, where $\psi(L) \equiv A(L)^{-1}B(L)$. Note that $\psi^*(L) = (1 - \alpha L)\psi(L)$ and that $\psi(0) = \psi^*(0) = 1$. The parameters θ and γ measure the magnitude of the possible trend break. For example, when there is a unit root, the intercept of $\{y_t\}$ is y_0 up to time T_b^c and $y_0 + \theta$ afterwards. When the series is stationary, the intercept is μ up to time T_b^c and $\mu + \theta$ afterwards.

Testing for a unit root in the AO framework consists of a two-step procedure. The first step involves detrending the series using the following regressions estimated by OLS,

$$y_t = \mu + \beta t + \theta DU_t + \tilde{y}_t^1, \qquad \text{(Model 1)} \qquad (2.1)$$

$$y_t = \mu + \beta t + \theta DU_t + \gamma DT_t + \tilde{y}_t^2, \qquad \text{(Model 2)} \qquad (2.2)$$

$$y_t = \mu + \beta t + \gamma DT_t + \tilde{y}_t^3, \qquad \text{(Model 3)} \qquad (2.3)$$

where $DU_t = 1(t > T_b)$ and $DT_t = 1(t > T_b)(t - T_b)$. In the second step, the unit root hypothesis is tested using the t-statistic for testing $\alpha = 1$ in the regressions:

$$\tilde{y}_t^j = \sum_{i=0}^{k} \omega_i D(T_b)_{t-i} + \alpha \tilde{y}_{t-1}^j + \sum_{i=1}^{k} c_i \Delta \tilde{y}_{t-i}^j + u_t, \quad (j = 1, 2) \qquad (3.1)$$

$$\tilde{y}_t^3 = \alpha \tilde{y}_{t-1}^3 + \sum_{i=1}^{k} c_i \Delta \tilde{y}_{t-i}^3 + u_t, \qquad (3.2)$$

where $D(T_b)_t = 1(t = T_b + 1)$. The inclusion of the $k+1$ dummy variables $D(T_b)_{t-i} (i = 0, \ldots, k)$ in (3.1) is necessary to ensure that the limiting distributions of the t-statistics on α are invariant to the correlation structure of the errors and the same as in the IO model below. The one-time dummies are not needed in Model 3 as the t-statistic on α is asymptotically invariant to the correlation structure of the data with the appropriate choice of k. Its limiting distribution is, however, different from the IO model (see Perron and Vogelsang 1993a for details). The t-statistics for testing $\alpha = 1$ using regressions (3.1) and (3.2) are denoted by $t_{\hat{\alpha}}(j,\ \text{AO},\ T_b,\ k)$, where j denotes the model ($j = 1, 2, 3$), T_b is the break date used, and k is the lag length of the autoregression.

2.2. *The innovational outlier model*

The IO model is applicable to cases where it is more reasonable to view the break as occurring more slowly over time. In principle, the dynamic path of adjustment of the shift could take any form. However, a natural and convenient way to model the dynamics is to assume that the series reacts to shocks to the trend function in the same way that it responds to shocks to the innovation process. This assumption can be captured using the following specification. Under the null hypothesis of a unit root, y_t is given by

$$y_t = y_{t-1} + \beta + \psi^*(L)(\theta D(T_b^c)_t + e_t), \qquad \text{(Model 1)} \quad (4.1\text{a})$$

$$y_t = y_{t-1} + \beta + \psi^*(L)(\theta D(T_b^c)_t + \gamma DU_t^c + e_t), \quad \text{(Model 2)} \quad (4.2\text{a})$$

$$y_t = y_{t-1} + \beta + \psi^*(L)(\gamma DU_t^c + e_t). \qquad \text{(Model 3)} \quad (4.3\text{a})$$

Note that the immediate impact of a shift in slope, say, for Model 3 is given by γ, while the long-run impact is given by $\psi^*(1)\gamma$. Under the alternative hypothesis, y_t is given by

$$y_t = \mu + \beta t + \psi(L)(\theta DU_t^c + e_t), \qquad \text{(Model 1)} \quad (4.1\text{b})$$

$$y_t = \mu + \beta t + \psi(L)(\theta DU_t^c + \gamma DT_t^c + e_t), \quad \text{(Model 2)} \quad (4.2\text{b})$$

$$y_t = \mu + \beta t + \psi(L)(\gamma DT_t^c + e_t). \qquad \text{(Model 3)} \quad (4.3\text{b})$$

Here, the immediate impact of a slope shift in Model 3 is given by γ, while the long-run impact is given by $\psi(1)\gamma$. Models (4.ja) and (4.jb) ($j = 1, 2$) can be nested, and the unit root hypothesis tested in the following Dickey–Fuller (1979) type regressions estimated by OLS:

$$y_t = \mu + \beta t + dD(T_b)_t + \theta DU_t + \alpha y_{t-1} + \sum_{i=1}^{k} c_i \Delta y_{t-i} + u_t, \quad (5.1)$$

$$y_t = \mu + \beta t + dD(T_b)_t + \theta DU_t + \gamma DT_t + \alpha y_{t-1} + \sum_{i=1}^{k} c_i \Delta y_{t-i} + u_t. \quad (5.2)$$

For Model 3, a nesting of (4.3a) and (4.3b) is not considered. Instead, the regression considered by Zivot and Andrews (1992) and Banerjee *et al.* (1992) is used,

$$y_t = \mu + \beta t + \gamma DT_t + \alpha y_{t-1} + \sum_{i=1}^{k} c_i \Delta y_{t-i} + u_t. \quad (5.3)$$

Under unit root hypothesis, $\alpha = 1$, $\theta = 0$, and $\gamma = 0$ in (5.1)–(5.3). Note that (5.3) does not permit a break in the trend under the null hypothesis as DU_t is absent. The null hypothesis can be tested using the t-statistics for testing $\alpha = 1$ in regressions (5.1)–(5.3). These t-statistics, for given values of T_b and k, are denoted by $t_{\hat{\alpha}}(j, IO, T_b, k)(j = 1, 2, 3)$.

2.3. *Procedures for selecting T_b and k*

To implement the above tests, some choice of T_b and k must be made. Two data-dependent methods for choosing T_b are considered in this paper. The first involves choosing T_b such that $t_{\hat{\alpha}}(j, m, T_b, k)$ is minimized. This choice of T_b corresponds to the break date that is most likely to reject the unit root hypothesis. Let $T_b(t_{\hat{\alpha}})$ denote the value of T_b chosen by this procedure and let $t_{\hat{\alpha}}(j, m, T_b(t_{\hat{\alpha}}), k) = \inf_{\lambda \in \Lambda} t_{\hat{\alpha}}(j, m, [\lambda T], k)$ denote the resulting t-statistic, where Λ is the interval $[0, 1]$ and $[\lambda T]$ is the integer part of λT. Therefore, all possible break dates are considered and trimming is not required. Limiting distributions are obtained without requiring any trimming by using results in Perron (1997).

The second approach involves choosing T_b to maximize some statistic that tests the significance of one or more of the break parameters (θ, γ). For Model 1, T_b is chosen using the maxima of $t_{\hat{\theta}}$ and $|t_{\hat{\theta}}|$ from regressions (2.1) and (5.1). The resulting value of T_b from these procedures is denoted by $T_b(t_{\hat{\theta}})$ and $T_b(|t_{\hat{\theta}}|)$. The maximum of $|t_{\hat{\theta}}|$ is used when the direction of the break is unknown while the maximum of $t_{\hat{\theta}}$ is used when the direction of the break is known *a priori* to be positive. If the direction of the break is known to be negative, then the minimum of $t_{\hat{\theta}}$ is used. When appropriate, the assumption that the direction of the break is known can lead to more powerful tests (see Perron, 1997; Perron and Vogelsang, 1992). For Model 2, T_b is chosen as the argmax of $t_{\hat{\gamma}}, |t_{\hat{\gamma}}|$ and $F_{\hat{\theta},\hat{\gamma}}$ (the argmin of $t_{\hat{\gamma}}$ if the break is negative) from regressions (2.2) and (5.2). $F_{\hat{\theta},\hat{\gamma}}$ is the F-statistic for testing the joint hypothesis that $\theta = \gamma = 0$. Again, the resulting values of T_b are denoted by $T_b(\cdot)$ using the appropriate argument. For Model 3, T_b is chosen as the argmax of $t_{\hat{\gamma}}$ and $|t_{\hat{\gamma}}|$ (or argmin of $t_{\hat{\gamma}}$ for negative breaks) from regressions (2.3) and (5.3). Note that some trimming is required to obtain limiting distributions in these cases as the results of Perron (1997) only apply to the statistics where T_b is chosen to minimize $t_{\hat{\alpha}}$.

The truncation lag parameter k is chosen using several data-dependent methods. The first procedure follows Perron (1989, 1990, 1997) and Perron and Vogelsang (1992) and is denoted by $k(t-\text{sig})$. For any given value of T_b, k is chosen so that the coefficient on the last included lagged first difference is significant at the 10% level, and the coefficient on the last included lagged first difference in higher-order autoregressions is insignificant up to some *a priori* specified maximum lag length denoted by *kmax*. Asymptotic normality of the t-statistic on these coefficients is used to carry out the inference. In the second procedure, the truncation lag is chosen using either the AIC or Schwartz (BIC) information criterion where k ranges from 0 to *kmax*. The choice of k based on these criteria are denoted by $k(\text{aic})$ and $k(\text{bic})$. See Hall (1994) and Ng and Perron (1995) for theoretical justifications of these procedures.

3. The Limiting Distributions of the Statistics

The discussion of the asymptotic distributions is divided into two cases. The first is where no break has occurred ($\theta = \gamma = 0$). For completeness, asymptotic results are stated for all models outlined in Section 2. New results pertain to AO Models 1, 2, and 3 for all choices of T_b and for IO Model 2 where T_b is chosen using $t_{\hat{\gamma}}$ and $F_{\hat{\theta},\hat{\gamma}}$. The second case is where a break has occurred under the unit root hypothesis ($\theta \neq 0$ and/or $\gamma \neq 0$). This case has not been considered elsewhere in the literature. Results are given for all the models outlined in Section 2. For simplicity, results are presented and proven assuming the errors are uncorrelated and no lagged first differences are included in regressions (3.1), (3.2), and (5.1)–(5.3). Thus $\psi(L) = 1$ and $k = 0$ are assumed. When no break has occurred under the null hypothesis, it is argued in Section 3.2 that the results obtained in this special case remain valid under more general error structures and when k is chosen using the data-dependent procedures. We set $y_0 = 0$ and $\beta = 0$, as the statistics are exactly invariant to y_0 and β under the null hypothesis.

3.1. *Limiting distributions when no break has occurred*

The results in this section are valid for $\theta = \gamma = 0$. Imposing $y_0 = \beta = 0$ and using the simplification that $\psi(L) = 1$, the DGP is the same in all models and can be written as:

$$y_t = y_{t-1} + e_t. \tag{6}$$

Consider first the limiting behavior of $t_{\hat{\alpha}}(j, m, T_b(t_{\hat{\alpha}}), k = 0)$ ($j = 1, 2, 3$; $m =$ AO, IO). Let $W_j(r, \lambda)$ denote the residuals from the projection of a standard Wiener process $W(r)$ onto the subspace generated by the functions $\{1, r, du(r, \lambda)\}$ ($j = 1$), $\{1, r, du(r, \lambda), dt(r, \lambda)\}$ ($j = 2$) and $\{1, r, dt(r, \lambda)\}$ ($j = 3$), where $du(r, \lambda) = 1(r > \lambda)$ and $dt(r, \lambda) = 1(r > \lambda)(r - \lambda)$ on $(0, 1)$. Under the null hypothesis (6), Perron (1997) proved that:

$$t_{\hat{\alpha}}(j, \mathrm{IO}, T_b(t_{\hat{\alpha}}), k = 0) \Rightarrow \inf_{\lambda \in \Lambda} \{R_j(\lambda)\} \quad (j = 1, 2, 3), \tag{7}$$

where $R_j(\lambda) = \int_0^1 W_j(r,\lambda)\, dW(r)(\int_0^1 W_j(r,\lambda)^2\, dr)^{-1/2}$. Results from the AO models are obtained from results in Perron and Vogelsang (1993b) and Perron (1997).

Under (6),

$$t_{\hat{\alpha}}(j, \mathrm{AO}, T_b(t_{\hat{\alpha}}), k=0) \Rightarrow \inf_{\lambda\in\Lambda}\{R_j(\lambda)\} \quad (j=1,2), \tag{8}$$

$$t_{\hat{\alpha}}(3, \mathrm{AO}, T_b(t_{\hat{\alpha}}), k=0) \Rightarrow \inf_{\lambda\in\Lambda}\{R_{3,\mathrm{AO}}(\lambda)\}, \tag{9}$$

where

$$\begin{aligned} R_{3,\mathrm{AO}}(\lambda) &\\ &= \left(\int_0^1 W_3(r,\lambda)^2 dr\right)^{-1/2} \times \left(\int_0^1 W_3(r,\lambda) dW(r)\right. \\ &\quad \left. - a_{22}^{-1}\int_\lambda^1 (r-\lambda)W^*(r)dr \int_\lambda^1 W_3(r,\lambda)dr\right), \end{aligned}$$

with $a_{22} = \lambda^3(1-\lambda)^3/3$ and $W^*(r)$ the residuals from a projection of $W(r)$ onto the subspace generated by $\{1, r\}$. The limiting distributions are identical in the AO and IO cases for Models 1 and 2. Without the one-time dummies, $D(T_b)_{t-i}(i=0,\ldots,k)$, in regression (3.1) this equivalence no longer holds. See Perron and Vogelsang (1993a) for details. For Model 3 the limiting distribution in the AO case is different from the IO case. Critical values for (7) and (8) can be found in Zivot and Andrews (1992) and are reproduced for Models 1 and 2 in Tables 1 and 2. Critical values for (9) are given by Perron and Vogelsang (1993a).

Now consider the limiting distributions of the tests when T_b is chosen using $t_{\hat{\theta}}, t_{\hat{\gamma}}$, or $F_{\hat{\theta},\hat{\gamma}}$. Note that in the AO models, $t_{\hat{\theta}}, t_{\hat{\gamma}}$, and $F_{\hat{\theta},\hat{\gamma}}$ diverge as $T \to \infty$. Nondegenerate limiting distributions are obtained by considering $T^{-1/2}t_{\hat{\theta}}, T^{-1/2}t_{\hat{\gamma}}$, and $T^{-1}F_{\hat{\theta},\hat{\gamma}}$. Choosing T_b to maximize these rescaled statistics will yield valid asymptotic testing procedures. Of course, in finite samples it makes no difference whether one maximizes the scaled or unscaled statistics. For these statistics some trimming of break dates is required. Let Λ^* denote a closed subset of $(0,1)$ that corresponds to the allowed (trimmed)

Table 1. Model (1, AO); distribution of $t_{\hat{\alpha}}(1, \mathrm{AO}, T_b(\cdot))$.

		1.0%	2.5%	5.0%	10.0%	50.0%	90.0%	95.0%	97.5%	99.0%
Panel (a): Choosing T_b Minimizing $t_{\hat{\alpha}}, t_{\hat{\alpha}}(1, \mathrm{AO}, T_b(t_{\hat{\alpha}}))$										
$T = 50$	$k = k(t-\mathrm{sig})$	−5.92	−5.49	−5.20	−4.83	−3.89	−3.08	−2.80	−2.60	−2.34
	$k = k(\mathrm{aic})$	−5.92	−5.58	−5.26	−4.90	−3.89	−3.06	−2.79	−2.57	−2.34
	$k = k(\mathrm{bic})$	−5.91	−5.51	−5.20	−4.80	−3.74	−2.87	−2.62	−2.43	−2.14
$T = 100$	$k = k(t-\mathrm{sig})$	−5.67	−5.34	−5.06	−4.73	−3.82	−3.02	−2.84	−2.65	−2.45
	$k = k(\mathrm{aic})$	−5.68	−5.42	−5.12	−4.77	−3.84	−3.04	−2.83	−2.63	−2.41
	$k = k(\mathrm{bic})$	−5.48	−5.25	−5.94	−4.66	−3.72	−2.96	−2.75	−2.54	−2.39
$T = 150$	$k = k(t-\mathrm{sig})$	−5.63	−5.33	−5.06	−4.77	−3.80	−3.01	−2.79	−2.64	−2.44
	$k = k(\mathrm{aic})$	−5.63	−5.32	−5.06	−4.79	−3.83	−3.01	−2.81	−2.61	−2.40
	$k = k(\mathrm{bic})$	−5.51	−5.25	−4.94	−4.64	−3.75	−2.97	−2.76	−2.56	−2.40
$T = \infty$		−5.41	−5.02	−4.80	−4.58	−3.75	−2.99	−2.77	−2.56	−2.32
Panel (b): Choosing T_b Maximizing $t_{\hat{\theta}}, t_{\hat{\alpha}}(1, \mathrm{AO}, T_b(t_{\hat{\theta}}))$										
$T = 50$	$k = k(t-\mathrm{sig})$	−5.24	−4.86	−4.53	−4.14	−3.04	−2.02	−1.73	−1.50	−1.20
	$k = k(\mathrm{aic})$	−5.25	−4.86	−4.54	−4.12	−2.97	−1.96	−1.65	−1.39	−1.09
	$k = k(\mathrm{bic})$	−5.26	−4.80	−4.43	−4.03	−2.88	−1.94	−1.64	−1.39	−1.09
$T = 100$	$k = k(t-\mathrm{sig})$	−4.95	−4.61	−4.32	−4.04	−2.93	−1.94	−1.70	−1.49	−1.29
	$k = k(\mathrm{aic})$	−5.09	−4.65	−4.33	−4.02	−2.86	−1.86	−1.64	−1.45	−1.25
	$k = k(\mathrm{bic})$	−4.83	−4.49	−4.18	−3.87	−2.82	−1.90	−1.66	−1.46	−1.21
$T = 150$	$k = k(t-\mathrm{sig})$	−4.98	−4.64	−4.31	−3.99	−2.90	−1.93	−1.67	−1.45	−1.09
	$k = k(\mathrm{aic})$	−4.85	−4.55	−4.30	−3.94	−2.85	−1.93	−1.66	−1.48	−1.14
	$k = k(\mathrm{bic})$	−4.81	−4.46	−4.21	−3.84	−2.83	−1.92	−1.66	−1.45	−1.17
$T = \infty$		−4.57	−4.24	−4.01	−3.74	−2.77	−1.85	−1.57	−1.34	−1.02

(Continued)

Table 1. (*Continued*)

		1.0%	2.5%	5.0%	10.0%	50.0%	90.0%	95.0%	97.5%	99.0%
Panel (c): Choosing T_b Maximizing $\|t_{\hat{\theta}}\|, t_{\hat{\alpha}}(1, \text{AO}, T_b(\|t_{\hat{\theta}}\|))$										
$T = 50$	$k = k(t - \text{sig})$	−5.35	−4.95	−4.73	−4.34	−3.29	−2.27	−1.96	−1.74	−1.43
	$k = k(\text{aic})$	−5.31	−4.98	−4.74	−4.37	−2.24	−2.22	−1.93	−1.69	−1.34
	$k = k(\text{bic})$	−5.35	−4.92	−4.58	−4.20	−3.10	−2.21	−1.93	−1.73	−1.39
$T = 100$	$k = k(t - \text{sig})$	−5.10	−4.77	−4.51	−4.19	−3.17	−2.19	−1.98	−1.74	−1.40
	$k = k(\text{aic})$	−5.11	−4.84	−4.49	−4.19	−3.13	−2.15	−1.87	−1.69	−1.40
	$k = k(\text{bic})$	−5.00	−4.69	−4.36	−4.05	−3.06	−2.14	−1.92	−1.74	−1.45
$T = 150$	$k = k(t - \text{sig})$	−5.07	−4.81	−4.52	−4.18	−3.13	−2.18	−1.96	−1.72	−1.40
	$k = k(\text{aic})$	−5.08	−4.75	−4.46	−4.11	−3.11	−2.16	−1.94	−1.68	−1.38
	$k = k(\text{bic})$	−5.04	−4.70	−4.38	−4.03	−3.06	−2.13	−1.93	−1.76	−1.44
$T = \infty$		−4.70	−4.40	−4.17	−3.90	−2.94	−2.05	−1.78	−1.52	−1.22

Table 2. Model (2, AO); Distribution of $t_{\hat{\alpha}}(2, \mathrm{AO}, T_b(\cdot))$.

		1.0%	2.5%	5.0%	10.0%	50.0%	90.0%	95.0%	97.5%	99.0%
Panel (a): Choosing T_b Minimizing $t_{\hat{\alpha}}, t_{\hat{\alpha}}(2, \mathrm{AO}, T_b(t_{\hat{\alpha}}))$										
$T = 50$	$k = k(t-\mathrm{sig})$	−6.17	−5.87	−5.56	−5.21	−4.21	−3.35	−3.12	−2.93	−2.77
	$k = k(\mathrm{aic})$	−6.18	−5.90	−5.57	−5.27	−4.21	−3.28	−3.02	−2.82	−2.60
	$k = k(\mathrm{bic})$	−6.15	−5.87	−5.53	−5.13	−4.07	−3.14	−2.88	−2.69	−2.44
$T = 100$	$k = k(t-\mathrm{sig})$	−5.82	−5.52	−5.13	−4.85	−3.87	−3.13	−2.98	−2.82	−2.61
	$k = k(\mathrm{aic})$	−5.91	−5.52	−5.20	−4.89	−3.88	−3.16	−2.98	−2.82	−2.64
	$k = k(\mathrm{bic})$	−5.74	−5.43	−5.11	−4.79	−3.81	−3.07	−2.86	−2.74	−2.56
$T = 150$	$k = k(t-\mathrm{sig})$	−5.57	−5.40	−5.16	−4.86	−4.01	−3.28	−3.08	−2.95	−2.81
	$k = k(\mathrm{aic})$	−5.59	−5.41	−5.16	−4.90	−4.03	−3.31	−3.15	−2.99	−2.81
	$k = k(\mathrm{bic})$	−5.57	−5.32	−5.08	−4.81	−3.95	−3.25	−3.07	−2.92	−2.76
$T = \infty$		−5.57	−5.30	−5.08	−4.82	−3.98	−3.25	−3.06	−2.91	−2.72
Panel (b): Choosing T_b Maximizing $t_{\hat{\gamma}}, t_{\hat{\alpha}}(2, \mathrm{AO}, T_b(t_{\hat{\gamma}}))$										
$T = 50$	$k = k(t-\mathrm{sig})$	−5.62	−5.30	−4.98	−4.60	−3.44	−2.39	−2.04	−1.78	−1.47
	$k = k(\mathrm{aic})$	−5.71	−5.30	−4.99	−4.62	−3.41	−2.29	−1.94	−1.65	−1.40
	$k = k(\mathrm{bic})$	−5.61	−5.16	−4.73	−4.44	−3.26	−2.20	−1.90	−1.65	−1.40
$T = 100$	$k = k(t-\mathrm{sig})$	−5.27	−4.90	−4.57	−4.23	−3.12	−2.08	−1.77	−1.49	−1.18
	$k = k(\mathrm{aic})$	−5.31	−4.89	−4.59	−4.19	−3.08	−2.01	−1.69	−1.40	−1.16
	$k = k(\mathrm{bic})$	−5.10	−4.74	−4.41	−4.07	−3.01	−1.97	−1.68	−1.40	−1.16
$T = 150$	$k = k(t-\mathrm{sig})$	−5.05	−4.69	−4.46	−4.13	−3.04	−1.91	−1.66	−1.39	−1.11
	$k = k(\mathrm{aic})$	−5.05	−4.66	−4.39	−4.09	−3.01	−1.89	−1.64	−1.34	−1.08
	$k = k(\mathrm{bic})$	−4.83	−4.57	−4.28	−3.99	−2.94	−1.89	−1.59	−1.34	−1.11
$T = \infty$		−4.86	−4.55	−4.28	−3.95	−2.84	−1.58	−1.22	−0.88	−0.51

(*Continued*)

Table 2. (*Continued*)

		1.0%	2.5%	5.0%	10.0%	50.0%	90.0%	95.0%	97.5%	99.0%
Panel (c): Choosing T_b Maximizing $\lvert t_{\hat{\gamma}}\rvert, t_{\hat{\alpha}}(2, \mathrm{AO}, T_b(\lvert t_{\hat{\gamma}}\rvert))$										
$T = 50$	$k = k(t - \mathrm{sig})$	−5.64	−5.31	−5.01	−4.65	−3.47	−2.46	−2.17	−1.95	−1.72
	$k = k(\mathrm{aic})$	−5.75	−5.33	−5.03	−4.65	−3.45	−2.37	−2.07	−1.81	−1.55
	$k = k(\mathrm{bic})$	−5.64	−5.19	−4.76	−4.50	−3.30	−2.31	−1.99	−1.77	−1.53
$T = 100$	$k = k(t - \mathrm{sig})$	−5.33	−4.98	−4.64	−4.30	−3.18	−2.24	−1.98	−1.73	−1.36
	$k = k(\mathrm{aic})$	−5.34	−5.00	−4.64	−4.25	−3.14	−2.17	−1.89	−1.63	−1.35
	$k = k(\mathrm{bic})$	−5.17	−4.83	−4.46	−4.13	−3.07	−2.10	−1.83	−1.58	−1.28
$T = 150$	$k = k(t - \mathrm{sig})$	−5.06	−4.75	−4.51	−4.23	−3.21	−2.21	−1.87	−1.66	−1.33
	$k = k(\mathrm{aic})$	−5.06	−4.77	−4.47	−4.19	−3.18	−2.17	−1.85	−1.63	−1.24
	$k = k(\mathrm{bic})$	−4.91	−4.66	−4.37	−4.09	−3.09	−2.16	−1.84	−1.51	−1.30
$T = \infty$		−5.04	−4.73	−4.50	−4.20	−3.20	−2.20	−1.90	−1.59	−1.32
Panel (d): Choosing T_b Maximizing $F_{\hat{\theta},\hat{\gamma}}, t_{\hat{\alpha}}(2, \mathrm{AO}, T_b(F_{\hat{\theta},\hat{\gamma}}))$										
$T = 50$	$k = k(t - \mathrm{sig})$	−5.71	−5.42	−5.11	−4.77	−3.59	−2.61	−2.33	−2.05	−1.77
	$k = k(\mathrm{aic})$	−5.73	−5.43	−5.12	−4.75	−3.58	−2.55	−2.28	−1.95	−1.64
	$k = k(\mathrm{bic})$	−5.70	−5.37	−4.95	−4.62	−3.49	−2.52	−2.27	−1.95	−1.66
$T = 100$	$k = k(t - \mathrm{sig})$	−5.44	−5.03	−4.66	−4.35	−3.27	−2.31	−2.07	−1.78	−1.42
	$k = k(\mathrm{aic})$	−5.47	−5.09	−4.68	−4.35	−3.21	−2.25	−2.02	−1.73	−1.45
	$k = k(\mathrm{bic})$	−5.35	−4.94	−4.56	−4.24	−3.18	−2.26	−2.02	−1.75	−1.42
$T = 150$	$k = k(t - \mathrm{sig})$	−5.14	−4.86	−4.62	−4.31	−3.35	−2.32	−2.03	−1.70	−1.39
	$k = k(\mathrm{aic})$	−5.09	−4.78	−4.57	−4.28	−3.32	−2.31	−2.02	−1.73	−1.50
	$k = k(\mathrm{bic})$	−5.04	−4.72	−4.52	−4.23	−3.28	−2.33	−2.05	−1.71	−1.47
$T = \infty$		−5.15	−4.86	−4.61	−4.31	−3.35	−2.46	−2.20	−1.94	−1.63

break dates. Limiting distributions for Model 1 follow from Banerjee *et al.* (1992). Under (6),

$$t_{\hat{\alpha}}(1, \text{IO}, T_b(t_{\hat{\theta}}), k=0) \Rightarrow R_1(\lambda_1^*), \tag{10}$$

$$t_{\hat{\alpha}}(1, \text{IO}, T_b(|t_{\hat{\theta}}|), k=0) \Rightarrow R_1(\lambda_2^*), \tag{11}$$

where $\lambda_1^* = \text{argmax}_{\lambda\in\Lambda^*} Q_{1,\text{IO}}(\lambda)$, $\lambda_2^* = \text{argmax}_{\lambda\in\Lambda^*} |Q_{1,\text{IO}}(\lambda)|$, $Q_{1,\text{IO}}(\lambda) = H_1(\lambda)K_1(\lambda)^{-1/2}$ with $H_1(\lambda) = B_1(\lambda) - \int_\lambda^1 W^*(r)dr \int_0^1 W^*(r)dW(r)/\int_0^1 W^*(r)^2dr$, $K_1(\lambda) = a_{11} - (\int_\lambda^1 W^*(r)dr)^2/\int_0^1 W^*(r)^2dr$, $B_1(\lambda) = \lambda(4-3\lambda)W(1) - W(\lambda) - 6\lambda(1-\lambda)\int_0^1 rdW(r)$, and $a_{11} = \lambda(1-\lambda)(3\lambda^2 - 3\lambda + 1)$. Critical values for (10) and (11) can be found in Perron (1997). In the case of the AO model, it is shown in the appendix that under (6),

$$t_{\hat{\alpha}}(1, \text{AO}, T_b(t_{\hat{\theta}}), k=0) \Rightarrow R_1(\lambda_1^*), \tag{12}$$

$$t_{\hat{\alpha}}(1, \text{AO}, T_b(|t_{\hat{\theta}}|), k=0) \Rightarrow R_1(\lambda_2^*), \tag{13}$$

where $\lambda_1^* = \arg\max_{\lambda\in\Lambda^*} Q_{1,\text{AO}}(\lambda)$, $\lambda_2^* = \arg\max_{\lambda\in\Lambda^*} |Q_{1,\text{AO}}(\lambda)|$, $Q_{1,\text{AO}}(\lambda) = [a_{11}L_1(\lambda)]^{-1/2}\int_\lambda^1 W^*(r)dr$, with $L_1(\lambda) = \int_0^1 W^*(r)^2dr - [\int_\lambda^1 W^*(r)dr]^2/a_{11}$. Critical values for (12) and (13) were calculated via simulation methods using i.i.d. $N(0,1)$ errors with 1,000 steps to approximate the Wiener processes with 10,000 replications. They are given in Table 1.

Limiting distributions for Model 2 are as follows. These results are new, and proofs are given in the Appendix. For the IO model under (6),

$$t_{\hat{\alpha}}(2, \text{IO}, T_b(t_{\hat{\gamma}}), k=0) \Rightarrow R_2(\lambda_1^*), \tag{14}$$

$$t_{\hat{\alpha}}(2, \text{IO}, T_b(|t_{\hat{\gamma}}|), k=0) \Rightarrow R_2(\lambda_2^*), \tag{15}$$

$$t_{\hat{\alpha}}(2, \text{IO}, T_b(F_{\hat{\theta},\hat{\gamma}}), k=0) \Rightarrow R_2(\lambda_3^*), \tag{16}$$

where

$$\lambda_1^* = \underset{\lambda\in\Lambda^*}{\text{argmax}}\, Q_{2,\text{IO}}^A(\lambda), \quad \lambda_2^* = \underset{\lambda\in\Lambda^*}{\text{argmax}}\, |Q_{2,\text{IO}}^A(\lambda)|,$$

$$\lambda_3^* = \underset{\lambda\in\Lambda^*}{\text{argmax}}\, Q_{2,\text{IO}}^B(\lambda),$$

$$Q^A_{2,\mathrm{IO}}(\lambda) = [H_2(\lambda) - K_1(\lambda)^{-1}K_3(\lambda)H_1(\lambda)][K_2(\lambda) - K_1(\lambda)^{-1}K_3(\lambda)^2]^{-1/2},$$

$$Q^B_{2,\mathrm{IO}}(\lambda) = [H_1(\lambda)^2K_2(\lambda) - 2H_1(\lambda)K_3(\lambda)H_2(\lambda) + H_2(\lambda)^2K_1(\lambda)] \times [K_1(\lambda)K_2(\lambda) - K_3(\lambda)^2]^{-1},$$

with

$$H_2(\lambda) = B_2(\lambda) - \int_\lambda^1 (r-\lambda)W^*(r)dr \int_0^1 W^*(r)dW(r) \Big/ \int_0^1 W^*(r)^2dr,$$

$$K_2(\lambda) = a_{22} - \left[\int_\lambda^1 (r-\lambda)W^*(r)dr\right]^2 \Big/ \int_0^1 W^*(r)^2dr,$$

$$K_3(\lambda) = a_{12} - \int_\lambda^1 W^*(r)dr \int_\lambda^1 (r-\lambda)W^*(r)dr \Big/ \int_0^1 W^*(r)^2dr,$$

$$B_2(\lambda) = \lambda(1-\lambda)^2W(1) + \int_\lambda^1 (r-\lambda)dW(r) - (1-\lambda)^2(1+2\lambda)\int_0^1 rdW(r),$$

and $a_{12} = \lambda^2(1-\lambda)^2(2\lambda-1)/2$. Critical values obtained by simulations are in Table 4 for (14) and (15) and in Table 3 for (16). In the AO model, under (6), it is shown in the appendix that,

$$t_{\hat{\alpha}}(2, \mathrm{AO}, T_b(t_{\hat{\gamma}}), k=0) \Rightarrow R_2(\lambda_1^*), \tag{17}$$

$$t_{\hat{\alpha}}(2, \mathrm{AO}, T_b(|t_{\hat{\gamma}}|), k=0) \Rightarrow R_2(\lambda_2^*), \tag{18}$$

$$t_{\hat{\alpha}}(2, \mathrm{AO}, T_b(F_{\hat{\theta},\hat{\gamma}}), k=0) \Rightarrow R_2(\lambda_3^*), \tag{19}$$

where

$$\lambda_1^* = \operatorname*{argmax}_{\lambda\in\Lambda^*} Q^A_{2,\mathrm{AO}}(\lambda), \quad \lambda_2^* = \operatorname*{argmax}_{\lambda\in\Lambda^*} |Q^A_{2,\mathrm{AO}}(\lambda)|,$$

$$\lambda_3^* = \operatorname*{argmax}_{\lambda\in\Lambda^*} Q^B_{2,\mathrm{AO}}(\lambda),$$

Table 3. Model (2, IO); Distribution of $t_{\hat{\alpha}}(2, \text{IO}, T_b(F_{\hat{\theta},\hat{\gamma}}))$; choosing T_b maximizing $F_{\hat{\theta},\hat{\gamma}}$.

		1.0%	2.5%	5.0%	10.0%	50.0%	90.0%	95.0%	97.5%	99.0%
$T = 50$	$k = k(t - \text{sig})$	−6.29	−5.95	−5.59	−5.23	−3.81	−2.16	−1.58	−0.85	−0.08
	$k = k(\text{aic})$	−6.34	−6.00	−5.62	−5.22	−3.83	−2.23	−1.64	−0.96	−0.40
	$k = k(\text{bic})$	−6.35	−6.08	−5.67	−5.23	−3.72	−2.35	−1.87	−1.43	−0.83
$T = 100$	$k = k(t - \text{sig})$	−5.93	−5.60	−5.26	−4.90	−3.65	−2.32	−1.94	−1.55	−0.93
	$k = k(\text{aic})$	−5.90	−5.63	−5.27	−4.95	−3.68	−2.38	−1.95	−1.60	−0.99
	$k = k(\text{bic})$	−5.81	−5.49	−5.15	−4.78	−3.56	−2.35	−2.04	−1.70	−1.38
$T = 150$	$k = k(t - \text{sig})$	−5.86	−5.50	−5.28	−4.90	−3.61	−2.35	−2.07	−1.71	−1.09
	$k = k(\text{aic})$	−5.81	−5.50	−5.22	−4.86	−3.64	−2.40	−2.09	−1.80	−1.28
	$k = k(\text{bic})$	−5.69	−5.39	−5.11	−4.75	−3.53	−2.42	−2.12	−1.91	−1.57
$T = \infty$		−5.71	−5.41	−5.16	−4.86	−3.91	−2.79	−2.44	−2.13	−1.67

$$Q^A_{2,\mathrm{AO}}(\lambda) = [(a_{22} - a_{12}^2/a_{11})L_1(\lambda) - L_3(\lambda)^2]^{-1/2} L_3(\lambda),$$

$$Q^B_{2,\mathrm{AO}}(\lambda) = [L_1(\lambda)(a_{11}a_{22} - a_{12}^2) - a_{11}L_3(\lambda)^2]^{-1} \times \left[a_{22}\left(\int_\lambda^1 W^*(r)dr\right)^2 - 2a_{12}\int_\lambda^1 W^*(r)dr \int_\lambda^1 \times (r-\lambda)W^*(r)dr + a_{11}\left(\int_\lambda^1 (r-\lambda)W^*(r)dr\right)^2\right]$$

with $L_3(\lambda) = \int_\lambda^1 (r-\lambda)W^*(r)dr - (a_{12}/a_{11})\int_\lambda^1 W^*(r)dr$. Critical values for (17)–(19) were simulated and are given in Table 2.

Limiting distributions for Model 3 are as follows. Under (6) Banerjee *et al.* (1992) showed that,

$$t_{\hat{\alpha}}(3, \mathrm{IO}, T_b(t_{\hat{\gamma}}), k = 0) \Rightarrow R_3(\lambda_1^*), \tag{20}$$

$$t_{\hat{\alpha}}(3, \mathrm{IO}, T_b(|t_{\hat{\gamma}}|), k = 0) \Rightarrow R_3(\lambda_2^*), \tag{21}$$

where

$$\lambda_1^* = \underset{\lambda \in \Lambda^*}{\operatorname{argmax}}\, Q_{3,\mathrm{IO}}(\lambda), \quad \lambda_2^* = \underset{\lambda \in \Lambda^*}{\operatorname{argmax}}\, |Q_{3,\mathrm{IO}}(\lambda)| \quad \text{and}$$

$$Q_{3,\mathrm{IO}}(\lambda) = K_2(\lambda)^{-1/2} H_2(\lambda).$$

Critical values for (21) can be found in Banerjee *et al.* (1992). Under (6) it is shown in the Appendix,

$$t_{\hat{\alpha}}(3, \mathrm{AO}, T_b(t_{\hat{\gamma}}), k = 0) \Rightarrow R_{3,\mathrm{AO}}(\lambda_1^*), \tag{22}$$

$$t_{\hat{\alpha}}(3, \mathrm{AO}, T_b(|t_{\hat{\gamma}}|), k = 0) \Rightarrow R_{3,\mathrm{AO}}(\lambda_2^*), \tag{23}$$

where $\lambda_1^* = \arg\max_{\lambda \in \Lambda} Q_{3,\mathrm{AO}}(\lambda)$, $\lambda_2^* = \arg\max_{\lambda \in \Lambda} |Q_{3,\mathrm{AO}}(\lambda)|$, $Q_{3,\mathrm{AO}}(\lambda) = [a_{22}L_2(\lambda)]^{-1/2}\int_\lambda^1 (r-\lambda)W^*(r)dr$ with $L_2(\lambda) = \int_0^1 W^*(r)^2 dr - a_{22}^{-1}\left[\int_\lambda^1 (r-\lambda)W^*(r)dr\right]^2$. Critical values for (22) and (23) were simulated and are given in Table 4.

Table 4. Additional asymptotic distributions.

	Statistic			
	$t_{\hat{\alpha}}(2, \mathrm{IO}, T_b(t_{\hat{\gamma}}))$	$t_{\hat{\alpha}}(2, \mathrm{IO}, T_b(\|t_{\hat{\gamma}}\|))$	$t_{\hat{\alpha}}(3, \mathrm{AO}, T_b(t_{\hat{\gamma}}))$	$t_{\hat{\alpha}}(3, \mathrm{AO}, T_b(\|t_{\hat{\gamma}}\|))$
1.0%	−5.28	−5.57	−4.67	−4.87
2.5%	−4.95	−5.20	−4.36	−4.58
5.0%	−4.62	−4.91	−4.08	−4.34
10.0%	−4.28	−4.59	−3.77	−4.04
50.0%	−2.94	−3.47	−2.65	−3.08
90.0%	−1.64	−2.15	−1.57	−2.14
95.0%	−1.33	−1.86	−1.22	−1.87
97.5%	−0.98	−1.59	−0.90	−1.61
99.0%	−0.59	−1.30	−0.49	−1.30

3.2. *Limiting distributions under more general error processes*

The previous asymptotic results were obtained assuming i.i.d. errors and using $k = 0$ in the various regressions. The results of Section 3.1 remain valid for more general error processes. When the errors follow an AR(p) process, it is straightforward to extend the results of Dickey and Fuller (1979) to show that the above distributions remain valid provided $k \geq p$. This is formally proved for Model (3, AO), when the break date is known by Vogelsang (1993). The distributions remain valid when the errors follow an ARMA(p, q) process, provided k increases at an appropriate rate as the sample size grows following Said and Dickey (1984). In the ARMA(p, q) case the limiting distributions using the data-dependent methods remain valid as long as $kmax$ increases at an appropriate rate as the sample size grows (or $kmax \geq p$ in the pure AR(p) case). See Hall (1994), and Ng and Perron (1995).

3.3. *Limiting distributions when a break occurs*

This subsection describes the limiting behavior of the statistics when θ and γ are different from zero. These results are new. Throughout, 'true break date' and 'incorrect break date' refer to λ_c and $\lambda\,(\lambda \neq \lambda_c)$, respectively. In finite samples, we adopt the conventions that $T_b^c = [\lambda_c T]$ and $T_b = [\lambda T]$. When regressions (2.1)–(2.3), (5.1), and

(5.2) are estimated using the true break date, $t_{\hat{\alpha}}$ is exactly invariant to θ under the null hypothesis. Similarly, when regressions (2.2), (2.3), and (5.2) are estimated using the true break date, $t_{\hat{\alpha}}$ is exactly invariant to γ under the null hypothesis. On the other hand, when regression (5.3) is estimated at the true break date, $t_{\hat{\alpha}}$ is not invariant to γ. Moreover, when any of the above regressions are estimated at the incorrect break date, $t_{\hat{\alpha}}$ is not invariant to θ, γ in finite samples. In fact, $t_{\hat{\alpha}}$ will no longer be invariant to γ asymptotically. Because the testing procedures involve estimation at incorrect break dates, the limiting results in Section 3.1 no longer apply when $\gamma \neq 0$. In many cases that follow results are established by showing the statistics are $O_p(1)$ for fixed λ, but note that most of the $O_p(1)$ results do not hold uniformly for $\lambda \in [0, 1]$.

Asymptotically, the dependence upon θ and γ can be made precise. It can be shown that all of the statistics are asymptotically invariant to θ, and we set $\theta = 0$ in the analysis that follows. Perron and Vogelsang (1992) prove asymptotic invariance with respect to θ in models that do not permit trending data. Proving invariance with respect to θ of the statistics in this paper involves simple extensions to the results by Perron and Vogelsang (1992) and a proof is omitted. When $\gamma \neq 0$, results are given for Models 2 and 3 as these are the models that permit a shift in slope. As before, results are derived assuming i.i.d. errors so that $\psi(L) = 1$. The results that follow also hold for more general errors provided $k = 0$. We do not provide results for $k > 0$ or for data-dependent lag length selection as the analysis becomes very complicated and is well beyond the scope of the paper. The complexity arises from collinearity in the lagged first differences due to the shift in slope.

Under the simplifying assumptions, the DGP can be written as,

$$y_t = \gamma DT_t^c + z_t, \quad z_t = z_{t-1} + e_t. \tag{24}$$

Consider estimating the models using the true break date ($\lambda = \lambda_c$). It is shown in Perron (1989) and Perron and Vogelsang (1993b) that under (24),

$$t_{\hat{\alpha}}(2, m, [\lambda_c T], k = 0) \Rightarrow R_2(\lambda_c) \quad (m = \mathrm{IO}, \mathrm{AO}), \tag{25}$$

$$t_{\hat{\alpha}}(3, \mathrm{AO}, [\lambda_c T], k = 0) \Rightarrow R_{3,\mathrm{AO}}(\lambda_c). \tag{26}$$

Things are different for Model (3, IO). It is shown in the appendix that under (24),

$$T^{-1/2}t_{\hat{\alpha}}(3, \mathrm{IO}, [\lambda_c T], k=0) \Rightarrow O_p(1). \tag{27}$$

It is easy to establish that $T^{-1/2}t_{\hat{\alpha}}(3, \mathrm{IO}, [\lambda_c T], k=0)$ converges to a random variable that has support over the entire real line. Thus, $t_{\hat{\alpha}}(3, \mathrm{IO}, [\lambda_c T], k=0)$ will diverge to minus infinity with nonzero probability. Therefore, even when the date of the break is known, the asymptotic size of a unit root test using $t_{\hat{\alpha}}(3, \mathrm{IO}, T_b^c, k=0)$ will be distorted if $\gamma \neq 0$.

When Models (2, IO), (2, AO), (3, IO), and (3, AO) are estimated at the incorrect break date, it is shown in the Appendix that under (24),

$$T^{-1/2}t_{\hat{\alpha}}(j, m, [\lambda T], k=0) = O_p(1) \quad (j = 2, 3;\ m = \mathrm{AO}, \mathrm{IO}). \tag{28}$$

It can be shown that the limits of $T^{-1/2}t_{\hat{\alpha}}(j, m, [\lambda T], k=0)$ ($j = 2, 3;\ m = \mathrm{AO}, \mathrm{IO}$) are nonrandom functions of γ, λ, and λ_c, and that the sign of these limits depends on λ and λ_c. It is tedious but straightforward to show that given λ_c, there exists λ such that the limit of $T^{-1/2}t_{\hat{\alpha}}(j, m, [\lambda T], k=0)$ ($j = 2, 3;\ m = \mathrm{AO}, \mathrm{IO}$) is negative. Thus, $t_{\hat{\alpha}}(j, m, [\lambda T], k=0)$ ($j = 2, 3;\ m = \mathrm{AO}, \mathrm{IO}$) diverges to $-\infty$ for some values of λ. Directly it follows when $\gamma \neq 0$:

$$t_{\hat{\alpha}}(j, m, T_b(t_{\hat{\alpha}}), k=0) \Rightarrow -\infty \quad (j = 2, 3;\ m = \mathrm{AO}, \mathrm{IO}). \tag{29}$$

The consequences of (29) is that the size of unit root tests using $t_{\hat{\alpha}}(j, m, T_b(t_{\hat{\alpha}}), k=0)$ ($j = 2, 3;\ m = \mathrm{AO}, \mathrm{IO}$) will approach one as the sample size grows when $\gamma \neq 0$. What is not evident from (29) is how large γ must be, given T, before size distortions become problematic. This question is examined in Section 5 using finite sample simulations.

Given γ, size distortions will occur using $t_{\hat{\alpha}}(j, m, T_b(t_{\hat{\alpha}}), k=0)$ ($j = 2, 3;\ m = \mathrm{AO}, \mathrm{IO}$) if T is large enough. One way to avoid this potential problem is to use procedures that pick the true break date asymptotically. In Model (3, IO) however, choosing the true break date will not yield a valid test as $t_{\hat{\alpha}}$ is not invariant to γ even when T_b^c is known.

Suppose that T_b is chosen using $t_{\hat{\gamma}}$ or $F_{\hat{\theta},\hat{\gamma}}$. In models (2, AO) and (3, AO) it is shown in the appendix that the correct break date is chosen asymptotically. Therefore,

$$t_{\hat{\alpha}}(2, \mathrm{AO}, m, k = 0) \Rightarrow R_2(\lambda_c) \quad (m = T_b(t_{\hat{\gamma}}), T_b(|t_{\hat{\gamma}}|), T_b(F_{\hat{\theta},\hat{\gamma}})), \tag{30}$$

$$t_{\hat{\alpha}}(3, \mathrm{AO}, m, k = 0) \Rightarrow R_{3,\mathrm{AO}}(\lambda_c) \quad (m = T_b(t_{\hat{\gamma}}), T_b(|t_{\hat{\gamma}}|)). \tag{31}$$

The limiting distributions given by (30) and (31) are equivalent to the case where the break date is known as in Perron (1989). Critical values for (30) can be found in Perron (1989), and critical values for (31) can be found in Perron and Vogelsang (1993b). The limiting distributions given by (30) and (31) are different from the case when $\gamma = 0$ and have larger (less negative) critical values. A conservative approach is to use the critical values from the $\gamma = 0$ distribution. When $\gamma = 0$, the tests have asymptotic size equal to the asymptotic level and when $\gamma \neq 0$ the tests have asymptotic size less than the asymptotic level. Since this approach is conservative, power is sacrificed when $\gamma \neq 0$. A way to avoid this potential power loss is to first perform a pretest for a shift in slope that is valid for both unit root and stationary errors. Perron (1991) and Vogelsang (1997a, 1997b) provide such tests. Of course, any time a pretest is conducted, the size of the ultimate test is likely to be distorted, and we leave a detailed investigation for future research.

When $t_{\hat{\gamma}}$ is used to choose T_b in Models (2, IO) and (3, IO), we show in the appendix that the true break is not chosen asymptotically. When $F_{\hat{\theta},\hat{\gamma}}$ is used to choose T_b in the Model (2, IO), we conjecture that the true break is not chosen asymptotically. We have not been able to prove this conjecture, but simulation results in Section 5 suggest $F_{\hat{\theta},\hat{\gamma}}$ does not choose the true break date. So, in Models (2, IO) and (3, IO), when T_b is chosen using $t_{\hat{\gamma}}$ or $F_{\hat{\theta},\hat{\gamma}}$ and $\gamma \neq 0$, size distortions are possible asymptotically.

4. Simulation of Finite Sample Critical Values

In this section, finite sample critical values are presented for the statistics $t_{\hat{\alpha}}(1, \mathrm{AO}, m, k)(m = T_b(t_{\hat{\alpha}}), T_b(t_{\hat{\theta}}), T_b(|t_{\hat{\theta}}|))$, $t_{\hat{\alpha}}(2, \mathrm{AO},$

$m, k)(m = T_b(t_{\hat{\alpha}}), T_b(t_{\hat{\gamma}}), T_b(|t_{\hat{\gamma}}|), T_b(F_{\hat{\theta},\hat{\gamma}}))$, and $t_{\hat{\alpha}}(2, \text{IO}, T_b(F_{\hat{\theta},\hat{\gamma}}), k)$ under the unit root null hypothesis. The purpose is to assess the adequacy of the asymptotic approximations and to provide finite sample critical values for the data-dependent methods for choosing k. Results are reported for the models that have not been examined elsewhere in the literature. Other finite sample critical values are given by Zivot and Andrews (1992), Perron (1997), and Banerjee *et al.* (1992).

The critical values were simulated using the DGP, $y_t = y_{t-1} + \gamma DU_t^c + e_t$, $y_0 = 0$ where e_t are i.i.d. $N(0, 1)$ random deviates. Setting $y_0 = 0$ and $\beta = 0$ is without loss of generality since the statistics are exactly invariant to y_0 and β. We set $\theta = 0$ since the statistics are asymptotically invariant θ. For most of the simulations $\gamma = 0$, however, for the $t_{\hat{\alpha}}(2, \text{AO}, T_b(t_{\hat{\alpha}}), k)$ statistic results are also reported for $\gamma \neq 0$. In the $\gamma \neq 0$ simulations $k = 0$ so that effects of γ on the distributions can be isolated from effects due to the choice of k. Even though the finite sample distributions of the statistics depend on correlation in the errors, they are asymptotically invariant to the correlation in the errors using the data-dependent methods to choose k. Additionally, i.i.d. errors helps to delineate the effects that γ has on the finite sample distributions. More general error processes are considered in Section 5.

Critical values were simulated for $T = 50$, 100, and 150, except for Model (2, AO), $\gamma \neq 0$, where $T = 100$, 250, 500. In all cases $kmax = 5$. Larger values of $kmax$ were prohibited by excessive computation time. Each simulation was based on 2,000 replications with the $N(0, 1)$ random deviates obtained from the routine RAN1 of Press *et al.* (1986) written in the C language. To minimize sampling variability across simulations, the same set of generated data was used for simulations with the same T. The initial seed used for the random number generator was $-T$ in all cases.

Results for Model (1, AO) are given in Table 1. When the sample size is small ($T = 50$), there are nontrivial differences between the finite sample and asymptotic critical values. In general, the finite sample critical values are smaller (more negative) than the asymptotic critical values. As T increases, the asymptotic approximation improves albeit slowly. Table 2 gives finite sample critical values for

Model (2, AO). As for Model 1, when $T = 50$ there are nontrivial differences between the finite sample and asymptotic critical values. However, as T increases, the asymptotic approximations become quite good. The approximation for $T = 150$ is excellent. In Table 3 critical values are reported for $t_{\hat{\alpha}}(2, \mathrm{IO}, T_b(F_{\hat{\theta},\hat{\gamma}}), k)$. For $T = 150$ the asymptotic approximation is quite good. Finite sample critical values for the other Model (2, IO) statistics can be found in Perron (1997).

The results in Section 3.3 showed that when T_b is chosen to minimize $t_{\hat{\alpha}}$ the asymptotic distributions are not invariant to γ. To illustrate how γ affects finite sample distributions, critical values for $t_{\hat{\alpha}}(2, \mathrm{AO}, T_b(t_{\hat{\alpha}}), k = 0)$ were simulated using $\gamma = 0.5$, 1.0, 2.0, and 5.0. Three sample sizes were used, $T = 100$, 250, 500, and three breaks dates were considered corresponding to $\lambda_c = 0.2$, 0.5, 0.8. The results are given in Table 5. For a given γ, the critical values shift to the left as T increases although the divergence is slow when γ is small. For example, with $\lambda_c = 0.5$ and $\gamma = 0.5$ the medians of the distributions range from -3.96 for $T = 100$ to -4.23 for $T = 500$. When γ is large, the divergence is faster. For example, with $\lambda_c = 0.5$ and $\gamma = 5.0$ the medians of the distributions range from -7.33 for $T = 100$ to -14.13 for $T = 500$. Because the distributions shift to the left as γ increases, use of the $\gamma = 0$ critical values can result in tests that are oversized when $\gamma \neq 0$. For small values of γ, size distortions are minimal. For example, suppose that the 5% asymptotic critical value of -5.08 were used when $\lambda_c = 0.5$. Comparing -5.08 to the critical values in Table 5 indicates that size will be close to 0.05 when $T = 100$ and $\gamma \leq 1.0$. As γ increases, size distortions can become a problem. With $\gamma = 5.0$ size is nearly 1.0. As T increases, size distortions increase if γ is held fixed. If $\gamma = 1.0$, size is about 0.05 when $T = 100$ but exceeds 0.5 when $T = 500$. This is consistent with the asymptotic results in Section 3.3. In practice, if a very large slope shift is suspected, the $t_{\hat{\alpha}}(2, \mathrm{AO}, T_b(t_{\hat{\alpha}}), k)$ statistic should not be used as there are potentially substantial size distortions.

5. Finite Sample Size and Power Simulations

This section presents finite sample size and power simulation results. The purpose is to determine how θ and γ effect size and power and

Table 5. Model (2, AO); Distribution of $t_{\hat{\alpha}}(2, \text{AO}, T_b(t_{\hat{\alpha}}), k = 0)$; Choosing T_b minimizing $t_{\hat{\alpha}}, \gamma \neq 0$.

		1.0%	2.5%	5.0%	10.0%	50.0%	90.0%	95.0%	97.5%	99.0%
					Panel (a): $T = 100$					
$T_b^c = 20$	$\gamma = 0.5$	−5.70	−5.38	−5.09	−4.83	−3.87	−3.09	−2.89	−2.75	−2.54
$(\lambda_c = 0.2)$	$\gamma = 1.0$	−5.84	−5.52	−5.22	−4.95	−3.97	−3.21	−3.02	−2.83	−2.61
	$\gamma = 2.0$	−6.93	−6.64	−6.38	−6.10	−5.12	−4.18	−3.93	−3.72	−3.53
	$\gamma = 5.0$	−9.82	−9.52	−9.19	−8.91	−7.94	−7.01	−6.76	−6.54	−6.29
$T_b^c = 50$	$\gamma = 0.5$	−5.68	−5.38	−5.12	−4.87	−3.96	−3.14	−2.92	−2.68	−2.49
$(\lambda_c = 0.5)$	$\gamma = 1.0$	−5.83	−5.58	−5.31	−5.03	−4.14	−3.24	−2.99	−2.77	−2.47
	$\gamma = 2.0$	−6.70	−6.40	−6.12	−5.84	−4.98	−4.08	−3.83	−3.57	−3.27
	$\gamma = 5.0$	−9.13	−8.81	−8.56	−8.28	−7.33	−6.49	−6.22	−6.02	−5.78
$T_b^c = 80$	$\gamma = 0.5$	−5.78	−5.30	−5.03	−4.70	−3.69	−2.85	−2.63	−2.43	−2.18
$(\lambda_c = 0.8)$	$\gamma = 1.0$	−5.45	−5.17	−4.87	−4.50	−3.50	−2.58	−2.35	−2.18	−1.96
	$\gamma = 2.0$	−5.85	−5.44	−5.17	−4.83	−3.79	−2.80	−2.48	−2.24	−2.04
	$\gamma = 5.0$	−7.34	−7.08	−6.72	−6.37	−5.33	−4.27	−3.98	−3.69	−3.48
					Panel (b): $T = 250$					
$T_b^c = 50$	$\gamma = 0.5$	−5.70	−5.38	−5.12	−4.81	−3.92	−3.18	−3.00	−2.85	−2.64
$(\lambda_c = 0.2)$	$\gamma = 1.0$	−6.51	−6.28	−5.99	−5.72	−4.81	−3.86	−3.58	−3.37	−3.11
	$\gamma = 2.0$	−8.70	−8.39	−8.16	−7.89	−7.06	−6.16	−5.90	−5.62	−5.28
	$\gamma = 5.0$	−13.30	−13.03	−12.83	−12.83	−11.67	−10.76	−10.50	−10.27	−9.95

(*Continued*)

Table 5. (*Continued*)

		1.0%	2.5%	5.0%	10.0%	50.0%	90.0%	95.0%	97.5%	99.0%
$T_b^c = 125$	$\gamma = 0.5$	−5.79	−5.37	−5.16	−4.94	−4.02	−3.19	−2.92	−2.70	−2.48
$(\lambda_c = 0.5)$	$\gamma = 1.0$	−6.38	−6.03	−5.81	−5.53	−4.67	−3.73	−3.41	−3.22	−2.99
	$\gamma = 2.0$	−8.04	−7.70	−7.47	−7.21	−6.34	−5.43	−5.16	−4.86	−4.72
	$\gamma = 5.0$	−12.20	−11.78	−11.54	−11.33	−10.43	−9.58	−9.34	−9.14	−8.91
$T_b^c = 200$	$\gamma = 0.5$	−5.36	−5.09	−4.81	−4.51	−3.51	−2.70	−2.48	−2.26	−2.08
$(\lambda_c = 0.8)$	$\gamma = 1.0$	−5.63	−5.28	−5.00	−4.71	−3.64	−2.65	−2.39	−2.17	−1.84
	$\gamma = 2.0$	−6.50	−6.13	−5.90	−5.60	−4.54	−3.47	−3.15	−2.84	−2.47
	$\gamma = 5.0$	−9.42	−9.06	−8.78	−8.47	−7.38	−6.29	−5.97	−5.66	−5.29
				Panel (c): $T = 500$						
$T_b^c = 100$	$\gamma = 0.5$	−5.93	−5.64	−5.45	−5.18	−4.17	−3.32	−3.09	−2.95	−2.77
$(\lambda_c = 0.2)$	$\gamma = 1.0$	−7.60	−7.35	−7.13	−6.84	−5.91	−4.93	−4.60	−4.38	−4.03
	$\gamma = 2.0$	−10.91	−10.63	−10.38	−10.14	−9.27	−8.40	−8.12	−7.81	−7.57
	$\gamma = 5.0$	−17.60	−17.30	−17.09	−16.89	−15.98	−15.07	−14.83	−14.64	−14.28
$T_b^c = 250$	$\gamma = 0.5$	−5.99	−5.60	−5.38	−5.11	−4.23	−3.35	−3.01	−2.76	−2.31
$(\lambda_c = 0.5)$	$\gamma = 1.0$	−7.11	−6.78	−6.51	−6.31	−5.42	−4.49	−4.20	−3.91	−3.54
	$\gamma = 2.0$	−9.69	−9.40	−9.19	−8.95	−8.08	−7.20	−6.90	−6.67	−6.42
	$\gamma = 5.0$	−15.76	−15.50	−15.24	−15.01	−14.13	−13.24	−13.00	−12.80	−12.55
$T_b^c = 400$	$\gamma = 0.5$	−5.31	−5.01	−4.81	−4.53	−3.52	−2.60	−2.35	−2.12	−1.94
$(\lambda_c = 0.8)$	$\gamma = 1.0$	−5.95	−5.61	−5.40	−5.10	−4.02	−2.97	−2.63	−2.30	−1.90
	$\gamma = 2.0$	−7.46	−7.22	−6.99	−6.64	−5.56	−4.45	−4.11	−3.82	−3.50
	$\gamma = 5.0$	−11.98	−11.66	−11.37	−11.09	−9.97	−8.80	−8.49	−8.11	−7.66

to see how the AO (IO) statistics perform when data is generated according to the IO (AO) models. Results are only reported for k chosen using $k(t-\text{sig})$ with $kmax = 5$. Also, we only report results for Models (2, AO) and (2, IO) as similar results were obtained for the other models. See the working paper by Vogelsang and Perron (1994) for results with Model 1, and Perron (1997) for results with Model 3.

Two DGPs were used for the simulations given by,

$$y_t = \theta DU_t^c + \gamma DT_t^c + z_t, \tag{32}$$

$$\begin{aligned} y_t &= \psi(L)[\bar{\theta}U_t^c + \bar{\gamma}DT_t^c] + z_t(\alpha < 1), \\ \Delta y_t &= \psi^*(L)[\bar{\theta}D(T_b^c)_t + \bar{\gamma}DU_t^c] + \Delta z_t(\alpha = 1), \end{aligned} \tag{33}$$

where $\psi(L) = (1-(\alpha+\rho)L+\rho L^2)^{-1}(1+\psi L)$, $\psi^*(L) = (1-\rho L)^{-1}(1+\psi L)$, $z_t = \alpha z_{t-1} + \rho\Delta z_{t-1} + e_t + \psi e_{t-1}$ and $e_t \sim$ i.i.d. $N(0,1)$. DGP (32) is an AO model, and DGP (33) is an IO model. For the size simulations we set $\alpha = 1$ and for power we set $\alpha = 0.8$. The sample size was $T = 100$, and 5% finite sample critical values with $\theta = \gamma = 0$ were used. The number of replications was 1,000. Five error specifications were used: (1) $\rho = 0$, $\psi = 0$, (2) $\rho = 0.6$, $\psi = 0$, (3) $\rho = -0.6$, $\psi = 0$, (4) $\rho = 0$, $\psi = 0.5$ and (5) $\rho = 0$, $\psi = -0.5$. Experiment (1) has i.i.d. errors. Experiment (2) has positively correlated errors and is common in economic data. Experiment (3) has negatively correlated errors. Experiments (4) and (5) have MA(1) errors to determine how the $k(t-\text{sig})$ procedure handles MA errors.

For the simulations where θ, $\gamma \neq 0$, the true break date was $T_b^c = 50$ ($\lambda_c = 0.5$). To allow direct comparison of results from both DGPs the magnitudes of $\bar{\theta}$ and $\bar{\gamma}$ in (33) were chosen so that the long run trend functions are the same in both DGPs. Several values of θ and γ were considered. In (32) we used $\theta = 0, 5, 10$ and $\gamma = 0, 1, 2$. The corresponding values of $\bar{\theta}$ and $\bar{\gamma}$ in (33) are $\bar{\gamma} = \gamma(1-\rho)/(1+\psi)$, $\bar{\theta} = [\theta(1-\rho)+\gamma\rho+\bar{\gamma}\psi]/(1+\psi)$ when $\alpha = 1$, and $\bar{\gamma} = \gamma(1-\alpha)/(1+\psi)$, $\bar{\theta} = [\theta(1-\alpha) + \gamma(\alpha-\rho) + \bar{\gamma}\psi]/(1+\psi)$ when $\alpha = 0.8$.

The simulation results are presented in Tables 6–9 for the AO and IO statistics, respectively. In each table the labels A and B correspond to data generated according to the AO and IO models,

Table 6. Finite sample size and power simulations; Model (2, AO), $t_{\hat{\alpha}}(2, \mathrm{AO}, T_b(\cdot), k(t-\mathrm{sig}))$.

			Size ($\alpha = 1$)					Power ($\alpha = 0.8$)				
			θ, γ	$\theta(\gamma = 0)$		$\gamma(\theta = 0)$		θ, γ	$\theta(\gamma = 0)$		$\gamma(\theta = 0)$	
ρ	ψ	T_b	0.0	5.0	10.0	1.0	2.0	0.0	5.0	10.0	1.0	2.0
0.0	0.0	$T_b(t_{\hat{\alpha}})$	0.040	0.108	0.507	0.044	0.076	0.295	0.435	0.861	0.236	0.386
		$T_b(\|t_{\hat{\gamma}}\|)$	0.049	0.048	0.032	0.036	0.032	0.301	0.098	0.042	0.239	0.230
		$T_b(t_{\hat{\gamma}})$	0.050	0.050	0.040	0.072	0.061	0.350	0.133	0.055	0.376	0.371
		$T_b(F_{\hat{\theta},\hat{\gamma}})$	0.052	0.050	0.020	0.027	0.021	0.339	0.194	0.163	0.193	0.184
0.6	0.0	$T_b(t_{\hat{\alpha}})$	0.033	0.036	0.072	0.030	0.029	0.832	0.731	0.804	0.696	0.714
		$T_b(\|t_{\hat{\gamma}}\|)$	0.036	0.036	0.023	0.032	0.031	0.738	0.473	0.118	0.726	0.425
		$T_b(t_{\hat{\gamma}})$	0.043	0.033	0.032	0.061	0.061	0.797	0.548	0.178	0.821	0.815
		$T_b(F_{\hat{\theta},\hat{\gamma}})$	0.038	0.037	0.030	0.035	0.024	0.740	0.657	0.646	0.650	0.659
−0.6	0.0	$T_b(t_{\hat{\alpha}})$	0.037	0.214	0.802	0.043	0.101	0.115	0.455	0.926	0.128	0.251
		$T_b(\|t_{\hat{\gamma}}\|)$	0.039	0.037	0.032	0.030	0.030	0.114	0.042	0.027	0.088	0.089
		$T_b(t_{\hat{\gamma}})$	0.045	0.046	0.037	0.057	0.053	0.138	0.064	0.004	0.165	0.164
		$T_b(F_{\hat{\theta},\hat{\gamma}})$	0.052	0.028	0.016	0.023	0.019	0.132	0.067	0.061	0.073	0.065
0.0	0.5	$T_b(t_{\hat{\alpha}})$	0.053	0.068	0.193	0.039	0.047	0.304	0.302	0.612	0.211	0.253
		$T_b(\|t_{\hat{\gamma}}\|)$	0.059	0.047	0.041	0.035	0.035	0.260	0.163	0.070	0.221	0.213
		$T_b(t_{\hat{\gamma}})$	0.061	0.057	0.043	0.067	0.069	0.303	0.216	0.101	0.332	0.325
		$T_b(F_{\hat{\theta},\hat{\gamma}})$	0.049	0.057	0.030	0.035	0.029	0.295	0.231	0.164	0.188	0.173
0.0	−0.5	$T_b(t_{\hat{\alpha}})$	0.500	0.622	0.906	0.408	0.574	0.888	0.911	0.983	0.786	0.854
		$T_b(\|t_{\hat{\gamma}}\|)$	0.298	0.175	0.112	0.265	0.242	0.685	0.150	0.013	0.640	0.626
		$T_b(t_{\hat{\gamma}})$	0.230	0.143	0.072	0.320	0.302	0.701	0.148	0.010	0.715	0.709
		$T_b(F_{\hat{\theta},\hat{\gamma}})$	0.391	0.239	0.209	0.254	0.230	0.723	0.591	0.591	0.613	0.612

Note: DGP(AO): $y_t = \theta DU_t^c + \gamma DT_t^c + z_t, z_t = \alpha z_{t-1} + \rho\Delta z_{t-1} + e_t + \psi e_{t-1},\ e_t \sim$ i.i.d. $N(0,1)$; $T = 100$, $T_b^c = 50$; 1,000 replications; 5% nominal size; kmax = 5.

Table 7. Finite sample size and power simulations; Model (2, AO), $t_{\hat{\alpha}}(2, \text{AO}, T_b(\cdot), k(t-\text{sig}))$.

			Size ($\alpha = 1$)					Power ($\alpha = 0.8$)				
			θ, γ	$\theta(\gamma = 0)$		$\gamma(\theta = 0)$		θ, γ	$\theta(\gamma = 0)$		$\gamma(\theta = 0)$	
ρ	ψ	T_b	0.0	5.0	10.0	1.0	2.0	0.0	5.0	10.0	1.0	2.0
0.0	0.0	$T_b(t_{\hat{\alpha}})$	0.040	0.098	0.526	0.040	0.072	0.290	0.171	0.155	0.225	0.???
		$T_b(\|t_{\hat{\gamma}}\|)$	0.045	0.043	0.025	0.036	0.031	0.285	0.109	0.027	0.219	0.216
		$T_b(t_{\hat{\gamma}})$	0.047	0.055	0.051	0.064	0.051	0.345	0.126	0.039	0.356	0.348
		$T_b(F_{\hat{\theta},\hat{\gamma}})$	0.056	0.039	0.016	0.026	0.022	0.331	0.218	0.115	0.176	0.166
0.6	0.0	$T_b(t_{\hat{\alpha}})$	0.024	0.022	0.041	0.023	0.027	0.821	0.772	0.789	0.713	0.750
		$T_b(\|t_{\hat{\gamma}}\|)$	0.022	0.039	0.025	0.023	0.023	0.743	0.554	0.223	0.725	0.722
		$T_b(t_{\hat{\gamma}})$	0.032	0.033	0.028	0.058	0.048	0.810	0.590	0.263	0.817	0.821
		$T_b(F_{\hat{\theta},\hat{\gamma}})$	0.030	0.028	0.022	0.021	0.020	0.727	0.692	0.632	0.657	0.653
−0.6	0.0	$T_b(t_{\hat{\alpha}})$	0.036	0.307	0.852	0.036	0.105	0.122	0.055	0.019	0.117	0.256
		$T_b(\|t_{\hat{\gamma}}\|)$	0.040	0.034	0.013	0.027	0.026	0.112	0.045	0.008	0.093	0.097
		$T_b(t_{\hat{\gamma}})$	0.048	0.052	0.024	0.065	0.058	0.143	0.057	0.019	0.162	0.166
		$T_b(F_{\hat{\theta},\hat{\gamma}})$	0.049	0.017	0.009	0.022	0.021	0.135	0.077	0.016	0.070	0.072

(*Continued*)

Table 7. (*Continued*)

			Size ($\alpha = 1$)					Power ($\alpha = 0.8$)				
			θ, γ	$\theta(\gamma = 0)$		$\gamma(\theta = 0)$		θ, γ	$\theta(\gamma = 0)$		$\gamma(\theta = 0)$	
ρ	ψ	T_b	0.0	5.0	10.0	1.0	2.0	0.0	5.0	10.0	1.0	2.0
0.0	0.5	$T_b(t_{\hat{\alpha}})$	0.051	0.058	0.151	0.034	0.049	0.292	0.238	0.155	0.194	0.249
		$T_b(\lvert t_{\hat{\gamma}}\rvert)$	0.050	0.058	0.038	0.040	0.037	0.248	0.144	0.046	0.220	0.216
		$T_b(t_{\hat{\gamma}})$	0.055	0.057	0.048	0.084	0.069	0.282	0.179	0.068	0.313	0.304
		$T_b(F_{\hat{\theta},\hat{\gamma}})$	0.055	0.061	0.035	0.035	0.030	0.271	0.221	0.147	0.198	0.187
0.0	−0.5	$T_b(t_{\hat{\alpha}})$	0.514	0.793	0.962	0.449	0.643	0.849	0.747	0.752	0.792	0.866
		$T_b(\lvert t_{\hat{\gamma}}\rvert)$	0.323	0.269	0.216	0.303	0.284	0.689	0.145	0.017	0.623	0.622
		$T_b(t_{\hat{\gamma}})$	0.265	0.216	0.125	0.329	0.322	0.689	0.131	0.023	0.699	0.697
		$T_b(F_{\hat{\theta},\hat{\gamma}})$	0.384	0.350	0.453	0.296	0.273	0.705	0.588	0.462	0.606	0.598

Note: DGP(IO): $y_t = \psi(L)[\bar{\theta}DU_t^c + \bar{\gamma}DT_t^c] + z_t, (\alpha < 1), \Delta y_t = \psi^*(L)[\bar{\theta}D(T_b^c)_t + \bar{\gamma}DU_t^c] + \Delta z_t(\alpha = 1)$, $e_t \sim$ i.i.d. $N(0,1)$, $\psi(L) = (1-(\alpha+\rho)L+\rho L^2)^{-1}(1+\psi L)$, $\psi^*(L) = (1-\rho L)^{-1}(1+\psi L)$, $z_t = \alpha z_{t-1} + \rho\Delta z_{t-1} + e_t + \psi e_{t-1}$, $T = 100$, $T_b^c = 50$. 1,000 replications, 5% nominal size, kmax = 5. When $\alpha = 1$, $\bar{\gamma} = \gamma(1-\rho)/(1+\psi)$ and $\bar{\theta} = (\theta(1-\rho) + \gamma\rho + \bar{\gamma}\psi)/(1+\psi)$. When $\alpha < 1$, $\bar{\gamma} = \gamma(1-\alpha)/(1+\psi)$ and $\bar{\theta} = (\theta(1-\alpha) + \gamma(\alpha-\rho) + \bar{\gamma}\psi)/(1+\psi)$.

Table 8. Finite sample size and power simulations; Model (2, IO), $t_{\hat{\alpha}}(2, \mathrm{IO}, T_b(\cdot), k(t-\mathrm{sig}))$.

			Size ($\alpha = 1$)					Power ($\alpha = 0.8$)				
			θ,γ	$\theta(\gamma=0)$		$\gamma(\theta=0)$		θ,γ	$\theta(\gamma=0)$		$\gamma(\theta=0)$	
ρ	ψ	T_b	0.0	5.0	10.0	1.0	2.0	0.0	5.0	10.0	1.0	2.0
0.0	0.0	$T_b(t_{\hat{\alpha}})$	0.034	0.119	0.637	0.065	0.314	0.211	0.442	0.976	0.233	0.688
		$T_b(\lvert t_{\hat{\gamma}}\rvert)$	0.047	0.119	0.471	0.126	0.494	0.214	0.158	0.339	0.418	0.851
		$T_b(t_{\hat{\gamma}})$	0.052	0.067	0.212	0.212	0.640	0.221	0.109	0.238	0.573	0.940
		$T_b(F_{\hat{\theta},\hat{\gamma}})$	0.067	0.189	0.707	0.106	0.233	0.309	0.564	0.989	0.345	0.447
0.6	0.0	$T_b(t_{\hat{\alpha}})$	0.034	0.043	0.115	0.038	0.006	0.756	0.639	0.841	0.623	0.683
		$T_b(\lvert t_{\hat{\gamma}}\rvert)$	0.048	0.051	0.104	0.055	0.116	0.651	0.158	0.382	0.759	0.843
		$T_b(t_{\hat{\gamma}})$	0.049	0.048	0.061	0.122	0.193	0.677	0.109	0.238	0.874	0.925
		$T_b(F_{\hat{\theta},\hat{\gamma}})$	0.062	0.074	0.151	0.061	0.093	0.747	0.715	0.893	0.730	0.773
−0.6	0.0	$T_b(t_{\hat{\alpha}})$	0.030	0.328	0.958	0.159	0.755	0.097	0.619	0.998	0.323	0.938
		$T_b(\lvert t_{\hat{\gamma}}\rvert)$	0.038	0.263	0.650	0.297	0.879	0.089	0.260	0.507	0.507	0.973
		$T_b(t_{\hat{\gamma}})$	0.040	0.136	0.328	0.438	0.950	0.103	0.159	0.339	0.662	0.995
		$T_b(F_{\hat{\theta},\hat{\gamma}})$	0.058	0.407	0.977	0.184	0.217	0.127	0.706	0.999	0.310	0.298
0.0	0.5	$T_b(t_{\hat{\alpha}})$	0.051	0.079	0.294	0.060	0.131	0.238	0.282	0.768	0.196	0.386
		$T_b(\lvert t_{\hat{\gamma}}\rvert)$	0.066	0.080	0.249	0.103	0.240	0.202	0.178	0.265	0.328	0.591
		$T_b(t_{\hat{\gamma}})$	0.072	0.067	0.116	0.172	0.361	0.219	0.145	0.157	0.459	0.740
		$T_b(F_{\hat{\theta},\hat{\gamma}})$	0.092	0.124	0.363	0.089	0.169	0.306	0.385	0.835	0.294	0.416
0.0	−0.5	$T_b(t_{\hat{\alpha}})$	0.447	0.656	0.999	0.435	0.896	0.843	0.955	1.00	0.857	0.994
		$T_b(\lvert t_{\hat{\gamma}}\rvert)$	0.373	0.457	0.679	0.561	0.964	0.699	0.361	0.375	0.914	0.998
		$T_b(t_{\hat{\gamma}})$	0.235	0.258	0.343	0.662	0.988	0.646	0.262	0.317	0.963	1.00
		$T_b(F_{\hat{\theta},\hat{\gamma}})$	0.443	0.708	0.999	0.456	0.469	0.750	0.968	1.00	0.849	0.833

Note: DGP(AO): $y_t = \theta DU_t^c + \gamma DT_t^c + z_t, z_t = \alpha z_{t-1} + \rho\Delta z_{t-1} + e_t + \psi e_{t-1}$ $e_t \sim$ i.i.d. $N(0,1)$; $T = 100$, $T_b^c = 50$; 1,000 replications; 5% nominal size; kmax = 5.

Table 9. Finite sample size and power simulations; Model (2, IO), $t_{\hat{\alpha}}(2, \text{IO}, T_b(\cdot), k(t-\text{sig}))$.

			Size ($\alpha = 1$)					Power ($\alpha = 0.8$)				
			θ, γ	$\theta(\gamma = 0)$		$\gamma(\theta = 0)$		θ, γ	$\theta(\gamma = 0)$		$\gamma(\theta = 0)$	
ρ	ψ	T_b	0.0	5.0	10.0	1.0	2.0	0.0	5.0	10.0	1.0	2.0
0.0	0.0	$T_b(t_{\hat{\alpha}})$	0.033	0.107	0.643	0.050	0.333	0.212	0.132	0.150	0.237	0.664
		$T_b(\lvert t_{\hat{\gamma}}\rvert)$	0.035	0.119	0.500	0.110	0.486	0.184	0.092	0.072	0.387	0.834
		$T_b(t_{\hat{\gamma}})$	0.048	0.083	0.238	0.183	0.612	0.228	0.103	0.109	0.546	0.928
		$T_b(F_{\hat{\theta},\hat{\gamma}})$	0.056	0.168	0.709	0.087	0.208	0.285	0.207	0.260	0.304	0.429
0.6	0.0	$T_b(t_{\hat{\alpha}})$	0.028	0.032	0.059	0.030	0.049	0.740	0.676	0.778	0.643	0.751
		$T_b(\lvert t_{\hat{\gamma}}\rvert)$	0.036	0.044	0.069	0.057	0.100	0.620	0.432	0.200	0.794	0.873
		$T_b(t_{\hat{\gamma}})$	0.041	0.052	0.059	0.100	0.180	0.653	0.429	0.173	0.877	0.940
		$T_b(F_{\hat{\theta},\hat{\gamma}})$	0.055	0.063	0.107	0.063	0.080	0.707	0.722	0.843	0.758	0.799
−0.6	0.0	$T_b(t_{\hat{\alpha}})$	0.036	0.397	0.963	0.161	0.755	0.103	0.044	0.046	0.294	0.919
		$T_b(\lvert t_{\hat{\gamma}}\rvert)$	0.031	0.323	0.735	0.281	0.854	0.086	0.048	0.045	0.467	0.975
		$T_b(t_{\hat{\gamma}})$	0.045	0.171	0.355	0.403	0.936	0.102	0.060	0.076	0.628	0.992
		$T_b(F_{\hat{\theta},\hat{\gamma}})$	0.059	0.471	0.974	0.149	0.189	0.158	0.093	0.085	0.268	0.269

0.0	0.5	$T_b(t_{\hat{\alpha}})$	0.045	0.070	0.273	0.037	0.121	0.252	0.184	0.144	0.181	0.366
		$T_b(\lvert t_{\hat{\gamma}}\rvert)$	0.052	0.087	0.238	0.075	0.230	0.208	0.127	0.085	0.310	0.574
		$T_b(t_{\hat{\gamma}})$	0.057	0.069	0.132	0.155	0.349	0.210	0.116	0.097	0.450	0.739
		$T_b(F_{\hat{\theta},\hat{\gamma}})$	0.085	0.117	0.348	0.068	0.154	0.308	0.246	0.223	0.281	0.409
0.0	−0.5	$T_b(t_{\hat{\alpha}})$	0.461	0.839	1.00	0.455	0.866	0.825	0.702	0.849	0.842	0.996
		$T_b(\lvert t_{\hat{\gamma}}\rvert)$	0.369	0.634	0.779	0.570	0.956	0.654	0.287	0.188	0.914	1.00
		$T_b(t_{\hat{\gamma}})$	0.268	0.329	0.386	0.685	0.987	0.634	0.259	0.286	0.950	1.00
		$T_b(F_{\hat{\theta},\hat{\gamma}})$	0.449	0.868	1.00	0.468	0.332	0.743	0.700	0.886	0.841	0.778

Note: DGP(IO): $y_t = \psi(L)[\bar{\theta}DU_t^c + \bar{\gamma}DT_t^c] + z_t(\alpha < 1), \Delta y_t = \psi^*(L)[\bar{\theta}D(T_b^c)_t + \bar{\gamma}DU_t^c] + \Delta z_t(\alpha = 1)$, $e_t \sim$ i.i.d. $N(0,1)$, $\psi(L) = (1-(\alpha+\rho)L+\rho L^2)^{-1}(1+\psi L)$, $\psi^*(L) = (1-\rho L)^{-1}(1+\psi L)$, $z_t = \alpha z_{t-1} + \rho\Delta z_{t-1} + e_t + \psi e_{t-1}$, $T = 100$, $T_b^c = 50$. 1,000 replications, 5% nominal size, kmax = 5. When $\alpha = 1$, $\bar{\gamma} = \gamma(1-\rho)/(1+\psi)$ and $\bar{\theta} = (\theta(1-\rho) + \gamma\rho + \bar{\gamma}\psi)/(1+\psi)$. When $\alpha < 1$, $\bar{\gamma} = \gamma(1-\alpha)/(1+\psi)$ and $\bar{\theta} = (\theta(1-\alpha) + \gamma(\alpha-\rho) + \bar{\gamma}\psi)/(1+\psi)$.

respectively. Consider cases where the model matches the DGP (Tables 6 and 9). Several pattern emerge. First, notice that the exact size of all the tests clearly depends on the correlation structure of the errors regardless of the values of θ and γ. When the first differences of errors are positively correlated as in experiment (2), all of the tests have exact size close to 0.05, especially if θ and γ are not too large. When the errors have an MA(1) component with $\psi = -0.5$ (experiment 5), all the tests are oversized. It is well known that unit root tests can have size distortions with such MA errors. Second, for $t_{\hat{\alpha}}(2, \mathrm{AO}, T_b(t_{\hat{\alpha}}), k(t-\mathrm{sig}))$ and $t_{\hat{\alpha}}(2, \mathrm{IO}, m, k(t-\mathrm{sig}))(m = T_b(t_{\hat{\alpha}}), T_b(|t_{\hat{\gamma}}|), T_b(t_{\hat{\gamma}}), T_b(F_{\hat{\theta},\hat{\gamma}}))$ the tests become oversized as θ and γ grow. For example, consider $t_{\hat{\alpha}}(2, \mathrm{AO}, T_b(t_{\hat{\alpha}}), k(t-\mathrm{sig}))$ (Table 6). In experiment (1) where the true process is a random walk, exact size is 0.04 when $\theta = 0$. Exact size is 0.108 for $\theta = 5.0$ and 0.507 for $\theta = 10$. When $\gamma = 2.0$, exact size is 0.076. For $t_{\hat{\alpha}}(2, \mathrm{IO}, T_b(t_{\hat{\gamma}}), k(t-\mathrm{sig}))$ (Table 9) exact size is 0.183 and 0.612 for $\gamma = 1.0$ and 2.0, respectively. These size distortions occur because the wrong break date is being chosen. Third, for $t_{\hat{\alpha}}(2, \mathrm{AO}, m, k(t-\mathrm{sig}))(m = T_b(|t_{\hat{\gamma}}|), T_b(t_{\hat{\gamma}}), T_b(F_{\hat{\theta},\hat{\gamma}}))$ exact size is close to 0.05 and does not increase as θ or γ increase (see Table 6) since as the break magnitude grows, the correct break date is chosen.

It is important to note that the magnitudes of θ and γ where size distortion become a problem are for θ, 5 to 10 times the standard deviation of the innovation errors, and for γ, 1 to 2 times the standard deviation of the innovation errors. For many macroeconomic time series, intercept shifts are often less than 5 standard deviations, and slope shifts are often less than 0.5 standard deviations. Also, such series often have first differences of the errors that are positively correlated. Thus, size distortions are not necessarily a problem in practice, but care must be used if a series is suspected of having large trend shifts or negative correlation in the first differences of the errors. In unreported calculations, θ and γ were estimated (imposing a unit root) for the long historical GNP series considered by Perron (1992) who applied the statistics from Model (2, IO). It was found that $\hat{\theta}$ and $\hat{\gamma}$ were generally no larger (in absolute value) than 8 and 0.7, respectively, relative to the standard deviation of the innovation errors. In addition, most of the series exhibited positive correlation in their first differences. Thus, experiment (2) most closely mimics the data considered by Perron (1992). In Table 9 size distortions are minimal for $\theta = 10$ and $\gamma = 1.0$ suggesting the tests used by Perron (1992) are not adversely affected by θ and γ.

Consider the finite sample power of the tests. Meaningful conclusions are most easily drawn for Model (2, AO) as size distortions are minimal. Consider the power of $t_{\hat{\alpha}}(2, \text{AO}, m, k(t - \text{sig}))(m = T_b(|t_{\hat{\gamma}}|), T_b(t_{\hat{\gamma}}), T_b(F_{\hat{\theta},\hat{\gamma}}))$ (Table 6). As θ or γ rise, power tends to fall. For example, the power of $t_{\hat{\alpha}}(2, \text{AO}, T_b(F_{\hat{\theta},\hat{\gamma}}), k(t - \text{sig}))$ in experiment (1) is 0.339 when $\theta = \gamma = 0.0$ but falls to 0.194, 0.163 for $\theta = 5, 10$, and 0.193, 0.184 for $\gamma = 1.0, 2.0$. Part of the reason for this decrease in power is that the true break date is chosen more often as θ and γ rise, and the critical values for the known break date case would be more appropriate. The conservative approach of using the critical values from the unknown break date case with $\gamma = 0$ controls size, but results in lower power. When it is reasonable to make the assumption that the direction of the break is known, power is higher.

Finally, what are the implications of using the AO model when the true process follows the IO model and vice versa? Compare Tables 6 with 7. Notice that size and power of the AO tests when data follows the IO model is very similar compared to when data follows the AO model. The same conclusion holds for the IO statistics by comparing Tables 8 and 9. Therefore, little is lost in terms of size and power if the AO model is incorrectly applied to IO data and vice versa. A practical implication is that the AO statistics should be used if large slope shifts are suspected regardless of whether the AO or IO model better approximates the true DGP since size is not inflated in either case.

6. Conclusions

This paper adds to the literature on unit root tests that allow for a shift in the trend function with the location of the break point unknown. Previous results focused mainly on the IO approach where the break is modeled as occurring slowly over time, as opposed to the AO approach where the break is sudden. Limiting distributions were derived for several models in the AO framework and asymptotic critical values tabulated. The limiting distributions of the statistics in both the IO and AO frameworks were also derived in the case where the shift in the trend function occurred under the null hypothesis of a unit root. It was shown that the limiting distributions are asymptotically invariant to a shift in intercept but not to a shift in slope. In fact, many of the tests currently proposed in the literature are no longer asymptotically valid testing procedures when a slope shift occurs under the null hypothesis. However, these asymptotic results were shown to provide poor approximations to finite sample distributions for trend breaks of the magnitudes typically encountered in practice. Indeed, for typical shifts in slope the asymptotic distributions obtained assuming no break under the unit root null hypothesis provide adequate approximations of finite sample distributions. In finite samples it was found that with the exception of the procedures in the AO framework where the break date is chosen using the significance of the trend break parameters, size increases as the magnitude of the shifts grow. Therefore, if the magnitude of the shift is suspected to be very large, the AO framework is preferred since it permits tests with less size distortion. Even if data is generated according to the IO model, the AO statistics provide reliable tests with size that is (practically speaking) invariant to large shifts in intercept or slope.

Appendix

The proofs of the limiting distributions stated in the text are given in this appendix. Weak convergence for a fixed λ is established for the various statistics. The results of Zivot and Andrews (1992) can then be applied to give the limiting results for the minimal and maximal statistics via the continuous mapping theorem. Recall that Zivot and Andrews (1992) require $\lambda \in \Lambda^*$. In the case of the statistics whereby the break is chosen to minimize $t_{\hat{\alpha}}$, results from Perron (1997) permit this condition to be relaxed so that $\lambda \in \Lambda$. The following notation is used throughout the appendix. A '$\tilde{\ }$' is used to denote the residuals from a projection of a variable onto the space spanned by $\{1, t\}$. A '$\hat{\ }$'

is used to denote the residuals from a projection of a variable onto the space spanned by $\{1, t, DU_t\}$. Let Σ be short for $\Sigma_{t=1}^{T}$, and let Σ' be short for $\Sigma_{t=T_b+1}^{T}$. Finally, given a matrix X, M_X is used to denote the matrix $I - X(X'X)^{-1}X'$. The data-generating process (DGP) in matrix form is,

$$Y = D_2^c\gamma + Z, \quad Z = Z_{-1} + e, \tag{A.1a}$$

where $Y \equiv \{y_t\}$, $Z \equiv \{z_t\}$, $Z_{-1} \equiv \{z_{t-1}\}$, $e \equiv \{e_t\}$, $D_2^c \equiv \{DT_t^c\}$. The small 'c' superscript on the 'D' variables means that these dummy variables are defined using the true break date T_b^c. A 'D' variable *without* a superscript 'c' (e.g., D_1, D_2) is used to denote dummy variables defined using the break date used in the estimation (T_b). Define the following matrices: $Y_{-1} \equiv \{y_{t-1}\}$, $D_1^c \equiv \{DU_t^c\}$, $D \equiv [D_1\ D_2]$, $D^c \equiv [D_1^c\ D_2^c]$, $N \equiv \{1, t\}$, $\tau_1 = \text{diag}(T^{-1/2}, T^{-3/2})$, $\tau_2 = \text{diag}(T^{-1/2}, T^{-1})$. It is useful to rewrite (A.1a) in several ways:

$$Y = Y_{-1} + D_1^c\gamma + e,$$
$$Y_{-1} = (D_2^c - D_1^c)\gamma + Z_{-1},\ D_2^c = (Y_{-1} - Z_{-1})/\gamma + D_1^c. \tag{A.1b}$$

Detrending both sides of (A.1a) and (A.1b) by $\{1,\ t\}$ yields,

$$\tilde{Y} = \tilde{D}_2^c\gamma + \tilde{Z}, \quad \tilde{Z} = \tilde{Z}_{-1} + \tilde{e} \tag{A.2a}$$
$$\tilde{Y} = \tilde{Y}_{-1} + \tilde{D}_1^c\gamma + \tilde{e},$$
$$\tilde{Y}_{-1} = (\tilde{D}_2^c - \tilde{D}_1^c)\gamma + \tilde{Z}_{-1}, \tilde{D}_2^c = (\tilde{Y}_{-1} - \tilde{Z}_{-1})/\gamma + \tilde{D}_1^c. \tag{A.2b}$$

Likewise, detrending both sides of (A.1a) and (A.1b) by $\{1, t, DU_t\}$ yields,

$$\hat{Y} = \hat{D}_2^c\gamma + \hat{Z}, \quad \hat{Z} = \hat{Z}_{-1} + \hat{e} \tag{A.3a}$$
$$\hat{Y} = \hat{Y}_{-1} + \hat{D}_1^c\gamma + \hat{e},$$
$$\hat{Y}_{-1} = (\hat{D}_2^c - \hat{D}_1^c)\gamma + \hat{Z}_{-1}, \hat{D}_2^c = (\hat{Y}_{-1} - \hat{Z}_{-1})/\gamma + \hat{D}_1^c. \tag{A.3b}$$

The next three lemmas establish convergence of sample moments that are used in the proofs. We state them without proof (see Vogelsang and Perron 1994 for proofs).

Lemma A1. *Let $\gamma = 0$. Under (A.1a) these limiting results hold:*

$$\text{i) } \tau_1\tilde{D}'\tilde{D}\tau_1 = \begin{bmatrix} T^{-1}\tilde{D}_1'\tilde{D}_1 & T^{-2}\tilde{D}_1'\tilde{D}_2 \\ T^{-2}\tilde{D}_2'\tilde{D}_1 & T^{-3}\tilde{D}_2'\tilde{D}_2 \end{bmatrix} \Rightarrow \begin{bmatrix} a_{11} & a_{12} \\ a_{12} & a_{22} \end{bmatrix},$$

$$\begin{aligned} &\text{ii) } T^{-1}\tilde{D}_1'M_{\tilde{Y}_{-1}}\tilde{e} \Rightarrow \sigma H_1(\lambda), && \text{iii) } T^{-3/2}\tilde{D}_2'M_{\tilde{Y}_{-1}}\tilde{e} \Rightarrow \sigma H_2(\lambda), \\ &\text{iv) } T^{-1}\tilde{D}_1'M_{\tilde{Y}_{-1}}\tilde{D}_1 \Rightarrow K_1(\lambda), && \text{v) } T^{-3}\tilde{D}_2'M_{\tilde{Y}_{-1}}\tilde{D}_2 \Rightarrow K_2(\lambda), \\ &\text{vi) } T^{-2}\tilde{D}_2'M_{\tilde{Y}_{-1}}\tilde{D}_1 \Rightarrow K_3(\lambda), && \text{vii) } T^{-2}\tilde{Y}'M_{\tilde{D}_1}\tilde{Y} \Rightarrow \sigma^2 L_1(\lambda), \\ &\text{viii) } T^{-2}\tilde{Y}'M_{\tilde{D}_2}\tilde{Y} \Rightarrow \sigma^2 L_2(\lambda), && \text{ix) } T^{-5/2}\tilde{D}_2M_{\tilde{D}_1}\tilde{Y} \Rightarrow \sigma L_3(\lambda), \\ &\text{x) } T^{-3}\tilde{D}_2'M_{\tilde{D}_1}\tilde{D}_2 \Rightarrow a_{22} - a_{12}^2/a_{11}. \end{aligned}$$

Lemma A2. *Let $\gamma \neq 0$ and suppose that $\lambda = \lambda_c$. Under (A.1a) these limiting results hold:*

$$\begin{aligned} &\text{i) } T^{-1/2}\tilde{Y}'M_{\tilde{Y}_{-1}}\tilde{D}\tau_2 = O_p(1), && \text{ii) } [\tau_2\tilde{D}'M_{\tilde{Y}_{-1}}\tilde{D}\tau_2]^{-1} = O_p(1), \\ &\text{iii) } T^{-1}\hat{D}_2'M_{\hat{Y}_{-1}}\hat{Y} = O_p(1), && \text{iv) } [T^{-2}\hat{D}_2'M_{\hat{Y}_{-1}}\hat{D}_2]^{-1} = O_p(1), \\ &\text{v) } T^{-3/2}\tilde{Y}_{-1}M_{\tilde{D}_2}(\tilde{Y} - \tilde{Y}_{-1}) = O_p(1), && \text{vi) } [T^{-2}\tilde{Y}_{-1}M_{\tilde{D}_2}\tilde{Y}_{-1}]^{-1} = O_p(1), \end{aligned}$$

Lemma A3. *Let $\gamma \neq 0$ and suppose that $\lambda \neq \lambda_c$. Under (A.1a) these limiting results hold:*

$$\begin{aligned} &\text{i) } T^{-1/2}\tilde{Y}'M_{\tilde{Y}_{-1}}\tilde{D}\tau_1 = O_p(1), && \text{ii) } [\tau_1\tilde{D}'M_{\tilde{Y}_{-1}}\tilde{D}\tau_1]^{-1} = O_p(1), \\ &\text{iii) } T^{-2}\hat{D}_2'M_{\hat{Y}_{-1}}\hat{Y} = O_p(1), && \text{iv) } [T^{-3}\hat{D}_2'M_{\hat{Y}_{-1}}\hat{D}_2]^{-1} = O_p(1), \\ &\text{v) } T^{-2}\tilde{Y}_{-1}M_{\tilde{D}_2}(\tilde{Y} - \tilde{Y}_{-1}) = O_p(1), && \text{vi) } [T^{-3}\tilde{Y}_{-1}M_{\tilde{D}_2}\tilde{Y}_{-1}]^{-1} = O_p(1), \\ &\text{vii) } T^{-2}\tilde{Y}_{-1}M_{\tilde{D}}(\tilde{Y} - \tilde{Y}_{-1}) = O_p(1), && \text{viii) } [T^{-3}\tilde{Y}_{-1}M_{\tilde{D}}\tilde{Y}_{-1}]^{-1} = O_p(1), \end{aligned}$$

Using the results from the lemmas, the proofs of the limiting distributions follow easily. Proofs are not given in the order as presented in the text but rather by model.

Proofs for Model (1, AO). To prove (12) and (13) all that needs to be shown is $T^{-1/2}t_{\hat{\theta}} \Rightarrow Q_{1,\text{AO}}(\lambda)$. Using regression (2.1) and (A.2a) with $\gamma = 0$, write

$$\begin{aligned} T^{-1/2}t_{\hat{\theta}} &= T^{-3/2}\tilde{Z}'\tilde{D}_1[T^{-1}s^2T^{-1}\tilde{D}_1'\tilde{D}_1]^{-1/2} \\ &= T^{-3/2}\tilde{Z}'\tilde{D}_1[T^{-2}\tilde{Z}'M_{\tilde{D}_1}\tilde{Z}T^{-1}\tilde{D}_1'\tilde{D}_1]^{-1/2} \\ &\Rightarrow \int_\lambda^1 W^*(r)dr/(a_{11}L_1(\lambda))^{1/2} \\ &= Q_{1,\mathrm{AO}}(\lambda) \text{ by Lemma A1 (i) and (vii).} \end{aligned}$$

Proofs for Model (2, IO). Proofs are first given for the case where $\gamma = 0$. Since the limiting results do not depend on the regressor $D(T_b)_t$ in (5.2), it is ignored. To prove (14)–(16) all that is needed are the limits of $t_{\hat{\gamma}}$ and $F_{\hat{\theta},\hat{\gamma}}$. Using (5.2) with $k = 0$ and (A.2a) and (A.2b) with $\gamma = 0$, we have,

$$\begin{aligned} t_{\hat{\gamma}} &= \left(T^{-3/2}\tilde{D}_2'M_{\tilde{Z}_{-1}}\tilde{e} - T^{-2}\tilde{D}_2'M_{\tilde{Z}_{-1}}\tilde{D}_1\right. \\ &\quad \times \left.\left(T^{-1}\tilde{D}_1'M_{\tilde{Z}_{-1}}\tilde{D}_1\right)^{-1}T^{-1/2}\tilde{D}_1'M_{\tilde{Z}_{-1}}\tilde{e}\right) \\ &\quad \times \left(s^2\{T^{-3}\tilde{D}_2'M_{\tilde{Z}_{-1}}\tilde{D}_2 - T^{-4}\left(\tilde{D}_2'M_{\tilde{Z}_{-1}}\tilde{D}_1\right)^2\right. \\ &\quad \times \left.\left(T^{-1}\tilde{D}_1'M_{\tilde{Z}_{-1}}\tilde{D}_1\right)^{-1}\}\right)^{-1/2} \\ &\Rightarrow [H_2(\lambda) - K_3(\lambda)H_1(\lambda)/K_1(\lambda)]/[K_2(\lambda) - K_3(\lambda)^2/K_1(\lambda)]^{-1/2} \\ &= Q^A_{2,\mathrm{IO}}(\lambda), \\ F_{\hat{\theta},\hat{\gamma}} &= \tilde{e}'M_{\tilde{Y}_{-1}}\tilde{D}\tau_1\left(\tau_1\tilde{D}'M_{\tilde{Y}_{-1}}\tilde{D}\tau_1\right)^{-1}\tau_1\tilde{D}M_{\tilde{Y}_{-1}}\tilde{e}/s^2 \\ &\Rightarrow [H_1(\lambda)^2K_2(\lambda) - 2H_1(\lambda)K_3(\lambda)H_2(\lambda) + H_2(\lambda)^2K_1(\lambda)] \\ &\quad \times[K_1(\lambda)K_2(\lambda) - K_3(\lambda)^2]^{-1} \\ &= Q^B_{2,\mathrm{IO}}(\lambda), \end{aligned}$$

by Lemma A1 (ii)–(vi) and the result $s^2 \to \sigma^2$ from Perron (1989). Suppose that $\gamma \neq 0$. Results for $t_{\hat{\alpha}}$ are established first. When $\lambda = \lambda_c$, $t_{\hat{\alpha}}$ is invariant to γ under the null hypothesis, and its limiting distribution is given by (25). Suppose that $\lambda \neq \lambda_c$. Two steps are used to establish (28) ($j = 2$, $m = \mathrm{IO}$).

Step 1. Using (5.2) (ignoring $D(T_b)_t$) write $T(\hat{\alpha}-1) = T^{-2}\tilde{Y}'_{-1}M_{\tilde{D}}(\tilde{Y}-\tilde{Y}_{-1})/[T^{-3}\tilde{Y}'_{-1}M_{\tilde{D}}\tilde{Y}_{-1}]$ which leads to

$$\begin{aligned}
s^2 &= T^{-1}\left[(\tilde{Y}-\tilde{Y}_{-1}(\hat{\alpha}-1)\tilde{Y}_{-1}\right]' M_{\tilde{D}}\left[(\tilde{Y}-\tilde{Y}_{-1}-(\hat{\alpha}-1)\tilde{Y}_{-1}\right] \\
&= T^{-1}(\tilde{Y}-\tilde{Y}_{-1})'M_{\tilde{D}}(\tilde{Y}-\tilde{Y}_{-1}) \\
&\quad -2T(\hat{\alpha}-1)T^{-2}\tilde{Y}'_{-1}M_{\tilde{D}}(\tilde{Y}-\tilde{Y}_{-1}) \\
&\quad +T^2(\hat{\alpha}-1)^2T^{-3}\tilde{Y}'_{-1}M_{\tilde{D}}\tilde{Y}_{-1} \\
&= T^{-1}(\tilde{D}_1^c\gamma+\tilde{e})'M_{\tilde{D}}(\tilde{D}_1^c\gamma+\tilde{e}) \\
&\quad -\gamma^2[T^{-2}\tilde{D}_1^cM_{\tilde{D}}\tilde{D}_2^c]^2[T^{-3}\tilde{D}_2^cM_{\tilde{D}}\tilde{D}_2^c]^{-1}+o_p(1) \\
&= T^{-1}\tilde{e}'M_{\tilde{D}}\tilde{e}+\gamma^2T^{-1}A'M_BA+o_p(1) \\
&\quad (\text{where } A = M_{\tilde{d}}\tilde{D}_1^c \text{ and } B = M_{\tilde{D}}\tilde{D}_2^c) \\
&\Rightarrow \sigma^2+\gamma^2\lim_{T\to\infty}T^{-1}A'M_BA.
\end{aligned}$$

Because the limit is strictly positive, we have $1/s^2 = O_p(1)$.

Step 2. Using (5.2), $T^{-1/2}t_{\hat{\alpha}}(2, \text{IO}, [\lambda T], k=0) = T^{-2}\tilde{Y}'_{-1}M_{\tilde{D}}(\tilde{Y}-\tilde{Y}_{-1})/[s^2T^{-3}\tilde{Y}'_{-1}M_{\tilde{D}}\tilde{Y}_{-1}]^{1/2} = O_p(1)$, which follows from Step 1 and Lemma A3 (vii) and (viii).

Consider $t_{\hat{\gamma}}$ and $F_{\hat{\theta},\hat{\gamma}}$ when $\gamma \neq 0$. Let $\lambda = \lambda_c$. Using Lemma A2(i–iv) and $s^2 \to \sigma^2$,

$$t_{\hat{\gamma}} = T^{-1}\hat{D}_2'M_{\hat{Y}_{-1}}\hat{Y}/(s^2T^{-2}\hat{D}_2'M_{\hat{Y}_{-1}}\hat{D}_2)^{1/2} = O_p(1), \tag{A.4}$$

$$\begin{aligned}
T^{-1}F_{\hat{\theta},\hat{\gamma}} &= T^{-1/2}\tilde{Y}'M_{\tilde{Y}_{-1}}\tilde{D}\tau_2(\tau_2\tilde{D}'M_{\tilde{Y}_{-1}}\tilde{D}\tau_2)^{-1}\tau_2\tilde{D}M_{\tilde{Y}_{-1}}\tilde{Y}T^{-1/2}/s^2 \\
&= O_p(1).
\end{aligned} \tag{A.5}$$

Suppose now that $\lambda \neq \lambda_c$. Using Lemma A2(i–iv) and $1/s^2 = O_p(1)$,

$$T^{-1/2}t_{\hat{\gamma}} = T^{-2}\hat{D}_2'M_{\hat{Y}_{-1}}\hat{Y}/(s^2T^{-3}\hat{D}_2'M_{\hat{Y}_{-1}}\hat{D}_2)^{1/2} = O_p(1), \tag{A.6}$$

$$\begin{aligned}
T^{-1}F_{\hat{\theta},\hat{\gamma}} &= T^{-1/2}\tilde{Y}'M_{\tilde{Y}_{-1}}\tilde{D}\tau_1(\tau_1\tilde{D}'M_{\tilde{Y}_{-1}}\tilde{D}\tau_1)^{-1}\tau_1\tilde{D}M_{\tilde{Y}_{-1}}\tilde{Y}T^{-1/2}/s^2 \\
&= O_p(1).
\end{aligned} \tag{A.7}$$

(A.4) and (A.6) establish that $t_{\hat{\gamma}}$ does not choose the correct break date. (A.5) and (A.7) establish $T^{-1}F_{\hat{\theta},\hat{\gamma}} = O_p(1)$ for all break dates.

Proofs for model (2, AO). Proofs are first given for the case where $\gamma = 0$. To establish (17)–(19) all that is needed are the limits of $T^{-1/2}t_{\hat{\gamma}}$ and $T^{-1}F_{\hat{\theta},\hat{\gamma}}$. Using regression (2.2) write,

$$\begin{aligned} T^{-1/2}t_{\hat{\gamma}} &= (T^{-5/2}\tilde{D}_2' M_{\tilde{D}_1}\tilde{Y})(T^{-1}s^2 T^{-3}\tilde{D}_2' M_{\tilde{D}_1}\tilde{D}_2)^{-1/2}, \\ T^{-1}F_{\hat{\theta},\hat{\gamma}} &= T^{-1}\tilde{Y}'\tilde{D}\tau_1(\tau_1\tilde{D}'\tilde{D}\tau_1)^{-1}\tau_1\tilde{D}'\tilde{Y}T^{-1}/(T^{-1}s^2), \end{aligned}$$

where $T^{-1}s^2 = T^{-2}\tilde{Y}'M_{\tilde{D}_1}\tilde{Y} - T^{-5/2}\tilde{Y}'M_{\tilde{D}_1}\tilde{D}_2(T^{-3}\tilde{D}_2'M_{\tilde{D}_1}\tilde{D}_2)^{-1}$ $T^{-5/2}\tilde{D}_2'M_{\tilde{D}_1}\tilde{Y}$.

$$T^{-1}s^2 \Rightarrow \sigma^2[L_1(\lambda) - L_3(\lambda)^2/(a_{22} - a_{12}^2/a_{11})], \qquad \text{(A.8)}$$

by Lemma A1 (vii), (ix) and (x). Using (A.8) and Lemma A1 it directly follows that,

$$\begin{aligned} T^{-1/2}t_{\hat{\gamma}} &\Rightarrow L_3(\lambda)\left[(a_{22} - a_{12}^2/a_{11})L_1(\lambda) - L_3(\lambda)^2\right]^{-1/2} = Q^A_{2,\mathrm{AO}}(\lambda), \\ T^{-1}F_{\hat{\theta},\hat{\gamma}} &\Rightarrow \left((a_{11}a_{22} - a_{12}^2)L_1(\lambda) - a_{11}L_3(\lambda)^2\right)^{-1} \\ &\quad \times \left(a_{22}\left(\int_\lambda^1 W^*(r)dr\right)^2 - 2a_{12}\int_\lambda^1 W^*(r)dr\right. \\ &\quad \left.\times \int_\lambda^1 (r-\lambda)W^*(r)dr + a_{11}\left(\int_\lambda^1 (r-\lambda)W^*(r)dr\right)^2\right) \\ &= Q^B_{2,\mathrm{AO}}(\lambda) \text{ as required.} \end{aligned}$$

Now consider the case where $\gamma \neq 0$. Using results in Perron and Vogelsang (1993a), the proof of (28) ($j = 2$, $m =$ AO) is analogous to the proof of (28) ($j = 2$, $m =$ IO) from above and is omitted. We now establish the result that $t_{\hat{\gamma}}$ and $F_{\hat{\theta},\hat{\gamma}}$ choose the correct break date asymptotically. Using (A2.a) and regression (2.2), we have,

$$T^{-1/2}t_{\hat{\gamma}} = T^{-1/2}\tilde{D}_2' M_{\tilde{D}_1}(\tilde{D}_2^c\gamma + \gamma\tilde{Z})(s^2\tilde{D}_2' M_{\tilde{D}_1}\tilde{D}_2)^{-1/2}, \quad \text{(A.9)}$$

$$T^{-1}F_{\hat{\theta},\hat{\gamma}} = T^{-1}(\tilde{D}_2^c\gamma + \tilde{Z})'\tilde{D}(\tilde{D}'\tilde{D})^{-1}\tilde{D}'(\tilde{D}_2^c\gamma + \tilde{Z})/s^2. \quad \text{(A.10)}$$

Consider the limiting behavior of (A.9) and (A.10) when (2.2) is estimated at the true break date ($\lambda = \lambda_c$). In this case s^2 is invariant to γ and $T^{-1}s^2 = O_p(1)$. Using this fact,

$$\begin{aligned} T^{-1/2}t_{\hat{\gamma}} &= T^{1/2}(T^{-3}\tilde{D}_2^{c'} M_{\tilde{D}_1^c}(\tilde{D}_2^c\gamma + \tilde{Z})(T^{-1}s^2T^{-3}\tilde{D}_2^{c'} M_{\tilde{D}_1^c}\tilde{D}_2^c)^{-1/2}) \\ &= \gamma T^{1/2}O_p(1) \qquad \text{(A.11)} \\ T^{-1}F_{\hat{\theta},\hat{\gamma}} &= T(T^{-3/2}(\tilde{D}_2^c\gamma + \tilde{Z})'\tilde{D}\tau_1(\tau_1\tilde{D}'\tilde{D}\tau_1)^{-1}\tau_1\tilde{D}' \\ &\quad \times (\tilde{D}_2^c\gamma + \tilde{Z})T^{-3/2})/(T^{-1}s^2) \\ &= \gamma^2 TO_p(1). \qquad \text{(A.12)} \end{aligned}$$

Next suppose that $\lambda \neq \lambda_c$. It is no longer true that s^2 is invariant to γ as,

$$\begin{aligned} [T^{-2}s^2]^{-1} &= [T^{-3}(\tilde{D}_2^c\gamma + \tilde{Z})'M_{\tilde{D}}(\tilde{D}_2^c\gamma + \tilde{Z})]^{-1} \\ &= [\gamma^2T^{-3}\tilde{D}_2^{c'} M_{\tilde{D}}\tilde{D}_2^c + o_p(1)]^{-1} = O_p(1). \qquad \text{(A.13)} \end{aligned}$$

Using (A.13), we have

$$\begin{aligned} T^{-1/2}t_{\hat{\gamma}} &= T^{-3}\tilde{D}_2' M_{\tilde{D}_1}(\tilde{D}_2^c + \gamma\tilde{Z})(T^{-2}s^2T^{-3}\tilde{D}_2' M_{\tilde{D}_1}\tilde{D}_2)^{-1/2} \\ &= O_p(1), \\ T^{-1}F_{\hat{\theta},\hat{\gamma}} &= T^{-3/2}(\tilde{D}_2^c\gamma + \tilde{Z})'\tilde{D}\tau_1(\tau_1\tilde{D}'\tilde{D}\tau_1)^{-1}\tau_1\tilde{D}' \\ &\quad \times (\tilde{D}_2^c\gamma + \tilde{Z})T^{-3/2}(T^{-2}s^2)^{-1} \\ &= O_p(1), \end{aligned}$$

which with (A.11) and (A.12) establishes that $t_{\hat{\gamma}}$ and $F_{\hat{\theta},\hat{\gamma}}$ diverge at a faster rate for $\lambda = \lambda_c$ than $\lambda \neq \lambda_c$. Therefore, λ_c is chosen asymptotically and (30) directly follows.

Proof for model (3, IO). The only results that need to be shown are (27) and (28) ($j = 3$, $m =$ IO) in the case $\gamma \neq 0$. Suppose that $\lambda = \lambda_c$. Two steps are used to establish (27).

Step 1. Using (5.3) write $T^{1/2}(\hat{\alpha} - 1) = T^{-3/2}\tilde{Y}'_{-1}M_{\tilde{D}_2}(\tilde{Y} - \tilde{Y}_{-1})/[T^{-2}\tilde{Y}'_{-1}M_{\tilde{D}_2}\tilde{Y}_{-1}]$ which gives

$$\begin{aligned}
s^2 &= T^{-1}(\tilde{Y} - \tilde{Y}_{-1})'M_{\tilde{D}_2}(\tilde{Y} - \tilde{Y}_{-1}) \\
&\quad - 2T^{1/2}(\hat{\alpha} - 1)T^{-3/2}\tilde{Y}'_{-1}M_{\tilde{D}_2}(\tilde{Y} - \tilde{Y}_{-1}) \\
&\quad + T(\hat{\alpha} - 1)^2 T^{-2}\tilde{Y}'_{-1}M_{\tilde{D}_2}\tilde{Y}_{-1} \\
&= T^{-1}(\tilde{D}_1^c\gamma + \tilde{e})'M_{\tilde{D}_2}(\tilde{D}_1^c\gamma + \tilde{e}) \\
&\quad - \gamma^2[T^{-2}\tilde{D}_1^c M_{\tilde{D}_2}\tilde{D}_2^c]^2[T^{-3}\tilde{D}_2^c M_{\tilde{D}_2}\tilde{D}_2^c]^{-1} + o_p(1) \\
&= T^{-1}\tilde{e}'M_{\tilde{D}_2}\tilde{e} + \gamma^2 T^{-1}A'M_B A + o_p(1) \\
&\quad (\text{where } A = M_{\tilde{D}_2}\tilde{D}_1^c \text{ and } B = M_{\tilde{D}_2}\tilde{D}_2^c) \\
&\Rightarrow \sigma^2 + \gamma^2 \lim_{T\to\infty} T^{-1}A'M_B A.
\end{aligned}$$

Since the limit is strictly positive, we have $1/s^2 = O_p(1)$.

Step 2. Again using (5.3) write

$$\begin{aligned}
&T^{-1/2}t_{\hat{\alpha}}(3, \text{IO}, T_b, k = 0) \\
&\quad = T^{-3/2}\tilde{Y}'_{-1}M_{\tilde{D}_2}(\tilde{Y} - \tilde{Y}_{-1})/[s^2 T^{-2}\tilde{Y}'_{-1}M_{\tilde{D}_2}\tilde{Y}_{-1}]^{1/2} = O_p(1)
\end{aligned}$$

which follows from Step 1 and Lemma A2 (v) and (vi). This proves (27). Suppose now that $\lambda \neq \lambda_c$. To establish (28) ($j = 3$, $m =$ IO) note that $1/s^2 = O_p(1)$ using arguments similar to those for Model (2, IO). Using (5.3) write

$$\begin{aligned}
&T^{-1/2}t_{\hat{\alpha}}(3, \text{IO}, T_b, k = 0) \\
&\quad = T^{-2}\tilde{Y}'_{-1}M_{\tilde{D}_2}(\tilde{Y} - \tilde{Y}_{-1})/[s^2 T^{-3}\tilde{Y}'_{-1}M_{\tilde{D}_2}\tilde{Y}_{-1}]^{1/2} = O_p(1),
\end{aligned}$$

which follows from $1/s^2 = O_p(1)$, Lemma A3 (v) and (vi). This proves (28) ($j = 3$, $m =$ IO).

Proofs for model (3, AO). As before, proofs are first given for the case where $\gamma = 0$, followed by proofs when $\gamma \neq 0$. Consider results (22) and (23). All that needs to be established is the limiting distribution of $T^{-1/2}t_{\hat{\gamma}}$. Using regression (2.3) write

$$\begin{aligned} T^{-1/2}t_{\hat{\gamma}} &= (T^{-5/2}\tilde{D}_2'\tilde{Y})(T^{-1}s^2T^{-3}\tilde{D}_2'\tilde{D}_2)^{-1/2} \\ &= (T^{-5/2}\tilde{D}_2'\tilde{Y})(T^{-2}\tilde{Y}M_{\tilde{D}_2}\tilde{Y}T^{-3}\tilde{D}_2'\tilde{D}_2)^{-1/2} \\ &\Rightarrow \int_{\lambda}^{1}(r-\lambda)W^*(r)dr/(a_{22}L_2(\lambda))^{1/2} \\ &= Q_{3,\mathrm{AO}}(\lambda), \text{ by Lemma A1 (i) and (viii).} \end{aligned}$$

Now suppose that $\gamma \neq 0$. Using results from Perron and Vogelsang (1993a), the proof of (28) ($j = 3$, $m = \mathrm{AO}$) is analogously to the proof of (28) ($j = 3$, $m = \mathrm{IO}$) from above and is omitted. Finally, using arguments similar to those for Model (2, AO), it is straightforward to show that $t_{\hat{\gamma}}$ diverges at a faster rate for $\lambda = \lambda_c$ than $\lambda \neq \lambda_c$ and details are omitted. Therefore λ_c is chosen asymptotically using $t_{\hat{\gamma}}$ and (31) directly follows.

Acknowledgments

We are grateful to an editor and two anonymous referees for helpful comments. The first author thanks the Center for Analytic Economics at Cornell University, USA. The second author acknowledges financial support from the Social Sciences and Humanities Research Council of Canada (SSHRC), the National Sciences and Engineering Research Council of Canada (NSERC) and the Fonds pour la Formation et l'Aide à la Recherche du Québec (FCAR), Canada.

References

Banerjee, A., Lumsdaine, R.L. and Stock, J.H. (1992). Recursive and sequential tests of the unit-root and trend-break hypothesis: Theory and international evidence. *Journal of Business and Economic Statistics* 10, 271–287.

Christiano, L.J. (1992). Searching for a break in GNP. *Journal of Business and Economic Statistics* 10, 237–250.

Dickey, D.A. and Fuller, W.A. (1979). Distribution of the estimators for autoregressive time series with a unit root. *Journal of the American Statistical Association* 74, 427–431.

Hall, A. (1994). Testing for a unit root in time series with pretest data based model selection. *Journal of Business and Economic Statistics* 12, 461–470.

Nelson, C.R. and Plosser, C.I. (1982). Trends and random walks in macroeconomic time series. *Journal of Monetary Economics* 10, 139–162.

Ng, S. and Perron, P. (1995). Unit root tests in ARMA models with data dependent methods for the truncation lag. *Journal of the American Statistical Association* 90, 268–281.

Perron, P. (1989). The great crash, the oil price shock and the unit root hypothesis. *Econometrica* 57, 1361–1401.

Perron, P. (1990). Testing for a unit root in a time series with a changing mean. *Journal of Business and Economic Statistics* 8, 153–162.

Perron, P. (1991). *A Test for Changes in a Polynomial Trend Function for a Dynamic Time Series.* Mimeo, Department of Economics, Université of Montréal.

Perron, P. (1992). Trend, Unit Root and Structural Change: A Multi-Country Study With Historical Data. In *Proceedings of the Business and Economic Statistics Section*, American Statistical Association, 144–149.

Perron, P. (1997). Further evidence on breaking trend functions in macroeconomic variables. *Journal of Econometrics* 80, 355–385.

Perron, P. and Vogelsang, T.J. (1992). Nonstationarity and level shifts with an application to purchasing power parity. *Journal of Business and Economic Statistics* 10, 301–320.

Perron, P. and Vogelsang, T.J. (1993a). A note on the asymptotic distributions of unit root tests in the additive outlier model with breaks. *Revista De Econometrica* 8, 181–202.

Perron, P. and Vogelsang, T.J. (1993b). The great crash, the oil price shock and the unit root hypothesis: Erratum. *Econometrica* 61, 248–249.

Press, W.H., Flannery, B.P., Teukolsky, S.A. and Vetterling, W.T. (1986). *Numerical Recipes: The Art of Scientific Computing*, Cambridge University Press: Cambridge, U.K.

Said, S.E. and Dickey, D.A. (1984). Testing for unit roots in autoregressive-moving average of unknown order. *Biometrika* 71, 599–608.

Vogelsang, T.J. (1993). Essays on testing for nonstationarities and structural change in time series models, Ph.D. Dissertation, Princeton University.

Vogelsang, T.J. (1997a). Wald-Type tests for detecting shifts in the trend function of a dynamic time series. *Econometric Theory* 13, 818–849.

Vogelsang, T.J. (1997b). Testing for a shift in trend when serial correlation is of unknown form, Working Paper 97-11, Center for Analytic Economics, Cornell University.

Vogelsang, T.J. and Perron, P. (1994). Additional tests for a unit root allowing for a break in the trend function at an unknown time, Working Paper 94-13, Center for Analytic Economics, Cornell University.

Zivot, E. and Andrews, D.W.K. (1992). Further evidence on the great crash, the oil price shock and the unit-root hypothesis. *Journal of Business and Economic Statistics* 10, 251–270.

Chapter 18

GLS Detrending, Efficient Unit Root Tests and Structural Change*

Pierre Perron[†,§] and Gabriel Rodríguez[‡,¶]

[†]*Department of Economics, Boston University*
270 Bay State Road, Boston, MA 02215, USA

[‡]*Département de Sciences Économiques*
Université d'Ottawa, 200 Wilbrod St., P.O. Box 450
Station A, Ottawa, Ont., Canada, K1N-6N5

[§]*perron@bu.edu*
[¶]*gabrielr@uottawa.ca*

We extend the class of M-tests for a unit root analyzed by Perron and Ng [*Rev. Econ. Studies* 63 (1996) 435] and Ng and Perron [*Econometrica* 69 (2001) 1519] to the case where a change in the trend function is allowed to occur at an unknown time. These tests (M^{GLS}) adopt the GLS detrending approach developed by Elliott *et al.* [*Econometrica* 64 (1996) 813] (ERS) following the results of Dufour and King [*J. Econometrics* 47 (1991) 115]. Following Perron [*Econometrica* 57 (1989) 1361], we consider two models: one allowing for a change in slope and the other for both a change in intercept and slope. We derive the asymptotic distributions of the tests as well as that of the feasible point optimal test (P_T^{GLS}) suggested by ERS. Also, we compute the noncentrality parameter used for the local GLS detrending that permits the test P_T^{GLS} to have 50% asymptotic power at that value. The asymptotic critical values of the tests are tabulated. We show that the M^{GLS} and P_T^{GLS} tests have an asymptotic power function close to the power envelope. A simulation study analyzes the size and power in finite samples

*Originally Published in *Journal of Econometrics*, 115, pp. 1–27.

under various methods to select the truncation lag for the autoregressive spectral density estimator. An empirical application is also provided.

Keywords: Integrated process; quasi-differencing; change-point; truncation lag; information criteria.

1. Introduction

Since the seminal paper of Nelson and Plosser (1982), the unit root hypothesis has received a lot of attention from both theoretical and empirical perspectives (e.g., Campbell and Perron, 1991; Stock, 1994 for surveys). Using tests developed by Dickey and Fuller (1979), Nelson and Plosser (1982) argued that current shocks have permanent effects on the level of most macroeconomic series. This finding was supported by other approaches which found that a typical shock has both important transitory and permanent components (see, e.g., Campbell and Mankiw, 1987a, 1987b; Shapiro and Watson, 1988; Clark, 1987; Cochrane, 1988).

In contrast to this literature, Perron (1989) argued, as an alternative to the unit root hypothesis, that macroeconomic fluctuations are most likely stationary if allowance is made for the trend function to exhibit occasional changes. Allowing for a single change in intercept and/or slope, he rejected the unit root hypothesis for 11 of the 14 series analyzed by Nelson and Plosser. As discussed in Banerjee *et al.* (1992) this finding may be important for the following reasons. First, it offers an alternative picture of the persistence in macroeconomics series. Second, this approach can provide a parsimonious model for a slowly changing trend component that may be useful as a data description. Third, the implications for inference in more complex models are very different.

Christiano (1992) criticized the results of Perron (1989) on the basis that the break points should not be treated as exogenous since the imposition of a given break date involves an issue of data mining. Accordingly, Zivot and Andrews (1992), Banerjee *et al.* (1992) and Perron (1997) considered unit root tests with unknown break points.

We continue to treat the potential break points as occurring at unknown times and contribute to this literature in two ways.

First, we use the M^{GLS} tests analyzed by Perron and Ng (1996) and extend them to permit a one time change in the trend function. Second, following Elliott *et al.* (1996) (hereafter ERS) and the prior work of Dufour and King (1991), we use local to unity GLS detrending of the data. We consider two specific models: one with a break in the slope of the trend function and one with a break in both the intercept and slope. In this setup, there is no need to analyze the case where only a change in the intercept is allowed since the tests then have the same asymptotic distribution as the case where the deterministic components include a constant and a time trend which was analyzed in ERS (since a change in intercept is a special case of what they refer to as a 'slowly evolving deterministic component').

The reasons for considering the M-tests, originally proposed by Stock (1999) and further analyzed by Perron and Ng (1996) is that these tests have much smaller size distortions than other classes of unit root tests when the errors have strong negative serial correlation. Also, using GLS detrending when constructing the M-tests allows substantial gains in power as showed by Ng and Perron (2001), similar to the DF^{GLS} test proposed by ERS.

Given that a uniformly most powerful test is not attainable, we follow ERS and derive a feasible point optimal test (P_T^{GLS}). The asymptotic power function of this test is derived and we use the associated power envelope to choose the noncentrality parameter ($\bar{c}$) to perform the GLS detrending such that the asymptotic power of the test is 50% against the local alternative $\bar{\alpha} = 1 + \bar{c}/T$. For our two models, we obtain $\bar{c} = -22.5$.

The rest of the paper is organized as follows. The model and some preliminary theoretical results are presented in Section 2. In Section 3, we derive the asymptotic distribution of the M^{GLS} and DF^{GLS} tests in both cases where the break point is known or unknown. Section 4 considers the asymptotic Gaussian power envelope and the limit distribution of the feasible point optimal test. The asymptotic critical values and the asymptotic power function of the various tests are presented in Section 5. Section 6 considers the size and power of the tests in finite samples using simulations.

Section 7 presents an empirical application and Section 8 briefly concludes. An appendix contains technical derivations.

2. GLS Detrending with Structural Change

The data generating process considered is of the form:

$$y_t = d_t + u_t, \quad t = 0, \ldots, T, \tag{1}$$

$$u_t = \alpha u_{t-1} + v_t, \tag{2}$$

where $\{v_t\}$ is an unobserved stationary mean-zero process. We use the assumption that $u_0 = 0$ throughout, though the results generally hold for the weaker requirement that $E(u_0^2) < \infty$. The noise function is $v_t = \sum_{i=0}^{\infty} \gamma_i \eta_{t-i}$ with $\sum_{i=0}^{\infty} i|\gamma_i| < \infty$ and where $\{\eta_t\}$ is a martingale difference sequence. The process v_t has a nonnormalized spectral density at frequency zero given by $\sigma^2 = \sigma_\eta^2 \gamma(1)^2$, where $\sigma_\eta^2 = \lim_{T\to\infty} T^{-1} \sum_{t=1}^{\infty} E(\eta_t^2)$. Furthermore, $T^{-1/2} \sum_{t=1}^{[rT]} v_t \Rightarrow \sigma W(r)$, where $\Rightarrow$ denotes weak convergence in distribution and $W(r)$ is the Wiener process defined on $C[0,\ 1]$ the space of continuous functions on the interval $[0,\ 1]$. In (1), $d_t = \psi' z_t$, where z_t is a set of deterministic components to be discussed below. For any series y_t, with deterministic components z_t, we define the transformed data $y_t^{\bar{\alpha}}$ and $z_t^{\bar{\alpha}}$ by

$$y_t^{\bar{\alpha}} = (y_0, (1 - \bar{\alpha}L)y_t), \quad z_t^{\bar{\alpha}} = (z_0, (1 - \bar{\alpha}L)z_t), \quad t = 0, \ldots, T.$$

We let $\hat{\psi}$ be the estimate that minimizes

$$S^*(\psi, \bar{\alpha}, \delta) = \sum_{t=0}^{T} (y_t^{\bar{\alpha}} - \psi' z_t^{\bar{\alpha}})^2 \tag{3}$$

and denote the minimized value by $S(\bar{\alpha}, \delta)$.

2.1. *The specifications of the deterministic components*

Model I. Structural change in the slope: For this model, the set of deterministic components, z_t in (1), is given by

$$z_t = \{1, t, 1(t > T_B)(t - T_B)\}, \tag{4}$$

where 1(.) is the indicator function and T_B is the time of the change. Without loss of generality, we assume that $T_B = T\delta$ for some $\delta \in (0,1)$. In this case, $\hat{\psi}(\delta) = (\hat{\mu}_1, \hat{\beta}_1, \hat{\beta}_2)'$ is the vector of estimates that minimizes (3).

Model II. Structural change in intercept and slope: For Model II,

$$z_t = \{1, 1(t > T_B), t, 1(t > T_B)(t - T_B)\}. \tag{5}$$

In this case, the vector of coefficient estimates is $\hat{\psi}(\delta) = (\hat{\mu}_1, \hat{\mu}_2, \hat{\beta}_1, \hat{\beta}_2)'$. In this model, we have the same results as with Model 1 since the effect of $\hat{\mu}_2 - \mu_2$ is negligible in large samples. This is because the change in intercept is a special case of a slowly evolving deterministic component in condition B of ERS.

3. The Tests and their Asymptotic Distributions

3.1. *The tests*

The M-tests, originally proposed by Stock (1999), and further analyzed by Perron and Ng (1996), exploit the feature that a series converges with different rates of normalization under the null and the alternative hypotheses. They are defined by

$$MZ_\alpha^{\mathrm{GLS}}(\delta) = (T^{-1}\tilde{y}_T^2 - s^2)\left(2T^{-2}\sum_{t=1}^{T}\tilde{y}_{t-1}^2\right)^{-1}, \tag{6}$$

$$MSB^{\mathrm{GLS}}(\delta) = \left(T^{-2}\sum_{t=1}^{T}\tilde{y}_{t-1}^2/s^2\right)^{1/2}, \tag{7}$$

$$MZ_t^{\mathrm{GLS}}(\delta) = (T^{-1}\tilde{y}_T^2 - s^2)\left(4s^2T^{-2}\sum_{t=1}^{T}\tilde{y}_{t-1}^2\right)^{-1/2} \tag{8}$$

with $\tilde{y}_t = y_t - \hat{\psi}' z_t$ where $\hat{\psi}$ minimizes (3). The term s^2 is the autoregressive estimate of the spectral density at frequency zero of v_t, defined as

$$s^2 = s_{ek}^2/(1 - \hat{b}(1))^2, \tag{9}$$

with $s^2_{ek} = (T-k)^{-1}\sum_{t=k+1}^{T}\hat{e}^2_{tk}$, $\hat{b}(1) = \sum_{j=1}^{k}\hat{b}_j$, and $\hat{b}_j$, $\{\hat{e}_{tk}\}$ obtained from the regression (see Perron and Ng (1998) for more details):

$$\Delta\tilde{y}_t = b_0\tilde{y}_{t-1} + \sum_{j=1}^{k} b_j\Delta\tilde{y}_{t-j} + e_{tk}. \tag{10}$$

The first statistic is a modified version of the Phillips and Perron (1988) Z_α test originally developed by Phillips (1987). The second statistic is a modified version of Bhargava's (1986) R_1 statistic which builds upon the work of Sargan and Bhargava (1983). The third statistic is a modified version of the Phillips and Perron (1988) Z_t test. As Perron and Ng (1996) showed, the MSB and Z_α tests are related by $Z_t \approx MSB \times Z_\alpha$. This relation suggests the MZ_t test defined by (8) since it satisfies the relation $MZ_t = MSB \times MZ_\alpha$. Another test of interest is the so-called ADF test which is the t-statistic for testing $b_0 = 0$ in the regression (10), see Dickey and Fuller (1979) and Said and Dickey (1984). We denote this test by $ADF^{\text{GLS}}(\delta)$. Our approach is an extension of Ng and Perron (2001) and Elliott *et al.* (1996) to the case where the trend function contains a structural change. In this case, the M^{GLS} tests will depend on the unknown break point δ.

3.2. *Asymptotic distributions of the tests*

We start with a statement of the limiting distribution of the various tests in the case where the break point is considered known.

Theorem 1. *Let y_t be generated by (1) with $\alpha = 1 + c/T$, MZ_α^{GLS}, MSB^{GLS} and MZ_t^{GLS} be defined by (6)–(8) with data obtained from local GLS detrending ($\tilde{y}_t$) at $\bar{\alpha} = 1 + \bar{c}/T$, and ADF^{GLS} be the t-statistic for testing $b_0 = 0$ in the regression (10). Also, s^2 is a consistent estimate of σ^2. For Models I and II, we have*

$$MZ_\alpha^{\text{GLS}}(\delta) \Rightarrow \frac{0.5K_1(c,\bar{c},\delta)}{K_2(c,\bar{c},\delta)} \equiv H^{MZ_\alpha^{\text{GLS}}}(c,\bar{c},\delta),$$

$$MSB_\alpha^{\text{GLS}}(\delta) \Rightarrow (K_2(c,\bar{c},\delta))^{1/2} \equiv H^{MSB^{\text{GLS}}}(c,\bar{c},\delta),$$

$$MZ_t^{\text{GLS}}(\delta) \Rightarrow \frac{0.5K_1(c,\bar{c},\delta)}{(K_2(c,\bar{c},\delta))^{1/2}} \equiv H^{MZ_t^{\text{GLS}}}(c,\bar{c},\delta),$$

$$ADF^{\text{GLS}}(\delta) \Rightarrow \frac{0.5K_1(c,\bar{c},\delta)}{(K_2(c,\bar{c},\delta))^{1/2}} \equiv H^{ADF^{\text{GLS}}}(c,\bar{c},\delta),$$

where

$$K_1(c,\bar{c},\delta) = V_{c\bar{c}}^{(1)}(1,\delta)^2 - 2V_{c\bar{c}}^{(2)}(1,\delta) - 1,$$

$$K_2(c,\bar{c},\delta) = \int_0^1 V_{c\bar{c}}^{(1)}(r,\delta)^2 dr - 2\int_\delta^1 V_{c\bar{c}}^{(2)}(r,\delta)\,dr,$$

and $V_{c\bar{c}}^{(1)}(r,\delta) = W_c(r) - rb_3$, $V_{c\bar{c}}^{(2)}(r,\delta) = b_4(r-\delta)[W_c(r) - rb_3 - (1/2)(r-\delta)b_4]$ *with* $W_c(r)$ *the Ornstein–Uhlenbeck process that is the solution to the stochastic differential equation* $\mathrm{d}W_c(r) = cW_c(r)\mathrm{d}r + \mathrm{d}W(r)$ *with* $W_c(0) = 0$. *Also,* b_3, b_4 *are defined by* $b_3 = (\lambda_1 b_1 + \lambda_2 b_2)$ *and* $b_4 = (\lambda_2 b_1 + \lambda_3 b_2)$ *where* $b_1 = (1-\bar{c})W_c(1) + \bar{c}^2\int_0^1 rW_c(r)\,dr$, $b_2 = (1-\bar{c}+\delta\bar{c})W_c(1) + \bar{c}^2\int_\delta^1 W_c(r)(r-\delta)\,dr - W_c(\delta)$, $\lambda_1 = d/\Theta$, $\lambda_2 = -m/\Theta$, $d = 1-\delta-\bar{c}+2\bar{c}\delta-\bar{c}\delta^2-\bar{c}^2\delta+\bar{c}^2\delta^2+(\bar{c}^2/3)(1-\delta^3)$, $m = 1-\delta-\bar{c}+\bar{c}\delta-(\bar{c}^2/2)\delta+(\bar{c}^2/2)\delta^3+(\bar{c}^2/3)(1-\delta^3)$, $a = 1-\bar{c}+\bar{c}^2/3$, $\Theta = ad - m^2$ *and* $\lambda_3 = a/\Theta$.

In practice, it is usually the case that an investigator wants to treat the break point as unknown. In this case, an estimate of the break point is needed. A method suggested by Zivot and Andrews (1992) is to consider estimating δ as the break point that yields the minimal value of the statistics, i.e., using $\inf_\delta J^{\text{GLS}}(\delta)$ where $J =$ *MZ*$_\alpha$, *MSB*, *MZ*$_t$, and *ADF*. Using the continuous mapping theorem and arguments as in Perron (1997), we have, assuming no shift in the trend function under the null hypothesis:

$$\inf_{\delta\in[0,1]} J^{\text{GLS}}(\delta) \Rightarrow \inf_{\delta\in[0,1]} H^{J^{\text{GLS}}}(c,\bar{c},\delta), \tag{11}$$

for $J =$ *MZ*$_\alpha$, *MSB*, *MZ*$_t$, and *ADF* with the functions $H(\cdot)$ defined in Theorem 1. Note that no truncation for the range of possible break points needs to be imposed. As discussed in Vogelsang and Perron (1998), the implied estimate of δ is not consistent for the true value of the break point when the data generating process contains a break.

These authors also note that the tests statistic are not invariant (even asymptotically) to values of the coefficients of the change in the trend. Nevertheless, they argue that, in typical sample sizes, this is not a problem unless the changes are extremely large. Thus, these tests can still be used with the critical values derived assuming no shift under the null hypothesis.

An alternative method to select the break date, as used in Perron (1997), is to choose it such that the absolute value of the t-statistic on the coefficient of the change in slope is maximized. This procedure has been used by many authors, e.g., Christiano (1992), Banerjee *et al.* (1992), Perron (1997) and Vogelsang and Perron (1998). Consider, for example Model I where the deterministic component is given by $d_t = \mu_1 + \beta_1 t + \beta_2(t - T_B)1(t > T_B)$. Let $\hat{\beta}_2(\delta)$ be the GLS estimate of β_2 and $t_{\hat{\beta}_2}(\delta)$ be its associated t-statistic. The break point can be selected using the estimate $\hat{\delta} = \arg\max_{\delta\in[\varepsilon,1-\varepsilon]} |t_{\hat{\beta}_2}(\delta)|$, where ε is some small number imposing a trimming on the possible values of the break dates. We shall use $\varepsilon = 0.15$ throughout. As discussed in Vogelsang and Perron (1998), if under the null hypothesis we have $\beta_2 \neq 0$ and the true break point given by $T_B^0/T = \delta^0$, then $\hat{\delta}$ is a consistent estimate of δ^0 and the limiting distributions of the test statistics correspond to those in the case where the break date is known, i.e., the limit distributions given in Theorem 1 evaluated at δ^0. In practice, one can simply evaluate these limit distributions at the estimated value $\hat{\delta}$.

When, under the null hypothesis, $\beta_2 = 0$ in which case there is no change in the slope of the trend function, it is easy to show (using the results of Lemma A.2 in Appendix A) that $t_{\hat{\beta}_2}(\delta) \Rightarrow b_4/(\lambda_3^{1/2})$, where b_4 and λ_3 are defined in Theorem 1. We then have

$$\hat{\delta} = \arg\max_{\delta\in[\varepsilon,1-\varepsilon]} |t_{\hat{\beta}_2}(\delta)| \Rightarrow \arg\max_{\delta\in[\varepsilon,1-\varepsilon]} |b_4/(\lambda_3^{1/2})| = \delta^*. \tag{12}$$

Hence, the limiting distributions of the statistics are given by

$$J^{\mathrm{GLS}}(\hat{\delta}) \Rightarrow H^{J^{\mathrm{GLS}}}(c, \bar{c}, \delta^*), \tag{13}$$

for $J = MZ_\alpha$, MSB, MZ_t, and ADF with the functions $H(\cdot)$ defined in Theorem 1.

In practice, it is difficult to know if there is a change in slope since any test of such hypothesis would depend on whether a unit root is present or not. Hence, a conservative procedure is to use the critical values corresponding to the case where it is assumed that no break is present, i.e., (13). This is the procedure we use in the following.

4. Feasible Point Optimal Test and the Power Envelope

Elliott *et al.* (1996), following Dufour and King (1991), have considered the issue of developing tests with optimality properties under Gaussian errors. The case where the break point is assumed known follows closely their analysis. While a uniformly most powerful test is not attainable, it is possible to define a point optimal test against the alternative $\alpha = \bar{\alpha}$. If v_t is i.i.d., this is provided by the likelihood ratio statistic, which simplifies, under normality, to $L(\delta) = S(\bar{\alpha}, \delta) - S(1, \delta)$, where $S(\bar{\alpha}, \delta)$ and $S(1, \delta)$ are the sums of squared errors from a GLS regression with $\alpha = \bar{\alpha}$ and $\alpha = 1$, respectively. Varying the value of $\bar{\alpha}$, gives a family of point optimal tests and the Gaussian power envelope for testing $\alpha = 1$. To allow for serial correlation in the errors v_t, ERS proposed a feasible point optimal test (P_T^{GLS}) defined by

$$P_T^{\text{GLS}}(c, \bar{c}, \delta) = \{S(\bar{\alpha}, \delta) - \bar{\alpha} S(1, \delta)\}/s^2. \tag{14}$$

The next theorem provides the limiting distribution of the P_T^{GLS} test.

Theorem 2. *Let y_t be generated by* (1) *with $\alpha = 1 + c/T$. Let P_T^{GLS} be defined by (14) with data obtained from local GLS detrending ($\tilde{y}_t$) at $\bar{\alpha} = 1 + \bar{c}/T$. Also, let s^2 be a consistent estimate of σ^2. The limit distribution of the P_T^{GLS} test under Models I and II is given by*

$$\begin{aligned} P_T^{\text{GLS}}(c, \bar{c}, \delta) &\Rightarrow M(c, 0, \delta) - M(c, \bar{c}, \delta) \\ &\quad - 2\bar{c} \int_0^1 W_c(r) dW(r) + (\bar{c}^2 - 2\bar{c}c) \int_0^1 W_c(r)^2 dr - \bar{c} \\ &\equiv H^{P_T^{\text{GLS}}}(c, \bar{c}, \delta). \end{aligned} \tag{15}$$

where $M(c,\bar{c},\delta) = A(c,\bar{c},\delta)B(\bar{c},\delta)^{-1}A(c,\bar{c},\delta)$ *with* $A(c,\bar{c},\delta)$ *a* 2×1 *vector defined by*

$$\begin{bmatrix} W(1) + (c-\bar{c})\displaystyle\int_0^1 W_c(r)dr \\ \qquad -\bar{c}\displaystyle\int_0^1 r\,dW(r) - (c-\bar{c})\bar{c}\int_0^1 rW_c(r)\,dr \\ (1+\delta\bar{c})\left([W(1)-W(\delta)] + (c-\bar{c})\displaystyle\int_\delta^1 W_c(r)\,dr\right) \\ \qquad -\bar{c}\displaystyle\int_\delta^1 r\,dW(r) - (c-\bar{c})\bar{c}\int_\delta^1 rW_c(r)\,dr \end{bmatrix}$$

and $B(\bar{c},\delta)$ *is a symmetric matrix with entries*

$$\begin{bmatrix} \bar{c}^2/3 - \bar{c} + 1 & (1-\delta)(1-\bar{c}) + \bar{c}^2(2+\delta^3-3\delta)/6 \\ & \bar{c}^2(1-\delta^3)/3 - \bar{c}(1-\delta^2)(1+\delta\bar{c}) + (1-\delta)(1+\delta\bar{c})^2 \end{bmatrix}.$$

The asymptotic expression (15) for the P_T^{GLS} test allows us to define the asymptotic power envelope for the two models. It is given by $\pi(c,\delta) = \Pr[H^{P_T^{\mathrm{GLS}}}(c,c,\delta) < b^{P_T^{\mathrm{GLS}}}(c,\delta)]$, where $b^{P_T^{\mathrm{GLS}}}(c,\delta)$ is such that $\Pr[H^{P_T^{\mathrm{GLS}}}(0,c,\delta) < b^{P_T^{\mathrm{GLS}}}(c,\delta)] = v$, with v the size of the test. Note that, in general, a different power envelope exists for each values of δ.

When δ is unknown, things are rather different. The principle is, however, the same. To maximize the likelihood function under the null and alternative hypotheses, the estimate of δ must be chosen to minimize the sum of squares residuals $S(1,\delta)$ and $S(\bar{\alpha},\delta)$, respectively. Hence, the corresponding asymptotic version of the feasible point optimal test is then

$$P_{T,*}^{\mathrm{GLS}}(c,\bar{c}) = \left\{\inf_{\delta\in[\varepsilon,1-\varepsilon]} S(\bar{\alpha},\delta) - \inf_{\delta\in[\varepsilon,1-\varepsilon]} \bar{\alpha}S(1,\delta)\right\} \Big/ s^2.$$

Note that a trimming ε is necessary otherwise the critical values become unbounded. The reason is similar to that encountered in the context of tests for structural change (see, e.g., Andrews, 1993). We use $\varepsilon = 0.15$ throughout. In the case of this feasible point optimal test, there is a problem of which method to choose to select the break date to construct the estimate s^2. Based on finite sample properties assessed via simulations we opted for evaluating s^2 at the break point $\hat{\delta}$, say, which minimizes the sum of squared residuals under the alternative, i.e., we select $\hat{\hat{\delta}} = \arg\min_{\delta\in[\varepsilon,1-\varepsilon]} S(\bar{\alpha}, \delta)$. Using Theorem 2, we have

$$
\begin{aligned}
P_{T,*}^{\mathrm{GLS}}(c,\bar{c}) \Rightarrow & \sup_{\delta\in[\varepsilon,1-\varepsilon]} M(c,0,\delta) - \sup_{\delta\in[\varepsilon,1-\varepsilon]} M(c,\bar{c},\delta) \\
& - 2\bar{c}\int_0^1 W_c(r)\mathrm{d}W(r) + (\bar{c}^2 - 2\bar{c}c)\int_0^1 W_c(r)^2\mathrm{d}r - \bar{c} \\
= & H_*^{P_T^{\mathrm{GLS}}}(c,\bar{c}). \qquad (16)
\end{aligned}
$$

The asymptotic Gaussian power envelope is then defined as $\pi^*(c) = \Pr[H_*^{P_T^{\mathrm{GLS}}}(c,c) < b_*^{P_T^{\mathrm{GLS}}}(c)]$, where, with v the size of the test, $b_*^{P_T^{\mathrm{GLS}}}(c)$ is such that $\Pr[H_*^{P_T^{\mathrm{GLS}}}(0,c) < b_*^{P_T^{\mathrm{GLS}}}(c)] = v$. Furthermore, the power envelope allows us to find the 'optimal' noncentrality parameter $\bar{c}$ for our models. ERS recommended to choose the value $\bar{c}$ such that the asymptotic power of the test is 50%, i.e., $\bar{c}$ is such that $\Pr[H_*^{P_T^{\mathrm{GLS}}}(\bar{c},\bar{c}) < b_*^{P_T^{\mathrm{GLS}}}(\bar{c})] = 0.5$. Using simulations, we found that $\bar{c} = -22.5$ and we use this value in the rest of the paper.

5. Critical Values and Asymptotic Power Functions

In this section, we obtain the asymptotic critical values for the tests assuming $\bar{c} = -22.5$ is used to detrend the data. We simulate directly the asymptotic distributions using 1000 steps to approximate the Wiener process on [0, 1] as the partial sums of i.i.d. N(0, 1) random variables. The limiting distributions are tabulated for the null hypothesis $c = 0$. For the finite sample distributions, we use $T = 100$ with data generated by a random walk with zero initial condition and i.i.d. N(0, 1) errors. Here k is set to 0 which is equivalent to using the true value of σ^2; the effects of selecting k are investigated in the next section. In all cases, 10,000 replications are used. The results are in the first three columns of Table 1a and Table 1b for, respectively, the case where the break point is selected by minimizing the tests and when the break point is selected

Table 1. (a) Critical values; P_T test, and M and ADF tests choosing T_B minimizing the statistics.

			$k=0$		Model I						Model II					
			$T=100$		$T=100$			$T=200$			$T=100$			$T=200$		
Test	Size	$T=\infty$	Model I	Model II	*MAIC*	*BIC*	*t*-sig	*MAIC*	*BIC*	*t*-sig	*MAIC*	*BIC*	*t*-sig	*MAIC*	*BIC*	*t*-sig
MZ_α	0.01	−40.54	−41.90	−44.97	−29.44	−41.3	−185.7	−33.2	−38.3	−69.6	−31.3	−65.2	−293.9	−33.5	−40.0	−82.1
	0.025	−35.08	−36.71	−39.99	−26.4	−34.2	−113.5	−29.4	−31.1	−51.7	−28.1	−39.1	−163.7	−30.5	−33.1	−62.5
	0.05	−31.04	−32.86	−35.99	−23.5	−28.6	−69.8	−25.4	−28.2	−42.8	−25.9	−33.7	−99.4	−27.1	−29.9	−49.9
	0.10	−26.57	−28.42	−31.38	−21.3	−24.2	−43.7	−22.2	−24.6	−34.9	−23.0	−26.9	−60.5	−23.7	−26.5	−39.5
	0.20	−22.07	−23.60	−26.48	−18.2	−20.2	−30.5	−18.7	−20.6	−26.3	−19.9	−22.6	−39.0	−20.2	−22.2	−30.1
MSB	0.01	0.110	0.109	0.105	0.130	0.110	0.052	0.122	0.114	0.085	0.125	0.087	0.041	0.121	0.110	0.078
	0.025	0.118	0.116	0.111	0.137	0.120	0.066	0.129	0.126	0.098	0.132	0.112	0.055	0.127	0.122	0.089
	0.05	0.126	0.123	0.117	0.145	0.130	0.084	0.139	0.132	0.107	0.137	0.121	0.072	0.135	0.129	0.099
	0.10	0.135	0.132	0.125	0.151	0.142	0.106	0.149	0.141	0.119	0.146	0.135	0.090	0.143	0.137	0.112
	0.20	0.148	0.144	0.136	0.163	0.155	0.127	0.161	0.154	0.137	0.157	0.146	0.113	0.155	0.148	0.128
MZ_t	0.01	−4.49	−4.56	−4.74	−3.83	−4.54	−9.59	−4.06	−4.36	−5.89	−3.93	−5.59	−12.11	−4.07	−4.44	−6.41
	0.025	−4.17	−4.27	−4.46	−3.63	−4.13	−7.53	−3.79	−3.94	−5.08	−3.75	−4.41	−9.03	−3.86	−4.06	−5.56
	0.05	−3.91	−4.04	−4.23	−3.42	−3.76	−5.89	−3.53	−3.74	−4.61	−3.56	−4.07	−7.01	−3.64	−3.85	−4.98
	0.10	−3.62	−3.75	−3.94	−3.24	−3.44	−4.65	−3.30	−3.49	−4.17	−3.36	−3.65	−5.48	−3.42	−3.60	−4.41
	0.20	−3.29	−3.41	−3.62	−2.99	−3.15	−3.87	−3.04	−3.19	−3.61	−3.13	−3.34	−4.40	−3.15	−3.30	−3.86
ADF	0.01	−4.49	−4.70	−4.87	−4.54	−5.00	−5.00	−4.47	−4.72	−4.89	−4.74	−5.19	−5.18	−4.48	−4.88	−4.99
	0.025	−4.17	−4.37	−4.54	−4.25	−4.67	−4.68	−4.11	−4.32	−4.46	−4.42	−4.77	−4.88	−4.22	−4.42	−4.49
	0.05	−3.91	−4.11	−4.27	−3.91	−4.32	−4.42	−3.78	−4.06	−4.19	−4.14	−4.44	−4.63	−3.90	−4.17	−4.31
	0.10	−3.62	−3.83	−3.99	−3.67	−3.91	−4.13	−3.49	−3.70	−3.88	−3.84	−4.19	−4.34	−3.62	−3.85	−4.07
	0.20	−3.29	−3.48	−3.65	−3.31	−3.54	−3.73	−3.18	−3.37	−3.51	−3.50	−3.72	−3.93	−3.29	−3.49	−3.69
P_T	0.01	6.71	7.86	8.87	9.47	6.96	1.85	8.11	7.23	4.05	9.94	8.18	2.12	9.16	8.80	4.59
	0.025	7.82	8.96	9.55	10.53	8.27	3.15	9.39	8.82	5.44	10.91	9.31	3.08	10.60	9.46	5.76
	0.05	8.84	9.77	10.53	11.68	9.57	4.65	10.60	9.49	6.90	12.04	10.35	5.28	11.50	10.42	7.66
	0.10	10.29	11.53	11.99	13.01	11.46	7.22	12.29	11.12	8.31	13.83	11.77	7.43	13.08	11.79	9.32
	0.20	12.28	13.50	14.08	15.12	13.67	10.08	14.62	13.23	10.89	15.79	14.06	10.20	15.06	14.04	11.50

Table 1. (b) Critical values; M and ADF tests choosing T_B maximizing $|t_{\hat{\beta}_2}|$.

			$k = 0$		Model I						Model II					
			$T = 100$		$T = 100$			$T = 200$			$T = 100$			$T = 200$		
Test	Size	$T = \infty$	Model I	Model II	$MAIC$	BIC	t-sig	$MAIC$	BIC	t-sig	$MAIC$	BIC	t-sig	$MAIC$	BIC	t-sig
MZ_α	0.01	−39.97	−32.03	−30.88	−27.2	−39.8	−139.1	−32.7	−37.1	−64.3	−27.0	−34.6	−130.5	−31.2	−32.8	−62.1
	0.025	−34.68	−28.79	−27.75	−25.7	−31.7	−87.0	−28.3	−30.5	−49.9	−24.9	−29.5	−86.5	−27.2	−30.1	−48.5
	0.05	−30.54	−26.12	−25.56	−22.6	−27.1	−58.2	−24.8	−27.5	−39.6	−22.9	−25.9	−55.3	−24.5	−27.5	−40.6
	0.10	−26.30	−23.23	−22.86	−20.6	−23.4	−37.0	−21.8	−24.3	−32.3	−20.7	−23.4	−36.6	−21.4	−23.8	−30.1
	0.20	−21.82	−19.91	−19.72	−17.4	−19.6	−26.5	−18.2	−20.4	−24.8	−17.4	−19.5	−25.8	−18.2	−20.2	−24.4
MSB	0.01	0.111	0.124	0.126	0.135	0.111	0.059	0.122	0.116	0.088	0.134	0.119	0.061	0.125	0.123	0.089
	0.025	0.119	0.131	0.132	0.138	0.124	0.075	0.132	0.127	0.099	0.140	0.128	0.076	0.133	0.128	0.101
	0.05	0.127	0.137	0.138	0.147	0.135	0.092	0.140	0.133	0.111	0.145	0.137	0.094	0.141	0.134	0.110
	0.10	0.137	0.145	0.146	0.154	0.144	0.115	0.150	0.143	0.124	0.154	0.145	0.116	0.150	0.143	0.128
	0.20	0.150	0.156	0.157	0.166	0.158	0.136	0.164	0.155	0.140	0.166	0.158	0.137	0.163	0.156	0.141
MZ_t	0.01	−4.46	−3.98	−3.91	−3.69	−4.44	−8.34	−4.02	−4.30	−5.66	−3.66	−4.15	−7.97	−3.91	−4.04	−5.56
	0.025	−4.14	−3.77	−3.70	−3.56	−3.96	−6.58	−3.74	−3.87	−4.99	−3.51	−3.83	−6.58	−3.67	−3.86	−4.88
	0.05	−3.89	−3.59	−3.54	−3.35	−3.67	−5.37	−3.49	−3.71	−4.44	−3.35	−3.59	−5.19	−3.47	−3.67	−4.47
	0.10	−3.59	−3.38	−3.35	−3.16	−3.39	−4.29	−3.27	−3.46	−3.98	−3.19	−3.37	−4.28	−3.24	−3.43	−3.86
	0.20	−3.27	−3.13	−3.11	−2.92	−3.09	−3.61	−2.99	−3.16	−3.49	−3.91	−3.08	−3.58	−2.99	−3.14	−3.46
ADF	0.01	−4.46	−4.67	−4.59	−4.40	−4.91	−4.87	−4.41	−4.67	−4.83	−4.31	−4.85	−4.86	−4.25	−4.44	−4.66
	0.025	−4.14	−4.33	−4.29	−4.18	−4.58	−4.57	−4.00	−4.23	−4.36	−4.11	−4.40	−4.48	−3.96	−4.20	−4.24
	0.05	−3.89	−4.06	−4.04	−3.83	−4.25	−4.33	−3.73	−4.00	−4.08	−3.83	−4.16	−4.25	−3.67	−3.94	−4.05
	0.10	−3.59	−3.78	−3.76	−3.55	−3.86	−3.99	−3.45	−3.67	−3.79	−3.59	−3.84	−4.02	−3.42	−3.63	−3.76
	0.20	−3.27	−3.44	−3.43	−3.23	−3.48	−3.64	−3.14	−3.34	−3.47	−3.24	−3.49	−3.65	−3.12	−3.31	−3.43

Note: $\bar{c} = -22.5$ when constructing the tests and s^2.

maximizing the absolute value of the t-statistic on the coefficient of the change in slope. In general, the approximation to the finite sample distribution is adequate but somewhat less good for Model II which contains a change in intercept that is asymptotically negligible.

The asymptotic power functions of the tests are defined by $\pi^*_{J\text{GLS}}(c,\bar{c}) = \Pr[\inf_{\delta\in[0,1]} H^{J^{\text{GLS}}}(c,\bar{c},\delta) < b^{J^{\text{GLS}}}(\bar{c})]$ or $\pi^*_{J\text{GLS}}(c,\bar{c}) = \Pr[H^{J^{\text{GLS}}}(c,\bar{c},\delta^*) < b_*^{J^{\text{GLS}}}(\bar{c})]$ for $J = MZ_\alpha$, MSB, MZ_t, ADF and with $H^i(c,\bar{c})$ defined in Theorem 1 and δ^* defined by (12). The constants $b^{J^{\text{GLS}}}(\bar{c})$ and $b_*^{J^{\text{GLS}}}(\bar{c})$ are such that $\Pr[\inf_{\delta\in[0,1]} H^{J^{\text{GLS}}}(0,\bar{c},\delta) < b^{J^{\text{GLS}}}(\bar{c})] = v$, and $\Pr[H^{J^{\text{GLS}}}(0,\bar{c},\delta^*) < b_*^{J^{\text{GLS}}}(\bar{c})] = v$, the size of the tests. The asymptotic power functions are shown in Fig. 1 where the solid line is the power envelope. The M^{GLS} tests, and especially the P_T^{GLS} test, have asymptotic power functions very close to the power envelope both when the break point is selected by minimizing the tests and when it is selected maximizing the absolute value of the t-statistic of the coefficient on the change in slope. This is also true of the ADF^{GLS} test since it is asymptotically equivalent to the MZ_t^{GLS} test. Hence, in terms of asymptotic power, all tests considered are basically equivalent.

6. Size and Power of the Tests in Finite Samples

6.1. *The size issue, the selection of k and information criteria*

All tests require the estimation of the augmented autoregression (10). Ng and Perron (2001) recommended using GLS detrended data with the same noncentrality parameter $\bar{c}$ for constructing s^2 and the tests. We follow their suggestion and, in subsequent results, $\bar{c} = -22.5$ is used to detrend the data when constructing the tests and when estimating the autoregression (10) to construct s^2.

In our simulations and empirical applications, we consider three data dependent methods to select the order of the autoregression. The first is the standard Bayesian Information Criterion (BIC). We follow the recommendation of Ng and Perron (2001) by confining the search for the best value of k in a range $[0, k_{\max}]$. Also, all regressions are estimated using the same number of

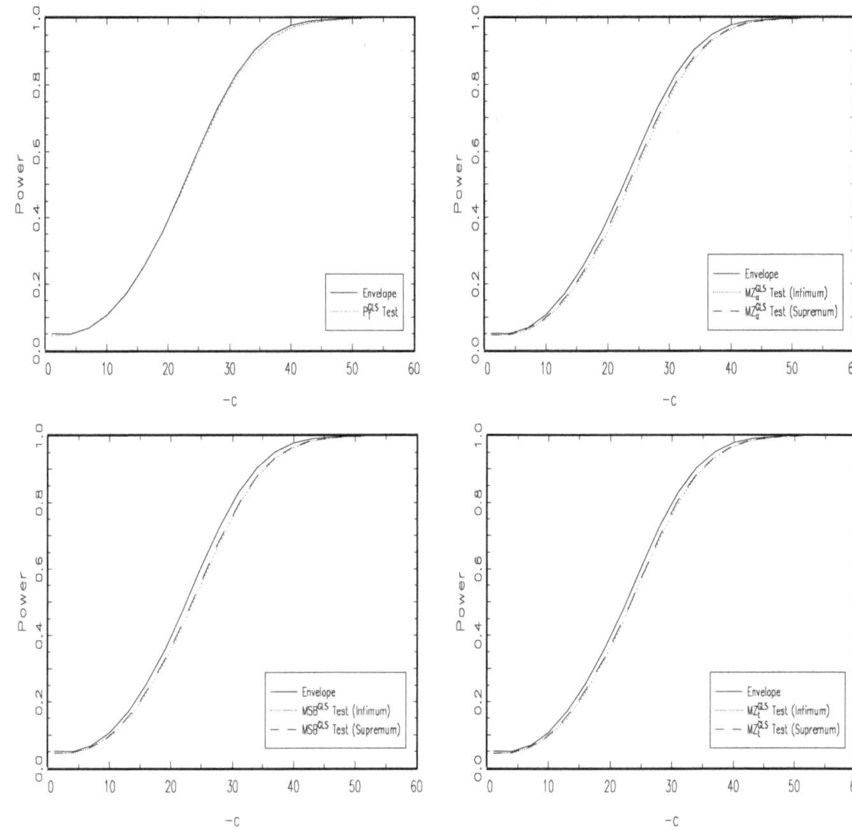

Fig. 1. Gaussian local power envelope and the local asymptotic power functions of the tests.

effective observations, $T^* = T - k_{\max}$. The *BIC* is then defined as $k_{\text{bic}} = \arg\min_{k\in[0,k_{\max}]}\{\log(s_{ek}^2) + (\ln(T^*)k)/T^*\}$ with $s_{ek}^2 = T^{*-1}\sum_{t=k_{\max}+1}^{T}\hat{e}_{tk}^2$ with $\hat{e}_{tk}$ obtained from (10) estimated from $t = k_{\max}+1$ to T. We also consider the Modified Akaike Information Criterion (*MAIC*), advocated by Ng and Perron (2001), defined by $k_{\text{maic}} = \arg\min_{k\in[0,k_{\max}]}\{\log(s_{ek}^2) + (2(\hat{\tau}_T(k) + k))/T^*\}$, where $\hat{\tau}_T(k) = (s_{ek}^2)^{-1}\hat{b}_0^2\sum_{t=k_{\max}+1}^{T}\tilde{y}_{t-1}^2$ with $\hat{b}_0$ obtained from (10). As shown in Ng and Perron (2001), the *MAIC* works as well as standard information criteria when the extent of correlation is mild but provides unit root tests having better finite sample size with a negative *MA* component. We also consider the sequential t-test, denoted t-sig, for the significance of the last lag, considered in Ng and Perron (1995), with a two-tailed 10% size. When using an information criterion, we set $k_{\max} = int[10 * (T/100)^\wedge(1/4)]$ and for the t-sig method, we set $k_{\max} = int[4 * (T/100)^\wedge(1/4)]$. Note that once the order k is selected, at k^* say, s^2 is constructed using all possible observations, i.e., estimating (10) from $t = k^* + 1$ to T.

6.2. *Critical values with data-dependent methods to select k*

While the asymptotic distribution is a good approximation to the finite sample distribution for any of the test considered when k is fixed, significant differences can occur when using a data-dependent method to select k, especially in the context of tests involving breaks at unknown dates (e.g., Perron, 1997). We present in Tables 2 and 3, for both Models I and II, the finite sample critical values of the tests when using either of the three data-dependent methods to select k. Two sample sizes are considered, $T = 100$ and 200. These were obtained from simulations with 1,000 replications from the data-generating process defined by (1) with $d_t = 0$, $\alpha = 1$ and $v_t \sim$ i.i.d. N(0, 1).

The results show substantial differences, especially when using the method t-sig which yields finite sample critical values much smaller than the asymptotic ones (hence, using the latter would imply liberal size distortions). When using the *MAIC*, the result is opposite, namely finite sample critical values that are higher than the asymptotic ones (hence, using the latter would imply conservative tests). With the *BIC*, the difference are not so large but still important. For these reasons, we recommend the use of these finite sample critical values 'adjusted for the effect of using a data-dependent method to select k'. The simulations and empirical applications below make use of these instead of the asymptotic ones.

6.3. *Size and power of the tests*

We now consider the size and power of the tests in finite samples using the various data-dependent methods to select the truncation

Table 2. Size and power; choosing T_B minimizing the tests; Model I; $T = 100$ ($\bar{c} = -22.5$ when constructing s^2 for the M^{GLS} and P_T^{GLS} tests; 5% nominal size tests).

		Size					Power				
	Criteria	MZ_α	*MSB*	MZ_t	*Pt*	*ADF*	MZ_α	*MSB*	MZ_t	*Pt*	*ADF*
i.i.d.	*BIC*	0.051	0.051	0.051	0.050	0.050	0.348	0.319	0.347	0.375	0.462
	MAIC	0.050	0.051	0.050	0.050	0.050	0.481	0.484	0.468	0.486	0.459
	t-sig	0.050	0.051	0.051	0.051	0.051	0.131	0.131	0.130	0.153	0.421
Moving-average processes											
$\theta = -0.8$	*BIC*	0.930	0.931	0.930	0.922	0.969	1.000	1.000	1.000	0.999	1.000
	MAIC	0.334	0.334	0.329	0.286	0.353	0.778	0.783	0.776	0.669	0.809
	t-sig	0.074	0.074	0.074	0.071	0.717	0.293	0.293	0.295	0.250	0.938
$\theta = -0.4$	*BIC*	0.391	0.385	0.385	0.368	0.447	0.874	0.873	0.874	0.865	0.903
	MAIC	0.145	0.146	0.141	0.138	0.124	0.506	0.505	0.497	0.475	0.461
	t-sig	0.025	0.025	0.025	0.026	0.228	0.096	0.096	0.096	0.100	0.683
$\theta = 0.4$	*BIC*	0.226	0.224	0.227	0.201	0.076	0.696	0.687	0.698	0.683	0.494
	MAIC	0.109	0.108	0.098	0.066	0.017	0.166	0.169	0.160	0.162	0.100
	t-sig	0.051	0.053	0.051	0.056	0.058	0.157	0.157	0.158	0.199	0.378
$\theta = 0.8$	*BIC*	0.481	0.482	0.478	0.364	0.107	0.703	0.699	0.707	0.679	0.400
	MAIC	0.185	0.191	0.176	0.091	0.011	0.336	0.337	0.327	0.301	0.084
	t-sig	0.074	0.076	0.074	0.081	0.052	0.233	0.231	0.232	0.269	0.215

(Continued)

Table 2. (*Continued*)

	Criteria	Size					Power				
		MZ_α	*MSB*	MZ_t	*Pt*	*ADF*	MZ_α	*MSB*	MZ_t	*Pt*	*ADF*
Autoregressive processes											
$\rho = -0.8$	*BIC*	0.010	0.010	0.010	0.007	0.041	0.030	0.029	0.031	0.026	0.410
	MAIC	0.003	0.002	0.003	0.002	0.043	0.015	0.017	0.015	0.016	0.359
	t-sig	0.016	0.016	0.015	0.016	0.043	0.054	0.054	0.054	0.067	0.384
$\rho = -0.4$	*BIC*	0.129	0.120	0.128	0.115	0.135	0.475	0.469	0.476	0.465	0.549
	MAIC	0.064	0.063	0.063	0.063	0.059	0.360	0.365	0.356	0.374	0.396
	t-sig	0.030	0.030	0.030	0.035	0.071	0.116	0.116	0.116	0.134	0.420
$\rho = 0.4$	BIC	0.146	0.146	0.146	0.112	0.038	0.496	0.477	0.499	0.502	0.275
	MAIC	0.122	0.131	0.113	0.083	0.022	0.122	0.124	0.116	0.083	0.058
	t-sig	0.057	0.058	0.057	0.057	0.047	0.147	0.146	0.147	0.162	0.273
$\rho = 0.8$	*BIC*	0.290	0.298	0.278	0.183	0.060	0.346	0.341	0.343	0.331	0.107
	MAIC	0.342	0.360	0.321	0.204	0.074	0.404	0.415	0.390	0.355	0.141
	t-sig	0.136	0.139	0.134	0.084	0.064	0.108	0.108	0.107	0.119	0.118

Table 3. Size and power; choosing T_B minimizing the tests; Model I; $T = 200$ ($\bar{c} = -22.5$ when constructing s^2 for the M^{GLS} and P_T^{GLS} tests; 5% nominal size tests).

		Size					Power				
	Criteria	MZ_α	*MSB*	MZ_t	*Pt*	*ADF*	MZ_α	*MSB*	MZ_t	*Pt*	*ADF*
i.i.d.	*BIC*	0.051	0.051	0.050	0.051	0.051	0.512	0.494	0.505	0.488	0.501
	MAIC	0.050	0.050	0.051	0.050	0.051	0.495	0.492	0.505	0.482	0.476
	t-sig	0.051	0.051	0.051	0.050	0.051	0.235	0.230	0.239	0.289	0.430
Moving-average processes											
$\theta = -0.8$	*BIC*	0.814	0.809	0.814	0.783	0.860	0.992	0.991	0.992	0.989	0.997
	MAIC	0.102	0.099	0.104	0.076	0.178	0.366	0.366	0.365	0.272	0.510
	t-sig	0.278	0.277	0.277	0.296	0.539	0.655	0.655	0.656	0.628	0.931
$\theta = -0.4$	*BIC*	0.239	0.230	0.235	0.218	0.225	0.817	0.808	0.815	0.797	0.820
	MAIC	0.091	0.091	0.094	0.074	0.074	0.453	0.458	0.460	0.418	0.427
	t-sig	0.034	0.034	0.035	0.043	0.110	0.297	0.287	0.295	0.342	0.605
$\theta = 0.4$	*BIC*	0.144	0.144	0.141	0.130	0.089	0.702	0.687	0.698	0.673	0.552
	MAIC	0.102	0.104	0.101	0.083	0.046	0.514	0.511	0.510	0.468	0.356
	t-sig	0.068	0.067	0.069	0.072	0.050	0.329	0.324	0.331	0.380	0.401
$\theta = 0.8$	*BIC*	0.296	0.290	0.291	0.241	0.076	0.694	0.683	0.693	0.665	0.392
	MAIC	0.181	0.182	0.181	0.120	0.027	0.447	0.445	0.453	0.382	0.155
	t-sig	0.149	0.151	0.146	0.156	0.064	0.515	0.509	0.517	0.550	0.313

(*Continued*)

Table 3. (*Continued*)

		Size					Power				
	Criteria	MZ_α	*MSB*	MZ_t	*Pt*	*ADF*	MZ_α	*MSB*	MZ_t	P_T	*ADF*
Autoregressive processes											
$\rho = -0.8$	*BIC*	0.006	0.006	0.006	0.006	0.045	0.045	0.037	0.044	0.040	0.463
	MAIC	0.001	0.001	0.001	0.000	0.047	0.033	0.033	0.035	0.032	0.395
	t-sig	0.016	0.016	0.016	0.014	0.046	0.061	0.059	0.061	0.068	0.411
$\rho = -0.4$	*BIC*	0.052	0.049	0.049	0.037	0.048	0.444	0.424	0.434	0.417	0.458
	MAIC	0.047	0.045	0.046	0.036	0.045	0.400	0.403	0.408	0.381	0.425
	t-sig	0.034	0.034	0.033	0.041	0.043	0.201	0.194	0.199	0.228	0.418
$\rho = 0.4$	*BIC*	0.106	0.105	0.104	0.090	0.041	0.585	0.562	0.582	0.560	0.411
	MAIC	0.105	0.103	0.108	0.083	0.054	0.494	0.493	0.506	0.465	0.356
	t-sig	0.068	0.068	0.067	0.065	0.044	0.267	0.263	0.268	0.320	0.367
$\rho = 0.8$	*BIC*	0.146	0.141	0.138	0.106	0.057	0.405	0.394	0.393	0.371	0.229
	MAIC	0.164	0.175	0.156	0.108	0.062	0.399	0.397	0.400	0.369	0.229
	t-sig	0.091	0.094	0.091	0.070	0.053	0.212	0.204	0.211	0.228	0.206

lag described above. Our simulations are based on 1,000 replications of the DGP defined by (1) with $d_t = 0$. We consider pure $MA(1)$ processes, i.e., with $v_t = (1 + \theta L)e_t$ and pure $AR(1)$ processes, i.e., $(1 - \rho L)v_t = e_t$, where $e_t \sim$ i.i.d. N(0, 1). For both the $MA(1)$ and $AR(1)$ cases, we consider θ and ρ in the range $[-0.8, 0.8]$. We consider the sample sizes $T = 100$ and 200. The power is evaluated at $\bar{\alpha} = 1 + \bar{c}/T$ for $\bar{c} = -22.5$ which implies that the asymptotic power is 50%. All results presented are for 5% nominal size tests.

The results for the case where the break point is chosen by minimizing the tests are presented in Table 2 for $T = 100$ and Table 3 for $T = 200$. With i.i.d. errors, as expected, the power of the tests when the sequential t-sig is used to select k is low. With the *MAIC* or *BIC*, the power is indeed close to the asymptotic value of 50%. For the ADF^{GLS} test, the power is high for all methods to choose k. Given these results, we shall not discuss further the behavior of the tests with the sequential t-sig method.

Consider now the case where the errors have a negative *MA* component. For all tests, the use of the *BIC* to select k implies tests with severe size distortions with exact sizes above 90% with $T = 100$ (80% with $T = 200$) when the *MA* component is -0.8 and about 40% with $T = 100$ (23% with $T = 200$) when it is -0.4. On the other hand, the *MAIC* allows the M^{GLS} and the P_T^{GLS} tests to have much less size distortions. The exact size is about 33% with $T = 100$ (10% with $T = 200$) when the coefficient is -0.8 and about 14% with $T = 100$ (9% with $T = 200$) when it is -0.4. But even when using the *MAIC*, the ADF^{GLS} test still suffers from high size distortions (with $\theta = -0.8$, the size is 18% when $T = 200$).

When the errors have a positive moving average coefficient, the M^{GLS} tests and the P_T^{GLS} tests are liberal, while the ADF^{GLS} test has the correct size (especially with $T = 200$). With a negative autoregressive coefficient, the M^{GLS} tests and the P_T^{GLS} test are very conservative and, hence, show basically no power. The *ADF* has the correct size and power is good. When the autoregressive coefficient is positive, the M^{GLS} tests and the P_T^{GLS} test are liberal. The ADF^{GLS} has better size but no power.

The results show that the ADF^{GLS} with k chosen using the $MAIC$ has better overall properties unless there is a negative MA component in the errors, in which case the M^{GLS} and the P_T^{GLS} tests are superior.

The results for the case where the break is selected by maximizing the absolute value of the t-statistic on the coefficient of the change in slope are in Table 4 for $T = 100$ and Table 5 for $T = 200$. They show properties with basically similar qualitative features with size distortions being slightly smaller.

7. Empirical Applications

Among the macroeconomic time series considered by Nelson and Plosser (1982), Perron (1989) argued that two of them were likely affected by a significant change in slope and intercept for the samples analyzed, namely the Real Wages and Stock Prices series. The series are presented in Figs. 2 and 3. We re-evaluate the claim made by Perron (1989) that these series are trend-stationary if allowance is made for such a change in slope and intercept using our new tests. We applied the MZ_t^{GLS}, P_T^{GLS}, and ADF^{GLS} tests using the BIC and $MAIC$ criteria to select the autoregressive order (imposing a minimal value of 1).

The results are presented in Table 6 for the case where the break date is selected minimizing the test. Using the BIC to select k, all tests point to a strong rejection at the 2.5% significance level for the Stock Prices series with the break date selected between 1931 and 1941 depending on the specification used. With the $MAIC$ to select k, there is a rejection at the 1% level (except with the test P_T^{GLS}). For the Real Wages series, there is a rejection at least at the 1% or 2.5% significance level using the criterion $MAIC$ to select k and at the 5% level using the BIC. The break date is selected at 1938 or 1940 depending on the specification used. The estimated trend function is plotted in Figs. 2 and 3 using $T_B = 1937$ for the Stock Prices series and $T_B = 1938$ for the Real Wages series.

Table 7 presents the results of the tests when the break date is selected by maximizing the absolute value of the t-statistic on the coefficient of the slope change. For Stock Prices, the break date

Table 4. Size and power; choosing T_B maximizing $|t_{\hat{\beta}_2}|$; Model I; $T = 100$ ($\bar{c} = -22.5$ when constructing s^2 for the M^{GLS} and P_T^{GLS} tests; 5% nominal size tests).

		Size				Power			
	Criteria	MZ_α	MSB	MZ_t	ADF	MZ_α	MSB	MZ_t	ADF
i.i.d.	*BIC*	0.050	0.051	0.051	0.051	0.411	0.414	0.402	0.478
	MAIC	0.051	0.050	0.050	0.050	0.503	0.475	0.499	0.465
	t-sig	0.051	0.051	0.051	0.051	0.150	0.151	0.154	0.447
Moving-average processes									
$\theta = -0.8$	*BIC*	0.928	0.930	0.926	0.966	1.000	1.000	1.000	1.000
	MAIC	0.289	0.286	0.289	0.304	0.669	0.669	0.669	0.698
	t-sig	0.070	0.070	0.070	0.689	0.250	0.250	0.250	0.925
$\theta = -0.4$	*BIC*	0.399	0.405	0.397	0.430	0.870	0.870	0.870	0.895
	MAIC	0.138	0.131	0.136	0.113	0.479	0.466	0.475	0.433
	t-sig	0.025	0.025	0.025	0.206	0.099	0.099	0.100	0.668
$\theta = 0.4$	*BIC*	0.216	0.225	0.211	0.072	0.707	0.711	0.703	0.494
	MAIC	0.064	0.065	0.066	0.010	0.156	0.139	0.161	0.107
	t-sig	0.053	0.053	0.054	0.055	0.195	0.199	0.195	0.389
$\theta = 0.8$	*BIC*	0.390	0.397	0.380	0.089	0.683	0.686	0.682	0.386
	MAIC	0.089	0.088	0.087	0.008	0.297	0.281	0.294	0.091
	t-sig	0.079	0.080	0.078	0.045	0.260	0.262	0.263	0.212
Autoregressive processes									
$\rho = -0.8$	*BIC*	0.006	0.006	0.006	0.041	0.028	0.028	0.027	0.398
	MAIC	0.002	0.003	0.002	0.041	0.015	0.017	0.016	0.335
	t-sig	0.016	0.016	0.016	0.040	0.062	0.062	0.062	0.377
$\rho = -0.4$	*BIC*	0.121	0.120	0.121	0.128	0.474	0.475	0.468	0.546
	MAIC	0.063	0.057	0.063	0.053	0.368	0.352	0.370	0.391
	t-sig	0.032	0.035	0.033	0.066	0.127	0.128	0.127	0.415
$\rho = 0.4$	*BIC*	0.132	0.136	0.127	0.033	0.525	0.530	0.520	0.287
	MAIC	0.078	0.084	0.080	0.014	0.080	0.077	0.079	0.047
	t-sig	0.055	0.056	0.054	0.040	0.156	0.158	0.157	0.281
$\rho = 0.8$	*BIC*	0.206	0.217	0.197	0.037	0.353	0.356	0.344	0.108
	MAIC	0.225	0.240	0.210	0.048	0.372	0.362	0.365	0.127
	t-sig	0.087	0.091	0.087	0.042	0.117	0.119	0.118	0.114

Table 5. Size and power; choosing T_B maximizing $|t_{\hat{\beta}_2}|$; Model I; $T = 200$ ($\bar{c} = -22.5$ when constructing s^2 for the M^{GLS} and P_T^{GLS} tests; 5% nominal size tests).

		Size				Power			
	Criteria	MZ_α	MSB	MZ_t	ADF	MZ_α	MSB	MZ_t	ADF
i.i.d.	*BIC*	0.051	0.051	0.051	0.051	0.518	0.504	0.513	0.520
	MAIC	0.051	0.050	0.051	0.050	0.505	0.484	0.515	0.478
	t-sig	0.051	0.050	0.051	0.051	0.275	0.267	0.279	0.483
Moving-average processes									
$\theta = -0.8$	*BIC*	0.796	0.793	0.796	0.856	0.991	0.989	0.991	0.997
	MAIC	0.076	0.072	0.079	0.139	0.277	0.272	0.281	0.422
	t-sig	0.290	0.286	0.288	0.538	0.625	0.620	0.624	0.915
$\theta = -0.4$	*BIC*	0.226	0.220	0.224	0.223	0.822	0.808	0.818	0.826
	MAIC	0.079	0.077	0.081	0.068	0.436	0.424	0.438	0.406
	t-sig	0.038	0.037	0.040	0.116	0.325	0.318	0.327	0.633
$\theta = 0.4$	*BIC*	0.142	0.138	0.138	0.088	0.703	0.683	0.696	0.567
	MAIC	0.094	0.092	0.094	0.042	0.479	0.472	0.481	0.342
	t-sig	0.068	0.070	0.069	0.057	0.365	0.351	0.364	0.422
$\theta = 0.8$	*BIC*	0.261	0.256	0.249	0.070	0.684	0.669	0.681	0.400
	MAIC	0.132	0.128	0.128	0.018	0.403	0.388	0.407	0.148
	t-sig	0.148	0.147	0.146	0.069	0.544	0.535	0.542	0.350
Autoregressive processes									
$\rho = -0.8$	*BIC*	0.006	0.005	0.006	0.046	0.045	0.041	0.045	0.471
	MAIC	0.000	0.000	0.000	0.037	0.034	0.031	0.035	0.376
	t-sig	0.014	0.014	0.015	0.051	0.061	0.062	0.062	0.428
$\rho = -0.4$	*BIC*	0.047	0.045	0.042	0.046	0.454	0.426	0.448	0.477
	MAIC	0.039	0.036	0.041	0.043	0.399	0.387	0.412	0.421
	t-sig	0.037	0.035	0.035	0.052	0.214	0.200	0.214	0.449
$\rho = 0.4$	*BIC*	0.101	0.095	0.096	0.041	0.594	0.563	0.584	0.427
	MAIC	0.087	0.083	0.088	0.048	0.480	0.462	0.490	0.353
	t-sig	0.065	0.064	0.063	0.043	0.308	0.298	0.308	0.407
$\rho = 0.8$	*BIC*	0.119	0.116	0.110	0.046	0.400	0.381	0.384	0.229
	MAIC	0.118	0.128	0.121	0.053	0.382	0.364	0.387	0.224
	t-sig	0.070	0.067	0.066	0.043	0.218	0.212	0.215	0.227

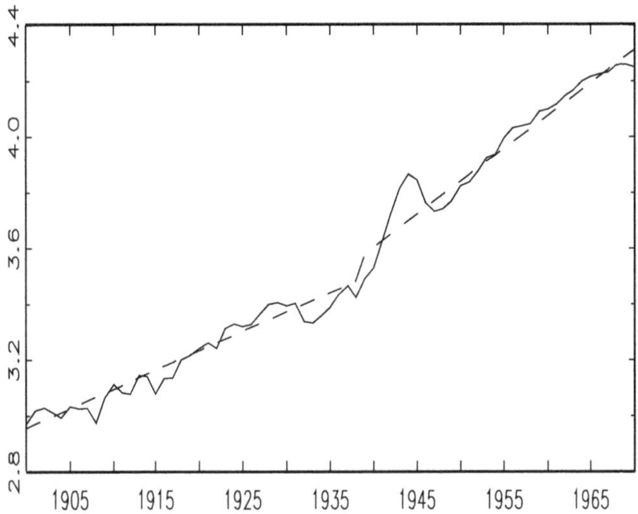

Fig. 2. Logarithm of real wages (1900–1970).

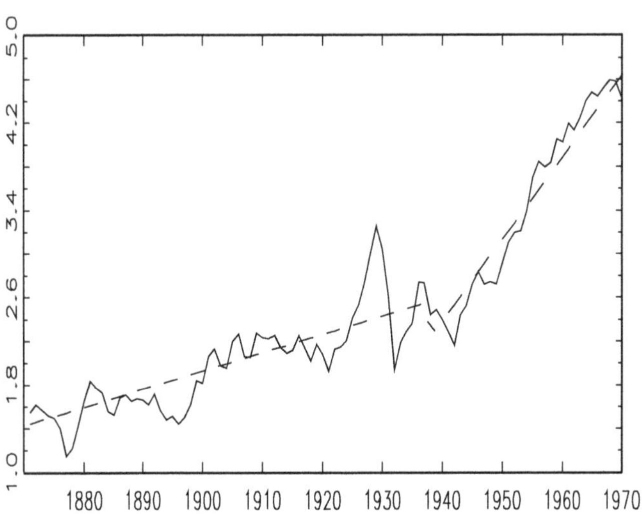

Fig. 3. Logarithm of common stock prices (1871–1970).

selected is 1931 and there is a rejection at the 2.5% or 5% significance level using *BIC* but there is little evidence against the unit root using the *MAIC*. For the Real Wages series, the break date selected is 1933. The tests show a rejection at least at the 5% significance level

Table 6. Empirical results for the real wages and stock prices series choosing the break point minimizing the tests.

	Series	
	Stock prices	Real wages
T	100	71
Criteria	BIC	BIC
	$MAIC$	$MAIC$
MZ_{α}	-48.4^{b}	-38.4^{c}
	-47.7^{a}	-38.4^{a}
k	1	1
	1	1
T_{B}	1941	1938
	1937	1938
MZ_{t}	-4.9^{b}	-4.3^{c}
	-4.8^{a}	-4.3^{a}
k	1	1
	1	1
T_{B}	1941	1938
	1937	1938
P_{T}	8.3^{b}	10.3^{c}
	12.5^{d}	10.3^{b}
k	1	1
	1	1
T_{B}	1931	1940
	1931	1940
ADF	-5.1^{b}	-4.6^{c}
	-5.1^{a}	-4.6^{b}
k	1	1
	1	1
T_{B}	1937	1938
	1937	1938
$\hat{\alpha}$	0.666	0.619
	0.666	0.619

Notes: (1) For the applications, we impose a minimal value $k = 1$; (2) the superscripts a, b, c and d denote significance levels at the 1.0%, 2.5%, 5.0%, and 10.0%, respectively.

Table 7. Empirical results for the real wages and stock prices series choosing the break point maximizing $|t_{\hat{\beta}_2}|$.

Series	T	Criteria	MZ_α	MZ_t	ADF	k	T_{B}	$\hat{a}$
Stock prices	100	*BIC*	-31.9^{b}	-3.9^{b}	-4.1^{c}	1	1931	0.753
		MAIC	-21.2^{d}	-3.2^{d}	-3.2	1	1931	0.793
Real wages	71	*BIC*	-27.7^{c}	-3.6^{c}	-3.8^{d}	1	1933	0.697
		MAIC	-27.7^{a}	-3.6^{a}	-3.8^{c}	1	1933	0.697

Notes: (1) For the applications, we impose a minimal value $k = 1$; (2) the superscripts a, b, c and d denote significance levels at the 1.0%, 2.5%, 5.0%, and 10.0%, respectively.

(except for the ADF^{GLS} test using *BIC* to select the autoregressive order).

8. Conclusions

We considered tests for the null hypothesis of a unit root in the presence of a one time change in the trend function and followed Elliott *et al.* (1996) and Dufour and King (1991) by detrending the data using a local to unity GLS approach. The extensions of the *ADF* and the P_T as well as of the various *M* tests suggested by Perron and Ng (1996) were studied. We also investigated the properties of the tests when the break point is selected either by minimizing the tests or by maximizing the absolute value of the t-statistic on the coefficient of the change in slope. All tests have a local asymptotic power function that lies close to the Gaussian power envelope though our simulations reveal that, in finite sample, the latter method yields tests with less power. Hence, for applications we recommend using either the ADF^{GLS} or the M^{GLS} or P_T^{GLS} tests with the break point selected by minimizing the tests. The main difference among the tests is that the ADF^{GLS} has worse size distortions in the negative *MA* case but better power in the negative *AR* case; the M^{GLS} and P_T^{GLS} tests have good size overall but very little power in the negative *AR* case. The choice between the two depends on the investigator's assessment of the likely importance of one or the other class of processes in the data considered. Our experiments also suggest that the use of the *MAIC* to select the autoregressive truncations lag leads to tests with better overall properties.

Appendix A

Throughout, we use the following lemma which is by now standard.

Lemma A.1. *Let* $\{u_t\}$ *be a near-integrated series generated by* (2). *Then, we have* (a) $T^{-1/2}u_{[Tr]} \Rightarrow \sigma Wc(r)$; (b) $T^{-3/2}\sum_{t=1}^{T} u_t \Rightarrow \sigma \int_0^1 W_c(r)dr$; (c) $T^{-2}\sum_{t=1}^{T} u_t^2 \Rightarrow \sigma^2 \int_0^1 W_c^2(r)dr$; (d) $T^{-1}\sum_{t=1}^{T} u_{t-1}v_t \Rightarrow \sigma^2\{\int_0^1 W_c(r)dW(r)+\gamma\}$ *with* $\gamma = (\sigma^2-\sigma_v^2)/2\sigma^2$.

We start with results concerning the limit of the estimates of the coefficients of the trend function obtained from (3). For Model I, we have

Lemma A.2. *Suppose that* y_t *is generated by* (1) *with* $\alpha = 1 + c/T$ *and* $\{z_t\}$ *is given by* (4). *Let* $\hat{\psi}(\delta)$ *be the GLS estimates, from minimizing* (3), *of the coefficients of the trend function obtained using* $\bar{\alpha} = 1 + \bar{c}/T$. *Then, with terms as defined in Theorem* 1:

$$\begin{aligned}
&\hat{\mu}_1 - \mu_1 \Rightarrow v_1,\\
&T^{1/2}(\hat{\beta}_1 - \beta_1) \Rightarrow \sigma(\lambda_1 b_1 + \lambda_2 b_2) \equiv \sigma b_3,\\
&T^{1/2}(\hat{\beta}_2 - \beta_2) \Rightarrow \sigma(\lambda_2 b_1 + \lambda_3 b_2) \equiv \sigma b_4,
\end{aligned}$$

Proof of Lemma A.2. In matrix notation, we have

$$\begin{aligned}
\hat{\psi}(\delta) - \psi = {} & [(\Delta z - \bar{c}T^{-1}z_{-1})'(\Delta z - \bar{c}T^{-1}z_{-1})]^{-1}\\
& \times [(\Delta z - \bar{c}T^{-1}z_{-1})'(\Delta u - \bar{c}T^{-1}u_{-1})], \qquad \text{(A.1)}
\end{aligned}$$

where

$$\begin{aligned}
\Delta z &= (z_1, z_2 - z_1, \ldots, z_T - z_{T-1}),\\
z_{-1} &= (0, z_1, z_2, \ldots z_{T-1}),\\
\Delta u &= (u_1, u_2 - u_1, \ldots, u_T - u_{T-1})\\
u_{-1} &= (0, u_1, u_2, \ldots u_{T-1}).
\end{aligned}$$

Now define the scaling matrix $\Upsilon_T = diag(1, T^{1/2}, T^{1/2})$, we can write expression (A.1) as

$$\Upsilon_T(\hat{\psi}(\delta) - \psi) = \Gamma_T(\delta)^{-1}\Psi_T(\delta), \tag{A.2}$$

where

$$\Gamma_T(\delta) = \Upsilon_T^{-1}[(\Delta z - \bar{c}T^{-1}z_{-1})'(\Delta z - \bar{c}T^{-1}z_{-1})]\Upsilon_T^{-1},$$

and

$$\Psi_T(\delta) = \Upsilon_T^{-1}[(\Delta z - \bar{c}T^{-1}z_{-1})'(\Delta u - \bar{c}T^{-1}u_{-1})].$$

We first consider the limit of each element of the matrix $\Gamma_T(\delta)$ denoted $\Gamma_{ij}(i, j = 1, 2, 3)$. We let $\Delta z_{(i)}$ and $z_{-1(i)}$ be the ith element of the vectors Δz and z_{-1}, respectively. We have

$$\begin{aligned}
\Gamma_{11} &= (\Delta z_{(1)} - \bar{c}T^{-1}z_{-1(1)})'(\Delta z_{(1)} - \bar{c}T^{-1}z_{-1(1)}) \Rightarrow 1,\\
\Gamma_{12} &= T^{-1/2}(\Delta z_{(1)} - \bar{c}T^{-1}z_{-1(1)})'(\Delta z_{(2)} - \bar{c}T^{-1}z_{-1(2)}) \Rightarrow 0,\\
\Gamma_{13} &= T^{-1/2}(\Delta z_{(1)} - \bar{c}T^{-1}z_{-1(1)})'(\Delta z_{(3)} - \bar{c}T^{-1}z_{-1(3)}) \Rightarrow 0,\\
\Gamma_{22} &= T^{-1}(\Delta z_{(2)} - \bar{c}T^{-1}z_{-1(2)})'(\Delta z_{(2)} - \bar{c}T^{-1}z_{-1(2)})\\
&\quad \Rightarrow 1 - \bar{c} + \bar{c}^2/3 \equiv a,\\
\Gamma_{23} &= T^{-1}(\Delta z_{(2)} - \bar{c}T^{-1}z_{-1(2)})'(\Delta z_{(3)} - \bar{c}T^{-1}z_{-1(3)})\\
&\quad \Rightarrow 1 - \delta - \bar{c} + \bar{c}\delta - (\bar{c}^2/2)\delta + (\bar{c}^2/2)\delta^3 + (\bar{c}^2/3)(1 - \delta^3) \equiv m,\\
\Gamma_{33} &= T^{-1}(\Delta z_{(3)} - \bar{c}T^{-1}z_{-1(3)})'(\Delta z_{(3)} - \bar{c}T^{-1}z_{-1(3)})\\
&\quad \Rightarrow 1 - \delta - \bar{c} + 2\bar{c}\delta - \bar{c}\delta^2 - \bar{c}^2\delta + \bar{c}^2\delta^2 + (\bar{c}^2/3)(1 - \delta^3) \equiv d.
\end{aligned}$$

We next consider the limit of each element of the vector $\Psi_T(\delta)$, denoted $\Psi_i(i = 1, 2, 3)$. We have

$$\begin{aligned}
\Psi_1 &= (\Delta z_{(1)} - \bar{c}T^{-1}z_{-1(1)})'(\Delta u - \bar{c}T^{-1}u_{-1}) \Rightarrow v_1,\\
\Psi_2 &= T^{-1/2}(\Delta z_{(2)} - \bar{c}T^{-1}z_{-1(2)})'(\Delta u - \bar{c}T^{-1}u_{-1})\\
&\quad \Rightarrow \sigma\left[W_c(1)(1 - \bar{c}) + \bar{c}^2\int_0^1 rW_c(r)dr\right] \equiv \sigma b_1,
\end{aligned}$$

$$\Psi_3 = T^{-1/2}(\Delta z_{(3)} - \bar{c}T^{-1}z_{-1(3)})'(\Delta u - \bar{c}T^{-1}u_{-1})$$
$$\Rightarrow \sigma \left[W_c(1)(1 - \bar{c} + \delta\bar{c}) + \bar{c}^2 \int_\delta^1 W_c(r)(r-\delta)dr - W_c(\delta) \right] \equiv \sigma b_2.$$

Hence, using the symmetry of $\Gamma_T(\delta)$,

$$\Upsilon_T(\hat{\Psi}(\delta) - \Psi) \Rightarrow \begin{bmatrix} 1 & 0 & 0 \\ 0 & a & m \\ 0 & m & d \end{bmatrix}^{-1} \begin{bmatrix} v_1 \\ \sigma b_1 \\ \sigma b_1 \end{bmatrix}.$$

□

The proof of the lemma follows upon solving for the inverse. For Model II, we have

Lemma A.3. *Suppose that y_t is generated by (1) with $\alpha = 1 + c/T$ and $\{z_t\}$ is given by (5). Let $\hat{\psi}(\delta)$ be the GLS estimates, from minimizing (3), of the coefficients of the trend function obtained using $\bar{\alpha} = 1 + \bar{c}/T$. Then, the result of Lemma A.2 still apply with the addition that $\hat{\mu}_2 - \mu_2 \Rightarrow \lim_{T\to\infty} v_{[T\delta]+1} \equiv v^*$.*

The proof of Lemma A.3 is basically the same as that of Lemma A.2 and, hence, omitted.

Proof of Theorem 1. The proof uses the results of Lemmas A.2 and A.3. We show the proof only for Model I and for the $MZ_\alpha^{\text{GLS}}(\delta)$ test, the proof for the other Model and tests follows analogously. We first have

$$T^{-1}\tilde{y}_T^2 = T^{-1}\{y_T - (\hat{\mu}_1 + \hat{\beta}_1 T + \hat{\beta}_2(T - T\delta))\}^2$$
$$= T^{-1}\{u_T - [(\hat{\mu}_1 - \mu_1) + (\hat{\beta}_1 - \beta_1)T + (\hat{\beta}_2 - \beta_2)(T - T\delta)]\}^2.$$

After some algebra, we obtain: $T^{-1}u_T^2 \Rightarrow \sigma^2 W_c(1)^2$; $2T^{-1}u_T(\hat{\mu}_1 - \mu_1) \Rightarrow 0$,

$$2T^{-1}u_T(\hat{\beta}_1 - \beta_1)T \Rightarrow 2\sigma^2 b_3 W_c(1),$$
$$2T^{-1}u_T(\hat{\beta}_2 - \beta_2)(T - T\delta) \Rightarrow 2\sigma^2 b_4 W_c(1)(1-\delta),$$
$$2T^{-1}(\hat{\mu}_1 - \mu_1)(\hat{\beta}_1 - \beta_1)T \Rightarrow 0,$$
$$T^{-1}(\hat{\beta}_1 - \beta_1)^2 T^2 \Rightarrow \sigma^2 b_3^2,$$

$$2T^{-1}(\hat{\mu}_1 - \mu_1)(\hat{\beta}_2 - \beta_2)(T - T\delta) \Rightarrow 0,$$
$$2T^{-1}(\hat{\beta}_1 - \beta_1)T(\hat{\beta}_2 - \beta_2)(T - T\delta) \Rightarrow 2\sigma^2 b_3 b_4(1-\delta),$$
$$T^{-1}(\hat{\beta}_2 - \beta_2)^2(T - T\delta)^2 \Rightarrow \sigma^2 b_4^2(1-\delta)^2.$$

Using these results, we have

$$T^{-1}\tilde{y}_T^2 \Rightarrow \sigma^2\{V_{c\bar{c}}^{(1)}(1,\delta)^2 - 2V_{c\bar{c}}^{(2)}(1,\delta)\}, \tag{A.3}$$

where $V_{c\bar{c}}^{(1)}(1,\delta) = W_c(1) - b_3$, and

$$V_{c\bar{c}}^{(2)}(1,\delta) = b_4(1-\delta)[W_c(1) - b_3 - (1/2)(1-\delta)b_4].$$

Consider now the term $2T^{-2}\sum_{t=1}^T \tilde{y}_t^2$, defined by

$$\begin{aligned} 2T^{-2}\sum_{t=1}^T \tilde{y}_t^2 &= 2T^{-2}\sum_{t=1}^T \{y_t - [\hat{\mu}_1 + \hat{\beta}_1 t + \hat{\beta}_2 1(t > T\delta)(t - T\delta)]\}^2 \\ &= 2T^{-2}\sum_{t=1}^T \{u_t - [(\hat{\mu}_1 - \mu_1) + (\hat{\beta}_1 - \beta_1)t \\ &\qquad + (\hat{\beta}_2 - \beta_2)1(t > T\delta)(t - T\delta)]\}^2. \end{aligned}$$

After some algebra, we obtain

$$2T^{-2}\sum_{t=1}^T u_t^2 \Rightarrow 2\sigma^2\int_0^1 W_c(r)^2 dr,$$

$$4T^{-2}(\hat{\mu}_1 - \mu_1)\sum_{t=1}^T u_t \Rightarrow 0,$$

$$4T^{-2}(\hat{\beta}_1 - \beta_1)\sum_{t=1}^T t u_t \Rightarrow 4\sigma^2\int_0^1 r b_3 W_c(r) dr,$$

$$4T^{-2}(\hat{\beta}_2 - \beta_2)\sum_{t=1}^T 1(t > T\delta)(t - T\delta)u_t \Rightarrow 4\sigma^2\int_\delta^1 b_4 W_c(r)(r-\delta) dr,$$

$$2T^{-1}(\hat{\mu}_1 - \mu_1)^2 \Rightarrow 0,$$

$$4T^{-2}(\hat{\mu}_1 - \mu_1)(\hat{\beta}_1 - \beta_1)\sum_{t=1}^{T} t \Rightarrow 0,$$

$$2T^{-2}(\hat{\beta}_1 - \beta_1)^2\sum_{t=1}^{T} t^2 \Rightarrow 2\sigma^2 \int_0^1 b_3^2 r^2 dr,$$

$$4T^{-2}(\hat{\mu}_1 - \mu_1)(\hat{\beta}_2 - \beta_2)\sum_{t=1}^{T} 1(t > T\delta)(t - T\delta) \Rightarrow 0,$$

$$4T^{-2}(\hat{\beta}_1 - \beta_1)(\hat{\beta}_2 - \beta_2)\sum_{t=1}^{T} t1(t > T\delta)(t - T\delta)$$

$$\Rightarrow 4\sigma^2 \int_\delta^1 b_3 b_4 r(r - \delta)dr,$$

$$2T^{-2}(\hat{\beta}_2 - \beta_2)^2\sum_{t=1}^{T} 1(t > T\delta)(t - T\delta)^2 \Rightarrow 2\sigma^2 \int_\delta^1 b_4^2(r - \delta)^2 dr.$$

Using these results we have

$$2T^{-2}\sum_{t=1}^{T} \tilde{y}_t^2 \Rightarrow 2\sigma^2 \left\{ \int_0^1 V_{c\bar{c}}^{(1)}(r,\delta)^2 dr - 2\int_\delta^1 V_{c\bar{c}}^{(2)}(r,\delta)dr \right\}. \quad \text{(A.4)}$$

Using (A.3), (A.4) and the fact that s^2 is a consistent estimate of σ^2, the proof is complete. □

Proof of Theorem 2. We first give the proof for Model I. Defining

$$M_T(c,\bar{c},\delta) = (u^{\alpha\prime} z^\alpha)(z^{\alpha\prime} z^\alpha)^{-1}(z^{\alpha\prime} u^\alpha),$$

we have $S(\bar{\alpha},\delta) = u^{\bar{\alpha}\prime} u^{\bar{\alpha}} - M_T(c,\bar{c},\delta)$ and $S(1) = u^{1\prime} u^1 - M_T(c,0,\delta)$. Using the fact that

$$u_t^{\bar{\alpha}} = v_t + T^{-1}(c - \bar{c})u_{t-1}$$

for $t = 2, \ldots, T$ and $u_1^{\bar{\alpha}} = v_1$, we have

$$s^2 P_T^{\mathrm{GLS}}(c, \bar{c}, \delta) = M_T(c, 0, \delta) - M_T(c, \bar{c}, \delta) - 2\bar{c}T^{-1}\sum_{t=2}^{T} u_{t-1}v_t + (\bar{c}^2 - 2\bar{c}c)T^{-2}\sum_{t=2}^{T} u_{t-1}^2 - cT^{-1}u^{1\prime}u^1 + o_p(1). \tag{A.5}$$

Note that $T^{-1}\sum_{t=2}^{T} u_{t-1}v_t \Rightarrow \sigma^2\int_0^1 W_c(r)dW(r)$ and $T^{-2}\sum_{t=2}^{T} u_{t-1}^2 \Rightarrow \sigma^2\int_0^1 W_c(r)^2 dr$. Consider now the limit of $M_T(c, \bar{c}, \delta)$. Using the scaling matrix $D_T = diag(1, T^{-1/2}, T^{-1/2})$, we have $M_T(c, \bar{c}, \delta) = (u^{\alpha\prime}z^\alpha D_T)(D_T z^{\alpha\prime}z^\alpha D_T)^{-1}(D_T z^{\alpha\prime}u^\alpha)$. The first term is given by

$$D_T u^{\alpha\prime}z^\alpha = \begin{bmatrix} v_1 + T^{-1}(c-\bar{c})u_0 - \bar{c}T^{-1}\sum_{t=2}^{T}[v_t + (c-\bar{c})T^{-1}u_{t-1}] \\ T^{-1/2}\sum_{t=2}^{T}(v_t + (c-\bar{c})T^{-1}u_{t-1})(-T^{-1}\bar{c}t + 1) + o_p(1) \\ T^{-1/2}\sum_{t=T_B+1}^{T}(v_t + (c-\bar{c})T^{-1}u_{t-1})(-T^{-1}\bar{c}t + 1 + \delta\bar{c}) \end{bmatrix}$$

and its limit is

$$\sigma\begin{bmatrix} v_1/\sigma \\ W(1) + (c-\bar{c})\int_0^1 W_c(r)\mathrm{d}r \\ \quad - \bar{c}\int_0^1 r\mathrm{d}W(r) - (c-\bar{c})\bar{c}\int_0^1 rW_c(r)\mathrm{d}r \\ (1+\delta\bar{c})\left([W(1) - W(\delta)] + (c-\bar{c})\int_\delta^1 W_c(r)\mathrm{d}r\right) \\ \quad - \bar{c}\int_\delta^1 r\mathrm{d}W(r) - (c-\bar{c})\bar{c}\int_\delta^1 rW_c(r)\mathrm{d}r \end{bmatrix}.$$

The term $D_T z^{\alpha'} z^{\alpha} D_T$ is given by

$$\begin{bmatrix} 1+\sum_{t=2}^{T}(-\bar{c}/T)^2 T^{-1/2}\left[1+\sum_{t=2}^{T}(-\bar{c}/T)(-\bar{c}t/T+1)\right] \\ T^{-1/2}\left[\sum_{t=T_B+1}^{T}(-\bar{c}/T)(-\bar{c}t/T+1+\delta\bar{c})\right] \\ T^{-1}\left[1+\sum_{t=2}^{T}(-\bar{c}t/T+1)^2\right] \\ T^{-1}\sum_{t=T_B+1}^{T}(-\bar{c}t/T+1)(-\bar{c}t/T+1+\delta\bar{c}) \\ T^{-1}\sum_{t=T_B+1}^{T}(-\bar{c}t/T+1+\delta\bar{c})^2 \end{bmatrix}$$

and its limit is

$$\begin{bmatrix} 1 & 0 & 0 \\ & \bar{c}^2/3-\bar{c}+1 & (1-\delta)(1-\bar{c})+\bar{c}^2(2+\delta^3-3\delta)/6 \\ & & \bar{c}^2(1-\delta^3)/3-\bar{c}(1-\delta^2)(1+\delta\bar{c})+(1-\delta)(1+\delta\bar{c})^2 \end{bmatrix}.$$

Simple algebra shows that $M_T(c,\bar{c},\delta) \Rightarrow v_1^2 + \sigma^2 M(c,\bar{c},\delta)$ and the result of the theorem follows using (A.5). The proof for Model II is entirely analogous and, hence, omitted. The result stated in (16) follows using the fact that $S(\bar{\alpha},\delta) = u^{\bar{\alpha}'}u^{\bar{\alpha}} - M_T(c,\bar{c},\delta)$ which depends on δ only through $M_T(c,\bar{c},\delta)$ which enters with a negative sign, hence taking the supremum instead of the infimum. □

Acknowledgments

This paper is drawn from Chapter 1 of Gabriel Rodríguez's Ph.D. Dissertation at the Université de Montréal (Rodríguez, 1999). We wish to thank Alain Guay for useful comments when this paper was presented at the 38th annual meeting of the Société Canadienne de Sciences Economiques. Perron acknowledges financial support from the Fonds pour la Formation de Chercheurs et l'Aide à la Recherche du Québec (FCAR).

References

Andrews, D.W.K. (1993). Tests for parameter instability and structural change with unknown change point. *Econometrica* 61, 821–856.

Banerjee, A., Lumsdaine, R. and Stock, J.H. (1992). Recursive and sequential tests of the unit root and trend break hypothesis: Theory and international evidence. *Journal of Business and Economics Statistics* 10, 271–287.

Bhargava, A. (1986). On the theory of testing for unit root in observed time series. *Review of Economic Studies* 53, 369–384.

Campbell, J.Y. and Mankiw, N.G. (1987a). Permanent and transitory components in macroeconomics fluctuations. *American Economic Review: Papers and Proceedings* 77, 111–117.

Campbell, J.Y. and Mankiw, N.G. (1987b). Are output fluctuations transitory? *Quarterly Journal of Economics* 102, 857–880.

Campbell, J.Y. and Perron, P. (1991). Pitfalls and Opportunities: What Macroeconomists Should Know About Unit Roots. In *NBER Macroeconomics Annual*, Blanchard, O.J., Fischer, S. (eds.), Vol. 6, pp. 141–201.

Christiano, L.J. (1992). Searching for a break in GNP. *Journal of Business and Economics Statistics* 10, 237–250.

Clark, P.K. (1987). The cyclical component of U.S. economic activity. *Quarterly Journal of Economics* 102, 798–814.

Cochrane, J.H. (1988). How big is the random walk in GNP? *Journal of Political Economy* 96, 893–920.

Dickey, D.A. and Fuller, W.A. (1979). Distribution of the estimator for autoregressive time series with a unit root. *Journal of the American Statistical Association* 74, 427–431.

Dufour, J.-M. and King, M. (1991). Optimal invariant tests for the autocorrelation coefficient in linear regressions with stationary or nonstationary errors. *Journal of Econometrics* 47, 115–143.

Elliott, G., Rothenberg, T. and Stock, J.H. (1996). Efficient tests for an autoregressive unit root. *Econometrica* 64, 813–839.

Nelson, C.R. and Plosser, C.I. (1982). Trends and random walks in macroeconomics time series: Some evidence and implications. *Journal of Monetary Economics* 10, 139–162.

Ng, S. and Perron, P. (1995). Unit root tests in ARMA models with data dependent methods for the selection of the truncation lag. *Journal of the American Statistical Association* 90, 268–281.

Ng, S. and Perron, P. (2001). Lag length selection and the construction of unit root tests with good size and power. *Econometrica* 69, 1519–1554.

Perron, P. (1989). The great crash, the oil price shock and the unit root hypothesis. *Econometrica* 57, 1361–1401.

Perron, P. (1997). Further evidence of breaking trend functions in macroeconomic variables. *Journal of Econometrics* 80, 355–385.

Perron, P. and Ng, S. (1996). Useful modifications to some unit root tests with dependent errors and their local asymptotic properties. *Review of Economics Studies* 63, 435–463.

Perron, P. and Ng, S. (1998). An autoregressive spectral density estimator at frequency zero for nonstationarity tests. *Econometric Theory* 14, 560–603.

Phillips, P.C.B. (1987). Time series regression with unit roots. *Econometrica* 55, 277–302.

Phillips, P.C.B. and Perron, P. (1988). Testing for a unit root in time series regression. *Biometrika* 75, 335–346.

Rodríguez, G. (1999). Unit root, outliers and cointegration analysis with macroeconomic applications. Unpublished Ph.D. Dissertation, Départment de Sciences Économiques, Université de Montréal.

Said, S.E. and Dickey, D.A. (1984). Testing for unit roots in autoregressive-moving average models of unknown order. *Biometrika* 71, 599–608.

Sargan, J.D. and Bhargava, A. (1983). Testing residuals from least squares regression for being generated by the Gaussian random walk. *Econometrica* 51, 153–174.

Shapiro, M. and Watson, M. (1988). Sources and Business Cycles Fluctuations. In *NBER Macroeconomics Annual*, Fischer, S. (ed.), Vol. 3, pp. 111–148.

Stock, J.H. (1994). Unit Roots, Structural Breaks and Trends. In *Handbook of Econometrics*, McFaden, D.L., Engle, R.F. (eds.), Vol. IV, Elsevier Science, Amsterdam, pp. 2740–2841.

Stock, J.H. (1999). A Class of Tests for Integration and Cointegration. In: *Cointegration, Causality and Forecasting. A Festschrift in Honour of Clive W.J. Granger*, Engle, R.F., White, H. (eds.), Oxford University Press, Oxford, pp. 137–167.

Vogelsang, T.J. and Perron, P. (1998). Additional tests for a unit root allowing the possibility of breaks in the trend function. *International Economic Review* 39, 1073–1100.

Zivot, E. and Andrews, D.W.K. (1992). Further evidence on the great crash, the oil-price shock and the unit root hypothesis. *Journal of Business and Economics Statistics* 10, 251–270.

Chapter 19

Unit Root Tests Allowing for a Break in the Trend Function at an Unknown Time Under Both the Null and Alternative Hypotheses*

Dukpa Kim[†,§] and Pierre Perron[‡,¶]

[†]*Department of Economics, University of Virginia, 2015 Ivy Road, Charlottesville, VA 22903, USA*
[‡]*Department of Economics, Boston University, 270 Bay State Road, Boston, MA 02215, USA*
[§]*dk4p@virginia.edu*
[¶]*perron@bu.edu*

Perron [Perron, P., 1989. The great crash, the oil price shock and the unit root hypothesis. *Econometrica* 57, 1361–1401] introduced a variety of unit root tests that are valid when a break in the trend function of a time series is present. The motivation was to devise testing procedures that were invariant to the magnitude of the shift in level and/or slope. In particular, if a change is present it is allowed under both the null and alternative hypotheses. This analysis was carried under the assumption of a known break date. The subsequent literature aimed to devise testing procedures valid in the case of an unknown break date. However, in doing so, most of the literature and, in particular the commonly used test of Zivot and Andrews [Zivot, E., Andrews, D.W.K., 1992. Further evidence on the great crash, the oil price shock and the unit root hypothesis. *Journal of Business and Economic Statistics* 10, 251–270], assumed that if a break occurs, it does so only under the alternative hypothesis of stationarity. This is undesirable since (a) it imposes an asymmetric

*Originally Published in *Journal of Econometrics*, 148, pp. 1–13.

treatment when allowing for a break, so that the test may reject when the noise is integrated but the trend is changing; (b) if a break is present, this information is not exploited to improve the power of the test. In this chapter, we propose a testing procedure that addresses both issues. It allows a break under both the null and alternative hypotheses and, when a break is present, the limit distribution of the test is the same as in the case of a known break date, thereby allowing increased power while maintaining the correct size. Simulation experiments confirm that our procedure offers an improvement over commonly used methods in small samples.

Keywords: Structural change; pretest; trend function; integrated processes; hypothesis testing.

1. Introduction

It is well known that a break in the deterministic trend affects the outcome of unit root tests. Perron (1989) showed that a Dickey and Fuller (1979) (DF) type unit root test is not consistent if the alternative is that of a stationary noise component with a break in the slope of the deterministic trend. His main point is that the existence of an exogenous shock which has a permanent effect will lead to a nonrejection of the unit root hypothesis even though it is not true.[1] Perron (1989, 1990) proposed alternative unit root tests which allow the possibility of a break under both the null and alternative hypotheses. They have less power than a DF test when there is no break but they have a correct size asymptotically and are consistent whether there is a break or not. Moreover, they are invariant to the break parameters and thus their performance does not depend on the magnitude of the break. However, the most controversial assumption is that its timing is known *a priori* (Christiano, 1992).

[1]See also Montañés and Reyes (1998,1999,2000) who examined the asymptotic behavior of the Augmented Dickey–Fuller test (Dickey and Fuller, 1979; Said and Dickey, 1984) and the Phillips-Perron (Phillips and Perron, 1988) test under the crash alternative hypothesis. Leybourne *et al.* (1998) and Leybourne and Newbold (2000) analyzed the effect of a break on a standard DF test under the unit root null hypothesis and showed that size distortions can occur, especially when the break is early in the sample.

In order for Perron's (1989 1990) test procedures to be valid, the break date should be chosen independently of the given data. Whenever a systematic search for a break is done, the limiting distributions in Perron (1989, 1990) are no longer appropriate. Historical facts can often be a good guidance in choosing a break date independently of the given data. Even in that case, it is very likely that an imprecise break date is used. Hecq and Urbain (1993) showed, by simulations, that the use of an incorrect break date in Perron's (1990) tests, applicable with nontrending data, causes size distortions and power loss, though as shown by Montañés (1997), this effect disappears asymptotically. Montañés and Olloqui (1999) extended the analysis to Perron's (1989) tests, applicable with trending data, and showed that a loss of power occurs even in large samples. Kim *et al.* (2000) examined the effect of using a wrong break date under the null hypothesis.

The work by Zivot and Andrews (1992), Perron (1997) and Vogelsang and Perron (1998), among others, provide methods that treat the occurrence of the break date as unknown. In this line of work, a break is not allowed under the null hypothesis, only under the alternative. This means, for example, that under the null hypothesis a level shift must be viewed as coming from the tail of the distribution of the data generating process, and a slope change involves errors with a different mean in some sub-samples. This framework is convenient since it allows them to establish various unit root testing procedures; for example by minimizing the t-statistic related to the sum of the autoregressive coefficients over each possible break date or computing this t-statistic at the break date that minimizes (or maximizes) the t-statistic associated with the break parameter (or maximizes its absolute value). This approach is, however, contrary to Perron's (1989) original motivation, which was to devise testing procedures that were invariant to the magnitude of the shift in level and/or slope. In particular, if a change is present it is allowed under both the null and alternative hypotheses.

The existence of a structural break in the trend function is a problem of long horizon data; it can happen whether the noise component is stationary or has a unit root. As Vogelsang and

Perron (1998) showed, if the noise component has a unit root and a break occurs in the trend function, Zivot and Andrews (1992) tests often diverge or are not invariant to break parameters (see also, Nunes *et al.*, 1997 and Lee and Strazicich, 2003). This is a natural consequence of not permitting a break under the null hypothesis. An added consequence is that this type of tests have substantially less power than Perron's (1989) tests, because they do not fully utilize the information about the break, when one is present.

It is clear that Perron's (1989, 1990) test has the most desirable property with respect to allowing a break under both the null and alternative hypotheses and having tests that are invariant to parameters of the trend function. This robustness is attained at the cost of the need to assume a known break date. This motivates us to develop new test procedures, which allow a break in the trend function at an unknown time under both the null and alternative hypotheses. Also, when a break is present, the limit distribution of the test is the same as in the case of a known break date, thereby allowing increased power while maintaining the correct size. Simulation experiments confirm that our procedure offers an improvement over commonly used methods in small samples.

It is important to remark that our work is not an extension of Zivot and Andrews (1992). On the contrary, we argue that the testing problem they considered is inadequate and can lead to misleading results. In order to allow the possibility of a break under both the null and alternative hypotheses, a completely different approach is required. This is now possible given recent advances related to structural change problems in models with trending variables, in particular those of Perron and Zhu (2005) and Perron and Yabu (2005).

The rest of the paper is organized as follows. In Section 2, we describe the models in detail. In Section 3, we analyze unit root tests which use an estimate of the break date instead of the true one. The estimates considered are based on minimizing the sum of squared residuals from the appropriate regression. We derive sufficient conditions for the limiting distributions of the test statistics to be the same as those occurring with a known break date. In Section 4, for the cases where the available estimate of the break fraction does not satisfy the sufficient condition, we propose a test procedure that uses trimming of the data around the estimated break date, which has the effect of increasing the rate of convergence of the estimate so that the limit distribution of the unit root test is the same as in the known break date case. In Section 5, we discuss an extension of the proposed test procedure that is robust to the case where no break occurs under the null hypothesis. Our approach is to use a pretest for a break that is valid whether the noise component is integrated or stationary (see Perron and Yabu, 2005). In Section 6, we present simulation results about the finite sample properties of the proposed test statistics. Section 7 considers an empirical application to real GDP series. Section 8 offers brief concluding remarks and an appendix some technical derivations. Note finally that our exposition pertains to the case of a single break, for ease of exposition. All results and procedures suggested are, however, still valid in the multiple break case with minor modifications.

2. The Data Generating Processes

We consider a univariate process $\{y_t\}$ generated by one of three additive outlier models or two innovational outlier models. For each model, the series is generated by the sum of a deterministic time trend and an error process. The deterministic time trend has a one time break at an unknown date in the intercept, slope, or both depending on the model. The Data Generating Processes (DGP) of the additive outlier (AO) models are given by

$$y_t = z(T_1)_t'\phi + u_t = z_{t,1}'\phi_1 + z(T_1)_{t,2}'\phi_2 + u_t,$$

where $z_{t,1} = (1, t)', \phi_1 = (\mu, \beta)'$,

$$\begin{aligned} z(T_1)_{t,2} &= \begin{cases} DU_t & \text{for Model A1} \\ B_t & \text{for Model A2} \\ (DU_t,\ B_t)' & \text{for Model A3,} \end{cases} \\ \phi_2 &= \begin{cases} \mu_b & \text{for Model A1} \\ \beta_b & \text{for Model A2} \\ (\mu_b, \beta_b)' & \text{for Model A3} \end{cases} \end{aligned} \tag{1}$$

with $DU_t = B_t = 0$ if $t \leq T_1$, and $DU_t = 1$, $B_t = t - T_1$, if $t > T_1$. Here $\mathrm{T}_1 = \lambda^c T$, with $0 < \lambda^c < 1$, denotes the true break date (and λ^c the true break fraction). Note that DU_t and B_t depend on T_1 and T but this dependency is omitted. The noise $\{u_t\}$ is such that $A(L)u_t = B(L)\varepsilon_t$ where $\varepsilon_t \sim$ i.i.d. $(0, \sigma_\varepsilon^2)$, and $A(L)$ and $B(L)$ are polynomials in L of order $p + 1$ and q, respectively. We factor $A(L)$ as $(1 - \alpha L)A^*(L)$ and it is assumed that $A^*(L)$ and $B(L)$ have roots strictly outside the unit circle. The null and alternative hypotheses are H_0: $\alpha = 1$ and $H_1 : |\alpha| < 1$, respectively. The ARMA specification could be relaxed to allow more general processes, but we shall use it to ease the presentation. The DGP of the innovational outlier (IO) models under the null hypothesis are given by:

$$y_t = y_{t-1} + \beta + \Psi^*(L)(d(T_1)'_{t,2}\phi_2 + \varepsilon_t),$$

where

$$d(T_1)_{t,2} = \begin{cases} D(T_1)_t & \text{for Model I1} \\ (D(T_1)_t, DU_t)' & \text{for Model I3} \end{cases}$$

and $D(T_1)_t = 1$ if $t = T_1 + 1$ and 0 otherwise. Under the alternative hypothesis, we have:

$$y_t = z'_{t,1}\phi_1 + \Psi(L)(z(T_1)'_{t,2}\phi_2 + \varepsilon_t),$$

where

$$Z(T_1)_{t,2} = \begin{cases} DU_t & \text{for Model I1} \\ (DU_t, B_t)' & \text{for Model I3,} \end{cases}$$

$$\phi_2 = \begin{cases} \mu_b, & \text{for Model I1} \\ (\mu_b, \beta_b)' & \text{for Model I3} \end{cases}$$

with $\Psi^*(L)$ and $\Psi(L)$ such that $\Psi^*(L) = A^*(L)^{-1}B(L)$ and $(1 - \alpha L)^{-1}\Psi^*(L) = \Psi(L)$.

Note that Models A1, A2, A3 (with $\mu_b = c + \beta_b T_1$) I1 and I3 (with $\mu_b = c + \beta_b T_1$) are the same as those in Perron (1989), except that we shall treat the break date as unknown. Models A1 and I1 are called 'level shift' models, A2 'changing growth' model, and A3 and I3 are 'mixed' models. A changing growth model of the IO type is not considered, because it is necessary to assume that no break occurs under the null hypothesis to obtain Perron's (1989) result, which is contrary to our stated objective. Perron's (1990) models with no trending component are also not considered explicitly in this paper. Since they correspond to Model A1 and I1 with $\beta = 0$, most of the results pertaining to Model A1 and I1 will apply to Perron's (1990) nontrending models as well.

3. The Limit Properties of Unit Root Tests With an Estimate of the Break Fraction

One notable feature of the models is that the regressors depend on an unknown parameter, T_1 Hence, one cannot even construct correct regressors when the break date is unknown. Instead, one can naturally think of using an estimate of the break date. We examine in this section a unit root test procedure which is the same as that of Perron (1989) except that an estimate of the break date is used instead of the true one.

Consider first the AO models. Perron's (1989) test procedure first detrends $\{y_t\}$ using the deterministic components $\{z(T_1)_t\}$ and then tests the unit root hypothesis using the t-statistic on the sum of the autoregressive coefficients α in the following regressions where $\{\tilde{y}_t\}$ denotes the detrended series:

$$\tilde{y}_t = \hat{\alpha}\tilde{y}_{t-1} + \sum_{j=0}^{k} \hat{w}_j D(T_1)_{t-j} + \sum_{j=1}^{k} \hat{d}_j \Delta\tilde{y}_{t-j} + \hat{u}_t \quad \text{(Models A1 and A3)}, \tag{2}$$

$$\tilde{y}_t = \hat{\alpha}\tilde{y}_{t-1} + \sum_{j=1}^{k} \hat{d}_j \Delta\tilde{y}_{t-j} + \hat{u}_t \quad \text{(Model A2)}.$$

For the IO models, Perron (1989) uses the t-statistic on the sum of the autoregressive coefficients α in the following regressions:

$$y_t = \hat{\alpha}y_{t-1} + \hat{\mu} + \hat{\beta}t + \hat{d}D(T_1)_t + \hat{\mu}_b DU_t + \sum_{i=1}^{k} \hat{d}_i \Delta y_{t-i} + \hat{u}_t \quad \text{(Model I1)}, \tag{3}$$

$$y_t = \hat{\alpha} y_{t-1} + \hat{\mu} + \hat{\beta} t + \hat{d} D(T_1)_t + \hat{\mu}_b DU_t + \hat{\beta}_b B_t$$
$$+ \sum_{i=1}^{k} \hat{d}_i \Delta y_{t-i} + \hat{u}_t \quad \text{(Model I3)}.$$

These equations nest the DGPs implied by the null and alternative hypotheses of each model. Note that in order to obtain the required asymptotic distribution, the truncation lag k is such that $k \to \infty$ and $k^3/T \to 0$, as $T \to \infty$ (see Ng and Perron, 1995). The limiting distributions of the unit root tests with a known break date, labeled $t_\alpha(\lambda^c)$, are given in Perron (1989) and Perron and Vogelsang (1994). Denote the limit result for any given model by $t_\alpha(\lambda^c) \Rightarrow R(\lambda^c)$, where '$\Rightarrow$' denotes weak convergence under the Skorohod metric. We use the notation $t_\alpha(\lambda^c)$ to indicate the fact that the correct break date, $T_1 = \lambda^c T$ is used. When using an estimate, say $\hat{\lambda}$, we denote the relevant t-statistic by $t_\alpha(\hat{\lambda})$. We are interested in conditions under which we get the same limiting distribution for the unit root test when using $\hat{\lambda}$ as we do when using the true break date λ^c, i.e., the conditions such that

$$t_\alpha(\hat{\lambda}) \Rightarrow R(\lambda^c). \tag{4}$$

When this condition holds, one can use the critical values corresponding to the case of a known break date conditioning on the break date being $\hat{\lambda}$, i.e., using the quantiles of $R(\hat{\lambda})$.

3.1. *Limit results pertaining to estimates of the break date when $\beta_b \neq 0$*

We first consider cases with a change in slope, i.e., $\beta_b \neq 0$. We shall consider in Section 6, cases with $\beta_b = 0$ and $\mu_b \neq 0$ so that only a level shift occurs. Given that we are concerned about the limit distributions of the test statistics under the null hypothesis of a unit root, a requirement for an estimate of the break fraction λ^c to be consistent is that $\lambda^c \in (0, 1)$, i.e., that a break actually occurs. If no break occurs, the limit distribution of estimate of the break fraction, say based on minimizing the sum of squared residuals of the relevant regression, will not converge to 0 or 1, but to a random variable having support on the interval [0, 1], see Nunes *et al.* (1995) and Bai (1998). Hence, we shall first make the following assumption:

Assumption 1. There exists a break, i.e., $\lambda^c \in (0,1)$, and $\beta_b \neq 0$.

Requiring that Assumption 1 be satisfied is an important limitation because Perron's (1989, 1990) tests do not require a break to occur, they only allow for such a possibility. Hence, this assumption will be relaxed latter. Several methods to select the break date have been considered in the literature; for example minimizing the relevant unit root test or using the t or F tests associated with the coefficients of the dummy variables. Little is known, however, about the consistency and the convergence rate of these estimates. On the other hand, such results are now established when the break date is estimated by minimizing the sum of squared residuals (SSR) of the relevant regression (see, in particular, Perron and Zhu, 2005; Hatanaka and Yamada, 1999). Hence, we shall concentrate on this method.

For Models A2 and A3, Perron and Zhu (2005) analyzed the properties of the estimate, say $\hat{\lambda}_1^{AO}$, of the break fraction obtained by minimizing the SSR of the following regression over all possible break dates such that $T_b = \lambda T \in \{\pi T, \ldots, (1-\pi)T\}$ [2]:

$$y_t = z(T_b)_t'\hat{\phi} + \hat{u}_t, \tag{5}$$

where $z(T_b)_t$ is the same as $z(T_1)_t$ in (1) except that a generic break date $T_b = \lambda T$ is used instead of $T_1 = \lambda^c T$. They showed that $(\hat{\lambda}_1^{AO} - \lambda^c) = O_p(T^{-1/2})$.

It is possible to use an approach similar to that of Hatanaka and Yamada (1999) and obtain estimates of the break fraction for the AO models with faster rates of convergence when based on minimizing the SSR of IO-type regressions. These include a lagged dependent

[2]The trimming specified by π is used to ensure the invertibility of the moment matrix of the regressors. Since it can be chosen arbitrarily small, this requirement has no practical impact.

variable and the relevant regressions are:

$$\begin{aligned} y_t &= \hat{\alpha} y_{t-1} + \hat{d}_1 D(T_b)_t + \hat{d}_2 1(t \geq T_b) \\ &\quad + z(T_b)_t' \hat{\phi} + \hat{u}_t \quad \text{(Model A2)}, \\ y_t &= \hat{\alpha} y_{t-1} + \hat{d} D(T_b)_t + z(T_b)_t' \hat{\phi} + \hat{u}_t \quad \text{(Models A1 and A3)}. \end{aligned} \tag{6}$$

The following proposition states the rate of convergence of the estimate $\hat{\lambda}_2^{AO}$ obtained by minimizing the SSR of regression (6) when the DGP is of the AO type.

Proposition 1. *Suppose that $\{y_t\}$ is generated under the null hypothesis of Model* A2 *or* A3 *and that Assumption* 1 *holds. For each model, let $\hat{\lambda}_2^{AO}$ denote the break fraction which minimizes the SSR of regression* (6) *over the possible break dates $T_b \in \{\pi T, \ldots, (1-\pi)T\}$. Then, $(\hat{\lambda}_2^{AO} - \lambda^c) = O_p(T^{-1})$, for Models* A2 *and* A3.

Remark 1. The estimate of the break date obtained by minimizing the SSR of the dynamic regression (6), instead of the static regression (5), has faster convergence rate in the case of Models A2 and A3. However, as discussed by Hatanaka and Yamada (1999), this estimate suffers from a severe downward bias in finite samples in the case of Model A3. It will, however, yield unit root tests with good properties for Models A1 and A2. For Model A3, a trimmed version of $\hat{\lambda}_1^{AO}$ will be shown to be adequate.

We now consider the innovational outlier models, with a break fraction estimator comparable to $\hat{\lambda}^{AO}$. As shown in the Appendix, under the null hypothesis, we can write:

$$y_t = z_{t,1}' \phi_1 + \sum_{i=0}^{k-1} \xi_i D(T_1 + i)_t + z(T_1 + k)_{t,2}' \zeta_2^* + u_t^s, \tag{7}$$

where $\xi_i = \mu_b \sum_{j=0}^{i} \psi_j^* + \beta_b \sum_{j=0}^{i} \psi_j^* (i - j + 1)$, $\zeta_2^* = \zeta_2 + (\sum_{i=1}^{\infty} \psi_{k+i}^*)$ ϕ_2, $\zeta_2' = (\mu_b \sum_{j=0}^{k} \psi_j^* + \beta_b \sum_{j=0}^{k-1} \psi_j^* (k-j),\ \beta_b \sum_{j=0}^{k} \psi_j^*)$, $u_t^s = u_t - s_t$ and $s_t = \sum_{i=1}^{\infty} \psi_{k+i}^* \sum_{j=0}^{i-1} d(T_1 + k + j)_{t,2}' \phi_2$. In fact, this is a more general representation of the IO models because the shock to the trend function and the shock to the error process do not have to evolve in the same way. Furthermore, with $\Psi^*(L) = 1$, i.e., $k = 0$, it reduces to the additive outlier models. We will denote by $\hat{\lambda}^{IO}$ the estimate of break fraction obtained by minimizing the SSR of regression (7). The relevant results are stated in the following Proposition.

Proposition 2. *Suppose that* $\{y_t\}$ *it generated under the null hypothesis of Model* I3 *and that Assumption* 1 *holds. Let* $\hat{\lambda}^{IO}$ *denote the estimate of the break fraction obtained by minimizing the SSR of regression* (7) *over the possible break dates* $T_b \in \{\pi T, \ldots, (1-\pi)T\}$. *Then, for any value of* k, $(\hat{\lambda}^{IO} - \lambda^c) = O_p(T^{-1/2})$.

Remark 2. We also considered an estimate of the break date obtained by minimizing the SSR of the regression Equation (3) say $\hat{\lambda}_2^{IO}$. In this case, $(\hat{\lambda}_2^{IO} - \lambda^c) = O_p(T^{-1})$, for Model I3 so that the rate of convergence is faster (see Kim (2007)). However, as for $\hat{\lambda}_2^{AO}$ discussed in Remark 1, simulations showed that the finite sample properties were poor and it did not lead to unit root tests with good properties. Hence, we shall no longer consider that approach.

3.2. *Unit root tests with an estimate of the break date*

We now consider the limit distributions of the unit root tests when using an estimate of the break date. Of concern are the conditions under which the limit distributions are the same as when the break date is known, so that (4) holds. We start by looking for conditions on the rate of convergence of any estimate of the break fraction for this to happen and then draw the implications when using one of the estimates discussed.

We suppose that we have a consistent estimate $\hat{\lambda} = \hat{T}_1/T$ such that $\hat{\lambda} - \lambda^c = O_p(T^{-a})$ for some $a \geq 0$. Our aim is to obtain conditions on a such that (4) holds. In general, consistency of the estimate of the break fraction is not enough. To gain some understanding, consider the following heuristic explanation of the main issue involved. The use of a break date different from the true one leads to a misspecified trend function, in which case the detrending procedure does not annihilate the deterministic components in the DGP and the limit distribution of $t_\alpha(\hat{\lambda})$ ends up with extra terms compared with $R(\lambda^c)$.

For concreteness, consider the additive outlier models. Let $Y = (y_1, \ldots, y_T)'$ and $U = (u_1, \ldots, u_T)'$ and let $Z(T_1)$ and $Z(T_1)_2$ be the matrices corresponding to $\{z(T_1)'_t\}$ and $\{z(T_1)'_{t,2}\}$, respectively (with $Z(\hat{T}_1)$ and $Z(\hat{T}_1)_2$ defined analogously). Then, we can write the detrended series as:

$$\tilde{Y} = \hat{M}_Z Y = \hat{M}_Z Z(T_1)\phi + \hat{M}_Z U = \hat{M}_Z Z(T_1)_2\phi_2 + \hat{M}_Z U \qquad (8)$$

where, $\hat{M}_Z = I - \hat{P}_Z = I - Z(\hat{T}_1)[Z(\hat{T}_1)'Z(\hat{T}_1)]^- Z(\hat{T}_1)'$. The test statistic $t_\alpha(\hat{\lambda})$ is constructed using $\tilde{Y}$ and the first-differences of $\tilde{Y}$. Note that, in (8) $\hat{M}_Z Z(T_1)_2\phi_2$ will be zero only if the correct break date is used. With an incorrect break date, this term will affect the test statistic. The conditions under which it becomes negligible and has no effect on the limit distribution of $t_\alpha(\hat{\lambda})$ depend on the specific Model.

It is instructive to note that as $T \to \infty$ (a) if $0 < a < 1$, $\hat{T}_1 - T_1 \to \infty$, (b) if $a > 1$, $\hat{T}_1 - T_1 \to 0$, and (c) if $a = 1$, $\hat{T}_1 - T_1 \to \tau$ (some random variable). A fixed false break fraction can be understood as an extreme case with $a = 0$, thus $\hat{\lambda} - \lambda^c \sim O_p(1)$. A break date known *a priori* corresponds to the other extreme case where $\hat{\lambda} - \lambda^c = 0$. The case with a consistent estimate is between these two extremes. In any finite sample, $\hat{\lambda} \neq \lambda^c$, but as the size of the sample increases, $\hat{\lambda} \xrightarrow{p} \lambda^c$. Hence, the faster $\hat{\lambda}$ converges to λ^c, the smaller will be the effect of $\hat{M}_z Z(T_1)_2\phi_2$ on $t_\alpha(\hat{\lambda})$, and the closer will be the limiting distribution of $t_\alpha(\hat{\lambda})$ to $R(\lambda^c)$. A similar argument holds for the innovational outlier models. The following proposition states the sufficient conditions for (4) to hold.

Proposition 3. *Suppose that* $\{y_t\}$ *is generated under the null hypothesis of each model, and that Assumption* 1 *holds. For Models* A2, A3, *and* I3, (4) *holds if* $(\hat{\lambda} - \lambda^c) = o_p(T^{-1/2})$.

Hence, a consistent estimate of the break date is not needed but the break fraction must be consistently estimated at some rate greater than $\sqrt{T}$. This will be the case when using $\hat{\lambda}_2^{AO}$ for Models A2 and A3. However, for Models A2, A3, and I3, if a unit root test is constructed with an estimate of the break date obtained by minimizing the SSR of the static regressions (5) or (7) the limiting distribution is different from the known break date case.[3]

4. Unit Root Tests with Trimmed Data

We first consider Models A2 and A3. When constructing the unit root test with $\hat{\lambda}_1^{AO}$, i.e., with the break date selected using the static regression, the rate of convergence is not fast enough to guarantee that the limit distribution is the same as in the known break date case. It is, however, possible to use a trimmed data set so that the rate of convergence is increased in such a way that it is the same.

The idea is to exclude data points in a window around the estimated break date. Consider an estimate of the break fraction such that $(\hat{\lambda} - \lambda^c) = O_p(T^{-a})$ for some $0 < a < 1$ (for example, in the case of $\hat{\lambda}_1^{AO}$, $a = 1/2$). The window has length $2w(T)$ with $w(T) \equiv \kappa T^{\delta}$, $\kappa > 0$, and $-1 < -a < \delta < 0$. Hence, it includes the data points from $T_l + 1 \equiv T(\hat{\lambda} - w(T)) + 1$ to $T_h \equiv T(\hat{\lambda} + w(T))$. With this specification, the length of the window increases slowly enough to be asymptotically negligible relative to the size of the entire sample, since $T_h/T = \hat{\lambda} + w(T) \xrightarrow{p} \lambda^c$, and $T_l/T = \hat{\lambda} - w(T) \xrightarrow{p} \lambda^c$. On the other hand, it increases fast enough to include the true break date, since

$$T_l - T_1 = T(\hat{\lambda} - w(T)) - T\lambda^c = [T^{-a-\delta}T^a(\hat{\lambda} - \lambda^c) - \kappa]T^{\delta+1} \xrightarrow{p} -\infty$$

and

$$T_1 - T_h = T\lambda^c - T(\hat{\lambda} + w(T)) = [T^{-a-\delta}T^a(\lambda^c - \hat{\lambda}) - \kappa]T^{\delta+1} \xrightarrow{p} -\infty.$$

Our modified unit root test uses this trimmed data. The method can be applied to Models A2 and A3 and yields a unit root test with the same limiting distribution in both cases. More precisely, the test procedure is: (1) Get the estimate of the break date, $\hat{T}_1 = \hat{\lambda}T$ and set the window, T_l and T_h; (2) Create a new data set by removing

[3]In this case, the test statistic does not diverge but converges to some other distribution, which involves many nuisance parameters such as the slope change and, thus, has no practical usefulness.

the data points in the window, i.e., the data points from $T_l + 1$ to T_h and shifting down the data after the window by $S(T) = y_{T_h} - y_{T_l}$. The resulting data set $\{y_t^n\}$, for $t = 1, \ldots, T_r \equiv T - 2w(T)T$, is then

$$y_t^n = \begin{cases} y_t & \text{if } t \leq T_l \\ y_{t+T_h-T_l} - S(T) & \text{if } t > T_l. \end{cases} \tag{9}$$

(3) Perform the unit root test corresponding to Model A2 using the break date T_l i.e., compute $t_\alpha(\hat{\lambda}_{tr})$, the t-statistic on $\hat{\alpha}$ from the following regression:

$$\tilde{y}_t^n = \hat{\alpha}\tilde{y}_{t-1}^n + \sum_{j=1}^{k} \hat{d}_i \Delta\tilde{y}_{t-j}^n + \tilde{e}_t \quad (t = k+1, \ldots, T_r), \tag{10}$$

where $\hat{\lambda}_{tr} = T_l/T_r$ and $\tilde{y}_t^n$ is the detrended value of y_t^n. If the window includes either end of the data in the previous step, simply apply to the new data set $\{y_t^n\}$ a standard unit root test with only a constant and time trend without dummy variables.

Because the reconnected trimmed data satisfy the properties of a series generated by Model A2, whether the actual DGP of the original series is that of Model A2 or A3, we can use a common limit distribution. The following proposition states the limiting distribution of this test statistic, which corresponds to the case of a known break date for Model A2.

Proposition 4. *Suppose that $\{y_t\}$ is generated under the null hypothesis of Model* A2 *or* A3 *and that Assumption* 1 *holds. Then the limiting distribution of $t_\alpha(\hat{\lambda}_{tr})$ is the same as that of $t_\alpha(\lambda^c)$ in Model* A2.

Note that T_l can be thought of as a consistent estimate of the break date for the new data set, i.e., the rate of convergence of the estimate of the break fraction is greater than T, fast enough for the result stated above to apply. Hence, its proof is trivial and omitted.

It is also possible to construct a similar unit root test for Model I3 based on $\hat{\lambda}^{IO}$ using the result of Proposition 2. We again construct T_h and T_l as above. Since, $T_l - T_1 \xrightarrow{p} -\infty$ and $T_1 - T_h \xrightarrow{p} -\infty$, we only need to consider the case $T_l \leq T_1 \leq T_h$. Let $h = T_h - 1 - T_1$, using the representation (7), we can write the DGP under the null hypothesis as:

$$y_t = z'_{t,1}\phi_1 + \sum_{i=0}^{h-1} \xi_i D(T_1 + i)_t + z(T_1 + h)'_{t,2}\zeta_2^* + u_t^s,$$

where $\xi_i = \mu_b \sum_{j=0}^{i} \psi_j^* + \beta_b \sum_{j=0}^{i} \psi_j^*(i - j + 1)$, $\zeta_2^* = \zeta_2 + (\sum_{i=1}^{\infty} \psi_{h+i}^*)\phi_2$, $\zeta_2' = (\mu_b \sum_{j=0}^{h} \psi_j^* + \beta_b \sum_{j=0}^{h-1} \psi_j^*(h - j), \beta_b \sum_{j=0}^{h} \psi_j^*)$, $u_t^s = u_t - s_t$ and $s_t = \sum_{i=1}^{\infty} \psi_{h+i}^* \sum_{j=0}^{i-1} d(T_1 + h + j)'_{t,2}\phi_2$. This shows that the process y_t outside of the trimming window is similar to Model A3, i.e.,

$$y_t = \begin{cases} z'_{t,1}\phi_1 + u_t & t \leq T_l, \\ z'_{t,1}\phi_1 + z(T_h - 1)'_{t,2}\zeta_2^* + u_t^s & t \geq T_h. \end{cases}$$

We create a new data set removing the data points from $T_l + 1$ to T_h and shifting down the data after T_h by

$$S(T) = y_{T_h} - y_{T_l} = (T_h - T_l)\beta + \mu_b^* + \beta_b^* - s_{T_h} + \sum_{i=T_l+1}^{T_h} e_i.$$

The resulting data set $\{y_t^n; t = 1, \ldots, T_r \equiv T - 2w(T)T\}$ is constructed as in (9) where now,

$$\begin{aligned} & y_{t+T_h-T_l} - S(T) \\ & \quad = z'_{t+T_h-T_l,1}\phi_1 + z(T_h - 1)'_{t+T_h-T_l,2}\zeta_2^* + u_{t+T_h-T_l}^s - S(T) \\ & \quad = \mu + (t + T_h - T_l)\beta + \mu_b^* + (t + T_h - T_l - T_h + 1)\beta_b^* + u_{t+T_h-T_l} \\ & \qquad - s_{t+T_h-T_l} - \left[(T_h - T_l)\beta + \mu_b^* + \beta_b^* - s_{T_h} + \sum_{i=T_l+1}^{T_h} e_i \right] \\ & \quad = \mu + \beta t + (t - T_l)\beta_b^* - s_{t+T_h-T_l} + s_{T_h} + u_t^n. \end{aligned}$$

The difference from Model A2 is the term $s_{t+T_h-T_l} - s_{T_h}$ and which is of a small order of magnitude because ψ_{h+i}^* decreases at a geometric rate and $h = O(T^{\delta+1})$. Hence, we can perform the unit root test as for Model A2 using the break date T_l i.e., compute the t statistic on $\hat{\alpha}$ from regression (10) where $\tilde{y}_t^n$ is the detrended value of y_t^n using

$\hat{\lambda}_{tr} = T_l/T_r$. The resulting test is denoted $t_\alpha(\hat{\lambda}_{tr}^{IO})$, and from these arguments, the following result follows:

Proposition 5. *Suppose that* $\{y_t\}$ *is generated under the null hypothesis of Model* I3, *and that Assumption* 1 *holds. Then the limiting distribution of* $t_\alpha(\hat{\lambda}_{tr}^{IO})$ *is the same as that of* $t_\alpha(\lambda^c)$ *in Model* A2.

5. Extension to the Case where a Break Need not Occur

Up to this point, we assumed that a break occurs under both the null and alternative hypotheses. With no break, the asymptotic results described in the previous sections do not hold because, under the null hypothesis of a unit root, the estimate of the break fraction has a nondegenerate limit distribution on the interval [0, 1] instead of converging to either 0 or 1. If there is no break in the trend, then the proper unit root test procedure is to simply apply a standard Dickey and Fuller (1979) type test with no break dummies. Hence, what is needed is a pretest to assess whether a break is present or not. It should have the correct size if the noise is integrated but should also be powerful whether the noise is stationary or has a unit root in order to ensure a specification that allows a unit root test procedure with good power. This testing problem has recently been addressed and we shall use the procedure of Perron and Yabu (2005). It is based on a quasi-GLS approach using an autoregression of order one for the noise component, with a truncation to 1 when α is in some neighborhood of 1, and a bias correction. For a given break date, one constructs the F-test for the null hypothesis of no structural change in the deterministic components. A correction for possible additional correlation is then applied. The final statistic uses the *Exp* functional of Andrews and Ploberger (1994). The test has virtually the same asymptotic size whether the noise component is stationary or integrated. We label this test as $Exp{-}W_{FS}$ and define the alternative estimate of the break fraction, $\tilde{\lambda}$, as $\tilde{\lambda} = \hat{\lambda} \cdot 1(Exp - W_{FS} > cv)$, where cv is the critical value for a test with, say, nominal size $p\%$. When there is a break, $\tilde{\lambda}$ consistently estimates λ^c, as long as $\hat{\lambda}$ consistently estimates λ^c, given that $Exp{-}W_{FS}$ is a consistent test. When there is no break, $\tilde{\lambda}$ will yield a nonzero estimate $p\%$ of times even with infinitely many observations, if cv is fixed. Now, suppose that $Exp - W_{FS} = O_p(T^{\varpi}), \varpi > 0$ under the alternative of a break. Let $cv = cT^{\varpi-\varepsilon}, 0 < \varepsilon < \varpi$. While such an increasing sequence of critical values does not harm the consistency of the test, the size converges to zero as T increases, which ensures the consistency of $\tilde{\lambda}$ for $\lambda^c \in [0,\ 1]$. Given the consistency of $\tilde{\lambda}$, the modified test procedure is to use a DF type test when $\tilde{\lambda} = 0$ and the procedure described above when $\tilde{\lambda} \neq 0$, in particular using the trimming procedure for Models A2, A3, and I3.

6. Models with a Level Shift

We now consider results pertaining to models involving a level shift so that $\mu_b \neq 0$. The first result pertains to Models A1 and I1 involving only a fixed level shift, in which case λ^c cannot be estimated consistently under H_0 because an integrated error process dominates an intercept shift and the break is asymptotically negligible (see Perron and Zhu, 2005). The proof follows easily from Perron and Vogelsang (1992a, 1992b) and is therefore omitted.

Proposition 6. *Suppose that* $\{y_t\}$ *is generated under the null hypothesis of Model* A1 *or* I1. *If* $\hat{\lambda} \xrightarrow{p} \lambda_1 \in [0,\ 1]$, *then* $t_\alpha(\hat{\lambda}) \Rightarrow R(\lambda_1)$.

This proposition states that, as long as $\hat{\lambda}$ converges to any constant $\lambda_1 \in [0,\ 1]$, $R(\lambda_1)$ is the corresponding limiting distribution. Using an incorrect break date does not matter asymptotically. This is because of the asymptotic negligibility of a level shift in the presence of an integrated error process. Perron and Vogelsang (1992a) and Lanne *et al.* (2003) exploit this asymptotic negligibility in deriving the limiting distribution of their test statistics. Since the boundary values 0 and 1 are also permitted, even a standard unit root test without break is asymptotically robust to a shift in the intercept (see Montañés and Reyes (1999), for details). However, it is important to use a consistent estimate of the break fraction because the test statistics will have better power when the correct break date is used. The effect of an incorrect break date in finite samples can substantially decrease power. It is also noteworthy that λ_1 should be a fixed number. When $\hat{\lambda}$ converges to a nondegenerate random variable, using the critical values corresponding to $R(\hat{\lambda})$ will result in incorrect inference. In such cases, a separate limiting distribution has to be derived, as done, for example, in Perron and Vogelsang (1992a).

We can obtain results of more interest by modeling the level shift as increasing with the sample size, i.e., $\mu_b = cT^{1/2+\eta}$ for some $\eta > 0$ and some constant $c > 0$. The models are then labeled as Models A1b, A3b, I1b, and I3b, respectively. This framework is useful to obtain better approximations of the properties of the tests in finite samples when the level shifts are not very small, as shown in Harvey *et al.* (2001)

For Model A3b, Perron and Zhu (2005) showed that $(\hat{\lambda}_1^{AO} - \lambda^c) = o_p(T^{-2-2\eta})$. In the next Proposition, we extend their results to Models A1b, I1b, and I3b.

Proposition 7. *Suppose that* $\{y_t\}$ *is generated under the null hypothesis of Model* A1b *or* A3b. *Let* $\hat{\lambda}_1^{AO}$ *and* $\hat{\lambda}_2^{AO}$ *denote the estimates of the break fraction for the AO models obtained by minimizing the SSR of the static regression (5) and the dynamic regression* (6), *respectively, over the possible break dates* T_b *such that* $T_b \in \{\pi T, \ldots, (1-\pi)T\}$; *then* (a) $(\hat{\lambda}_1^{AO} - \lambda^c) = o_p(T^{-2-2\eta})$ *and* (b) $(\hat{\lambda}_2^{AO} - \lambda^c) = o_p(T^{-1})$. *Suppose that* $\{y_t\}$ *is generated under the null hypothesis of Model* I1b *or* I3b. *For each model, let* $\hat{\lambda}^{IO}$ *denote the estimate of the break fraction obtained by minimizing the SSR of regression* (7) *over the possible break dates* $T_b \in \{\pi T, \ldots, (1-\pi)T\}$. *Then, for any value of k,* (c) $(\hat{\lambda}^{IO} - \lambda^c) = o_p(T^{-1})$.

This result shows that when the size of the level shift is an increasing function of the sample size, the rate of convergence of the estimate of the break fraction is faster than in the case of a change in slope. The next proposition states the required condition for the weak convergence result in (4) to hold, so that the unit root tests evaluated using the estimate of the break fraction have the same limit distributions as in the known break date case.

Proposition 8. *Suppose that* $\{y_t\}$ *is generated under the null hypothesis of each model. Then* (4) *holds if* $(\hat{\lambda} - \lambda^c) = o_p(T^{-1})$, *for Models* A1b, A3b, I1b, *and* I3b.

This states that with large level shifts, a consistent estimate of the break fraction is not enough and a consistent estimate of the break date itself is needed. However, in light of the results of Proposition 7, this is guaranteed if one uses either $\hat{\lambda}_1^{AO}$ or $\hat{\lambda}_2^{AO}$ for Models A1b and A3b, and $\hat{\lambda}^{IO}$ for Models I1b and I3b. Hence, no trimming is needed with large level shifts.

7. Simulations of the Size and Power in Finite Samples

We now present simulation results about the size and power of the tests. The DGP used for the simulation is $y_t = d_t + u_t$ with $u_t = \alpha u_{t-1} + e_t$, $e_t \sim$ i.i.d. $N(0,1)$, and

$$d_t = \begin{cases} \mu_b DU_t & \text{for Models A1, I1} \\ \beta_b B_t & \text{for Model A2} \\ \mu_b DU_t + \beta_b B_t & \text{for Models A3, I3.} \end{cases}$$

For Model A1, the value of μ_b ranges from -10 to 10 in increments of 0.5, and for Models A2 and A3, the value of β_b ranges from -4 to 4 in increments of 0.2. For Model A3, we set $\mu_b = 0$. The number of replications is 2,000 and the sample size is $T = 150$. The value of the true break fraction λ^c is set to 0.5 (we also considered $\lambda^c = 0.3$, 0.7 and the general conclusions are the same, see Kim (2007)). The same set of random errors has been used across values of μ_b and β_b. One sided tests with nominal size 5 were constructed. For power, α is set to 0.8. For all test statistics, the asymptotic critical values are used.[4]

[4]The critical values for Models A1, A3, I1, and I3 are taken from Perron (1989, Tables IV.B and VI.B) and for Model A2 from Perron and Vogelsang (1993, Table 1).

We simulated the following statistics: $t_\alpha(\hat{\lambda}_1^{AO})$, $t_\alpha(\hat{\lambda}_2^{AO})$, t_α $(\hat{\lambda}^{IO})$, for Models A2 and A3, $t_\alpha(\hat{\lambda}_{tr})$, and for Model I3, $t_\alpha(\hat{\lambda}_{tr}^{IO})$ for which we use a trimming window of 6 observations (the results are not sensitive to other choices of window length). For the pretest, we used the *Exp*–W_{FS} test of Perron and Yabu (2005) with a nominal size 5%. The statistics with a pretest are denoted by $t_\alpha^p(\hat{\lambda}_1^{AO})$, $t_\alpha^p(\hat{\lambda}_2^{AO})$, $t_\alpha^p(\hat{\lambda}^{IO})$, and $t_\alpha^p(\hat{\lambda}_{tr})$. The results are presented only for the version of the statistic with a pretest given that without it the tests show considerable liberal size distortions when the break is zero or small. Three benchmark tests were also simulated for comparisons: a standard Dickey–Fuller t-test with a constant and a time trend (DF), Perron's (1989) t-test, $t_\alpha(\lambda^c)$ using the true break date, and the $\min_{\lambda\in\Lambda} t_\alpha(\lambda)$ test of Zivot and Andrews (1992) with $\Lambda = [0.15, 0.85]$. For all tests, we set the value of k at zero in order to better assess the relative properties of the tests without confounding issues related to the choice of the autoregressive order.

The results for Model A1 are presented in Fig. 1. The test $t_\alpha^p(\hat{\lambda}_1^{AO})$suffers from some liberal size distortions unless $|\mu_b|$ is large. This is due to the fact that the estimate of the break date is consistent only as $|\mu_b|$ increases with the sample size. This translates in finite samples into estimates that are quite variable when $|\mu_b|$ is not large enough, even though the break is large enough to be detected by the pretest. This randomness translates into a distribution of $t_\alpha^p(\hat{\lambda}_1^{AO})$ that is far from the fixed break date case and liberal size distortions occur. The test $t_\alpha^p(\hat{\lambda}_2^{AO})$, which uses the estimate of the break date from the dynamic regression is much less affected by this problem, though some distortions remain. As expected, the DF test is conservative for all nonzero values of μ_b and its power seriously decreases as $|\mu_b|$ increases. The often used $\min_{\lambda\in\Lambda} t_\alpha(\lambda)$ test suffers from large liberal size distortions when $|\mu_b|$ gets large, and when μ_b is near 0 its power is strictly smaller than that of $t_\alpha^p(\hat{\lambda}_1^{AO})$ and $t_\alpha^p(\hat{\lambda}_2^{AO})$. The power of $t_\alpha^p(\hat{\lambda}_1^{AO})$ and $t_\alpha^p(\hat{\lambda}_2^{AO})$ is no less than that of $t_\alpha(\lambda^c)$ and, hence, the cost of relaxing the assumption of a known break date is almost null in Model A1. Overall, in Model A1, the test $t_\alpha^p(\hat{\lambda}_2^{AO})$ has the best properties.

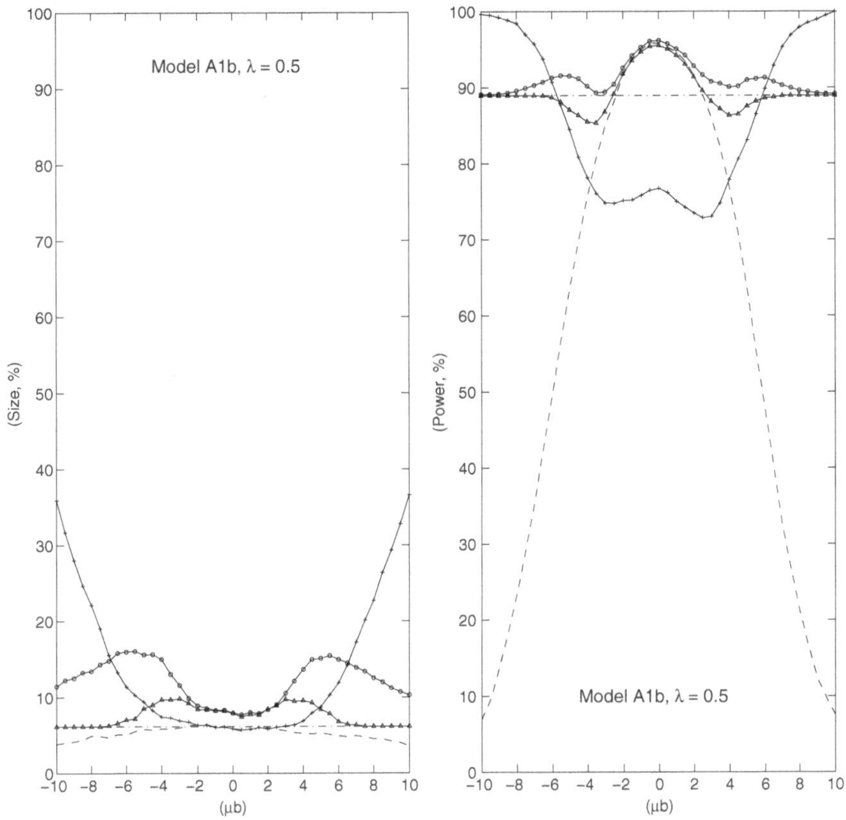

Fig. 1. Null rejection probabilities in model A1, $T = 150$, $\lambda = 0.5$ [$- \circ -$: $t_\alpha^p(\hat{\lambda}_1^{AO})$, $-\Delta-$: $t_\alpha^p(\hat{\lambda}_2^{AO})$, $--$: DF, $- \cdot -$: $t_\alpha(\lambda^c)$, $-+-$: $\min t_\alpha(\lambda)$].

The results for Model A2 are presented in Fig. 2. With a pretest, the test $t_\alpha^p(\hat{\lambda}_1^{AO})$ shows liberal size distortions, while the properties of $t_\alpha^p(\hat{\lambda}_2^{AO})$ and $t_\alpha^p(\hat{\lambda}_{tr})$ are closer to those of $t_\alpha(\lambda^c)$. However, $t_\alpha^p(\hat{\lambda}_2^{AO})$ is more conservative and has strictly less power than $t_\alpha^p(\hat{\lambda}_{tr})$, showing the advantage of using the trimmed version even though $t_\alpha^p(\hat{\lambda}_2^{AO})$ has asymptotically the right size since it is constructed with an estimate of the break fraction that is consistent at rate T. The power of $t_\alpha^p(\hat{\lambda}_{tr})$ is strictly less than that of $t_\alpha(\lambda^c)$ unless there is no break, which highlights the cost of relaxing the assumption of a known break date. However, it is strictly larger than that of DF and the commonly used $\min_{\lambda\in\Lambda} t_\alpha(\lambda)$. Hence, the simulations show that our trimmed

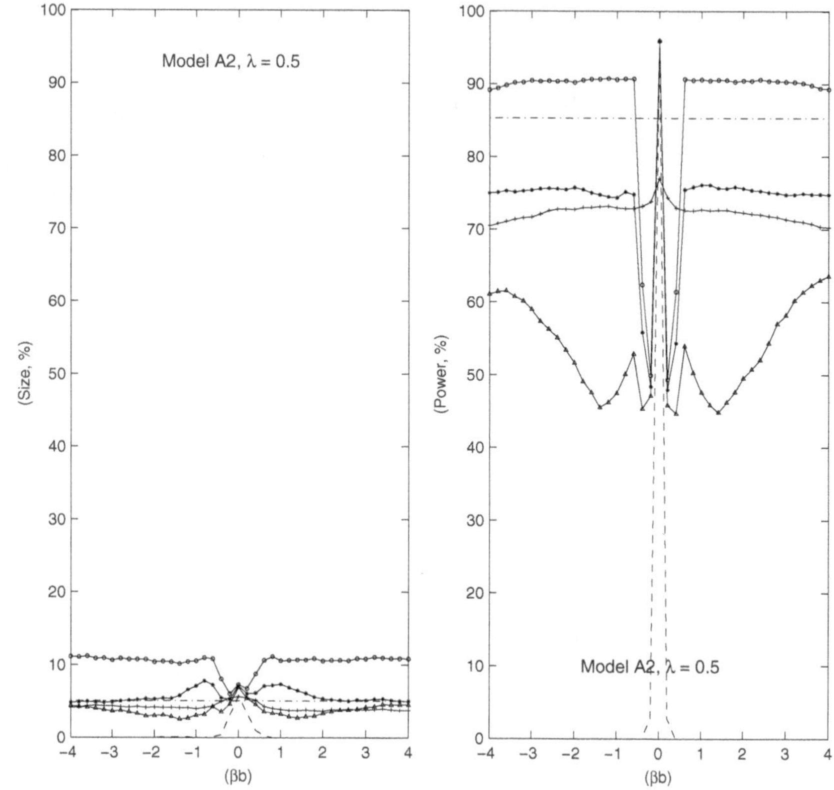

Fig. 2. Null rejection probabilities in model A2, $T = 150$, $\lambda = 0.5$ [$-\circ-$: $t^p_\alpha(\hat{\lambda}_1^{AO})$, $-\Delta-$: $t^p_\alpha(\hat{\lambda}_2^{AO})$, $-*-$: $t^p_\alpha(\hat{\lambda}_{tr})$, $--$: DF, $-\cdot-$: $t_\alpha(\lambda^c)$, $-+-$: $\min t_\alpha(\lambda)$].

version with a pretest offers an improvement in power over available alternative tests, while ensuring an adequate size whether a break is present or not.

The results for Model A3 are presented in Fig. 3. They show a rather different picture. The test $t^p_\alpha(\hat{\lambda}_2^{AO})$ now shows liberal size distortions but the test $t^p_\alpha(\hat{\lambda}_1^{AO})$ has good size and power, and overall properties similar to those of $t^p_\alpha(\hat{\lambda}_{tr})$. This points to the fact that the fast rate of convergence of the estimate of the break date when using the dynamic regression does not always translate into a test with good finite sample properties. The reason for this poor performance is that the estimate $\hat{\lambda}_2^{AO}$ from the dynamic regression

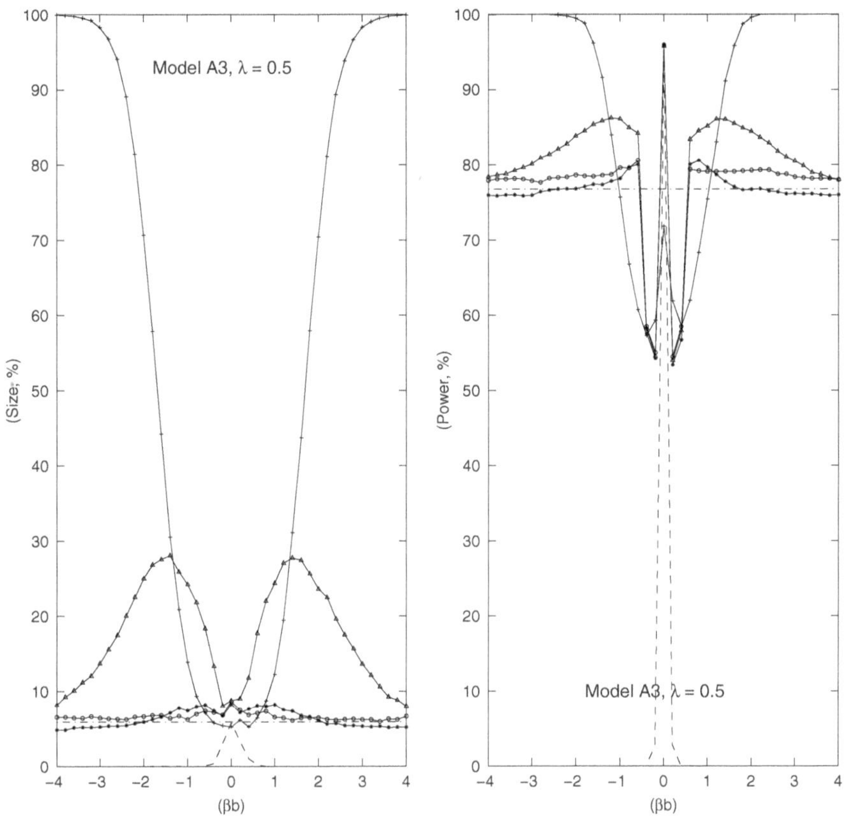

Fig. 3. Null rejection probabilities in model A3, $T = 150$, $\lambda = 0.5$ [$-\circ-$: $t^p_\alpha(\hat{\lambda}^{AO}_1)$, $-\Delta-$: $t^p_\alpha(\hat{\lambda}^{AO}_2)$, $-*-$: $t^p_\alpha(\hat{\lambda}_{tr})$, $--$: DF, $-\cdot-$: $t_\alpha(\lambda^c)$, $-+-$: $\min t_\alpha(\lambda)$].

suffers from a peculiar finite sample bias due to the joint presence of the regressors $D(T_b)_t$ and DU_t (see Lee and Strazicich (2001) and Harvey *et al.* (2001)). The results again clearly show that the test $t^p_\alpha(\hat{\lambda}_{tr})$ has better properties than the DF test and the commonly used $\min_{\lambda\in\Lambda} t_\alpha(\lambda)$ test and is, accordingly, also the preferred one in this case.

Consider now the results for the IO models. For Model I1, they are very similar to those for the AO case presented in Fig. 1 and the general conclusions still apply. Hence, we do not explicitly report them. The results for Model I3 are presented in Fig. 4. They show

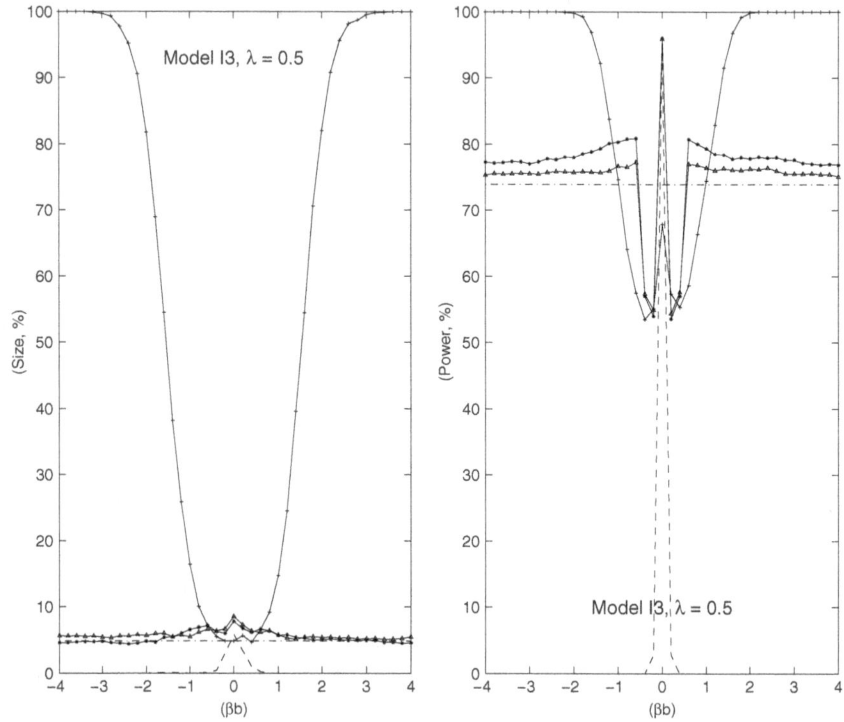

Fig. 4. Null rejection probabilities in model I3, $T = 150$, $\lambda = 0.5$ $[-\Delta- : t^p_\alpha(\hat{\lambda}^{IO})$, $-*- : t^p_\alpha(\hat{\lambda}^{IO}_{tr})$, $-- : DF$, $-\cdot- : t_\alpha(\lambda^c)$, $-+- : \min t_\alpha(\lambda)]$.

that the tests $t^p_\alpha(\hat{\lambda}^{IO})$ and $t^p_\alpha(\hat{\lambda}^{IO}_{tr})$ have very little size distortions, especially $t^p_\alpha(\hat{\lambda}^{IO}_{tr})$. The power function of $t^p_\alpha(\hat{\lambda}^{IO})$ is close to what can be achieved assuming a known break date, unless the magnitude is near zero. When it is close to zero, its power is actually higher (because of the use of the pretest which makes the test equivalent to the DF and, hence, more powerful). When the break is moderate, the power is lower because the pretest implies the use of the DF test, whose power decreases as the break gets larger. When the break is large enough to be detected by the pretest, the test $t^p_\alpha(\hat{\lambda}^{IO})$has power comparable to that in the known break date case. What is more interesting is the fact that, when the break is large enough, the power of the test $t^p_\alpha(\hat{\lambda}^{IO}_{tr})$ is actually higher than what can be achieved if the break date was known. This is because the trimming reduces Model I3 to Model A2 for which power is higher. Hence, for Model I3, the test $t^p_\alpha(\hat{\lambda}^{IO}_{tr})$ is clearly the best.

8. Empirical Applications

We analyze two sets of multi-country GDP data. The first includes postwar quarterly (log) seasonally adjusted real GDP series for the G7 countries. This data set is from the OECD. For France, we included a dummy in the second quarter of 1968 in order to account for the effect of the general strike in May 1968. For Japan, we also consider a subsample that excludes data after 1994 in order to account for a second likely break in the 1990's. The second one is a historical data set of (log) real GDP series and their per capita counterparts from 1870 to 1986 for 10 countries: Australia, Canada, Denmark, France, Germany, Italy, Norway, Sweden, the UK, and the US.[5]

The results for the postwar quarterly real GDP series for the G7 countries are reported in Table 1. Given the nature of these series, we considered the AO version of the test corresponding to both Models A2 and A3. For Model A2, the test $t^p_\alpha(\hat{\lambda}_{tr})$ is the preferred one, while for Model A3, both $t^p_\alpha(\hat{\lambda}^{AO}_1)$ and $t^p_\alpha(\hat{\lambda}_{tr})$ perform well. Hence, we report both. We also report the commonly used $\min_{\lambda\in\Lambda} t_\alpha(\lambda)$ test for comparisons. The first thing to note is that the *Exp*-W_{FS} break test of Perron and Yabu (2005) rejects the null hypothesis of a constant deterministic trend function in all cases

[5]This data set is the same as used by Kormendi and Meguire (1990), Perron (1992), Perron and Zhu (2005), Perron and Yabu (2005) and Vogelsang (1997). It was obtained through the *Journal of Money, Credit and Banking* editorial office. All series are real GDP except for the US for which real GNP is used. For the US, the series is real GNP from the National Income and Products Accounts for the period 1929–1986, spliced to Romer's (1989) estimates for the period 1870–1928. For the UK, the series is real GDP from Feinstein (1972) for the period 1870–1947 spliced to the International Financial Statistics (IFS) series of the IMF for the period 1948–1986. For the remaining countries, the series are indices of annual real GDP from Maddison (1982) spliced to the postwar IFS data. The population series used are from the same sources. All series are analyzed with a logarithmic transformation.

Table 1. Empirical results for the postwar real GDP series of the G7 countries.

	Span		T	DF	Model	$Exp - W_{FS}$	Break	$t_\alpha(\hat{\lambda}_1^{AO})$	$t_\alpha(\hat{\lambda}_{tr})$	$\min t_\alpha(\lambda)$
Canada	1947.1	1997.2	202	−0.19(1)	A2	2.69*	1978.3 [0.63]	−3.70**(1)	−3.64	−3.70
					A3	3.04**	1978.3 [0.63]	−3.66(1)	−3.64	−3.83
France	1963.1	2004.4	168	−2.66(3)	A2	17.16*	1973.4 [0.26]	−2.80(1)	−2.73	−3.70
					A3	22.47*	1973.3 [0.25]	−2.75(1)	−2.74	−5.10*
Germany	1960.1	2004.4	180	−1.51(0)	A2	1.56**	1971.3 [0.26]	−1.78(0)	−1.69	−2.33
					A3	74.59*	1974.3 [0.33]	−1.67(0)	−1.67	−3.42
Italy	1960.1	2004.4	180	−1.89(1)	A2	7.38*	1977.2 [0.39]	−3.23(1)	−3.06	−3.35
					A3	16.10*	1977.1 [0.38]	−3.55(1)	−3.07	−3.55
Japan	1960.1	2004.4	180	−2.04(3)	A2	42.33*	1973.2 [0.30]	−1.59(3)	−1.41	−2.66
					A3	58.15*	1973.2 [0.30]	−1.78(3)	−1.41	−3.98
	1960.1	1994.4	140	−1.56(3)	A2	27.46*	1972.1 [0.35]	−4.01*(3)	−2.82	−4.09
					A3	42.96*	1971.4 [0.34]	−3.65(3)	−2.85	−4.40
UK	1960.1	2004.4	180	−2.19(0)	A2	−0.10	1966.1 [0.14]	NA	NA	NA
					A3	7.79*	1980.1 [0.45]	−2.91(0)	−2.83	−4.46
US	1947.1	2004.2	230	−2.85(1)	A2	−0.08	1969.4 [0.40]	NA	NA	NA
					A3	2.77**	1965.2 [0.32]	−4.49*(1)	−4.50*	−5.61*

Notes: (a) The numbers in parentheses are the values of the autoregressive order k selected by the BIC; (b) the number in brackets in the 'break' column denote the implied break fraction.

*Denote a statistic significant at the 5% significance level.

**Denote a statistic significant at the 10% significance level.

with the exception of the UK and the US when Model A2 is used. In these two cases, the relevant unit root test is the DF test, which does not allow a rejection. For Canada (Model A3), France (Model A2), Germany (both A2 and A3), Italy (both A2 and A3), Japan (both A2 and A3, 1960–2004; and A3 for 1960–1994) and UK (Model A3) all three tests $t^p_\alpha(\hat{\lambda}^{AO}_1), t^p_\alpha(\hat{\lambda}_{tr})$ and $\min_{\lambda\in\Lambda} t_\alpha(\lambda)$ lead to a nonrejection of the unit root at the 10% significance level. Differences across tests occur for the following cases. For Canada and Japan (1960–1994) in the case of Model A2, the test $t^p_\alpha(\hat{\lambda}^{AO}_1)$ leads to a rejection, while the test $t^p_\alpha(\hat{\lambda}_{tr})$ does not. Given the simulation results reported above, which show the test $t^p_\alpha(\hat{\lambda}^{AO}_1)$ to suffer from liberal size distortions in Model A2, and the test $t^p_\alpha(\hat{\lambda}_{tr})$ to have better size and power, one must conclude that a nonrejection of the null of a unit root is warranted. The same conclusion, for the same reasons, apply to the case of France with Model A3 where the test $\min_{\lambda\in\Lambda} t_\alpha(\lambda)$ shows a rejection, while the tests $t^p_\alpha(\hat{\lambda}^{AO}_1)$ and $t^p_\alpha(\hat{\lambda}_{tr})$ do not. The only case where a rejection of the null hypothesis of a unit root is warranted is for the United Stated with Model A3.

Table 2 reports the results pertaining to the historical data set. For both the GDP and Per Capita GDP series, the *Exp*–W_{FS} break test of Perron and Yabu (2005) rejects the null hypothesis of a constant deterministic trend function in all cases with the exception of Canada, in which case the DF test does not allow a rejection of the unit root hypothesis. Given the gradual nature of the change in the trend, we used Model I3 to test the unit root hypothesis, in which case the test $t_\alpha(\hat{\lambda}^{IO}_{tr})$ is the appropriate one based on size and power properties. For the computation of $t_\alpha(\hat{\lambda}^{IO}_{tr})$, there is no easy way to select k using a data dependent method given the fact that the order pertains to the number of pulse dummies. But, by the same reason, any value of k is likely to yield similar estimates of the break date. We estimated the break date with $k = 5, 6$, and 7 (as reported in Table 3) and used the most frequently chosen date for the unit root test. As can be seen, the break date is most of the time the same

Table 2. Empirical results for the historical annual GDP series (1870–1986).

	GDP					Per capita GDP				
	DF	$Exp - W_{FS}$	Break	$t^p_\alpha(\hat{\lambda}^{IO}_{tr})$	$\min t_\alpha(\lambda)$	DF	$Exp - W_{FS}$	Break	$t^p_\alpha(\hat{\lambda}^{IO}_{tr})$	$\min t_\alpha(\lambda)$
Australia	−1.13(0)	3.84*	1927 [0.50]	−3.15(0)	−4.87**	−0.96(0)	3.18*	1939 [0.60]	−3.54(0)	−4.86**
Canada	−2.12(0)	1.51	1930 [0.52]	NA	NA	−2.00(1)	1.46	1930 [0.52]	NA	NA
Denmark	−2.34(0)	8.97*	1939 [0.60]	−2.94(0)	−4.98**	−1.64(0)	9.42*	1938 [0.59]	−2.49(0)	−6.11*
France	−1.54(1)	30.78*	1942 [0.62]	−2.17(1)	−6.66*	−2.00(1)	28.66*	1943 [0.63]	−2.26(0)	−7.06*
Germany	−2.67(1)	33.32*	1945 [0.65]	−1.58(1)	−4.87**	−2.50(1)	58.00*	1945 [0.65]	−1.71(1)	−5.56*
Italy	−1.68(1)	15.92*	1943 [0.63]	−1.96(0)	−4.56	−1.55(1)	17.08*	1943 [0.63]	−2.16(0)	−4.68
Norway	−1.02(0)	7.34*	1937 [0.58]	−3.24(0)	−3.14	−0.81(0)	7.72*	1937 [0.58]	−3.63(0)	−3.07
Sweden	−2.38(1)	13.57*	1924 [0.47]	−2.28(1)	−3.47	−2.21(0)	14.31*	1924 [0.47]	−2.54(1)	−4.01
UK	−2.44(1)	6.34*	1914 [0.38]	−4.46*(2)	−6.8*	−1.94(1)	12.92*	1914 [0.38]	−4.46*(2)	−6.76*
US	−3.66*(1)	6.84*	1929 [0.51]	−4.35*(1)	−4.82*	−4.14*(1)	6.33*	1930 [0.52]	−3.9**(1)	−5.01**

Notes: (a) The numbers in parentheses are the values of the autoregressive order k selected by the BIC; b) the number in brackets in the 'break' column denote the implied break fraction.
*Denote a statistic significant at the 5% significance level.
**Denote a statistic significant at the 10% significance level.

Table 3. Estimated break dates for the historical annual GDP series (1870–1986).

	GDP			Per capita GDP		
k	5	6	7	5	6	7
Australia	1927	1927	1927	1939	1939	1938
Canada	1930	1930	1930	1930	1930	1930
Denmark	1938	1939	1939	1938	1939	1938
France	1943	1942	1942	1943	1942	1941
Germany	1945	1945	1945	1945	1945	1945
Italy	1943	1943	1943	1943	1943	1943
Norway	1937	1936	1935	1937	1936	1935
Sweden	1925	1924	1924	1925	1924	1924
UK	1915	1914	1914	1915	1914	1914
US	1930	1929	1929	1930	1930	1931

and differs by at most two periods. For both series of Norway and the Per Capita GDP series of France, each value of k gave a different estimate of the break date estimate. However, they lead to the same conclusion for the unit root test and, in Table 2 we simply report the results obtained with $k = 5$. The size of the trimming window was set to 14 to cover all pulse dummies whatever the value of k is used. The statistic $t_\alpha(\hat{\lambda}_{tr}^{IO})$ shows a rejection for both the GDP and Per Capita GDP series only in the case of UK and the US. There are eight cases for which the $t_\alpha(\hat{\lambda}_{tr}^{IO})$ and $\min_{\lambda \in \Lambda} t_\alpha(\lambda)$ tests show conflicting results, namely the GDP and per Capita GDP series of Australia, Denmark, France, and Germany, where the former does not reject while the latter test does. This is well explained by the simulation results reported, which showed the $\min_{\lambda \in \Lambda} t_\alpha(\lambda)$ test to exhibit liberal size distortions in the presence of a change in the trend function. Hence, in these three cases the evidence does not favor a rejection. These results illustrate the pitfalls associated with the approach of Zivot and Andrews (1992), namely that a rejection can be due to a change in the trend rather than a stationary noise function. Our tests are immune to these problems and, hence, more reliable.

9. Conclusions

Perron (1989) introduced a variety of unit root tests that are valid when a break in the trend function of a time series is present. The motivation was to devise testing procedures that were invariant to the magnitude of the shift in level and/or slope. In particular, if a change is present it is allowed under both the null and alternative hypotheses. The subsequent literature, which aimed to devise testing procedures valid in the case of an unknown break date, assumed that if a break occurs, it does so only under the alternative hypothesis of stationarity. This is undesirable for several reasons. First, it imposes an asymmetric treatment when allowing for a break, so that the test may reject when the noise is integrated but the trend is changing. Second, if a break is present, this information is not exploited to improve the power of the test. The test procedures discussed in this paper remedy these problems. They allow a break under both the null and alternative hypotheses and, when a break is present, the limit distribution of the test is the same as in the case of a known break date, thereby allowing increased power while maintaining the correct size. Simulation experiments confirm that our procedure offers an improvement over commonly used methods in small samples. Indeed, in many cases, the power of our tests is close to what can be achieved assuming knowledge of the true break date. Hence, our procedure offers a clear improvement over existing tests and should be used in practical applications.

Appendix

Proof of Propositions 1 and 7(b). We first introduce some notation. $Y(a,b)$ refers to a subvector of Y which includes the ath to the bth element of Y; Let $e_t = A^*(L)^{-1}B(L)\varepsilon_t$, $e = (e_1,\dots,e_T)'$, and $d(T_b)_2 = (d(T_b)_{1,2}\dots d(T_b)_{T,2})'$, where $d(T_b)_{t,2}$ is 1 $(t \geq T_b)$ for Model A2 and, for the other models, the same as in the corresponding IO models. As before $d(T_b,a,b)_2$ and $e(a,b)$ define a subvector where the first argument T_b of $d(\cdot)_2$ denotes a specified break date. $P(T_b,2,T)$ is the projection matrix associated with the regressors in Equation (6). Again, write the sum of squared residuals of regression

(6) as $SSR_T(\lambda)$, $\hat{\lambda}_2^{AO} = \arg\min_{\lambda\in\{\pi,\ldots,1-\pi\}} SSR_T(\lambda)$. We have, with $\lambda = T_b/T$,

$$\begin{aligned}
SSR_T(\lambda) - SSR_T(\lambda^c) &= Y(2,T)'[P(T_1,2,T) - P(T_b,2,T)]Y(2,T) \\
&= \phi_2' d(T_1,2,T)_2'[P(T_1,2,T) - P(T_b,2,T)]d(T_1,2,T)_2\phi_2 \\
&\quad + 2\phi_2' d(T_1,2,T)_2'[P(T_1,2,T) - P(T_b,2,T)]e(2,T) \\
&\quad + e(2,T)'[P(T_1,2,T) - P(T_b,2,T)]e(2,T) \\
&= \phi_2'[d(T_1,2,T)_2 - d(T_b,2,T)_2]'[I - P(T_b,2,T)] \\
&\quad \times [d(T_1,2,T)_2 - d(T_b,2,T)_2]\phi_2 \qquad (A.1) \\
&\quad + 2\phi_2'[d(T_1,2,T)_2 - d(T_b,2,T)_2]' \\
&\quad \times [I - P(T_b,2,T)]e(2,T) \qquad (A.2) \\
&\quad + e(2,T)'[P(T_1,2,T) - P(T_b,2,T)]e(2,T). \qquad (A.3)
\end{aligned}$$

Denote the indicator function, $1(T_b - T_1 \neq 0)$ by $\tilde{I}$. The orders of magnitude of the three terms are, uniformly over all generic break date, $T_b \in \{\pi T, (1-\pi)T\}$, for Models A1b and A3b, $(XX) = \tilde{I}\cdot O_p(T^{1+2\eta})$, $(XU) = \tilde{I}\cdot O_p(T^{1/2+\eta})$, $(UU) = \tilde{I}\cdot O_p(1)$. For Models A2 and A3, we have $(XX) = |T_b - T_1|O_p(1)$, $(XU) = |T_b - T_1|^{1/2}O_p(1)$, $(UU) = \tilde{I}\cdot O_p(1)$. Given the orders of magnitude of (XX), (XU), and (UU), the rate of convergence results follow using the arguments of Perron and Zhu (2005). □

Proof of Propositions 2 and 7 (c). In order to derive (7), note that

$$\begin{aligned}
&\Psi^*(L)z(T_1)_{t,2}'\phi_2 \\
&\quad = DU(T_1)_t\mu_b + DT(T_1)_t\beta_b + \sum_{i=1}^{\infty}\psi_i^* z(T_1+i)_{t,2}'\phi_2 \\
&\quad = D(T_1)_t(\mu_b+\beta_b) + DU(T_1+1)_t(\mu_b+\beta_b) \\
&\qquad + DT(T_1+1)_t\beta_b + \sum_{i=1}^{\infty}\psi_i^* z(T_1+i)_{t,2}'\phi_2
\end{aligned}$$

$$= \sum_{i=0}^{k-1} D(T_1+i)_t \left[\mu_b \sum_{j=0}^{i} \psi_j^* + \beta_b \sum_{j=0}^{i} \psi_j^*(i-j+1)\right]$$

$$+ DU(T_1+k)_t \left[\mu_b \sum_{j=0}^{k} \psi_j^* + \beta_b \sum_{j=0}^{k-1} \psi_j^*(k-j)\right]$$

$$+ DT(T+k)_t \left(\beta_b \sum_{j=0}^{k} \psi_j^*\right) + \sum_{i=k+1}^{\infty} \psi_i^* z(T_1+i)'_{t,2}\phi_2$$

$$= \sum_{i=0}^{k-1} D(T_1+i)_t \xi_i + z(T_1+k)'_{t,2}\zeta_2 + \sum_{i=k+1}^{\infty} \psi_i^* z(T_1+i)'_{t,2}\phi_2$$

where $\xi_i = \mu_b \sum_{j=0}^{i} \psi_j^* + \beta_b \sum_{j=0}^{i} \psi_j^*(i-j+1)$ and $\zeta_2' = (\mu_b \sum_{j=0}^{k} \psi_j^* + \beta_b \sum_{j=0}^{k-1} \psi_j^*(k-j),\ \beta_b \sum_{j=0}^{k} \psi_j^*)$. Now, write the DGP under the null hypothesis as:

$$y_t = z'_{t,1}\phi_1 + \Psi^*(L) z(T_1)'_{t,2}\phi_2 + u_t$$

$$= z'_{t,1}\phi_1 + \sum_{i=0}^{k-1} D(T_1+i)_t \xi_i + z(T_1+k)'_{t,2}\zeta_2$$

$$+ \sum_{i=k+1}^{\infty} \psi_i^* z(T_1+i)'_{t,2}\phi_2 + u_t$$

$$= z'_{t,1}\phi_1 + \sum_{i=0}^{k-1} \xi_i D(T_1+i)_t + z(T_1+k)'_{t,2}\zeta_2$$

$$+ \sum_{i=1}^{\infty} \psi_{k+i}^* z(T_1+k+i)'_{t,2}\phi_2 + u_t$$

$$= z'_{t,1}\phi_1 + \sum_{i=0}^{k-1} \xi_i D(T_1+i)_t + z(T_1+k)'_{t,2}\zeta_2$$

$$+\sum_{i=1}^{\infty}\psi_{k+i}^{*}\left[z(T_1+k)_{t,2}-\sum_{j=0}^{i-1}d(T_1+k+j)_{t,2}\right]'\phi_2+u_t$$

$$=z_{t,1}'\phi_1+\sum_{i=0}^{k-1}\xi_i D(T_1+i)_t+z(T_1+k)_{t,2}'\zeta_2^{*}+u_t^{s} \tag{A.4}$$

by using the fact that $z(T_1+k+i)_{t,2}=z(T_1+k)_{t,2}-\sum_{j=0}^{i-1}d(T_1+k+j)_{t,2}$, where $\zeta_2^{*}=\zeta_2+(\sum_{i=1}^{\infty}\psi_{k+i}^{*})\phi_2$, $u_t^{s}=u_t-s_t$ and $s_t=\sum_{i=1}^{\infty}\psi_{k+i}^{*}\sum_{j=0}^{i-1}d(T_1+k+j)_{t,2}'\phi_2$. Let $\sup_t$ abbreviates $\sup_{T_1+k+1\le t\le T}$. In Model I3, $|s_t|$ is bounded uniformly in $1\le t\le T$, because

$$\begin{aligned}\sup_{1\le t\le T}|s_t|&=\sup_t\left|\sum_{i=1}^{\infty}\psi_{k+i}^{*}\sum_{j=0}^{i-1}d(T_1+k+j)_{t,2}'\phi_2\right|\\&\le\sup_t\left|\sum_{i=1}^{\infty}\psi_{k+i}^{*}\sum_{j=0}^{i-1}D(T_1+k+j)_{t,2}\mu_b\right|\\&\quad+\sup_t\left|\sum_{i=1}^{\infty}\psi_{k+i}^{*}\sum_{j=0}^{i-1}DU(T_1+k+j)_{t,2}\beta_b\right|\end{aligned}$$

and each term is such that

$$\begin{aligned}&\sup_t\left|\sum_{i=1}^{\infty}\psi_{k+i}^{*}\sum_{j=0}^{i-1}D(T_1+k+j)_{t,2}\mu_b\right|\\&\le|\mu_b|\sup_t\sum_{i=1}^{\infty}|\psi_{k+i}^{*}|\sum_{j=0}^{i-1}D(T_1+k+j)_{t,2}\\&=|\mu_b|\sup_t\sum_{i=1}^{\infty}\sum_{j=0}^{i-1}|\psi_{k+i}^{*}|D(T_1+k+j)_{t,2}\end{aligned}$$

$$= |\mu_b| \sup_t \sum_{j=0}^{\infty} \left(\sum_{i=j+1}^{\infty} |\psi_{k+i}^*| \right) D(T_1 + k + j)_{t,2}$$

$$= |\mu_b| \sup_t \left(\sum_{i=t-T_1-k}^{\infty} |\psi_{k+i}^*| \right) D(t-1)_{t,2} \leq |\mu_b| \sum_{i=1}^{\infty} |\psi_i^*| < \infty,$$

and, similarly,

$$\sup_t \left| \sum_{i=1}^{\infty} \psi_{k+i}^* \sum_{j=0}^{i-1} DU(T_1 + k + j)_{t,2} \beta_b \right| \leq |\beta_b| \sum_{i=1}^{\infty} i |\psi_i^*| < \infty.$$

Hence, s_t is of strictly smaller order of magnitude than u_t and thus asymptotically negligible. In Models I1b and I3b, s_t is also uniformly bounded by $|\mu_b|$. Let $U^s = (u_1^s, \ldots, u_T^s)', \underline{D}(T_b)_t = (D(T_b)_t, \ldots, D(T_b + k - 1)_t)', \underline{D}(T_b) = (\underline{D}(T_b)_1, \ldots, \underline{D}(T_b)_T)'$ and $\xi = (\xi_0, \ldots \xi_{k-1})'$. Denote by $P(T_b)$ the projection matrix corresponding to the regressors $Z(T_b + k)$ and $\underline{D}(T_b)$. Writing the sum of squared residuals of regression (7) as $SSR_T(\lambda)$, the estimate of the break fraction is $\hat{\lambda}_2^{IO} = \arg\min_{\lambda \in \{\pi, \ldots, 1-\pi\}} SSR_T(\lambda)$. With $\lambda = T_1/T$, we have

$$\begin{aligned}
&SSR_T(\lambda) - SSR_T(\lambda^c) \\
&\quad = Y'[P(T_1) - P(T_b)]Y \\
&\quad = \zeta_2^{*\prime}[Z(T_1 + k)_2 - Z(T_b + k)_2]'[I - P(T_b)] \\
&\qquad \times [Z(T_1 + k)_2 - Z(T_b + k)_2]\zeta_2^* && (XX) \\
&\qquad + \xi'[\underline{D}(T_1) - \underline{D}(T_b)]'[I - P(T_b)][\underline{D}(T_1) - \underline{D}(T_b)]\xi && (DD) \\
&\qquad + 2\zeta_2^{*\prime}[Z(T_1 + k)_2 - Z(T_b + k)_2]' \\
&\qquad \times [I - P(T_b)][\underline{D}(T_1) - \underline{D}(T_b)]\xi && (XD) \\
&\qquad + 2\zeta_2^{*\prime}[Z(T_1 + k)_2 - Z(T_b + k)_2]'[I - P(T_b)]U^s && (XU) \\
&\qquad + 2\xi'[\underline{D}(T_1) - \underline{D}(T_b)]'[I - P(T_b)]U^s && (DU) \\
&\qquad + U^{s\prime}[P(T_1) - P(T_b)]U^s. && (UU)
\end{aligned}$$

As explained above, we can substitute U^s by U, and the orders of magnitude of (XX), (XU), and (UU) are the same as in the additive outlier models, as given in the proof of Proposition 1 and in Perron and Zhu (2005). They are, uniformly over all generic break dates $T_b \in \{\pi T, (1-\pi)T\}$, for Models I1b and I3b, $(XX) = |T_b - T_1|O_p(T^{1+2\eta})$, $(XU) = |T_b - T_1|O_p(T^{1+\eta})$, $(UU) = |T_b - T_1|O_p(T)$, and for Model I3, $(XX) = |T_b - T_1|^3 O_p(1)$, $(XU) = |T_b - T_1|^2 O_p(T^{1/2})$, $(UU) = |T_b - T_1|O_p(T)$. The order of magnitude of (DD) is, uniformly over all generic break dates $T_b \in \{\pi T, (1-\pi)T\}$: $(DD) = \tilde{I} \cdot O_p(kT^{1+2\eta})$, for Models I1b and I3b and $\tilde{I} \cdot O_p(k)$ for Model I3. Given these results, the rest of the proof is immediate using the arguments of Perron and Zhu (2005) and Schwarz's inequality. □

Proof of Propositions 3 and 8. When the estimate of the break date is consistent, the proof is trivial and omitted. Hence, we focus on the case of a consistent estimate of the break fraction at some rate T^a for $0 < a \leq 1$. Consider first the cases corresponding to the AO Models. The vector of the detrended dependent variable is $\tilde{Y}(k+2, T)$ and the vectors of regressors other than dummy variables are $\tilde{Y}(k+1, T-1)$, $\Delta\tilde{Y}(k+1, T-1), \ldots, \Delta\tilde{Y}(2, T-k)$. The t-statistic consists of inner products of these vectors. Let the matrix of regressors be $\hat{W}$. The t-statistic is then given by:

$$t_\alpha(\hat{\lambda}) = \frac{R[\hat{W}'\hat{W}]^{-1}\hat{W}'\tilde{Y}(k+2,T) - 1}{\sqrt{s^2[\hat{W}'\hat{W}]_{11}^{-1}}}$$

where $s^2 = T^{-1}\tilde{Y}(k+2,T)'[I - \hat{W}(\hat{W}'\hat{W})^{-1}\hat{W}']\tilde{Y}(k+2,T)$ and $R = [1, 0, \ldots, 0]$. We need to show that the orders of magnitude of terms associated with the deterministic trend are smaller than those of the terms associated with the error process. For example, from (8)

$$\tilde{Y} = \hat{M}_z Z(T_1)\phi + \hat{M}_z U = \hat{M}_z Z(T_1)_2\phi_2 + \hat{M}_z U = \tilde{M} + \tilde{U}.$$

Again let $\tilde{M}(a, b)$ and $\tilde{U}(a, b)$ be subvectors of $\tilde{M}$ and $\tilde{U}$ which include the ath to the bth elements. Then, for $j = 0, \ldots, k$, $\tilde{Y}(j+2, T) =$

$\tilde{M}(j+2,T)+\tilde{U}(j+2,T)$ and

$$\tilde{Y}(j+2,T)'\tilde{Y}(j+2,T)=\tilde{M}(j+2,T)'\tilde{M}(j+2,T)\\ +2\tilde{M}(j+2,T)'\tilde{U}(j+2,T)+\tilde{U}(j+2,T)'\tilde{U}(j+2,T).$$

We need to show that the last term dominates the first two under the conditions stated in the proposition. There are two vectors in level ($\tilde{Y}(k+2,T)$ and ($\tilde{Y}(k+1,T-1)$) and k vectors in differences ($\Delta\tilde{Y}(k+1,T-1),\ldots,\Delta\tilde{Y}(2,T-k)$). Note that all the inner products of the level vectors share the same order of magnitude with $\tilde{Y}'\tilde{Y}=\tilde{Y}(1,T)'\tilde{Y}(1,T)$, because k is controlled to be small relative to T. Hence, we only need to investigate the order of magnitude of $\tilde{Y}'\tilde{Y}$. Similarly, we can consider only $\Delta\tilde{Y}(2,T)'\Delta\tilde{Y}(2,T)$ for the inner products of vectors in differences, and $\tilde{Y}(2,T)'\Delta\tilde{Y}(2,T)$ for the inner products of a vector in level and a vector in difference. Write

$$\begin{aligned}\tilde{Y}(1,T)'\tilde{Y}(1,T)&=\tilde{M}(1,T)'\tilde{M}(1,T)\\&\quad+2\tilde{M}(1,T)'\tilde{U}+\tilde{U}(1,T)'\tilde{U}(1,T)\\ \Delta\tilde{Y}(2,T)'\Delta\tilde{Y}(2,T)&=\Delta\tilde{M}(2,T)'\Delta\tilde{M}(2,T)\\&\quad+2\Delta\tilde{M}(2,T)'\Delta\tilde{U}(2,T)+\Delta\tilde{U}(2,T)'\Delta\tilde{U}(2,T)\\ \tilde{Y}(2,T)'\Delta\tilde{Y}(2,T)&=\tilde{M}(2,T)'\Delta\tilde{M}(2,T)+\tilde{U}(2,T)'\Delta\tilde{U}(2,T)\\&\quad+\tilde{M}(2,T)'\Delta\tilde{U}(2,T)+\Delta\tilde{M}(2,T)'\tilde{U}(2,T).\end{aligned}$$

It is easy to show that, in all models, uniformly over all generic break dates $T_b\in\{\pi T,(1-\pi)T\}$, $\tilde{U}(1,T)'\tilde{U}(1,T)=O_p(T^2)$, $\tilde{U}(2,T)'\Delta\tilde{U}(2,T)=O_p(T)$, $\Delta\tilde{U}(2,T)'\Delta\tilde{U}(2,T)=O_p(T)$. When $\tilde{M}(1,T)'\tilde{M}(1,T)$ is of smaller order of magnitude than $\tilde{U}(1,T)'\tilde{U}(1,T)$, so is $\tilde{M}(1,T)'\tilde{U}$ by Schwarz's inequality. Similarly, when $\Delta\tilde{M}(2,T)'\Delta\tilde{M}(2,T)$ is of smaller order than $\Delta\tilde{U}(2,T)'\Delta\tilde{U}(2,T)$, so is $\Delta\tilde{M}(2,T)'\Delta\tilde{U}(2,T)$. Also, the order of magnitude of $\Delta\tilde{M}(2,T)'\Delta\tilde{M}(2,T)$ is no greater than that of $\tilde{M}(2,T)'\Delta\tilde{M}(2,T)$. Given Assumption 1, we have, uniformly over all generic break dates $T_b\in\{\pi T,(1-\pi)T\}$ for the various models,

Model	A1b	A2	A3
$\tilde{M}(1,T)'\tilde{M}(1,T)$	$O_p(T^{2-a+2\eta})$	$O_p(T^{3-2a})$	$O_p(T^{3-3a})$
$\tilde{M}(2,T)'\Delta\tilde{M}(2,T)$	$O_p(T^{1+2\eta})$	$O_p(T^{2-2a})$	$O_p(T^{2-2a})$
$\tilde{M}(2,T)'\Delta\tilde{U}(2,T)$	$O_p(T^{1-a/2+\eta})$	$O_p(T^{3/2-a})$	$O_p(T^{3(1-a)/2})$
$\Delta\tilde{M}(2,T)'\tilde{U}(2,T)$	$O_p(T^{1+\eta})$	$O_p(T^{3/2-a})$	$O_p(T^{3/2-a})$

For Model A3b, the order of magnitude of each of these terms is the maximum of the orders in Models A1b and A3. From the given orders of magnitude, the rest of the proof for the additive outlier models is straightforward.

Consider now the innovational outlier models and write the DGP as:

$$y_t = y_{t-1} + \beta' + d(\hat{T}_1)'_{t,2}\phi_2 + \sum_{i=1}^{k} d_i \Delta y_{t-i} + \hat{\varepsilon}_t^k$$

where $\hat{\varepsilon}_t^k = \varepsilon_t^k + [d(T_1)_{t,2} - d(\hat{T}_1)_{t,2}]'\phi_2$. Denote the matrix of regressors in regression (3) constructed with $\hat{T}_1$ as $W(\hat{T}_1)$. The t-statistic is:

$$\begin{aligned} t_\alpha(\hat{\lambda}) &= \frac{R[W(\hat{T}_1)'W(\hat{T}_1)]^{-1}W(\hat{T}_1)'Y(k+2,T) - 1}{\sqrt{s^2[W(\hat{T}_1)'W(\hat{T}_1)]_{11}^{-1}}} \\ &= \frac{RD[DW(\hat{T}_1)'W(\hat{T}_1)D]^{-1}DW(\hat{T}_1)'\hat{\varepsilon}^k(k+2,T)}{\sqrt{s^2[W(\hat{T}_1)'W(\hat{T}_1)]_{11}^{-1}}} \end{aligned}$$

and

$$\begin{aligned} s^2 &= T^{-1}Y(k+2,T)'[I - W(\hat{T}_1)[W(\hat{T}_1)'W(\hat{T}_1)]^{-1} \\ &\quad \times W(\hat{T}_1)']Y(k+2,T) \\ &= T^{-1}\hat{\varepsilon}^k(k+2,T)'[I - W(\hat{T}_1)D[DW(\hat{T}_1)'W(\hat{T}_1)D]^{-1} \\ &\quad \times DW(\hat{T}_1)']\hat{\varepsilon}^k(k+2,T) \end{aligned}$$

where $R = [1, 0, \ldots 0]$, D is a diagonal scaling matrix, $\hat{\varepsilon}^k = (\hat{\varepsilon}_1^k, \ldots, \hat{\varepsilon}_T^k)'$ and $\hat{\varepsilon}(a, b)$ is a subvector of $\hat{\varepsilon}$. Given the consistency of $\hat{\lambda}$, using $W(\hat{T}_1)$ or $W(T_1)$ makes no difference as long as $[d(T_1)_{t,2} - d(\hat{T}_1)_{t,2}]'\phi_2$ is asymptotically negligible relative to ε_t^k. Write

$$
\begin{aligned}
&\hat{\varepsilon}^k(k+2, T)'\hat{\varepsilon}^k(k+2, T) \\
&\quad = \varepsilon^k(k+2, T)'\varepsilon^k(k+2, T) \\
&\qquad + 2\phi_2'[d(T_1, k+2, T)_2 - d(\hat{T}_1, k+2, T)_2]'\varepsilon^k \\
&\qquad + \phi_2'[d(T_1, k+2, T)_2 - d(\hat{T}_1, k+2, T)_2]' \\
&\qquad \times [d(T_1, k+2, T)_2 - d(\hat{T}_1, k+2, T)_2]\phi_2
\end{aligned}
$$

and

$$
\begin{aligned}
&DW(\hat{T}_1)'\hat{\varepsilon}^k(k+2, T) \\
&\quad = DW(\hat{T}_1)'\varepsilon^k(k+2, T) \\
&\qquad + DW(\hat{T}_1)'[d(T_1, k+2, T)_2 - d(\hat{T}_1, k+2, T)_2]\phi_2.
\end{aligned}
$$

We need to show that the order of magnitude of $\|\varepsilon^k\|$ is strictly bigger than that of $\|[d(T_1, 1, T) - d(\hat{T}_1, 1, T)]\phi_2\|$ when the sufficient condition stated in the Proposition holds. We have $\|\varepsilon^k\| = O_p(T^{1/2})$ and

$$
\begin{aligned}
&\|[d(T_1, 1, T) - d(\hat{T}_1, 1, T)]\phi_2\| \\
&\quad = \begin{cases} \tilde{I} \cdot O_p(T^{1/2+\eta}), & \text{for Models I1b and I3b,} \\ \tilde{I} \cdot O_p(T^{1/2-a/2}), & \text{for Model I3.} \end{cases}
\end{aligned}
$$

To show the asymptotic negligibility of $\phi_2'[d(T_1, 1, T) - d(\hat{T}_1, 1, T)]'\varepsilon^k$, Schwarz's inequality can be applied again. For $DW(\hat{T}_1)'\hat{\varepsilon}^k$, D is such that $\|DW(\hat{T}_1)'\varepsilon^k\| = O_p(k^{1/2})$, and then

$$
\begin{aligned}
&\|DW(\hat{T}_1)'[d(T_1, 1, T) - d(\hat{T}_1, 1, T)]\phi_2\| \\
&\quad = \begin{cases} \tilde{I} \cdot O_p(T^{\eta}k^{1/2}), & \text{for Models I1b and I3b,} \\ \tilde{I} \cdot O_p(T^{1/2-a}k^{1/2}), & \text{for Model I3.} \end{cases}
\end{aligned}
$$

The rest of the proof is immediate. □

Proof of Propositions 7(a). We follow the arguments in Perron and Zhu (2005). Let $P(T_b)$ be the projection matrix constructed with $Z(T_b)$. Writing the sum of squared residuals of regression (5) as $SSR_t\ (\lambda)$, the estimate of the break fraction is $\hat{\lambda}_1^{AO} = \arg\min_{\lambda\in\{\pi,\ldots,1-\pi\}} SSR_T(\lambda)$. Obviously, we have for all T,

$$SSR_T(\hat{\lambda}_1^{AO}) - SSR_T(\lambda^c) \leq 0. \tag{A.5}$$

Note that, with $\lambda = T_b/T$,

$$\begin{aligned}
SSR_T(\lambda) - SSR_T(\lambda^c) &= Y'[P(T_1) - P(T_b)]Y \\
&= \phi_2' Z(T_1)_2'[P(T_1) - P(T_b)]Z(T_1)_2\phi_2 \\
&\quad + 2\phi_2' Z(T_1)_2'[P(T_1) - P(T_b)]U + U'[P(T_1) - P(T_b)]U \qquad \text{(A.6)} \\
&= \phi_2'[Z(T_1)_2 - Z(T_b)_2]'[I - P(T_b)][Z(T_1)_2 - Z(T_b)_2]\phi_2 \qquad (XX) \\
&\quad + 2\phi_2'[Z(T_1)_2 - Z(T_b)_2]'[I - P(T_b)]U \qquad (XU) \\
&\quad + U'[P(T_1) - P(T_b)]U. \qquad (UU)
\end{aligned}$$

The order of magnitude of each term is given by, uniformly over all generic break dates $T_b \in \{\pi T, (1-\pi)T\}$, $(XX) = |T_b - T_1|O_p(T^{1+2\eta})$, $(XU) = |T_b - T_1|O_p(T^{1+\eta})$, $(UU) = |T_b - T_1|O_p(T)$. We use the same contradiction argument as in Perron and Zhu (2005). Suppose that $|\hat{T}_1 - T_1| = O_p(T^s), -1-2\eta \leq s \leq 1$, the first term (XX) is then strictly positive and, unless $\hat{T}_1 = T_1$, dominates the others. This implies that the inequality in (A.5) will not hold with probability one, a contradiction, and we conclude that $(\hat{\lambda}_1^{AO} - \lambda^c) = o_p(T^{-2-2\eta})$. □

Acknowledgments

Perron acknowledges financial support for this work from the National Science Foundation under Grant SES-0649350.

References

Andrews, D.W.K. and Ploberger, W. (1994). Optimal tests when a nuisance parameter is present only under the alternative. *Econometrica* 62, 1383–1414.

Bai, J. (1998). A note on spurious break. *Econometric Theory* 14, 663–669.

Christiano, L.J. (1992). Searching for a break in GNP. *Journal of Business and Economic Statistics* 10, 237–250.

Dickey, D.A. and Fuller, W.A. (1979). Distribution of the estimators for autoregressive time series with a unit root. *Journal of the American Statistical Association* 74, 427–431.

Feinstein, C.H. (1972). National Income Expenditure and Output of the United Kingdom. In *Studies in the National Income and Expenditure of the United Kingdom*, Stone, Richard (ed.), vol. 6. Cambridge University Press, London.

Hatanaka, M. and Yamada, K. (1999). A Unit Root Test in the Presence of Structural Changes in I(1) and I(0) Models. In *Cointegration, Causality, and Forecasting: A Festschrift in Honour of Clive* W. J. Granger. Engle, R.F., White, H. (eds.), Oxford University Press.

Harvey, D.I., Leybourne, S.J. and Newbold, P. (2001). Innovational outlier unit root tests with an endogenously determined break in level. *Oxford Bulletin of Economics and Statistics* 63, 559–575.

Hecq, A. and Urbain, J.P. (1993). Misspecification tests, unit roots and level shifts. *Economics Letters* 43, 129–135.

Kim, D. (2007). Unit root, cointegration and structural changes: Theoretical analyses and improved testing procedures. Ph.D. Dissertation. Department of Economics, Boston University (Unpublished).

Kim, T., Leybourne, S.J. and Newbold, P. (2000). Spurious rejections by Perron tests in the presence of a break. *Oxford Bulletin of Economics and Statistics* 62, 433–444.

Kormendi, R.C. and Meguire, P. (1990). A multicountry characterization of the nonstationarity of aggregate output. *Journal of Money, Credit and Banking* 22, 77–93.

Lanne, M., Lütkepohl, H. and Saikkonen, P. (2003). Test procedures for unit roots in time series with level shifts at unknown time. *Oxford Bulletin of Economics and Statistics* 65, 91–115.

Lee, J. and Strazicich, M.C. (2001). Break point estimation and spurious rejections with endogenous unit root tests. *Oxford Bulletin of Economics and Statistics* 63, 535–558.

Lee, J. and Strazicich, M.C. (2003). Minimum Lagrange multiplier unit root test with two structural breaks. *The Review of Economics and Statistics* 85, 1082–1089.

Leybourne, S.J., Mills, T.C. and Newbold, P. (1998). Spurious rejections by Dickey–Fuller tests in the presence of a break under the null. *Journal of Econometrics* 87, 191–203.

Leybourne, S.J. and Newbold, P. (2000). Behavior of the standard and symmetric Dickey–Fuller-type tests when there is a break under the null hypothesis. *Econometrics Journal* 3, 1–15.

Maddison, A. (1982). *Phases of Capitalist Development.* Oxford University Press, London.

Montañés, A. (1997). Level shifts, unit roots and misspecification of the breaking date. *Economics Letters* 54, 7–13.

Montañés, A. and Olloqui, I. (1999). Misspecification of the breaking date in segmented trend variables: Effect on the unit root tests. *Economics Letters* 65, 301–307.

Montañés, A. and Reyes, M. (1998). Effect of a shift in the trend function on Dickey–Fuller unit root tests. *Econometric Theory* 14, 355–363.

Montañés, A. and Reyes, M. (1999). The asymptotic behaviour of the Dickey–Fuller tests under the crash hypothesis. *Statistics & Probability Letters* 42, 81–89.

Montañés, A. and Reyes, M. (2000). Structural breaks, unit roots and methods for removing the autocorrelation pattern. *Statistics & Probability Letters* 48, 401–409.

Ng, S. and Perron, P. (1995). Unit root tests in ARMA models with data dependent methods for the selection of the truncation lag. *Journal of the American Statistical Association* 90, 268–281.

Nunes, L.C., Kuan, C.-M. and Newbold, P. (1995). Spurious break. *Econometric Theory* 11, 736–749.

Nunes, L.C., Newbold, P. and Kuan, C. (1997). Testing for unit root with breaks: Evidence on the great crash and the unit root hypothesis reconsidered. *Oxford Bulletin of Economics and Statistics* 59, 435–448.

Perron, P. (1989). The great crash, the oil price shock and the unit root hypothesis. *Econometrica* 57, 1361–1401.

Perron, P. (1990). Testing for a unit root in a time series with a changing mean. *Journal of Business and Economic Statistics* 8, 153–162.

Perron, P. (1992). Trend, Unit Root and Structural Change: A Multi-country Study with Historical Data. In *Proceedings of the Business and Economic Statistics Section*. American Statistical Association, pp. 144–149.

Perron, P. (1997). Further evidence on breaking trend functions in macroeconomic variables. *Journal of Econometrics* 80, 355–385.

Perron, P. and Vogelsang, T.J. (1992a). Nonstationarity and level shifts with an application to purchasing power parity. *Journal of Business and Economic Statistics* 10, 301–320.

Perron, P. and Vogelsang, T.J. (1992b). Testing for a unit root in a time series with a changing mean: Correction and extensions. *Journal of Business and Economic Statistics* 10, 467–470.

Perron, P. and Vogelsang, T.J. (1993). The great crash, the oil price shock and the unit root hypothesis: Erratum. *Econometrica* 61, 248–249.

Perron, P. and Vogelsang, T.J. (1994). A note on the asymptotic distributions of unit root tests in the additive outlier model with breaks. *Revista de Econometria* 13, 181–201.

Perron, P. and Yabu, T. (2005). Testing for shifts in trend with an integrated or stationary noise component. *Journal of Business and Economic Statistics* (in press).

Perron, P. and Zhu, X. (2005). Structural breaks with deterministic and stochastic trends. *Journal of Econometrics* 129, 65–119.

Phillips, P.C.B. and Perron, P. (1988). Testing for a unit root in time series regression. *Biometrika* 75, 335–346.

Romer, C. (1989). The prewar business cycle reconsidered: New estimates of gross national product, 1869–1928. *Journal of Political Economy* 97, 1–37.

Said, S.E. and Dickey, D.A. (1984). Testing for unit roots in autoregressive moving average models with unknown order. *Biometrika* 71, 599–607.

Vogelsang, T.J. (1997). Wald-type tests for detecting breaks in the trend function of a dynamic time series. *Econometric Theory* 13, 818–849.

Vogelsang, T.J. and Perron, P. (1998). Additional tests for a unit root allowing for a break in the trend function at an unknown time. *International Economic Review* 39, 1073–1100.

Zivot, E. and Andrews, D.W.K. (1992). Further evidence on the great crash, the oil price shock and the unit root hypothesis. *Journal of Business and Economic Statistics* 10, 251–270.

Chapter 20

GLS-based Unit Root Tests with Multiple Structural Breaks Under Both the Null and the Alternative Hypotheses*

Josep Lluís Carrion-i-Silvestre[†,¶], Dukpa Kim[‡] and Pierre Perron[§]

[†] *University of Barcelona, Gran Via de les Corts, Catalanes, 585, 08007 Barcelona, Spain*

[‡] *University of Virginia, Charlotlesville, VA, USA*

[§] *Boston University, Boston, MA, USA*

[¶] *carrion@ub.edu*

Perron (1989, *Econometrica* 57, 1361–1401) introduced unit root tests valid when a break at a known date in the trend function of a time series is present. In particular, they allow a break under both the null and alternative hypotheses and are invariant to the magnitude of the shift in level and/or slope. The subsequent literature devised procedures valid in the case of an unknown break date. However, in doing so most research, in particular the commonly used test of Zivot and Andrews (1992, *Journal of Business & Economic Statistics* 10, 251–270), assumed that if a break occurs it does so only under the alternative hypothesis of stationarity. This is undesirable for several reasons. Kim and Perron (2009, *Journal of Econometrics* 148, 1–13) developed a methodology that allows a break at an unknown time under both the null and alternative hypotheses. When a break is present, the limit distribution of the test is the same as in the case of a known break date, allowing increased power while maintaining the correct size. We extend their work in several directions: (1) we allow for an arbitrary number of changes

*Originally Published in *Econometric Theory*, 25, pp. 1754–1792.

This chapter was prepared for the conference in honor of Professor Paul Newbold at the University of Nottingham, 21–22 September 2007.

in both the level and slope of the trend function; (2) we adopt the quasi-generalized least squares detrending method advocated by Elliott *et al.* (1996, *Econometrica* 64, 813–836) that permits tests that have local asymptotic power functions close to the local asymptotic Gaussian power envelope; (3) we consider a variety of tests, in particular the class of M-tests introduced in Stock (1999, *Cointegration, Causality, and Forecasting: A Festschrift for Clive W.J. Granger*) and analyzed in Ng and Perron (2001, *Econometrica* 69, 1519–1554).

Keywords: Multiple structural breaks, unit root, GLS detrending, point-optimal tests, trend breaks.

1. Introduction

Professor Paul Newbold has made important contributions in econometrics, in particular in the area of the analysis of nonstationary time series. I (Perron) have been honored and privileged to have him participate in a research agenda that I put forward in Perron (1989). This chapter is, in part, a product motivated and made possible by some of his work.

It is well known that a break in the deterministic trend affects the outcome of unit root tests. Perron (1989) showed that a standard Dickey and Fuller (1979) (DF) type unit root test is not consistent if the alternative is a stationary noise component with a break in the slope of the deterministic trend. His main point is that the existence of an exogenous shock that has a permanent effect will lead to a nonrejection of the unit root hypothesis even though it is not true. Of interest also is the fact that Leybourne *et al.* (1998) and Leybourne and Newbold (2000) analyzed the effect of a break on a standard DF test under the unit root null hypothesis and showed that size distortions can occur, especially when the break is early in the sample. Perron (1989, 1990) proposed alternative unit root tests that allow the possibility of a break under both the null and alternative hypotheses. These tests have less power than a standard DF type test when there is no break. Nonetheless, they have a correct size asymptotically and are consistent whether there is a break or not. Moreover, they are invariant to the break parameters, and thus their performance does not depend on the magnitude of the break. The most controversial assumption, however, is that its timing is known *a priori* (see Christiano, 1992).

For Perron (1989, 1990) test procedures to be valid, the break date should be chosen independently of the given data. Whenever a systematic search for a break is done, the limiting distributions in Perron (1989, 1990) are no longer appropriate. Historical facts can often provide good guidance in choosing a break date independently of the given data. Even in that case, it is very likely that an imprecise break date is used. Hecq and Urbain (1993) showed, by simulations, that the use of an incorrect break date in the Perron (1990) test, applicable with nontrending data, causes size distortions and power loss, though as shown by Montañés (1997) this effect disappears asymptotically. Montañés and Olloqui (1999) extended the analysis to the Perron (1989) tests, applicable with trending data, and showed that a loss of power occurs even in large samples. Kim *et al.* (2000) examined the effect of using a wrong break date under the null hypothesis.

Zivot and Andrews (1992) provided methods that treat the occurrence of the break date as unknown and that have become quite popular (see also Perron and Vogelsang, 1992a, 1992b; parts of Perron, 1997). However, Professor Newbold and his co-authors have clearly recognized the potential pitfalls of their approach. In this line of work, a break is not allowed under the null hypothesis, only under the alternative, mostly because of the theoretical apparatus adopted. This means, e.g., that under the null hypothesis a level shift must be viewed as coming from the tail of the distribution of the data generating process, and a slope change involves errors with a different mean in some subsamples. This framework is convenient because it allows establishing various unit root testing procedures; e.g., by minimizing the t-statistic related to the sum of the autoregressive coefficients over each possible break date. This approach is, however, contrary to the original motivation of Perron (1989), which was to devise procedures that are invariant to the magnitude of the shift in level and/or slope. In particular, if a change is present it is allowed under both the null and alternative hypotheses.

The existence of a structural break in the trend function is a problem of long horizon data; it can happen whether the noise component is stationary or has a unit root. As argued in Nunes *et al.* (1997) and Harvey *et al.* (2001) (see also Vogelsang and Perron, 1998; Lee and Strazicich, 2001), if the noise component has a unit root and a break occurs in the trend function, the Zivot and Andrews (1992) type test statistics often diverge or are not invariant to break parameters. This is a natural consequence of not permitting a break under the null hypothesis. An added consequence is that these types of tests have substantially less power than the Perron (1989) tests because they do not fully utilize the information about the break when one is present.

Despite these shortcomings, the method of Zivot and Andrews (1992) has remained popular in empirical work, probably in part because of the lack of sound statistical methods that could tackle the problem of allowing for changes in the trend function at unknown times under both the null and alternative hypotheses. Using recent developments on structural change problems related to nonstationary data by Perron and Zhu (2005) and Perron and Yabu (2009). Kim and Perron (2009) developed a methodology to devise new test procedures that allow a break in the trend function at an unknown time under both the null and alternative hypotheses. Also, when a break is present, the limit distribution of the test is the same as in the case of a known break date, thereby allowing increased power while maintaining the correct size. Simulation experiments confirm that it offers an improvement over commonly used methods in small samples.

In this chapter, we extend the work of Kim and Perron (2009) in several directions: (1) we allow for an arbitrary number of changes in both the level and slope of the trend function; (2) we adopt the so-called quasi-generalized least squares (quasi-GLS) detrending method advocated by Elliott *et al.* (1996) that permits tests that have local asymptotic power functions close to the local asymptotic

Gaussian power envelope[1]; (3) we consider a variety of tests, in particular the class of M-tests introduced in Stock (1999) and analyzed in Ng and Perron (2001). On the other hand, we restrict our analysis to the case of the so-called additive outlier (AO) models as defined in Perron (1989). Hatanaka and Yamada (1999) is related to the approach of Kim and Perron (2009) and deals with two breaks for the so-called innovational outlier (IO) type model. In independent work, Harris *et al.* (2009) consider a similar problem. Their analysis is, however, restricted to a single break, and they propose using a DF type test with different methods to obtain tests valid with or without a break. Our studies are therefore complementary.

The chapter is organized as follows. In Section 2, we present the model with multiple structural breaks in the trend function. Section 3 discusses the feasible point optimal test assuming the break dates to be known and provides details on the methods to construct the relevant noncentrality parameter arising in the quasi-differencing procedures used. Section 4 analyzes the M-class and related unit root tests allowing for multiple breaks. Section 5 considers the case with unknown break dates. It shows that if these are estimated by minimizing the sum of squared residuals from the appropriate GLS regression, the limit distributions of the tests are the same as in the known break date case, provided breaks are present. It also provides details on how to compute the estimate of the break dates. Section 6 provides preliminary simulations showing that the tests perform well but that they exhibit important size distortions when no break occurs. A solution to this problem is offered in Section 7 based on a pretest for changes in the slope of a trend function allowing the noise component to be stationary or integrated based on the work of Perron and Yabu (2009). We show that the resulting procedure involving the pretest has good size and

[1]Perron and Rodríguez (2003) also consider GLS-detrended type procedures allowing a single break. The approach is, however, quite different. Also, although their treatment yields 'optimal' tests when a break is present, this is not the case if a break is absent.

power superior to that of alternative methods. Section 8 offers brief concluding remarks and an Appendix the proofs of various theoretical results.

2. The Model

Let y_t be a stochastic process generated according to

$$y_t = d_t + u_t, \tag{1}$$

$$u_t = \alpha u_{t-1} + v_t, \quad t = 0, \dots, T, \tag{2}$$

where $\{u_t\}$ is an unobserved mean-zero process. We assume that $u_0 = 0$, although the results generally hold for the weaker requirement that $\mathrm{E}(u_0^2) < \infty$. The disturbance term v_t is defined by $v_t = \sum_{i=0}^{\infty} \gamma_i \eta_{t-i}$ with $\sum_{i=0}^{\infty} i\,|\gamma_i| < \infty$ and $\{\eta_t\}$ a martingale difference sequence adapted to the filtration $F_t = \sigma{-}\mathit{field}\{\eta_{t-i}; i \geq 0\}$. We define the long-run and short-run variance as $\sigma^2 = \sigma_\eta^2 \gamma(1)^2$ and $\sigma_\eta^2 = \lim_{T\to\infty} T^{-1} \Sigma_{t=1}^T E(\eta_t^2)$, respectively.

We consider three models: Model 0 ('level shift' or 'crash'), Model I ('slope change' or 'changing growth'), and Model II ('mixed change').[2] Let $DU_t(T_j^0) = 1$ and $DT_t^*(T_j^0) = (t - T_j^0)$ for $t > T_j^0$ and 0 elsewhere, with $T_j^0 = [T\lambda_j^0]$ denoting the jth break date, with $[\cdot]$ the integer part and $\lambda_j^0 \equiv T_j^0/T \in (0, 1)$ the break fraction parameter. As a matter of notation, the true break fractions and break dates are denoted with a superscript zero, and their estimates are denoted with a hat. We also use the convention that $T_0^0 = 0$ and $T_{m+1}^0 = T$. We collect the m break fraction parameters in the vector $\lambda^0 = (\lambda_1^0, \dots, \lambda_m^0)'$. For now it is assumed that the break dates are known; this will be relaxed later.

The deterministic component in (1) is given by

$$d_t = z_t'(T_0^0)\psi_0 + z_t'(T_1^0)\psi_1 + \cdots + z_t'(T_m^0)\psi_m \equiv z_t'(\lambda^0)\psi, \tag{3}$$

where $z_t(\lambda^0) = [z_t'(T_0^0), \dots, z_t'(T_m^0)]'$ and $\psi = (\psi_0', \dots, \psi_m')'$. The deterministic components and associated coefficients are defined by

[2] Perron and Rodríguez (2003) also considered Models I and II but only with one break.

$z_t(T_0^0) \equiv z_t(0) = (1, t)'$ with $\psi_0 = (\mu_0, \beta_0)'$ and

$$z_t(T_j^0) = \begin{cases} DU_t(T_j^0), & \text{in Model 0,} \\ DT_t^*(T_j^0), & \text{in Model I,} \\ (DU_t(T_j^0), DT_t^*(T_j^0))', & \text{in Model II,} \end{cases}$$

for $1 \leq j \leq m$, with $\psi_j = \mu_j$ in Model 0, $\psi_j = \beta_j$ in Model I, and $\psi_j = (\mu_j, \beta_j)'$ in Model II.

For Models 0 and II, we also consider the case where the magnitude of level shifts gets larger as the sample size grows, i.e., $(\mu_1, \ldots, \mu_m) = T^{1/2+\eta}(\kappa_1, \ldots, \kappa_m)$, with $\eta > 0$. The models are then labeled as Models 0b and IIb, respectively. These models are useful to capture important properties of the tests in finite samples. In Models 0 and II, the level shifts belong to the class of 'slowly evolving trend' defined by Elliott *et al.* (1996) and have no effect on the asymptotic size and power of the tests. When the magnitude of the shifts is nonnegligible, this typically implies that the derived asymptotic distribution is a bad approximation to the finite-sample distribution. In Models 0b and IIb, the level shifts do not belong to the class of 'slowly evolving trend'. As shown in Harvey *et al.* (2001) this framework provides better approximations. A second feature of importance is related to the estimation of the unknown break fractions. As shown by Perron and Zhu (2005), the rate of convergence increases when the level shifts are modeled to increase as the sample size grows. This has important implications for the properties of the unit root tests, as will be shown in the following sections.

Remark 1. It is possible to assume that there is no trending deterministic component in Model 0. We do not explicitly consider this case in the subsequent analysis, but it should be understood that most of our results pertaining to Models 0 and 0b can readily be applied to the level shift model with no time trend.

The so-called GLS-detrended unit root test statistics are based on the use of the quasi-differenced variables $y_t^{\bar{\alpha}}$ and $z_t^{\bar{\alpha}}(\lambda^0)$ defined by $y_1^{\bar{\alpha}} = y_1, z_t^{\bar{\alpha}}(\lambda^0) = z_1(\lambda^0)$, and

$$y_t^{\bar{\alpha}} = (1 - \bar{\alpha}L)y_t,$$
$$z_t^{\bar{\alpha}}(\lambda^0) = (1 - \bar{\alpha}L)z_t(\lambda^0)$$

for $t = 2, \ldots, T$ with $\bar{\alpha} = 1 + \bar{c}/T$ where $\bar{c}$ is a noncentrality parameter to be defined later. Once the data have been transformed, the parameters ψ, associated with the deterministic components, can be estimated by minimizing the following objective function:

$$S^*(\psi, \bar{\alpha}, \lambda^0) = \sum_{t=1}^{T} (y_t^{\bar{\alpha}} - \psi' z_t^{\bar{\alpha}}(\lambda^0))^2. \tag{4}$$

We denote the minimum of this function by $S(\bar{\alpha}, \lambda^0)$.

3. Feasible Point Optimal Test with Multiple Structural Breaks

The choice of the noncentrality parameter $\bar{c}$ is related to the Gaussian point optimal statistic to test the null hypothesis of $\alpha = 1$ in (2) against the alternative hypothesis that $\alpha = \bar{\alpha}$, as suggested by Elliott *et al.* (1996). Following their analysis and Perron and Rodríguez (2003), the feasible point optimal statistic is given by

$$P_T^{GLS}(\lambda^0) = \left\{S(\bar{\alpha}, \lambda^0) - \bar{\alpha}S(1, \lambda^0)\right\} / s^2(\lambda^0), \tag{5}$$

where $s^2(\lambda^0)$ is an estimate of the spectral density at frequency zero of v_t. Following Ng and Perron (2001) and Perron and Ng (1998), we use an autoregressive estimate defined by

$$s(\lambda^0)^2 = s_{ek}^2 \Big/ \left(1 - \sum_{j=1}^{k} \hat{b}_j\right)^2, \tag{6}$$

where $s_{ek}^2 = (T-k)^{-1}\Sigma_{t=k+1}^{T}\hat{e}_{t,k}^2$ and $\{\hat{b}_j, \hat{e}_{t,k}\}$ obtained from the ordinary least squares (OLS) regression

$$\Delta\tilde{y}_t = b_0\tilde{y}_{t-1} + \sum_{j=1}^{k} b_j \Delta\tilde{y}_{t-j} + e_{t,k}, \tag{7}$$

with $\tilde{y}_t = y_t - \hat{\psi}' z_t(\lambda^0)$ where $\hat{\psi}$ minimizes (4). The order of the autoregression k is selected using the modified information criteria suggested by Ng and Perron (2001) with the modification proposed by Perron and Qu (2007).

Let $W_c(r)$ be an Ornstein–Uhlenbeck process, i.e., the solution to the stochastic differential equation $dW_c(r) = cW_C(r)dr + dW(r)$, with $W_c(0) = 0$ where $W(r)$ is the standard Brownian motion. Denoting by $\Longrightarrow$ weak convergence of the associated measure of probability, the limiting distribution of the test $P_T^{GLS}(\lambda^0)$ is given in the following theorem.

Theorem 1. *Let $\{y_t\}_{t=1}^T$ be the stochastic process generated according to (1) and (2) with $\alpha = 1 + c/T$ and $s^2(\lambda^0)$ be a consistent estimate of σ^2. Then,*

(i) *For Models 0 and 0b:*

$$P_T^{GLS}(\lambda^0) \Rightarrow \bar{c}^2 \int_0^1 V_{c,\bar{c}}^2(r)dr + (1-\bar{c})V_{c,\bar{c}}^2(1) \equiv K^{P_T^{GLS}}(c,\bar{c}),$$

where $V_{c,\bar{c}}^2(r) = W_c(r) - r[bW_c(1) + 3(1-b)\int_0^1 sW_c(s)ds]$ and $b = (1-\bar{c})/(1-\bar{c}+\bar{c}^2/3)$.

(ii) *For Models* I, II, *and* II*b:*

$$P_T^{GLS}(\lambda^0) \Rightarrow M(c,0,\lambda^0) - M(c,\bar{c},\lambda^0) - 2\bar{c}\int_0^1 W_c(r)dW(r)$$

$$+ (\bar{c}^2 - 2\bar{c}c)\int_0^1 W_c(r)^2 dr - \bar{c} \equiv H^{P_T^{GLS}}(c,\bar{c},\lambda^0),$$

where $M(c,\bar{c},\lambda^0) = \bar{V}(\lambda^0)' A(\lambda^0)^{-1}\bar{V}(\lambda^0)$, $\bar{V}(\lambda^0) = (V(\lambda_0^0),\ldots, V(\lambda_m^0))'$ with

$$V(\lambda_i^0) = (1+\bar{c}\lambda_i^0)[W(1) - W(\lambda_i^0) + (c-\bar{c})\int_{\lambda_i^0}^1 W_c(r)dr]$$

$$-\bar{c}\int_{\lambda_i^0}^1 r dW(r) - (c-\bar{c})\bar{c}\int_{\lambda_i^0}^1 rW_c(r)dr$$

and $A(\lambda^0)$ *a symmetric matrix defined by*

$$A(\lambda^0) = \begin{bmatrix} a(\lambda_0^0, \lambda_0^0) & a(\lambda_0^0, \lambda_1^0) & \cdots & a(\lambda_0^0, \lambda_m^0) \\ & a(\lambda_1^0, \lambda_1^0) & \cdots & a(\lambda_1^0, \lambda_m^0) \\ & & \ddots & \vdots \\ & & & a(\lambda_m^0, \lambda_m^0) \end{bmatrix} \tag{8}$$

with

$$a(\lambda_i^0, \lambda_j^0) = (1/6)(1 - \lambda_j) \\ \times [6\bar{c}(\lambda_i - 1) + \bar{c}^2\{\lambda_j(3\lambda_i - 1) - 3\lambda_i - \lambda_j^2 + 2\} + 6].$$

The proof for Model II is given in the appendix. Because Models 0 and I can be viewed as a special case of Model II, no separate proof is provided. The limiting distribution in (i) is the same as that of the linear time trend model with no break, which can be found in Elliott *et al.* (1996). Because the break dates are assumed known here, the test statistic $P_T^{GLS}(\lambda^0)$ is exactly invariant to the value of the coefficients associated with all regressors including those pertaining to the change in the trend. Hence, there is no distinction between Models 0 and 0b and between Models II and IIb. The limiting distributions of the tests for Models I, II, and IIb depend on the number of breaks and the vector of break fractions.

We now turn to the choice of the noncentrality parameter $\bar{c}$. Using the limiting distributions in Theorem 1, we can obtain the local Gaussian power envelope for the various cases. For Models I, II, and IIb, it is defined by $\pi^*(c) = \Pr[H^{P_T^{GLS}}(c, c, \lambda^0) < b^{P_T^{GLS}}(c, \lambda^0)]$, where, with υ the size of the test, $b^{P_T^{GLS}}(c, \lambda^0)$ is such that $\Pr[H^{P_T^{GLS}}(0, c, \lambda^0) < b^{P_T^{GLS}}(c, \lambda^0)] = \upsilon$. This power envelope allows us to obtain the 'optimal' noncentrality parameter $\bar{c}$ that Elliott *et al.* (1996) recommended choosing as the value that yields tests with a 50% asymptotic power, i.e., $\bar{c}$ is such that $\Pr[H^{P_T^{GLS}}(\bar{c}, \bar{c}) < b^{P_T^{GLS}}(\bar{c})] = 0.5$.

For Models I, II, and IIb the parameter $\bar{c}$ depends on the number of structural breaks and their positions. Table 1 presents the relevant values for $m = 1$ and 2 breaks for all possible combinations of break

Table 1. Values for the noncentrality parameter $\bar{c}$.

	$m = 1$								
λ_1	0.1	0.2	0.3	0.4	0.5	0.6	0.7	0.8	0.9
$\bar{c}$	−16.0	−17.6	−18.2	−18.2	−17.8	−17.2	−16.4	−15.6	−14.8

	$m = 2$								
$\lambda_1 \backslash \lambda_2$	0.1	0.2	0.3	0.4	0.5	0.6	0.7	0.8	0.9
0.1		−18.0	−19.7	−20.6	−21.0	−20.9	−20.5	−19.6	−18.1
0.2			−19.5	−20.9	−21.5	−21.6	−21.2	−20.4	−19.2
0.3				−20.3	−21.5	−21.8	−21.5	−20.8	−19.6
0.4					−20.5	−21.3	−21.4	−20.8	−19.6
0.5						−20.1	−20.6	−20.3	−19.3
0.6							−19.2	−19.4	−18.7
0.7								−18.0	−17.8
0.8									−16.5

fraction vectors $\lambda^0 = (\lambda_1^0, \lambda_2^0)'$, with $\lambda_i^0 = \{0.1, 0.2, \ldots, 0.9\}$. These were obtained by simulations using 1,000 steps to approximate the Wiener process and 10,000 replications. This should be sufficient for most applications, but we also constructed a response surface analysis that can provide the relevant values for up to 5 breaks and any combinations of break fractions. The details of its construction and the results are available on the corresponding author's web page, www.eco.ub.es/∼carrion. Note that for Models 0 and 0b, a similar analysis holds, though much simpler because the limit distribution is invariant to any break parameters. The results are described in Elliott *et al.* (1996), and $\bar{c} = -13.5$ (if no trend is present $\bar{c} = -7$).

4. The M-Class and Related Unit Root Tests

Following Perron and Rodriguez (2003), we use the M-class of tests analyzed in Ng and Perron (2001) allowing for multiple structural breaks, defined by

$$MZ_\alpha^{GLS}(\lambda^0) = (T^{-1}\tilde{y}_T^2 - s(\lambda^0)^2)\left(2T^{-2}\sum_{t=1}^{T}\tilde{y}_{t-1}^2\right)^{-1}, \qquad (9)$$

$$MSB^{GLS}(\lambda^0) = \left(s(\lambda^0)^{-2} T^{-2} \sum_{t=1}^{T} \tilde{y}_{t-1}^2 \right)^{1/2}, \tag{10}$$

$$MZ_t^{GLS}(\lambda^0) = (T^{-1}\tilde{y}_T^2 - s(\lambda^0)^2) \left(4s(\lambda^0)^2 T^{-2} \sum_{t=1}^{T} \tilde{y}_{t-1}^2 \right)^{-1/2}, \tag{11}$$

with $\tilde{y}_t = y_t - \hat{\psi}' z_t(\lambda^0)$, where $\hat{\psi}$ minimizes (4) and $s(\lambda^0)^2$ is defined in (6). Another statistic, following Ng and Perron (2001), is a modified feasible point optimal test defined by

$$MP_T^{GLS}(\lambda^0) = \left[\bar{c}^2 T^{-2} \sum_{t=1}^{T} \tilde{y}_{t-1}^2 + (1-\bar{c}) T^{-1} \tilde{y}_T^2 \right] \Big/ s(\lambda^0)^2.$$

This test is based on the same motivation that leads to the definition of the M-tests in Stock (1999), namely, to provide functionals of sample moments that have the same asymptotic distributions as well-known unit root tests. The $MP_T^{GLS}(\lambda^0)$ is important because its limiting distribution coincides with that of the feasible point optimal test.

The following theorem provides the limit null distribution of the various tests considered.

Theorem 2. *Let $\{y_t\}_{t=1}^T$ be the stochastic process generated according to (1) and (2) with $\alpha = 1 + c/T$ and $s^2(\lambda^0)$ be a consistent estimate of σ^2. Then,*

(i) *For Models 0 and 0b:*

$$MZ_\alpha^{GLS}(\lambda^0) \Rightarrow 0.5(V_{c,\bar{c}}(1)^2 - 1) \left(\int_0^1 V_{c,\bar{c}}(r)^2 dr \right)^{-1},$$

$$MSB^{GLS}(\lambda^0) \Rightarrow \left(\int_0^1 V_{c,\bar{c}}(r)^2 dr \right)^{1/2},$$

where $V_{c,\bar{c}}^2(r) = W_c(r) - r[bW_c(1) + 3(1-b)\int_0^1 sW_c(s)ds]$ and $b = (1-\bar{c})/(1-\bar{c}+\bar{c}^2/3)$.

(ii) *For Models* I, II, *and* II*b:*

$$MZ_\alpha^{GLS}(\lambda^0) \Rightarrow 0.5(V_{c,\bar{c}}(1,\lambda^0)^2 - 1)\left(\int_0^1 V_{c,\bar{c}}(r,\lambda^0)^2 dr\right)^{-1}$$
$$\equiv H^{MZ_\alpha^{GLS}}(c,\bar{c},\lambda^0),$$
$$MSB^{GLS}(\lambda^0) \Rightarrow \left(\int_0^1 V_{c,\bar{c}}(r,\lambda^0)^2 dr\right)^{1/2} \equiv H^{MSB^{GLS}}(c,\bar{c},\lambda^0),$$

where $V_{c,\bar{c}}(r,\lambda^0) = W_c(r) - z_2(r)A(\lambda^0)^{-1}\bar{V}(\lambda^0)$ *with* $A(\lambda^0)$ *and* $\bar{V}(\lambda^0)$ *as defined in Theorem* 1, *and* $z_2(r) = (r, (r-\lambda_1^0)1(r > \lambda_1^0), \ldots, (r-\lambda_m^0)1(r > \lambda_m^0))$, *where* $1(\cdot)$ *is the indicator function.*

(iii) *The limiting distribution of* $MZ_t^{GLS}(\lambda^0)$ *in all models can be obtained in view of the fact that* $MZ_t^{GLS}(\lambda^0) = MZ_\alpha^{GLS}(\lambda^0) \cdot MSB^{GLS}(\lambda^0)$, *denoted by* $H^{MZ_t^{GLS}}(c,\bar{c},\lambda^0)$.

Again, the limiting distribution in (i) is the same as that of the linear time trend model with no break given in Ng and Perron (2001) because of the invariance of the tests to the break parameters in these models. This is not the case for Models I, II, and IIb, for which the limiting distribution depends on the number and location of the break points. The proof of Model II is given in the appendix, and the proof for the other models easily follows.

For Models I, II, and IIb, the limit distributions depend on the number of breaks and their positions. Tables 2–5 present the 1, 2.5, 5, and 10% percentiles of the limit distributions of the tests $m = 1$ and 2 breaks for all possible combinations of break fraction vectors $\lambda^0 = (\lambda_1^0, \lambda_2^0)'$ with $\lambda_i^0 = \{0.1, 0.2, \ldots, 0.9\}$ (again, obtained via simulations using 1,000 steps to approximate the Wiener process and 10,000 replications). We also constructed a response surface analysis that can provide the relevant values for up to 5 breaks. The details of its construction and the results are available on the corresponding author's web page, www.eco.ub.es/~carrion.

The asymptotic power functions of the tests are defined by $\pi^*_{J^{GLS}}(c,\bar{c},\lambda^0) = \Pr[H^{J^{GLS}}(c,\bar{c},\lambda^0) < b^{J^{GLS}}(\bar{c},\lambda^0)]$ for $J = MZ_\alpha$, MSB, MZ_t with $H^{J^{GLS}}(c,\bar{c},\lambda^0)$ defined in Theorem 2. The constants $b^{J^{GLS}}(\bar{c},\lambda^0)$ are such that $\Pr[H^{J^{GLS}}(0,\bar{c},\lambda^0) < b^{J^{GLS}}(\bar{c},\lambda^0)] = \upsilon$, the size of the tests. To assess the efficiency in terms of local asymptotic

Table 2. Asymptotic critical values for the $P_T^{GLS}(\lambda^0)$ and $MP_T^{GLS}(\lambda^0)$ tests.

	$m = 1$								
λ_1	0.1	0.2	0.3	0.4	0.5	0.6	0.7	0.8	0.9
1%	4.354	4.807	5.042	5.079	4.949	4.704	4.423	4.191	4.075
2.5%	5.235	5.732	5.978	6.000	5.837	5.556	5.250	5.016	4.933
5%	6.238	6.759	6.992	6.986	6.789	6.476	6.146	5.910	5.865
10%	7.492	8.076	8.344	8.335	8.105	7.744	7.373	7.118	7.086
$m = 2$	1% (top triangle), 2.5% (bottom triangle)								
0.1		4.723	5.213	5.536	5.71	5.724	5.573	5.266	4.810
0.2	5.520		5.134	5.585	5.819	5.866	5.743	5.475	5.087
0.3	6.095	5.948		5.471	5.829	5.932	5.841	5.596	5.241
0.4	6.471	6.496	6.328		5.617	5.848	5.812	5.595	5.252
0.5	6.670	6.768	6.772	6.502		5.524	5.622	5.461	5.135
0.6	6.686	6.817	6.892	6.798	6.403		5.214	5.197	4.923
0.7	6.514	6.678	6.783	6.759	6.548	6.063		4.772	4.654
0.8	6.155	6.384	6.516	6.516	6.373	6.081	5.596		4.328
0.9	5.593	5.954	6.143	6.157	6.025	5.795	5.511	5.172	
$m = 2$	5% (top triangle), 10% bottom triangle								
0.1		6.478	7.034	7.398	7.592	7.607	7.434	7.072	6.511
0.2	7.791		6.897	7.431	7.708	7.765	7.624	7.310	6.844
0.3	8.376	8.210		7.264	7.712	7.850	7.749	7.461	7.037
0.4	8.733	8.757	8.568		7.435	7.747	7.727	7.475	7.070
0.5	8.903	9.013	9.012	8.719		7.338	7.498	7.328	6.951
0.6	8.887	9.039	9.123	9.019	8.600		6.997	7.015	6.719
0.7	8.697	8.873	8.997	8.978	8.754	8.245		6.524	6.424
0.8	8.361	8.563	8.706	8.722	8.582	8.276	7.772		6.094
0.9	7.902	8.159	8.322	8.345	8.226	7.998	7.705	7.358	

Note: For 1% and 5% (resp., 2.5% and 10%), the value of λ_1 is on the horizontal (resp., vertical) axis and that of λ_2 on the vertical (resp., horizontal) axis.

power of the various tests, we consider the case with a single break occurring at $\lambda^0 = 0.3$, 0.5, and 0.7. The asymptotic power functions are shown in Fig. 1, where the solid line is the power envelope. As can be seen, the local power functions are nearly identical and indeed very close to the Gaussian power envelope. So from this local asymptotic power perspective, all tests are nearly efficient. For the case of Models 0 and 0b, we already know from Ng and Perron (2001) that the local power functions of the various tests considered are very close to the power envelope.

Table 3. Asymptotic critical values for the $MZ_t^{GLS}(\lambda^0)$ test.

	$m = 1$								
λ_1	0.1	0.2	0.3	0.4	0.5	0.6	0.7	0.8	0.9
1%	−3.800	−3.942	−3.981	−3.969	−3.938	−3.897	−3.843	−3.763	−3.639
2.5%	−3.485	−3.650	−3.702	−3.695	−3.662	−3.618	−3.561	−3.477	−3.344
5%	−3.232	−3.396	−3.448	−3.445	−3.419	−3.382	−3.328	−3.240	−3.094
10%	−2.950	−3.118	−3.171	−3.166	−3.137	−3.096	−3.039	−2.949	−2.801
$m = 2$	1% (top triangle), 2.5% (bottom triangle)								
0.1	−4.131	−4.276	−4.338	−4.347	−4.326	−4.285	−4.222	−4.124	
0.2	−3.812		−4.288	−4.384	−4.415	−4.403	−4.363	−4.298	−4.203
0.3	−3.980	−3.965		−4.337	−4.403	−4.415	−4.385	−4.321	−4.226
0.4	−4.048	−4.084	−4.019		−4.325	−4.371	−4.365	−4.313	−4.223
0.5	−4.059	−4.118	−4.105	−4.011		−4.278	−4.304	−4.277	−4.202
0.6	−4.040	−4.105	−4.116	−4.074	−3.964		−4.203	−4.206	−4.158
0.7	−4.001	−4.066	−4.083	−4.066	−4.007	−3.887		−4.100	−4.079
0.8	−3.933	−4.003	−4.022	−4.012	−3.977	−3.905	−3.775		−3.960
0.9	−3.807	−3.902	−3.927	−3.918	−3.892	−3.845	−3.761	−3.619	
$m = 2$	5% (top triangle), 10% bottom triangle								
0.1		−3.566	−3.729	−3.797	−3.812	−3.801	−3.768	−3.699	−3.558
0.2	−3.266		−3.723	−3.836	−3.872	−3.865	−3.834	−3.775	−3.665
0.3	−3.425	−3.437		−3.780	−3.861	−3.875	−3.849	−3.792	−3.693
0.4	−3.499	−3.553	−3.502		−3.777	−3.836	−3.832	−3.781	−3.684
0.5	−3.524	−3.598	−3.588	−3.502		−3.737	−3.777	−3.747	−3.657
0.6	−3.519	−3.600	−3.611	−3.566	−3.463		−3.665	−3.678	−3.609
0.7	−3.489	−3.571	−3.589	−3.565	−3.503	−3.388		−3.551	−3.524
0.8	−3.417	−3.507	−3.528	−3.512	−3.470	−3.395	−3.269		−3.377
0.9	−3.272	−3.387	−3.417	−3.405	−3.372	−3.319	−3.232	−3.089	

Note: For 1% and 5% (resp., 2.5% and 10%), the value of λ_1 is on the horizontal (resp., vertical) axis and that of λ_2 on the vertical (resp., horizontal) axis.

5. The Case with Unknown Break Dates

The analysis so far has assumed that the timing of the structural breaks is known. We need to establish a procedure to estimate them and deduce the effect on the limit distribution of the various unit root tests. Section 5.1 presents the method of estimation, the rate of convergence of the estimates, and the properties of the tests when evaluated at these estimates. Section 5.2 presents an efficient method to compute the estimates when $m > 1$.

Table 4. Asymptotic critical values for the $MSB^{GLS}(\lambda^0)$ test.

	$m = 1$								
λ_1	0.1	0.2	0.3	0.4	0.5	0.6	0.7	0.8	0.9
1%	0.130	0.125	0.123	0.124	0.125	0.127	0.129	0.132	0.137
2.5%	0.142	0.135	0.133	0.133	0.135	0.137	0.139	0.143	0.149
5%	0.152	0.144	0.142	0.143	0.144	0.147	0.149	0.154	0.161
10%	0.166	0.157	0.154	0.155	0.157	0.159	0.163	0.169	0.178
$m = 2$	1% (top triangle), 2.5% (bottom triangle)								
0.1		0.120	0.115	0.114	0.114	0.115	0.116	0.118	0.121
0.2	0.129		0.115	0.113	0.112	0.113	0.114	0.116	0.119
0.3	0.124	0.124		0.114	0.112	0.112	0.113	0.115	0.118
0.4	0.122	0.121	0.123		0.115	0.114	0.114	0.115	0.118
0.5	0.122	0.120	0.120	0.123		0.116	0.115	0.116	0.119
0.6	0.122	0.121	0.120	0.121	0.125		0.118	0.118	0.120
0.7	0.124	0.122	0.122	0.122	0.124	0.127		0.121	0.122
0.8	0.126	0.124	0.124	0.124	0.125	0.127	0.132		0.126
0.9	0.131	0.128	0.127	0.127	0.128	0.130	0.133	0.138	
$m = 2$	5% (top triangle), 10% bottom triangle								
0.1		0.138	0.132	0.130	0.129	0.130	0.131	0.134	0.140
0.2	0.150		0.132	0.128	0.127	0.128	0.129	0.132	0.136
0.3	0.143	0.143		0.131	0.128	0.128	0.129	0.131	0.135
0.4	0.140	0.138	0.140		0.131	0.129	0.130	0.132	0.135
0.5	0.140	0.137	0.137	0.141		0.133	0.131	0.133	0.137
0.6	0.140	0.137	0.137	0.138	0.143		0.135	0.135	0.138
0.7	0.142	0.139	0.138	0.139	0.141	0.146		0.140	0.142
0.8	0.145	0.141	0.141	0.141	0.143	0.146	0.152		0.148
0.9	0.153	0.147	0.146	0.146	0.148	0.150	0.154	0.162	

Note: For 1% and 5% (resp., 2.5% and 10%), the value of λ_1 is on the horizontal (resp., vertical) axis and that of λ_2 on the vertical (resp., horizontal) axis.

5.1. *The limit distributions of the tests when using estimates of the break dates*

We propose to estimate the break dates via a global minimization of the sum of squared residuals (SSR) of the GLS-detrended model,[3]

[3]Note that this approach is different from the one adopted by Perron and Rodríguez (2003), who estimated the break date through a minimization of the SSR under the null and alternative hypotheses.

Table 5. Asymptotic critical values for the $MZ_{\alpha}^{GLS}(\lambda^0)$ test.

	$m = 1$								
λ_1	0.1	0.2	0.3	0.4	0.5	0.6	0.7	0.8	0.9
1%	−29.315	−31.442	−32.024	−31.828	−31.333	−30.713	−29.873	−28.539	−26.400
2.5%	−24.730	−26.811	−27.544	−27.518	−27.131	−26.544	−25.698	−24.372	−22.293
5%	−21.244	−23.322	−24.018	−23.969	−23.612	−23.113	−22.393	−21.188	−19.126
10%	−17.815	−19.793	−20.433	−20.350	−19.965	−19.450	−18.743	−17.599	−15.657
$m = 2$	1% (top triangle), 2.5% (bottom triangle)								
0.1		−34.600	−36.957	−38.023	−38.147	−37.698	−36.884	−35.727	−34.161
0.2	−29.635		−37.208	−38.881	−39.455	−39.195	−38.368	−37.097	−35.403
0.3	−32.072	−31.932		−38.001	−39.208	−39.418	−38.798	−37.548	−35.792
0.4	−33.127	−33.725	−32.784		−37.720	−38.551	−38.408	−37.405	−35.764
0.5	−33.283	−34.279	−34.104	−32.643		−36.806	−37.238	−36.726	−35.428
0.6	−32.935	−34.038	−34.279	−33.577	−31.806		−35.447	−35.441	−34.674
0.7	−32.231	−33.319	−33.683	−33.400	−32.350	−30.436		−33.666	−33.314
0.8	−31.049	−32.180	−32.551	−32.418	−31.792	−30.539	−28.591		−31.352
0.9	−29.068	−30.458	−30.903	−30.834	−30.415	−29.604	−28.244	−26.244	

(Continued)

Table 5. (*Continued*)

	$m = 1$								
λ_1	0.1	0.2	0.3	0.4	0.5	0.6	0.7	0.8	0.9
$m = 2$	5% (top triangle), 10% bottom triangle								
0.1		−25.856	−28.150	−29.163	−29.379	−29.144	−28.555	−27.445	−25.411
0.2	−21.932		−28.123	−29.825	−30.383	−30.243	−29.663	−28.640	−26.924
0.3	−23.932	−24.159		−28.948	−30.194	−30.399	−29.929	−28.935	−27.329
0.4	−24.888	−25.683	−24.966		−28.848	−29.736	−29.612	−28.740	−27.186
0.5	−25.178	−26.270	−26.113	−24.890		−28.176	−28.715	−28.187	−26.768
0.6	−25.044	−26.245	−26.385	−25.715	−24.275		−27.045	−27.150	−26.073
0.7	−24.528	−25.760	−26.014	−25.642	−24.744	−23.210		−25.350	−24.879
0.8	−23.466	−24.763	−25.061	−24.811	−24.210	−23.204	−21.564		−22.848
0.9	−21.514	−23.021	−23.405	−23.218	−22.756	−22.054	−20.930	−19.082	

Note: For 1% and 5% (resp., 2.5% and 10%), the value of λ_1 is on the horizontal (resp., vertical) axis and that of λ_2 on the vertical (resp., horizontal) axis.

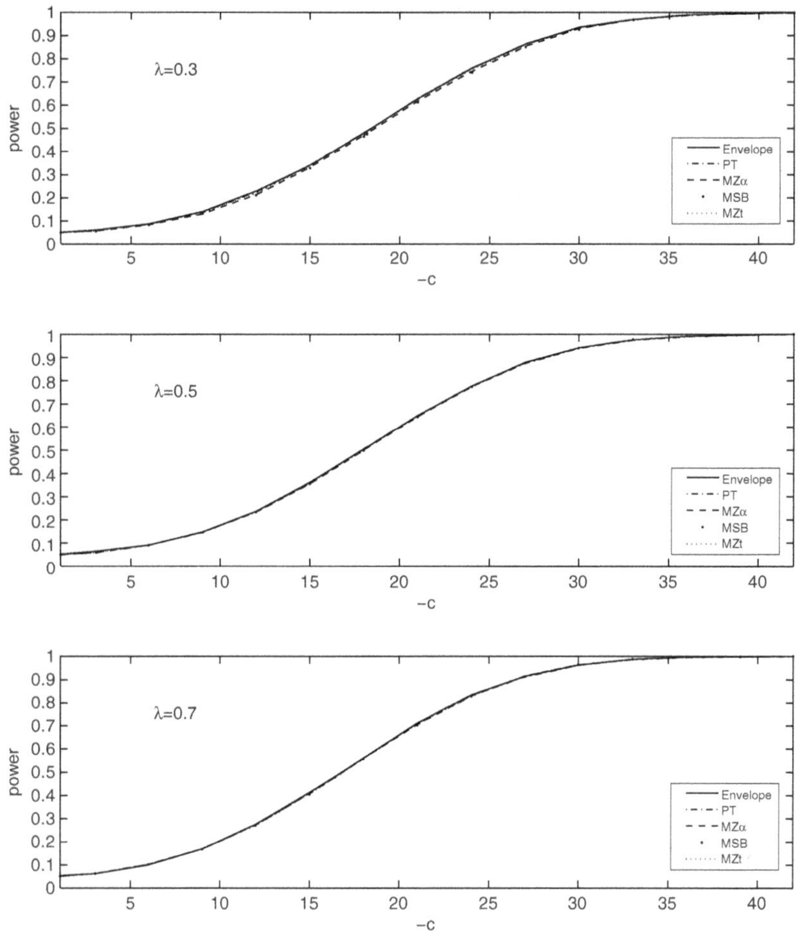

Fig. 1. Gaussian local power envelope and the asymptotic local power functions.

i.e., $\hat{\lambda} = \text{argmin}_{\lambda \in \Lambda(\varepsilon)} S(\bar{\alpha}, \lambda)$, so that

$$S(\bar{\alpha}, \hat{\lambda}) = \min_{\lambda \in \Lambda(\varepsilon)} S(\bar{\alpha}, \lambda), \tag{12}$$

where the infimum is taken over all possible break fractions defined on the set

$$\Lambda(\varepsilon) = \{(\lambda_1, \ldots, \lambda_m); |\lambda_{i+1} - \lambda_i| \\ \geq \varepsilon(i = 1, \ldots, m-1), \lambda_1 \geq \varepsilon, \lambda_m \leq 1 - \varepsilon\},$$

with ε some trimming parameter that dictates the minimal length of a segment. A common value in the related literature is $\varepsilon = 0.15$. The following proposition establishes the consistency and rate of convergence of the estimate of the vector of break fractions λ^0.

Proposition 1. *Let $\{y_t\}_{t=1}^T$ be the stochastic process generated according to (1) and (2) with $\alpha = 1$. Assume that $m > 0$ and $\psi_j \neq 0 (j = 1, \ldots, m)$, so that there are structural breaks affecting y_t under the null hypothesis. Let $\hat{\lambda} = \arg\min_{\lambda \in \Lambda(\varepsilon)} S(\bar{\alpha}, \lambda)$; then, as $T \to \infty$: (i) in Models I and II: $\|\hat{\lambda} - \lambda^0\| = O_p(T^{-1})$ and (ii) in Models 0b and IIb: $\|\hat{\lambda} - \lambda^0\| = o_p(T^{-1})$.*

The proof is given in the appendix. Following Kim and Perron (2009), the next step is to derive the limit distribution of the unit root tests when this estimate of the break fractions is used instead of the true values. The next proposition, proved in the appendix, demonstrates that the rate of convergence is fast enough to guarantee that we recover the same limit distribution as in the known break date case.

Proposition 2. *Let $\{y_t\}_{t=1}^T$ be the stochastic process generated according to (1) and (2) with $\alpha = 1$. Assume that $m > 0$ and $\psi_j \neq 0 (j = 1, \ldots, m)$ and that $s(\hat{\lambda})^2$ is a consistent estimate of σ^2. Let $\hat{\lambda} = \arg\min_{\lambda \in \Lambda(\varepsilon)} S(\bar{\alpha}, \lambda)$; then the limit distributions of $P_T^{GLS}(\hat{\lambda})$, $MP_T^{GLS}(\hat{\lambda})$, $MZ_\alpha^{GLS}(\hat{\lambda})$, $MSB^{GLS}(\hat{\lambda})$, and $MZ_t^{GLS}(\hat{\lambda})$ are the same as those of $P_T^{GLS}(\lambda^0)$, $MP_T^{GLS}(\lambda^0)$, $MZ_\alpha^{GLS}(\lambda^0)$, $MSB^{GLS}(\lambda^0)$, and $MZ_t^{GLS}(\lambda^0)$ respectively, for all models.*

This result is important because even if the break dates are unknown, the use of the particular estimate $\hat{\lambda}$ considered allows us to obtain unit root tests with the same limit distribution as in the known break date case. Because the latter have a local asymptotic power function close to the power envelope, this implies that whether the break dates are known or not the same optimality properties hold and that in the case of Gaussian errors we cannot do better in terms of local asymptotic power. Note that the result is different from that of Kim and Perron (2009), who considered a framework using

OLS based on a regression involving the raw variables (not quasi-detrended). For Models I and II, the rate of convergence was not fast enough to obtain such an equivalence, and to solve the problem they proposed a procedure involving data trimmed around the estimate of the break date. No such device is needed here with the GLS type procedure.

5.2. *Computation of the estimates of the break dates*

In practice, the computation of the estimates of the break dates defined in (12) is computationally prohibitive using a regular grid search when $m > 2$. Following Bai and Perron (2003), we propose to use a dynamic programming approach. The procedure involves, however, an additional layer of difficulty because the quasi-differencing used in (12) destroys the block-diagonality of the matrix of regressors. We can, nevertheless, recover a block diagonal matrix provided appropriate restrictions are imposed on the coefficients. We provide details for Model II and summarize the relevant conditions for the others at the end of the section. We start by noting that the matrix of regressors is given by (assuming only two breaks):

$$\begin{bmatrix}
1 & 1 & 0 & 0 & 0 & 0 \\
(1-\bar{\alpha}) & (1-\bar{\alpha}L)2 & 0 & 0 & 0 & 0 \\
\vdots & \vdots & \vdots & \vdots & \vdots & \vdots \\
(1-\bar{\alpha}) & (1-\bar{\alpha}L)T_1^0 & 0 & 0 & 0 & 0 \\
(1-\bar{\alpha}) & (1-\bar{\alpha}L)(T_1^0+1) & 1 & 1 & 0 & 0 \\
(1-\bar{\alpha}) & (1-\bar{\alpha}L)(T_1^0+2) & (1-\bar{\alpha}) & (1-\bar{\alpha}L)2 & 0 & 0 \\
\vdots & \vdots & \vdots & \vdots & \vdots & \vdots \\
(1-\bar{\alpha}) & (1-\bar{\alpha}L)T_2^0 & (1-\bar{\alpha}) & (1-\bar{\alpha}L)(T_2^0-T_1^0) & 0 & 0 \\
(1-\bar{\alpha}) & (1-\bar{\alpha}L)(T_2^0+1) & (1-\bar{\alpha}) & (1-\bar{\alpha}L)(T_2^0-T_1^0+1) & 1 & 1 \\
(1-\bar{\alpha}) & (1-\bar{\alpha}L)(T_2^0+2) & (1-\bar{\alpha}) & (1-\bar{\alpha}L)(T_2^0-T_1^0+2) & (1-\bar{\alpha}) & (1-\bar{\alpha}L)2 \\
\vdots & \vdots & \vdots & \vdots & \vdots & \vdots \\
(1-\bar{\alpha}) & (1-\bar{\alpha}L)T & (1-\bar{\alpha}) & (1-\bar{\alpha}L)(T-T_1^0) & (1-\bar{\alpha}) & (1-\bar{\alpha}L)(T-T_2^0)
\end{bmatrix}$$

For $t = 1$, we have

$$y_1^{\bar{\alpha}} = \mu_0 + \beta_0 + u_1,$$

which can be expressed as

$$\begin{aligned} y_1^{\bar{\alpha}} &= \mu_0 + \beta_0 \pm \mu_0(1-\bar{\alpha}) \pm \beta_0(1-\bar{\alpha}L)1 + u_1 \\ &= \mu_0(1-\bar{\alpha}) + \beta_0(1-\bar{\alpha}L)1 \\ &\quad + [\mu_0 + \beta_0 - \mu_0(1-\bar{\alpha}) - \beta_0(1-\bar{\alpha}L)1] + u_1 \\ &= \mu_0(1-\bar{\alpha}) + \beta_0(1-\bar{\alpha}L)1 \\ &\quad + [\mu_0 + \beta_0 - \mu_0(1-\bar{\alpha}) - \beta_0(1-\bar{\alpha}L)1]D_1(T_0) + u_1, \end{aligned}$$

where $D_t(T_j) = 1$ for $t = T_j + 1$ and zero otherwise, with the convention that $T_0 = 0$. Therefore, the model for $1 \le t \le T_1^0$ can be written as

$$y_1^{\bar{\alpha}} = \mu_0 1^{\bar{\alpha}} + \beta_0 t^{\bar{\alpha}} + \gamma_0 D_t(T_0^0) + u_t^{\bar{\alpha}},$$

with $\gamma_0 = \mu_0\bar{\alpha} + \beta_0 - \beta_0(1-\bar{\alpha}L)\cdot 1 = \mu_0\bar{\alpha}$, using the fact that $(1-\bar{\alpha}L)t = t - (1+\bar{c}/T)(t-1) = 1 - (\bar{c}/T)(t-1)$. When $t = T_1^0 + 1$ the quasi-differenced variable is given by

$$\begin{aligned} y_{T_1^0+1}^{\bar{\alpha}} &= \mu_0(1-\bar{\alpha}) + \beta_0(1-\bar{\alpha}L)(T_1^0+1) + \mu_1 + \beta_1 + u_{T_1^0+1}^{\bar{\alpha}} \\ &= \mu_0(1-\bar{\alpha}) + \beta_0(1-\bar{\alpha}L)(T_1^0+1) \\ &\quad \pm \mu_1(1-\bar{\alpha}) \pm \beta_1(1-\bar{\alpha}L)(T_1^0+1) + \mu_1 + \beta_1 + u_{T_1^0+1}^{\bar{\alpha}} \\ &= \mu_0(1-\bar{\alpha}) + \beta_0(1-\bar{\alpha}L)(T_1^0+1) + \mu_1(1-\bar{\alpha}) \\ &\quad + \beta_1(1-\bar{\alpha}L)(T_1^0+1) + [\mu_1 + \beta_1 - \mu_1(1-\bar{\alpha}) \\ &\quad - \beta_1(1-\bar{\alpha}L)(T_1^0+1)]D_{T_1^0+1}(T_1^0) + u_{T_1^0+1}^{\bar{\alpha}} \\ &= (\mu_0 + \mu_1 - \beta_1 T_1^0)(1-\bar{\alpha}) + (\beta_0+\beta_1)(1-\bar{\alpha}L)(T_1^0+1) \\ &\quad + \bar{\alpha}\mu_1 D_{T_1^0+1}(T_1^0) + u_{T_1^0+1}^{\bar{\alpha}}, \end{aligned}$$

and for $T_1^0 < t \le T_2^0$ it is given by

$$\begin{aligned} y_t^{\bar{\alpha}} &= (\mu_0 + \mu_1 - \beta_1 T_1^0)(1-\bar{\alpha}) + (\beta_0+\beta_1)(1-\bar{\alpha}L)t \\ &\quad + \gamma_1 D_t(T_1^0) + u_t^{\bar{\alpha}}, \end{aligned}$$

where $\gamma_1 = \bar{\alpha}\mu_1$. A similar expression can be obtained for $T_2^0 < t \leq T$, namely,

$$\begin{aligned} y_t^{\bar{\alpha}} &= (\mu_0 + \mu_1 + \mu_2 - \beta_1 T_1^0 - \beta_2 T_2^0)(1 - \bar{\alpha}) \\ &\quad + (\beta_0 + \beta_1 + \beta_2)(1 - \bar{\alpha}L)t + \gamma_2 D_t(T_2^0) + u_t^{\bar{\alpha}}, \end{aligned}$$

with $\gamma_2 = \bar{\alpha}\mu_2$. In general, $y_t^{\bar{\alpha}}$ in the jth regime $(T_{j-1}^0 < t \leq T_j^0), j = 1, \ldots, m+1$, can therefore be written as

$$y_t^{\bar{\alpha}} = \mu^*(1 - \bar{\alpha}) + \beta^*(1 - \bar{\alpha}L)t + \gamma_{j-1} D_t(T_{j-1}^0) + u_t^{\bar{\alpha}}, \tag{13}$$

where the coefficients satisfy the restrictions $\mu^* = (\sum_{i=0}^{j-1} \mu_i - \sum_{i=0}^{j-1} \beta_i T_i^0)$, $\beta^* = (\sum_{i=0}^{j-1} \beta_i)$, and $\gamma_{j-1} = \bar{\alpha}\mu_{j-1}$. Then, the moment matrix of the regressors can be expressed via the block diagonal matrix,

$$\left[\begin{array}{ccccccccc}
(1-\bar{\alpha}) & (1-\bar{\alpha}L)1 & 1 & & & & & & \\
\vdots & \vdots & \vdots & & & & & & 0 \\
(1-\bar{\alpha}) & (1-\bar{\alpha}L)T_1^0 & 0 & & & & & & \\
 & & & (1-\bar{\alpha}) & (1-\bar{\alpha}L)(T_1^0+1) & 1 & & & \\
 & & & \vdots & \vdots & \vdots & & & \\
 & & & (1-\bar{\alpha}) & (1-\bar{\alpha}L)T_2^0 & 0 & & & \\
 & & & & & & (1-\bar{\alpha}) & (1-\bar{\alpha}L)(T_2^0+1) & 1 \\
0 & & & & & & \vdots & \vdots & \vdots \\
 & & & & & & (1-\bar{\alpha}) & (1-\bar{\alpha}L)T & 0
\end{array}\right],$$

and the dynamic programming algorithm can be used provided the appropriate restrictions on the coefficients of the system (13) are imposed. With this specification, the estimation of the break dates can be done using an iterative procedure similar to that of Perron and Qu (2006). The exact steps are as follows.

1. Compute initial estimates of the break dates and associated break fractions $\hat{\lambda} = (\hat{\lambda}_1, \ldots, \hat{\lambda}_m)$ and coefficients $\hat{\psi} = (\hat{\psi}_0', \hat{\psi}_1', \ldots, \hat{\psi}_m')'$ using an OLS method applied to (1). This involves a standard application of the algorithm described in Bai and Perron (2003).
2. For a given set of preliminary estimates of the break dates, obtain an initial value of $\bar{c}$.

3. Let $T^*(\psi, r, n) = (T_1^*(\psi, r, n), \ldots, T_r^*(\psi, r, n))$ be the vector of the optimal r break dates using the first n observations for a given vector of coefficients ψ and let $RSSR(T^*(\psi, r, n))$ be the associated restricted sum of squared residuals. Then, compute the restricted sum of squared residuals $RSSR(T^*(\psi, 1, n))$ for $2h \leq n \leq T - (m-1)h$ by

$$\begin{aligned} &RSSR(T^*(\hat{\psi}, 1, n)) \\ &\quad = \min_{h \leq j \leq n-h} [RSSR^1(1, j) + RSSR^2(j+1, n)] \end{aligned}$$

and

$$T^*(\hat{\psi}, 1, n) = \arg \min_{h \leq j \leq n-h} [RSSR^1(1, j) + RSSR^2(j+1, n)],$$

where

$$RSSR^1(1, j) = \sum_{t=1}^{j} (y_t^{\bar{\alpha}} - z_t^{\bar{\alpha}}(T_0)' \hat{\phi}_0 - \hat{\gamma}_0 D_t(T_0))^2,$$

with $\hat{\phi}_0 = \hat{\psi}_0$ and $\hat{\gamma}_0 = \bar{\alpha}\hat{\mu}_0$, and

$$RSSR^2(j+1, n) = \sum_{t=j+1}^{n} (y_t^{\bar{\alpha}} - z_t^{\bar{\alpha}}(T_0)' \hat{\phi}_1(j) - \hat{\gamma}_1 D_t(j))^2,$$

with $\hat{\phi}_1(j) = (\hat{\mu}_0 + \hat{\mu}_1 - \hat{\beta}_1 j, \hat{\beta}_0 + \hat{\beta}_1)'$ and $\hat{\gamma}_1 = \bar{\alpha}\hat{\mu}_1$. Then, sequentially compute and store $RSSR(T^*(\hat{\psi}, r, n))$ for $r = 2, \ldots, m-1$, with n ranging from $(r+1)h$ to $T - (m-r)h$. This is done solving

$$\begin{aligned} RSSR(T^*(\hat{\psi}, r, n)) = \min_{rh \leq j \leq n-h} &[RSSR(T^*(\hat{\psi}, r-1, n)) \\ &+ RSSR^{r+1}(j+1, n)]. \end{aligned}$$

The last rth element is

$$\begin{aligned} T^*(\hat{\psi}, r, n) = \arg \min_{rh \leq j \leq n-h} &[RSSR(T^*(\hat{\psi}, r-1, n)) \\ &+ RSSR^{r+1}(j+1, n)], \end{aligned}$$

where

$$RSSR^{r+1}(j+1,n) = \sum_{t=j+1}^{n} (y_t^{\bar{\alpha}} - z_t^{\bar{\alpha}}(T_0)'\hat{\phi}_r(j) - \hat{\gamma}_r D_t(j))^2, \quad (14)$$

with $\hat{\phi}_r(j) = (\sum_{i=0}^{r} \hat{\mu}_i - \sum_{i=1}^{r-1} \hat{\beta}_i T_i^*(\hat{\psi}, r-1, j) - \beta_r j, \sum_{i=0}^{r} \hat{\beta}_i)'$ and $\hat{\gamma}_r = \bar{\alpha}\mu_r$. Finally compute

$$RSSR(T^*(\hat{\psi}, m-1, n)) = \min_{mh \leq j \leq T-h} [RSSR(T^*(\hat{\psi}, m-1, n)) + RSSR^{m+1}(j+1, T)],$$

where $RSSR^{m+1}(j+1,T)$ is computed as in (14). Then store the estimated break dates and update $\bar{c}$ accordingly.

4. Repeat steps 2 and 3 until convergence.

Note that for Models 0 and I, the constraints to be imposed are slightly different. For the first regime ($j = 0$) they are the same as for Model II, but for the other regimes, $T_{j-1}^0 < t \leq T_j^0 (j = 1, \ldots, m)$, we have $\mu^* = \sum_{i=1}^{j-1} \mu_i, \beta^* = \beta_0$, and $\gamma_{j-1} = \bar{\alpha}\mu_{j-1}$ for Model 0 and $\mu^* = \mu_0 - \sum_{i=1}^{j-1} \beta_i T_i^0, \beta^* = \sum_{i=0}^{j-1} \beta_i$, and $\gamma_{j-1} = 0$ for Model I.

6. Preliminary Simulations

Even though we have assumed the existence of a break in the trend function so far, it is instructive to analyze the properties of the tests. We consider cases with a single break without loss of generality. The data generating process used in the simulations is given by

$$y_t = d_t + u_t; \quad d_t = \mu_b DU_t(T_1^0) + \beta_b DT_t^*(T_1^0); \quad u_t = \alpha u_{t-1} + v_t,$$

$v_t \sim i.i.d.\ N(0,1), u_0 = 0$. We specified four values for the magnitude of the level shift $\mu_b = \{0, 0.5, 1, 5\}$. For each value of μ_b, we considered values of β_b ranging from 0 to 3 in increments of 0.2. The sample size is set at $T = \{100, 200, 300\}$, and the results are based on 1,000 replications. The empirical size is analyzed setting $\alpha = 1$, whereas the power is evaluated with $\alpha = \bar{\alpha} = 1 + \bar{c}/T$, where the parameter $\bar{c}$ depends on the break fractions. Hence, in large samples, the power is 50% in all cases. In this chapter, we only report results for $\lambda^0 = 0.5$

(the results were similar for $\lambda^0 = 0.3$ and 0.7) and for the tests $P_T^{GLS}(\hat{\lambda})$ (the properties of the other tests are qualitatively similar).

Figure 2 presents the results of the empirical size using the asymptotic critical values at the 5% level of significance drawn from the estimated response surfaces, and Fig. 3 presents results for power. For size, the following features are worthy of note. First, the exact size is close to the nominal size as $|\beta_b|$ increases and more so the larger the sample. This is in accordance with our theoretical results. When $|\beta_b|$ and $|\mu_b|$ are small, the tests show liberal size distortions and more so as the sample size increases. This is due to the fact that our theoretical results so far assumed $\beta_b \neq 0$ so that a break is present. The limit null distributions when no break occurs are indeed different because the estimate of the break fraction has a nondegenerate limit distribution on the interval [0, 1] instead of converging to either zero or one (see Nunes *et al.* 1995; Bai, 1998). What is of special interest is the fact that when $|\mu_b|$ is large, the exact sizes are close to the nominal 5% level irrespective of the value of β_b. This suggests that a correction will be needed to account for cases with both $|\beta_b|$ and $|\mu_b|$ small. With respect to power, the following features emerge. The power quickly approaches the 50% limit value suggested by the local asymptotic power analysis unless T is small. This is so even if $|\beta_b|$ is not very large. For very small values of $|\beta_b|$, the power is above 50% but only because of the presence of important size distortions.

7. Extension to the Case Where a Break Need Not Occur

Up to this point, we have assumed that a break occurs under both the null and alternative hypotheses. When no break occurs, the asymptotic results described in the previous sections do not hold because, under the null hypothesis of a unit root, the estimates of the break fractions have a nondegenerate limit distribution on the interval [0, 1] instead of converging to either zero or one. If there is no break in the trend, then the proper unit root test procedure is to simply apply a standard DF type test with no break dummies.

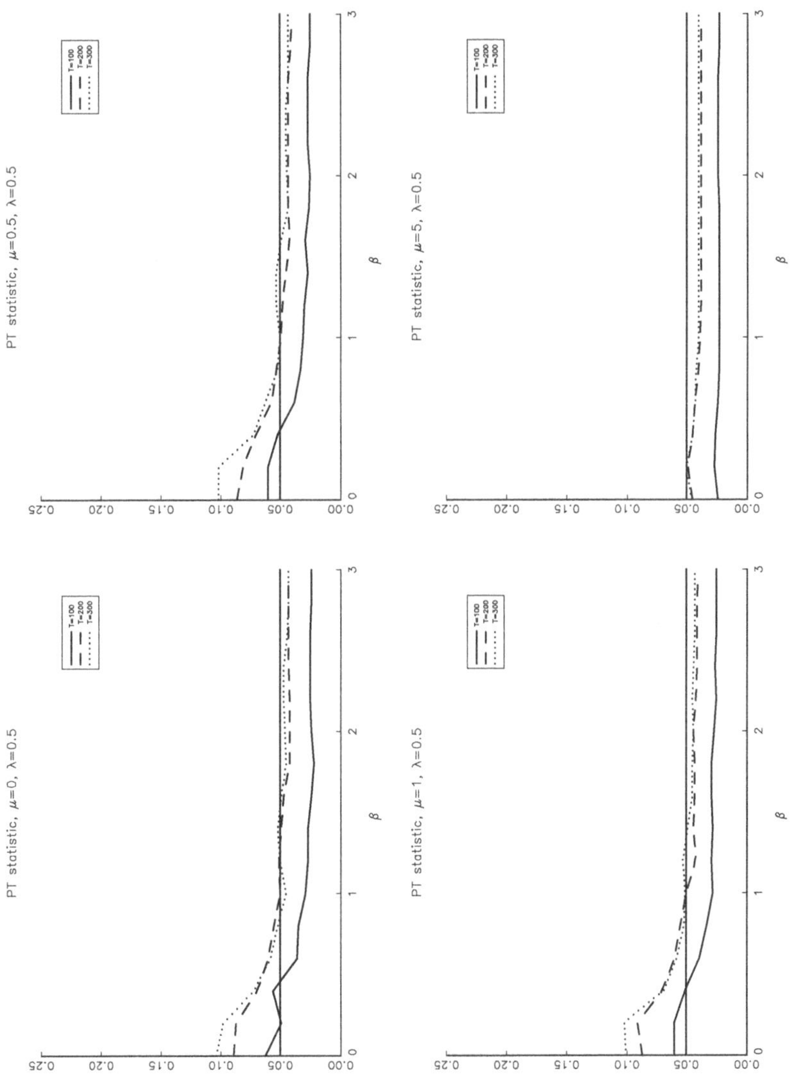

Fig. 2. Empirical size of the $P_T^{GLS}(\hat{\lambda})$ test, $\lambda^0 = 0.5$.

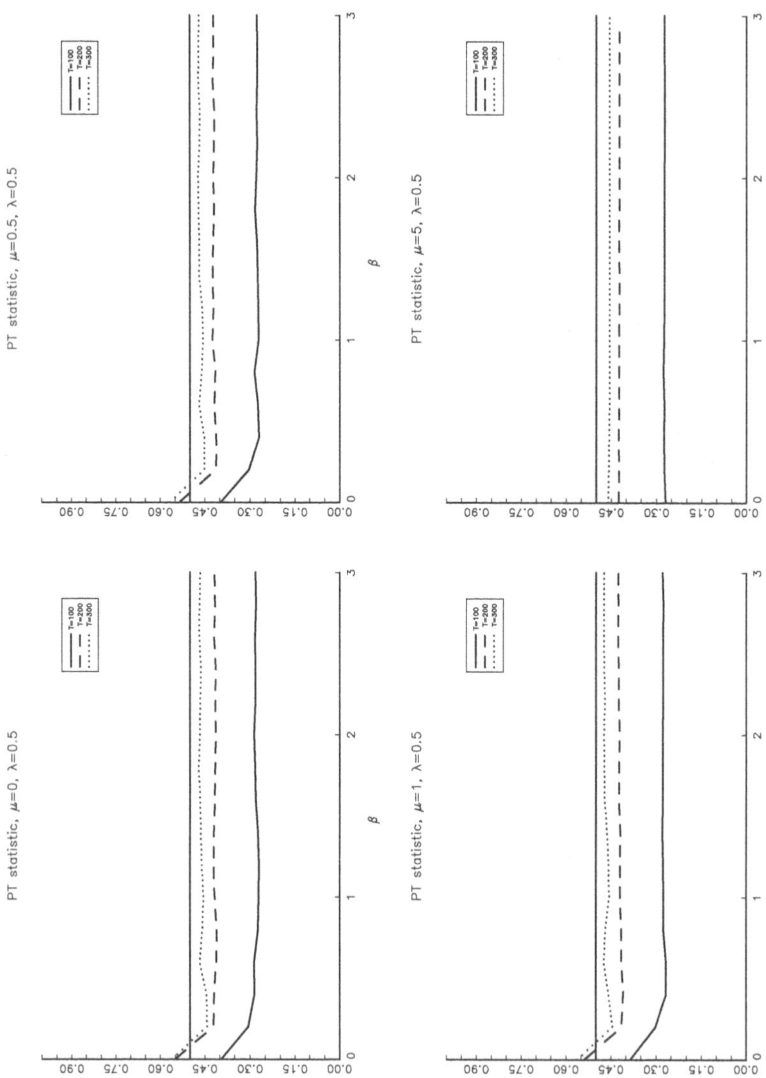

Fig. 3. Empirical power of the $P_T^{GLS}(\hat{\lambda})$ test, $\lambda^0 = 0.5$.

Hence, what is needed is a pretest to assess whether a break is present or not. This pretest for a break should have the correct size if the noise is integrated but should also be powerful whether the noise is stationary or has a unit root, to ensure a specification that allows a unit root test procedure with good power. This testing problem has recently been addressed by Vogelsang (2001) and Perron and Yabu (2009). Because the procedure of Perron and Yabu (2009) has better size and power, we shall use it as the pretest. It is based on a quasi-GLS approach using an autoregression for the noise component, with a truncation to one when the sum of the autoregressive coefficients is in some neighborhood of 1, along with a bias correction. For given break dates, one constructs the F-test for the null hypothesis of no structural change in the deterministic components. The final statistic uses the *Exp* functional of Andrews and Ploberger (1994). The test has virtually the same asymptotic size with a stationary or integrated noise component. We label this test as $Exp\text{-}W_{FS}$ and define the alternative estimate of the break fraction, $\tilde{\lambda}$, as $\tilde{\lambda} = \hat{\lambda} \cdot 1(Exp\text{-}W_{FS} > cv)$ where cv is the critical value for a test with, say, nominal size $p\%$. When there is a break, $\tilde{\lambda}$ consistently estimates λ^0, as long as $\hat{\lambda}$ consistently estimates λ^0, given that $Exp\text{-}W_{FS}$ is a consistent test. When there is no break, $\tilde{\lambda}$ will yield a nonzero estimate $p\%$ of times even with infinitely many observations, if cv is fixed. Now, suppose that $Exp\text{-}W_{FS} = O_p(T^{\bar{\omega}}), \bar{\omega} > 0$ under the alternative of a break. Let $cv = cT^{\bar{\omega}-\varepsilon}, 0 < \varepsilon < \bar{\omega}$. Such an increasing sequence of critical values does not harm the consistency of the test, and with the size converging to zero as $T \to \infty$, it ensures the consistency of $\tilde{\lambda}$ for $\lambda^0 \in [0, 1]$. Hence, the modified test procedure is to use a DF type test when $\tilde{\lambda} = 0$ and the procedure described earlier when $\tilde{\lambda} \neq 0$.

We now present simulation evidence about the size and power properties of our testing procedures using the pretest for the presence of breaks. We start with the case of a single break and then consider the case with an unknown number of breaks up to two. This will allow us, among other things, to discuss issues related to the estimation of the number of breaks.

The Case With One Break. We performed simulations with the same data generating process and design as used in Section 6 but this time for the version of our statistics involving the pretest. We only report results for $\lambda^0 = 0.5$ and for the test $P_T^{GLS}(\hat{\lambda})$ (the properties of the other tests are similar). The results are presented in Figs. 4 (for size) and 5 (for power). With respect to the size of the test, one can see that the size distortions for values of $|\beta_b|$ near zero disappear to a large extent. The test $P_T^{GLS}(\hat{\lambda})$ with the pretest for break in trend has an exact size close to the nominal level in all cases.

The power functions with the pretest for break approach the target 50% level quickly as T increases when $|\beta_b|$ is moderate to large. When $|\beta_b|$ is near zero, the power is actually higher than 50%. This is due to the fact that with $\beta_b = 0$ the test performed without allowing for a change in trend has a local power function that is actually higher than the test that does allow for breaks. So there is improvement in terms of power also. The only drawback is for values of $|\beta_b|$ neither moderate nor small. Here, the power dips down. This is due to the fact that the pretest for break is not powerful enough for this range of values of $|\beta_b|$ so that a standard test without breaks is applied. Yet, the break is large enough to affect the power of the test, which decreases for the reasons explained in Perron (1989).

Overall, the performance of the $P_T^{GLS}(\hat{\lambda})$ tests (and the others presented in the chapter whose performance is similar) is quite satisfactory and offers an improvement in terms of both size and power over existing procedures. It avoids the large size distortions of the Zivot and Andrews (1992) procedure while still allowing for much improved power by taking advantage of the information about the presence or absence of a break whether it be under the null or the alternative hypothesis.

The Case With an Unknown Number of Breaks up to Two. We now consider the case with a maximum of up to two breaks. When dealing with the possibility of multiple breaks, one needs a method to estimate the number of breaks present. The application of the test of Perron and Yabu (2009) for two breaks is not appropriate because it would indicate a rejection when a single break is present;

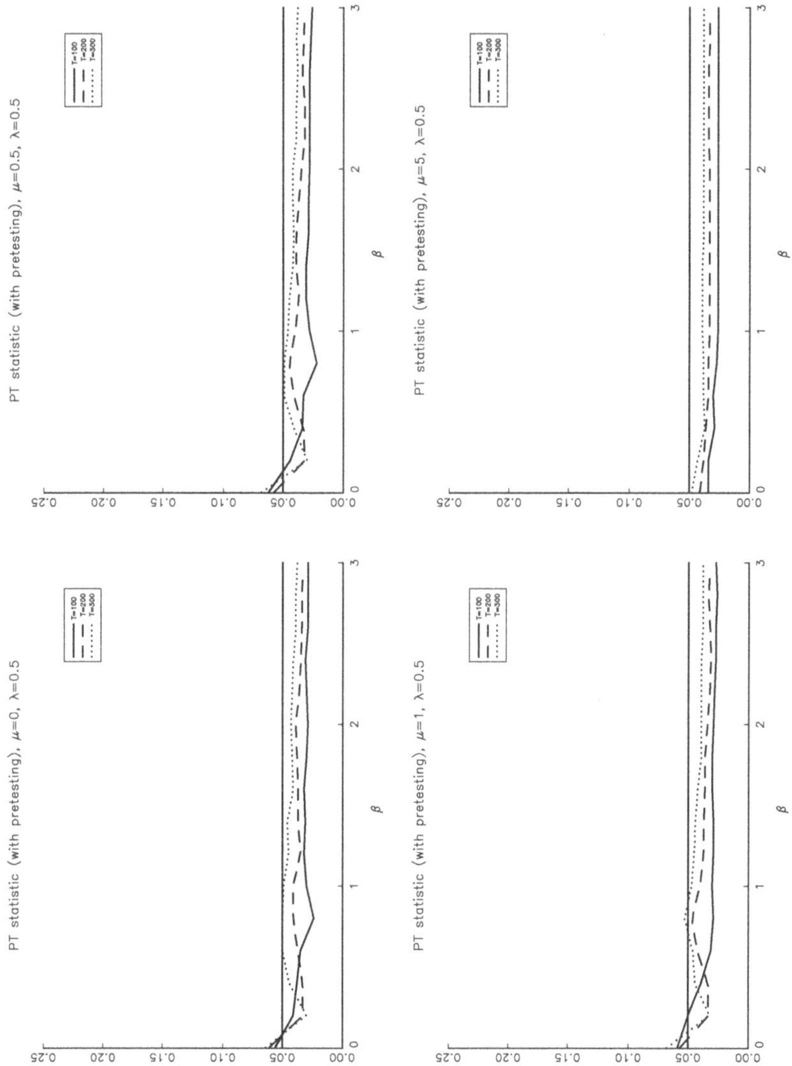

Fig. 4. Empirical size of the $P_T^{GLS}(\hat{\lambda})$ test, $\lambda^0 = 0.5$, with pretesting.

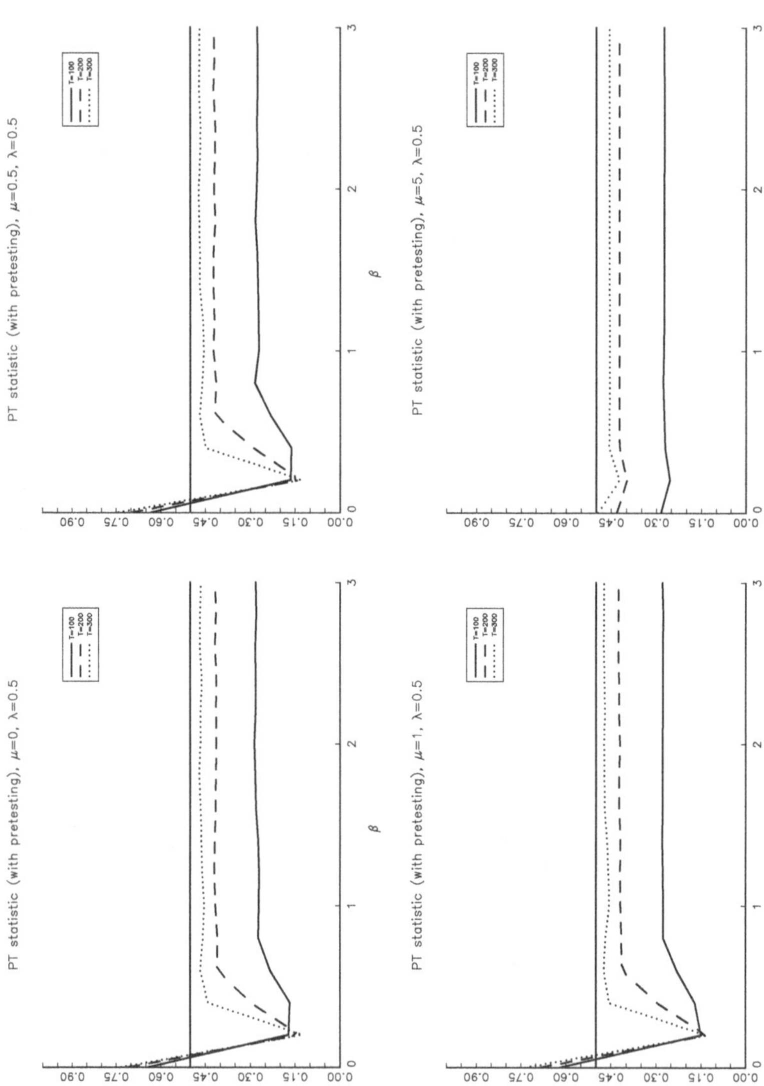

Fig. 5. Empirical power of the $P_T^{GLS}(\hat{\lambda})$ test, $\lambda^0 = 0.5$, with pretesting.

i.e., it does not allow us to estimate the number of breaks. Kejriwal and Perron (2008) have extended the Perron–Yabu test so that it is applicable in a specific to general sequential manner, thereby allowing the estimation of the number of breaks. The procedure first applies the Perron–Yabu test for a single break. If the test does not reject, a unit root test with no break is applied. If the test does reject, the sample is split at the estimate of the break date obtained using the method described in Section 5.2. The Perron–Yabu test is then applied to each subsample (with a 15% trimming applied at each end), and the maximum value of the two tests is compared to the critical values derived in Kejriwal and Perron (2008). If no rejection occurs, a unit root test with a single break is used. Upon a rejection, a unit root test with two breaks is applied with the estimates again obtained using the method laid out in Section 5.2.

To illustrate the size and power properties of the unit root tests using this method to determine the number and location of the breaks, we consider simulations using Model I as the data generating process so that

$$y_t = \beta_1 DT_t^*(T_1^0) + \beta_2 DT_t^*(T_2^0) + u_t,$$
$$u_t = \alpha u_{t-1} + v_t,$$

with $v_t \sim i.i.d.\ N(0,1)$, $u_0 = 0$, and $T_i^0 = [T\lambda_i^0]$ for $i = 1, 2$. We set $\lambda_1^0 = 0.3$ and $\lambda_2^0 = 0.7$. We consider three sample sizes, $T = 200, 300$, and 500. The nominal size of the pretest and the unit root test is again 5%, and the results are based on 1,000 replications. For the size simulation, $\alpha = 1$; for the power simulation, $\alpha = \bar{\alpha} = 1 + \bar{c}/T$, with $\bar{c}$ the value of the noncentrality parameter corresponding to the number and location of the breaks detected. Hence, for two nonzero breaks, the power is 50% in large samples. The magnitudes of the break coefficients β_1 and β_2 take values $\{0, 1, 2, 3\}$. When $\beta_1 = \beta_2 = 0$, no break occurs; if either β_1 or β_2 is zero (but not both), we have a one break model, whereas if both β_1 and β_2 are nonzero, we have a two break model. In all cases, the method searches for up to a maximum of two breaks.

The results for size are presented in Fig. 6, again only for the test $P_T^{GLS}(\hat{\lambda})$, the results for the other tests being qualitatively similar.

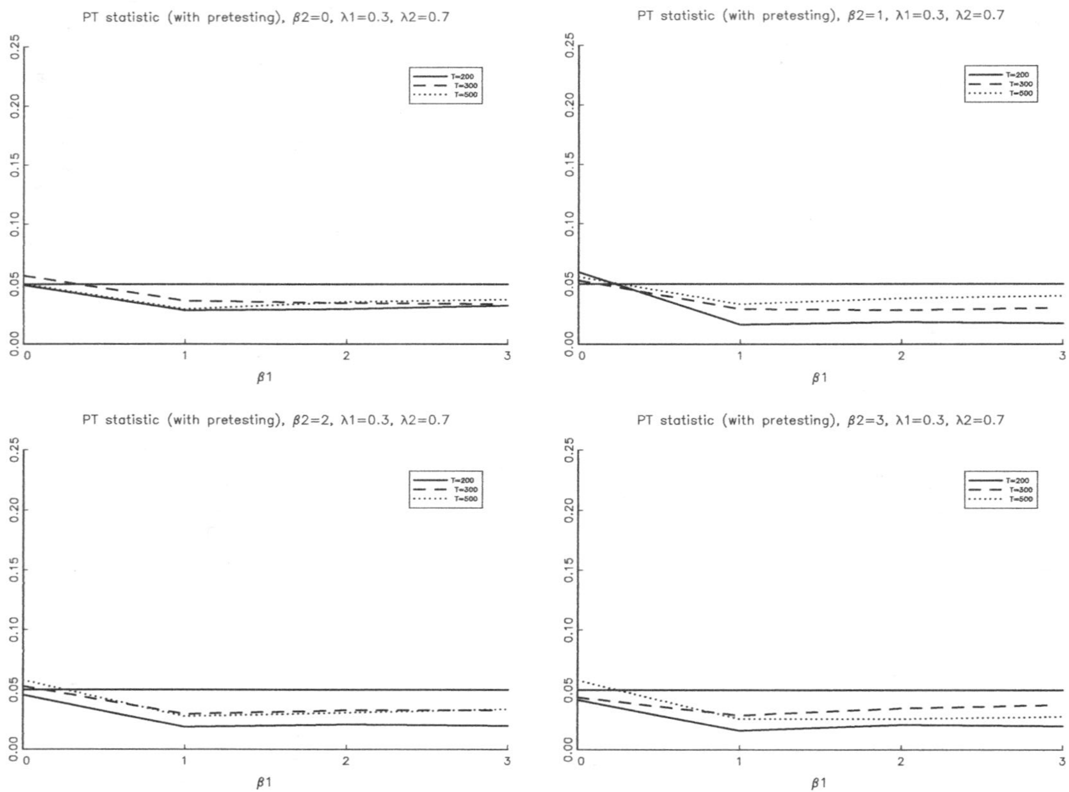

Fig. 6. Empirical size of the $P_T^{GLS}(\hat{\lambda})$ test, $\lambda^0 = (0.3, 0.7)'$, with pretesting.

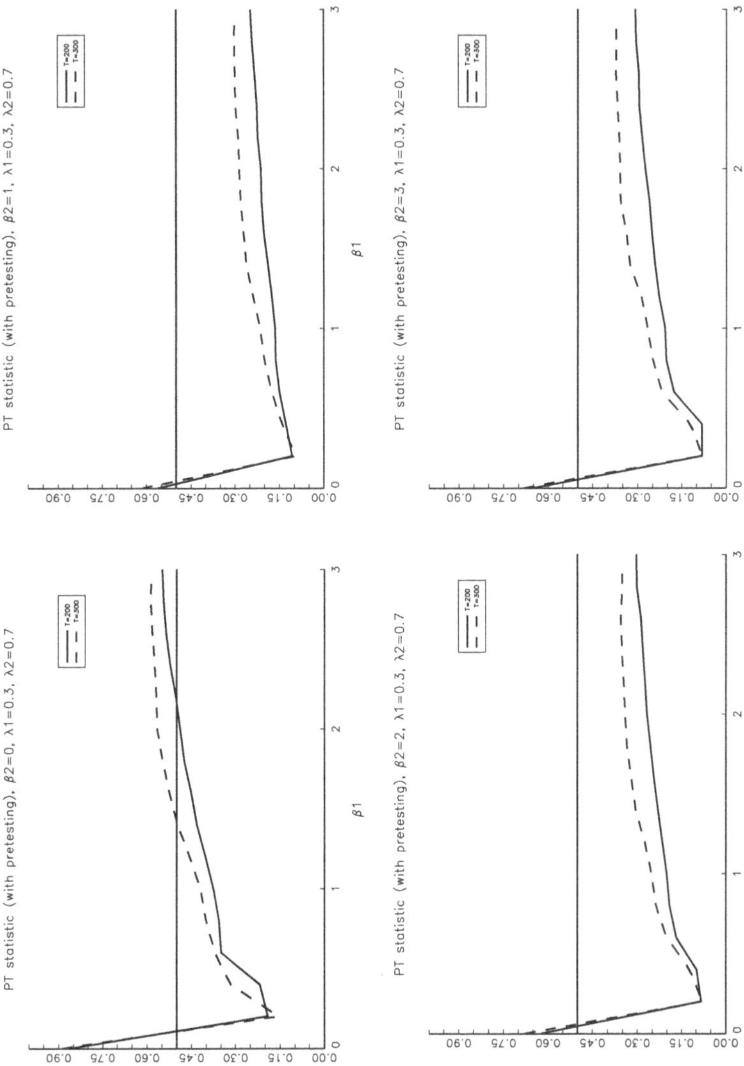

Fig. 7. Empirical power of the $P_T^{GLS}(\hat{\lambda})$ test, $\lambda^0 = (0.3, 0.7)'$, with pretesting.

One can see that for small values of the break coefficients, the exact size is close to the 5% nominal level. There is, however, a tendency for the unit root test to be slightly conservative when at least one of the breaks is large, though the size generally approaches 5% quickly unless both breaks are large.

The results for power are presented in Fig. 7. When $\beta_1 = \beta_2 = 0$, the test has power well above 50%. This indicates that the method to estimate the number of breaks correctly selects no break, in which case power is larger than 50% in large samples. The power is also above 50% when there is a single break of large enough magnitude, as illustrated in the top left panel. When one of the breaks is small we have, as in the single break case, power below the target 50% level. The power, however, increases and approaches the target 50% level as the magnitudes of the breaks or the sample size increase. Hence, we can conclude from this limited set of simulations that the method is adequate in delivering tests with good size and power in the multiple break case with an unknown number of breaks.

8. Conclusions

Following the work of Kim and Perron (2009), the procedures suggested in this chapter solve many of the problems raised in the Introduction that plague most existing unit root tests designed for series with a breaking trend function. First, the tests allow for a break under both the null and the alternative hypotheses. This is desirable for several reasons. It imposes a symmetric treatment when allowing for a break, so that the tests do not reject when the noise is integrated but the trend is changing. Also, if a break is present, this information is exploited to improve power. Second, when a break is present, the limit distributions of the tests are the same as in the case of a known break date, thereby allowing increased power while maintaining the correct size. Our chapter used the quasi-GLS procedure suggested by Elliott *et al.* (1996) to obtain tests that have a local asymptotic power close to the power envelope, except perhaps in a small neighborhood where the change in slope is small but nonzero. We have also extended the analysis to the multiple break case made possible by a modification of the dynamic programming algorithm as described in Bai and Perron (2003) and Perron and Qu (2006). Simulation experiments confirm that our procedures offer improvements over commonly used methods in small samples. Our tests should therefore be useful in empirical applications.

Appendix

We start by presenting some results that will be used throughout. We have, by definition

$$DT^{*\bar{\alpha}}(T_i) = \begin{cases} 0, & t \leq T_i \\ 1 - \bar{c}(t - T_i - 1)/T, & t \geq T_i + 1 \end{cases} \quad \text{and}$$

$$DU^{\bar{\alpha}}(T_i) = \begin{cases} 0, & t \leq T_i \\ 1, & t = T_i + 1 \\ -\bar{c}/T, & t \geq T_i + 2. \end{cases}$$

Also, for $T_j > T_i$,

$$DT^{*\bar{\alpha}}(T_j) - DT^{*\bar{\alpha}}(T_i) = \begin{cases} -1 + \bar{c}(t - T_i - 1)/T, & T_i + 1 \leq t \leq T_j \\ \bar{c}(\lambda_j - \lambda_i), & T_j + 1 \leq t \end{cases}$$

and zero otherwise,

$$DU^{\bar{\alpha}}(T_j) - DU^{\bar{\alpha}}(T_i) = \begin{cases} -1, & t = T_i + 1 \\ \bar{c}/T, & T_i + 2 \leq t \leq T_j \\ 1 + \bar{c}/T & t = T_j + 1 \end{cases}$$

and zero otherwise. We can then show that, with $T_i \leq T_j$,

$$\begin{aligned} T^{-1}DT^{*\bar{\alpha}}(T_i)'DT^{*\bar{\alpha}}(T_j) &= (1/6)(1 - \lambda_j) \\ &\quad \times [6\bar{c}(\lambda_i - 1) + \bar{c}^2\{\lambda_j(3\lambda_i - 1) - 3\lambda_i - \lambda_j^2 + 2\} + 6] + o(1), \\ &\equiv a(\lambda_i, \lambda_j) + o(1), \end{aligned} \tag{A.1}$$

$$DU^{\bar{\alpha}}(T_i)'DU^{\bar{\alpha}}(T_j) = 1(|T_i - T_j| = 0) + o(1), \tag{A.2}$$

where $1(\cdot)$ denotes the indicator function. Also,

$$\begin{cases} DU^{\bar{\alpha}}(T_i)' DT^{*\bar{\alpha}}(T_j) = 1(|T_i - T_j| = 0) - \bar{c}(1-\lambda_j) \\ \qquad\qquad\qquad\qquad + \bar{c}^2(1-\lambda_j)^2/2 + o(1), \\ DU^{\bar{\alpha}}(T_j)' DT^{*\bar{\alpha}}(T_i) = 1 - \bar{c}(\lambda_j - \lambda_i) - \bar{c}(1-\lambda_j) \\ \qquad\qquad\qquad\qquad + \bar{c}^2(1+\lambda_j - 2\lambda_i)(1-\lambda_j)/2 + o(1). \end{cases} \tag{A.3}$$

From (A.1)–(A.3), we have, with $T_i \le T_j$,

$$\begin{cases} DU^{\bar{\alpha}}(T_j)'[DU^{\bar{\alpha}}(T_j) - DU^{\bar{\alpha}}(T_i)] = 1(|T_i - T_j| \ne 0) + o(1), \\ DU^{\bar{\alpha}}(T_i)'[DU^{\bar{\alpha}}(T_j) - DU^{\bar{\alpha}}(T_i)] = -1(|T_i - T_j| \ne 0) + o(1), \end{cases} \tag{A.4}$$

$$\begin{cases} DU^{\bar{\alpha}}(T_j)'[DT^{*\bar{\alpha}}(T_j) - DT^{*\bar{\alpha}}(T_i)] \\ \quad = \bar{c}(\lambda_j - \lambda_i) - \bar{c}^2(\lambda_j - \lambda_i)(1-\lambda_j) + o(1), \\ DU^{\bar{\alpha}}(T_i)'[DT^{*\bar{\alpha}}(T_j) - DT^{*\bar{\alpha}}(T_i)] \\ \quad = -1(|T_i - T_j| \ne 0) + \bar{c}(\lambda_j - \lambda_i) \\ \qquad -\bar{c}^2(\lambda_j - \lambda_i)^2/2 - \bar{c}^2(\lambda_j - \lambda_i)(1-\lambda_j) + o(1), \end{cases} \tag{A.5}$$

$$\begin{cases} DT^{*\bar{\alpha}}(T_j)'[DU^{\bar{\alpha}}(T_j) - DU^{\bar{\alpha}}(T_i)] = 1(|T_i - T_j| \ne 0) + o(1), \\ DT^{*\bar{\alpha}}(T_i)'[DU^{\bar{\alpha}}(T_j) - DU^{\bar{\alpha}}(T_i)] = -\bar{c}^2(\lambda_j - \lambda_i)^2/2 + o(1), \end{cases} \tag{A.6}$$

$$\begin{cases} T^{-1}DT^{*\bar{\alpha}}(T_j)'[DT^{*\bar{\alpha}}(T_j) - DT^{*\bar{\alpha}}(T_i)] \\ \quad = \bar{c}(\lambda_j - \lambda_i)[(1-\lambda_j) - \bar{c}(1-\lambda_j)^2/2] + o(1), \\ T^{-1}DT^{*\bar{\alpha}}(T_i)'[DT^{*\bar{\alpha}}(T_j) - DT^{*\bar{\alpha}}(T_i)] \\ \quad = -(\lambda_j - \lambda_i) + \bar{c}(\lambda_j - \lambda_i)(1-\lambda_j) \\ \qquad -\bar{c}^2(\lambda_j - \lambda_i)\{3(1-\lambda_j)^2 - (\lambda_j - \lambda_i)^2\}/6 + o(1). \end{cases} \tag{A.7}$$

From (A.4)–(A.7), we have (with $T_j \ge T_i$)

$$\|DU^{\bar{\alpha}}(T_j) - DU^{\bar{\alpha}}(T_i)\|^2 = 2 \cdot 1(|T_j - T_i| \ne 0) + o(1), \tag{A.8}$$

$$
\begin{aligned}
&T^{-1}\|DT^{*\bar{\alpha}}(T_j) - DT^{*\bar{\alpha}}(T_i)\|^2 \\
&\quad = (\lambda_j - \lambda_i) - \bar{c}(\lambda_j - \lambda_i)^2 + \bar{c}^2(\lambda_j - \lambda_i)^3/3 \\
&\qquad + \bar{c}^2(\lambda_j - \lambda_i)^2(1 - \lambda_j) + o(1), \qquad \text{(A.9)} \\
&(DU^{\bar{\alpha}}(T_j) - DU^{\bar{\alpha}}(T_i))'(DT^{*\bar{\alpha}}(T_j) - DT^{*\bar{\alpha}}(T_i)) \\
&\quad = 1(|T_i - T_j| \neq 0) + \bar{c}(\lambda_j - \lambda_i)^2/2 + o(1). \qquad \text{(A.10)}
\end{aligned}
$$

Noting that $u_1^{\bar{\alpha}} = \upsilon_1$ and $u_t^{\bar{\alpha}} = \upsilon_t + T^{-1}(c - \bar{c})u_{t-1} (t = 2, \ldots, T)$, we deduce that

$$
\begin{aligned}
(DU^{\bar{\alpha}}(T_i)'u^{\bar{\alpha}} &= \upsilon_{T_i+1} + T^{-1}(c - \bar{c})u_{T_i} + \bar{c}T^{-1} \\
&\quad \times \sum_{t=T_i+2}^{T} (\upsilon_t + T^{-1}(c - \bar{c})u_{t-1}) = \upsilon_{T_i+1} + o_p(1),
\end{aligned} \tag{A.11}
$$

$$
\begin{aligned}
&T^{-1/2}DT^{*\bar{\alpha}}(T_i)'u^{\bar{\alpha}} \\
&\quad = T^{-1/2}\sum_{T_i+1}^{T}(\upsilon_t + T^{-1}(c - \bar{c})u_{t-1})(1 - \bar{c}(t - T_i - 1)/T) \\
&\quad \Rightarrow \sigma V(\lambda_i),
\end{aligned} \tag{A.12}
$$

where

$$
\begin{aligned}
V(\lambda_i) = (1 + \bar{c}\lambda_i)&\left[W(1) - W(\lambda_i) + (c - \bar{c})\int_{\lambda_i}^{1} W_c(r)dr\right] \\
&- \bar{c}\int_{\lambda_i}^{1} rdW(r) - (c - \bar{c})\bar{c}\int_{\lambda_i}^{1} rW_c(r)dr.
\end{aligned} \tag{A.13}
$$

Proof of Theorem 1. Define the quadratic form $M_T(c, \bar{c}, \lambda^0) = (u^{\bar{\alpha}'}z^{\bar{\alpha}})(z^{\bar{\alpha}'}z^{\bar{\alpha}})^{-1} \times (z^{\bar{\alpha}'}u^{\bar{\alpha}})$, where $z^{\bar{\alpha}}$ is the vector of the quasi-differences of $z_t(\lambda^0)$ defined in (3). From Elliott *et al.* (1996) and

Perron and Rodriguez (2003), we have

$$\begin{aligned} s^2(\lambda^0)P_T^{GLS}(c,\bar{c},\lambda^0) \\ &= M_T(c,0,\lambda^0) - M_T(c,\bar{c},\lambda^0) - 2\bar{c}T^{-1}\sum_{t=2}^{T} u_{t-1}v_t \\ &\quad + (\bar{c}^2 - 2\bar{c}c)T^{-2}\sum_{t=2}^{T} u_{t-1}^2 - \bar{c}T^{-1}u^{1'}u^1 + o_p(1). \end{aligned} \tag{A.14}$$

From the invariance principle and the continuous mapping theorem (CMT), $T^{-1}\Sigma_{t=2}^{T}u_{t-1}v_t \Rightarrow \sigma^2[\int_0^1 W_c(r)dW(r) + \gamma]$ and $T^{-2}\Sigma_{t=2}^{T}u_{t-1}^2 \Rightarrow \sigma^2\int_0^1 W_c(r)^2dr$ where $\gamma = (\sigma^2 - \sigma_v^2)/(2\sigma^2)$ and $\sigma_v^2 = Ev_t^2$. By the law of large numbers, $p\lim T^{-1}u^{1'}u^1 = \sigma_v^2$. Note that the expression in (A.14) is similar to that in Elliott *et al.* (1996) and Perron and Rodríguez (2003), the only difference coming from the definition of the quadratic forms $M_T(c,0,\lambda^0)$ and $M_T(c,\bar{c},\lambda^0)$. Let $z_t' = (z_{t,1}', z_{t,2}')$ be a rearranged version of $z_t(\lambda^0)$ so that $z_{t,1}$ collects the $m+1$ regressors that correspond to the constant and the impulse dummy variables and $z_{t,2}$ collects the $m+1$ trending regressors. Also, let $z_t^{\bar{\alpha}'} = (z_{t,1}^{\bar{\alpha}'}, z_{t,2}^{\bar{\alpha}'})$ be the quasi-differenced z_t. Then, the scaled matrix $M_T(c,\bar{c},\lambda^0)$ can be expressed as

$$M_T(c,\bar{c},\lambda^0) = (u^{\bar{\alpha}'}z^{\bar{\alpha}}D_T)(D_Tz^{\bar{\alpha}'}z^{\bar{\alpha}}D_T)^{-1}(D_Tz^{\bar{\alpha}'}u^{\bar{\alpha}}),$$

where $D_T = \text{diag}\{D_{1,T},\ D_{2,T}\} = \text{diag}(1,\ldots,1,T^{-1/2},\ldots,T^{-1/2})$. From (A.11) and (A.12),

$$\begin{aligned} D_Tz^{\bar{\alpha}'}u^{\bar{\alpha}} &= \sigma(v_1/\sigma,\ldots,v_{Tj+1}/\sigma,\ldots,v_{Tm+1}/\sigma, \\ &\quad V(\lambda_0^0),\ldots,V(\lambda_j^0),\ldots,V(\lambda_m^0))' + o_p(1). \end{aligned} \tag{A.15}$$

Using (A.1)–(A.3), the limit of $D_Tz^{\bar{\alpha}'}z^{\bar{\alpha}}D_T$ is given by the following block diagonal matrix:

$$D_Tz^{\bar{\alpha}'}z^{\bar{\alpha}}D_T \to \begin{bmatrix} I_{m+1} & 0 \\ 0 & A(\lambda^0) \end{bmatrix}, \tag{A.16}$$

where I_{m+1} is the identity matrix of order $m+1$ and $A(\lambda^0)$ is a symmetric matrix defined by (8) with $a(\lambda_i^0,\lambda_j^0)$ as given in (A.1).

Therefore, we have

$$M_T(c,\bar{c},\lambda^0) = \upsilon_1^2 + \sum_{j=1}^{m} \upsilon_{T_j+1}^2 + \sigma^2 M(c,\bar{c},\lambda^0) + o_p(1),$$

where $M(c,\bar{c},\lambda^0) = \bar{V}(\lambda^0)' A(\lambda^0)^{-1} \bar{V}(\lambda^0)$, with $\bar{V}(\lambda^0) = (V(\lambda_0^0),\ldots, V(\lambda_m^0))'$ and $V(\lambda_j^0)$ as defined in (A.13). This weak convergence result also holds when $\bar{c} = 0$, and, thus,

$$M_T(c,0,\lambda^0) = \upsilon_1^2 + \sum_{j=1}^{m} \upsilon_{T_j+1}^2 + \sigma^2 M(c,0,\lambda^0) + o_p(1).$$

It follows that the limiting distribution is as stated in Theorem 1.□

Proof of Theorem 2. Note that the scaled detrended variable $\tilde{y}_t$ is given by

$$T^{-1/2}\tilde{y}_t = T^{-1/2}u_t - T^{-1/2}z_t' D_T (D_T z^{\bar{\alpha}'} z^{\bar{\alpha}} D_T)^{-1} D_T z^{\bar{\alpha}'} u^{\bar{\alpha}}.$$

Using (A.15), (A.16), and the asymptotic block diagonality of $(D_T z^{\bar{\alpha}'} z^{\bar{\alpha}} D_T)^{-1}$, we obtain

$$\begin{aligned} T^{-1/2}\tilde{y}_t &= T^{-1/2}u_t - T^{-1/2}z'_{t,1}(z_1^{\bar{\alpha}'} z_2^{\bar{\alpha}})^{-1} z_1^{\bar{\alpha}'} u^{\bar{\alpha}} \\ &\quad - T^{-1/2}z'_{t,2} D_{2,T}(D_{2,T} z_2^{\bar{\alpha}'} z_2^{\bar{\alpha}} D_{2,T})^{-1} D_{2,T} z_2^{\bar{\alpha}'} u^{\bar{\alpha}} + o_p(1). \end{aligned}$$

Note that $T^{-1/2}z'_{[Tr],1}(z_1^{\bar{\alpha}'} z_1^{\bar{\alpha}})^{-1} z_1^{\bar{\alpha}'} u^{\bar{\alpha}} \xrightarrow{p} 0$, provided that $T^{-1/2}\upsilon_{T_{b,j}+1}/\sigma \xrightarrow{p} 0$, $\forall j = 0,\ldots,m$. Also, $T^{-1/2}D_{2,T}z_{[Tr],2} \to z_2(r)$ uniformly in $r \in [0,1]$, where $z_2(r) = (r,(r-\lambda_1^0)1(r>\lambda_1^0),\ldots,(r-\lambda_m^0)1(r>\lambda_{m1}^0))$. Therefore,

$$T^{-1/2}\tilde{y}_{[Tr]} \Rightarrow \sigma[W_c(r) - z_2(r)A(\lambda^0)^{-1}\bar{V}(\lambda^0)] = \sigma V_{c,\bar{c}}(r,\lambda^0),$$

with $A(\lambda^0)$ and $\bar{V}(\lambda^0)$ as defined before. Using this weak convergence result, the CMT and the fact that $s(\lambda^0)^2 \xrightarrow{p} \sigma^2$, the results stated in the theorem for Models I, II, and IIb follow. □

Proof of Proposition 1. Let λ^0 denote the true break fraction and λ a generic one. Except when indicated, $\bar{\alpha}$ is computed using λ. Note that

$$\begin{aligned} S(\bar{\alpha}, \lambda) &= (z^{\bar{\alpha}}(\lambda^0)\psi + u^{\bar{\alpha}})' M_\lambda^{\bar{\alpha}} (z^{\bar{\alpha}}(\lambda^0)\psi + u^{\bar{\alpha}}) \\ &= (d^{\bar{\alpha}}(\lambda)\psi + u^{\bar{\alpha}})' M_\lambda^{\bar{\alpha}} (d^{\bar{\alpha}}(\lambda)\psi + u^{\bar{\alpha}}) \\ &= \psi' d^{\bar{\alpha}'}(\lambda) M_\lambda^{\bar{\alpha}} d^{\bar{\alpha}}(\lambda)\psi - 2u^{\bar{\alpha}'} M_\lambda^{\bar{\alpha}} d^{\bar{\alpha}}(\lambda)\psi + u^{\bar{\alpha}'} M_\lambda^{\bar{\alpha}} u^{\bar{\alpha}} \\ &\equiv Q^{\bar{\alpha}}(\lambda) - 2G^{\bar{\alpha}}(\lambda) + u^{\bar{\alpha}'} M_\lambda^{\bar{\alpha}} u^{\bar{\alpha}}, \end{aligned} \tag{A.17}$$

where $d^{\bar{\alpha}}(\lambda) = z^{\bar{\alpha}}(\lambda) - z^{\bar{\alpha}}(\lambda^0)$, $M_\lambda^{\bar{\alpha}} = I - P_\lambda^{\bar{\alpha}}$, and $P_\lambda^{\bar{\alpha}} = z^{\bar{\alpha}}(\lambda) D_T (D_T z^{\bar{\alpha}'}(\lambda) z^{\bar{\alpha}}(\lambda) D_T)^{-1} D_T z^{\bar{\alpha}'}(\lambda)$. When the break fraction is correctly specified, $\lambda = \lambda^0$ and $S(\bar{\alpha}_0, \lambda^0) = u^{\bar{\alpha}'_0} M_{\lambda^0}^{\bar{\alpha}_0} u^{\bar{\alpha}_0}$. Note that the preceding expressions are functions of $u^{\bar{\alpha}_0}$ and $z^{\bar{\alpha}_0}(\lambda^0)$, which are quasi-differenced with $\bar{\alpha}_0$, i.e., the parameter $\bar{\alpha}$ computed using the true break fraction λ^0. The difference between these two sums of squared residuals is given by

$$S(\bar{\alpha}, \lambda) - S(\bar{\alpha}_0, \lambda^0) = Q^{\bar{\alpha}}(\lambda) - 2G^{\bar{\alpha}}(\lambda) + u^{\bar{\alpha}'} M_\lambda^{\bar{\alpha}} u^{\bar{\alpha}} - u^{\bar{\alpha}'_0} M_{\lambda^0}^{\bar{\alpha}_0} u^{\bar{\alpha}_0}.$$

Consider first

$$Q^{\bar{\alpha}}(\lambda) = \psi' d^{\bar{\alpha}'}(\lambda) d^{\bar{\alpha}}(\lambda)\psi - \psi' d^{\bar{\alpha}'}(\lambda) P_\lambda^{\bar{\alpha}} d^{\bar{\alpha}}(\lambda)\psi.$$

For simplicity, assume that there is only one break. In Models II and IIb,

$$\begin{aligned} &\psi' d^{\bar{\alpha}'}(\lambda) d^{\bar{\alpha}}(\lambda)\psi \\ &\quad = \mu_1^2 \|DU^{\bar{\alpha}}(T_1) - DU^{\bar{\alpha}}(T_1^0)\|^2 + \beta_1^2 \|DT^{*\bar{\alpha}}(T_1) - DT^{*\bar{\alpha}}(T_1^0)\|^2 \\ &\qquad + 2\mu_1\beta_1 [DU^{\bar{\alpha}}(T_1) - DU^{\bar{\alpha}}(T_1^0)]' [DT^{*\bar{\alpha}}(T_1) - DT^{*\bar{\alpha}}(T_1^0)]. \end{aligned}$$

From (A.8)–(A.10), we have, in Model II (if $\lambda > \lambda^0$),

$$\begin{aligned} T^{-1}\psi' d^{\bar{\alpha}'}(\lambda) d^{\bar{\alpha}}(\lambda)\psi = \beta_1^2 [&(\lambda - \lambda^0) - \bar{c}(\lambda - \lambda^0)^2 + \bar{c}^2(\lambda - \lambda^0)^3/3 \\ &+ \bar{c}^2(\lambda - \lambda^0)^2(1 - \lambda)] + o(1) \end{aligned}$$

and, in Model IIb,

$$T^{-1-2\eta}\psi' d^{\bar{\alpha}'}(\lambda) d^{\bar{\alpha}}(\lambda)\psi = 2k_1^2 \cdot 1(|T_1 - T_1^0| \neq 0) + o(1).$$

Also,

$$
\begin{aligned}
& d^{\bar{\alpha}'}(\lambda)P_{\lambda}^{\bar{\alpha}}d^{\bar{\alpha}}(\lambda) \\
& \quad = d^{\bar{\alpha}'}(\lambda)z^{\bar{\alpha}}(\lambda)D_T(D_T z^{\bar{\alpha}'}(\lambda)z^{\bar{\alpha}}(\lambda)D_T)^{-1}D_T z^{\bar{\alpha}'}(\lambda)d^{\bar{\alpha}}(\lambda).
\end{aligned}
$$

We have shown that $(D_T z^{\bar{\alpha}'}(\lambda)z^{\bar{\alpha}}(\lambda)D_T)^{-1} = O(1)$ with the limit being a block diagonal matrix. It also follows from (A.2)–(A.7) that

$$
\begin{aligned}
& D_T z^{\bar{\alpha}'}(\lambda)d^{\bar{\alpha}}(\lambda)\psi \\
& = [\iota^{\bar{\alpha}}, DU^{\bar{\alpha}}(T_1), T^{-1/2}\tau^{\bar{\alpha}}, T^{-1/2}DT^{*\bar{\alpha}}(T_1)]' \\
& \quad \times [DU^{\bar{\alpha}}(T_1) - DU^{\bar{\alpha}}(T_1^0), DT^{*\bar{\alpha}}(T_1) - DT^{*\bar{\alpha}}(T_1^0)](\mu_1, \beta_1)' \\
& = \begin{pmatrix} o(1) & O(T^{-1}|T_1 - T_1^0|) \\ 1 + o(1) & 1(T_1 < T_1^0) + O(T^{-1}|T_1 - T_1^0|) \\ o(1) & O(T^{-1/2}|T_1 - T_1^0|) \\ o(1) & O(T^{-1/2}|T_1 - T_1^0|) \end{pmatrix} \begin{pmatrix} \mu_1 \\ \beta_1 \end{pmatrix}.
\end{aligned}
\tag{A.18}
$$

Hence, we have in Model II

$$
\begin{aligned}
& T^{-1}\psi' d^{\bar{\alpha}'}(\lambda)P_{\lambda}^{\bar{\alpha}}d^{\bar{\alpha}}(\lambda)\psi \\
& = \beta_1^2 \begin{pmatrix} a(0,\lambda) - a(0,\lambda^0) \\ a(\lambda,\lambda) - a(\lambda,\lambda^0) \end{pmatrix}' A(\lambda)^{-1} \begin{pmatrix} a(0,\lambda) - a(0,\lambda^0) \\ a(\lambda,\lambda) - a(\lambda,\lambda^0) \end{pmatrix} + o(1) \\
& = |T_1 - T_1^0|^2 O(T^{-2})
\end{aligned}
$$

with $a(\lambda_i, \lambda_j)$ as defined in (A.1) and, in Model IIb, $T^{-1-2\eta}\psi' d^{\bar{\alpha}'}(\lambda)P_{\lambda}^{\bar{\alpha}}d^{\bar{\alpha}}(\lambda)\psi = o(1)$. An entirely analogous argument applies to Models 0b and I. Therefore, collecting terms,

$$
Q^{\bar{\alpha}}(\lambda) = \begin{cases} O(T^{1+2\eta}) \cdot \tilde{I}, & \text{in Models 0b and IIb,} \\ |T_1 - T_1^0| O(1), & \text{in Models I and II,} \end{cases} \tag{A.19}
$$

where $\tilde{I} = 1(|T_1 - T_1^0| \neq 0)$. Consider now

$$
G^{\bar{\alpha}}(\lambda) = u^{\bar{\alpha}'}d^{\bar{\alpha}}(\lambda)\psi - u^{\bar{\alpha}'}P_{\lambda}^{\bar{\alpha}}d^{\bar{\alpha}}(\lambda)\psi.
$$

As before, we first analyze the order of magnitude of $u^{\bar{\alpha}\prime} d^{\bar{\alpha}}(\lambda)\psi$. Note first that

$$
\begin{aligned}
&u^{\bar{\alpha}\prime}(DU^{\bar{\alpha}}(T_1) - DU^{\bar{\alpha}}(T_1^0)) \\
&\quad = -u^{\bar{\alpha}}_{\lambda^0 T+1} + \bar{c}T^{-1} \sum_{t=\lambda^0 T+2}^{\lambda T} u^{\bar{\alpha}}_t + (1+\bar{c}T^{-1})u^{\bar{\alpha}}_{\lambda T+1} \\
&\quad = -(v_{\lambda^0 T+1} + T^{-1}(c-\bar{c})u_{\lambda^0 T}) + \bar{c}T^{-1} \\
&\qquad \times \sum_{t=\lambda^0 T+2}^{\lambda T} (v_t + T^{-1}(c-\bar{c})u_{t-1}) \\
&\qquad + (1+\bar{c}T^{-1})(v_{\lambda T+1} + T^{-1}(c-\bar{c})u_{\lambda T}) \\
&\quad = -v_{\lambda^0 T+1} + v_{\lambda T+1} + o_p(1) = O_p(1) \cdot \tilde{I}
\end{aligned}
$$

and

$$
\begin{aligned}
&u^{\bar{\alpha}\prime}(DT^{*\bar{\alpha}}(T_1) - DT^{*\bar{\alpha}}(T_1^0)) \\
&\quad = -u^{\bar{\alpha}}_{\lambda^0 T+1} - \sum_{t=\lambda^0 T+2}^{\lambda T+1} (-\bar{c}T^{-1}t + 1 + \lambda^0\bar{c} + \bar{c}T^{-1})u^{\bar{\alpha}}_t + u^{\bar{\alpha}}_{\lambda T+1} \\
&\qquad + \sum_{t=\lambda T+2}^{\lambda T} ((-\bar{c}T^{-1}t + 1 + \lambda\bar{c} + \bar{c}T^{-1}) \\
&\qquad - (-\bar{c}T^{-1}t + 1 + \lambda^0\bar{c} + \bar{c}T^{-1}))u^{\bar{\alpha}}_t \\
&\quad = u^{\bar{\alpha}}_{\lambda T+1} - u^{\bar{\alpha}}_{\lambda^0 T+1} - \sum_{t=\lambda^0 T+2}^{\lambda T+1} ((-\bar{c}T^{-1}t + 1 + \lambda^0\bar{c})u^{\bar{\alpha}}_t + (\lambda - \lambda^0)\bar{c} \\
&\qquad \times \sum_{t=\lambda T+2}^{T} (v_t + \frac{c-\bar{c}}{T} u_{t-1}) + o_p(1) = |T_1 - T_1^0| O_p(T^{-1/2}).
\end{aligned}
$$

Then,

$$
u^{\bar{\alpha}\prime} d^{\bar{\alpha}}(\lambda)\psi = \begin{cases} O_p(T^{1/2+\eta}) \cdot \tilde{I}, & \text{in Models 0b and IIb,} \\ |T_1 - T_1^0| O_p(T^{-1/2}), & \text{in Models I and II.} \end{cases}
$$

Now,

$$\begin{aligned}
u^{\bar{\alpha}'} P_{\lambda}^{\bar{\alpha}} d^{\bar{\alpha}}(\lambda)\psi &= u^{\bar{\alpha}'} z^{\bar{\alpha}}(\lambda) D_T (D_T z^{\bar{\alpha}'}(\lambda) z^{\bar{\alpha}}(\lambda) D_T)^{-1} D_T z^{\bar{\alpha}'}(\lambda) d^{\bar{\alpha}}(\lambda)\psi \\
&= O_p(1) D_T z^{\bar{\alpha}'}(\lambda) d^{\bar{\alpha}}(\lambda)\psi \\
&= \begin{cases} o_p(T^{1/2+\eta}) \cdot \tilde{I}, & \text{in Models 0b and IIb,} \\ |T_1 - T_1^0| O_p(T^{-1/2}), & \text{in Models I and II.} \end{cases}
\end{aligned}$$

Therefore, we have

$$G^{\bar{\alpha}}(\lambda) = \begin{cases} O_p(T^{1/2+\eta}) \cdot \tilde{I}, & \text{in Models 0b and IIb,} \\ |T_1 - T_1^0| O_p(T^{-1/2}), & \text{in Models I and II.} \end{cases} \tag{A.20}$$

For the last term, it follows from previous results that

$$u^{\bar{\alpha}'} M_{\lambda}^{\bar{\alpha}} u^{\bar{\alpha}} - u^{\bar{\alpha}_0'} M_{\lambda^0}^{\bar{\alpha}_0} u^{\bar{\alpha}_0} \leq O_p(1) \cdot \tilde{I}.$$

Therefore, collecting terms,

$$T^{-1}(S(\bar{\alpha}, \lambda) - S(\bar{\alpha}_0, \lambda^0)) = T^{-1} Q^{\bar{\alpha}}(\lambda) + o_p(1).$$

Furthermore, note that for a given estimate of the break fraction vector, $\hat{\lambda}$, the inequality $T^{-1}S(\bar{\alpha}, \hat{\lambda}) \leq T^{-1}S(\bar{\alpha}_0, \lambda^0)$ is always satisfied. Now suppose that $\hat{\lambda} \nrightarrow_p \lambda^0$. Then, according to this inequality, we will need that, for large T, $T^{-1}Q^{\bar{\alpha}}(\hat{\lambda}) \leq 0$, but we have shown that $Q^{\bar{\alpha}}(\lambda) > 0$ when $\lambda \neq \lambda^0$. Hence a contradiction, and the only way that the inequality is satisfied is when $\hat{\lambda} \rightarrow_p \lambda^0$. Thus, the minimization of $S(\bar{\alpha}, \lambda)$ over all possible values of the break fraction vector λ results in a consistent estimate of the break fractions. Although our derivations have considered only one structural break, the arguments are also valid for the multiple break case.

To establish the rate of convergence for Models I and II, we first define the sets $V_\epsilon = \{T_k : |T_k - T_1^0| < \epsilon T\}$ for $\epsilon \in (0, 1)$ and $V_\epsilon(C) = \{T_k : |T_k - T_1^0| < \epsilon T, |T_k - T_1^0| > C\}$ for $C > 0$, so that $V_\epsilon\ (C) \subset V_\epsilon$. Note that $S(\bar{\alpha}, \hat{\lambda}) \leq S(\bar{\alpha}_0, \lambda^0)$ with probability one and $\Pr(\hat{T}_1 \in V\epsilon) \rightarrow 1$, as $T \rightarrow \infty$. Given the previous results, there is a constant

$C > 0$ such that

$$\Pr\left(\min_{\lambda T\in V\epsilon(C)} \frac{S(\bar{\alpha},\lambda)-S(\bar{\alpha}_0,\lambda^0)}{|\lambda-\lambda^0|T} \leq 0\right) < \zeta$$

for some small $\zeta > 0$ because, when C is properly chosen,

$$\begin{aligned}
&\Pr\left(\min_{\lambda T\in V\epsilon(C)} \frac{S(\bar{\alpha},\lambda)-S(\bar{\alpha}_0,\lambda^0)}{|\lambda-\lambda^0|T} \leq 0\right)\\
&\quad = \Pr\left(\min_{\lambda T\in V\epsilon(C)} \frac{Q^{\bar{\alpha}}(\lambda)-2G^{\bar{\alpha}}(\lambda)+u^{\bar{\alpha}'}M_{\lambda}^{\bar{\alpha}}u^{\bar{\alpha}}-u^{\bar{\alpha}_0'}M_{\lambda^0}^{\bar{\alpha}_0}u^{\bar{\alpha}_0}}{|\lambda-\lambda^0|T} \leq 0\right)\\
&\quad < \zeta
\end{aligned}$$

because

$$\begin{aligned}
&\frac{Q^{\bar{\alpha}}(\lambda)-2G^{\bar{\alpha}}(\lambda)+u^{\bar{\alpha}'}M_{\lambda}^{\bar{\alpha}}u^{\bar{\alpha}}-u^{\bar{\alpha}_0'}M_{\lambda^0}^{\bar{\alpha}_0}u^{\bar{\alpha}_0}}{|\lambda-\lambda^0|T}\\
&\quad = O(1)-2O_p(T^{-1/2})+\frac{Op(1)\cdot\tilde{I}}{C}
\end{aligned}$$

with the $O(1)$ and $O_p(1)$ terms positive. Hence, $[S(\bar{\alpha},\lambda) - S(\bar{\alpha}_0,\lambda^0)]/(|\lambda-\lambda^0|T) > 0$ on $V_\epsilon(C)$ with large probability. This implies that the minimum cannot be achieved on $V_\epsilon(C)$ and, thus, $\Pr(T|\hat{\lambda}-\lambda^0| \geq C) \leq \zeta$, so that $(\hat{\lambda}-\lambda^0) = O_p(T^{-1})$. For Models 0b and IIb, a similar argument can be applied to show that $(\hat{\lambda}-\lambda^0) = O_p(T^{-1})$. □

Proof of Proposition 2. Consider first the statistic $P_T^{GLS}(c,\bar{c},\hat{\lambda})$ in Models I and II.

$$\begin{aligned}
s(\hat{\lambda})^2 P_T^{GLS}(c,\bar{c},\hat{\lambda}) &= S(\bar{\alpha},\lambda)-\bar{\alpha}S(1,\hat{\lambda})\\
&= Q^{\bar{\alpha}}(\hat{\lambda})-2G^{\bar{\alpha}}(\hat{\lambda})+u^{\bar{\alpha}'}M_{\hat{\lambda}}^{\bar{\alpha}}u^{\bar{\alpha}}-(1+\bar{c}/T)\\
&\quad\times[Q^1(\hat{\lambda})-2G^1(\hat{\lambda})+u^{1'}M_{\hat{\lambda}}^1u^1]\\
&= [Q^{\bar{\alpha}}(\hat{\lambda})-Q^1(\hat{\lambda})]-\bar{c}T^{-1}Q^1(\hat{\lambda})-2[G^{\bar{\alpha}}(\hat{\lambda})-(1+\bar{c}/T)G^1(\hat{\lambda})]\\
&\quad+[u^{\bar{\alpha}'}M_{\hat{\lambda}}^{\bar{\alpha}}u^{\bar{\alpha}}-(1+\bar{c}/T)u^{1'}M_{\hat{\lambda}}^1u^1]\\
&= [Q^{\bar{\alpha}}(\hat{\lambda})-Q^1(\hat{\lambda})]+[u^{\bar{\alpha}'}M_{\hat{\lambda}}^{\bar{\alpha}}u^{\bar{\alpha}}-(1+\bar{c}/T)u^{1'}M_{\hat{\lambda}}^1u^1]+o_p(1).
\end{aligned}$$

The last equality follows from (A.19), (A.20), and Proposition 1. From this expression, we can see that if $[Q^{\bar{\alpha}}(\hat{\lambda}) - Q^1(\hat{\lambda})]$ is asymptotically negligible, the limit distribution of $s(\hat{\lambda})^2 P_T^{GLS}(c, \bar{c}, \hat{\lambda})$ will be the same as that of $s(\lambda^0)^2 P_T^{GLS}(c, \bar{c}, \lambda^0)$. Now,

$$\begin{aligned} Q^{\bar{\alpha}}(\hat{\lambda}) - Q^1(\hat{\lambda}) &= \psi' d^{\bar{\alpha}'}(\hat{\lambda}) d^{\bar{\alpha}}(\hat{\lambda})\psi - \psi' d^{\bar{\alpha}'}(\hat{\lambda}) P_{\hat{\lambda}}^{\bar{\alpha}} d^{\bar{\alpha}}(\hat{\lambda})\psi \\ &\quad - \psi' d^{1'}(\hat{\lambda}) d^1(\hat{\lambda})\psi + \psi' d^{1\prime}(\hat{\lambda}) P_{\hat{\lambda}}^1 d^1(\hat{\lambda})\psi. \end{aligned}$$

From (A.8)–(A.10),

$$\psi' d^{\bar{\alpha}'}(\hat{\lambda}) d^{\bar{\alpha}}(\hat{\lambda})\psi - \psi' d^{1'}(\hat{\lambda}) d^1(\hat{\lambda})\psi = o_p(1),$$

and we recall that

$$\begin{aligned} & d^{\bar{\alpha}'}(\hat{\lambda}) P_{\hat{\lambda}}^{\bar{\alpha}} d^{\bar{\alpha}}(\hat{\lambda}) \\ &\quad = d^{\bar{\alpha}'}(\hat{\lambda}) z^{\bar{\alpha}}(\hat{\lambda}) D_T (D_T z^{\bar{\alpha}'}(\hat{\lambda}) z^{\bar{\alpha}}(\hat{\lambda}) D_T)^{-1} D_T z^{\bar{\alpha}'}(\hat{\lambda}) d^{\bar{\alpha}}(\hat{\lambda}). \end{aligned}$$

Then, from (A.18) and Proposition 1, we have (assuming for simplicity a single break)

$$\begin{aligned} & D_T z^{\bar{\alpha}'}(\hat{\lambda}) d^{\bar{\alpha}}(\hat{\lambda})\psi \\ &\quad = [\iota^{\bar{\alpha}}, DU^{\bar{\alpha}}(\hat{T}_1), T^{-1/2}\tau^{\bar{\alpha}}, T^{-1/2} DT^{*\bar{\alpha}}(\hat{T}_1)]' \\ &\qquad \times [DU^{\bar{\alpha}}(\hat{T}_1) - DU^{\bar{\alpha}}(T_1^0), DT^{*\bar{\alpha}}(\hat{T}_1) - DT^{*\bar{\alpha}}(T_1^0)](\mu_1, \beta_1)' \\ &\quad = \begin{pmatrix} o_p(1) & o_p(1) \\ 1 + o_p(1) & 1(T_1 < T_1^0) + o_p(1) \\ o_p(1) & o_p(1) \\ o_p(1) & o_p(1) \end{pmatrix} \begin{pmatrix} \mu_1 \\ \beta_1 \end{pmatrix} \end{aligned}$$

and, similarly,

$$D_T z^{1'}(\hat{\lambda}) d^1(\hat{\lambda})\psi = \begin{pmatrix} o_p(1) & o_p(1) \\ 1 + o_p(1) & 1(T_1 < T_1^0) + o_p(1) \\ o_p(1) & o_p(1) \\ o_p(1) & o_p(1) \end{pmatrix} \begin{pmatrix} \mu_1 \\ \beta_1 \end{pmatrix}.$$

Given the asymptotic block diagonality of $D_T z^{\bar{\alpha}'}(\hat{\lambda}) z^{\bar{\alpha}}(\hat{\lambda}) D_T$ and $D_T z^{1'}(\hat{\lambda}) z^1(\hat{\lambda}) D_T$,we have

$$\psi' d^{\bar{\alpha}'}(\hat{\lambda}) P_{\hat{\lambda}}^{\bar{\alpha}} d^{\bar{\alpha}}(\hat{\lambda})\psi - \psi' d^{1'}(\hat{\lambda}) P_{\hat{\lambda}}^{1} d^{1}(\hat{\lambda})\psi = o_p(1).$$

We focus on $MSB^{GLS}(\lambda)$ defined by (10). Note that $\tilde{y}_{-1}'\tilde{y}_{-1}$ is of the same order of magnitude as $\tilde{y}'\tilde{y}$, and so we concentrate on $\tilde{y}'\tilde{y}$. First, note that, in Models I and II,

$$\begin{aligned}\hat{\psi} &= (z^{\bar{\alpha}}(\hat{\lambda})' z^{\bar{\alpha}}(\hat{\lambda}))^{-1} z^{\bar{\alpha}}(\hat{\lambda})' y^{\bar{\alpha}} = (z^{\bar{\alpha}}(\hat{\lambda})' z^{\bar{\alpha}}(\hat{\lambda}))^{-1} z^{\bar{\alpha}}(\hat{\lambda})' z^{\bar{\alpha}}(\lambda^0)\psi \\ &\quad + (z^{\bar{\alpha}}(\hat{\lambda})' z^{\bar{\alpha}}(\hat{\lambda}))^{-1} z^{\bar{\alpha}}(\hat{\lambda})' u^{\bar{\alpha}} = \psi + (z^{\bar{\alpha}}(\hat{\lambda})' z^{\bar{\alpha}}(\hat{\lambda}))^{-1} z^{\bar{\alpha}}(\hat{\lambda})' u^{\bar{\alpha}} \\ &\quad + (z^{\bar{\alpha}}(\hat{\lambda})' z^{\bar{\alpha}}(\hat{\lambda}))^{-1} z^{\bar{\alpha}}(\hat{\lambda})' [z^{\bar{\alpha}}(\lambda^0) - z^{\bar{\alpha}}(\hat{\lambda})]\psi\end{aligned}$$

and

$$\begin{aligned}\tilde{y} &= y - z(\hat{\lambda})\hat{\psi} = z(\lambda^0)(\psi - \hat{\psi}) + u + [z(\lambda^0) - z(\tilde{\lambda})]\hat{\psi} \\ &= u - z(\lambda^0)(z^{\bar{\alpha}}(\hat{\lambda})' z^{\bar{\alpha}}(\hat{\lambda}))^{-1} z^{\bar{\alpha}}(\hat{\lambda})' u^{\bar{\alpha}} - z(\lambda^0)(z^{\bar{\alpha}}(\hat{\lambda})' z^{\bar{\alpha}}(\hat{\lambda}))^{-1} z^{\bar{\alpha}}(\hat{\lambda})' \\ &\quad \times [z^{\bar{\alpha}}(\lambda^0) - z^{\bar{\alpha}}(\hat{\lambda})]\psi + [z(\lambda^0) - z(\hat{\lambda})]\hat{\psi}.\end{aligned}$$

We need to show that the last two terms are asymptotically negligible compared to the first two. Note that the first two terms are such that

$$\|u - z(\lambda^0)(z^{\bar{\alpha}}(\hat{\lambda})' z^{\bar{\alpha}}(\hat{\lambda}))^{-1} z^{\bar{\alpha}}(\hat{\lambda})' u^{\bar{\alpha}}\| = O_p(T).$$

For the third term,

$$\begin{aligned}&\|z(\lambda^0)(z^{\bar{\alpha}}(\hat{\lambda})' z^{\bar{\alpha}}(\hat{\lambda}))^{-1} z^{\bar{\alpha}}(\hat{\lambda})' [z^{\bar{\alpha}}(\lambda^0) - z^{\bar{\alpha}}(\hat{\lambda})]\psi\| \\ &= (\psi[z^{\bar{\alpha}}(\lambda^0) - z^{\bar{\alpha}}(\hat{\lambda})] z^{\bar{\alpha}}(\hat{\lambda}) D_T S_T D_T z(\lambda^0)' \\ &\quad \times z(\lambda^0) D_T S_T D_T z^{\bar{\alpha}}(\hat{\lambda})' [z^{\bar{\alpha}}(\lambda^0) - z^{\bar{\alpha}}(\hat{\lambda})]\psi)^{1/2} \\ &= (\psi[z^{\bar{\alpha}}(\lambda^0) - z^{\bar{\alpha}}(\hat{\lambda})] z^{\bar{\alpha}}(\hat{\lambda}) D_T S_T D_T z(\lambda^0)' \\ &\quad \times z(\lambda^0) D_T S_T D_T z^{\bar{\alpha}}(\hat{\lambda})' [z^{\bar{\alpha}}(\lambda^0) - z^{\bar{\alpha}}(\hat{\lambda})]\psi)^{1/2} \\ &= O_p(T^{1/2}),\end{aligned}$$

where $S_T = (D_T z^{\bar{\alpha}}(\hat{\lambda})' z^{\bar{\alpha}}(\hat{\lambda}) D_T)^{-1}$. The last equality follows from the fact that $S_T = O_p(1)$, $D_T z(\lambda^0)' z(\lambda^0) D_T = O_p(T^2)$ and, from the previous proof, $D_T z^{\bar{\alpha}}(\lambda)'[z^{\bar{\alpha}}(\lambda^0) - z^{\bar{\alpha}}(\hat{\lambda})]\psi = O_p(T^{-1/2})$. For the last term,

$$\begin{aligned}\|[z(\lambda^0) - z(\hat{\lambda})]\hat{\psi}\| &= (\hat{\psi}'[z(\lambda^0) - z(\hat{\lambda})]'[z(\lambda^0) - z(\hat{\lambda})]\hat{\psi})^{1/2} \\ &= |\hat{\lambda} - \lambda^0|O_p(T^{3/2}) = O_p(T^{1/2}).\end{aligned}$$

The CMT with the consistency of $s^2(\lambda)$ completes the proof for $MSB^{GLS}(\lambda)$. An analogous argument applies to $\tilde{y}_T$, and the proofs for $MZ_\alpha^{GLS}(\lambda)$ and $MZ_t^{GLS}(\lambda)$ directly follow. □

Acknowledgments

Perron acknowledges financial support for this work from the National Science Foundation under grant SES-0649350. Carrion-i-Silvestre acknowledges financial support from the Ministerio de Ciencia y Tecnología under grant SEJ2005-08646/ECON. We thank two referees for useful comments.

References

Andrews, D.W.K. and Ploberger, W. (1994). Optimal tests when a nuisance parameter is present only under the alternative. *Econometrica* 62, 1383–1414.

Bai, J. (1998). A note on spurious break. *Econometric Theory* 14, 663–669.

Bai, J. and Perron, P. (2003). Computation and analysis of multiple structural change models. *Journal of Applied Econometrics* 18, 1–22.

Christiano, L.J. (1992). Searching for a break in GNP. *Journal of Business & Economic Statistics* 10, 237–250.

Dickey, D.A. and Fuller, W.A. (1979). Distribution of the estimators for autoregressive time series with a unit root. *Journal of the American Statistical Association* 74, 427–431.

Elliott, G., Rothenberg, T.J. and Stock, J.H. (1996). Efficient tests for an autoregressive unit root. *Econometrica* 64, 813–836.

Harris, D., Harvey, D.I., Leybourne, S.J. and Taylor, A.M.R. (2009). Testing for a unit root in the presence of a possible break in trend. *Econometric Theory* 25, 1545–1588.

Harvey, D.I., Leybourne, S.J. and Newbold, P. (2001). Innovational outlier unit root tests with an endogenously determined break in level. *Oxford Bulletin of Economics and Statistics* 63, 559–575.

Hatanaka, M. and Yamada, K. (1999). A unit root test in the presence of structural changes in I(1) and I(0) models. In *Cointegration, Causality, and*

Forecasting: A Festschrift in Honour of Clive W. J. Granger, Engle R.F. and White H. (eds.), Oxford University Press, pp. 256–282.

Hecq, A. and Urbain, J.P. (1993). Misspecification tests, unit roots and level shifts. *Economics Letters* 43,129–135.

Kejriwal, M. and Perron, P. (2008). A Note on the Sequential Application of the Perron–Yabu Test for Shifts in Trend with an Integrated or Stationary Noise Component. Manuscript, Department of Economics, Boston University.

Kim, D. and Perron, P. (2009). Unit root tests allowing for a break in the trend function under both the null and alternative hypotheses. *Journal of Econometrics* 148, 1–13.

Kim, T., Leybourne, S.J. and Newbold, P. (2000). Spurious rejections by Perron tests in the presence of a break. *Oxford Bulletin of Economics and Statistics* 62, 433–444.

Lee, J. and Strazicich, M.C. (2001). Breakpoint estimation and spurious rejections with endogenous unit root tests. *Oxford Bulletin of Economics and Statistics* 63, 535–558.

Leybourne, S.J., Mills, T.C. and Newbold, P. (1998). Spurious rejections by Dickey-Fuller tests in the presence of a break under the null. *Journal of Econometrics* 87, 191–203.

Leybourne, S.J. and Newbold, P. (2000). Behavior of the standard and symmetric Dickey-Fuller-type tests when there is a break under the null hypothesis. *Econometrics Journal* 3, 1–15.

Montañés, A. (1997). Level shifts, unit roots and misspecification of the breaking date. *Economics Letters* 54, 7–13.

Montañés, A. and Olloqui, I. (1999). Misspecification of the breaking date in segmented trend variables: Effect on the unit root tests. *Economics Letters* 65, 301–307.

Ng, S. and Perron, P. (2001). Lag length selection and the construction of unit root tests with good size and power. *Econometrica* 69, 1519–1554.

Nunes, L.C., Kuan, C.-M. and Newbold, P. (1995). Spurious break. *Econometric Theory* 11, 736–749.

Nunes, L.C., Newbold, P. and Kuan, C.-M. (1997). Testing for unit root with breaks: Evidence on the Great Crash and the unit root hypothesis reconsidered. *Oxford Bulletin of Economics and Statistics* 59, 435–448.

Perron, P. (1989). The Great Crash, the oil price shock and the unit root hypothesis. *Econometrica* 57, 1361–1401.

Perron, P. (1990). Testing for a unit root in a time series with a changing mean. *Journal of Business & Economic Statistics* 8, 153–162.

Perron, P. (1997). Further evidence on breaking trend functions in macroeconomic variables. *Journal of Econometrics* 80, 355–385.

Perron, P. and Ng, S. (1998). An autoregressive spectral density estimator at frequency zero for nonstationarity tests. *Econometric Theory* 14, 560–603.

Perron, P. and Qu, Z. (2006). Estimating restricted structural change models. *Journal of Econometrics* 134, 373–399.

Perron, P. and Qu, Z. (2007). A simple modification to improve the finite sample properties of Ng and Perron's unit root tests. *Economics Letters* 94, 12–19.

Perron, P. and Rodriguez, G.H. (2003). GLS detrending, efficient unit root tests and structural change. *Journal of Econometrics* 115, 1–27.

Perron, P. and Vogelsang, T.J. (1992a). Nonstationarity and level shifts with an application to purchasing power parity. *Journal of Business & Economic Statistics* 10, 301–320.

Perron, P. and Vogelsang, T.J. (1992b). Testing for a unit root in a time series with a changing mean: Correction and extensions. *Journal of Business & Economic Statistics* 10, 467–470.

Perron, P. and Yabu, T. (2009). Testing for shifts in trend with an integrated or stationary noise component. *Journal of Business & Economic Statistics* 27, 369–396.

Perron, P. and Zhu, X. (2005). Structural breaks with deterministic and stochastic trends. *Journal of Econometrics* 129, 65–119.

Stock, J.H. (1999). A Class of Tests for Integration and Cointegration. In *Cointegration, Causality, and Forecasting: A Festschrift for Clive W.J. Granger*, Oxford University Press, pp. 135–167.

Vogelsang, T.J. (2001). Testing for a shift in trend when serial correlation is of unknown form. Manuscript, Department of Economics, Cornell University.

Vogelsang, T.J. and Perron, P. (1998). Additional tests for a unit root allowing for a break in the trend function at an unknown time. *International Economic Review* 39, 1073–1100.

Zivot, E. and Andrews, D.W.K. (1992). Further evidence on the Great Crash, the oil price shock and the unit root hypothesis. *Journal of Business & Economic Statistics* 10, 251–270.

www.ingramcontent.com/pod-product-compliance
Lightning Source LLC
LaVergne TN
LVHW010539300926
849162LV00002B/181

* 9 7 8 9 8 1 3 2 3 7 8 6 5 *